Problem Books in Mathematics

Series Editor:
Peter Winkler
Department of Mathematics
Dartmouth College
Hanover, NH 03755
USA

More information about this series at http://www.springer.com/series/714

Antonio Caminha Muniz Neto

An Excursion through Elementary Mathematics, Volume I

Real Numbers and Functions

 Springer

Antonio Caminha Muniz Neto
Mathematics
Universidade Federal do Ceará
Fortaleza, Ceará, Brazil

ISSN 0941-3502   ISSN 2197-8506   (electronic)
Problem Books in Mathematics
ISBN 978-3-319-85261-4   ISBN 978-3-319-53871-6   (eBook)
DOI 10.1007/978-3-319-53871-6

Printed on acid-free paper

This Springer imprint is published by Springer Nature
The registered company is Springer International Publishing AG
The registered company address is: Gewerbestrasse 11, 6330 Cham, Switzerland

To Gabriel and Isabela, my most beautiful theorems.
To my teacher Valdenísio Bezerra, in memorian

Preface

This is the first of a series of three volumes (the other ones being [4] and [5]) devoted to the mathematics of mathematical olympiads. Generally speaking, they are somewhat expanded versions of a collection of six volumes, first published in Portuguese by the Brazilian Mathematical Society in 2012 and currently in its second edition.

The material collected here and in the other two volumes is based on course notes that evolved over the years since 1991, when I first began coaching students of Fortaleza to the Brazilian Mathematical Olympiad and to the International Mathematical Olympiad. Some 10 years ago, preliminary versions of the Portuguese texts also served as textbooks for several editions of summer courses delivered at UFC to math teachers of the Cape Verde Republic.

All volumes were carefully planned to be a balanced mixture of a smooth and self-contained introduction to the fascinating world of mathematical competitions, as well as to serve as textbooks for students and instructors involved with math clubs for gifted high school students.

Upon writing the books, I have stuck myself to an invaluable advice of the eminent Hungarian-American mathematician George Pólya, who used to say that one cannot learn mathematics without *getting one's hands dirty*. That's why, in several points throughout the text, I left to the reader the task of checking minor aspects of more general developments. These appear either as small omitted details in proofs or as subsidiary extensions of the theory. In this last case, I sometimes refer the reader to specific problems along the book, which are marked with an * and whose solutions are considered to be an essential part of the text. In general, in each section, I collect a list of problems, carefully chosen in the direction of applying the material and ideas presented in the text. Dozens of them are taken from former editions of mathematical competitions and range from the almost immediate to real challenging ones. Regardless of their level of difficulty, we provide generous hints, or even complete solutions, to virtually all of them.

This first volume concentrates on real numbers, elementary algebra, and real functions. The book starts with a non-axiomatic discussion of the most elementary properties of real numbers, followed by a detailed study of basic algebraic identities, equations and systems of equations, elementary sequences, mathematical induction, and the binomial theorem. These pave the way for an initial presentation of algebraic inequalities like that between the arithmetic and geometric means, as well as those of Cauchy, Chebyshev, and Abel. We then run through an exhaustive elementary study of functions that culminates with a first look at implicitly defined functions. This is followed by a second look on real numbers, focusing on the concept of convergence for sequences and series of reals. We then return to functions, this time to successively develop, in detail, the basics of continuity, differentiability, and integrability. Along the way, the text stays somewhere between a thorough calculus course and an introductory analysis one. Lots of interesting examples and important applications are presented throughout. Whenever possible (or desirable), the examples are taken from mathematical competitions, whereas the applications vary from the proof and several applications of Jensen's convexity inequality to Lambert's theorem on the irrationality of π and Stirling's formula on the asymptotic behavior of $n!$. The text ends with a chapter on sequences and series of functions, where, among other interesting topics, we construct an example of a continuous and nowhere differentiable function, develop the rudiments of the generating function method, and discuss Weierstrass' approximation theorem and the rudiments of the theory of Fourier series.

Several people and institutions contributed throughout the years for my effort of turning a bunch of handwritten notes into these books. The State of Ceará Mathematical Olympiad, created by the Mathematics Department of the Federal University of Ceará (UFC) back in 1980 and now in its 36th edition, has since then motivated hundreds of youngsters of Fortaleza to deepen their studies of mathematics. I was one such student in the late 1980s, and my involvement with this competition and with the Brazilian Mathematical Olympiad a few years later had a decisive influence on my choice of career. Throughout the 1990s, I had the honor of coaching several brilliant students of Fortaleza to the Brazilian Mathematical Olympiad. Some of them entered Brazilian teams to the IMO or other international competitions, and their doubts, comments, and criticisms were of great help in shaping my view on mathematical competitions. In this sense, sincere thanks go to João Luiz Falcão, Roney Castro, Marcelo Oliveira, Marcondes França Jr., Marcelo C. de Souza, Eduardo Balreira, Breno Falcão, Fabrício Benevides, Rui Vigelis, Daniel Sobreira, Samuel Feitosa, Davi Máximo Nogueira, and Yuri Lima.

Professor João Lucas Barbosa, upon inviting me to write the textbooks to the Amílcar Cabral Educational Cooperation Project with Cape Verde Republic, had unconsciously provided me with the motivation to complete the Portuguese version of these books. The continuous support of Professor Hilário Alencar, president of the Brazilian Mathematical Society when the Portuguese edition was first published, was also of great importance for me. Special thanks go to professors Abdênago Barros and Fernanda Camargo, my colleagues at the Mathematics Department of UFC, who had made quite useful comments on the Portuguese editions, which were incorporated in the text in a way or another; they had also read the entire English version and helped me in improving it in a number of ways. If it weren't for my editor at Springer-Verlag, Mr. Robinson dos Santos, I almost surely would not have had the courage to embrace the task of translating more than 1500 pages from Portuguese into English. I acknowledge all the staff of Springer involved with this project in his name.

Finally, and mostly, I would like to express my deepest gratitude to my parents Antonio and Rosemary, my wife Monica, and our kids Gabriel and Isabela. From early childhood, my parents have always called my attention to the importance of a solid education, having done their best for me and my brothers to attend the best possible schools. My wife and kids filled our home with the harmony and softness I needed to get to endure on several months of solitary nights of work while translating this book.

Fortaleza, Brazil   Antonio Caminha Muniz Neto
December 2016

Contents

Chapter 1
The Set of Real Numbers

This first chapter recalls some definitions and results which are essential to all further developments. We assume from the reader a modest acquaintance with the most basic concepts of set theory; we also assume that he or she is familiar with the sets of **naturals**,

$$\mathbb{N} = \{1, 2, 3, 4, \ldots\},$$

integers,

$$\mathbb{Z} = \{0, \pm 1, \pm 2, \pm 3, \ldots\},$$

and **rationals**,

$$\mathbb{Q} = \left\{ \frac{a}{b}; \ a, b \in \mathbb{Z}, \ b \neq 0 \right\},$$

as well as with the elementary arithmetic operations within these sets.

In what concerns the integers[1], given $a, b \in \mathbb{Z}$, with $a \neq 0$, we say that *a divides* b if there exists an integer c such that $b = ac$. Equivalently, to say that a divides b is the same as to say that the rational number $\frac{b}{a}$ is an integer; for example, 13 divides 52, since $\frac{52}{13} = 4$.

If a divides b, we also say that a is a *divisor* of b, or that b is *divisible* by a; in such a case, we denote $a \mid b$. If a doesn't divide b (or, which is the same, if $\frac{b}{a} \notin \mathbb{Z}$), we denote $a \nmid b$. An integer n is *even* if $2 \mid n$; otherwise, n is said to be *odd*. Hence, $0, \pm 2, \pm 4, \pm 6, \ldots$ are the even integers, while $\pm 1, \pm 3, \pm 5, \ldots$ are the odd ones.

---

[1] We refer the reader to Chap. 6 of [5] for a systematic discussion of what follows.

© Springer International Publishing AG 2017
A. Caminha Muniz Neto, *An Excursion through Elementary Mathematics, Volume I*,
Problem Books in Mathematics, DOI 10.1007/978-3-319-53871-6_1

Given natural numbers a and b, it is well known that there exist *unique* integers q and r satisfying the following conditions:

$$b = aq + r \text{ and } 0 \leq r < a. \tag{1.1}$$

The above relation is known as the *division algorithm*, and the integers q and r are respectively called the *quotient* and the *remainder* of the division of b by a. Conditions (1.1) are usually condensed in the diagram

$$
\begin{array}{c|c}
b & a \\
\hline
r & q
\end{array}
$$

In particular, for given naturals a and b and in the above notations, $a \mid b$ is the same as $r = 0$ and $q = \frac{b}{a}$.

Two nonzero integers a and b always have a *greatest common divisor*, which will be denoted $\gcd(a, b)$; Moreover, a and b are said to be *relatively prime* if $\gcd(a, b) = 1$; in particular, if $b = ka + 1$, then a and b are relatively prime.

If $r = \frac{m}{n}$, with m and n integers, is a representation of the nonzero rational r, then, by *cancelling out factors common to m and n* (i.e., cancelling $\gcd(m, n)$ out of m and n), we get an *irreducible* representation for r. For example, since $\gcd(-12, 18) = 6$, the rational number $\frac{-12}{18}$ has $\frac{-2}{3}$ as an irreducible representation, which was obtained by cancelling out a factor 6 from both -12 and 18.

An integer $p > 1$ is *prime* if 1 and p are its only positive divisors; in another way, an integer $p > 1$ is prime if, for $a \in \mathbb{N}$, we have that

$$\frac{p}{a} \in \mathbb{N} \Rightarrow a = 1 \text{ or } p.$$

It is a well known fact (cf. Chap. 6 of [5]) that the set of prime numbers is infinite. Below, we list all prime numbers less than 100:

$$2, 3, 5, 7, 11, 13, 17, 19, 23, 29, 31, 37, 41, 43, 47, 53,$$

$$59, 61, 67, 71, 73, 79, 83, 89, 97.$$

An integer greater than 1 and which is not prime is said to be *composite*. It is also a well known fact (cf. Chap. 6 of [5]) that every natural number $n > 1$ can be written (or *decomposed*) as a product of a finite number of powers[2] of prime numbers (its *prime factors*); also, such a representation of n is unique, up to the order of the powers. For instance, $9000 = 2^3 \cdot 3^2 \cdot 5^3$ is the decomposition of 9000 as a product of powers of primes. The existence and uniqueness of such a decomposition of $n > 1$ is known as the **Fundamental Theorem of Arithmetic**.

---

[2]To recall the definition and the main properties of powers of numbers, we refer the reader to Section 1.2.

1.1 Arithmetic in $\mathbb{R}$

We use to represent rational numbers in *decimal notation*. For the rational number $\frac{1}{8}$, for example, we write $\frac{1}{8} = 0.125$ as a shorthand for the equality

$$\frac{1}{8} = \frac{1}{10} + \frac{2}{10^2} + \frac{5}{10^3},$$

and say that 0.125 is the **decimal representation** of $\frac{1}{8}$.

Some rational numbers have more complicated decimal representations. We take as an example the rational $\frac{1}{12}$, for which we usually write

$$\frac{1}{12} = 0.08333\ldots.$$

What does this equality mean? Arguing as in the case of $\frac{1}{8}$, we are tempted to say that this equality is a shorthand for

$$\frac{1}{12} = \frac{8}{10^2} + \frac{3}{10^3} + \frac{3}{10^4} + \frac{3}{10^5} + \cdots. \tag{1.2}$$

This is actually so, provided we correctly interpret the sum with an infinite number of summands at the right hand side. Rigorously, (1.2) means that, if we fix the *a priori* maximum error $\frac{1}{10^n} = 0.\underbrace{00\ldots01}_{n}$, then we have

$$0 < \frac{1}{12} - \left(\frac{8}{10^2} + \frac{3}{10^3} + \frac{3}{10^4} + \cdots + \frac{3}{10^k} \right) \leq \frac{1}{10^n}$$

for every natural number $k \geq n$; yet in another way, (1.2) means that all of the numbers $\frac{8}{10^2} + \frac{3}{10^3} + \frac{3}{10^4} + \cdots + \frac{3}{10^k}$, with $k \geq n$, **approximate** $\frac{1}{12}$ **by defect** with error less than or equal to $0.\underbrace{00\ldots01}_{n}$. In fact, it will follow from Proposition 3.12 that

$$\frac{1}{12} - \left(\frac{8}{10^2} + \frac{3}{10^3} + \frac{3}{10^4} + \cdots + \frac{3}{10^k} \right) = \frac{1}{3 \cdot 10^k},$$

so that the error in the defect approximation $\frac{8}{10^2} + \frac{3}{10^3} + \frac{3}{10^4} + \cdots + \frac{3}{10^k}$ of $\frac{1}{12}$ equals $\frac{1}{3 \cdot 10^k}$, which is always less than or equal to the maximum error $\frac{1}{10^n}$, whenever $k \geq n$. It is precisely in this sense that we should think of the equality $\frac{1}{12} = 0.08333\ldots.$

In the light of the above discussion, one question suggests itself naturally: for an arbitrary *sequence*[3] $(a_1, a_2, a_3, \ldots)$ of decimal digits, can we think of

$$0.a_1 a_2 a_3 \ldots$$

as the decimal representation of some rational number?

It is possible to prove (and we will do so in Problem 2, page 7) that the answer to this question is *yes* if and only if the list $(a_1, a_2, a_3, \ldots)$ is *periodic* from a certain point on, i.e., if and only if it is of the form

$$(a_1, a_2, \ldots, a_l, \underbrace{b_1, b_2, \ldots, b_p}_{p}, \underbrace{b_1, b_2, \ldots, b_p}_{p}, \underbrace{b_1, b_2, \ldots, b_p}_{p}, \ldots). \qquad (1.3)$$

(In particular, for the rational number $\frac{1}{12}$ the list is $(0, 8, 3, 3, 3, \ldots)$, which is clearly periodic.) Therefore, if we are able to exhibit a list of digits that is not periodic from any point on, we will conclude that the general answer to the question posed above is *no*! We show an example of such a list now.

Example 1.1 The sequence of digits $(0, 1, 0, 1, 1, 0, 1, 1, 1, 0, \ldots)$, with infinitely many 0's and such that the quantity of digits 1's after each digit 0 equals the previous quantity of digits 1 plus one, is not periodic from any point on.

Proof By the sake of contradiction, suppose that the list in the statement is periodic from some point on, with, say, a block of p digits that repeats itself. The way we defined the sequence assures that, from some point on, each occurrence of a block of digits 1 would bring more than p digits 1. Therefore, the block of digits that repeats itself should be composed only by digits 1, so that the list should be, from some point on, equal to $(1, 1, 1, 1, 1, 1, 1, \ldots)$. However, if this was so, then we could not have an infinite number of digits 0, which is a contradiction. $\square$

In short, the previous discussion points to the following deficiency of the set of rationals: every rational number admits a decimal representation, but not every *decimal representation* represents a rational number. At this point, in order to fulfill this deficiency, we **postulate**[4] the existence of a set $\mathbb{R}$, containing $\mathbb{Q}$ and having the following properties:

---

[3]As we shall see in Section 6.1, a sequence of real numbers is just a function $f : \mathbb{N} \to \mathbb{R}$; however, for our purposes here, we can think of a sequence just as an *ordered list* of numbers, i.e., a list of numbers in which we specify which is the first number, which is the second, third, and so on. We will sistematically discuss some important elementary sequences of real numbers in Chap. 3.

[4]An **axiom** or **postulate** in a certain theory is a property imposed as true. One of the fundamental characteristics of Mathematics as a branch of human knowledge is the use of the **axiomatic method**, i.e., the acceptance of the fact that not every mathematical property can be logically deduced from previously established properties, being necessary the adoption of an adequate set of axioms.

(I) Addition, subtraction, multiplication and division in $\mathbb{Q}$ extend to similar operations on $\mathbb{R}$, in the sense that they have, on $\mathbb{R}$, properties similar to those on $\mathbb{Q}$.

(II) The order relation within $\mathbb{Q}$ extends to an order relation within $\mathbb{R}$, so that it has, in $\mathbb{R}$, the same properties its restriction has in $\mathbb{Q}$; in particular, every element of $\mathbb{R}$ is *negative*, zero or *positive*.

(III) In the sense of our previous discussions, to every sequence $(a_1, a_2, a_3, \ldots)$ of digits there corresponds a unique element $x \in [0, 1]$, which will be denoted by $x = 0.a_1 a_2 a_3 \ldots$. Conversely, to every $x \in [0, 1]$, there corresponds a (not necessarily unique) sequence $(a_1, a_2, a_3, \ldots)$ of digits, such that $x = 0.a_1 a_2 a_3 \ldots$.

The elements of $\mathbb{R}$ are called **real numbers**, and the set $\mathbb{R}$ as a whole is the **set of real numbers**[5].

In the rest of this section and in the next two section we discuss in detail each of the items (I), (II) and (III) above, showing what they really mean.

Beginning with (I), we postulate that $\mathbb{R}$ is furnished with two operations, respectively denoted $+$ and $\cdot$ and called (by analogy with the corresponding operations on $\mathbb{Q}$) *addition* and *multiplication*. Such operations satisfy axioms (1) to (7), quoted below:

(1) *Consistency*: for $a, b \in \mathbb{Q}$, the result $a + b$ of the addition of a and b is the same, whether we consider the usual addition on $\mathbb{Q}$ or the corresponding operation on $\mathbb{R}$. Analogously, the result $a \cdot b$ of the multiplication of a and b is the same, whether we consider the usual multiplication on $\mathbb{Q}$ or the corresponding operation on $\mathbb{R}$.

(2) *Commutativity*: the operations $+$ and $\cdot$ are commutative, i.e., they are such that $a + b = b + a$ and $a \cdot b = b \cdot a$, for all $a, b \in \mathbb{R}$.

(3) *Associativity*: the operations $+$ and $\cdot$ are associative, i.e., they are such that $a + (b + c) = (a + b) + c$ and $a \cdot (b \cdot c) = (a \cdot b) \cdot c$, for all $a, b, c \in \mathbb{R}$.

(4) *Distributivity*: Multiplication is distributive with respect to addition, i.e., it is such that $a \cdot (b + c) = (a \cdot b) + (a \cdot c)$, for all $a, b, c \in \mathbb{R}$.

(5) *The roles of 0 and 1 in $\mathbb{R}$*: the rational numbers 0 and 1 are such that $0 + a = a$ and $1 \cdot a = a$, for every $a \in \mathbb{R}$.

(6) *Law of cancellation*: if $a, b \in \mathbb{R}$ are such that $a \cdot b = 0$, then $a = 0$ or $b = 0$.

(7) *Existence of inverses*: for every $a \in \mathbb{R}$, there exists $b \in \mathbb{R}$ such that $a + b = 0$. For every $a \in \mathbb{R} \setminus \{0\}$, there exists $b \in \mathbb{R}$ such that $a \cdot b = 1$.

The properties above have the following important consequences:

(i) Uniqueness of the additive inverse: for a given $a \in \mathbb{R}$, if $b, b' \in \mathbb{R}$ are such that $a + b = 0$ and $a + b' = 0$, the associativity and commutativity of addition

---

[5]There are more *construtive* (and rigorous) ways of introducing the real numbers, as the reader can find in [6], for example. However, in these notes we chose to follow an approach that was as close as possible to the previous experience of the medium reader, sacrificing rigor in the name of understanding.

give us

$$b = b + 0 = b + (a + b')$$
$$= (b + a) + b' = (a + b) + b'$$
$$= 0 + b' = b'.$$

Hence, the real number a has a unique additive inverse, which from now on will be denoted by $-a$, as is usually done within $\mathbb{Q}$. It follows from $a + (-a) = 0$ that a is the additive inverse of $-a$; therefore, according to the notation just established for additive inverses, we have $-(-a) = a$.

(ii) Uniqueness of the multiplicative inverse: for a given $a \in \mathbb{R} \setminus \{0\}$, if $b, b' \in \mathbb{R}$ are such that $a \cdot b = 1$ and $a \cdot b' = 1$, then $b = b'$. The proof of this fact is completely analogous to that of the previous item (cf. Problem 3, page 7). From now on, we will denote the multiplicative inverse of $a \in \mathbb{R} \setminus \{0\}$ by a^{-1}, as is also usually done within $\mathbb{Q}$.

(iii) For $a \in \mathbb{R}$, we have $a \cdot 0 = 0$: in order to check this, let $a \cdot 0 = e$. The distributivity of multiplication with respect to addition gives

$$e = a \cdot 0 = a \cdot (0 + 0) = a \cdot 0 + a \cdot 0 = e + e.$$

Hence,

$$e = e + 0 = e + (e + (-e))$$
$$= (e + e) + (-e)$$
$$= e + (-e) = 0,$$

so that $a \cdot 0 = e = 0$.

In view of the above properties, we adopt the convention (as is usual within $\mathbb{Q}$) to omit the sign $\cdot$ of multiplication, thus writing simply ab instead of $a \cdot b$. Now, observe that the associativity and commutativity of addition and multiplication in $\mathbb{R}$ allow us to add or multiply an arbitrary finite number of real numbers without worrying with which summands or factors should be initially operated; the final result will always be the same[6].

We also *define* the operations of *subtraction* $(-)$ and *division* $(\div)$ in $\mathbb{R}$ as is usually done in $\mathbb{Q}$: for $a, b \in \mathbb{R}$, we set

$$a - b = a + (-b) \quad \text{and} \quad a \div b = ab^{-1},$$

---

[6]Rigorously speaking, the validity of such a statement should be proved as a theorem, what can be done with the aid of the principle of mathematical induction (cf. Chap. 5). However, in order to have a less terse reading, we chose not to give a formal proof of this fact, just relying on the previous experience of the reader.

with $b \neq 0$ in the last case; yet in the case of division, and whenever there is no danger of confusion, we write a/b or $\frac{a}{b}$ as synonymous for $a \div b$.

Problems: Section 1.1

1. * Establish the following *properties of proportions*: if a, b, c and d are nonzero reals, such that $\frac{a}{b} = \frac{c}{d}$, then

$$\frac{a}{b} = \frac{c}{d} = \frac{a \pm c}{b \pm d}.$$

2. * Given a sequence $(a_1, a_2, a_3, \ldots)$ of digits, prove that the real number $0.a_1a_2a_3 \ldots$ represents the decimal expansion of a rational if and only if the sequence $(a_1, a_2, a_3, \ldots)$ is periodic from some point on, in the sense of (1.3).
3. * Prove the uniqueness of multiplicative inverses in ℝ. More precisely, prove that if $a \neq 0$ is a real number and $b, b' \in ℝ$ are such that $a \cdot b = a \cdot b' = 1$, then $b = b'$.
4. Prove, from the axioms for addition and multiplication of reals, that $-a = (-1)a$, for every $a \in ℝ$.
5. In each of the following items, decide whether the real number in question is rational or not. In doing so, assume that the pattern of digits suggested up to the dots is actually followed. Moreover, in case the number is rational, write it as an irreducible fraction:

   (a) $2.324444\ldots$
   (b) $0.12121212\ldots$
   (c) $2.1345454545\ldots$
   (d) $0.12345678910111213 14\ldots$

1.2 The Order Relation in ℝ

We postulate the existence of an **order relation** on ℝ, which amounts to a *way of comparing* real numbers. By the sake of analogy with the corresponding order relation on ℚ, we denote this order relation by $\geq$, and also read it as *greater than or equal to*. The order relation on ℝ satisfies axioms (1') to (5') below:

(1') *Consistency*: for $a, b \in ℚ$ such that $a \geq b$ in ℚ, we have $a \geq b$ in ℝ.
(2') *Reflexivity*: $a \geq a$, for every $a \in ℝ$.
(3') *Antisymmetry*: if $a, b \in ℝ$ are such that $a \geq b$ and $b \geq a$, then $a = b$.
(4') *Transitivity*: if $a, b, c \in ℝ$ are such that $a \geq b$ and $b \geq c$, then $a \geq c$.
(5') *Dichotomy*: for all $a, b \in ℝ$, one has either $a \geq b$ or $b \geq a$.

In all that follows, if $a, b \in \mathbb{R}$ are such that $a \geq b$ and $a \neq b$, we write $a > b$ and read *a is greater than b*. We also write $a \leq b$ (read *a is less than or equal to b*) as a synonymous for $b \geq a$, and $a < b$ (read *a is less than b*) as a synonymous for $b > a$. If $a \in \mathbb{R}$ is such that $a > 0$, we say that a is **positive**; if $a < 0$, we say that a is **negative**.

With respect to the order relation $\geq$ on $\mathbb{R}$, we also impose axioms (6') and (7') below, which guarantee – as they do in $\mathbb{Q}$ – its *compatibility* with the operations of addition and multiplication:

(6') $a > b \Leftrightarrow a - b > 0$.
(7') $a, b > 0 \Rightarrow a + b, ab > 0$.

The next result collects some other useful properties of the order relation on $\mathbb{R}$, which can be deduced from axioms (1') to (7'). From now on, we say that two nonzero real numbers *have equal signs* if they are both positive or both negative.

Proposition 1.2 *Let $a, b, c, d \in \mathbb{R}$.*

(a) *If $a > 0$, then $-a < 0$, and vice-versa.*
(b) *If $a > 0$, then* $\begin{cases} b > 0 \Rightarrow ab > 0 \\ b < 0 \Rightarrow ab < 0 \end{cases}$.
(c) *If $a < 0$, then* $\begin{cases} b > 0 \Rightarrow ab < 0 \\ b < 0 \Rightarrow ab > 0 \end{cases}$.
(d) $a > b \Rightarrow a + c > b + c$.
(e) $a > b, c \geq d \Rightarrow a + c > b + d$.
(f) *If $a > b$, then* $\begin{cases} c > 0 \Rightarrow ac > bc \\ c < 0 \Rightarrow ac < bc \end{cases}$.
(g) $a \neq 0 \Rightarrow a^2 > 0$.
(h) $a > 0 \Leftrightarrow \frac{1}{a} > 0$.
(i) *If a and b have equal signs and $a > b$, then $\frac{1}{a} < \frac{1}{b}$.*

Proof We prove just a few of the above items, letting the others as exercises (see Problem 1, page 10).

(a) This follows from axiom (6'):

$$0 > -a \Leftrightarrow 0 - (-a) > 0 \Leftrightarrow a > 0.$$

(d) Again by axiom (6'), we have

$$a > b \Rightarrow a - b > 0 \Rightarrow (a + c) - (b + c) > 0 \Rightarrow a + c > b + c.$$

(e) We use (d) and the transitivity of $\geq$:

$$\left. \begin{array}{l} a > b \Rightarrow a + c > b + c \\ c \geq d \Rightarrow b + c \geq b + d \end{array} \right\} \Rightarrow a + c > b + d.$$

(f) Suppose $c > 0$ (the case $c < 0$ can be treated analogously). It follows from (6')
and (7') that

$$a > b \Rightarrow a - b > 0 \Rightarrow (a - b)c > 0 \Rightarrow ac - bc > 0 \Rightarrow ac > bc.$$

(h) Suppose $a > 0$. If we had $\frac{1}{a} < 0$, it would follow from (b) that $1 = a \cdot \frac{1}{a} < 0$,
which is an absurd (for the order relation on $\mathbb{R}$ extends that on $\mathbb{Q}$).

(i) Since a and b have equal signs, it follows from (b) and (c) that $ab > 0$. Thus,
item (h) gives $\frac{1}{ab} > 0$. Since $b - a < 0$, we get from either (b) or (c) that

$$\frac{1}{a} - \frac{1}{b} = \frac{b - a}{ab} = (b - a) \cdot \frac{1}{ab} < 0.$$

Finally, (6') assures that this is equivalent to $\frac{1}{a} < \frac{1}{b}$.

$\square$

Items (b) and (c) of the former proposition are known as the **sign rules** for
multiplication of real numbers.

Next, let $r \in \mathbb{R}$ be given. The **square** of r, denoted r^2, is the real number $r^2 = r \cdot r$;
the **cube** of r, denoted r^3, is the real number $r^3 = r \cdot r \cdot r$. More generally, for a given
$n \in \mathbb{N}$, we define the **n–th power** of r, denoted r^n, as being r, if $n = 1$, or the real
number obtained by multiplying r by itself n times, if $n > 1$:

$$r^n = \underbrace{r \cdot r \cdot \cdots \cdot r}_{n}.$$

Once more we call the attention of the reader to the fact that the associativity of the
multiplication of reals, together with the principle of mathematical induction (cf.
Chap. 5) allows us to prove that the result of the right hand side of the equality above
does not depend on the order in which we multiply the n copies of r. Therefore, r^n
is a well defined real number. Problem 3, page 10, lists some useful properties of
the powers of real numbers.

We collect below an important consequence of the properties of the order relation
on $\mathbb{R}$, listed in the last proposition.

Corollary 1.3 *Let r be a positive real number and m and n be natural numbers,
with $m > n$. Then:*

(a) $0 < r < 1 \Rightarrow r^m < r^n$.
(b) $r > 1 \Rightarrow r^m > r^n$.

Proof

(a) Since r is positive, multiplying both sides of the inequality $r < 1$ by r, we get
$r^2 < r$. Multipliying both sides of this last inequality by r once more, it follows
that $r^3 < r^2$ and, hence, that $r^3 < r^2 < r$. Proceeding this way, we arrive at the
desired result, i.e.,

$$\cdots < r^4 < r^3 < r^2 < r.$$

(b) The proof of this item is essentially equal to that of (a), with the only difference that, in this case, we have $r > 1$.

$\square$

We next illustrate, in an example, how the above corollary can be useful in comparing certain real numbers.

Example 1.4 In order to compare $2^{100} + 3^{100}$ and 4^{100}, for instance, it suffices to see that

$$2^{100} + 3^{100} < 3^{100} + 3^{100} = 2 \cdot 3^{100}$$
$$= 2 \cdot 3^3 \cdot 3^{97} = 54 \cdot 3^{97}$$
$$< 64 \cdot 4^{97} = 4^3 \cdot 4^{97} = 4^{100}.$$

In Chapter 5, we will sistematically study some important inequalities involving real numbers. For the time being, the following corollary – in spite of its simplicity – will play an important role.

Corollary 1.5 *For $a, b \in \mathbb{R}$, we have*

$$a^2 + b^2 \geq 0, \tag{1.4}$$

with equality if and only if $a = b = 0$.

Proof Item (g) of Proposition 1.2 gives $a^2, b^2 \geq 0$. Therefore, item (d) of that proposition gives $a^2 + b^2 \geq 0$. Now, let $a \neq 0$. Then, it follows from item (g) of the above mentioned proposition that $a^2 > 0$. On the other hand, since we still have $b^2 \geq 0$, item (e) of that proposition guarantees that $a^2 + b^2 > 0$. $\square$

Problems: Section 1.2

1. * Prove items (b), (c) and (g) of Proposition 1.2.
2. * Generalize Corollary 1.5, showing that, if $a, b, c \in \mathbb{R}$, then

$$a^2 + b^2 + c^2 \geq 0,$$

with equality if and only if $a = b = c = 0$.
3. * Given $r, s \in \mathbb{R}$ and $m, n \in \mathbb{N}$, prove that[7]:

(a) $(rs)^n = r^n s^n$.

---

[7]In order to prove properties (a) to (d) in a rigorous way, we have to rely on the principle of mathematical induction (cf. Chap. 5). Thus, our intention here is simply to make the reader give some arguments on their plausibility.

(b) $r^{m+n} = r^m r^n$.

(c) $(r^m)^n = r^{mn}$.

(d) $\left(\frac{r}{s}\right)^n = \frac{r^n}{s^n}$, if $s \neq 0$.

4. * For $r \in \mathbb{R} \setminus \{0\}$ and $n \in \mathbb{N}$, we extend the notion of powers with natural exponents to powers with integer exponents by defining $r^{-n} = \frac{1}{r^n}$. For example, $r^{-1} = \frac{1}{r}$, $r^{-2} = \frac{1}{r^2}$ etc. If we also set $r^0 = 1$, prove that, for all $m, n \in \mathbb{N}$, one has $\frac{r^m}{r^n} = r^{m-n}$.

5. Let a and b be nonzero integers, with $b > 1$. If the only prime divisors of b are 2 or 5, prove that the decimal representation of $\frac{a}{b}$ is finite.

6. * If $x \neq 0$ is a real number and $n \in \mathbb{N}$, prove that x^n is positive if n is even, and has the same sign of x if n is odd.

7. Prove that $\frac{1}{2} - \frac{1}{3} + \frac{1}{4} - \frac{1}{5} + \cdots - \frac{1}{99} + \frac{1}{100} > \frac{1}{5}$.

8. Given positive real numbers a and b such that $a < b$, compare (i.e., decide which is the greatest of) the numbers $\frac{a+1}{b+1}$ and $\frac{a+2}{b+2}$.

9. (TT) We are given ten real numbers such that the sum of any four of them is positive. Show that the sum of all ten numbers is also positive.

10. Decide which of the numbers 31^{11} or 17^{14} is the greatest one.

11. * Let $n \in \mathbb{N}$ and a, b be positive reals. Prove that:

(a) $a < b$ if and only if $a^2 < b^2$.

(b) $a < b$ if and only if $a^n < b^n$.

(c) $a^n + b^n < (a+b)^n$.

12. (EKMC - adapted) Let a, b and c be the lenghts of the sides of a right triangle, c being the hypotenuse. Which one is the greatest: $a^3 + b^3$ or c^3? Justify your answer.

13. Show that, for every $n \in \mathbb{N}$, we have $1^{2n} + 2^{2n} + 3^{2n} \geq 2 \cdot 7^n$.

14. * Find all natural numbers a, b and c such that $a \leq b \leq c$ and $\frac{1}{a} + \frac{1}{b} + \frac{1}{c}$ is an integer.

15. (IMO) Explain how to write 100 as a sum of naturals whose product is as large as possible.

16. (Russia)[8] The leftmost digit of the decimal representations of the natural numbers 2^n and 5^n is the same. Prove that such digit is equal to 3.

17. (Russia) Let a, b, c and d be positive real numbers. Show that, among the inequalities

$$a + b < c + d, \quad (a+b)(c+d) < ab + cd \text{ and } (a+b)cd < ab(c+d),$$

at least one is false.

---

[8]For a converse to this problem, see Example 10.60.

1.3   Completeness of the Real Number System

For the time being, we postulate that to every sequence $(a_1, a_2, a_3, \ldots)$ of digits, there corresponds a unique $x \in \mathbb{R}$, in the following sense: for a fixed maximum error $\frac{1}{10^n}$, with $n \in \mathbb{N}$, we have

$$0 \leq x - \left(\frac{a_1}{10} + \frac{a_2}{10^2} + \cdots + \frac{a_k}{10^k} \right) \leq \frac{1}{10^n},$$

for every natural $k \geq n$. In particular, it follows from this inequality that

$$x \leq \frac{a_1}{10} + \frac{a_2}{10^2} + \cdots + \frac{a_k}{10^k} + \frac{1}{10^n},$$

for every $k \geq n$. Taking $k = n$ and recalling that $a_j \leq 9$ for every j, we get

$$
\begin{aligned}
x &\leq \frac{a_1}{10} + \frac{a_2}{10^2} + \cdots + \frac{a_{n-1}}{10^{n-1}} + \frac{a_n}{10^n} + \frac{1}{10^n} \\
&\leq \frac{9}{10} + \frac{9}{10^2} + \cdots + \frac{9}{10^{n-1}} + \frac{9}{10^n} + \frac{1}{10^n} \\
&= \frac{9}{10} + \frac{9}{10^2} + \cdots + \frac{9}{10^{n-1}} + \frac{1}{10^{n-1}} \\
&= \frac{9}{10} + \frac{9}{10^2} + \cdots + \frac{1}{10^{n-2}} \\
&= \cdots = 1,
\end{aligned}
$$

Thus, $0 \leq x \leq 1$.

As a shorthand to the above postulate, we say that the set $\mathbb{R}$ of real numbers is **complete**. In this respect, we refer the reader to Section 7.1, as well as to Problem 9, page 242, where the completeness of the real number system will be more rigorously discussed.

Conversely, we also postulate that to every positive real number x there corresponds a nonnegative integer m and a sequence $(a_1, a_2, a_3, \ldots)$ of digits, such that $x = m + 0.a_1 a_2 a_3 \ldots$, in the sense of the previous paragraph. If $m > 0$ and $m = b_n \ldots b_1 b_0$, with the b_i's being their digits, we write

$$x = b_n \ldots b_1 b_0 . a_1 a_2 a_3 \ldots$$

and say that $b_n \ldots b_1 b_0 . a_1 a_2 a_3 \ldots$ is the **decimal representation** or **expansion** of x.

As was seen in Problem 2, page 7, a real number is rational exactly when its decimal representation is finite or infinite and periodic. On the other hand, a real number which is not rational is said to be **irrational**. Thus, irrational numbers are those real numbers which cannot be written as quotients of two integers, or, in another way, those reals whose decimal representations are **infinite and nonperiodic**.

Up to the present, the only irrational number we have met was the number $0,0101101110\ldots$ (cf. Example 1.1). In a sense, such an example is pretty dissatisfying, for it is difficult to manipulate this number (i.e., it is difficult to *make calculations* with it). We will remedy this situation in what follows.

From an arithmetic point of view, one great advantage of the set of reals, in comparison to the set of rationals, is the possibility of making *root extractions* of positive real numbers. More precisely, given a real number $x > 0$ and a natural number n, it is possible to prove (and we will do so, in two different ways, in sections 7.1 and 8.3) that there exists a unique positive real number y such that $y^n = x$. From now on, we shall call this real number y the **n−th root** of x, and write $y = \sqrt[n]{x}$; the natural n is the **index** of the root. In short,

$$y = \sqrt[n]{x} \Leftrightarrow x = y^n.$$

Indices $n = 2$ and $n = 3$ occur so frequently that deserve special names and notations. When $n = 2$ (and $x > 0$), we write simply $\sqrt{x}$, instead of $\sqrt[2]{x}$, and say that $\sqrt{x}$ is the **square root** of x; when $n = 3$ (and $x > 0$), we say that $\sqrt[3]{x}$ is the **cubic root** of x.

The argument in the next example allows us to heuristically understand why roots of positive reals do exist.

Example 1.6 By definition, we have $\sqrt{2}^2 = 2$. Since $1^2 < 2 < 2^2$, the result of Problem 11, page 11, gives $1 < \sqrt{2} < 2$. Now, since $1.4^2 < 2 < 1.5^2$, it follows once again from that problem that $1.4 < \sqrt{2} < 1.5$. Analogously, since $1.41^2 < 2 < 1.42^2$, we have $1.41 < \sqrt{2} < 1.42$. Continuing this way, we obtain a unique list $(a_1 = 4, a_2 = 1, a_3 = 4, \ldots)$ of digits, such that $1.a_1 a_2 \ldots a_{n-1} a_n < \sqrt{2} < 1.a_1 a_2 \ldots a_{n-1} a'_n$ for all $n \geq 1$, where $a'_n = a_n + 1$. Thus, we have no option but to conclude that $\sqrt{2} = 1.a_1 a_2 a_3 \ldots = 1.414\ldots$.

At this point, we urge the reader to at least read the statements of Problems 1 and 2, page 14, to get an idea on how to (partially) extend the concept of n−th root to negative reals, as well as to take a look at the main properties of root extraction.

We now turn to powers of natural numbers. A **perfect square** is a natural number which can be written in the form m^2, for some $m \in \mathbb{N}$; hence, the perfect squares are the natural numbers $1^2 = 1, 2^2 = 4, 3^2 = 9, 4^2 = 16$ etc. A **perfect cube** is a natural number which can be written in the form m^3, for some $m \in \mathbb{N}$; the perfect cubes are, then, the naturals $1^3 = 1, 2^3 = 4, 3^3 = 27, 4^3 = 64$ etc. More generally, a natural n is a **perfect power** if there exist natural numbers n and k, with $k > 1$, such that $n = m^k$. If this is the case, we say that n is a **k-th perfect power**, i.e., is equal to one of the natural numbers $1^k, 2^k, 3^k, 4^k$ etc. Equivalently, to say that $n \in \mathbb{N}$ is a k−th perfect power is the same as to say that its k−th root, $\sqrt[k]{n}$, is a natural number.

The following proposition, which by now we take for granted, gives an infinite supply of examples of irrational numbers, which, as we will see in a moment, are more or less easy to handle. For a proof of it, we refer the reader to Chap. 6 of [5]. (Nevertheless, see Problems 7 and 8.)

Proposition 1.7 *Given natural numbers n and k, with k > 1, either n is a k-th perfect power or $\sqrt[k]{n}$ is an irrational number.*

According to the above proposition, numbers like $\sqrt{2}$, $\sqrt[3]{3}$, $\sqrt[5]{10}$ etc are all examples of irrationals (for 2 is not a perfect square, 3 is not a perfect cube and 10 is not a perfect 5–th power).

Now we can, at least *formally* (i.e., without worrying with *approximations*), operate with several irrational numbers. Let us see an example where we (indirectly) apply the last proposition to explain why a certain real number is irrational.

Example 1.8 The number $\sqrt{2} + \sqrt{3}$ is irrational. In fact, if we let $r = \sqrt{2} + \sqrt{3}$, there are two possibilities: $r \in \mathbb{Q}$ or $r \notin \mathbb{Q}$. Suppose, for the sake of contradiction, that $r \in \mathbb{Q}$. Then, since the set of rationals is closed with respect to multiplication, we would have $r^2 \in \mathbb{Q}$. On the other hand, the distributivity of multiplication with respect to addition gives

$$r^2 = (\sqrt{2} + \sqrt{3})(\sqrt{2} + \sqrt{3})$$
$$= \sqrt{2}(\sqrt{2} + \sqrt{3}) + \sqrt{3}(\sqrt{2} + \sqrt{3})$$
$$= (2 + \sqrt{6}) + (\sqrt{6} + 3)$$
$$= 5 + 2\sqrt{6}$$

so that $\sqrt{6} = \frac{r^2 - 5}{2}$. Therefore, $\sqrt{6}$ would be the quotient of the rational numbers $r^2 - 5$ and 2 and, as such, $\sqrt{6}$ would be itself a rational number. This contradicts the result of Proposition 1.7, so that $r \notin \mathbb{Q}$.

In order to finish our discussion on rational and irrational numbers, note that the set of irrationals is not *closed* with respect to the ordinary arithmetic operations. For instance, for a given irrational number r, although $-r$ is also irrational, we have $r + (-r) = 0$, which is a rational. On the other hand, if we set $r = \sqrt{2}$, we get $r \cdot r = r^2 = 2$, which is also a rational. Finally, if $r \neq 0$, then the quotient of r by itself is equal to 1, again a rational number.

Problems: Section 1.3

1. * Given $x < 0$ real and $n \in \mathbb{N}$ odd, let $y = -\sqrt[n]{-x}$. Prove that $y^n = x$ (so that the real number y is also called the **n–th root** of x).
2. * Given $m, n \in \mathbb{N}$ and $x, y > 0$, prove that:
   (a) $\sqrt[n]{xy} = \sqrt[n]{x}\sqrt[n]{y}$.
   (b) $\sqrt[mn]{x} = \sqrt[m]{\sqrt[n]{x}}$.
   (c) $\sqrt[n]{\frac{x}{y}} = \frac{\sqrt[n]{x}}{\sqrt[n]{y}}$.

Then, extend the above properties to all nonzero reals x and y, provided m and n are odd naturals.

3. * Let a and b be rational numbers, and let r be an irrational one. If $a + br = 0$, prove that $a = b = 0$.

4. Let a, b, c and d be rational numbers, and let r be an irrational one. If $a + br = c + dr$, prove that $a = c$ and $b = d$.

5. Let r be a positive real and let k be an integer greater than 1. If r is irrational, prove that the numbers $\frac{1}{r}$ and $\sqrt[k]{r}$ are also irrational.

6. (Canada) Let a, b and c be rational numbers, such that $a + b\sqrt{2} + c\sqrt{3} = 0$. Prove that a, b and c are all equal to zero.

7. * Assuming the validity of the Fundamental Theorem of Arithmetic (according to the last paragraph of the introduction to this chapter), prove that $\sqrt{2}$ is an irrational number.

8. Let p be a prime number, and $k > 1$ be natural. Prove that $\sqrt[k]{p}$ is irrational.

1.4 The Geometric Representation

A quite useful way of thinking *geometrically* on the set of rational numbers is the following: we take a line r, and mark on it a point O; then, we choose one of the half-lines that O determines on r, call it **positive** (the other being called **negative**) and set a line segment ℓ as unit of measure. Then, we associate to each rational number a point of r, in the following way: first, we associate 0 to the point O; then (according to Figure 1.1), given a rational number $\frac{a}{b}$, with $a, b \in \mathbb{N}$, we mark, starting from O and on the positive half-line, a line segment OA of length $a\ell$ (i.e., OA is obtained by juxtaposition of a segments equal to ℓ). If $b = 1$, we associate $\frac{a}{1} = a$ to the point A. If $b > 1$, we divide OA into b equal line segments, by plotting $b - 1$ points on OA; from these $b - 1$ points, we let B denote the one closest to O, and associate $\frac{a}{b}$ to the point B. It is not difficult to show (cf. Problem 1) that this geometric construction is consistent, in the sense that, by changing $\frac{a}{b}$ for another equivalent fraction, we get the same point B on r. Moreover, an analogous construction can obviously be made for the negative rationals, marked on the negative half-line.

By proceeding as described in the previous paragraph, it happens that there are lots of points on r which are not associated to any rational number. In order to give a simple example, let A be the point associated to the number 1, and construct a square

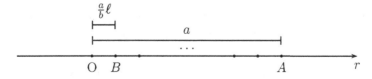

Fig. 1.1 plotting rationals on a real line.

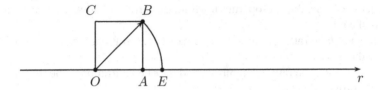

Fig. 1.2 a point that doesn't represent any rational number.

OABC, as shown in Figure 1.2. By using a compass, mark on the positive half-line a point *E* such that $\overline{OE} = \overline{OB}$. Since $\overline{OA} = 1$, it follows from Pythagoras[9] Theorem[10] that $\overline{OE} = \overline{OB} = \sqrt{2}$. However, since $\sqrt{2}$ is irrational (cf. Proposition 1.7 or Problem 7, page 15), we conclude that *E* is not associated to any rational number.

At this point, two natural questions pose themselves: is it possible to mark on *r* all of the real numbers? If we assume that the answer to the previous question is *yes*, then, after we mark all reals on *r*, will there be unmarked points on *r*? One of the axioms of Plane Euclidean Geometry[11], stated below, assures that the answers to these two questions are, respectively, *yes* and *no*.

Axiom 1.9 *There is a one-to-one correspondence between the points on an Euclidean line r onto the set of real numbers, which is completely determined by the following choices:*

(a) A point O on r, to represent the real number 0.
(b) A half-line, among those that O determines on r, where we mark the positive reals.
(c) A point on the half-line of item (b), to represent the number 1.

If we fix choices on a line *r* as specified by the above axiom, we say that *r* is the **real line** (cf. Figure 1.3).

For further use, we need the following definition.

Definition 1.10 For given real numbers $a < b$, we set[12]:

(i) $[a, b] = \{x \in \mathbb{R};\ a \leq x \leq b\}$.

---

[9]Pythagoras of Samos was one of the greatest mathematicians of classical antiquity. The theorem that bears his name was already known to babylonians, at least two thousand years before he was born; nevertheless, Pythagoras was the first one to prove it. It is also attributed to him the first proof of the irrationality of $\sqrt{2}$.

[10]We recall that Pythagoras' Theorem, one of the most celebrated (and important) results of Plane Euclidean Geometry, says that, in every right triangle, the square of the length of the hypotenuse equals the sum of the squares of the lengths of the legs. For two different proofs of it, see chapters 4 and 5 of [4].

[11]For an axiomatic construction of plane Euclidean Geometry, we refer the reader to [16].

[12]We call the reader's attention to the less common notations $[a, b[$ instead of $[a, b)$, $]a, b]$ instead of $(a, b]$, $]a, b[$ instead of (a, b), $[a, +\infty[$ instead of $[a, +\infty)$ and $]-\infty, a[$ instead of $(-\infty, a)$.

(ii) $[a, b) = \{x \in \mathbb{R}; a \le x < b\}$.
(iii) $(a, b] = \{x \in \mathbb{R}; a < x \le b\}$.
(iv) $(a, b) = \{x \in \mathbb{R}; a < x < b\}$.
(v) $[a, +\infty) = \{x \in \mathbb{R}; a \le x\}$.
(vi) $(a, +\infty) = \{x \in \mathbb{R}; a < x\}$.
(vii) $(-\infty, a] = \{x \in \mathbb{R}; x \le a\}$.
(viii) $(-\infty, a) = \{x \in \mathbb{R}; x < a\}$.

An **interval** in $\mathbb{R}$ is the set $\mathbb{R}$ itself, or a subset of $\mathbb{R}$ of any of the seven types above. Observe that, in the real line, an interval corresponds to a line segment (perhaps with the exclusion of one or both of its end points), to a half-line (perhaps with the exclusion of its real end point), or even to the whole real line.

Remarks 1.11

i. We stress that the symbols $+\infty$ and $-\infty$ (one respectively reads *plus infinite* and *minus infinite*) *do not represent* real numbers. They merely serve to point out that, in each of the items (v), (vi), (vii), (viii) above, the corresponding intervals do contain all reals greater than or equal to (resp. greater than), or less than or equal to (resp. less than) a.

ii. According to the previous definition, we shall denote $\mathbb{R} = (-\infty, +\infty)$.

Given real numbers $a < b$, we say that a and b are the **endpoints** and that $b - a$ is the **length** of each one of the intervals of items (i) to (iv), in the previous definition. In this case, we also say that those intervals have *finite* length. Analogously, the real number a is the (only) endpoint of each of the intervals of items (v) to (viii), which have *infinite* (i.e., not finite) lengths. An interval in $\mathbb{R}$ is **bounded** provided it has finite length; otherwise, the interval is said to be **unbounded**. In particular, the bounded intervals of $\mathbb{R}$ are precisely those of items (i) to (iv), in the previous definition.

If a bounded interval has endpoints $a < b$, we shall say that it is **closed, closed on the left, closed on the right** or **open** when such an interval is respectively equal to $[a, b]$, $[a, b)$, $(a, b]$ or (a, b). Alternatively, we say that $[a, b)$ is **open on the right** and $(a, b]$ is **open on the left**. Finally, infinite intervals are named accordingly, by using obvious extensions of the above terminologies. Figure 1.3 shows the interval $[a, b)$,

Fig. 1.3 the open on the right interval $[a, b)$.

Problems: Section 1.4

1. * With respect to the geometric interpretation of rational numbers, discussed at section 1.4, let the fractions $\frac{a}{b}$ and $\frac{c}{d}$ be given, where $a, b, c, d \in \mathbb{N}$. If $\frac{a}{b} = \frac{c}{d}$, explain why the construction given in the text associates both these fractions to the same point of the real line.

Chapter 2
Algebraic Identities, Equations and Systems

The rest of this volume, up to Chap. 5, approaches and develops several tools necessary for an adequate presentation of the material in volumes 2 and 3. We start by studying, in this chapter, some important algebraic identities, equations and systems of equations.

2.1 Algebraic Identities

Through the rest of these notes, we refer to a varying real number as a **real variable**.[1] In general, real variables will be denoted by lower case Latin letters, for example a, b, c, x, y, z etc (an important exception to this usage is mentioned in the next paragraph).

An **algebraic expression**, or simply an *expression*, is a real number formed from a finite number of real variables, possibly with the aid of one or more **algebraic operations**, i.e., additions, subtractions, multiplications, divisions, power computations and root extractions (whenever the results of these operations make sense in $\mathbb{R}$). In particular, every real variable can be seen as an algebraic expression. For another example,

$$\frac{x + \sqrt{y} - x^2 z}{yz} + 3\sqrt[5]{x^2 y z^3 - x^4}$$

is an algebraic expression which makes sense for all reals x, y, z, such that $y > 0$ and $z \neq 0$ (recall that Problem 1, page 14, assures that we can extract roots of odd index of any real number). We shall denote algebraic expressions by upper case latin letters, as E, F etc.

---

[1]In [5], we shall have the opportunity to consider *complex* variables.

© Springer International Publishing AG 2017
A. Caminha Muniz Neto, *An Excursion through Elementary Mathematics, Volume I*,
Problem Books in Mathematics, DOI 10.1007/978-3-319-53871-6_2

We say that an algebraic expression E is a **monomial** if E is a product of a given nonzero real number by powers of its variables, each of which having nonnegative integer exponents. Thus, the monomials in the real variables x and y are the expressions of the form $ax^k y^l$, where $a \neq 0$ is a given real number and $k, l \geq 0$ are nonnegative integers (here, we adopt the convention that $x^k = 1$ whenever $k = 0$, and $y^l = 1$ whenever $l = 0$—see Problem 4, page 11). For an arbitrary monomial, the given nonzero real number that plays the role of a in $ax^k y^l$ is called its **coefficient**. Hence, the monomials in x, y with coefficient 2 are those of one of the forms

$$2, \ 2x \ 2y, \ 2x^2, \ 2xy, \ 2y^2, \ 2x^3, \ 2x^2 y, \ 2xy^2, \ 2y^3 \ \text{etc.}$$

A **polynomial expression** or simply a **polynomial** is (an expression that is) a finite sum of monomials, as, for instance,

$$2 + 3xy - \sqrt{5}x^2 yz.$$

The **coefficients** of a polynomial are the coefficients of its monomials.

Let E and F be algebraic expressions. We say that equality $E = F$ is an **algebraic identity** provided it is true for all possible values of the involved real variables. In order to give a relevant example, let us consider the algebraic expression $E = (x + y)^2$. The elementary properties of the operations of addition and multiplication of real numbers (i.e., commutativity and associativity of addition and multiplication, as well as distributivity of multiplication with respect to addition) give

$$\begin{aligned} E = (x + y)(x + y) &= x(x + y) + y(x + y) \\ &= (x^2 + xy) + (yx + y^2) \\ &= x^2 + 2xy + y^2, \end{aligned}$$

for all values of the real variables x and y. Therefore, by setting $F = x^2 + 2xy + y^2$, we obtain the algebraic identity $E = F$, i.e.,

$$(x + y)^2 = x^2 + 2xy + y^2, \tag{2.1}$$

to which we refer, from now on, as the *formula* for the square of a sum of two real numbers.

The following proposition collects some important algebraic identities, which the reader must keep for future use.

Proposition 2.1 *For all* $x, y, z \in \mathbb{R}$, *we have:*

(a) $x^2 - y^2 = (x - y)(x + y)$.
(b) $(x \pm y)^2 = x^2 \pm 2xy + y^2$.
(c) $x^3 \pm y^3 = (x \pm y)(x^2 \mp xy + y^2)$.

(d) $(x \pm y)^3 = x^3 \pm y^3 \pm 3xy(x \pm y)$.

(e) $(x + y + z)^2 = x^2 + y^2 + z^2 + 2xy + 2xz + 2yz$.

Proof We let the proofs of items (a)–(c) as exercises (see Problem 1), observing that the identity of item (b), with the $+$ sign, was established in (2.1). In item (d), let us prove the identity for $(x + y)^3$; that for $(x - y)^3$ is totally analogous: by invoking the distributivity of the multiplication with respect to addition, as well as identity (2.1), we get

$$
\begin{aligned}
(x + y)^3 &= (x + y)(x + y)^2 = x(x + y)^2 + y(x + y)^2 \\
&= x(x^2 + 2xy + y^2) + y(x^2 + 2xy + y^2) \\
&= (x^3 + 2x^2y + xy^2) + (x^2y + 2xy^2 + y^3) \\
&= x^3 + y^3 + 3x^2y + 3xy^2 \\
&= x^3 + y^3 + 3xy(x + y).
\end{aligned}
$$

In order to get the result of item (e), we apply that of item (b), with $x + y$ in the place of x and z in the place of y:

$$
\begin{aligned}
(x + y + z)^2 &= [(x + y) + z]^2 \\
&= (x + y)^2 + 2(x + y)z + z^2 \\
&= (x^2 + 2xy + y^2) + 2(xz + yz) + z^2 \\
&= x^2 + y^2 + z^2 + 2xy + 2xz + 2yz.
\end{aligned}
$$

$\square$

The reader has certainly noticed that, in the previous proposition, one either has: (i) an identity of the form $E = F$, where E is a product of (at least two) polynomials and F is the sum of monomials we get from expanding the products in E (this is the case of the identities of items (b), (d) and (e)); or else (ii) an identity of the form $E = F$, where E is a polynomial and F is a product of (at least two) polynomials (as in items (a) and (c) of the previous proposition). In case (ii), we shall sometimes say that F is a **factorisation** of E, or that it is obtained by *factoring out* expression E.

The coming examples will give us an idea on how to apply the identities collected in the previous proposition to solve several interesting problems.

Example 2.2 Let x, y, z be real numbers, not all zero, such that $x + y + z = 0$. Explain why $xy + xz + yz \neq 0$ and, then, compute all possible values of the expression

$$
\frac{x^2 + y^2 + z^2}{xy + yz + zx}.
$$

Solution Squaring both sides of $x + y + z = 0$, it follows from item (e) of Proposition 2.1 that $x^2 + y^2 + z^2 + 2(xy + xz + yz) = 0$. If $xy + xz + yz = 0$, we would have $x^2 + y^2 + z^2 = 0$, and a simple extension of Corollary 1.5 (cf. Problem 2, page 10) would give us $x = 0$, $y = 0$ and $z = 0$, contradicting our hypotheses. Therefore, $xy + xz + yz \neq 0$, and it follows from $x^2 + y^2 + z^2 = -2(xy + xz + yz)$ that

$$\frac{x^2 + y^2 + z^2}{xy + yz + zx} = -2.$$

□

Our next example shows how to use the algebraic identities we know so far to prove *inequalities*.[2]

Example 2.3 (Poland) For given positive real numbers a and b, prove that $4(a^3 + b^3) \geq (a + b)^3$.

Proof By expanding the right hand side with the aid of item (d) of Proposition 2.1, it is immediate to see that the inequality we want to prove is equivalent to $a^3 + b^3 \geq a^2 b + ab^2$. It now suffices to see that

$$a^3 + b^3 - a^2 b - ab^2 = a^3 - a^2 b + b^3 - ab^2 = a^2(a - b) - b^2(a - b)$$

$$= (a^2 - b^2)(a - b) = (a + b)(a - b)(a - b)$$

$$= (a + b)(a - b)^2 \geq 0,$$

for $a + b > 0$ and $(a - b)^2 \geq 0$.

□

We now generalize Example 1.8.

Example 2.4 (Austria) Let a and b be positive rationals, such that $\sqrt{ab}$ is irrational. Prove that $\sqrt{a} + \sqrt{b}$ is also irrational.

Proof By contraposition, suppose that $r = \sqrt{a} + \sqrt{b}$ were a rational number. Then, $r^2 = a + b + 2\sqrt{ab}$ would also be rational. However, in such a case, we would have

$$\sqrt{ab} = \frac{r^2 - a - b}{2},$$

which would be a rational number too, for, in the right hand side of the above equality, both the numerator and the denominator are rational numbers.

□

Example 2.5 (Canada) For each natural number n, prove that

$$n(n + 1)(n + 2)(n + 3)$$

is never a perfect square.

---

[2]We will undertake a thorough discussion of inequalities in Chap. 5 and Sects. 9.7 and 10.8.

Proof Letting $p = n(n+1)(n+2)(n+3)$, we have

$$p = [n(n+3)][(n+1)(n+2)]$$
$$= (n^2 + 3n)[(n^2 + 3n) + 2]$$
$$= (n^2 + 3n)^2 + 2(n^2 + 3n)$$
$$= [(n^2 + 3n)^2 + 2(n^2 + 3n) + 1] - 1$$
$$= [(n^2 + 3n) + 1]^2 - 1.$$

If we set $m = n^2 + 3n + 1$, we have $m > 1$ and, hence,

$$p = m^2 - 1 > m^2 - 2m + 1 = (m-1)^2.$$

Therefore, p is situated between the consecutive perfect squares $(m-1)^2$ and m^2, so that it cannot be, itself, a perfect square. □

Apart from the algebraic identities collected in Proposition 2.1, another frequently useful one is that given by the equality

$$(x - y)(x - z) = x^2 - (y + z)x + yz. \tag{2.2}$$

Observe that, at the right hand side of the above expression, both the sum $S = y + z$ and the product $P = yz$ of y and z do appear. An expression of the form $x^2 - Sx + P$, where S and P represent the sum and the product of two numbers or expressions, is called a **second degree trinomial** in x. Hence, writing (2.2) backwards, we can also see it as giving a factorisation for the second degree trinomial $x^2 - Sx + P$, where $S = y + z$ and $P = yz$:

$$x^2 - Sx + P = (x - y)(x - z). \tag{2.3}$$

The above factorisation is sometimes called **Viète's formula**, in honor of the French mathematician François Viète.[3]

The following example shows us how to apply Viète's formula.

Example 2.6 (Soviet Union) Let a, b and c be pairwise distinct real numbers. Show that the number

$$a^2(c - b) + b^2(a - c) + c^2(b - a)$$

is always different from zero.

---

[3]François Viète, French mathematician of the XVI century. By his pioneerism in the usage of letters to represent variables, Viète is sometimes called the father of modern Algebra.

Proof Letting S denote the given number, we have

$$
\begin{aligned}
S &= a^2(c-b) + b^2a - b^2c + c^2b - c^2a \\
&= a^2(c-b) + (b^2a - c^2a) + (c^2b - b^2c) \\
&= a^2(c-b) + a(b+c)(b-c) + bc(c-b) \\
&= (c-b)[a^2 - a(b+c) + bc] \\
&= (c-b)(a-b)(a-c),
\end{aligned}
$$

where we used (2.3) in the last equality. Now, it follows from $a \neq b$, $b \neq c$ and $c \neq a$ that $a - b, c - b, a - c \neq 0$, so that $S \neq 0$. □

A useful variant of Viète's formula is the factorisation for the expression $x^2 + Sx + P$, where, as before, $S = y + z$ and $P = yz$:

$$
x^2 + Sx + P = (x+y)(x+z). \tag{2.4}
$$

If we change S, y and z in (2.3) respectively by $-S$, $-y$ and $-z$, we immediately see that (2.4) is indeed equivalent to that factorisation.

The next example uses (2.4) to get yet another algebraic identity, which will be further applied in a number of places, both in this volume as well as in [4] and [5].

Example 2.7 For all $x, y, z \in \mathbb{R}$, we have

$$
(x+y+z)^3 = x^3 + y^3 + z^3 + 3(x+y)(x+z)(y+z). \tag{2.5}
$$

Proof Applying item (d) of Proposition 2.1 twice, first with $x + y$ in place of x and z in place of y, we successively get

$$
\begin{aligned}
(x+y+z)^3 &= [(x+y)+z]^3 \\
&= (x+y)^3 + z^3 + 3(x+y)z[(x+y)+z] \\
&= x^3 + y^3 + 3xy(x+y) + z^3 + 3(x+y)[(x+y)z + z^2) \\
&= x^3 + y^3 + z^3 + 3(x+y)[xy + (x+y)z + z^2] \\
&= x^3 + y^3 + z^3 + 3(x+y)(y+z)(x+z),
\end{aligned}
$$

where, in the last equality, we have used the variant (2.4) of Viète's formula. □

Problems: Section 2.1

1. * Prove the other items of Proposition 2.1.

2. If $m + n + p = 6$, $mnp = 2$ and $mn + mp + np = 11$, compute all possible values of $\frac{m}{np} + \frac{n}{mp} + \frac{p}{mn}$.

3. Let a and b be nonzero real numbers, such that $a \neq b, 1$. If $\left(\frac{b}{a}\right)^2 = \left(\frac{1-b}{1-a}\right)^2$, compute all possible values of $\frac{1}{a} + \frac{1}{b}$.

4. Given positive real numbers x and y, simplify the expression

$$\frac{1 - \left(\frac{x}{y}\right)^{-2}}{(\sqrt{x} - \sqrt{y})^2 + 2\sqrt{xy}}.$$

5. For $x, y, z \neq 0$, such that $y + z \neq 0$, simplify the expression

$$\frac{(x^3 + y^3 + z^3)^2 - (x^3 - y^3 - z^3)^2}{y + z}.$$

6. Let a and b be real numbers such that $ab = 1$ and $a \neq b$. Simplify the expression

$$\frac{\left(a - \frac{1}{a}\right)\left(b + \frac{1}{b}\right)}{a^2 - b^2}.$$

7. Let x and y be natural numbers such that $x^2 + 361 = y^2$. Find all possible values of x.

8. Real numbers a and b are such that $a + b = m$ and $ab = n$. Compute the value of $a^4 + b^4$ in terms of m and n.

9. If $a^2 + b^2 = 1$, find all possible values of $\frac{1 - 3(ab)^2}{a^6 + b^6}$.

10. (EKMC) Let a, b, c and d be real numbers such that $a^2 + b^2 = 1$ and $c^2 + d^2 = 1$. If $ac + bd = \frac{\sqrt{3}}{2}$, compute the value of $ad - bc$, provided it is a positive number.

11. (Brazil) Find all natural numbers x and y such that $x + y + xy = 120$.

12. * Given positive distinct real numbers x and y, prove that the following rationalisations[4] are valid:

   (a) $\dfrac{1}{\sqrt{x} \pm \sqrt{y}} = \dfrac{\sqrt{x} \mp \sqrt{y}}{x - y}$.

   (b) $\dfrac{1}{\sqrt[3]{x} \pm \sqrt[3]{y}} = \dfrac{\sqrt[3]{x^2} \mp \sqrt[3]{xy} + \sqrt[3]{y^2}}{x \pm y}$.

   (c) $\dfrac{1}{\sqrt[3]{x^2} \mp \sqrt[3]{xy} + \sqrt[3]{y^2}} = \dfrac{\sqrt[3]{x} \pm \sqrt[3]{y}}{x \pm y}$.

---

[4]In an informal way, one can think of a *rationalisation* as a way of *clearing roots* from denominators.

13. * For a natural number $n > 1$, show that

$$2\left(\sqrt{n+1} - \sqrt{n}\right) < \frac{1}{\sqrt{n}} < 2\left(\sqrt{n} - \sqrt{n-1}\right).$$

14. Rationalise $\frac{1}{2+\sqrt{2}+\sqrt{3}}$.

15. Rationalise $\frac{1}{\sqrt{2}+\sqrt[3]{3}}$. More precisely, obtain integers a, b, c, d, e, f and g such that

$$\frac{1}{\sqrt{2} + \sqrt[3]{3}} = \frac{1}{g}[(a\sqrt{2} + b) + (c\sqrt{2} + d)\sqrt[3]{3} + (e\sqrt{2} + f)\sqrt[3]{9}].$$

16. Let x, y and z be nonzero real numbers, such that $x + y + z = 0$. Explain why the sum of any two of them is also nonzero, and compute all possible values of each of the following expressions:

    (a) $\frac{x^2}{(y+z)^2} + \frac{y^2}{(x+z)^2} + \frac{z^2}{(x+y)^2}$.

    (b) $\frac{x^3}{(y+z)^3} + \frac{y^3}{(x+z)^3} + \frac{z^3}{(x+y)^3}$.

17. Let a and b be distinct integers. Find, in terms of a and b, the quotient of the division of $a^{64} - b^{64}$ by $(a + b)(a^2 + b^2)(a^4 + b^4)(a^8 + b^8)(a^{16} + b^{16})$.

18. * Given $n > 1$ integer and $a, b \in \mathbb{R}$, prove that the following factorizations are valid:

    (a) $a^n - b^n = (a - b)(a^{n-1} + a^{n-2}b + a^{n-3}b^2 + \cdots + b^{n-1})$.

    (b) $a^n + b^n = (a + b)(a^{n-1} - a^{n-2}b + a^{n-3}b^2 - \cdots + b^{n-1})$, provided n is odd.

19. Write $x^4 + 4y^4$ as a product of two non constant polynomials in x and y, both having integer coefficients.

20. (Canada) Let $a, b, c \in \mathbb{Z}$. Prove that 6 divides $a + b + c$ if and only if 6 divides $a^3 + b^3 + c^3$.

21. (Canada) If a, b and c are real numbers for which $a + b + c = 0$, show that $a^3 + b^3 + c^3 = 3abc$.

22. Prove the *double radical formula*, also known as *Bhaskara's formula*[5]: for all positive real numbers a and b, such that $a^2 \geq b$, one has

$$\sqrt{a \pm \sqrt{b}} = \sqrt{\frac{a + \sqrt{a^2 - b}}{2}} \pm \sqrt{\frac{a - \sqrt{a^2 - b}}{2}}.$$

---

[5] In honor of the Indian mathematician of the XII century Bhaskara II, also known as Bhaskaracharya (Bhaskara, the professor). The idea behind Bhaskara's formula is that, if a and b are naturals for which $a^2 - b$ is a perfect square, then his formula provides a simpler expression for $\sqrt{a \pm \sqrt{b}}$.

23. Show that there do not exist nonzero real numbers x, y and z such that $x + y + z \neq 0$ and

$$\frac{1}{x+y+z} = \frac{1}{x} + \frac{1}{y} + \frac{1}{z}.$$

24. (Soviet Union) Let a, b and c be pairwise distinct rationals. Prove that

$$\frac{1}{(b-c)^2} + \frac{1}{(c-a)^2} + \frac{1}{(a-b)^2}$$

is the square of a rational.

25. (TT) Let a, b and c be distinct rationals. If $\sqrt[3]{a} + \sqrt[3]{b} \in \mathbb{Q}$, prove that $\sqrt[3]{a}$, $\sqrt[3]{b} \in \mathbb{Q}$.

26. (TT) Let a, b, c, d, e and f be real numbers such that $a + b + c + d + e + f = 0$ and $a^3 + b^3 + c^3 + d^3 + e^3 + f^3 = 0$. If no two of them are opposite to each other, prove that

$$(a+c)(a+d)(a+e)(a+f) = (b+c)(b+d)(b+e)(b+f).$$

27. (Poland) For positive integers $a \leq b$, do the following items:

(a) Show that $b^3 < b^3 + 6ab + 1 < (b+2)^3$.

(b) Find all such a and b for which both $a^3 + 6ab + 1$ and $b^3 + 6ab + 1$ are perfect cubes.

2.2   The Modulus of a Real Number

We start this section by recalling the definition of modulus of a real number, a concept which will be important in a number of places hereafter.

Definition 2.8 For $x \in \mathbb{R}$, the **modulus** of x, denoted $|x|$, is defined as

$$|x| = \begin{cases} x, & \text{if } x \geq 0 \\ -x, & \text{if } x < 0 \end{cases}.$$

As an example, since $-5 < 0$, we have $|-5| = -(-5) = 5$; analogously, $|-\sqrt{3}| = -(-\sqrt{3}) = \sqrt{3}$ etc. More generally, an immediate consequence of the definition is that $|x| \geq 0$ for all $x \in \mathbb{R}$, with equality if and only if $x = 0$. Moreover, one always has

$$x \leq |x| = |-x|,$$

with equality if and only if $x \geq 0$. Note also that

$$|x| = \sqrt{x^2} = \max\{x, -x\}. \tag{2.6}$$

The simplest **modular equation** is the equation

$$|x - a| = b,$$

where a and b are given real numbers. Since $|x - a| \geq 0$, such an equation does not admit roots when $b < 0$. On the other hand, when $b \geq 0$, it follows from the definition of modulus that one must have either $x - a = b$ or $x - a = -b$, from where we get the roots

$$x = a + b, a - b.$$

The coming example shows how to solve a more elaborate equation in a single variable, involving the concept of modulus of a real number.

Example 2.9 Solve equation $|x + 1| + |x - 2| + |x - 5| = 7$.

Solution First of all, note that

$$|x + 1| = \begin{cases} x + 1, \text{ if } x \geq -1 \\ -x - 1, \text{ if } x < -1 \end{cases},$$

$$|x - 2| = \begin{cases} x - 2, \text{ if } x \geq 2 \\ -x + 2, \text{ if } x < 2 \end{cases}$$

and

$$|x - 5| = \begin{cases} x - 5, \text{ if } x \geq 5 \\ -x + 5, \text{ if } x < 5 \end{cases}.$$

Now, the conjunction of the conditions $x < -1$ or $x \geq -1$, $x < 2$ or $x \geq 2$, $x < 5$ or $x \geq 5$ partitions the real line into the intervals $(-\infty, -1)$, $[-1, 2)$, $[2, 5)$ e $[5, +\infty)$. Hence, in order to simplify the left hand side of the given equation, we separately consider x as varying in each one of these intervals. We thus obtain

$$|x + 1| + |x - 2| + |x - 5| = \begin{cases} -3x + 6, \text{ if } x < -1 \\ -x + 8, \text{ if } -1 \leq x < 2 \\ x + 4, \text{ if } 2 \leq x < 5 \\ 3x - 6, \text{ if } x \geq 5 \end{cases}.$$

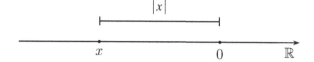

Finally, note that

- $-3x + 6 = 7 \Leftrightarrow x = -\frac{1}{3}$; however, since the condition $-\frac{1}{3} < -1$ is not satisfied, there are no roots in this case.
- $-x + 8 = 7 \Leftrightarrow x = 1$; since the condition $-1 \le 1 < 2$ is satisfied, $x = 1$ is a root of the equation.
- $x + 4 = 7 \Leftrightarrow x = 3$; since the condition $2 \le 3 < 5$ is satisfied, $x = 3$ is also a root of the equation.
- $3x - 6 = 7 \Leftrightarrow x = \frac{13}{3}$; since the condition $\frac{13}{3} \ge 5$ is not satisfied, there are no roots in this case.

Therefore, the solution set of the given equation is $S = \{1, 3\}$. □

Back to the study of the properties of modulus, let us represent the real numbers as points in the real line. It is easy to see that $|x|$ is simply the distance from (the point that represents) x to (the one representing) 0 (cf. Fig. 2.1). More generally, given $x, y \in \mathbb{R}$, we can look at $|x - y|$ as the distance from the points x and y in the real line. In fact, since $|x - y| = |y - x|$, we can suppose that $x \le y$. Then,

$$|x - y| = y - x = \text{ distance from } x \text{ to } y \text{ in the real line.}$$

In the above reasoning, if we do not wish to consider which of x and y is the greatest one, we can write

$$|x - y| = \max\{x, y\} - \min\{x, y\}, \tag{2.7}$$

for

$$\{x, y\} = \{\max\{x, y\}, \min\{x, y\}\}. \tag{2.8}$$

There simple remarks suffice to consider the following hard example.

Example 2.10 (Yugoslavia) Let $n \in \mathbb{N}$ and $M = \{1, 2, 3, \ldots, 2n\}$. Also, let $M_1 = \{a_1, a_2, \ldots, a_n\}$ and $M_2 = \{b_1, b_2, \ldots, b_n\}$ be subsets of M such that $a_1 < a_2 < \cdots < a_n$ and $b_1 > b_2 > \cdots > b_n$. If $M_1 \cup M_2 = M$ and $M_1 \cap M_2 = \emptyset$, prove that

$$|a_1 - b_1| + |a_2 - b_2| + \cdots + |a_n - b_n| = n^2.$$

Proof It follows from (2.8) that

$$\bigcup_{i=1}^{n} \{\max\{a_i, b_i\}, \min\{a_i, b_i\}\} = \bigcup_{i=1}^{n} \{a_i, b_i\} = \{1, 2, 3, \ldots, 2n\}.$$

Also, (2.7) gives

$$|a_1 - b_1| + |a_2 - b_2| + \cdots + |a_n - b_n| =$$
$$= (\max\{a_1, b_1\} + \max\{a_2, b_2\} + \cdots + \max\{a_n, b_n\})$$
$$- (\min\{a_1, b_1\} + \min\{a_2, b_2\} + \cdots + \min\{a_n, b_n\}).$$

On the other hand, given integers $1 \le k, l \le n$, with $k \ne l$, we have

$$k > l \Rightarrow \max\{a_k, b_k\} \ge a_k > a_l \ge \min\{a_l, b_l\}$$

and

$$k < l \Rightarrow \max\{a_k, b_k\} \ge b_k > b_l \ge \min\{a_l, b_l\}.$$

Therefore,

$$\{\max\{a_1, b_1\}, \max\{a_2, b_2\}, \ldots, \max\{a_n, b_n\}\} = \{n + 1, n + 2, \ldots, 2n\}$$

and

$$\{\min\{a_1, b_1\}, \min\{a_2, b_2\}, \ldots, \min\{a_n, b_n\}\} = \{1, 2, \ldots, n\}.$$

Finally, the above relations give

$$|a_1 - b_1| + |a_2 - b_2| + \cdots + |a_n - b_n| =$$
$$= ((n + 1) + (n + 2) + \cdots + 2n) - (1 + 2 + \cdots + n)$$
$$= n^2 + (1 + 2 + \cdots + n) - (1 + 2 + \cdots + n) = n^2.$$

$$\square$$

We continue our study of the concept of modulus with the following important result, which is known in mathematical literature as the **triangle inequality**.[6]

Proposition 2.11 *For all real numbers a and b, we have*

$$|a + b| \le |a| + |b|. \tag{2.9}$$

Moreover, if $a, b \ne 0$, then equality holds if and only if a and b have the same sign.

Proof Since $|a + b|$ and $|a| + |b|$ are both nonnegative, we have

$$|a + b| \leq |a| + |b| \Leftrightarrow |a + b|^2 \leq (|a| + |b|)^2$$
$$\Leftrightarrow (a + b)^2 \leq |a|^2 + |b|^2 + 2|ab|$$
$$\Leftrightarrow 2ab \leq 2|ab|,$$

which is clearly true. From the computations above it also follows that $|a + b| = |a| + |b|$ if and only if $ab = |ab|$; in turn, this happens if and only if $ab \geq 0$. Finally, if $a, b \neq 0$, then we have the equality if and only if $ab > 0$. □

Corollary 2.12 *For all real numbers a and b, we have*

$$||a| - |b|| \leq |a - b|.$$

Moreover, if $a, b \neq 0$, then the equality holds if and only if a and b have the same sign.

Proof Applying the triangle inequality to $a - b$ in place of a, we get

$$|a| = |(a - b) + b| \leq |a - b| + |b|$$

and, hence, $|a| - |b| \leq |a - b|$. Repeating the above argument with the roles of a and b interchanged, it follows that $|b| - |a| \leq |a - b|$.

Now, since $|a - b| \geq |a| - |b|, |b| - |a|$, we get

$$|a - b| \geq \max\{|a| - |b|, |b| - |a|\} = ||a| - |b||,$$

where we used (2.6) in the last equality.

Equality happens if and only if we have equality in at least one of the triangular inequalities $|a| \leq |a - b| + |b|$ or $|b| \leq |b - a| + |a|$. If $a, b \neq 0$ and, say, equality holds in the first one of theses inequalities, i.e., if $|a| = |a - b| + |b|$, then the condition for equality in Proposition 2.11 assures that we must have $(a - b)b \geq 0$, or, which is the same $ab \geq b^2$. In particular, we must have $ab > 0$.

Conversely, suppose that $ab > 0$, and let us show that equality holds. There are two possibilities: $a, b > 0$ or $a, b < 0$. Suppose that $a, b < 0$ (the other case can be treated in a similar way). Then, $|a| = -a$ and $|b| = -b$, so that $||a| - |b|| = |(-a) - (-b)| = |b - a| = |a - b|$. □

Given real numbers a, b and c and applying triangle inequality twice, we get

$$|a + b + c| \leq |a + b| + |c| \leq |a| + |b| + |c|, \qquad (2.10)$$

Hence, we have the inequality

$$|a + b + c| \leq |a| + |b| + |c|, \qquad (2.11)$$

which is the analogous of (2.9) for three real numbers, instead of two. Therefore, we shall also refer to this last inequality as the *triangle inequality*.

If $a, b, c \neq 0$ and we have equality in (2.11), then we should also have equality in all inequalities in (2.10). In particular, we have $|a + b| \leq |a| + |b|$, and it follows from Proposition 2.11 that a and b have equal signs. Since we can also reach (2.11) by writing

$$|a + b + c| \leq |a| + |b + c| \leq |a| + |b| + |c|,$$

we conclude, analogously, that b and c should also have equal signs.

Conversely, if a, b and c all have equal signs, say $a, b, c < 0$ (the case $a, b, c > 0$ is completely analogous), then $a + b + c < 0$, so that

$$|a + b + c| = -(a + b + c) = (-a) + (-b) + (-c) = |a| + |b| + |c|.$$

Hence, we have just shown that equality holds in (2.11) if and only if a, b and c have equal signs.

As we shall see in Sect. 4.1 (cf. Problem 7, page 96), inequalities (2.9) and (2.10) can be easily generalized for n real numbers. For the time being, we end this section with the following

Example 2.13 Prove that, for every $x \in \mathbb{R}$, we have

$$|x - 1| + |x - 2| + |x - 3| + \cdots + |x - 10| \geq 25.$$

Proof It follows from the triangle inequality that

$$|x - a| + |x - b| = |x - a| + |b - x| \geq |(x - a) + (b - x)| = |b - a|.$$

Hence, grouping the summands at the left hand side in pairs, we get,

$$|x - 1| + |x - 10| \geq |10 - 1| = 9;$$
$$|x - 2| + |x - 9| \geq |9 - 2| = 7;$$
$$|x - 3| + |x - 8| \geq |8 - 3| = 5;$$
$$|x - 4| + |x - 7| \geq |7 - 4| = 3;$$
$$|x - 5| + |x - 6| \geq |6 - 5| = 1.$$

Adding these inequalities, we obtain that of the statement.   □

Problems: Section 2.2

1. * Given real numbers a and b, show that

$$\{x \in \mathbb{R};\ |x - a| < b\} = \begin{cases} \varnothing, & \text{if } b < 0 \\ \{a\}, & \text{if } b = 0 \\ (a - b, a + b), & \text{if } b > 0 \end{cases}.$$

   Do the same for $|x - a| \leq b$, $|x - a| > b$ and $|x - a| \geq b$.

2. * Prove that, for all $x, y \in \mathbb{R}$, one has $|xy| = |x| \cdot |y|$.

3. Solve, for $x \in \mathbb{R}$, the following equations:

   (a) $|x| = x - 6$.
   (b) $|x + 1| + |x - 2| + |x - 5| = 4$.

4. Solve, for $x \in \mathbb{R} \setminus \{0, 1\}$, equation $\frac{|x|}{x} = \frac{|x-1|}{x-1}$.

5. Let a, b and c be given real numbers, with $a < b$. Discuss, in terms of a, b and c, the number of solutions of the equation

$$|x - a| + |x - b| = c.$$

6. (Mexico) Let r be a nonnegative rational number. Prove that

$$\left| \frac{r + 2}{r + 1} - \sqrt{2} \right| < \frac{1}{2} |r - \sqrt{2}|.$$

   (This inequality shows that the rational number $\frac{r+2}{r+1}$ approximates $\sqrt{2}$ twice as better as r does it.)

7. Prove that:

   (a) If $0 \leq x \leq y$, then $\frac{x}{1+x} \leq \frac{y}{1+y}$.
   (b) If $a, b \in \mathbb{R}$, then $\frac{|a|}{1+|a|} + \frac{|b|}{1+|b|} \geq \frac{|a+b|}{1+|a+b|}$.

8. Let $n > 1$ be an integer. Prove that

$$|x - 1| + |x - 2| + \cdots + |x - 2n| \geq n^2$$

   for every real x, with equality for infinitely many values of x.

2.3   A First Look at Polynomial Equations

In this section we study some particular types of *polynomial equations*, postponing a much deeper look to [5].

In general, a **polynomial equation of degree n** is an equation of the form

$$a_n x^n + a_{n-1} x^{n-1} + \cdots + a_1 x + a_0 = 0, \qquad (2.12)$$

where $n \geq 1$ is an integer and $a_0, a_1, \ldots, a_n$ are given real numbers,[7] with $a_n \neq 0$.

The simplest kind of such an equation is the **first degree equation** $ax + b = 0$, where a and b are given real numbers and $a \neq 0$. As the reader certainly knows, we have

$$ax + b = 0 \Leftrightarrow ax = -b \Leftrightarrow x = -\frac{b}{a},$$

so that $-\frac{b}{a}$ is its only root.

The second simplest kind of polynomial equation is the **second degree equation**

$$ax^2 + bx + c = 0, \qquad (2.13)$$

where a, b and c are given real numbers, with $a \neq 0$. For reasons that will soon be clear, the left hand side of (2.13) is also known as the **second degree trinomial** associated to Eq. (2.13).

In order to solve (2.13), we let Δ (one reads *delta*) denote the real number

$$\Delta = b^2 - 4ac,$$

and call it the **discriminant** of the equation (or of the associated trinomial). As we shall see in a moment, the sign of Δ *discriminates* whether or not the equation has real roots. To this end, we need the following auxiliary result.

Lemma 2.14 *Given $a, b, c \in \mathbb{R}$, with $a \neq 0$, one has*

$$ax^2 + bx + c = a \left[\left(x + \frac{b}{2a} \right)^2 - \frac{\Delta}{4a^2} \right]. \qquad (2.14)$$

*This algebraic identity is called the **canonical form** of the second degree trinomial* $ax^2 + bx + c$.

---

[7] Here, we are using the concept of a *sequence* of real numbers. For further details in this respect, we refer the reader to Chap. 3.

Proof It suffices to see that

$$
ax^2 + bx + c = a\left(x^2 + \frac{b}{a}x + \frac{c}{a}\right)
$$

$$
= a\left(x^2 + \frac{b}{a}x + \frac{b^2}{4a^2} - \frac{b^2}{4a^2} + \frac{c}{a}\right)
$$

$$
= a\left[\left(x^2 + \frac{b}{a}x + \frac{b^2}{4a^2}\right) - \frac{b^2}{4a^2} + \frac{4ac}{4a^2}\right]
$$

$$
= a\left[\left(x + \frac{b}{2a}\right)^2 - \frac{\Delta}{4a^2}\right].
$$

□

Remark 2.15 The idea of adding and subtracting a certain summand out of a given algebraic expression in order to *complete a square*, as was done right after the second equality in the proof of the previous lemma, is very important and should be learned as a kind of *algebraic trick* that will be useful in a number of places, here as well as in [4] and [5]. Later in this chapter, we shall see other interesting applications of such a technique.

Proposition 2.16 *Let a, b and c be given real numbers, with a ≠ 0.*

(a) *The equation $ax^2 + bx + c = 0$ has real roots if and only if $\Delta \geq 0$. Moreover, if this is so, then its roots are $\frac{-b-\sqrt{\Delta}}{2a}$ and $\frac{-b+\sqrt{\Delta}}{2a}$.*

(b) *If $\Delta \geq 0$, then the sum S and the product P of the roots of $ax^2 + bx + c = 0$ are given by $S = -\frac{b}{a}$ and $P = \frac{c}{a}$.*

Proof

(a) It follows from (2.14) that

$$
ax^2 + bx + c = 0 \Leftrightarrow \left(x + \frac{b}{2a}\right)^2 = \frac{\Delta}{4a^2}. \tag{2.15}
$$

Since $\left(x + \frac{b}{2a}\right)^2 \geq 0$ for all $x \in \mathbb{R}$, if the equation has real roots, then one must have $\Delta \geq 0$. In this case, it transpires from (2.15) that $x + \frac{b}{2a} = \pm\frac{\sqrt{\Delta}}{2a}$, and item (a) follows.

(b) It suffices to compute

$$
\frac{-b - \sqrt{\Delta}}{2a} + \frac{-b + \sqrt{\Delta}}{2a} = -\frac{b}{a}
$$

and

$$\left(\frac{-b-\sqrt{\Delta}}{2a}\right)\left(\frac{-b+\sqrt{\Delta}}{2a}\right) = \frac{(-b)^2 - \Delta}{4a^2} = \frac{c}{a}.$$

□

Remarks 2.17

i. When $\Delta \geq 0$, formulae $\frac{-b\pm\sqrt{\Delta}}{2a}$ for the roots of $ax^2 + bx + c = 0$ are known as **Bhaskara's formulae**.
ii. The formulas of item (b) are also known as **Viète's formulae**.
iii. In the notations of item (a), if $\Delta = 0$ we say that $ax^2 + bx + c = 0$ has *two equal roots*.

The coming example shows how one can reduce a seemingly complicated equation to a simpler one by means of a suitable **substitution of variable**.

Example 2.18 (Brazil) Find all real numbers x such that $x^2 + x + 1 = \frac{156}{x^2+x}$.

Solution By performing the substitution $y = x^2+x$, we get the equation $y+1 = \frac{156}{y}$ or, which is the same, $y^2 + y - 156 = 0$. For this last equation, since $\Delta = 1^2 - 4(-156) = 625 = 25^2$, it follows that $y = \frac{-1\pm25}{2} = -13$ or 12. Thus, we have reduced the original equation to the second degree equations $x^2 + x = -13$ and $x^2 + x = 12$. For the first one, we have $\Delta = -51 < 0$, so that there are no real roots. For the second, $\Delta = 49$ and, hence, $x = \frac{-1\pm7}{2} = -4$ or 3.   □

Our next example shows that it is sometimes more useful to *algebraically manipulate* a second degree equation than to solve it explicitly.

Example 2.19 Find the numerical value of $(3+\sqrt{2})^5+(3-\sqrt{2})^5$ without expanding the powers involved.

Proof Letting $u = 3 + \sqrt{2}$ and $v = 3 - \sqrt{2}$, we have $u + v = 6$ and $uv = 7$, so that u and v are the roots of the equation $x^2 - 6x + 7 = 0$. Therefore, making $x = u$ and $x = v$ in this equation gives us $u^2 - 6u + 7 = 0$ and $v^2 - 6v + 7 = 0$, or, equivalently, $u^2 = 6u - 7$ and $v^2 = 6v - 7$. Multiplying the first equality by u^k and the second one by v^k, where $k \geq 0$ is an integer, we get

$$u^{k+2} = 6u^{k+1} - 7u^k \text{ and } v^{k+2} = 6v^{k+1} - 7v^k;$$

adding both of these, we finally arrive at

$$u^{k+2} + v^{k+2} = 6(u^{k+1} + v^{k+1}) - 7(u^k + v^k).$$

Writing the previous relation for k respectively equal to 0, 1, 2 and 3, we successively get

$$u^2 + v^2 = 6(u + v) - 7 \cdot 2 = 6 \cdot 6 - 14 = 22;$$

$$u^3 + v^3 = 6(u^2 + v^2) - 7(u + v) = 6 \cdot 22 - 7 \cdot 6 = 90;$$

$$u^4 + v^4 = 6(u^3 + v^3) - 7(u^2 + v^2) = 6 \cdot 90 - 7 \cdot 22 = 386;$$

$$u^5 + v^5 = 6(u^4 + v^4) - 7(u^3 + v^3) = 6 \cdot 386 - 7 \cdot 90 = 1686.$$

□

For the next example, recall that if the sum and the product of two real numbers are positive, then both numbers are also positive.

Example 2.20 Let p and q be given real numbers. If the equation $x^2 + px + q = 0$ has real, positive and distinct real roots, show that the same is true for the equation $qx^2 + (p - 2q)x + (1 - p) = 0$.

Proof Note initially that $q \neq 0$, for otherwise the first equation would reduce to $x^2 + px = 0$, which has 0 as one of its roots, thus contradicting our hypotheses.

Now, let Δ and Δ' be, respectively, the discriminants of $x^2 + px + q = 0$ and $qx^2 + (p - 2q)x + (1 - p)$. Let us first show that $\Delta' > 0$, which will guarantee that the second equation has distinct real roots. Since $x^2 + px + q = 0$ has distinct real roots, we have $\Delta = p^2 - 4q > 0$. Therefore,

$$\Delta' = (p - 2q)^2 - 4q(1 - p)$$

$$= p^2 - 4q + 4q^2$$

$$= \Delta + 4q^2 > 0.$$

Finally, according to the paragraph that immediately precedes this example, in order to show that the roots of $qx^2 + (p - 2q)x + (1 - p) = 0$ are positive, it suffices to show that the sum S' and the product P' of them are both positive. In order to do this, we recall that the roots of $x^2 + px + q = 0$ are known to be positive, so that (by Viète's formulae) $-p > 0$ and $q > 0$. Hence, again by Viète's formulae, we have

$$S' = \frac{2q - p}{q} = 2 + \frac{-p}{q} > 0 \text{ and } P' = \frac{1 - p}{q} = \frac{1}{q} + \frac{-p}{q} > 0,$$

as we wished to show.

□

We finish our discussion of second degree equations with the following important remark: if $a \neq 0$ and $ax^2 + bx + c = 0$ has real roots α and β (not necessarily $\alpha \neq \beta$), then we have the factorisation

$$ax^2 + bx + c = a(x - \alpha)(x - \beta). \tag{2.16}$$

In fact, it follows from item (b) of Proposition 2.16 that, for every real x,

$$a(x - \alpha)(x - \beta) = a[x^2 - (\alpha + \beta)x + \alpha\beta]$$
$$= a\left[x^2 - \left(-\frac{b}{a}\right)x + \frac{c}{a}\right]$$
$$= ax^2 + bx + c.$$

It is instructive to compare (2.16) to (2.2). The right hand side of (2.16) is called the **factorised form** of the second degree trinomial $ax^2 + bx + c$.

In turning to more general polynomial equations, it is natural to try to look at those of degrees $n = 3$ and $n = 4$. In these cases, formulas have been built, in terms of the coefficients of the equations, to compute their real roots, if any.[8] As professor I. Stewart teaches us in Chap. 4 of his very interesting book [25], such formulas derive from the works of the Italian mathematicians Scipione del Ferro, Girolamo Cardano, Niccolò Fontana (conhecido como Tartaglia) and Lodovico Ferrari. However, they are too complicated to be useful, and for this reason we shall not discuss them here. In order to help convincing the reader, let us just mention that Cardano's formula for the roots of the third degree polynomial equation $ax^3 + bx^2 + cx + d = 0$ is the following[9]:

$$\sqrt[3]{q + \sqrt{q^2 + (r - p^2)^3}} + \sqrt[3]{q - \sqrt{q^2 + (r - p^2)^3}} + p,$$

where $p = -\frac{b}{3a}$, $q = p^3 + \frac{bc - 3ad}{6a^2}$ and $r = \frac{c}{3a}$ (however, see Problems 16, 21 and 22).

For polynomial Eq. (2.12) of degree $n \geq 5$, the Norwegian mathematician Niels H. Abel[10] and the French mathematician Évariste Galois,[11] both of the XIX century, proved that there exists no similar formula, built on the coefficients of the given polynomial equation, that gives its real (or even complex!) roots. Well understood, it doesn't matter how smart someone is; they proved that *it is impossible to find* such a formula, simply because it doesn't exist! For a beautiful and elementary account of the ideas involved, we refer the reader to [14].

Some particular kinds of polynomial equations of degrees 4 and 6, however, are sufficiently simple to deserve some attention, specially because appropriate

---

[8]As it happens, these formulas also give the *complex* roots of the corresponding equations. However, we shall postpone any considerations involving complex numbers to [5].

[9]Cf. http://www.math.vanderbilt.edu/~schectex/courses/cubic.

[10]In his 27 years of life, Abel made several deep contributions to Mathematics, among which the most famous one is perhaps the impossibility of solving general polynomial equations of degree 5.

[11]In spite of his premature death, when he was only 21 years old, Galois is considered to be one of the greatest mathematicians the world has ever seen. His work on the connection between the solvability of polynomial equations of degrees $n \geq 5$ and Group Theory constitute the foundations of what is known today as Galois' Theory, a branch of modern Algebra with applications to several distinct areas of Mathematics.

variable substitutions immediately reduce them to second degree equations. Let us first examine **biquadratic**, i.e., equations of the form

$$ax^4 + bx^2 + c = 0, \tag{2.17}$$

with $a, b, c \in \mathbb{R}$ and $a \neq 0$. The variable substitution $y = x^2$ transforms it into the second degree equation $ay^2 + by + c = 0$, which we already know how to solve in $\mathbb{R}$. Therefore, for each nonnegative root $y = \alpha$ of this last equation, solving equation $x^2 = \alpha$ gives us the pair of real roots $x = \pm\sqrt{\alpha}$ of the original biquadratic equation. Conversely, if $x = \beta$ is a real root of the given biquadratic equation, it is immediate to see that $y = \beta^2$ is a nonnegative root of the second degree equation $ay^2 + by + c = 0$. We have, thus, proved the following result.

Proposition 2.21 *Given real numbers a, b and c, with a $\neq$ 0, the real roots of the biquadratic equation $ax^4 + bx^2 + c = 0$ are the reals of the form $\pm\sqrt{\alpha}$, where α is a nonnegative root of the second degree equation $ay^2 + by + c = 0$.*

In order to actually compute the real roots of a specific biquadratic equation, instead of invoking the last proposition, it is usually much better to recall our previous discussion, remembering that the variable substitution $y = x^2$ does the job of reducing it to a second degree equation.

Example 2.22 Find the real roots of the biquadratic equation $x^4 + 5x^2 - 7 = 0$.

Solution The variable substitution $y = x^2$ leads us to the second degree equation $y^2 + 5y - 7 = 0$, for which $\Delta = 53$. Hence, the roots of this last equation are $y = \frac{-5\pm\sqrt{53}}{2}$. Since $\frac{-5-\sqrt{53}}{2} < 0$, the real roots of the original biquadratic equation are the solutions of $x^2 = \frac{-5+\sqrt{53}}{2}$, i.e., are the real numbers $\pm\sqrt{\frac{-5+\sqrt{53}}{2}}$.   □

Given $n \in \mathbb{N}$, we point out that we can discuss the problem of finding the roots of a polynomial equation of the form

$$ax^{2n} + bx^n + c = 0 \tag{2.18}$$

in a way quite similar to that used to study biquadratic equations. We refer the reader to Problem 16 for the corresponding details.

Let us now examine **reciprocal polynomial equations** of degree 4, i.e., polynomial equations of degree 4 having the form

$$ax^4 + bx^3 + cx^2 + bx + a = 0,$$

where a, b and c are given real numbers, with $a \neq 0$. As Problem 28 shows, the name *reciprocal* applies to a larger class of polynomial equations, and comes from the fact that $x \in \mathbb{R} \setminus \{0\}$ is a real root of it if and only if $\frac{1}{x}$ also is.

Initially, note that 0 is not a root of the equation above, for $a \neq 0$. Therefore, a real number x is a root of it if and only if it is a root of the equation

$$ax^2 + bx + c + \frac{b}{x} + \frac{a}{x^2} = 0, \qquad (2.19)$$

which is obtained from the original equation by dividing both sides of the equality by x^2. Rewrite the left hand side of the last equation above as

$$a\left(x^2 + \frac{1}{x^2}\right) + b\left(x + \frac{1}{x}\right) + c = 0.$$

Now, the idea is to perform the variable substitution $y = x + \frac{1}{x}$. In order to implement it, let us first of all note that, according to (2.1), one has

$$y^2 = \left(x + \frac{1}{x}\right)^2 = x^2 + 2 + \frac{1}{x^2}.$$

Hence, $x^2 + \frac{1}{x^2} = y^2 - 2$, so that solving (2.19) amounts to solving the second degree equation

$$a(y^2 - 2) + by + c = 0. \qquad (2.20)$$

However, the above discussion hides a subtlety: it is sure that every real root $x = \alpha$ of (2.19) generates the real root $\beta = \alpha + \frac{1}{\alpha}$ of $a(y^2 - 2) + by + c = 0$. Nevertheless, the converse statement is not true: not every real root $y = \beta$ of this last equation does generate real roots $x = \alpha$ of the initial reciprocal equation. In fact, once we get a real root $y = \beta$ of $a(y^2 - 2) + by + c = 0$, in order to obtain the possible real roots of the reciprocal equation corresponding to β, we have to solve in $\mathbb{R}$ the equation

$$x + \frac{1}{x} = \beta,$$

or, which is the same, $x^2 - \beta x + 1 = 0$. Since the discriminant of this last equation is

$$\Delta = \beta^2 - 4,$$

it will have real root only if $\beta^2 - 4 \geq 0$, i.e., only if $|\beta| \geq 2$.

As was the case with biquadratic equations, in order to actually find the real roots of a given reciprocal equation of degree 4, instead of memorizing the result of the variable substitution $y = x + \frac{1}{x}$, it is more profitable to follow the steps that led us from (2.19) to (2.20). Let us see an example in this respect.

Example 2.23 Find all real roots of the reciprocal equation

$$2x^4 + 5x^3 + 6x^2 + 5x + 2 = 0.$$

Solution Dividing both sides by x^2 and grouping summands, we obtain

$$2\left(x^2 + \frac{1}{x^2}\right) + 5\left(x + \frac{1}{x}\right) + 6 = 0.$$

Making the variable substitution $y = x + \frac{1}{x}$, it follows that $y^2 = x^2 + \frac{1}{x^2} + 2$, so that the given equation is equivalent to

$$2(y^2 - 2) + 5y + 6 = 0.$$

Since this second degree equation has real roots $y = -2$ and $y = -\frac{1}{2}$, it follows that the real roots of the original equation are the real roots of the equations $x + \frac{1}{x} = -2$ and $x + \frac{1}{x} = -\frac{1}{2}$. The first of these equations is equivalent to $x^2 + 2x + 1 = 0$ and, hence, has two real roots, both equal to -1. The second is equivalent to $2x^2 + x + 2 = 0$, which has discriminant $\Delta = -15 < 0$; therefore, it has no real roots.   □

Problems: Section 2.3

1. Let b and c be given real numbers, such that the equation $x^2 + b|x| + c = 0$ has real roots. Prove that the sum of these roots is always equal to 0.
2. Solve, for $x \in \mathbb{R}$, the following equations:

    (a) $x + \sqrt{x + 2} = 10$.
    (b) $\sqrt{x + 10} - \sqrt{2x + 3} = \sqrt{1 - 3x}$.
    (c) $x^2 + 18x + 30 = 2\sqrt{x^2 + 18x + 45}$.

3. (IMO) In each of the cases (a) $A = \sqrt{2}$, (b) $A = 1$ and (c) $A = 2$, find all $x \in \mathbb{R}$ for which we have

$$\sqrt{x + \sqrt{2x - 1}} + \sqrt{x - \sqrt{2x - 1}} = A.$$

4. A math teacher composed three different quizzes. In the first one, he put a second degree equation. In the second, he put almost the same equation, changing just the coefficient of the monomial of second degree. Finally, in the third quiz, once more he put almost the same equation of the first, this time changing just the constant coefficient. It is known that the roots of the equation of the second quiz are 2 and 3, and that those of the third one are 2 and -7. Decide whether the second degree equation of the first quiz has real roots and, if this is so, compute them.
5. Let a and b be two distinct, nonzero real numbers. If a and b are the roots of the equation $x^2 + ax + b = 0$, find all possible values of $a - b$.

6. Let α and β be the roots of $x^2 - 13x + 9 = 0$, and a and b be real numbers such that the equation $x^2 + ax + b = 0$ has roots α^2 and β^2. Compute the value of $a + b$.

7. Equation $x^2 + x - 1 = 0$ has roots u and v. Find a second degree equation whose roots are u^3 and v^3.

8. The roots of the equation $x^2 - Sx + P = 0$ are the real numbers α and β. Find a second degree trinomial whose coefficients are expressions built on S and P and whose roots are the real numbers $\alpha S + P$ and $\beta S + P$.

9. Use the theory of second degree equations developed in this section to compute the value of the sum $(7 + 4\sqrt{3})^5 + (7 - 4\sqrt{3})^5$.

10. If α is a root of $x^2 - x - 1 = 0$, find all possible values of $\alpha^5 - 5\alpha$.

11. For which integer values of m does the equation $x^2 + mx + 5 = 0$ have integer roots?

12. Show that, for every $a, b, c \in \mathbb{R}$, with $a \neq 0$, the equation

$$\frac{1}{x - b} + \frac{1}{x - c} = \frac{1}{a^2}$$

has exactly two distinct real roots.

13. Solve equation $x = \sqrt{x - \frac{1}{x}} + \sqrt{1 - \frac{1}{x}}$ in the set of real numbers.

14. Show that, for every real number $a \neq 0$, the equations

$$ax^3 - x^2 - x - (a + 1) = 0 \quad \text{and} \quad ax^2 - x - (a + 1) = 0$$

have a common root.

15. (Soviet Union) Do the following items:

   (a) For real x, write the number $x^3 - 3x^2 + 5x$ in the form

   $$a(x - 1)^3 + b(x - 1)^2 + c(x - 1) + d,$$

   with $a, b, c, d \in \mathbb{Z}$.

   (b) If x and y are real numbers such that $x^3 - 3x^2 + 5x = 1$ and $y^3 - 3y^2 + 5y = 5$, compute all possible values of $x + y$.

16. * Let $n \in \mathbb{N}$ and $a, b, c \in \mathbb{R}$, with $a \neq 0$. Elaborate, for the equation $ax^{2n} + bx^n + c = 0$, a discussion analogous to that made on the text for biquadratic equations, and which led us to Proposition 2.21.

17. * Consider the polynomial equation of third degree $x^3 + ax^2 + bx + c = 0$, where a, b and c are given real numbers, with $c \neq 0$. If α is a (nonzero) real root of it, prove that there exist real numbers b' and c' such that we have the factorisation

$$x^3 + ax^2 + bx + c = (x - \alpha)(x^2 + b'x + c').$$

Then, conclude that the original polynomial equation has at most three (not necessarily distinct) real roots. Moreover, if this is the case, and α, β and γ are its three real roots, show that

$$x^3 + ax^2 + bx + c = (x - \alpha)(x - \beta)(x - \gamma).$$

This result generalizes the factorised form (2.16) of a second degree trinomial and is a particular case of the *division algorithm*.[12].

18. (a) Show that the real number $\alpha = \sqrt[3]{2 + \sqrt{5}} + \sqrt[3]{2 - \sqrt{5}}$ is a root of the equation $x^3 + 3x - 4 = 0$.
    (b) Conclude that α is a rational number.

19. * Establish the following version of **Girard's relations**[13] between roots and coefficients of a polynomial equation of third degree: if the real numbers a_0, a_1, a_2 and a_3 ($a_3 \neq 0$) are such that the equation

$$a_3 x^3 + a_2 x^2 + a_1 x + a_0 = 0$$

has (not necessarily distinct) real roots x_1, x_2 and x_3, then

$$\begin{cases} x_1 + x_2 + x_3 = -\frac{a_2}{a_3} \\ x_1 x_2 + x_1 x_3 + x_2 x_3 = \frac{a_1}{a_3} \\ x_1 x_2 x_3 = -\frac{a_0}{a_3} \end{cases}. \qquad (2.21)$$

20. Assume that the equation $x^3 - 3x + 1 = 0$ has three real roots, say α, β and γ. Compute the values of $\alpha^2 + \beta^2 + \gamma^2$, $\alpha^3 + \beta^3 + \gamma^3$ and $\alpha^4 + \beta^4 + \gamma^4$.

21. Still concerning the third degree polynomial equation $x^3 + ax^2 + bx + c = 0$, with $a, b, c \in \mathbb{R}$, prove that there exists a real number d such that the variable substitution $y = x - d$ transforms the given equation into one of the form $y^3 + py + q = 0$, for certain real numbers p and q.

22. Concerning the equation $x^3 - 11x + 16 = 0$, do the following items:

    (a) Substitute $x = u + v$ and get an equivalent equation in the two real variables u and v.
    (b) Impose that $uv = \frac{11}{3}$ (i.e., make uv equals $-\frac{1}{3}$ times the coefficient of x in the original equation) and conclude that the equation in u and v of item (a) is equivalent to $u^6 + 16u^3 + \left(\frac{11}{3}\right)^3 = 0$.
    (c) Find u and v, and conclude that one of the roots of the given equation is

$$\sqrt[3]{-8 + \frac{\sqrt{1191}}{9}} + \sqrt[3]{-8 - \frac{\sqrt{1191}}{9}}.$$

---

[12]For more details concerning this point, as well as for the generalization of the result of Problem 19, we refer the reader to [5].

[13]After Albert Girard, French mathematician of the XVII century.

The items above describe, by means of a specific example, the ideas behind Cardano's formula for the roots of a polynomial equation of third degree.

23. * Let x be a nonzero real number, such that $\left(x + \frac{1}{x}\right)^2 = 3$. Compute all possible values of $x^3 + \frac{1}{x^3}$.

24. If x is a nonzero real number such that $x + \frac{1}{x} = 4$, calculate $x^4 + \frac{1}{x^4}$.

25. If $x^2 - x - 1 = 0$, compute all possible values of $\left(x - \frac{1}{x}\right)^2 + \left(x^3 - \frac{1}{x^3}\right)^2$.

26. (Singapore) If $x^2 - 4x + 1 = 0$, compute all possible values of

$$\frac{x^6 + \frac{1}{x^6} - \left(x + \frac{1}{x}\right)^6 + 2}{x^3 + \frac{1}{x^3} - \left(x + \frac{1}{x}\right)^3}.$$

27. Solve the reciprocal equation $x^4 - 7x^3 + 14x^2 - 7x + 1 = 0$.

28. Given real numbers $a_0, a_1, \ldots, a_n$, with $a_n \neq 0$, the polynomial equation

$$a_n x^n + a_{n-1} x^{n-1} + \cdots + a_1 x + a_0 = 0,$$

is said to be **reciprocal** if $a_k = a_{n-k}$, for $0 \leq k \leq n$. If α is a real root of such an equation, show that $\alpha \neq 0$ and that $\frac{1}{\alpha}$ is also a root of it. (Hence, the name *reciprocal* justifies itself by the fact that reciprocal polynomial equations have real roots which are pairwise reciprocal.[14] In this case, one also says that $a_n x^n + a_{n-1} x^{n-1} + \cdots + a_1 x + a_0$ is a **reciprocal polynomial**.).

29. Do the following items:

   (a) * Given a real number $x \neq 0$, write $x^3 + \frac{1}{x^3}$ in terms of $y = x + \frac{1}{x}$.
   (b) Use the result of (a) to reduce the reciprocal equation $ax^6 + bx^5 + cx^4 + dx^3 + cx^2 + bx + a = 0$, of degree 6, to a polynomial equation of degree 3.

30. (Brazil—adapted) We are given nonzero integers $a \geq b$, such that the quadratic equation $x^2 + ax + b = 0$ has nonzero integer roots $c \geq d$. Then we form the quadratic equation $x^2 + cx + d = 0$ and check if it also has nonzero integer roots. If this is so, we let $e \geq f$ be those roots and form the quadratic equation $x^2 + ex + f = 0$. We proceed in a likewise manner until we reach a quadratic equation with nonzero integer coefficients but with no integer roots. The purpose of this problem is to find all a and b for which this process continues indefinitely. To this end, do the following items:

   (a) Show that if the process is to continue indefinitely, then we can assume that $a > 0 > b$ and that every subsequent equation $x^2 + \alpha x + \beta = 0$ is also such that $\alpha > 0 > \beta$.
   (b) Under the choices of (a), let $x^2 + mx + n = 0$ and $x^2 + px + q = 0$ be two consecutive equations (i.e., such that $p > 0 > q$ are the roots of the first),

---

[14]We shall see in [5] that the same holds for the *complex* roots of this equation.

and let Δ and Δ' be their discriminants, respectively. Show that $\Delta' < \Delta$, unless $n = -1$ or $m = 1, n = -2$.

(c) Conclude that $a = 1$ and $b = -2$ is the only possible choice to begin with.

2.4 Linear Systems and Elimination

Let E and F be algebraic expressions in the real variables $x_1, \ldots, x_n$. By the **equation $E = F$ in the real variables or unknowns** $x_1, \ldots, x_n$, we mean the problem of finding all sequences[15] $(a_1, \ldots, a_n)$ of real numbers, such that E and F make sense and the equality $E = F$ holds when $x_1 = a_1, \ldots, x_n = a_n$. In this case, each such sequence $(a_1, \ldots, a_n)$ is said to be a **solution** of the equation $E = F$.

Now, let $E_1, \ldots, E_m, F_1, \ldots, F_m$ be algebraic expressions in the real variables $x_1, \ldots, x_n$. The **system of equations**

$$\begin{cases} E_1 = F_1 \\ E_2 = F_2 \\ \quad \vdots \\ E_m = F_m \end{cases} \tag{2.22}$$

is the problem of finding all sequences $(a_1, \ldots, a_n)$ of real numbers, such that the substitutions $x_1 = a_1, \ldots, x_n = a_n$ solve all of the equations $E_j = F_j$. As above, each such sequence $(a_1, \ldots, a_n)$ is said to be a **solution** of the system (2.22). To *solve* a system of equations as (2.22) means to find all of its solutions.

In this section and the next one, we shall learn how to solve some simple (though useful) systems of equations. We are also going to see that, under certain conditions, an equation $E = F$ (in several real variables) is equivalent to a system of equations like (2.22), and this will be a source of a number of interesting examples.

The simplest—and, for our purposes, also the most useful—systems of equations are the **linear systems** with two (resp. three) equations in two (resp. three) real unknowns, i.e., systems of equations of one of the forms

$$\begin{cases} a_1x + b_1y = c_1 \\ a_2x + b_2y = c_2 \end{cases} \quad \text{or} \quad \begin{cases} a_1x + b_1y + c_1z = d_1 \\ a_2x + b_2y + c_2z = d_2 \\ a_3x + b_3y + c_3z = d_3 \end{cases}, \tag{2.23}$$

respectively. Here, the real numbers a_i, b_i, c_i, d_i are given and not all zero, being called the **coefficients** of the linear system.

---

[15] Although the reader is probably acquainted with the concept of sequence, we refer to Sect. 6.1 for a rigorous definition.

The most efficient method for solving linear systems like those above is the **elimination** algorithm,[16] also known as **gaussian elimination**,[17] which is based in the following result.

Lemma 2.24 *Let E and F be given algebraic expressions in the real variables x and y (or x, y and z). For reals a, b and c, the systems of equations*

$$\begin{cases} E = a \\ F = b \end{cases} \quad and \quad \begin{cases} E + cF = a + cb \\ F = b \end{cases}$$

have the same solutions.

Proof We shall prove the lemma in the case in which E and F are algebraic expressions in the real variables x and y; the other case is completely analogous. Suppose that $x = x_0$ and $y = y_0$ solve the system on the left, so that, when we substitute x by x_0 and y by y_0 into E and F, both equalities $E = a$ and $F = b$ hold; we shall denote such a situation by writing $E(x_0, y_0) = a$ and $F(x_0, y_0) = b$. Then, substituting x by x_0 and y by y_0 into $E + cF$ gives us

$$(E + cF)(x_0, y_0) = E(x_0, y_0) + cF(x_0, y_0) = a + cb,$$

so that $x = x_0$ and $y = y_0$ also solve the system on the right. Conversely, if $x = x_0$ and $y = y_0$ solve the system on the right, then, since $E = (E+cF)-cF$, an argument entirely analogous to the above shows that $x = x_0$ and $y = y_0$ do solve the system on the left. Therefore, both systems have the same solutions.   □

Back to the analysis of the linear systems (2.23), let us show that a number of careful applications of the previous lemma lead us to quick solutions of them (the *elimination algorithm* consists exactly of this).

We start by the system on the left, which, for simplicity, we write as

$$\begin{cases} ax + by = e \\ cx + dy = f \end{cases}.$$

Since at least one of the coefficients a, b, c, d is nonzero, we can suppose, without loss of generality, that $a \neq 0$ (otherwise, it suffices to rewrite the system in one of the forms

---

[16] An **algorithm** is a finite sequence of precise procedures that, once followed, give an expected result, also known as the **output** of the algorithm. Although algorithms will play a very modest role in this volume, [5] brings a number of very interesting ones.

[17] After Joanne Carl Friedrich Gauss, German mathematician of the XVIII and XIX centuries. Gauss is generally considered to be the greatest mathematician of all times. In the several different areas of Mathematics and Physics in which he worked, like Number Theory, Analysis, Differential Geometry and Electromagnetism, there are always very important and deep results or methods that bear his name. We refer the reader to [26] for an interesting biography of Gauss.

$$\begin{cases} cx + dy = f \\ ax + by = e \end{cases}, \quad \begin{cases} by + ax = e \\ dy + cx = f \end{cases} \quad \text{or} \quad \begin{cases} dy + cx = f \\ by + ax = e \end{cases},$$

according to whether c, b or d is nonzero, and change a by this number in the following discussion).

Then, let $a \neq 0$ and $E = ax + by$, $F = cx + dy$. Changing the equation $F = f$ by the equation

$$F - \frac{c}{a}E = f - \frac{c}{a}e,$$

we get the system

$$\begin{cases} E = e \\ F - \frac{c}{a}E = f - \frac{c}{a}e \end{cases},$$

and Lemma 2.24 immediately assures that the solutions of this new system coincide with those of the original one. Hence, in order to solve that system, it suffices to solve this last one. On the other hand, since

$$F - \frac{c}{a}E = (cx + dy) - \frac{c}{a}(ax + by) = \left(d - \frac{bc}{a} \right) y, \tag{2.24}$$

the last system reduces to

$$\begin{cases} ax + by = e \\ d'y = f' \end{cases},$$

where $d' = d - \frac{bc}{a}$ and $f' = f - \frac{c}{a}e$. Now, we shall consider three different cases:

- If $d' \neq 0$ (or, equivalently, $ad - bc \neq 0$), then the second equation above gives $y = \frac{f'}{d'}$, and the substitution of this value into the first equation gives $x = \frac{1}{a}(e - by) = \frac{1}{a}\left(e - \frac{bf'}{d'} \right)$. In this case, the system is said to be **determined**, for it has a unique solution.
- If $d' = 0$ (or, equivalently, $ad - bc = 0$) and $f' \neq 0$, the system is **impossible**, for the second equation reduces to $0y = f'$, which has no roots at all.
- If $d' = f' = 0$, then the second equation reduces to $0y = 0$, an equality which is true for all real values of y. Therefore, the system as a whole consists only of the equation $ax + by = 0$, which has infinitely many solutions (making $y = \alpha$, with $\alpha \in \mathbb{R}$, we get $x = -\frac{b\alpha}{a}$). For this reason, the system is said to be **undetermined**.

As was the case in the previous chapter, we would like to stress that, in specific examples, rather than memorizing the formulas obtained through the above discussion, one should just execute the elimination process. The previous discussion (more precisely, (2.24)) makes it clear that this process consists of subtracting an appropriate multiple of the first equation from the second, in order to eliminate the variable x from it (and that's where the name *elimination* comes from).

Example 2.25 Use Gaussian elimination to find all real values of a for which the system equations

$$\begin{cases} 2x + ay = 3 \\ ax + 2y = \frac{3}{2} \end{cases}$$

is impossible, undetermined or determined.

Solution Subtracting $-\frac{a}{2}$ times the first equation from the second, we get the equivalent system

$$\begin{cases} 2x + ay = 3 \\ \left(2 - \frac{a^2}{2}\right)y = \frac{3}{2}(1 - a) \end{cases}.$$

If $2 - \frac{a^2}{2} \neq 0$, which is the same as $a \neq \pm 2$, then the second equation above gives us $y = \frac{3(1-a)}{4-a^2}$, so that the first equation furnishes a single value for x, namely,

$$x = \frac{1}{2}(3 - ay) = \frac{3(4 - a)}{2(4 - a^2)}.$$

Therefore, the system is determined whenever $a \neq \pm 2$.

If $a = \pm 2$, then the second equation reduces to $0y = \frac{3}{2}(1 \mp 2)$, which represents an impossible equality. Therefore, the system is impossible.   □

For a *geometric* interpretation of Gaussian elimination for linear systems of two equations in two real variables, see the problems of Sect. 6.2 of [4].

Let us now turn our attention to the linear system on the right, in (2.23). In order to analyse it, let $E_i = a_ix + b_iy + c_iz$, for $1 \leq i \leq 3$, so that it reduces to

$$\begin{cases} E_1 = d_1 \\ E_2 = d_2 \\ E_3 = d_3 \end{cases}.$$

As was done for linear systems in two variables, we can suppose that $a_1 \neq 0$ (the other cases being totally analogous). Changing equations $E_2 = d_2$ and $E_3 = d_3$ respectively by

$$E_2 - \frac{a_2}{a_1}E_1 = d_2 - \frac{a_2}{a_1}d_1 \text{ and } E_3 - \frac{a_3}{a_1}E_1 = d_3 - \frac{a_3}{a_1}d_1$$

(in order to eliminate the variable x from the second and third equations), we get the equivalent system

$$\begin{cases} E_1 = d_1 \\ E_2 - \frac{a_2}{a_1}E_1 = d_2 - \frac{a_2}{a_1}d_1 \\ E_3 - \frac{a_3}{a_1}E_1 = d_3 - \frac{a_3}{a_1}d_1 \end{cases}.$$

Since

$$E_2 - \frac{a_2}{a_1}E_1 = (a_2x + b_2y + c_2z) - \frac{a_2}{a_1}(a_1x + b_1y + c_1z)$$

$$= \left(b_2 - \frac{a_2b_1}{a_1} \right) y + \left(c_2 - \frac{a_2c_1}{a_1} \right) z$$

$$= b_2'y + c_2'z,$$

and, analogously, $E_3 - \frac{a_3}{a_1}E_1 = b_3'y + c_3'z$, it suffices to solve the system

$$\begin{cases} a_1x + b_1y + c_1z = d_1 \\ b_2'y + c_2'z = d_2' \\ b_3'y + c_3'z = d_3' \end{cases},$$

where $d_2' = d_2 - \frac{a_2}{a_1}d_1$ and $d_3' = d_3 - \frac{a_3}{a_1}d_1$.

In case all of the numbers b_2', c_2', b_3', c_3' are equal to 0, the last system above reduces to

$$\begin{cases} a_1x + b_1y + c_1z = d_1 \\ 0 = d_2' \\ 0 = d_3' \end{cases}.$$

If $d_2' \neq 0$ or $d_3' \neq 0$, the system is *impossible*; if $d_2' = d_3' = 0$, the system is *undetermined*, for it is equivalent to the single equation $a_1x + b_1y + c_1z = 0$, which obviously has infinitely many solutions.

Suppose, then, that at least one of the numbers b_2', c_2', b_3' or c_3' is nonzero, say $b_2' \neq 0$ (as before, the other cases can be treated quite analogously). Then, applying Gaussian elimination to the system

$$\begin{cases} b_2'y + c_2'z = d_2' \\ b_3'y + c_3'z = d_3' \end{cases},$$

we obtain a system of the form

$$\begin{cases} a_1x + b_1y + c_1z = d_1 \\ b_2'y + c_2'z = d_2' \\ c_3''z = d_3'' \end{cases},$$

which is also equivalent to the original system. The discussion, then, goes on as before, and we concentrate our attention in the third equation, $c_3'' z = d_3''$. Also as before, we have to distinguish three distinct cases:

- If $c_3'' \neq 0$, then the third equation above gives a single value for z, say $z = \gamma$; since $b_2' \neq 0$, the substitution of this value into the second equation furnishes $y = \frac{1}{b_2'}(d_2' - c_2' z) = \frac{1}{b_2'}(d_2' - c_2' \gamma)$. Finally, since $a_1 \neq 0$, the substitution of the values thus obtained for y and z into the first equation give us a single value for x. Then, we conclude that the system is *determined*.
- If $c_3'' = 0$ and $d_3'' \neq 0$, the system is *impossible*, for the third equation reduces to $0z = d_3''$.
- If $c_3'' = d_3'' = 0$, then the third equation reduces to the equality $0z = 0$, and the system as a whole reduces to

$$\begin{cases} a_1 x + b_1 y + c_1 z = d_1 \\ \quad\quad\quad b_2' y + c_2' z = d_2' \end{cases}$$

or, which is the same,

$$\begin{cases} a_1 x + b_1 y = d_1 - c_1 z \\ \quad\quad\quad b_2' y = d_2' - c_2' z \end{cases}.$$

Since a_1 and b_2' are both nonzero, for each real value $z = \gamma$ the system corresponding to

$$\begin{cases} a_1 x + b_1 y = d_1 - c_1 \gamma \\ \quad\quad\quad b_2' y = d_2' - c_2' \gamma \end{cases}$$

is determined; therefore, the original system in x, y, z is *undetermined*.

We can summarize the above discussion by saying that the elimination algorithm for linear systems of the form

$$\begin{cases} a_1 x + b_1 y + c_1 z = d_1 \\ a_2 x + b_2 y + c_2 z = d_2 \\ a_3 x + b_3 y + c_3 z = d_3 \end{cases}$$

consists in performing, one by one, the following three steps:

1st. *Eliminate* the variable x from the second and third equations, by adding to these equations appropriate multiples of the first one, thus obtaining a system of the form

$$\begin{cases} a_1 x + b_1 y + c_1 z = d_1 \\ \quad\quad\quad b_2' y + c_2' z = d_2' \\ \quad\quad\quad b_3' y + c_3' z = d_3' \end{cases}.$$

2nd. *Eliminate* the variable y (in case $b_2' \neq 0$ in the last system above) from the third equation, by adding to it an adequate multiple of the second equation, thus obtaining a system of the form

$$\begin{cases} a_1 x + b_1 y + c_1 z = d_1 \\ \quad\quad b_2' y + c_2' z = d_2' \\ \quad\quad\quad\quad c_3'' z = d_3'' \end{cases}$$

3rd. Analyse equation $c_3'' z = d_3''$ and, after this, the other two equations in succession, in order to decide whether the original system is *determined*, *undetermined* or *impossible*.

As before, it is much more useful to keep the general steps above in mind than to try to memorize any of the expressions obtained through the calculations in the previous discussions. Let us see one more example to illustrate this point.

Example 2.26 Find all real values of a for which the system of equations

$$\begin{cases} x + y - az = -1 \\ 2x + ay + z = 1 \\ ax + y - z = 2 \end{cases}$$

is impossible.

Solution Multiplying the first equation respectively by 2 and by a, and subtracting (also respectively) the results from the second and third equations, we get the equivalent system

$$\begin{cases} x + y - az \quad\quad\quad\quad = -1 \\ (a-2)y + (1+2a)z = 3 \\ (1-a)y - (1-a^2)z = 2+a \end{cases}$$

If $a = 1$, the last equation reduces to the impossible equality $0 = 3$, and the original system is impossible. If $a \neq 1$, add to the second equation $-\frac{a-2}{1-a}$ times the third one, thus obtaining the equivalent system

$$\begin{cases} x + y - az \quad\quad\quad\quad = -1 \\ (a^2 + a - 1)z \quad\quad = \frac{-a^2 - 3a + 7}{1-a} \\ (1-a)y - (1-a^2)z = 2+a \end{cases}$$

(Observe that we have slightly changed the second step, so that the roles of the second and third equations of the last system are interchanged—it is the second equation that, now, has just one variable. Obviously, this is perfectly right, and shows that the elimination algorithm is quite a flexible one.)

Now, if $a^2 + a - 1 = 0$, i.e., if $a = \frac{-1 \pm \sqrt{5}}{2}$, then the system is impossible, for $-a^2 - 3a + 7 \neq 0$ for these values of a and, hence, the second equation reduces to an impossible equality. If $a \neq \frac{-1 \pm \sqrt{5}}{2}$, then the second equation gives us a single possible value for z; in turn, upon substitution of this value of z into the third equation, we find a single possible value for y; finally, putting these values for y and z into the first equation, we find a single possible value for x, so that the system is determined.

We conclude $a = 1$ and $a = \frac{-1 \pm \sqrt{5}}{2}$ are the values of a for which the original system is impossible. $\square$

We finish our discussion of linear systems by observing that Gaussian elimination algorithm can easily be put to work to the analysis of the general linear system in m equations and n unknowns

$$\begin{cases} a_{11}x_1 + a_{12}x_2 + \cdots + a_{1n}x_n = b_1 \\ a_{21}x_1 + a_{22}x_2 + \cdots + a_{2n}x_n = b_2 \\ \qquad \vdots \qquad\qquad \vdots \\ a_{m1}x_1 + a_{m2}x_2 + \cdots + a_{mn}x_n = b_m \end{cases} ; \qquad (2.25)$$

here, the a_{ij} and b_i are given real numbers, such that at least one of the a_{ij} is nonzero.

Apart from an important particular case of general linear system, which will make its appearance in Sect. 18.1 of [5] (and will be analysed there, by other methods), we shall not make a systematic use of such systems along these notes; hence, we shall not develop the general analysis of the application of the elimination algorithm to them. If it is the case we have to solve such a linear system (as in the example below), some clever reasoning, perhaps with the aid of Lemma 2.24, will suffice.

Example 2.27 (OCM) Find all real solutions $x_1, x_2, \ldots, x_{100}$ of the linear system

$$\begin{cases} x_1 + x_2 + x_3 = 0 \\ x_2 + x_3 + x_4 = 0 \\ \qquad \vdots \qquad \vdots \\ x_{98} + x_{99} + x_{100} = 0 \\ x_{99} + x_{100} + x_1 = 0 \\ x_{100} + x_1 + x_2 = 0 \end{cases} .$$

Solution Observe that each x_j appears in exactly three of the given equations. Therefore, adding all of them and dividing both sides by 3, we get

$$x_1 + x_2 + x_3 + \cdots + x_{100} = 0.$$

In order to find $x_1 = 0$, just note that

$$0 = x_1 + (x_2 + x_3 + x_4) + \cdots + (x_{98} + x_{99} + x_{100})$$
$$= x_1 + 0 + 0 + \cdots + 0 = x_1.$$

For $x_2 = 0$, write

$$0 = x_2 + (x_3 + x_4 + x_5) + \cdots + (x_{99} + x_{100} + x_1)$$
$$= x_2 + 0 + 0 + \cdots + 0 = x_2.$$

Now, $x_1 + x_2 + x_3 = 0$ implies $x_3 = 0$. Then, $x_2 + x_3 + x_4 = 0$ implies $x_4 = 0$, and so on, so that all of the x_i's are equal to 0.   □

We refer the interested reader to Chap. 1 of [18] for quite a detailed exposition of the elimination algorithm for general linear systems.

Problems: Section 2.4

1. Assume that, in the linear system $\begin{cases} ax + by = e \\ cx + dy = f \end{cases}$, we have $a, b, c, d, e, f \neq 0$.
   Prove that:

   (a) $\frac{a}{c} \neq \frac{b}{d} \Leftrightarrow$ the system is determined.
   (b) $\frac{a}{c} = \frac{b}{d} = \frac{e}{f} \Leftrightarrow$ the system is undetermined.
   (c) $\frac{a}{c} = \frac{b}{d} \neq \frac{e}{f} \Leftrightarrow$ the system is impossible.

2. Find all real values of a for which the system of equations

$$\begin{cases} x + 2y - 3z = 4 \\ 3x - y + 5z = 2 \\ 4x + y + (a^2 - 14)z = a + 2 \end{cases}$$

   is impossible.

3. Solve the system of equations

$$\begin{cases} \frac{2}{x} + \frac{3}{y} - \frac{1}{z} = 8 \\ \frac{3}{x} - \frac{5}{y} + \frac{2}{z} = -1 \\ \frac{7}{x} - \frac{6}{y} + \frac{3}{z} = 5 \end{cases}.$$

4. For $1 \leq i, j \leq 3$, let a_{ij} be given real numbers, such that $a_{i1}^2 + a_{i2}^2 + a_{i3}^2 = 1$ for $1 \leq i \leq 3$ and $a_{i1}a_{j1} + a_{i2}a_{j2} + a_{i3}a_{j3} = 0$, for $1 \leq i, j \leq 3$ with $i \neq j$. Given real numbers b_1, b_2, b_3, solve the linear system

$$\begin{cases} a_{11}x_1 + a_{12}x_2 + a_{13}x_3 = b_1 \\ a_{21}x_1 + a_{22}x_2 + a_{23}x_3 = b_2 \ , \\ a_{31}x_1 + a_{32}x_2 + a_{33}x_3 = b_3 \end{cases}$$

writing x_1, x_2 and x_3 in terms of the a_{ij} and b_i.
5. With respect to the linear system (2.25), do the following items:

   (a) If $b_1 = b_2 = \cdots = b_m = 0$, then the system always has at least one solution.
   (b) If the system has at least two different solutions, then it has infinitely many solutions.
   (c) If the system has only one solution when $b_1 = b_2 = \cdots = b_m = 0$, then, for general values of $b_1, b_2, \ldots, b_m$, it has at most one solution.

2.5   Miscellaneous

Let us now turn our attention to the next simplest type of system of equations, namely, **second degree systems**. Ultimately, such a system is just a rephrasing of a second degree equation in terms of a system of two equations in two unknowns. Nevertheless, the reader will be amazed on how they provide a number of interesting applications.

Proposition 2.28 *Let S and P be given real numbers. The system of equations*

$$\begin{cases} x + y = S \\ \quad xy = P \end{cases} \tag{2.26}$$

has real solutions if and only if $S^2 \geq 4P$. Moreover, in this case, the solutions are given by $x = \alpha$, $y = \beta$ or vice-versa, where α and β are the roots of the second degree equation $u^2 - Su + P = 0$.

Proof First of all, if α and β are the roots of $u^2 - Su + P = 0$, then we know from Viète's formulas (see item (b) of Proposition 2.16) that $\alpha + \beta = S$ and $\alpha\beta = P$. Therefore, the pairs $x = \alpha$, $y = \beta$ and $x = \beta$, $y = \alpha$ do satisfy the system of Eq. (2.26).

Conversely, let $x = x_0$, $y = y_0$ be any solution of that system. Then, the first equation gives $y_0 = S - x_0$, and the second equation then gives $P = x_0 y_0 = x_0(S - x_0)$, or, which is the same, $x_0^2 - Sx_0 + P = 0$. Therefore, x_0 is a root of $u^2 - Su + P = 0$, from where we get $x_0 = \alpha$ or $x_0 = \beta$. Since $\alpha + \beta = S$, we have two possibilities:

- If $x_0 = \alpha$, then $y_0 = S - x_0 = S - \alpha = \beta$.
- If $x_0 = \beta$, then $y_0 = S - x_0 = S - \beta = \alpha$.

$\qquad\qquad\qquad\qquad\qquad\qquad\qquad\qquad\qquad\qquad\qquad\qquad\qquad\qquad\qquad\qquad\qquad\qquad\quad\square$

The following examples show how we can sometimes reduce the search for solutions of more complex systems of equations to that of simpler systems, of one of the types we met so far.

Example 2.29 Find all real solutions of the system of equations

$$\begin{cases} (x^2 + 1)(y^2 + 1) = 10 \\ (x + y)(xy - 1) = 3 \end{cases}.$$

Solution First of all, rewriting the left hand side of the first equation as

$$(x^2 + 1)(y^2 + 1) = x^2 y^2 + x^2 + y^2 + 1$$
$$= (xy)^2 + [(x + y)^2 - 2xy] + 1,$$

and letting $x + y = a$ and $xy = b$, we get the system

$$\begin{cases} b^2 + a^2 - 2b = 9 \\ a(b - 1) = 3 \end{cases}.$$

Now, writing

$$b^2 + a^2 - 2b = a^2 + (b^2 - 2b + 1) - 1$$
$$= a^2 + (b - 1)^2 - 1$$

and making the substitution $b - 1 = c$, we reach the system

$$\begin{cases} a^2 + c^2 = 10 \\ ac = 3 \end{cases}.$$

By squaring the second equation, we transform this last system into one of the form (2.26), whose unknowns are a^2 and c^2, and such that $S = 10$ and $P = 9$. Since the roots of the second degree equation $u^2 - 10u + 9 = 0$ are 1 and 9, it follows that $a^2 = 1$ or 9 and, hence, $a = \pm 1$ or ± 3. Then, we have the possibilities

$$(a, c) = (1, 3), (-1, -3), (3, 1) \text{ or } (-3, -1),$$

from where

$$(a, b) = (1, 4), (-1, -2), (3, 2) \text{ or } (-3, 0).$$

Finally, each of these pairs (a, b) give us another system of the form (2.26), with unknowns x and y. Solving the four systems thus obtained, we arrive at the solutions of the original system:

$$(x, y) = (1, -2), (-2, 1), (1, 2), (2, 1), (0, -3) \text{ or } (-3, 0).$$

$\square$

Some seemingly complicated equations can be easily solved, provided we find a way to transform them into systems of equations. Since there is no general procedure that tells us when or how this can be done, each equation should be analysed separately. In this respect, the following example shows that a frequently useful algebraic trick is the introduction of new variables.

Example 2.30 (Israel)   Find all real solutions of the equation

$$\sqrt[4]{13 + x} + \sqrt[4]{4 - x} = 3.$$

Solution   Letting $a = \sqrt[4]{13 + x}$ and $b = \sqrt[4]{4 - x}$, we get $a + b = 3$ and $a^4 + b^4 = (13 + x) + (4 - x) = 17$. Hence, we have reduced the problem of solving the giving equation to that of solving the system of equations

$$\begin{cases} a + b = 3 \\ a^4 + b^4 = 17 \end{cases}.$$

Applying formula (2.1) for the square of a sum twice, we get

$$\begin{aligned} 17 = a^4 + b^4 &= (a^2 + b^2)^2 - 2a^2b^2 \\ &= [(a + b)^2 - 2ab]^2 - 2(ab)^2 \\ &= (9 - 2ab)^2 - 2(ab)^2 \\ &= 81 - 36ab + 2(ab)^2, \end{aligned}$$

so that

$$(ab)^2 - 18(ab) + 32 = 0.$$

Solving for ab the second degree equation above, we find $ab = 2$ or $ab = 16$. Therefore, there are two distinct possibilities:

$$(i)\ \begin{cases} a + b = 3 \\ ab = 2 \end{cases} \quad \text{or } (ii)\ \begin{cases} a + b = 3 \\ ab = 16 \end{cases}.$$

Possibility (i) gives $a = 1$ and $b = 2$, or vice-versa. If $a = 1$, then $13 + x = 1$ and, hence, $x = -12$. If $a = 2$, then $13 + x = 16$ and, hence, $x = 3$. Possibility (ii) does not give any real solutions, for, in this case, a and b would be real roots of the second degree equation $u^2 - 3u + 16 = 0$, which has none of them. $\square$

The following lemma states a relatively easy sufficient condition to transform the search for the roots of an equation in one unknown into that of a system of equations.

Lemma 2.31 *If $E_1, E_2, \ldots, E_n$ are expressions involving one or more real variables, then*

$$E_1^2 + E_2^2 + \cdots + E_n^2 = 0 \Leftrightarrow \begin{cases} E_1 = 0 \\ E_2 = 0 \\ \vdots \\ E_n = 0 \end{cases} . \tag{2.27}$$

Proof This is an easy generalization of Corollary 1.5 and of Problem 2, page 10.

$\square$

Example 2.32 Find all real roots of the equation

$$x^4 y^2 + y^2 + 2 = 2y + 2x^2 y.$$

Solution We can write the given equation as

$$(x^4 y^2 - 2x^2 y + 1) + (y^2 - 2y + 1) = 0,$$

or, which is the same,

$$(x^2 y - 1)^2 + (y - 1)^2 = 0.$$

Therefore, by the previous lemma, the equation is equivalent to the system of equations

$$\begin{cases} x^2 y - 1 = 0 \\ y - 1 \quad = 0 \end{cases},$$

whose solutions $y = 1, x = \pm 1$ can be obtained without difficulty. $\square$

Problems: Section 2.5

1. (IMO) Find all real numbers x, y and z, such that the sum of any of them with the product of the other two is always equal to 2.
2. Let a be a nonzero real constant. Find, in terms of a, all real numbers x and y that solve the system of equations

$$\begin{cases} \frac{1}{x+y} + x = a + 1 \\ \frac{x}{x+y} \quad = a \end{cases}.$$

3. Solve, for $x, y \in \mathbb{R}$, the system of equations

$$\begin{cases} \frac{x+y}{xy} + \frac{x-y}{xy} = 5 \\ \frac{xy}{x+y} + \frac{xy}{x-y} = \frac{5}{6} \end{cases}.$$

4. (IMO—adapted) Consider the system of equations

$$\begin{cases} ax_1^2 + bx_1 + c = x_2 \\ ax_2^2 + bx_2 + c = x_3 \\ ax_3^2 + bx_3 + c = x_1 \end{cases},$$

whose unknowns are x_1, x_2, x_3, where a, b and c are given real numbers, with $a \neq 0$. If $\Delta = (b-1)^2 - 4ac$, do the following items:

(a) If $\Delta < 0$, then there is no real solution.
(b) If $\Delta = 0$, then there is exactly one real solution.

5. (TT) Find all real solutions of the system of equations

$$\begin{cases} x^3 = 2y - 1 \\ y^3 = 2z - 1 \\ z^3 = 2x - 1 \end{cases}.$$

6. Solve, for $x, y \in \mathbb{R}$, the equation $x^2 + 2xy + 3y^2 + 2x + 6y + 3 = 0$.
7. (IMO) Find all real values of a for which the system of equations

$$\begin{cases} x^2 + y^2 = 4z \\ 3x + 4y + z = a \end{cases},$$

with unknowns x, y and z, has a single solution.
8. (NMC) Find all real numbers x, y, z greater than 1, such that

$$x + y + z + \frac{3}{x-1} + \frac{3}{y-1} + \frac{3}{z-1} = 2\left(\sqrt{x+2} + \sqrt{y+2} + \sqrt{z+2}\right).$$

9. Find all real roots of the equation $\sqrt[3]{x+5} + \sqrt[3]{11-x} = 6$.
10. Find all real roots of the equation $\sqrt{5 - \sqrt{5-x}} = x$.
11. (Canada) Find all real roots of the equation $x^2 + \frac{x^2}{(x+1)^2} = 3$.
12. (Canada) Solve the system of equations

$$\begin{cases} \frac{4x^2}{1+4x^2} = y \\ \frac{4y^2}{1+4y^2} = z \\ \frac{4z^2}{1+4z^2} = x \end{cases}$$

13. (Romania) Let a, b, c and d be given real numbers, such that

$$\begin{cases} a + b + c \le 3d \\ b + c + d \le 3a \\ c + d + a \le 3b \\ d + a + b \le 3c \end{cases}.$$

   Prove that $a = b = c = d$.

14. * Let $E_1, E_2, \ldots, E_n$ and $F_1, F_2, \ldots, F_n$ be given expressions in one or more real variables, such that $E_1 \le F_1$, $E_2 \le F_2$, $\ldots$, $E_n \le F_n$. Prove that the equation

$$E_1 + \cdots + E_n = F_1 + \cdots + F_n$$

   is equivalent to the system of equations

$$\begin{cases} E_1 = F_1 \\ E_2 = F_2 \\ \quad\vdots \\ E_n = F_n \end{cases}.$$

15. (Romania) Find all real roots of the equation

$$\sqrt{4x^2 - x^4 - 3} + \sqrt{4y^2 - y^4} + \sqrt{4z^2 - z^4 + 5} = 6.$$

16. Solve, in the set of positive reals, the system of equations

$$\begin{cases} x + \frac{4}{y} = \frac{5y}{4} \\ y + \frac{4}{z} = \frac{5z}{4} \\ z + \frac{4}{x} = \frac{5x}{4} \end{cases}.$$

17. (Soviet Union—adapted)

   (a) For $x > 0$, prove that $x + \frac{2}{x} \ge 2\sqrt{2}$, with equality if and only if $x = \sqrt{2}$.

   (b) Solve the system of equations

$$\begin{cases} 2y = x + \frac{2}{x} \\ 2z = y + \frac{2}{y} \\ 2x = z + \frac{2}{z} \end{cases}.$$

Chapter 3
Elementary Sequences

An (**infinite**) **sequence** of real numbers is an *infinite ordered list* $(a_1, a_2, a_3, \ldots)$ of real numbers, i.e., an infinite list of real numbers, in which we specify who is the first number of the list, who is the second, third and so on. It is customary to denote an infinite sequence as above simply by $(a_k)_{k \geq 1}$. A (**finite**) **sequence** of real numbers is a *finite ordered list* $(a_1, a_2, \ldots, a_n)$ of real numbers, i.e., a finite list of real numbers, in which, as was the case with infinite sequences, we specify who is the first number of the list, who is the second, the third, ..., the n-th. Is it customary to denote a finite sequence as above simply by $(a_k)_{1 \leq k \leq n}$. In any of the cases above, we say that a_k is the **k-th term** of the sequence.[1]

Our aim in this chapter is to study some types of elementary sequences that occur quite frequently in elementary Mathematics. Along this process, we shall introduce several definitions and properties that will apply to general sequences as well. Whenever there is no danger of confusion, we shall concentrate our discussion on infinite sequences, letting to the reader the task of adapting all that comes to finite sequences.

3.1 Progressions

We say that a sequence $(a_k)_{k \geq 1}$ is defined by a **positional formula** if the values $a_k \in \mathbb{R}$ are given by means of an expression that depends on k.

---

[1] As we will see in Chap. 6, an infinite (resp. finite) sequence of real numbers is, rigorously speaking, a function $f : \mathbb{N} \to \mathbb{R}$ (resp. $f : \{1, 2, \ldots, n\} \to \mathbb{R}$, for some $n \in \mathbb{N}$). Thus, the above notation comes from the shortcut of letting $a_k = f(k)$, for every $k \in \mathbb{N}$ (resp., every $1 \leq k \leq n$). However, for the time being, the above definition will be harmless and sufficient for all of our purposes.

© Springer International Publishing AG 2017
A. Caminha Muniz Neto, *An Excursion through Elementary Mathematics, Volume I*,
Problem Books in Mathematics, DOI 10.1007/978-3-319-53871-6_3

Example 3.1 The sequence $(a_k)_{k\geq1}$ of the perfect squares is the infinite sequence $(1^2, 2^2, 3^2, \ldots)$. Therefore, we have $a_1 = 1^2$, $a_2 = 2^2$, $a_3 = 3^2$ and, more generally, $a_k = k^2$, for every integer $k \geq 1$.

An alternative to defining a sequence by means of a positional formula is to give a **recursive definition** of it. This means that we should specify a (finite) number of initial terms of the sequence, as well as a *recipe* to calculate the other terms by *recurring* to those that precede it. Let us see an example.

Example 3.2 Consider the sequence $(a_k)_{k\geq1}$, defined recursively by $a_1 = 2$, $a_2 = 5$ and

$$a_k = 2a_{k-1} - a_{k-2}, \ \forall \ k \geq 3. \tag{3.1}$$

If we take $k = 3$ in the relation above, we get $a_3 = 2a_2 - a_1 = 2 \cdot 5 - 2 = 8$; if we now take $k = 4$, we find $a_4 = 2a_3 - a_2 = 2 \cdot 8 - 5 = 11$, and so on. In the above, relation (3.1) is the **recurrence relation**, or simply the **recurrence** satisfied by the sequence $(a_k)_{k\geq1}$.

In the previous example, it is important to notice that we were only capable to compute the value of a_3 because we knew, beforehand, both the recurrence relation (3.1) as well as the values of a_1 and a_2. Knowing just the value of a_1 would not suffice, for (3.1) computes each term as a function of the two ones that immediately precede it. On the other hand, if we changed the values of a_1 and a_2 (but maintained the recurrence relation), we would generally change the corresponding values of subsequent terms (just make $a_1 = 1$ and $a_2 = 2$ in the given recurrence relation, and compute the new value of the a_3).

It is also worth observing that there are other equivalent ways of writing the recurrence relation of the former example. For instance, letting $k - 2 = j$, we get $k = j + 2$ and, hence, $a_{j+2} = 2a_{j+1} - a_j$, $\forall \ j \geq 1$ (since $k \geq 3 \Rightarrow j \geq 1$). In other words, the *name* we give to the **index** of a sequence (i.e., j, k etc.) is not relevant to its definition; we could equally well write it as $a_{k+2} = 2a_{k+1} - a_k$, $\forall \ k \geq 1$.

Sometimes, we will also need to list the terms of a sequence starting from zero, denoting it as $(a_k)_{k\geq0}$. At first, this could look a little odd, for the first term of the sequence would be a_0, the second would be a_1 etc. Nevertheless, since this can be a natural choice in a number of contexts, we urge the reader to get used to it. For example, the sequence $(1, 4, 9, 16, \ldots)$ of the perfect squares could be written as $a_k = (k + 1)^2$, for all $k \geq 0$.

At this point, a natural question poses itself: if a certain sequence is defined recursively, how can we find a positional formula for its terms? Unfortunately, this question does not admit a simple answer. Instead of trying to attack this general problem right now, we shall first discuss some simple examples, postponing a deeper analysis to Chaps. 3 and 21 of [5].

We begin by looking at *arithmetic progressions*.

Definition 3.3 A sequence $(a_k)_{k\geq1}$ of real numbers is an **arithmetic progression** (we abbreviate **AP**) if there exists a real number r such that

$$a_{k+1} = a_k + r, \tag{3.2}$$

for every $k \geq 1$.

In the previous definition, we refer to the real number r as the **common difference** of the AP. We also observe that, for an AP $(a_k)_{k\geq 1}$ to be completely determined, one has to give both its first term a_1 and its common difference r. For example, the sequence $(a_k)_{k\geq 1}$ such that $a_1 = 2$ and $a_{k+1} = a_k + 3$ for $k \geq 1$, is the AP with initial term 2 and common difference 3, and it is completely determined by the recurrence relation it satisfies and by the value of its first term. However, if we only knew that $a_{k+1} = a_k + 3$ for $k \geq 1$, we would not have just one AP, for we would not know how to start it.

Example 3.4 Let $(a_k)_{k\geq 1}$ be an AP of common difference r. Prove that the sequences $(b_k)_{k\geq 1}$ and $(c_k)_{k\geq 1}$, defined by $b_k = a_{2k}$ and $c_k = a_{2k-1}$ for $k \geq 1$, are also AP's, with common differences equal to $2r$.

Proof We look at the sequence $(b_k)_{k\geq 1}$, the case of the sequence $(c_k)_{k\geq 1}$ being totally analogous. By the definition of an AP, it suffices to show that $b_{k+1} - b_k = 2r$, for $k \geq 1$. Since $(a_k)_{k\geq 1}$ is an AP with common difference r, it follows from the definition of b_k that

$$\begin{aligned} b_{k+1} - b_k &= a_{2(k+1)} - a_{2k} = a_{2k+2} - a_{2k} \\ &= (a_{2k+2} - a_{2k+1}) + (a_{2k+1} - a_{2k}) \\ &= r + r = 2r, \end{aligned}$$

as we wished to prove. □

Another useful recursive characterization of AP's is the one given in the following result.

Proposition 3.5 *A sequence $(a_k)_{k\geq 1}$ of real numbers is an AP if and only if*

$$a_{k+2} + a_k = 2a_{k+1}, \forall\, k \geq 1. \tag{3.3}$$

Proof By definition, a sequence is an AP if and only if $a_2 - a_1 = a_3 - a_2 = \cdots$, i.e., if and only if we have $a_{k+2} - a_{k+1} = a_{k+1} - a_k$ for every integer $k \geq 1$. This is an equivalent way of writing (3.3). □

The next result collects two useful formulas for AP's.

Proposition 3.6 *If $(a_k)_{k\geq 1}$ is an AP with common difference r, then*

(a) $a_k = a_1 + (k-1)r$, *for every $k \geq 1$.*
(b) $a_1 + a_2 + \cdots + a_n = \frac{n(a_1+a_n)}{2}$, *for every $n \geq 1$.*

Proof

(a) Diagram

$$a_1 \xrightarrow{\ +r\ } a_2 \xrightarrow{\ +r\ } a_3 \xrightarrow{\ +r\ } \cdots \xrightarrow{\quad} a_{k-1} \xrightarrow{\ +r\ } a_k$$

makes it clear that, in order to reach a_k from a_1, it is necessary $k-1$ steps, where each step consists of adding r to a term. Therefore, in order to obtain a_k from a_1, we have to add $(k-1)r$ to a_1, so that $a_k = a_1 + (k-1)r$.

(b) From diagram

$$a_1 \xrightarrow{\ +r\ } a_2 \xrightarrow{\ +r\ } a_3 \xrightarrow{\ +r\ } \cdots \xleftarrow{\ -r\ } a_{k-2} \xleftarrow{\ -r\ } a_{n-1} \xleftarrow{\ -r\ } a_n \ ,$$

we conclude that

$$a_1 + a_n = (a_2 - r) + (a_{n-1} + r) = a_2 + a_{n-1},$$
$$a_2 + a_{n-1} = (a_3 - r) + (a_{n-2} + r) = a_3 + a_{n-2}$$

etc. Therefore, letting $S = a_1 + a_2 + \cdots + a_n$, we have

$$
\begin{aligned}
2S &= 2(a_1 + a_2 + a_3 + \cdots + a_{n-2} + a_{n-1} + a_n) \\
&= (a_1 + a_n) + (a_2 + a_{n-1}) + (a_3 + a_{n-2}) + \cdots + (a_n + a_1) \\
&= \underbrace{(a_1 + a_n) + (a_1 + a_n) + (a_1 + a_n) + \cdots + (a_1 + a_n)}_{n \text{ summands}} \\
&= n(a_1 + a_n),
\end{aligned}
$$

and the formula follows.

$\square$

The formulas of items (a) and (b) of the previous proposition are respectively known as the formulas for the **general term** and for the sum of the first n terms of an AP. Let us see two simple applications of them.

Example 3.7 Compute, in terms of n, the sums of the first n positive integers and the first n odd positive integers.

Solution The positive integers form the AP $1, 2, 3, 4, \ldots$, of common difference 1. Since its n-th term is obviously equal to n, item (b) of the previous proposition gives

$$1 + 2 + 3 + \cdots + n = \frac{n(n+1)}{2}.$$

The odd positive integers form the AP $1, 3, 5, 7, \ldots$, of common difference 2. Its n-th term (i.e., the n-th odd positive integer) is, by the formula for the general term of an AP, equal to $1 + (n - 1) \cdot 2 = 2n - 1$. Thus, again by item (b) of the previous proposition, the sum of the first n odd positive integers is

$$1 + 3 + 5 + \cdots + (2n - 1) = \frac{n[1 + (2n - 1)]}{2} = n^2. \qquad \square$$

Example 3.8 Let $(a_k)_{k \geq 1}$ be the sequence given by $a_1 = 1$ and

$$a_{k+1} = \frac{a_k}{1 + 2a_k},$$

for every integer $k \geq 1$. Compute a_k as a function of k.

Solution It is clear that all terms of the sequence $(a_k)_{k \geq 1}$ are positive, so that we can define the sequence $(b_k)_{k \geq 1}$ by letting $b_k = \frac{1}{a_k}$, for every $k \geq 1$. Then, the recurrence for $(a_k)_{k \geq 1}$ gives us

$$b_{k+1} = \frac{1}{a_{k+1}} = \frac{1 + 2a_k}{a_k} = \frac{1}{a_k} + 2 = b_k + 2,$$

so that $(b_k)_{k \geq 1}$ is the AP of initial term $b_1 = \frac{1}{a_1} = 1$ and common difference 2. Therefore, this AP coincides with that of the odd positive integers, so that the previous example gives $b_k = 2k - 1$, for every $k \geq 1$. Hence,

$$a_k = \frac{1}{b_k} = \frac{1}{2k - 1}.$$

$\qquad \square$

Another useful class of sequences is the one of *geometric progressions*, according to the next definition.

Definition 3.9 A sequence $(a_k)_{k \geq 1}$ of real numbers is a **geometric progression** (we shall abbreviate **GP**) if there exists a real number q such that

$$a_{k+1} = q \cdot a_k, \qquad (3.4)$$

for every $k \geq 1$.

The real number q that appears in the definition of a GP is its **common ratio**. Observe that, if $q = 0$, then $a_k = 0$ for every $k > 1$. On the other hand, if $q = 1$, then $a_k = a_1$ for every $k \geq 1$. As it happens with AP's, a GP $(a_k)_{k \geq 1}$ is completely determined provided we know its first term a_1 and its common ratio q.

Example 3.10 For a fixed real number $q \neq 0$, the sequence $(a_k)_{k \geq 1}$ such that $a_k = q^k$ for every $k \geq 1$ (i.e., the sequence formed by the powers of q, with natural exponents) is a GP of common ratio q. If $q < 0$, then Problem 6, page 11, shows that a_k is positive if and only if k is even; if $0 < q < 1$, then Corollary 1.3 gives $a_1 > a_2 > a_3 > \cdots > 0$; if $q > 1$, then, again by that result, we have $0 < a_1 < a_2 < a_3 < \cdots$.

Another useful recursive characterization of (almost all) GP's is the one given by the following result.

Proposition 3.11 *A sequence* $(a_k)_{k \geq 1}$ *of nonzero real numbers is a GP if and only if*

$$a_{k+2} a_k = a_{k+1}^2, \tag{3.5}$$

for every $k \geq 1$.

Proof By definition, $(a_k)_{k \geq 1}$ is a GP (of common ratio q) if and only if

$$\frac{a_2}{a_1} = \frac{a_3}{a_2} = \frac{a_4}{a_3} = \cdots = q.$$

In other words, $(a_k)_{k \geq 1}$ is a GP if and only if for every integer $k \geq 1$, we have $\frac{a_{k+2}}{a_{k+1}} = \frac{a_{k+1}}{a_k}$, and this is an equivalent way of writing (3.5). $\square$

As we have done with AP's, our next result brings formulas for the **general term** and for the sum of the first k terms of a GP.

Proposition 3.12 *If* $(a_k)_{k \geq 1}$ *is a GP of common ratio* q, *then:*

(a) $a_k = a_1 \cdot q^{k-1}$, *for every* $k \geq 1$.
(b) *If* $q \neq 1$, *then* $a_1 + a_2 + \cdots + a_n = \frac{a_{n+1} - a_1}{q - 1}$, *for every* $n \geq 1$.

Proof

(a) The proof we present parallels that for the general term of an AP (see the proof of item (a) of Proposition 3.6): diagram

$$a_1 \xrightarrow{\cdot q} a_2 \xrightarrow{\cdot q} a_3 \xrightarrow{\cdot q} \cdots \xrightarrow{\cdot q} a_{k-1} \xrightarrow{\cdot q} a_k$$

makes it clear that, in order to reach a_k starting from a_1, it is necessary $k - 1$ steps, where each step consists of multiplying a term of the sequence by q. Hence, to reach a_k we have to multiply a_1 by q exactly $k - 1$ times, so that $a_k = a_1 \cdot q^{k-1}$.

(b) Letting $S_n = a_1 + a_2 + \cdots + a_n$, it follows from (3.4) that

$$\begin{aligned} q S_n &= q(a_1 + a_2 + \cdots + a_{n-1} + a_n) \\ &= q a_1 + q a_2 + \cdots + q a_{n-1} + q a_n \\ &= a_2 + a_3 + \cdots + a_n + a_{n+1}. \end{aligned}$$

Therefore,

$$(q-1)S_n = qS_n - S_n$$
$$= (a_2 + a_3 + \cdots + a_n + a_{n+1}) - (a_1 + a_2 + \cdots + a_n)$$
$$= (a_2 + a_3 + \cdots + a_n) + a_{n+1} - a_1 - (a_2 + \cdots + a_n)$$
$$= a_{n+1} - a_1,$$

where, in the last passage, we cancelled out the summand $a_2 + a_3 + \cdots + a_n$. It now suffices to divide both members of the equality $(q-1)S = a_{n+1} - a_1$ by $q - 1$.

□

Example 3.13 Let $(a_k)_{k\geq 1}$ be an AP of natural numbers, with common difference $r > 0$, and $(b_k)_{k\geq 1}$ be a GP of nonzero real numbers, with common ratio q. Consider the sequence $(c_k)_{k\geq 1}$, such that $c_k = b_{a_k}$ for every integer $k \geq 1$. Prove that $(c_k)_{k\geq 1}$ is a GP of common ratio q^r.

Proof It suffices to show that the ratio of any two consecutive terms of the sequence $(c_k)_{k\geq 1}$ is always equal to q^r. For this, we use the formulas for the general terms of AP's and GP's, as well as the result of Problem 4, page 11:

$$\frac{c_{k+1}}{c_k} = \frac{b_{a_{k+1}}}{b_{a_k}} = \frac{b_1 q^{a_{k+1}-1}}{b_1 q^{a_k-1}} = q^{a_{k+1}-a_k} = q^r.$$

□

The purpose of the next example is to call the reader's attention to the fact that the idea used in the proof of item (b) of Proposition 3.12 is as important as the actual result it collects. Hence, the reader should keep this idea in his/her mind.

Example 3.14 Compute the value of the sum

$$2 \cdot 1 + 7 \cdot 3 + 12 \cdot 3^2 + 17 \cdot 3^3 + \cdots + 497 \cdot 3^{99} + 502 \cdot 3^{100},$$

where, from left to right, the k-th summand equals the product of the k-th term of the AP $2, 7, 12, \ldots, 502$ with the k-th term of the GP $1, 3, 3^2, \ldots, 3^{100}$.

Solution Arguing exactly as in the proof of item (b) of Proposition 3.12, let S denote the desired sum, and compute $3S$ (we use the factor 3, for it is the common ratio of the GP $1, 3, 3^2, \ldots, 3^{100}$):

$$3S = 2 \cdot 3 + 7 \cdot 3^2 + 12 \cdot 3^3 + 17 \cdot 3^4 + \cdots + 497 \cdot 3^{100} + 502 \cdot 3^{101}.$$

Hence,

$$2S = 3S - S$$

$$= 2 \cdot 3 + 7 \cdot 3^2 + 12 \cdot 3^3 + 17 \cdot 3^4 + \cdots + 497 \cdot 3^{100} + 502 \cdot 3^{101}$$

$$- 2 \cdot 1 - 7 \cdot 3 - 12 \cdot 3^2 - 17 \cdot 3^3 - \cdots - 497 \cdot 3^{99} - 502 \cdot 3^{100}$$

$$= (502 \cdot 3^{101} - 2) - 5(3 + 3^2 + 3^3 + 3^4 + \cdots + 3^{100})$$

$$= (502 \cdot 3^{101} - 2) - \frac{5}{2}(3^{101} - 3),$$

where we grouped multiples of equal powers of 3 in the third equality, and used the formula of item (b) of Proposition 3.12 in the fourth one. Finally, grouping together the summands of the last expression above, we get $S = \frac{1}{4}(999 \cdot 3^{101} + 11)$. □

Problems: Section 3.1

1. Compute the first four terms of the sequence $(a_n)_{n \geq 1}$, defined by $a_1 = 1$ and $a_{k+1} = a_k^2 + a_k + 1$, for $k \geq 1$.

2. Write a positional formula for each of the following sequences:

   (a) $(1, -2, 3, -4, 5, -6, 7, -8, \ldots)$.
   (b) $\left(\frac{1}{2}, \frac{2}{3}, \frac{3}{4}, \frac{4}{5}, \frac{5}{6}, \frac{6}{7}, \frac{7}{8}, \frac{8}{9}, \ldots\right)$.
   (c) $\left(1, \frac{1}{2}, 3, \frac{1}{4}, 5, \frac{1}{6}, 7, \frac{1}{8}, \ldots\right)$.

3. Let $(a_n)_{n \geq 1}$ be a sequence of positive real numbers satisfying the recurrence relation $a_{k+1} = \frac{a_k}{3a_k + 1}$, for $k \geq 1$. Find a recurrence relation satisfied by the sequence $(b_n)_{n \geq 1}$, defined for $n \geq 1$ by $b_n = \frac{1}{a_n}$.

4. Write a recurrence relation satisfied for each of the following sequences:

   (a) $(1, 1, 1, 3, 5, 9, 17, 31, 57, 105, 183, \ldots)$.
   (b) $(1, 2, 2^2, 2^{2^2}, 2^{2^{2^2}}, \ldots)$.

5. Below, we show the first four lines of an infinite array of natural numbers, such that, for $i > 1$, the i-th line starts with the number i and has two more entries than the $(i-1)$-th line. Compute the sum of the entries in the n-th line.

$$
\begin{array}{ccccccc}
1 & & & & & & \\
2 & 3 & 4 & & & & \\
3 & 4 & 5 & 6 & 7 & & \\
4 & 5 & 6 & 7 & 8 & 9 & 10 \\
\cdots & \cdots & \cdots & \cdots & \cdots & \cdots & \cdots
\end{array}
$$

6. Show that the number $11 \ldots 1$ (n digits 1) is equal to $\frac{10^n - 1}{9}$.

7. Compute the sum $1 + 11 + 111 + \cdots + \underbrace{11 \ldots 1}_{n}$ as a function of n.

8. If $(a_k)_{k \geq 1}$ is an AP with common difference r, prove that the sequence $(b_k)_{k \geq 1}$, defined by $b_k = a_{k+1}^2 - a_k^2$ for every $k \geq 1$, is also an AP, and compute its common difference.

9. Let $(a_k)_{k \geq 1}$ be an AP of nonzero common difference, and p, q, u, v be given naturals. Prove that

$$a_p + a_q = a_u + a_v \Leftrightarrow p + q = u + v$$

10. Let $(a_k)_{k \geq 1}$ be an AP of common difference $r \neq 0$. If $\frac{a_1}{r}$ is a nonnegative integer, prove that the sum of any two terms of the AP is also a term of it.

11. Let $(a_k)_{k \geq 1}$ be an AP such that $a_p = \alpha$ and $a_q = \beta$, with $p \neq q$. Compute a_{p+q} in terms of p, q, α, β.

12. (Romania) The sum of some (more than one) consecutive odd integers equals 7^3. Find these numbers.

13. The AP $(a_k)_{k \geq 1}$ is formed by pairwise distinct naturals. Prove that it contains infinitely many composite naturals[2] among its terms.

14. (Canada) Let a_n be the sum of the first n terms of the sequence

$$(0, 1, 1, 2, 2, 3, 3, 4, 4, \ldots, r, r, r + 1, r + 1, \ldots).$$

    (a) Find a formula for a_n as a function of n.
    (b) Prove that $a_{m+n} - a_{m-n} = mn$ for every natural numbers m and n, with $m > n$.

15. (Canada) Show that the numbers $\sqrt{2}$, $\sqrt{3}$ and $\sqrt{5}$ cannot be terms of a single AP.

16. Let $(a_k)_{k \geq 1}$ be a GP with common ratio q. Prove that, for an arbitrary integer $n \geq 1$, we have

$$a_1 a_2 \ldots a_n = a_1^n q^{\frac{n(n-1)}{2}}.$$

17. Compute the value of the sum

$$\frac{1}{2} + \frac{3}{2^2} + \frac{5}{2^3} + \cdots + \frac{99}{2^{50}},$$

in which the k-th summand from left to right equals the quotient between the k-th term of the AP $1, 3, 5, \ldots, 99$ and the k-th term of the GP $2, 2^2, 2^3, \ldots, 2^{50}$.

---

[2] According to the introduction to Chap. 1, we recall that a natural $n > 1$ is *composite* if we can write $n = ab$, for some naturals $a, b > 1$.

18. The sequence $(a_k)_{k \geq 1}$ is an *arithmetic-geometric progression* if, for every integer $k \geq 1$, we have $a_k = b_k c_k$, where the sequences $(b_k)_{k \geq 1}$ and $(c_k)_{k \geq 1}$ are respectively an AP and a GP. Let r be the common difference of the AP and q be the common ratio of the GP. Compute, as a function of n, b_1, c_1, r and q, the value of the sum of the first n terms of $(a_k)_{k \geq 1}$.

19. (Macedonia) In a nonconstant AP of real numbers, the quotient between the first term and the common difference is an irrational number. Prove that no three distinct terms of this AP form a GP.

20. Compute, as a function of n, the n-th term of the sequence $(a_k)_{k \geq 1}$, given by $a_1 = 2$ and $a_{k+1} = 2a_k - 1$, for every integer $k \geq 1$.

21. The sequence $(a_n)_{n \geq 1}$ satisfies $a_1 = 1$ and $a_{k+1} = 3a_k - 1$, for every $k \geq 1$. Do the following items:

    (a) If $b_n = a_n - \frac{1}{2}$, prove that $b_{k+1} = 3b_k$ for every $k \geq 1$.
    (b) Write down the first five terms of the sequence $(b_n)_{n \geq 1}$ and, then obtain a general positional formula for it.
    (c) Find a positional formula for a_n.

22. Prove that there does not exist a GP having the numbers 2, 3 and 5 as three of its terms.

3.2   Linear Recurrences of Orders 2 and 3

At the end of last section, we saw that a sequence $(a_n)_{n \geq 1}$ is a GP if and only if it satisfies a recurrence relation of the form (3.4), i.e., if and only if

$$a_{k+1} - qa_k = 0, \tag{3.6}$$

for every integer $k \geq 1$, where q is a real constant. For reasons that will be clear in a few moments, we say that (3.6) is a **first order linear recurrence relation with constant coefficients**. There, we also saw that a sequence $(a_n)_{n \geq 1}$ is an AP if and only if it satisfies a recurrence relation like (3.3).

In this section, our task is to study the more general class of sequences $(a_n)_{n \geq 1}$ that satisfy recurrence relations like

$$a_{k+2} + ra_{k+1} + sa_k = 0, \tag{3.7}$$

where r and s are given real constants, not both zero. If $s = 0$, (3.7) essentially reduces to (3.6); if $s \neq 0$, it is said to be a **second order linear recurrence relation with constant coefficients**.

Problem 5 explains what we mean by the property of *linearity* of sequences that satisfy recurrence relations like (3.6) and, more generally, (3.7). For such a sequence $(a_n)_{n \geq 1}$, Theorem 3.16 will explain how to compute a_n as a function of n. Before we present it, it is instructive to show the idea behind its proof in a concrete example, and we start by doing this.

Example 3.15 Let $(a_n)_{n\geq 1}$ be the sequence such that $a_1 = 1$, $a_2 = 7$ and $a_{k+2} = 8a_{k+1} - 15a_k$, for every integer $k \geq 1$. Find a positional formula for a_n.

Solution For every integer $k \geq 1$, we have

$$a_{k+2} - 3a_{k+1} = 5(a_{k+1} - 3a_k).$$

Thus, the sequence $(b_k)_{k\geq 1}$, defined by $b_k = a_{k+1} - 3a_k$ for every $k \geq 1$, is a GP of common ratio 5 and initial term $b_1 = a_2 - 3a_1 = 4$. Hence, $b_k = b_1 \cdot 5^{k-1} = 4 \cdot 5^{k-1}$. Analogously, for every integer $k \geq 1$, we have

$$a_{k+2} - 5a_{k+1} = 3(a_{k+1} - 5a_k),$$

so that the sequence $(c_k)_{k\geq 1}$, given by $c_k = a_{k+1} - 5a_k$ for every $k \geq 1$, is a GP of common ratio 3. Since its initial term is $c_1 = a_2 - 5a_1 = 2$, it follows that $c_k = c_1 \cdot 3^{k-1} = 2 \cdot 3^{k-1}$.

Therefore, for every integer $k \geq 1$, we have the system of linear equations

$$\begin{cases} a_{k+1} - 3a_k = 4 \cdot 5^{k-1} \\ a_{k+1} - 5a_k = 2 \cdot 3^{k-1} \end{cases},$$

which easily gives

$$a_k = 2 \cdot 5^{k-1} - 3^{k-1}, \ \forall \, k \geq 1. \qquad \square$$

The reasoning presented in the solution to the previous example can be easily generalized to deal with a general second order linear recurrence relation with constant coefficients. We do this now, referring the reader to Sect. 11.4 for a different approach.

Theorem 3.16 *Let $(a_n)_{n\geq 1}$ be a sequence of real numbers such that*

$$a_{k+2} + ra_{k+1} + sa_k = 0$$

for every integer $k \geq 1$, where r and s are given real constants, not both zero. If the second degree equation $x^2 + rx + s = 0$ has real roots[3] α and β, then there exist real constants A and B, completely determined by the values of a_1 and a_2, such that:

(a) If $\alpha \neq \beta$, then $a_n = A\alpha^{n-1} + B\beta^{n-1}$, for every integer $n \geq 1$.
(b) If $\alpha = \beta$, then $a_n = (A + B(n - 1))\alpha^{n-1}$, for every integer $n \geq 1$.

Proof Recall from Proposition 2.16 that $\alpha + \beta = -r$ and $\alpha\beta = s$. Thus, (3.7) can be rewritten as

$$a_{k+2} - (\alpha + \beta)a_{k+1} + (\alpha\beta)a_k = 0$$

---

[3]The case of complex roots is dealt with, in a much more general setting, in Chap. 21 of [5].

or, which is the same, as

$$a_{k+2} - \alpha a_{k+1} = \beta(a_{k+1} - \alpha a_k)$$

or

$$a_{k+2} - \beta a_{k+1} = \alpha(a_{k+1} - \beta a_k),$$

for every integer $k \geq 1$.

Letting $b_k = a_{k+1} - \alpha a_k$ and $c_k = a_{k+1} - \beta a_k$, it follows from the above relations that $(b_k)_{k \geq 1}$ and $(c_k)_{k \geq 1}$ are GP's of common ratios respectively equal to β and α, and initial terms respectively equal to $b_1 = a_2 - \alpha a_1$ and $c_1 = a_2 - \beta a_1$. Therefore, the formula for the general term of a GP gives

$$b_k = (a_2 - \alpha a_1)\beta^{k-1} \quad \text{and} \quad c_k = (a_2 - \beta a_1)\alpha^{k-1}$$

or, which is the same,

$$\begin{cases} a_{k+1} - \alpha a_k = (a_2 - \alpha a_1)\beta^{k-1} \\ a_{k+1} - \beta a_k = (a_2 - \beta a_1)\alpha^{k-1} \end{cases}. \tag{3.8}$$

Now, let us first consider the case $\alpha \neq \beta$. In (3.8), subtracting the first relation from the second gives

$$a_k = \frac{a_2 - \beta a_1}{\alpha - \beta}\alpha^{k-1} - \frac{a_2 - \alpha a_1}{\alpha - \beta}\beta^{k-1};$$

letting

$$A = \frac{a_2 - \beta a_1}{\alpha - \beta} \quad \text{and} \quad B = -\frac{a_2 - \alpha a_1}{\alpha - \beta}$$

we get the formula of item (a).

In case $\alpha = \beta$, relations (3.8) are equal and, at a first glance, it seems that we do not have enough information to compute a_k. Nevertheless, we can use the following trick: it is immediate to verify that the sequences $(u_k)_{k \geq 1}$ and $(v_k)_{k \geq 1}$, given by $u_k = \alpha^{k-1}$ and $v_k = (k-1)\alpha^{k-1}$ (the same α as before, root of the second degree equation $x^2 + rx + s = 0$), are such that

$$\begin{cases} u_{k+2} + ru_{k+1} + su_k = 0 \\ v_{k+2} + rv_{k+1} + sv_k = 0 \end{cases},$$

for every integer $k \geq 1$ (see Problem 4); therefore, for fixed $A, B \in \mathbb{R}$, the sequence $z_k = Au_k + Bv_k$ is also such that

$$z_{k+2} + rz_{k+1} + sz_k = 0$$

(see Problem 5). Thus, the idea is to search for real numbers A and B such that, for every integer $k \geq 1$, we have $a_k = z_k$. Since both sequences $(a_k)_{k \geq 1}$ and $(z_k)_{k \geq 1}$ satisfy identical linear second order recurrence relations, Problem 13, page 96, will assure that it suffices to find real numbers A e B for which $a_1 = z_1$ and $a_2 = z_2$, i.e., such that

$$\begin{cases} a_1 = A \\ a_2 = (A + B)\alpha \end{cases}.$$

This can obviously be done, for $\alpha = -\frac{r}{2} \neq 0$. □

In practice, we employ the formulas of the previous result in the following way: let $(a_n)_{n \geq 1}$ be a sequence of real numbers satisfying the recurrence relation (3.7) for every integer $k \geq 1$, where r and s are given real constants, with $r \neq 0$. We start by computing the real roots (if any) of the second degree equation

$$x^2 + rx + s = 0,$$

which is called the **characteristic equation** of (3.7). If the characteristic equation actually has real roots α and β, we then check whether $\alpha \neq \beta$ or $\alpha = \beta$, and use the formulas of items (a) or (b) of Theorem 3.16, according to the case at hand. In order to find the real constants A and B in case $\alpha \neq \beta$, we solve the system of equations

$$\begin{cases} a_1 = A + B \\ a_2 = A\alpha + B\beta \end{cases};$$

in case $\alpha = \beta$, we solve the system

$$\begin{cases} a_1 = A \\ a_2 = (A + B)\alpha \end{cases}.$$

In both of these cases, the systems we have to solve are obtained by taking k respectively equal to 1 and 2 in the formulas of items (a) or (b) of the theorem.

Example 3.17 Let us execute the above procedure to get a positional formula for the n-th term of the celebrated **Fibonacci sequence**,[4] i.e., the sequence $(F_n)_{n \geq 1}$ given by $F_1 = 1, F_2 = 1$ and

$$F_{k+2} = F_{k+1} + F_k, \tag{3.9}$$

---

[4]After Leonardo di Pisa, also known as *Fibonacci*, Italian mathematician of the XII and XIII centuries. As can be seen in Chap. 1 of [5], the Fibonacci sequence plays a relevant role in Combinatorics.

for every integer $k \geq 1$. More precisely, let us show that

$$F_n = \frac{1}{\sqrt{5}}\left\{\left(\frac{1+\sqrt{5}}{2}\right)^n - \left(\frac{1-\sqrt{5}}{2}\right)^n\right\}, \quad \forall\, n \geq 1. \tag{3.10}$$

Solution The recurrence relation (3.9) can be written as $F_{k+2} - F_{k+1} - F_k = 0$ and, hence, has characteristic equation $x^2 - x - 1 = 0$. Letting $\alpha = \frac{1+\sqrt{5}}{2}$ and $\beta = \frac{1-\sqrt{5}}{2}$ be its roots, it follows from item (a) of Theorem 3.16 that

$$F_n = A\alpha^{n-1} + B\beta^{n-1},$$

with A and B chosen so that $F_1 = F_2 = 1$.

Making $n = 1$ and $n = 2$ in the above formula for F_n, we get the linear system

$$\begin{cases} A + B = 1 \\ \alpha A + \beta B = 1 \end{cases}.$$

Multiplying the first equation by α and subtracting the second equation from the result, we obtain $(\alpha - \beta)B = \alpha - 1$. Now, since $\alpha + \beta = 1$ and $\alpha - \beta = \sqrt{5}$, we can write this last equation as $\sqrt{5}B = -\beta$, so that $B = -\frac{\beta}{\sqrt{5}}$. Analogously, multiplying the first equation by β and subtracting the result from the second equation, we obtain $A = \frac{\alpha}{\sqrt{5}}$, so that

$$F_n = \frac{\alpha}{\sqrt{5}} \cdot \alpha^{n-1} - \frac{\beta}{\sqrt{5}} \cdot \beta^{n-1} = \frac{1}{\sqrt{5}}(\alpha^n - \beta^n).$$

$\square$

Our next example shows a situation where Theorem 3.16 can be applied, albeit not straightforwardly.

Example 3.18 Let $(a_n)_{n\geq1}$ be a sequence of real numbers satisfying the recurrence relation $a_{k+1} = ra_k + s$ for every integer $k \geq 1$, where r and s are given real constants, with $r \neq 0, 1$ (we need not consider the case $r = 1$, for in this case $(a_n)_{n\geq1}$ is an AP). For an integer $k \geq 1$, we have

$$a_{k+2} - ra_{k+1} = s = a_{k+1} - ra_k$$

or, which is the same,

$$a_{k+2} - (r + 1)a_{k+1} + ra_k = 0.$$

This is a second order linear recurrence relation with constant coefficients, whose characteristic equation is

$$x^2 - (r + 1)x + r = 0.$$

Since its roots are r and 1, it follows from item (a) of Theorem 3.16 that $a_n = Ar^{n-1} + B$. If we now recall that $a_2 = ra_1 + s$, we see that A and B can be found by solving the linear system

$$\begin{cases} A + B = a_1 \\ Ar + B = ra_1 + s \end{cases}.$$

Therefore, $A = a_1$ and $B = s$, so that

$$a_n = a_1 r^{n-1} + s.$$

Let us now turn our attention to third order linear recurrence relations with constant coefficients. As the reader has probably guessed by now, in this case we have a sequence $(a_n)_{n\geq 1}$ satisfying a recurrence relation of the form

$$a_{k+3} + ra_{k+2} + sa_{k+1} + ta_k = 0$$

for every integer $k \geq 1$, where r, s and t are given real constants, not all zero.

As in the case of second order linear recurrence relations with constant coefficients, we define the **characteristic equation** of (3.11) as the third degree polynomial equation

$$x^3 + rx^2 + sx + t = 0.$$

According to Problem 21, page 43, such an equation admits, at most, three real roots. In what follows, we shall assume that it indeed has three such roots, let us say α, β and γ. Then, we have the following result.

Theorem 3.19 *Let $(a_n)_{n\geq 1}$ be a sequence of real numbers such that*

$$a_{k+3} + ra_{k+2} + sa_{k+1} + ta_k = 0 \qquad (3.11)$$

for every integer $k \geq 1$, where r, s and t are given real constants, not all zero. If the characteristic equation $x^3 + rx^2 + sx + t = 0$ of (3.11) has real roots α, β and γ, then there exist real constants A, B and C, completely determined by the values of a_1, a_2 and a_3, such that:

(a) If $\alpha \neq \beta \neq \gamma \neq \alpha$, then $a_n = A\alpha^{n-1} + B\beta^{n-1} + C\gamma^{n-1}$, for every $n \geq 1$.
(b) If $\alpha = \beta \neq \gamma$, then $a_n = (A + B(n-1))\alpha^{n-1} + C\gamma^{n-1}$, for every $n \geq 1$.
(c) If $\alpha = \beta = \gamma$, then $a_n = (A + B(n-1) + C(n-1)^2)\alpha^{n-1}$, for every $n \geq 1$.

Proof We shall only sketch the proof, leaving the details to the reader (see Problem 6).

Let $(b_n)_{n\geq 1}$ be the sequence given by $b_n = \alpha^{n-1}$, for every integer $n \geq 1$. Since $\alpha^3 + r\alpha^2 + s\alpha + t = 0$, we also have $\alpha^{k+2} + r\alpha^{k+1} + s\alpha^k + t\alpha^{k-1} = 0$ or, which is the same,

$$b_{k+3} + rb_{k+2} + sb_{k+1} + tb_k = 0,$$

for every integer $k \geq 1$. Hence, $(b_n)_{n\geq 1}$ satisfies the same recurrence relation as $(a_n)_{n\geq 1}$ does. Analogously, the sequences $(c_n)_{n\geq 1}$ and $(d_n)_{n\geq 1}$, given for an integer $n \geq 1$ by $c_n = \beta^{n-1}$ and $d_n = \gamma^{n-1}$, also satisfy the same recurrence relation as $(a_n)_{n\geq 1}$.

We now turn our attention to the cases (a)–(c):

(a) For all $A, B, C \in \mathbb{R}$, the sequence $(u_n)_{n\geq 1}$ such that $u_n = A\alpha^{n-1} + B\beta^{n-1} + C\gamma^{n-1}$ for every integer $n \geq 1$ satisfies the same recurrence relation as $(a_n)_{n\geq 1}$ does. On the other hand, since $\alpha \neq \beta \neq \gamma \neq \alpha$, we can choose the constants A, B and C in such a way that $u_1 = a_1$, $u_2 = a_2$ and $u_3 = a_3$. Therefore, by invoking once more the result of Problem 13, page 96, we conclude that $u_n = a_n$, for every integer $n \geq 1$.

(b) The result of Problem 21, page 43, gives us

$$x^3 + rx^2 + sx + t = (x - \alpha)^2 (x - \gamma).$$

With the aid of this identity, it is easy to show that $b_n = (n-1)\alpha^{n-1}$ also satisfies the same recurrence relation as the sequence $(a_n)_{n\geq 1}$. Then, as in item (a), the sequence $(u_n)_{n\geq 1}$ such that $u_n = (A + B(n-1))\alpha^{n-1} + C\gamma^{n-1}$ for $n \geq 1$ satisfies that same recurrence relation, for all $A, B, C \in \mathbb{R}$. Moreover, we can again choose the real constants A, B and C such that $u_1 = a_1$, $u_2 = a_2$ and $u_3 = a_3$. Thus, once more by the result of Problem 13, page 96, we conclude that $u_n = a_n$ for every integer $n \geq 1$.

(c) This time, the result of Problem 21, page 43, gives

$$x^3 + rx^2 + sx + t = (x - \alpha)^3.$$

It is now easy to show that both sequences $b_n = (n-1)\alpha^{n-1}$ and $c_n = (n-1)^2\alpha^{n-1}$ do satisfy the same recurrence relation as the sequence $(a_n)_{n\geq 1}$. Hence, the same is true for the sequence $u_n = (A + B(n-1) + C(n-1)^2)\alpha^{n-1}$, for all $A, B, C \in \mathbb{R}$. The rest of the argument proceeds exactly as in items (a) and (b).

□

Example 3.20 Let $(a_n)_{n\geq 1}$ be the sequence of real numbers given by $a_1 = 1$, $a_2 = 4$, $a_3 = 14$ and

$$a_{k+3} - 6a_{k+2} + 12a_{k+1} - 8a_k = 0,$$

for every integer $k \geq 1$. Compute a_n as a function of n.

Solution The characteristic equation of the given recurrence relation is $x^3 - 6x^2 + 12x - 8 = 0$. Since $x^3 - 6x^2 + 12x - 8 = (x - 2)^3$, item (c) of the previous result gives

$$a_n = (A + B(n - 1) + C(n - 1)^2) \cdot 2^{n-1},$$

for every integer $n \geq 1$, where A, B and C are real constants chosen so that $a_1 = 1$, $a_2 = 4$ and $a_3 = 14$. These initial conditions, in turn, give us the linear system of equations

$$\begin{cases} A = 1 \\ A + B + C = 2 \\ A + 2B + 4C = 7/2 \end{cases},$$

whose solution is $A = 1$, $B = 3/4$, $C = 1/4$. Then, we obtain

$$a_n = \left(1 + \frac{3}{4}(n-1) + \frac{1}{4}(n-1)^2\right) \cdot 2^{n-1}$$

$$= (n^2 + n + 2) \cdot 2^{n-3}.$$

$\square$

Problems: Section 3.2

1. Let $(a_n)_{n \geq 1}$ be the sequence given by $a_1 = 1$, $a_2 = 4$ and $a_{k+2} = 5a_{k+1} - 6a_k$, for every integer $k \geq 1$. Compute a_n as a function of n.
2. Let $(a_n)_{n \geq 1}$ be the sequence given by $a_1 = 3$, $a_2 = 5$ and $a_{k+2} = 3a_{k+1} - 2a_k$, for every integer $k \geq 1$. Prove that $a_n = 2^n + 1$, for every $n \in \mathbb{N}$.
3. Let $(a_n)_{n \geq 1}$ be the sequence given by $a_1 = 3$ and $a_{k+1} = 2a_k - 1$, for every integer $k \geq 1$. Compute a_n as a function of n.
4. * If the second degree equation $x^2 + rx + s = 0$ has two real roots equal to α, prove that the sequences $u_k = \alpha^{k-1}$ and $v_k = (k-1)\alpha^{k-1}$ satisfy the recurrence relations

$$\begin{cases} u_{k+2} + ru_{k+1} + su_k = 0 \\ v_{k+2} + rv_{k+1} + sv_k = 0 \end{cases}. \tag{3.12}$$

5. * The sequences $(u_n)_{n \geq 1}$ and $(v_n)_{n \geq 1}$ satisfy the recurrence relations (3.12), for every integer $k \geq 1$. For any fixed real constants A and B, prove that the sequence $(z_n)_{n \geq 1}$, given by $z_k = Au_k + Bv_k$, satisfies an analogous recurrence relation.
6. * Fulfill the details in the proof of Theorem 3.19.
7. * The **Lucas sequence**[5] is the sequence $(L_n)_{n \geq 1}$ given by $L_1 = 1$, $L_2 = 3$ and $L_{k+2} = L_{k+1} + L_k$, for every integer $k \geq 1$. Show that, for every integer $n \geq 1$, we have:

   (a) $L_n = \alpha^n + \beta^n$, where $\alpha = \frac{1+\sqrt{5}}{2}$ and $\beta = \frac{1-\sqrt{5}}{2}$.
   (b) $L_{2n} = L_n^2 + 2(-1)^{n-1}$.

---

[5] After François Édouard Anatole Lucas, French mathematician of the XIX century.

(c) $L_n^2 - 5F_n^2 = 4(-1)^n$, where $(F_n)_{n\geq 1}$ is the Fibonacci sequence (cf. Example 3.17).

8. (IMO shortlist—adapted) With respect to the sequence $(a_n)_{n\geq 1}$, such that $a_1 = 1$ and

$$a_{n+1} = \frac{1}{16}(1 + 4a_n + \sqrt{1 + 24a_n})$$

for every integer $n \geq 1$, do the following items:

(a) If $b_n = \sqrt{1 + 24a_n}$, show that $2b_{n+1} = b_n + 3$ for every $n \geq 1$.
(b) Show that $b_n = 3 + \frac{4}{2^n}$ for every $n \geq 1$.
(c) Conclude that, for every $n \geq 1$, we have

$$a_n = \frac{1}{3}\left(1 + \frac{1}{2^{n-1}}\right)\left(1 + \frac{1}{2^n}\right).$$

3.3 The Σ and Π Notations

In this section, we introduce the notations $\sum$ (one reads **sigma**) for sums and $\prod$ (one reads **pi**) for products, which will reveal themselves quite useful in the context of sequences.

Definition 3.21 Given a sequence $(a_k)_{k\geq 1}$, we write $\sum_{j=1}^n a_j$ to denote the sum $a_1 + a_2 + \cdots + a_n$, and read it as **the sum of the a_j's, from $j = 1$ to n**. Hence,

$$\sum_{j=1}^n a_j = \begin{cases} a_1 & , \text{ if } n = 1 \\ a_1 + a_2 + \cdots + a_n, & \text{ if } n > 1 \end{cases}.$$

As a particular case of the above definition, if $(a_k)_{k\geq 1}$ is a **constant sequence** with, say, $a_k = c$ for every $k \geq 1$, we clearly have

$$\sum_{j=1}^n a_j = \sum_{j=1}^n c = nc.$$

The main reason for the success of the Σ notation owes to the fact that it turns it quite easy to manipulate sums with large numbers of summands, specially when each such summand is itself a sum. For example, given sequences $(a_k)_{k\geq 1}$ and $(b_k)_{k\geq 1}$ of real numbers, the associativity and commutativity of addition of reals give

$$(a_1 + a_2 + \cdots + a_n) \pm (b_1 + b_2 + \cdots + b_n) = (a_1 \pm b_1) + (a_2 \pm b_2) + \cdots + (a_n \pm b_n);$$

with the aid of the Σ notation, the equality above can be written in the much more compact form

$$\sum_{j=1}^{n} a_j \pm \sum_{j=1}^{n} b_j = \sum_{j=1}^{n}(a_j \pm b_j). \tag{3.13}$$

On the other hand, given $c \in \mathbb{R}$, the distributivity of the multiplication of reals with respect to addition gives

$$c(a_1 + a_2 + \cdots + a_n) = ca_1 + ca_2 + \cdots + ca_n,$$

an equality that can be written, again with the aid of the Σ notation, as

$$c\sum_{j=1}^{n} a_j = \sum_{j=1}^{n} ca_j. \tag{3.14}$$

The following example shows, in a simple case, how one could apply the two identities above in order to simplify the task of calculating the value of a sum.

Example 3.22 Compute the value of $\sum_{k=1}^{n}(2k + 1)$ in terms of $n \in \mathbb{N}$.

Solution Successively applying (3.13), (3.14) and the first part of Example 3.7, we get

$$\sum_{k=1}^{n}(2k + 1) = \sum_{k=1}^{n} 2k + \sum_{k=1}^{n} 1 = 2\sum_{k=1}^{n} k + n$$

$$= 2 \cdot \frac{n(n + 1)}{2} + n = n^2 + 2n.$$

$\square$

The Σ notation is particularly useful to perform cancellations in sums. More precisely, given a sequence $(a_k)_{k \geq 1}$, and performing the possible cancellations in the sum

$$(a_2 - a_1) + (a_3 - a_2) + (a_4 - a_3) + \cdots + (a_{n-1} - a_{n-2}) + (a_n - a_{n-1}),$$

we get $a_n - a_1$ as result. With the aid of the Σ notation, we can write the equality just get as

$$\sum_{j=1}^{n-1}(a_{j+1} - a_j) = a_n - a_1. \tag{3.15}$$

An equivalent formula (obtained from the above by writing $n + 1$ in place of n), which will sometimes be used in the place of (3.15), is

$$\sum_{j=1}^{n} (a_{j+1} - a_j) = a_{n+1} - a_1. \tag{3.16}$$

Any one of (3.15) or (3.16) is known as a **telescoping sum**. The idea behind the name is the following: as a telescope shortens the immense distance between a celestial body and our eyes, the above formulas shorten the way between certain given sums and their results.

Telescoping sums are the greatest advantage of having the $\sum$ notation at our disposal. Let us justify this claim by examining two interesting examples.

Example 3.23 Deduce the formula for the general term of an AP with the aid of telescoping sums.

Solution Let the sequence $(a_k)_{k\geq 1}$ be an AP of common difference r. Then, $a_{j+1} - a_j = r$ for each integer $j \geq 1$, so that (3.16) gives

$$a_n - a_1 = \sum_{j=1}^{n-1} (a_{j+1} - a_j) = \sum_{j=1}^{n-1} r = (n-1)r$$

or, which is the same, $a_n = a_1 + (n-1)r$.   □

To the next example, we say that a sequence $(a_k)_{k\geq 1}$ is a **second order AP** provided the sequence $(b_k)_{k\geq 1}$, given for $k \geq 1$ by $b_k = a_{k+1} - a_k$, is a nonconstant AP. In order to build a second order AP $(a_k)_{k\geq 1}$, we can start with a nonconstant AP, as

$$(3, 7, 11, 15, 19, 23, 27, \ldots),$$

and stipulate the initial term of the second order AP, say $a_1 = 2$; then, we successively compute the values of $a_2, a_3, \ldots$ from the relations $a_2 - a_1 = 3$, $a_3 - a_2 = 7$, $a_4 - a_3 = 11$ etc. Proceeding this way we obtain, from the AP given above, the second order AP

$$(2, 5, 12, 23, 38, 57, 80, \ldots).$$

The discussion of the previous paragraph makes it clear that a second order AP is completely determined only if we know its first three terms. In fact, it is only by knowing its first three terms that we will know the first two terms of the nonconstant AP formed by the differences of the consecutive terms of the second order AP.

We are now in position to use telescoping sums to deduce a formula for the general term of a general second order AP.

Example 3.24 Given a second order AP $(a_k)_{k\geq 1}$, prove that

$$a_n = a_1 + (a_2 - a_1)(n - 1) + \frac{(n - 1)(n - 2)r}{2}, \qquad (3.17)$$

where $r = a_3 - 2a_2 + a_1$ is the common difference of the nonconstant AP formed by the differences of the consecutive terms of $(a_k)_{k\geq 1}$.

Solution Let $b_k = a_{k+1} - a_k$ for every integer $k \geq 1$. The formulas for telescoping sums and for the sum of the terms of a finite AP give us

$$a_n - a_1 = \sum_{j=1}^{n-1}(a_{j+1} - a_j) = \sum_{j=1}^{n-1} b_j = \frac{(n - 1)(b_1 + b_{n-1})}{2}.$$

On the other hand, by applying the formula for the general term of an AP, we get

$$b_{n-1} = b_1 + (n - 2)r = (a_2 - a_1) + (n - 2)r$$

and, thus,

$$\begin{aligned}
a_n &= a_1 + \frac{(n - 1)(b_1 + b_{n-1})}{2} \\
&= a_1 + \frac{(n - 1)(2(a_2 - a_1) + (n - 2)r)}{2} \\
&= a_1 + (n - 1)(a_2 - a_1) + \frac{(n - 1)(n - 2)r}{2}.
\end{aligned}$$

Finally, it suffices to observe that

$$r = b_2 - b_1 = (a_3 - a_2) - (a_2 - a_1) = a_3 - 2a_2 + a_1.$$

$\square$

In spite of the previous examples, if we wish to use the formula for telescoping sums to compute the sum $\sum_{j=1}^{n} b_j$ of the first n terms of a given sequence $(b_k)_{k\geq 1}$, we first need to find a way to write the summands b_j as the differences $a_{j+1} - a_j$, between consecutive terms of some other sequence $(a_k)_{k\geq 1}$. The difficulty with this reasoning relies upon the fact that we do not know, beforehand, who the sequence $(a_k)_{k\geq 1}$ is. The discussion of some additional examples will help to clarify this point.

Example 3.25 Consider the sequence $(a_k)_{k\geq 1}$ given by $a_1 = 1$ and

$$a_{k+1} = \frac{a_k}{1 + ka_k},$$

for every integer $k \geq 1$. Compute a_n as a function of n.

Solution Since the terms of the sequence $(a_k)_{k\geq 1}$ are all positive (why?), we can define the sequence $(b_k)_{k\geq 1}$ by setting $b_k = \frac{1}{a_k}$. The given recurrence relation, then, gives,

$$b_{k+1} = \frac{1}{a_{k+1}} = \frac{1 + ka_k}{a_k} = \frac{1}{a_k} + k = b_k + k.$$

Thus, by applying the formula for telescoping sums, together with the formula for the sum of the terms of a finite AP, we get

$$b_n - b_1 = \sum_{k=1}^{n-1}(b_{k+1} - b_k) = \sum_{k=1}^{n-1} k = \frac{n(n-1)}{2}.$$

Therefore,

$$b_n = b_1 + \frac{n(n-1)}{2} = 1 + \frac{n(n-1)}{2} = \frac{n^2 - n + 2}{2},$$

and, hence, $a_n = \frac{1}{b_n} = \frac{2}{n^2 - n + 2}$. $\square$

For the next example define, for each positive integer n, the **factorial** of n, denoted $n!$, as the product of all natural numbers from 1 to n (by convention, $1! = 1$), i.e., $2! = 2$, $3! = 6$, $4! = 24$, $5! = 120$ etc. Observe that, in general, we have $(k+1)! = (k+1) \cdot k!$.

Example 3.26 (Canada) Compute the value of the sum $\sum_{j=1}^{n} j(j!)$ in terms of n.

Solution In order to use the formula for telescoping sums, we have first to succeed in writing $j(j!)$ as a difference $a_{j+1} - a_j$, of consecutive terms of a single sequence $(a_k)_{k\geq 1}$. We do this by observing that

$$j(j!) = [(j+1) - 1]j! = (j+1) \cdot j! - j! = (j+1)! - j!.$$

Hence, by defining $a_k = k!$ for $k \geq 1$, we get

$$\sum_{j=1}^{n} j(j!) = \sum_{j=1}^{n} [(j+1)! - j!] = \sum_{j=1}^{n}(a_{j+1} - a_j)$$

$$= a_{n+1} - a_1 = (n+1)! - 1!.$$ $\square$

As anticipated at the beginning of this section, we now present a very useful notation for products of real numbers.

Definition 3.27 Given a sequence $(a_k)_{k\geq 1}$, we write $\prod_{j=1}^{n} a_j$ to denote the product $a_1 a_2 \ldots a_n$, and read **the product of the a_j's, from $j = 1$ to $j = n$**. Thus,

$$\prod_{j=1}^{n} a_j = \begin{cases} a_1 & , \text{ if } n = 1 \\ a_1 a_2 \ldots a_n & , \text{ if } n > 1 \end{cases}.$$

With the aid of the Π notation, we can write the factorial of $n \in \mathbb{N}$ (see the paragraph that precedes Example 3.26) by writing

$$n! = \prod_{j=1}^{n} j. \tag{3.18}$$

As was the case with the $\sum$ notation, the usefulness of the Π notation comes from the fact that it *formally commutes* with multiplications and divisions. In fact, given a real number c and sequences $(a_k)_{k\geq 1}$ and $(b_k)_{k\geq 1}$, we have

$$(a_1 a_2 \ldots a_n)(b_1 b_2 \ldots b_n) = (a_1 b_1)(a_2 b_2) \ldots (a_n b_n),$$

$$\frac{a_1 a_2 \ldots a_n}{b_1 b_2 \ldots b_n} = \frac{a_1}{b_1} \frac{a_2}{b_2} \cdots \frac{a_n}{b_n}$$

and

$$c^n(a_1 a_2 \ldots a_n) = (ca_1)(ca_2) \ldots (ca_n)$$

(provided all of the b_j's are nonzero, in the second equality above). Writing both sides of the identities above by means of the Π notation, we obtain the relations

$$\left(\prod_{j=1}^{n} a_j \right) \left(\prod_{j=1}^{n} b_j \right) = \prod_{j=1}^{n} (a_j b_j),$$

$$\frac{\prod_{j=1}^{n} a_j}{\prod_{j=1}^{n} b_j} = \prod_{j=1}^{n} \frac{a_j}{b_j} \quad \text{and} \quad c^n \prod_{j=1}^{n} a_j = \prod_{j=1}^{n} (ca_j).$$

Below, we present an example of application of such formulas.

Example 3.28 Compute, in terms of n, the value of $\prod_{k=1}^{n} \left(2 + \frac{2}{k} \right)$.

Solution We apply the properties of the Π notation listed above:

$$\prod_{k=1}^{n} \left(2 + \frac{2}{k} \right) = \prod_{k=1}^{n} 2 \left(\frac{k+1}{k} \right) = 2^n \frac{\prod_{k=1}^{n} (k+1)}{\prod_{k=1}^{n} k}$$

$$= 2^n \frac{(n+1)!}{n!} = 2^n \frac{(n+1) \cdot n!}{n!}$$

$$= 2^n (n+1).$$

$\square$

Analogously to the case of the $\sum$ notation, the $\prod$ notation is particularly useful when performing cancellations in certain products. This is the content of the formula for **telescoping products**, collected in the following result.

Proposition 3.29 *If* $(a_k)_{k \geq 1}$ *is a sequence of nonzero real numbers, then*

$$\prod_{j=1}^{n} \frac{a_{j+1}}{a_j} = \frac{a_{n+1}}{a_1}. \tag{3.19}$$

Proof As with telescoping sums, it suffices to observe that the intermediate factors of the product at the left hand side all cancel. In symbols,

$$\prod_{j=1}^{n} \frac{a_{j+1}}{a_j} = \frac{a_2}{a_1} \cdot \frac{a_3}{a_2} \cdot \frac{a_4}{a_3} \cdots \frac{a_n}{a_{n-1}} \cdot \frac{a_{n+1}}{a_n} = \frac{a_{n+1}}{a_1}.$$

$\square$

Example 3.30 We take another look at the previous example, this time with the formula for telescoping products at our disposal. To this end, we first of all note that

$$\prod_{k=1}^{n} \left(2 + \frac{2}{k} \right) = 2^n \prod_{k=1}^{n} \left(\frac{k+1}{k} \right).$$

Now, defining the sequence $(a_k)_{k \geq 1}$ by $a_k = k$, it follows from (3.19) that

$$\prod_{k=1}^{n} \left(2 + \frac{2}{k} \right) = 2^n \prod_{k=1}^{n} \left(\frac{a_{k+1}}{a_k} \right) = 2^n \cdot \frac{a_{n+1}}{a_1} = 2^n(n + 1).$$

Problems: Section 3.3

1. Prove that a sequence $(a_k)_{k \geq 1}$ is a second order AP if and only if $a_{k+2} - 2a_{k+1} + a_k \neq 0$ and $a_{k+3} - 3a_{k+2} + 3a_{k+1} - a_k = 0$, for every integer $k \geq 1$.
2. Let $(a_k)_{k \geq 1}$ be the sequence defined by $a_1 = 1$ and $a_{n+1} = a_n + 3n - 1$, for every positive integer n. Compute the n-th term of this sequence in terms of n.
3. The sequence $(a_n)_{n \geq 1}$ is given by $a_1 = 1$ and $a_{n+1} = a_n + 8n$, for $n \geq 1$. Compute a_n in terms of n.
4. * Compute the value of the sum $\sum_{k=2}^{n} \frac{1}{(k-1)k}$ as a function of $n \in \mathbb{N}$.
5. The sequence $(a_k)_{k \geq 1}$ is an AP. Prove that, for every positive integer n, we have

$$\sum_{k=1}^{n-1} \frac{1}{a_k a_{k+1}} = \frac{n-1}{a_1 a_n}.$$

6. Prove that, for every positive integer n, we have

$$\frac{1}{1^2} + \frac{1}{2^2} + \frac{1}{3^2} + \cdots + \frac{1}{n^2} < 2 - \frac{1}{n}.$$

7. Compute the value of the sum $\sum_{k=1}^{n-1} \frac{1}{(4k-1)(4k+3)}$ as a function of $n \in \mathbb{N}$.
8. (Romania) Let k and n be positive integers. Prove that:

   (a) $(2k+1)^3 - (2k-1)^3$ can always be written as a sum of three perfect squares.
   (b) $(2n+1)^3 - 2$ can always be written as a sum of $3n - 1$ perfect squares.

9. Compute the value of the sum $\sum_{k=1}^{n} \frac{k}{(k+1)!}$ as a function of the integer $n > 1$.
10. (Brazil—adapted) Do the following items:

   (a) For $k \in \mathbb{N}$, write $(k+1)^2 + k^2 + k^2(k+1)^2$ as a perfect square.
   (b) Compute the value of the sum $\sum_{k=1}^{99} \sqrt{\frac{1}{k^2} + \frac{1}{(k+1)^2}} + 1$.

11. * (Brazil) Let $(F_k)_{k\geq 1}$ be the Fibonacci sequence, i.e., the sequence given by $F_1 = 1$, $F_2 = 1$ and $F_{k+2} = F_{k+1} + F_k$, for every integer $k \geq 1$. Compute the value of the sum $\sum_{k=1}^{n} \frac{F_{k+1}}{F_k F_{k+2}}$ in terms of $n \in \mathbb{N}$.
12. The sequence $(a_k)_{k\geq 1}$ is an AP. Prove that, for every positive integer n, we have

$$\sum_{k=1}^{n-1} \frac{1}{\sqrt{a_k} + \sqrt{a_{k+1}}} = \frac{n-1}{\sqrt{a_1} + \sqrt{a_n}}.$$

13. (a) Factorise the expression $x^4 + x^2 + 1$.
    (b) Compute the value of the sum $\sum_{k=1}^{n} \frac{k}{k^4+k^2+1}$ in terms of $n \in \mathbb{N}$.
14. (Australia) Compute the value of the sum

$$\sum_{k=1}^{999,999} \frac{1}{\sqrt[3]{(k+1)^2} + \sqrt[3]{k^2 - 1} + \sqrt[3]{(k-1)^2}}.$$

15. Compute, in terms of $n \in \mathbb{N}$, the value of the sum

$$\sum_{k=1}^{n} \frac{1}{(k+1)\sqrt{k} + k\sqrt{k+1}}.$$

16. (Germany) Given a natural number $n > 1$, compute, as a function of n, the value of the product $\prod_{j=2}^{n} \left(1 - \frac{1}{j^2}\right)$.
17. For $0 \leq k \leq 101$, let $x_k = \frac{k}{101}$. Compute the value of the sum

$$\sum_{k=0}^{101} \frac{x_k^3}{1 - 3x_k + 3x_k^2}.$$

18. (Leningrad) Compute, as a function of $n \in \mathbb{N}$, the value of the product

$$\prod_{k=2}^{n-1} \frac{k^3 + 1}{k^3 - 1}.$$

For the next problem, we recall that the **arithmetic mean** A of a finite collection $a_1, a_2, \ldots, a_k$ of real numbers is the real number

$$A = \frac{a_1 + a_2 + \cdots + a_k}{k}.$$

19. (Macedonia) Find all values of $n \in \mathbb{N}$ for which we can write the set $A = \{1, 2, 3, \ldots, 4n\}$ as the union of n pairwise disjoint 4-element subsets, such that, in each of them, one element is the arithmetic mean of the other three.

20. (China—adapted) Use the result of Problem 13, page 26, to compute the value of the greatest integer that is less than or equal to

$$S = \frac{1}{\sqrt{1}} + \frac{1}{\sqrt{2}} + \frac{1}{\sqrt{3}} + \cdots + \frac{1}{\sqrt{10,000}}.$$

21. * Given positive integers n and p, prove that:

(a) $k^p < \frac{(k+1)^{p+1} - k^{p+1}}{p+1} < (k+1)^p$, for every $k \in \mathbb{N}$.

(b) $\sum_{k=1}^{n-1} k^p < \frac{n^{p+1}}{p+1} < \sum_{k=1}^{n} k^p$.

Chapter 4
Induction and the Binomial Formula

With the algebraic background of the previous chapters at our disposal, we devote the first section of this chapter to the development of a very important topic in elementary Mathematics, namely, the *principle of mathematical induction*. It will considerably improve our ability of elaborating proofs of mathematical statements that depend on natural numbers. In the other two sections, we shall apply this principle to deduce Newton's formula for binomial expansion, as well as some important related results.

4.1 The Principle of Mathematical Induction

Generally speaking, there are several ways of proving something. For instance, we can make a direct proof, or a proof by contradiction (for more on Logic and proof techniques, see the first chapters of [21]). In this sense, the principle of mathematical induction will provide us yet another way of proving certain types of mathematical statements.

To understand how this principle works, let's consider a subset $A \subset \mathbb{N}$ such that $1 \in A$. Also, suppose we know that, whenever a certain natural number k is in A, then, so $k + 1$ also is in A. Therefore, $1 \in A$ assures that $2 = 1 + 1 \in A$; accordingly, $2 \in A$ allows us to conclude that $3 = 2 + 1 \in A$. Continuing this way, we conclude that A contains all natural numbers, or, which is the same, that $A = \mathbb{N}$. This intuitive discussion can be formalized as the following axiom, known as the **principle of mathematical induction**.

Axiom 4.1 *Let $A \subset \mathbb{N}$ be a set satisfying the following conditions:*

(a) $1 \in A$.
(b) $k \in A \Rightarrow k + 1 \in A$.

Then $A = \mathbb{N}$.

© Springer International Publishing AG 2017
A. Caminha Muniz Neto, *An Excursion through Elementary Mathematics, Volume I*,
Problem Books in Mathematics, DOI 10.1007/978-3-319-53871-6_4

At this point, a natural question is: how can we apply the principle of mathematical induction to prove something in Mathematics? In order to answer it, suppose there is given a property $P(n)$ of the natural number n, which we wish to prove to be true for every $n \in \mathbb{N}$. Then, we let

$$A = \{k \in \mathbb{N}; \ P(k) \text{ is true}\}$$

and observe that

$$A = \mathbb{N} \Leftrightarrow (P(n) \text{ is true for every } n \in \mathbb{N}).$$

Thus, in order to prove that $P(n)$ is true for every $n \in \mathbb{N}$, it suffices to prove that $A = \mathbb{N}$; yet in another way, by invoking the principle of mathematical induction, it suffices to prove that:

- $1 \in A$;
- $k \in A \Rightarrow k + 1 \in A$.

In turn, the definition of the set A assures that, showing the validity of the two items above is the same as showing that

- $P(1)$ is true;
- $P(k)$ true $\Rightarrow P(k + 1)$ true.

The previous discussion can be summarized as the following *recipe* for a **proof by induction**.

Proposition 4.2 *If $P(n)$ is a property of the natural number n, then $P(n)$ is true for every $n \in \mathbb{N}$ if and only if the following two conditions are satisfied:*

(a) $P(1)$ is true;
(b) $P(k)$ true $\Rightarrow P(k + 1)$ true.

To understand how a proof by induction works in practice, let's start by discussing two examples, the first of which was already considered in the previous chapter.

Example 4.3 Prove that, for every $n \in \mathbb{N}$, the sum of the first n odd natural numbers equals n^2.

Proof Since the k–th odd natural number is $2k - 1$, the property $P(n)$ is, in this case, the same as

$$P(n): \ \sum_{j=1}^{n} (2j - 1) = n^2.$$

As was said before, in order to prove by induction that $P(n)$ is true for every $n \in \mathbb{N}$, we have to verify that:

i. $P(1)$ is true.

ii. $P(k)$ true $\Rightarrow P(k+1)$ true.

Checking i. is immediate: the first odd natural number is 1, which is the same as 1^2. In order to verify ii., we have to assume that $P(k)$ is true, i.e., that

$$\sum_{j=1}^{k}(2j-1) = k^2, \tag{4.1}$$

and find a way to deduce from this that $P(k+1)$ is also true, i.e., that

$$\sum_{j=1}^{k+1}(2j-1) = (k+1)^2. \tag{4.2}$$

Since we are assuming that $P(k)$ is true, we can substitute (4.1) into the left hand side of (4.2) to get

$$\sum_{j=1}^{k+1}(2j-1) = \sum_{j=1}^{k}(2j-1) + (2k+1) = k^2 + (2k+1) = (k+1)^2.$$

Hence, it follows by induction that $P(n)$ is true for every $n \in \mathbb{N}$. $\qquad\square$

Example 4.4 Prove that, for every $n \in \mathbb{N}$, the sum of the first n perfect squares is equal to

$$\frac{1}{6}n(n+1)(2n+1).$$

Proof Since the k−th perfect square is the number k^2, the property $P(n)$ is, in this case,

$$P(n): \quad \sum_{j=1}^{n}j^2 = \frac{1}{6}n(n+1)(2n+1).$$

As before, in order to compose a proof by induction, we have to verify that:

i. $P(1)$ is true.

ii. $P(k)$ true $\Rightarrow P(k+1)$ true.

Again, verifying i. is immediate: $1^2 = \frac{1(1+1)(2\cdot 1+1)}{6}$. To verify ii. let's assume that $P(k)$ is actually true, i.e., that

$$\sum_{j=1}^{k}j^2 = \frac{1}{6}k(k+1)(2k+1), \tag{4.3}$$

and find a way to deduce that $P(k+1)$ is also true, i.e., that

$$\sum_{j=1}^{k+1} j^2 = \frac{1}{6}(k+1)[(k+1)+1][2(k+1)+1]. \tag{4.4}$$

Substituting (4.3) into the left hand side of (4.4), we obtain

$$\sum_{j=1}^{k+1} j^2 = \sum_{j=1}^{k} j^2 + (k+1)^2$$

$$= \frac{1}{6}k(k+1)(2k+1) + (k+1)^2$$

$$= \frac{1}{6}(k+1)[k(2k+1) + 6(k+1)]$$

$$= \frac{1}{6}(k+1)(k+2)(2k+3).$$

Hence, it follows by induction that $P(n)$ is true for every $n \in \mathbb{N}$. □

Below, we state (also as an axiom) a slightly more general form of the principle of mathematical induction, which gives it greater flexibility in applications.

Axiom 4.5 *Let $a \in \mathbb{N}$ and $A \subset \{a, a+1, a+2, \ldots\}$ be a set satisfying the following conditions:*

(a) $a \in A$.
(b) $k \in A \Rightarrow k+1 \in A$.

Then, $A = \{a, a+1, a+2, \ldots\}$.

As the reader may be guessing right now, the prototype application of this form of the principle of mathematical induction is to prove that a certain property $P(n)$ is true for every natural number $n \geq a$, where a is a natural number given in advance. This can be done by letting

$$A = \{k \in \mathbb{N};\ P(k) \text{ is true}\}$$

and observing that

$$A = \{a, a+1, a+2, \ldots\}$$
$$\Updownarrow$$
$$P(n) \text{ is true for every natural number } n \geq a.$$

This way, we obtain the following more general recipe for proofs by induction.

Proposition 4.6 *Let $a \in \mathbb{N}$. If $P(n)$ is a property of the natural number n, then $P(n)$ is true for every natural number $n \geq a$ if and only if the following two conditions are satisfied:*

(a) $P(a)$ is true;
(b) $P(k)$ true $\Rightarrow P(k+1)$ true.

The next example illustrates the fact that, sometimes, this more general form of proof by induction is really necessary.

Example 4.7 Prove that $n! > 2^n$, for every natural number $n \geq 4$.

Proof First of all, observe that we really need to start from $n = 4$, for the inequality $n! > 2^n$ is not true when $n = 1, 2$ or 3.
The property $P(n)$ we wish to prove is

$$P(n) : \quad n! > 2^n.$$

In order to prove it by means of mathematical induction, we have to prove that $P(4)$ is true and that $P(k)$ true $\Rightarrow P(k+1)$ true. The validity of $P(4)$ follows from $4! = 24 > 16 = 2^4$. Let's then suppose that, for some $k \in \mathbb{N}$, the property $P(k)$ is actually true, i.e., that

$$k! > 2^k.$$

From this, we wish to deduce the trueness of $P(k+1)$, i.e., that $(k+1)! > 2^{k+1}$.
To this end, we first use the trueness of $P(k)$ to get

$$(k+1)! = (k+1) \cdot k! > (k+1) \cdot 2^k;$$

subsequently, we note that

$$(k+1) \cdot 2^k \geq 2^{k+1}$$

for every integer $k \geq 1$. Hence, by combining the two inequalities got above, we arrive at $(k+1)! > 2^{k+1}$, thus concluding that $P(k+1)$ is indeed true.
Therefore, it follows by induction that the property $P(n)$ is true for every integer $n \geq 4$. □

Before we present another example, it is appropriate to make some comments on terminology: in a proof by induction, the step $P(k) \Rightarrow P(k+1)$ is generally called the **induction step**. In order to execute it, we assume that the property $P(k)$ is indeed true (and this assumption constitutes our **induction hypothesis**) and, then, use this trueness, possibly together with other arguments, to deduce that the property $P(k+1)$ is also true. Thus, a proof by mathematical induction, as stated in Proposition 4.6, is usually done by executing the following steps:

- *identification*: isolate the property $P(n)$ to be proved;
- *initial case:* verify the validity of $P(a)$;
- *induction hypothesis:* assume the trueness of $P(k)$.
- *induction step:* deduce the validity of $P(k + 1)$, with the aid of the induction hypothesis and (perhaps) other arguments.

The statement of each problem generally makes it quite clear what the property $P(n)$ is. Therefore, a proof by induction is usually centered around the three last steps of the above scheme. Moreover, it is also customary not to explicitly state either the property $P(k)$ or the induction step $P(k) \Rightarrow P(k + 1)$. The discussion of the next example is performed in this shortened way.

Example 4.8 (Brazil) For every integer $n > 2$, show that there exist n pairwise distinct natural numbers whose sum of inverses equals 1.

Proof Let's make induction on $n \geq 3$, the initial case $n = 3$ following from

$$\frac{1}{2} + \frac{1}{3} + \frac{1}{6} = 1.$$

Now suppose, by induction hypothesis, that, for a certain natural $k \geq 3$, there exist natural numbers $x_1 < x_2 < \cdots < x_k$ such that

$$\frac{1}{x_1} + \frac{1}{x_2} + \cdots + \frac{1}{x_k} = 1.$$

Multiplying both sides of the equality above by $\frac{1}{2}$ and adding $\frac{1}{2}$ to both sides of the resulting equality, we obtain

$$\frac{1}{2} + \frac{1}{2x_1} + \frac{1}{2x_2} + \cdots + \frac{1}{2x_k} = 1.$$

Since $2 < 2x_1 < 2x_2 < \cdots < 2x_k$, we have got $k + 1$ pairwise distinct natural numbers whose sum of inverses is also equal to 1, and this completes the induction step.   □

When applying the principle of mathematical induction to prove something, one must be very careful in the execution of the induction step, to avoid absurd conclusions. This point is better illustrated with the following classical example and the subsequent discussion on the reasoning we shall present.

Example 4.9 In a farm, if one horse is white, show that all horses are white.

Proof Let's "*prove*" the statement above by induction on the number n of horses.

To the initial step, note that if a farm has just one horse and at least one of its horses is white, then surely all horses in the farm are white.

By induction hypothesis, assume that in a farm with k horses, if at least one of them is white, then all are white. Then, consider a farm with $k + 1$ horses, also with at least a white one. Take one horse from the farm, one that is not that we knew *a priori* to be white. Since we are left with k horses in the farm, at least one of which is white, it follows from the induction hypothesis that all of the k horses are white. Now, bring back the horse who was taken from the farm, and take from it one that was left at the first time. Again, we are left with k horses, at least one of which is white, so that, applying the induction hypothesis one more time, we conclude that all of these k horses are also white. However, since the horse we took from the farm this second time was already white, we conclude that all of the $k + 1$ horses are white. □

It is quite evident that there is some absurd in the "proof" presented above, for the statement of the example does not reflect the reality. The point is that we did not succeed in executing the induction step, for our reasoning does not work in a farm with two horses, as you can readily verify.

There is yet another important form of the principle of mathematical induction, generally referred to as the **strong principle of mathematical induction**, or simply as **strong induction**. We describe it next.

Axiom 4.10 *Let $A \subset \mathbb{N}$ be a set satisfying the following conditions:*

(a) $1 \in A$.
(b) $\{1, \ldots, k\} \subset A \Rightarrow k + 1 \in A$.

Then, $A = \mathbb{N}$.

At this point, the reader should probably find it clear how to use of the strong principle of induction to compose proofs. Nevertheless, in order to further illustrate it, we take a look at a couple of examples.

Example 4.11 (OCS - adapted) Show that, for every $n \in \mathbb{N}$, the number $(7 + 4\sqrt{3})^n + (7 - 4\sqrt{3})^n$ is a positive even integer.

Proof Let $u = 7 + 4\sqrt{3}$ and $v = 7 - 4\sqrt{3}$. Then, $u + v = 14$ and $uv = 1$, from where it follows that u and v are the roots of the second degree equation $x^2 - 14x + 1 = 0$. It follows from this that $u^2 = 14u - 1$ and $v^2 = 14v - 1$, and hence, for every integer $k \geq 2$,

$$u^k = 14u^{k-1} - u^{k-2} \text{ and } v^k = 14v^{k-1} - v^{k-2}.$$

Thus, letting $s_j = u^j + v^j$ and adding both relations above, we get, for every integer $k \geq 2$, that

$$s_k = 14s_{k-1} - s_{k-2}.$$

Now, observe that $s_0 = 2$ and $s_1 = u + v = 14$, both integers. Suppose, by induction hypothesis, that $s_k \in \mathbb{Z}$ for every integer $1 \leq k < n$. Then, the recurrence relation above gives

$$s_n = 14s_{n-1} - s_{n-2},$$

from where we conclude that s_n, being the sum of two integers, is also integer.

To what was left to prove, note that $u, v > 0$ guarantee that $s_n = u^n + v^n$ is positive for every $n \in \mathbb{N}$. On the other hand, again from the recurrence relation for the sequence $(s_k)_{k \geq 1}$, we conclude that s_k and s_{k-2} always have *the same parity*, i.e., they are either both even or both odd. However, since s_0 and s_1 are both even, it follows again by induction that s_n is an even integer, for every natural n.   □

Example 4.12 Show that every natural number n can be written, in a unique way, as a sum of powers of 2 with nonnegative and pairwise distinct integer exponents. This way of writing n is called its **binary representation** or **expansion**.

Proof Let's make a proof by strong induction. For $n = 1$, we have $1 = 2^0$, and this is obviously the only possible binary representation of 1. Now, suppose that the desired result is true for every natural number less than a certain natural $n > 1$.

We first show that n has *at least one* binary representation. To this end, take the greatest power of 2 which is less than or equal to n, say 2^k. Then,

$$2^k \leq n < 2^{k+1},$$

so that $0 \leq n - 2^k < 2^k$. If $n - 2^k = 0$, there is nothing left to do. Otherwise, $1 \leq n - 2^k < n$ and, by induction hypothesis (here, we are assuming that *every* natural number less than n *has* a binary representation; therefore, we are making a strong induction argument), there exist nonnegative integers $0 \leq a_0 < a_1 < \cdots < a_l$ such that

$$n - 2^k = 2^{a_0} + 2^{a_1} + \cdots + 2^{a_l}.$$

Using once more the inequality $n - 2^k < 2^k$, it follows that $2^{a_0} + 2^{a_1} + \cdots + 2^{a_l} < 2^k$ and, thus, $a_l < k$. Hence,

$$n = 2^{a_0} + 2^{a_1} + \cdots + 2^{a_l} + 2^k,$$

with $0 \leq a_0 < a_1 < \cdots < a_l < k$.

We then show that the binary representation of n is unique. So, let's assume that

$$n = 2^{a_0} + 2^{a_1} + \cdots + 2^{a_j} = 2^{b_0} + 2^{b_1} + \cdots + 2^{b_l},$$

with $0 \leq a_0 < a_1 < \cdots < a_j$ and $0 \leq b_0 < b_1 < \cdots < b_l$. Then,

$$2^{a_j} \leq 2^{a_0} + 2^{a_1} + \cdots + 2^{a_j}$$
$$= n = 2^{b_0} + 2^{b_1} + \cdots + 2^{b_l}$$
$$\leq 2^0 + 2^1 + \cdots + 2^{b_l}$$
$$= 2^{b_l+1} - 1,$$

where in the last passage we used the formula for the sum of the terms of a finite GP. It follows from the above that $2^{a_j} < 2^{b_l+1}$ and, hence, $a_j < b_l + 1$, i.e., $a_j \leq b_l$. Changing the roles of a_j and b_l in the above computations, we analogously get $b_l \leq a_j$ and, thus, $a_j = b_l$. Letting $a_j = b_l = k$, say, we obtain

$$n - 2^k = 2^{a_0} + 2^{a_1} + \cdots + 2^{a_{j-1}} = 2^{b_0} + 2^{b_1} + \cdots + 2^{b_{l-1}}. \tag{4.5}$$

Since $n - 2^k < n$, if we now invoke the uniqueness part of the induction hypothesis (i.e., if we assume that the binary representation of every natural number less than n is *unique*), then (4.5) gives $j - 1 = l - 1$ and $a_0 = b_0, a_1 = b_1, \ldots, a_{j-1} = b_{l-1}$, as we wished to prove. □

Before you go on and try the proposed problems, there are two remarks that ought to be made. Firstly, if we want to prove the validity of a certain property $P(n)$ of the natural number n, then using induction will not always be the best possible choice. For example, try Problem 1 and, after you are done, compare your proof by induction to the one given in Example 3.7. Secondly, apart from Problems 8 to 12, we do not present any applications of induction (and there are lots of them) to either Combinatorics or Number Theory. Instead, we refer the interested reader to [5].

Problems – Section 4.1

1. Use the principle of induction to prove that the sum of the first n natural numbers equals $\frac{n(n+1)}{2}$.

2. Prove that, for every natural number n, one has

$$1^3 + 2^3 + \cdots + n^3 = \left(\frac{n(n+1)}{2}\right)^2.$$

3. Prove that, for every natural number n, one has

$$1 - \frac{1}{2} + \frac{1}{3} - \cdots + \frac{1}{2n-1} = \frac{1}{n} + \frac{1}{n+1} + \cdots + \frac{1}{2n-1}.$$

4. (Canada) For $n \in \mathbb{N}$, let $h(n) = 1 + \frac{1}{2} + \frac{1}{3} + \cdots + \frac{1}{n}$. Prove that

$$n + h(1) + h(2) + h(3) + \cdots + h(n-1) = nh(n).$$

5. Show that, for every integer $n > 1$, one has

$$1 \cdot 2 + 2 \cdot 3 + \cdots + (n-1)n = \frac{1}{3}(n-1)n(n+1).$$

6. Prove that, for every integer $n > 1$, one has

$$1^2 + 3^2 + 5^2 + \cdots + (2n-1)^2 = \frac{1}{6}(2n-1)2n(2n+1).$$

7. Given $n \in \mathbb{N}$ and real numbers $a_1, a_2, \ldots, a_n$, prove the general version of the **triangle inequality**:

$$|a_1 + a_2 + \cdots + a_n| \leq |a_1| + |a_2| + \cdots + |a_n|. \qquad (4.6)$$

Also, show that if $a_1, a_2, \ldots, a_n \neq 0$, then equality holds in (4.6) if and only if $a_1, a_2, \ldots, a_n$ have equal signs.

8. Prove that 9 divides $4^n + 15n - 1$, for each natural n.

9. Prove that 3 divides $n^3 - n$, for each natural number n.

10. Prove that 3^n divides $4^{3^{n-1}} - 1$, for every integer $n \geq 1$.

    The result of the next problem is known as the **fundamental principle of counting** or as the **multiplicative principle**.

11. * Show that, if we have n_1 ways of choosing an object of type 1, n_2 ways of choosing an object of type 2, $\ldots$, n_k ways of choosing an object of type k, then the number of ways of choosing simultaneously one object of each of the types from 1 to k is $n_1 n_2 \ldots n_k$.

12. * Prove that a set with n elements has exactly 2^n subsets.

13. * Let $(a_n)_{n \geq 1}$ and $(b_n)_{n \geq 1}$ be two sequences satisfying the second order linear recurrence relation with constant coefficients $u_{k+2} + r u_{k+1} + s u_k = 0$, for every $k \geq 1$. If $a_1 = b_1$ and $a_2 = b_2$, prove that $a_n = b_n$, for every $n \geq 1$. Then, extend this result to the case in which both sequences $(a_n)_{n \geq 1}$ and $(b_n)_{n \geq 1}$ satisfy identical third order linear recurrence relations with constant coefficients.

14. Let $(a_n)_{n \geq 1}$ be a sequence of nonzero real numbers such that, for every $n \geq 2$, we have

$$\sum_{j=1}^{n-1} \frac{1}{a_j a_{j+1}} = \frac{n-1}{a_1 a_n}.$$

Prove that the sequence is an *AP*.

15. (Bulgaria - adapted) * The sequence $(a_n)_{n \geq 1}$ is defined by $a_1 = 1$ and, for $k \geq 1$ integer, $a_{k+1} = a_k^2 - a_k + 1$. Prove that, for every integer $n \geq 1$, we have:

    (a) $a_{n+1} = a_1 \ldots a_n + 1$.
    (b) $\sum_{j=1}^{n} \frac{1}{a_j} = 2 - \frac{1}{a_1 a_2 \ldots a_n}$.

16. Prove that, for every natural number n, we have $2^{2^n} > n^n$.

17. (Macedonia) Let x be a nonzero real number, such that $x + x^{-1} \in \mathbb{Z}$. Prove that $x^n + x^{-n} \in \mathbb{Z}$, for every integer n.

18. (Brazil) Let $(x_n)_{n \geq 1}$ and $(y_n)_{n \geq 1}$ be sequences of real numbers such that, for every integer $k \geq 1$, we have $x_{k+1} = x_k^3 - 3x_k$ e $y_{k+1} = y_k^3 - 3y_k$. If $x_1^2 = y_1 + 2$, show that $x_n^2 = y_n + 2$ for every integer $n \geq 1$.

19. (Putnam) Let $(x_n)_{n \geq 0}$ be a sequence of nonzero real numbers satisfying, for every integer $n \geq 1$, the recurrence relation $x_n^2 - x_{n-1} x_{n+1} = 1$. Prove that there exists a real number α such that $x_{n+1} = \alpha x_n - x_{n-1}$, for every $n \in \mathbb{N}$.

The next four problems concern the Fibonacci sequence. We refer the reader to Example 3.17, for a review of its definition.

20. Let $(F_n)_{n\geq 1}$ be the Fibonacci sequence. Prove that, for every $n \in \mathbb{N}$, the following identities are true:

   (a) $F_1 + F_2 + \cdots + F_n = F_{n+2} - 1$.
   (b) $F_1^2 + F_2^2 + \cdots + F_n^2 = F_n F_{n+1}$.
   (c) $F_1 + F_3 + \cdots + F_{2n-1} = F_{2n}$.
   (d) $F_2 + F_4 + \cdots + F_{2n} = F_{2n+1} - 1$.
   (e) $F_{n+1}^2 - F_n F_{n+2} = (-1)^n$.

21. Let $(F_n)_{n\geq 1}$ be the Fibonacci sequence. Prove that, for every $m, n \in \mathbb{N}$, with $m > 1$, we have $F_{m+n} = F_m F_{n+1} + F_{m-1} F_n$.

22. Let $(F_n)_{n\geq 1}$ be the Fibonacci sequence and $(L_n)_{n\geq 1}$ be the Lucas sequence (cf. Problem 7, page 77). Given $m, n \in \mathbb{N}$, prove that:

   (a) $2F_{m+n} = F_m L_n + F_n L_m$.
   (b) $2L_{m+n} = 5F_m F_n + L_m L_n$.

23. Let $(F_n)_{n\geq 1}$ be the Fibonacci sequence.

   (a) Prove that $F_n = n^2$ if and only if $n = 1$ or 12.[1]
   (b) If $\alpha = \frac{1+\sqrt 5}{2}$, do the following items:

      i. Prove that $\alpha^n = F_n \alpha + F_{n-1}$, for every integer $n \geq 1$.
      ii. Find all $n \in \mathbb{N}$ such that $\alpha^n - n^2 \alpha$ is an integer.

24. (TT) Let $(a_n)_{n\geq 1}$ be a sequence of pairwise distinct natural numbers. Prove that there are infinitely many naturals k such that $a_k > k$.

25. (France) Let $(a_n)_{n\geq 1}$ be a sequence of positive real numbers such that $a_1 = 1$ and

$$a_1^3 + a_2^3 + \cdots + a_n^3 = (a_1 + a_2 + \cdots + a_n)^2,$$

for every integer $n \geq 1$. Show that $a_n = n$, for every integer $n \geq 1$.

26. * For a fixed real number $a > 1$, let $(x_n)_{n\geq 1}$ be a sequence of real numbers such that $\sqrt a < x_1 < \sqrt a + 1$ and $x_{k+1} = \frac{1}{2}\left(x_k + \frac{a}{x_k}\right)$, for each integer $k \geq 1$. Prove that, for every integer $n \geq 1$, we have

$$\sqrt a < x_n \leq \sqrt a + \frac{1}{2^{n-1}}.$$

27. (France) Prove that, for every integer $n > 5$, there exist n positive integers such that the sum of the inverses of their squares is equal to 1.

---

[1] More generally, it can be shown that F_n is a perfect square if and only if $n = 1$ or 12. For a proof of this fact, see the problems of Section 12.3 of [5].

28. Given $m, n \in \mathbb{N}$, with $m > 1$, prove that there exist unique nonnegative integers $k, a_0, a_1, \ldots, a_k$ such that $0 \leq a_i < m$ for $0 \leq i \leq k$, $a_k \neq 0$ and

$$n = a_k m^k + a_{k-1} m^{k-1} + \cdots + a_2 m^2 + a_1 m + a_0.$$

In this case, we write $n = (a_k a_{k-1} \ldots a_1 a_0)_m$ and say that the right hand side above is the *representation of n in base m*, or the *m−adic representation* of n.

29. (Sweden) Prove that, for every natural number n, there exists a unique sequence $(a_j)_{j \geq 1}$ of integers, such that $0 \leq a_j \leq j$ for every $j \geq 1$ and

$$n = a_1 \cdot 1! + a_2 \cdot 2! + a_3 \cdot 3! + \cdots.$$

30. * Prove **Zeckendorf's theorem**[2]: every natural number can be uniquely written as a sum of Fibonacci numbers of nonconsecutive and greater than 1 indices.

4.2   Binomial Numbers

We start by recalling the definition of fatorial, extended to the nonnegative integers.

Definition 4.13 The **fatorial** of a given nonnegative integer n is the number $n!$, given by

$$n! = \begin{cases} 1, & \text{if } n = 0 \\ \prod_{j=1}^{n} j, & \text{if } n \geq 1 \end{cases}.$$

One could guess that it would be much more reasonable to define 0! as being equal to 0. Nevertheless, the reasons for letting 0! be equal to 1 will soon become evident.

Definition 4.14 Given integers n and k, with $0 \leq k \leq n$, we define the **binomial number** $\binom{n}{k}$ by

$$\binom{n}{k} = \frac{n!}{k!(n-k)!}.$$

It is easy to verify, directly from the definition, that

$$\binom{n}{0} = 1, \quad \binom{n}{1} = n, \quad \binom{n}{2} = \frac{n(n-1)}{2}$$

---

[2] After Édouard Zeckendorf, belgian mathematician of the XX century.

for every nonnegative integers n for which the binomial numbers above are defined. On the other hand, for every integers n and k such that $0 \leq k \leq n$, we have

$$\binom{n}{k} = \binom{n}{n-k}. \tag{4.7}$$

Indeed, this follows at once from

$$\binom{n}{k} = \frac{n!}{k!(n-k)!} = \frac{n!}{(n-k)!(n-(n-k))!} = \binom{n}{n-k}.$$

Notice that the binomial numbers $\binom{n}{0}$, $\binom{n}{1}$ and $\binom{n}{2}$, whenever defined, are all naturals (in the case of $\binom{n}{2} = \frac{n(n-1)}{2}$, this is due to the fact that the product of any two consecutive integers is an even number). This is also true in the case of $\binom{n}{3} = \frac{n(n-1)(n-2)}{6}$, for, among any three consecutive integers, there is always a multiple of 3 and an even integer, so that $n(n-1)(n-3)$ is always a multiple of 6. On the other hand, (4.7) assures that $\binom{n}{n} = \binom{n}{0}$, $\binom{n}{n-1} = \binom{n}{1}$, $\binom{n}{n-2} = \binom{n}{2}$ and $\binom{n}{n-3} = \binom{n}{3}$, whenever defined, are also natural numbers.

The discussion of the previous paragraph makes it natural to ask whether $\binom{n}{k}$ is a natural number for every choice of integers n and k such that $0 \leq k \leq n$. This is indeed the case, and will be a straightforward consequence of relation (4.8) below,[3] which is known as **Stifel's relation**.[4]

Proposition 4.15 *Given integers n and k such that $1 \leq k < n$, we have*

$$\binom{n}{k} = \binom{n-1}{k} + \binom{n-1}{k-1}. \tag{4.8}$$

Proof It suffices to apply the definition of binomial number to the right hand side of the equality above, together with some algebraic manipulations:

$$\binom{n-1}{k} + \binom{n-1}{k-1} = \frac{(n-1)!}{k!(n-1-k)!} + \frac{(n-1)!}{(k-1)!(n-k)!}$$

$$= \frac{(n-1)!}{(k-1)!(n-1-k)!} \left(\frac{1}{k} + \frac{1}{n-k} \right)$$

$$= \frac{(n-1)!}{(k-1)!(n-1-k)!} \frac{n}{k(n-k)}$$

$$= \frac{n!}{k!(n-k)!} = \binom{n}{k}.$$

$\square$

---

[3] Another proof is the object of Problem 8.

[4] After Michael Stifel, german mathematician of the XVI century.

With the binomial numbers we construct an infinite triangular numerical table, known as **Pascal's triangle**,[5] in the following way: we count lines and columns starting from 0, lines being labelled from top to bottom and columns from left to right; for $0 \leq k \leq n$, the $(n, k)-th$ entry, i.e., the number written on the crossing of the $n-$th line and the $k-$th column, is the binomial number $\binom{n}{k}$. More precisely:

- The entries of column 0, read from top to bottom, are respectively equal to the binomial numbers $\binom{0}{0}, \binom{1}{0}, \binom{2}{0}, \binom{3}{0}, \ldots$. As we have already seen, all such numbers are equal to 1.
- Line 0 is formed solely by the binomial number $\binom{0}{0} = 1$. Line 1 is composed by the binomial numbers $\binom{1}{0}$ and $\binom{1}{1}$, both of which are also equal to 1.
- In general, the entries of line n, read from left to right, are respectively equal to the binomial numbers $\binom{n}{0}, \binom{n}{1}, \binom{n}{2}, \ldots, \binom{n}{n}$.

According to the above recipe, we show below the first few lines of Pascal's triangle:

Pascal's Triangle

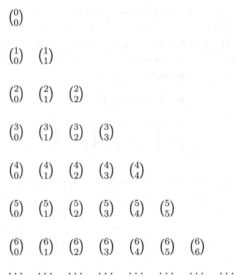

With respect to Pascal's triangle, Stifel's relation says that, whenever we add, in line $n - 1$, the entries at columns $k - 1$ and k, we get the entry situated at line n and column k. This is more difficult to say than to understand and verify, and allows us to *recursively get* the numerical values of the binomial numbers $\binom{n}{k}$. The table below shows the numerical values of the binomial numbers $\binom{n}{k}$ for $0 \leq n \leq 6$, obtained with the aid of Stifel's relation.

---

[5] After Blaise Pascal, french mathematician of the XVII century. Besides the *triangle* that bears his name, there is an important *Pascal's theorem* in the theory of conics, which has greatly motivated the developments of Projective and Algebraic Geometry.

```
1
1   1
1   2   1
1   3   3   1
1   4   6   4   1
1   5   10  10  5   1
1   6   15  20  15  6   1
...  ...  ...  ...  ...  ...  ...  ...
```

More gerally, since $\binom{n}{0} = 1$ and $\binom{n}{n} = 1$ whenever these numbers are defined, it is not difficult for the reader to convince himself or herself that $\binom{n}{k} \in \mathbb{N}$ for all integers n and k such that $0 \le k \le n$. We give a formal proof of this fact in the corollary below.

Corollary 4.16 *For all integers n and k such that $0 \le k \le n$, we have $\binom{n}{k} \in \mathbb{N}$.*

Proof Let's make induction on $n \ge 0$, the case $n = 0$ being obvious, for the only such binomial number is $\binom{0}{0} = 1$.

Now suppose, by induction hypothesis, that $\binom{n-1}{j}$ is a natural number for every $0 \le j \le n-1$, and consider a binomial number of the form $\binom{n}{k}$. There are two cases to consider:

(i) If $k = 0$ or $k = n$, then we have already observed that $\binom{n}{0} = \binom{n}{n} = 1$.
(ii) If $1 \le k \le n-1$, then it follows from the induction hypothesis that both $\binom{n-1}{k}$ and $\binom{n-1}{k-1}$ are natural numbers. Hence, Stifel's relation gives

$$\binom{n}{k} = \binom{n-1}{k} + \binom{n-1}{k-1} \in \mathbb{N}.$$

□

Later (cf. Theorem 4.20), we shall see that one of the main utilities of Pascal's triangle is that it provides an easy and quick way to write down the expansion of the binomial $(x + y)^n$ for small values of n. For the time being, let's establish some useful identities relating the entries of a row, column or diagonal of it.

The formula of the following proposition is known as the **columns' theorem** of Pascal's triangle.

Proposition 4.17 *In the column n of Pascal's triangle, the sum of the entries in rows $n, n + 1, \ldots, n + k - 1$ is equal to the entry located at column $n + 1$ and row $n + k$. In symbols,*

$$\sum_{j=0}^{k-1} \binom{n+j}{n} = \binom{n+k}{n+1}. \tag{4.9}$$

Proof Let's make a proof by induction on $k \geq 1$. The initial case $k = 1$ reduces to verifying that $\binom{n}{n} = \binom{n+1}{n+1}$, which is immediate.

By induction hypothesis, suppose that, when $k = l \geq 1$, (4.9) is true for all nonnegative integer values of n. Then, for $k = l + 1$ and every integer $n \geq 0$, we have

$$\sum_{j=0}^{l} \binom{n+j}{n} = \sum_{j=0}^{l-1} \binom{n+j}{n} + \binom{n+l}{n}$$

$$= \binom{n+l}{n+1} + \binom{n+l}{n}$$

$$= \binom{n+l+1}{n+1},$$

where we used Stifel's relation in the last equality above.

Therefore, it follows by induction that (4.9) is true for every $k \in \mathbb{N}$ and every integer $n \geq 0$.   □

In the coming example, we employ the columns' theorem to compute the sum of the first n perfect squares.

Example 4.18 Given $n \in \mathbb{N}$, compute the value of $1^2 + 2^2 + \cdots + n^2$ in terms of n.

Solution Letting $S = \sum_{j=1}^{n} j^2$, we have

$$S = \sum_{j=1}^{n} j(j-1) + \sum_{j=1}^{n} j = 2 \sum_{j=2}^{n} \binom{j}{2} + \sum_{j=1}^{n} \binom{j}{1}.$$

Therefore, it follows from (4.9) that

$$S = 2 \binom{n+1}{3} + \binom{n+1}{2} = \frac{1}{6} n(n+1)(2n+1).$$

□

Given an integer $n \geq 0$, the *diagonal n* of Pascal's triangle is formed by the binomial numbers

$$\binom{n}{0}, \binom{n+1}{1}, \binom{n+2}{2}, \binom{n+3}{3}, \ldots$$

With respect to them, the following corollary is the **diagonals' theorem** of Pascal's triangle.

Corollary 4.19 *In the diagonal n of Pascal's triangle, the sum of the entries in rows* $0, 1, \ldots, k-1$ *is equal to the entry located at row $n+k$ and column $k-1$. In symbols,*

$$\sum_{j=0}^{k-1} \binom{n+j}{j} = \binom{n+k}{k-1}. \qquad (4.10)$$

Proof Since $\binom{n+j}{j} = \binom{n+j}{n}$, it follows from the columns' theorem that

$$\sum_{j=0}^{k-1} \binom{n+j}{j} = \sum_{j=0}^{k-1} \binom{n+j}{n} = \binom{n+k}{n+1} = \binom{n+k}{k-1}.$$

□

Problems – Section 4.2

1. Prove that $\binom{2n}{n}$ is even, for every $n \in \mathbb{N}$.
2. Use the results of this section to compute the value of $1^3 + 2^3 + \cdots + n^3$, for every $n \in \mathbb{N}$.
3. * For $n \in \mathbb{N}$, prove that

$$\binom{n}{0} < \binom{n}{1} < \cdots < \binom{n}{\frac{n}{2}} > \binom{n}{\frac{n}{2}+1} > \cdots > \binom{n}{n}$$

if n is even, and

$$\binom{n}{0} < \binom{n}{1} < \cdots < \binom{n}{\frac{n-1}{2}} = \binom{n}{\frac{n+1}{2}} > \cdots > \binom{n}{n}$$

if n is odd.
4. Prove that, for every integer $n \geq 2$, one has $2^{\frac{5n}{4}} < \binom{2n}{n}$.
5. Given natural numbers k and m, with $m > k$, prove that

$$\sum_{n=k}^{m} \frac{\binom{n}{k-1}}{\binom{n}{k}\binom{n+1}{k}} = 1 - \frac{1}{\binom{m+1}{k}}.$$

6. Let n and k be given integers, such that $0 \leq k \leq n$. Prove that

$$\binom{n}{0} - \binom{n}{1} + \binom{n}{2} - \cdots + (-1)^k \binom{n}{k} = (-1)^k \binom{n-1}{k}.$$

7. Let $(F_k)_{k \geq 1}$ be the Fibonacci sequence (cf. Example 3.17). Show that, for every $n \in \mathbb{N}$, one has

$$F_n = \sum_{j=0}^{\lfloor \frac{n-1}{2} \rfloor} \binom{n-j-1}{j},$$

where $\lfloor \frac{n-1}{2} \rfloor$ denotes the greatest integer less than or equal to $\frac{n-1}{2}$, i.e.,

$$\left\lfloor \frac{n-1}{2} \right\rfloor = \begin{cases} \frac{n-1}{2}, & \text{if } n \text{ is odd} \\ \frac{n-3}{2}, & \text{if } n \text{ is even} \end{cases}.$$

8. * Given integers n and k such that $0 \leq k \leq n$, prove that the set $\{1, 2, \ldots, n\}$ has exactly $\binom{n}{k}$ subsets of k elements each.

4.3   The Binomial Formula

We shall now obtain **Newton's binomial formula**,[6] i.e., we shall explicitly write $(x + y)^n$ as a sum of monomials of the form $x^k y^l$.

Theorem 4.20 *Given $n \in \mathbb{N}$, we have*

$$(x + y)^n = \sum_{j=0}^{n} \binom{n}{j} x^{n-j} y^j. \tag{4.11}$$

Proof For $n = 1$, we have

$$(x + y)^1 = x + y = \binom{1}{0} x^1 + \binom{1}{1} y^1.$$

Suppose, by induction hypothesis, that (4.11) is true for $n = k$, i.e., that

$$(x + y)^k = \sum_{j=0}^{k} \binom{k}{j} x^{k-j} y^j.$$

---

[6]Sir Isaac Newton, english mathematician and physicist of the XVII century, is considered to be one of the greatest scientists ever. Actually, it is difficult to properly address Newton's contribution to the development of science. Known as the father of modern Physics, Newton also created, together with G. W. Leibniz, the Differential and Integral Calculus. His masterpiece, *Philosophiae Naturalis Principia Mathematica*, is one of the most influential books ever written and contains the cornerstones of both Calculus and Physics.

For $n = k + 1$, we thus have

$$(x + y)^{k+1} = (x + y)(x + y)^k = (x + y) \sum_{j=0}^{k} \binom{k}{j} x^{k-j} y^j$$

$$= \sum_{j=0}^{k} \binom{k}{j} x^{k+1-j} y^j + \sum_{j=0}^{k} \binom{k}{j} x^{k-j} y^{j+1}$$

$$= x^{k+1} + \sum_{j=1}^{k} \binom{k}{j} x^{k+1-j} y^j + \sum_{j=0}^{k-1} \binom{k}{j} x^{k-j} y^{j+1} + y^{k+1}.$$

In the last line above, let's perform the following changes of indices: in the first $\sum$, change j per l and, in the second $\sum$, change $j + 1$ per l; in this way, in the second $\sum$ we have $j = l - 1$ and $0 \leq j \leq k - 1 \Leftrightarrow 1 \leq l \leq k$. Then, we get

$$(x + y)^{k+1} = x^{k+1} + \sum_{l=1}^{k} \left[\binom{k}{l} + \binom{k}{l-1} \right] x^{k+1-l} y^l + y^{k+1}$$

$$= x^{k+1} + \sum_{l=1}^{k} \binom{k+1}{l} x^{k+1-l} y^l + y^{k+1},$$

where we used Stifel's relation in the last equality.

Finally, since $\binom{k+1}{0} = \binom{k+1}{k+1} = 1$, we can write the last line above as

$$\binom{k+1}{0} x^{k+1} + \sum_{l=1}^{k} \binom{k+1}{l} x^{k+1-l} y^l + \binom{k+1}{k+1} y^{k+1}$$

or, which is the same,

$$\sum_{l=1}^{k+1} \binom{k+1}{l} x^{k+1-l} y^l.$$

This is exactly the expression we would like to reach, so that, by induction, we conclude that (4.11) is true for every $n \in \mathbb{N}$. $\qquad \square$

Corollary 4.21 *Given $n \in \mathbb{N}$, we have*

$$(x - y)^n = \sum_{j=0}^{n} (-1)^j \binom{n}{j} x^{n-j} y^j.$$

Proof It suffices to apply (4.11), writing $-y$ in place of y. $\qquad \square$

It is customary to write

$$T_j = \binom{n}{j} x^{n-j} y^j$$

and say that such a monomial T_j is the **general term** of the expansion of $(x + y)^n$. Hence,

$$(x \pm y)^n = T_0 \pm T_1 \mp \cdots (\pm 1)^n T_n.$$

In what follows, we collect some examples of application of the binomial formula, as well as some important consequences of it.

Example 4.22 Find the least $n \in \mathbb{N}$ for which the expansion of $\left(x\sqrt{x} + \frac{1}{x^4}\right)^n$ has a summand not depending on x. For such an n, compute this summand.

Solution In expanding the given expression, the general term is

$$\binom{n}{k}(x\sqrt{x})^{n-k}\left(\frac{1}{x^4}\right)^k = \binom{n}{k} x^{\frac{3}{2}(n-k)} x^{-4k} = \binom{n}{k} x^{\frac{3n-11k}{2}}.$$

Therefore, there is a summand not depending on x if and only if there exists $0 \le k \le n$ such that $3n - 11k = 0$. Hence, $n = \frac{11k}{3}$, so that the least possible $n \in \mathbb{N}$ is $n = 11$, obtained for $k = 3$. In this case, the summand that doesn't depend on x is $\binom{11}{3} = 165$. □

Example 4.23 Let $k \in \mathbb{N}$ and $a, b, r \in \mathbb{Q}$, with $r > 0$ such that $\sqrt{r}$ is irrational. Prove that:

(a) There exist $c, d \in \mathbb{Q}$ such that $(a + b\sqrt{r})^k = c + d\sqrt{r}$.
(b) If $(a + b\sqrt{r})^k = c + d\sqrt{r}$, with $c, d \in \mathbb{Q}$, then $(a - b\sqrt{r})^k = c - d\sqrt{r}$.

Proof

(a) By expanding $(a + b\sqrt{r})^k$, we get

$$(a + b\sqrt{r})^k = \sum_{\substack{0 \le j \le k \\ 2|j}} \binom{k}{j} a^{k-j} b^j \sqrt{r^j} + \sqrt{r} \sum_{\substack{0 \le j \le k \\ 2|j}} \binom{k}{j} a^{k-j} b^j \sqrt{r^{j-1}}.$$

Making

$$c = \sum_{\substack{0 \le j \le k \\ 2|j}} \binom{k}{j} a^{k-j} b^j \sqrt{r^j} \quad \text{and} \quad d = \sum_{\substack{0 \le j \le k \\ 2|j}} \binom{k}{j} a^{k-j} b^j \sqrt{r^{j-1}}, \qquad (4.12)$$

it is immediate that $c, d \in \mathbb{Q}$ (since $\sqrt{r^j} \in \mathbb{Q}$ for an even j, and $\sqrt{r^{j-1}} \in \mathbb{Q}$ for an odd j).

(b) It suffices to note that the expansion of $(a - b\sqrt{r})^k$ is essentially equal to that of $(a + b\sqrt{r})^k$, the difference lying on the signs of half of the summands. More precisely, in the notations of (4.12), we get

$$(a - b\sqrt{r})^k = \sum_{\substack{0 \le j \le k \\ 2|j}} \binom{k}{j} a^{k-j} b^j \sqrt{r^j} - \sqrt{r} \sum_{\substack{0 \le j \le k \\ 2 \nmid j}} \binom{k}{j} a^{k-j} b^j \sqrt{r^{j-1}}$$

$$= c - d\sqrt{r}.$$

□

Example 4.24 Use Newton's binomial formula to prove **Lagrange's identity:**[7]

$$\sum_{j=0}^{n} \binom{n}{j}^2 = \binom{2n}{n}. \tag{4.13}$$

Proof Writing $(1+x)^{2n} = (1+x)^n(1+x)^n$ and applying Newton's binomial formula to both sides, we get

$$\sum_{k=0}^{2n} \binom{2n}{k} x^k = \left(\sum_{i=0}^{n} \binom{n}{i} x^i \right) \left(\sum_{j=0}^{n} \binom{n}{j} x^j \right) = \sum_{i,j=0}^{n} \binom{n}{i} \binom{n}{j} x^{i+j}.$$

Comparing the coefficients of x^n in the first and last expressions above and using (4.7), we get

$$\binom{2n}{n} = \sum_{i+j=n} \binom{n}{i} \binom{n}{j} = \sum_{i=0}^{n} \binom{n}{i} \binom{n}{n-i} = \sum_{i=0}^{n} \binom{n}{i}^2.$$

□

Item (a) of the following corollary is known as the **lines' theorem** of the Pascal's triangle.

Corollary 4.25 *Given $n \in \mathbb{N}$, we have:*

(a) $\sum_{j=0}^{n} \binom{n}{j} = 2^n$.
(b) $\sum_{\substack{0 \le j \le n \\ 2|j}} \binom{n}{j} = \sum_{\substack{0 \le j \le n \\ 2 \nmid j}} \binom{n}{j} = 2^{n-1}$.

Proof For item (a), it suffices to set $x = y = 1$ in the formula for $(x + y)^n$.

---

[7] Joseph Louis Lagrange, french physicist and mathematician of the XVIII century. Lagrange was one of the greatest scientists of his time, with notable contributions to Physics and Mathematics. In particular, he was a pioneer in the fields of Calculus of Variations and Celestial Mechanics.

In what concerns item (b), we start by setting $x = 1$ and $y = -1$ in the binomial formula to get (check it!)

$$0 = \sum_{\substack{0 \leq j \leq n \\ 2|j}} \binom{n}{j} - \sum_{\substack{0 \leq j \leq n \\ 2 \nmid j}} \binom{n}{j}.$$

Now, letting $A = \sum_{\substack{0 \leq j \leq n \\ 2|j}} \binom{n}{j}$ and $B = \sum_{\substack{0 \leq j \leq n \\ 2 \nmid j}} \binom{n}{j}$, it follows from (a) and the above relation that

$$\begin{cases} A + B = 2^n \\ A - B = 0 \end{cases}.$$

Hence, we have $A = B = 2^{n-1}$.   □

Remark 4.26 It is possible to generalize the formulas of item (b) of the previous corollary in the following way: given integers $0 \leq r < k < n$, we can compute the value of the sum

$$\binom{n}{r} + \binom{n}{k+r} + \binom{n}{2k+r} + \cdots$$

in terms of n, k and r. For instance, for $k = 3$ we can compute, in terms of $n \in \mathbb{N}$, the values of the sums $\binom{n}{0} + \binom{n}{3} + \binom{n}{6} + \cdots$, $\binom{n}{1} + \binom{n}{4} + \binom{n}{7} + \cdots$ and $\binom{n}{2} + \binom{n}{5} + \binom{n}{8} + \cdots$. The corresponding deduction uses the *multisection formula*, which will be obtained in Chapter 15 of [5], with the aid of complex numbers.

We finish this small chapter with an example that illustrates the use of the lines' theorem of Pascal's triangle to the computation of sums.

Example 4.27 Given $n \in \mathbb{N}$, compute the value of

$$\binom{n}{1} + 2\binom{n}{2} + 3\binom{n}{3} + \cdots + n\binom{n}{n}.$$

Solution First of all, note that for $j \in \mathbb{N}$ we have

$$j\binom{n}{j} = j\frac{n!}{j!(n-j)!} = \frac{n!}{(j-1)!(n-j)!}$$

$$= n\frac{(n-1)!}{(j-1)!(n-j)!} = n\binom{n-1}{j-1}.$$

Hence, it follows from the lines' theorem that

$$\sum_{j=1}^{n} j\binom{n}{j} = n\sum_{j=1}^{n}\binom{n-1}{j-1} = n \cdot 2^{n-1}.$$

$\square$

Problems – Section 4.3

1. Given $n \in \mathbb{N}$, compute the value of the sum $\sum_{j=0}^{n}\binom{n}{j}3^j$.
2. Let $A = \sum_{k=0}^{n}\binom{n}{k}3^k$ and $B = \sum_{k=0}^{n-1}\binom{n-1}{k}11^k$. If $\frac{B}{A} = \frac{3^8}{4}$, compute the value of n.
3. Show that $(1,1)^n \geq \frac{1}{200}(n^2 + 19n + 200)$.
4. (OCM) Without directly evaluating all summands, compute the value of the sum

$$\frac{1}{10!} + \frac{1}{3!8!} + \frac{1}{5!6!} + \frac{1}{7!4!} + \frac{1}{9!2!} + \frac{1}{11!}.$$

5. Find the maximal term in the expansion of $\left(1 + \frac{1}{3}\right)^{65}$.
6. (Baltic Way). Let a, b, c and d be real numbers for which $a^2 + b^2 + (a+b)^2 = c^2 + d^2 + (c+d)^2$. Prove that $a^4 + b^4 + (a+b)^4 = c^4 + d^4 + (c+d)^4$.
7. Given $n \in \mathbb{N}$, prove that $\binom{2n}{n} > \frac{2^{2n-1}}{n}$.
8. In the expansion of $x(1+x)^n$, divide the coefficient of each term by the exponent of x in that term. Prove that the sum of all these numbers equals $\frac{2^{n+1}-1}{n}$.
9. Given $a \in \mathbb{Z} \setminus \{0\}$ and $n \in \mathbb{N}$, compute the value of the sums $\sum_{j=0}^{n}\binom{n}{j}ja^j$ and $\sum_{j=0}^{n}\binom{n}{j}j^2$ in terms of a and n.
10. The sequence $(a_k)_{k\geq 1}$ is an AP. Prove that, for every integer $n > 1$, the following identities are valid:

   (a) $\sum_{j=0}^{n}\binom{n}{j}(-1)^{j+1}a_j = 0$.
   (b) $\sum_{j=0}^{n}\binom{n}{j}(-1)^{j+1}a_{j+1}^2 = 0$.

11. * If $0 < q < 1$ and $n \in \mathbb{N}$, prove that $q^n < \frac{q}{q+n(1-q)}$.
12. Given $a, b, n \in \mathbb{N}$, with $n > 1$, prove that the number $\sqrt{a} + \sqrt[n]{b}$ is the root of a polynomial equation with integer coefficients and degree $2n$.
13. (Croatia) Let x, y and z be nonzero real numbers, such that $x + y + z = 0$. Prove that the value of

$$\frac{x^5 + y^5 + z^5}{xyz(xy + yz + zx)}$$

does not depend on the particular values of x, y and z.

14. * Given $n \in \mathbb{N}$ and j, k, l nonnegative integers such that $j + k + l = n$, we define the **trinomial number** $\binom{n}{j,k,l}$ by

$$\binom{n}{j, k, l} = \frac{n!}{j!k!l!}.$$

Prove the **trinomial expansion formula**

$$(x + y + z)^n = \sum_{j+k+l=n} \binom{n}{j, k, l} x^j y^k z^l, \qquad (4.14)$$

where the sum in the right hand side extends over all possible choices of nonnegative integers j, k, l such that $j + k + l = n$.

15. Use the result of the previous problem to expand the trinomials $(x+y+z)^3$ and $(x+y+z)^4$.

16. Compute, in the expansion of $\left(1 + x + \frac{6}{x}\right)^{10}$, the term that doesn't depend on x.

17. Prove the following identities involving trinomial numbers:

    (a) $\sum_{j+k+l=n} \binom{n}{j,k,l} = 3^n$.

    (b) $\sum_{j+k+l=n} (-1)^l \binom{n}{j,k,l} = 1$.

18. * Prove that, for every $n \in \mathbb{N}$, we have

$$\sum_{j=1}^{n} \binom{n}{j} F_{2n+1-j} = F_{2n+1} - 1,$$

where F_k is the k-th Fibonacci number (cf. Example 3.17).

19. (IMO shortlist) Prove that

$$\sum_{k=0}^{995} \frac{(-1)^k}{1991 - k} \binom{1991 - k}{k} = \frac{1}{1991}.$$

Chapter 5
Elementary Inequalities

This chapter is an invitation to the systematic study of algebraic inequalities. More precisely, our main purpose here is to discuss some interesting examples of inequalities, for whose derivation we can use the simple mathematics we developed so far. Later, when we have the tools of Calculus at our disposal, we shall return to the study of algebraic inequalities, largely generalizing some of those we will study here.

5.1 The AM-GM Inequality

As will be clear in this section, what allows us to develop a systematic study of inequalities is the basic fact that the square of any real number is nonnegative, and it is zero if and only if the number in question is also equal to zero.

To start with, for $x, y \in \mathbb{R}$ we know that $(|x| - |y|)^2 \geq 0$, with equality if and only if $|x| = |y|$. If we expand the left hand side, we get the inequality $|x|^2 + |y|^2 \geq 2|xy|$ or, which is the same,

$$\frac{x^2 + y^2}{2} \geq |xy|, \tag{5.1}$$

with equality if and only if $|x| = |y|$.

If we now start with two positive real numbers a and b, and make $x = \sqrt{a} \geq 0$ and $y = \sqrt{b} \geq 0$, it follows from (5.1) that

$$\frac{a + b}{2} \geq \sqrt{ab}, \tag{5.2}$$

with equality if and only if $\sqrt{a} = \sqrt{b}$, i.e., if and only if $a = b$.

© Springer International Publishing AG 2017
A. Caminha Muniz Neto, *An Excursion through Elementary Mathematics, Volume I*,
Problem Books in Mathematics, DOI 10.1007/978-3-319-53871-6_5

The simplicity of the previous reasoning hides its importance. Actually, (5.2) is a particular case of a much more general inequality, known as the *arithmetic-geometric mean inequality* (see Theorem 5.7). For the time being, let's see how to deduce other interesting inequalities from (5.2).

Example 5.1 Given positive real numbers x and y, we have:

(a) $x + \frac{1}{x} \geq 2$, with equality if and only if $x = 1$.
(b) $\frac{1}{x} + \frac{1}{y} \geq \frac{4}{x+y}$, with equality if and only if $x = y$.

Proof

(a) Applying (5.2) with $a = x$ and $b = \frac{1}{x}$, we get

$$x + \frac{1}{x} \geq 2\sqrt{x \cdot \frac{1}{x}} = 2,$$

with equality if and only if $x = \frac{1}{x}$, i.e., if and only if $x = 1$ (for, $x > 0$ by hypothesis).

(b) Since $x + y > 0$, we have $\frac{1}{x} + \frac{1}{y} \geq \frac{4}{x+y}$ if and only if $(x + y)\left(\frac{1}{x} + \frac{1}{y}\right) \geq 4$ or, which is the same, if and only if $\frac{x}{y} + \frac{y}{x} \geq 2$. In turn, this last inequality follows directly from that of item (a), with $\frac{x}{y}$ in the place of x. It also follows from (a) that equality holds if and only if $\frac{x}{y} = \frac{y}{x}$, i.e., if and only if $x^2 = y^2$. However, since $x, y > 0$, such condition is clearly equivalent to $x = y$.

For another proof of item (b), apply (5.2) twice and multiply the results:

$$(x + y)\left(\frac{1}{x} + \frac{1}{y}\right) \geq 2\sqrt{xy} \cdot 2\sqrt{\frac{1}{x} \cdot \frac{1}{y}} = 4.$$

It follows directly from (5.2) that we have equality if and only if $x = y$.   □

Example 5.2 Given positive real numbers x, y and z, prove that

$$x^2 + y^2 + z^2 \geq xy + xz + yz, \tag{5.3}$$

with equality if and only if $x = y = z$.

Proof Applying (5.1) three times, we get the inequalities

$$\frac{x^2 + y^2}{2} \geq xy, \quad \frac{x^2 + z^2}{2} \geq xz, \quad \frac{y^2 + z^2}{2} \geq yz.$$

It now suffices to add these three inequalities to reach (5.3).

If $x = y = z$, then (5.3) clearly becomes an equality. Conversely, if $x \neq y$ (the cases $x \neq z$ and $y \neq z$ can be treated analogously), then $\frac{x^2+y^2}{2} > xy$. Therefore, an obvious generalization of item (e) of Proposition 1.2 assures that, adding the inequalities

$$\frac{x^2 + y^2}{2} > xy, \quad \frac{x^2 + z^2}{2} \geq xz, \quad \frac{y^2 + z^2}{2} \geq yz,$$

we get $x^2 + y^2 + z^2 > xy + xz + yz$. Hence, in order to have equality in (5.3), we must have $x = y = z$. $\square$

The next two examples extend (5.2) to three and four positive reals. We start with the case of four positive numbers.

Example 5.3 Given positive real numbers a, b, c and d, we have

$$\frac{a + b + c + d}{4} \geq \sqrt[4]{abcd}, \tag{5.4}$$

with equality if and only if $a = b = c = d$.

Proof We already know that $\frac{a+b}{2} \geq \sqrt{ab}$ and $\frac{c+d}{2} \geq \sqrt{cd}$. Therefore,

$$\frac{a + b + c + d}{4} = \frac{\frac{a+b}{2} + \frac{c+d}{2}}{2} \geq \frac{\sqrt{ab} + \sqrt{cd}}{2} \geq \sqrt{\sqrt{ab}\sqrt{cd}} = \sqrt[4]{abcd},$$

where the last inequality above follows from yet another application of (5.2).

Equality happens if and only if we have equality in all of the inequalities above. For the first one of them, we must have $\frac{a+b}{2} = \sqrt{ab}$ and $\frac{c+d}{2} = \sqrt{cd}$, and, hence, $a = b$ and $c = d$. For the second one, we must have $\sqrt{ab} = \sqrt{cd}$; however, since $a = b$ and $c = d$, this is the same as $\sqrt{a^2} = \sqrt{c^2}$, i.e., $a = c$. Therefore, we have equality if and only if $a = b = c = d$. $\square$

The previous example raises the natural question of whether a similar inequality is true for three positive real numbers. As we have already anticipated, this is indeed the case, albeit its deduction will not be as immediate as the one just discussed (nevertheless, see Problem 11).

Example 5.4 Given positive real numbers a, b and c, we have

$$\frac{a + b + c}{3} \geq \sqrt[3]{abc}, \tag{5.5}$$

with equality if and only if $a = b = c$.

Proof We apply the inequality of Example 5.4 to the four positive reals a, b, c and $d = \sqrt[3]{abc}$ to get

$$\frac{a + b + c + \sqrt[3]{abc}}{4} \geq \sqrt[4]{abc\sqrt[3]{abc}} = \sqrt[4]{d^3 d} = d = \sqrt[3]{abc}.$$

Then, $a + b + c + \sqrt[3]{abc} \geq 4\sqrt[3]{abc}$ or, which is the same, $\frac{a+b+c}{3} \geq \sqrt[3]{abc}$.

Fig. 5.1 Rectangular parallelepiped with dimensions x, y and z

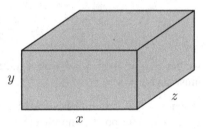

From the above computations it is clear that equality holds if and only if it also happens in the inequality

$$\frac{a+b+c+\sqrt[3]{abc}}{4} \geq \sqrt[4]{abc\sqrt[3]{abc}}.$$

Therefore, it follows from Example 5.3 that equality holds if and only if $a = b = c = \sqrt[3]{abc}$, i.e., if and only if $a = b = c$.   □

At this point, the reader is probably formulating in his/her mind a natural generalization of inequalities (5.2), (5.4) and (5.5), which we anticipated to be true. Before we formally state and prove it, let's see an interesting application of (5.5).

Example 5.5 A box (i.e., a rectangular parallelepiped[1]) is such that the sum of its edge lengths equals 48 cm. If the box encloses the largest possible volume, find its dimensions.

Solution Figure 5.1 shows a rectangular parallelepiped with dimensions x, y and z (measured in centimeters), such that the sum of the lengths of its twelve edges equals 48 cm. Since each dimension comprises four equal and parallel edges, this is the same as saying that $x + y + z = 12$.

It is a well known fact that the volume V of such a parallelepiped is given by the formula $V = xyz$. Hence, algebraically our problem reduces to the one of *maximizing* the product xyz, under the restrictions that $x, y, z > 0$ and $x+y+z = 12$. To this end, we apply inequality (5.5) to get

$$V = xyz \leq \left(\frac{x+y+z}{3}\right)^3 = \left(\frac{12}{3}\right)^3 = 64,$$

thus concluding that V is at most $64\,\text{cm}^3$. Example 5.4 also shows that the volume equals $64\,\text{cm}^3$ (i.e., that equality holds in the above inequality) if and only if $x = y = z$ or, which is the same, if and only if $x = y = z = 4\,\text{cm}$. Therefore, the box of largest possible volume is a cube with edge length equal to $4\,\text{cm}$.   □

---

[1]For the elementary facts on the notion of volume of solids, we refer the reader to [4].

Before we state the promised generalization of (5.2), (5.4) and (5.5), we need some terminology.

Definition 5.6 Given $n > 1$ positive real numbers $a_1, a_2, \ldots, a_n$, we define their:

(a) **Arithmetic mean** as the number $\frac{1}{n}(a_1 + a_2 + \cdots + a_n)$.
(b) **Geometric mean**[2] as the number $\sqrt[n]{a_1 a_2 \ldots a_n}$.

In the context of the previous definition, what we did in (5.2) and in Examples 5.3 and 5.4 was to show that the arithmetic means of two, three or four positive reals are always greater than or equal to the respective geometric means, with equality in each case if and only if the given numbers are all equal. We establish the general case in the coming result, whose proof can be omitted in a first reading. As the title of this section suggests, inequality (5.6) below is known as the **arithmetic mean-geometric mean inequality**.

Theorem 5.7 *Given $n > 1$ positive reals $a_1, a_2, \ldots, a_n$, their arithmetic mean is always greater than or equal to their geometric mean. In symbols:*

$$\frac{a_1 + a_2 + \cdots + a_n}{n} \geq \sqrt[n]{a_1 a_2 \ldots, a_n}, \tag{5.6}$$

with equality if and only if $a_1 = a_2 = \cdots = a_n$.

Proof Firstly, let's prove by induction that the desired inequality is true whenever n is a power of 2, with equality if and only if $a_1 = a_2 = \cdots = a_n$. To this end, we have to verify the initial case $n = 2$ (which was already done along the discussion that established (5.2)), formulate the induction hypothesis (for $n = 2^j$, say) and execute the induction step (i.e., deduce the case $n = 2^{j+1}$ from the case $n = 2^j$). However, since $2^{j+1} = 2 \cdot 2^j$, it suffices to suppose the inequality to be true for any k positive reals (with equality if and only if these k numbers are all equal) and, from this, to deduce that it is also true for any $2k$ positive reals (with equality if and only if these $2k$ positive reals are also all equal). Hence, to perform the induction step, consider $2k$ positive reals $a_1, a_2, \ldots, a_{2k}$. The induction hypothesis, together with the case $n = 2$, give

$$\frac{1}{2k} \sum_{j=1}^{2k} a_j = \frac{1}{2}\left(\frac{1}{k}\sum_{j=1}^{k} a_j + \frac{1}{k}\sum_{j=1}^{k} a_{k+j}\right)$$

$$\geq \frac{1}{2}\left(\sqrt[k]{a_1 \ldots a_k} + \sqrt[k]{a_{k+1} \ldots a_{2k}}\right)$$

$$\geq \sqrt{\sqrt[k]{a_1 \ldots a_k}\sqrt[k]{a_{k+1} \ldots a_{2k}}}$$

$$= \sqrt[2k]{a_1 \ldots a_k a_{k+1} \ldots a_{2k}}.$$

---

[2]The reader maybe find it useful if we observe that the adjective *geometric* attached to this number comes from the case $n = 2$. In this case, inequality (5.2) has a simple geometric interpretation, for which we refer to the problems of Sect. 4.2 of [4].

To have equality, we must have it in all passages above. Then, it must be that

$$\frac{a_1 + \cdots + a_k}{k} = \sqrt[k]{a_1 \ldots a_k}, \quad \frac{a_{k+1} + \cdots + a_{2k}}{k} = \sqrt[k]{a_{k+1} \ldots a_{2k}}$$

and

$$\frac{\sqrt[k]{a_1 \ldots a_k} + \sqrt[k]{a_{k+1} \ldots a_{2k}}}{2} = \sqrt{\sqrt[k]{a_1 \ldots a_k} \sqrt[k]{a_{k+1} \ldots a_{2k}}}.$$

For the first two equalities, it follows from the induction hypothesis that $a_1 = \cdots = a_k$ and $a_{k+1} = \cdots = a_{2k}$. For the third one, we must have $\sqrt[k]{a_1 \ldots a_k} = \sqrt[k]{a_{k+1} \ldots a_{2k}}$; this last condition, together with the two former ones, implies that we must have $a_1 = \cdots = a_k = a_{k+1} = \cdots = a_{2k}$. It is also evident (verify, anyhow!) that, if the $2k$ positive given numbers are equal, then equality must happen. Therefore, (5.6) is true by induction, with the stated condition for equality, whenever n is a power of 2.

Let's now prove, by strong induction, that the inequality is true in general, with equality holding if and only if all of the numbers are equal. To this end, let $n > 1$ be natural and $a_1, a_2, \ldots, a_n$ be given positive reals. Take $k \in \mathbb{N}$ such that $2^k > n$, and apply the arithmetic mean-geometric mean inequality to the numbers $a_1, a_2, \ldots, a_n$, together with $2^k - n$ copies of the number $a = \sqrt[n]{a_1 a_2 \ldots a_n}$ (thus, to a total of $n + (2^k - n) = 2^k$ numbers, in which case we already know that the inequality is true). We get

$$\frac{a_1 + \cdots + a_n + a + \cdots + a}{2^k} \geq \sqrt[2^k]{a_1 \ldots a_n \cdot a^{2^k - n}}$$

$$= \sqrt[2^k]{a^n a^{2^k - n}} = \sqrt[2^k]{a^{2^k}} = a.$$

Therefore, it follows that $a_1 + a_2 + \cdots + a_n + (2^k - n)a \geq 2^k a$ and, hence,

$$\frac{a_1 + a_2 + \cdots + a_n}{n} \geq a = \sqrt[n]{a_1 a_2 \ldots a_n}.$$

For the equality to hold, we already know that we must have $a_1 = a_2 = \cdots = a_n = a = \cdots = a$; in particular, all of the numbers $a_1, a_2, \ldots, a_n$ must be equal. Finally, it's immediate to see that, if all of the numbers $a_1, a_2, \ldots, a_n$ are equal, then equality holds. $\square$

The following corollary generalizes item (b) of Example 5.1. For an alternative proof, see Problem 21.

Corollary 5.8 *Given $n > 1$ positive reals $a_1, a_2, \ldots, a_n$, we have*

$$(a_1 + a_2 + \cdots + a_n) \left(\frac{1}{a_1} + \frac{1}{a_2} + \cdots + \frac{1}{a_n} \right) \geq n^2, \tag{5.7}$$

with equality if and only if $a_1 = a_2 = \cdots = a_n$.

Proof Applying (5.6) twice, we get

$$(a_1 + a_2 + \cdots + a_n) \left(\frac{1}{a_1} + \frac{1}{a_2} + \cdots + \frac{1}{a_n} \right) \geq$$

$$\geq \left(n \sqrt[n]{a_1 a_2 \ldots a_n} \right) \left(n \sqrt[n]{\frac{1}{a_1} \cdot \frac{1}{a_2} \cdots \frac{1}{a_n}} \right) = n^2.$$

In order to have equality in (5.7), we must have equality in (5.6), and we know that this implies $a_1 = a_2 = \cdots = a_n$. Conversely, it is immediate to verify that, if all of the n given numbers are equal, then we have equality in (5.7). $\qquad\square$

Remark 5.9 The inequality (5.7) is sometimes referred to as the **arithmetic mean-harmonic mean inequality**. This is due to the fact that it can be written as

$$\frac{a_1 + a_2 + \cdots + a_n}{n} \geq \left(\frac{1/a_1 + 1/a_2 + \cdots + 1/a_n}{n} \right)^{-1},$$

and that the right hand side above (i.e., the inverse of the arithmetic mean of the inverses of $a_1, a_2, \ldots, a_n$) is known as the **harmonic mean** of $a_1, a_2, \ldots, a_n$.

We now illustrate the use of the inequality on the arithmetic and geometric means in three examples.

Example 5.10 (Israel-Hungary) Let k and n be positive integers, with $n > 1$. Prove that

$$\frac{1}{kn} + \frac{1}{kn + 1} + \cdots + \frac{1}{kn + n - 1} > n \left(\sqrt[n]{\frac{k + 1}{k}} - 1 \right).$$

Proof It suffices to see that

$$\sum_{j=0}^{n-1} \frac{1}{kn + j} + n = \sum_{j=0}^{n-1} \left(\frac{1}{kn + j} + 1 \right) = \sum_{j=0}^{n-1} \frac{kn + j + 1}{kn + j}$$

$$> n \sqrt[n]{\prod_{j=0}^{n-1} \frac{kn + j + 1}{kn + j}} = n \sqrt[n]{\frac{k + 1}{k}},$$

where we've applied the inequality on the arithmetic and geometric means once. Note that, since the numbers $\frac{kn+j+1}{kn+j}$ are pairwise distinct, equality never holds. $\qquad\square$

Example 5.11 (APMO) If a, b and c are positive reals, prove that

$$\left(1+\frac{a}{b}\right)\left(1+\frac{b}{c}\right)\left(1+\frac{c}{a}\right) \geq 2\left(1+\frac{a+b+c}{\sqrt[3]{abc}}\right).$$

Proof Expanding the left hand side, we obtain

$$\left(1+\frac{a}{b}\right)\left(1+\frac{b}{c}\right)\left(1+\frac{c}{a}\right) = 2+\frac{a+c}{b}+\frac{b+c}{a}+\frac{a+b}{c},$$

and it suffices to prove that

$$\frac{a+c}{b}+\frac{b+c}{a}+\frac{a+b}{c} \geq \frac{2(a+b+c)}{\sqrt[3]{abc}}.$$

Letting S denote the left hand side of the last expression above, it follows from (5.6) and (5.7) that

$$
\begin{aligned}
S &= (a+b+c)\left(\frac{1}{a}+\frac{1}{b}+\frac{1}{c}\right) - 3 \\
&= \frac{2}{3}(a+b+c)\left(\frac{1}{a}+\frac{1}{b}+\frac{1}{c}\right) + \frac{1}{3}(a+b+c)\left(\frac{1}{a}+\frac{1}{b}+\frac{1}{c}\right) - 3 \\
&\geq \frac{2}{3}(a+b+c)\left(\frac{3}{\sqrt[3]{abc}}\right) + \frac{1}{3}\cdot 9 - 3 \\
&= \frac{2(a+b+c)}{\sqrt[3]{abc}}.
\end{aligned}
$$

□

Example 5.12 Gabriel has a sheet of cardboard of 2 m by 3 m. In order to assemble an open box, he cuts four equal squares from the corners of the sheet, folds it along the cuts and glues the lateral faces of the box along their common edges. If the box is to have the largest possible volume, what should be the length of the sides of the squares he cut? Justify your answer.

Solution If x is the common length of the sides of the cut squares, then one must clearly have $0 < x < 1$. Since the box has height x and its bottom is a rectangle of side lengths $2-2x$ and $3-2x$, the volume Gabriel wants to maximize depends on x and equals $(2-2x)(3-2x)x$.

One possibility for him is to try to apply the inequality between the arithmetic and geometric means to get rid of x and, then, see what the condition for equality says about the size of x. However, he cannot do this directly, for, although

$$(2-2x)(3-2x)x \leq \left(\frac{(2-2x)+(3-2x)+x}{3}\right)^3,$$

the expression at the right hand side still depends on x.

Nevertheless, the following trick does the job: he starts by choosing positive reals a, b and c such that $a(2-2x)+b(3-2x)+cx$ doesn't depend on x and such that there exists at least one value of $x \in (0, 1)$ for which $a(2 - 2x) = b(3 - 2x) = cx$. This amounts to finding a positive solution (a, b, c) for the linear equation $2a+2b-c = 0$, such that this solution, in turn, gives equal solutions for the first degree equations $2(a-b)x = 2a - 3b$ and $(2a+c)x = 2a$, which should belong to the interval $(0, 1)$. Hence, we should have

$$2a + 2b = c \text{ and } \frac{2a - 3b}{2(a - b)} = \frac{2a}{2a + c} \in (0, 1).$$

If we succeed in finding $a, b, c > 0$ satisfying the given equations, we will automatically have $\frac{2a}{2a+c} \in (0, 1)$. To what is left to do, substitute $c = 2a + 2b$ into the second equation to get $\frac{2a-3b}{2(a-b)} = \frac{a}{2a+b}$ or, which is the same, $2a^2 - 2ab - 3b^2 = 0$. Therefore, $\frac{a}{b}$ is a positive solution of the second degree equation $2u^2 - 2u - 3 = 0$, so that $\frac{a}{b} = \frac{1+\sqrt{7}}{2}$. It thus suffices to choose $b = 2$, $a = 1 + \sqrt{7}$ and $c = 2a+2b = 6 + 2\sqrt{7}$.

With these choices of a, b, c at hand, Gabriel can successfully implement the heuristic reasoning of the second paragraph of the proof. Writing V for the volume and recalling that $2a + 2b = c$, we have

$$abc\,V = a(2 - 2x)b(3 - 2x)cx$$
$$\leq \left(\frac{a(2 - 2x) + b(3 - 2x) + cx}{3}\right)^3$$
$$= \left(\frac{2a + 3b}{3}\right)^3.$$

Equality holds if and only if $a(2-2x) = b(3-2x) = cx$. However, we already know that these two equations have $x = \frac{2a}{2a+c}$ as common solution, so that the maximal possible volume is attained only for $x = x_0 = \frac{2a}{2a+c} = \frac{5-\sqrt{7}}{6}$, and equals

$$\frac{1}{abc}\left(\frac{2a + 3b}{3}\right)^3 = x_0(2 - 2x_0)(3 - 2x_0) = \frac{10 + 7\sqrt{7}}{27}.$$

□

Later, when we have the methods of Calculus at our disposal, the previous example will fall into the general framework of the analysis of the *first* and *second variations* of the function $V(x) = (2 - 2x)(3 - 2x)x$; as such, it will have a straightforward solution. Nevertheless, the solution we presented above, of choosing adequate *weights* prior to applying the inequality between the arithmetic and geometric means, albeit quite tricky, is an instructive one for several other situations.

Problems: Section 5.1

1. * Generalize item (a) of Example 5.1. More precisely, prove that, if x is a nonzero real number, then $\left|x + \frac{1}{x}\right| \geq 2$, with equality if and only if $|x| = 1$.

2. Given positive reals a and b, prove that $\frac{a^3}{b} + \frac{b^3}{a} \geq a^2 + b^2$. When does equality hold?

3. Given real numbers $a < b < c$, prove that the equation

$$\frac{1}{x - a} + \frac{1}{x - b} + \frac{1}{x - c} = 0$$

has exactly two distinct real roots.

4. (Brazil) Let a, b and c be positive real numbers. Prove that

$$(a + b)(a + c) \geq 2\sqrt{abc(a + b + c)}.$$

5. (USA) Prove that, for every positive real numbers a, b and c, one has

$$\frac{1}{a^3 + b^3 + abc} + \frac{1}{b^3 + c^3 + abc} + \frac{1}{c^3 + a^3 + abc} \leq \frac{1}{abc}.$$

For the next two problems, the reader shall need some Euclidean Geometry, which we review next.[3] More precisely (cf. Fig. 5.2), if $a = \overline{BC}, b = \overline{AC}$ and $c = \overline{AB}$ are the lengths of the sides of a triangle ABC, then there exist $x, y, z > 0$ such that $a = y + z$, $b = x + z$ and $c = x + y$. In fact, it suffices to take x, y and z as being equal to the lengths of the line segments determined on the sides of ABC by the points of tangency of its incircle. In the context of inequalities involving the lengths of the sides of a triangle, the substitution of them by $y + z$, $x + z$ and $x + y$ is frequently referred to as **Ravi's transformation**.

6. (IMO) If a, b and c are the lengths of the sides of a triangle, prove that

$$abc \geq (a + b - c)(b + c - a)(c + a - b).$$

Fig. 5.2 Ravi's transformation

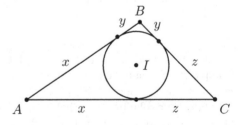

---

[3]For a thorough discussion of the facts that follow, see Sect. 3.4 of [4], for instance.

7. Let a, b and c be the lengths of the sides of a triangle. Prove that

$$\frac{a}{b+c-a} + \frac{b}{c+a-b} + \frac{c}{a+b-c} \geq 3.$$

8. (Baltic Way) Let a, b, c and d be given positive reals. Prove that

$$\frac{a+c}{a+b} + \frac{b+d}{b+c} + \frac{c+a}{c+d} + \frac{d+b}{d+a} \geq 4.$$

9. If $0 < x \neq 1$ and n is a positive integer, prove that

$$\frac{1 - x^{2n+1}}{1 - x} \geq (2n + 1)x^n.$$

10. (England) Prove that $3a^4 + b^4 \geq 4a^3b$, for all nonzero real numbers a and b, with equality if and only if $|a| = |b|$ and $ab > 0$.

11. * Prove directly (i.e., without appealing to (5.5)) that $a^3 + b^3 + c^3 \geq 3abc$.

12. (Soviet Union) Let a, b and c be positive reals. Prove that

$$(ab + ac + bc)^2 \geq 3abc(a + b + c).$$

13. (Soviet Union) If $x, y, z > 0$, prove that

$$\frac{x^2}{y^2} + \frac{y^2}{z^2} + \frac{z^2}{x^2} \geq \frac{y}{x} + \frac{z}{y} + \frac{x}{z}.$$

14. Let a and b be given positive reals. Prove that

$$9(a^3 + b^3 + c^3) \geq (a + b + c)^3.$$

15. Given positive reals a, b and c, prove that

$$a^4(1 + b^4) + b^4(1 + c^4) + c^4(1 + a^4) \geq 6a^2b^2c^2,$$

with equality if and only if $a = b = c = 1$.

16. Let $a_1, a_2, \ldots, a_n$ be given positive reals. Prove that

$$\frac{a_1}{a_2} + \frac{a_2}{a_3} + \frac{a_3}{a_4} + \cdots + \frac{a_{n-1}}{a_n} + \frac{a_n}{a_1} \geq n.$$

17. Let $n > 1$ be an odd integer and $a_1, a_2, \ldots, a_n$ be negative reals. Show that

$$\frac{a_1 + a_2 + \cdots + a_n}{n} \leq \sqrt[n]{a_1 a_2 \ldots a_n},$$

with equality if and only if all of the a_i's are equal.

18. (BMO) Prove that, for every natural n, one has:

   (a) $(n+1)^n \geq 2^n n!$.
   (b) $(n+1)^n (2n+1)^n \geq 6^n (n!)^2$.

19. (Slovenia) Let x be a positive real number and m be a natural number. Prove that

$$x(x+1)(x+2)\ldots(x+m-1) \geq m!\, x^{1+\frac{1}{2}+\frac{1}{3}+\cdots+\frac{1}{m}}.$$

20. (Poland) If $x_1, x_2, \ldots, x_n$ are positive reals whose sum equals S, prove that

$$\frac{S}{S-x_1} + \frac{S}{S-x_2} + \cdots + \frac{S}{S-x_n} \geq \frac{n^2}{n-1},$$

   with equality if and only if all of the x_i's are equal.

21. The purpose of this problem is to present an alternative proof of inequality (5.7), one that doesn't make use of (5.6). To this end, do the following items:

   (a) Show that $\left(\sum_{i=1}^n a_i\right)\left(\sum_{i=1}^n \frac{1}{a_i}\right) = n + \sum_{i<j}\left(\frac{a_i}{a_j} + \frac{a_j}{a_i}\right)$.
   (b) Apply item (a) of Example 5.1 to each sum $\frac{a_i}{a_j} + \frac{a_j}{a_i}$, with $i < j$, to get (5.7).

22. Prove the **weighted arithmetic-geometric mean inequality**: let $a_1, a_2, \ldots, a_n$ be positive reals and $k_1, k_2, \ldots, k_n$ be positive integers whose sum of inverses equals 1. Prove that

$$\frac{a_1^{k_1}}{k_1} + \frac{a_2^{k_2}}{k_2} + \cdots + \frac{a_n^{k_n}}{k_n} \geq a_1 a_2 \ldots a_n,$$

   with equality if and only if $a_1 = a_2 = \cdots = a_n$. (Note that the case $k_1 = k_2 = \cdots = k_n = \frac{1}{n}$ corresponds to the usual inequality between the arithmetic and geometric means.)

23. (Romania) Let $n > 1$ be an integer and $0 < a_1 < a_2 < \cdots < a_n$ be given real numbers. Prove that

$$\frac{1^2}{a_1} + \frac{2^2}{a_2} + \cdots + \frac{n^2}{a_n} \leq \frac{n}{a_1} + \frac{n-1}{a_2 - a_1} + \frac{n-2}{a_3 - a_2} + \cdots + \frac{1}{a_n - a_{n-1}}.$$

   Under what conditions does equality occur?

24. (China) Given positive reals a, b and c, prove that

$$\frac{a + \sqrt{ab} + \sqrt[3]{abc}}{3} \leq \sqrt[3]{a\left(\frac{a+b}{2}\right)\left(\frac{a+b+c}{3}\right)}.$$

5.2 Cauchy's Inequality

As a further application of the ideas of the previous section, let $n > 1$ be an integer and $a_1, a_2, a_3, b_1, b_2, b_3$ be real numbers such that $a_1^2 + a_2^2 + a_3^2 = 1$ and $b_1^2 + b_2^2 + b_3^2 = 1$. Since $x^2 + y^2 \geq 2|xy|$ for all $x, y \in \mathbb{R}$, with equality if and only if $|x| = |y|$, we have

$$
\begin{aligned}
a_1^2 + b_1^2 &\geq |a_1 b_1|, \\
a_2^2 + b_2^2 &\geq |a_2 b_2|, \\
a_3^2 + b_3^2 &\geq |a_3 b_3|,
\end{aligned}
\tag{5.8}
$$

with equality if and only if $|a_1| = |b_1|$, $|a_2| = |b_2|$ and $|a_3| = |b_3|$. Adding the left and right hand sides of the above inequalities, we get

$$
\begin{aligned}
(a_1^2 + a_2^2 + a_3^2) + (b_1^2 + b_2^2 + b_3^2) &= (a_1^2 + b_1^2) + (a_2^2 + b_2^2) + (a_3^2 + b_3^2) \\
&\geq 2(|a_1 b_1| + |a_2 b_2| + |a_3 b_3|) \\
&\geq 2|a_1 b_1 + a_2 b_2 + a_3 b_3|,
\end{aligned}
$$

where in the last step we applied the triangle inequality for three real numbers, (2.11).

Hence, it follows from $a_1^2 + a_2^2 + a_3^2 = 1$ and $b_1^2 + b_2^2 + b_3^2 = 1$ that

$$
|a_1 b_1 + a_2 b_2 + a_3 b_3| \leq 1.
$$

Equality holds if and only if it holds in the three inequalities (5.8), as well as in the triangle inequality. Therefore, equality holds if and only if $|a_1| = |b_1|$, $|a_2| = |b_2|$, $|a_3| = |b_3|$ and either $a_1 b_1, a_2 b_2, a_3 b_3 \geq 0$ or $a_1 b_1, a_2 b_2, a_3 b_3 \leq 0$. However, it is immediate to check that such conditions are equivalent to $a_1 = b_1$, $a_2 = b_2$ and $a_3 = b_3$.

Now, consider arbitrary real numbers a_1, a_2, a_3 and b_1, b_2, b_3, such that at least one of a_1, a_2, a_3 and at least one of b_1, b_2, b_3 are nonzero. For a positive real number c, let $x_i = \frac{a_i}{c}$ for $1 \leq i \leq 3$. Since

$$
x_1^2 + x_2^2 + x_3^2 = \frac{a_1^2 + a_2^2 + a_3^2}{c^2},
$$

we have

$$
x_1^2 + x_2^2 + x_3^2 = 1 \Leftrightarrow c = \sqrt{a_1^2 + a_2^2 + a_3^2}.
$$

Analogously, letting $y_i = \frac{b_i}{d}$ for $1 \leq i \leq 3$, with $d = \sqrt{b_1^2 + b_2^2 + b_3^2}$, we have $y_1^2 + y_2^2 + y_3^2 = 1$. Therefore, it follows from our previous discussion that

$$|x_1y_1 + x_2y_2 + x_3y_3| \leq 1,$$

with equality if and only if $x_i = y_i$ for $1 \leq i \leq 3$.

Substituting the definitions of x_i and y_i in the above inequality and recalling that $c, d > 0$, we conclude that the last inequality above is equivalent to

$$\frac{|a_1b_1 + a_2b_2 + a_3b_3|}{cd} \leq 1,$$

with $c = \sqrt{a_1^2 + a_2^2 + a_3^2}$ and $d = \sqrt{b_1^2 + b_2^2 + b_3^2}$. Moreover, equality holds if and only if $a_i = \frac{c}{d}b_i$ for $1 \leq i \leq 3$. Finally, note that the last inequality above is equivalent to

$$|a_1b_1 + a_2b_2 + a_3b_3| \leq cd = \sqrt{a_1^2 + a_2^2 + a_3^2}\sqrt{b_1^2 + b_2^2 + b_3^2}.$$

Up to this point, what we have done was to establish, for $n = 3$, the famous **Cauchy's inequality**.[4] We now turn to the general case.

Theorem 5.13 (Cauchy) *Let $n > 1$ be an integer and $a_1, a_2, \ldots, a_n, b_1, b_2, \ldots, b_n$ given real numbers. Then,*

$$\left| \sum_{j=1}^{n} a_j b_j \right| \leq \sqrt{\sum_{j=1}^{n} a_j^2} \cdot \sqrt{\sum_{j=1}^{n} b_j^2}, \tag{5.9}$$

with equality if and only if the a_i's and b_i's are respectively proportional, i.e., if and only if there exists a nonzero real number λ such that

$$a_1 = \lambda b_1, \ a_2 = \lambda b_2, \ \ldots, \ a_n = \lambda b_n.$$

Proof If all of the a_i's or all of the b_i's are equal to zero, there is nothing to do. Otherwise, in order to establish (5.9), it suffices to follow the steps of the particular case $n = 3$ discussed above. The only difference is that, whenever convenient, we have to use (4.6) instead of (2.11) (see Problem 2, page 127). $\qquad\square$

Later (see Example 6.18), we shall give another proof of Cauchy's inequality as an application of the theory of maxima and minima of quadratic functions. For a geometric interpretation of Cauchy's inequality for $n = 2$, see the problems of Sect. 6.3 of [4].

The coming two examples illustrate how one can apply Cauchy's inequality.

---

[4]Augustin Louis Cauchy, one of the greatest mathematicians of the XIX century, and maybe of History. Cauchy was one of the precursors of Mathematical Analysis, an extremely important area of higher Mathematics. He also has his name attached to several important results in Differential Equations and Mathematical Physics.

Example 5.14 Let a, b and c be given real numbers. Show that the system of equations

$$\begin{cases} 3(x^2 + y^2 + z^2) + a^2 + b^2 + c^2 = 6 \\ ax + by + cz = 2 \end{cases}$$

doesn't have any real solutions x, y, z.

Proof By contradiction, if there existed a real solution x, y, z, we would have, by Cauchy's inequality,

$$4 = (ax + by + cz)^2 \leq (a^2 + b^2 + c^2)(x^2 + y^2 + z^2).$$

Thus, letting $u = a^2 + b^2 + c^2$ and $v = x^2 + y^2 + z^2$, we would have $u + 3v = 6$ and $uv \geq 4$. At this point, the inequality on the arithmetic and geometric means would give us

$$6 = u + 3v \geq 2\sqrt{u \cdot 3v} = 2\sqrt{3uv} = 2\sqrt{3 \cdot 4} = 4\sqrt{3},$$

which is an absurd. □

Example 5.15 (Romania) Let $x_1, x_2, \ldots, x_{n+1}$ be positive reals satisfying $x_1 + x_2 + \cdots + x_n = x_{n+1}$. Prove that

$$\sqrt{x_1(x_{n+1} - x_1)} + \cdots + \sqrt{x_n(x_{n+1} - x_n)} \leq$$

$$\leq \sqrt{x_{n+1}(x_{n+1} - x_1) + \cdots + x_{n+1}(x_{n+1} - x_n)}.$$

Proof For $1 \leq j \leq n$, let $y_j = x_{n+1} - x_j$. By Cauchy's inequality, we have

$$\sqrt{x_1 y_1} + \cdots + \sqrt{x_n y_n} \leq \sqrt{x_1 + \cdots + x_n}\sqrt{y_1 + \cdots + y_n}$$

$$= \sqrt{x_{n+1}}\sqrt{(x_{n+1} - x_1) + \cdots + (x_{n+1} - x_n)}.$$

□

For future use, we collect the following corollary of Cauchy's inequality.

Corollary 5.16 *Given real numbers $a_1, \ldots, a_n$ and $b_1, \ldots, b_n$, we have*

$$\sqrt{\sum_{j=1}^{n}(a_j + b_j)^2} \leq \sqrt{\sum_{j=1}^{n} a_j^2} + \sqrt{\sum_{j=1}^{n} b_j^2}, \tag{5.10}$$

with equality holding if and only if $a_1, \ldots, a_n$ and $b_1, \ldots, b_n$ are positively proportional, i.e., if and only if there exists a positive real number λ such that $a_i = \lambda b_i$ for $1 \leq i \leq n$.

Proof For the sake of clarity, let's prove the corollary for $n = 3$, the general case being totally analogous. Since both sides of (5.10) are nonnegative real numbers, it suffices to show that the square of the left hand side is less than or equal to the square of the right hand side. In symbols,

$$(a_1 + b_1)^2 + (a_2 + b_2)^2 + (a_3 + b_3)^2 \leq \left(\sqrt{a_1^2 + a_2^2 + a_3^2} + \sqrt{b_1^2 + b_2^2 + b_3^2} \right)^2.$$

Expanding $(a_i + b_i)^2$, it follows that the square of the left hand side equals

$$(a_1^2 + 2a_1b_1 + b_1^2) + (a_2^2 + 2a_2b_2 + b_2^2) + (a_3^2 + 2a_3b_3 + b_3^2).$$

Analogously, the square of the right hand side equals

$$(a_1^2 + a_2^2 + a_3^2) + 2\sqrt{a_1^2 + a_2^2 + a_3^2}\sqrt{b_1^2 + b_2^2 + b_3^2} + (b_1^2 + b_2^2 + b_3^2).$$

Cancelling the summand $(a_1^2 + a_2^2 + a_3^2) + (b_1^2 + b_2^2 + b_3^2)$ from both sides, we conclude that (5.10) is equivalent to

$$2(a_1b_1 + a_2b_2 + a_3b_3) \leq 2\sqrt{a_1^2 + a_2^2 + a_3^2}\sqrt{b_1^2 + b_2^2 + b_3^2},$$

which is precisely Cauchy's inequality.

The analysis of the conditions for equality will be left as an exercise to the reader.

$\Box$

As will be seen in Chaps. 8 and 13 of [4], for $n = 2$ and 3 inequality (5.10) has the following *geometric interpretation*: in a cartesian coordinate system, if $O = (0, \ldots, 0)$, $A = (a_1, \ldots, a_n)$ and $B = (b_1, \ldots, b_n)$, and we set $C = A + B$, then $C = (a_1 + b_1, \ldots, a_n + b_n)$ and (5.10) amounts to the inequality

$$\overline{OC} \leq \overline{OA} + \overline{OB},$$

where $\overline{XY}$ stands for the length of the line segment XY. This is the same as the (geometric) triangle inequality for the (possibly degenerated) triangle OAC, and for this reason (5.10) is also known as the **triangle inequality**.

Note also that, if $n = 1$ and we make $a_1 = a$ and $b_1 = b$, then (5.10) reduces to $\sqrt{(a + b)^2} \leq \sqrt{a^2} + \sqrt{b^2}$ or, which is the same, $|a+b| \leq |a|+|b|$. This is the reason why (5.10), as well as its generalization (4.6), are known as *triangle inequality*.

Problems: Section 5.2

1. Given real numbers x and y such that $3x + 4y = 12$, compute the minimum possible value of $x^2 + y^2$.
2. Prove the general case (5.9) of Cauchy's inequality.
3. Given positive reals $a_1, a_2, \ldots, a_n$, we define its *quadratic mean* as the number

$$\sqrt{\frac{a_1^2 + a_2^2 + \cdots + a_n^2}{n}}.$$

Prove the inequality between the quadratic and arithmetic means:

$$\sqrt{\frac{a_1^2 + a_2^2 + \cdots + a_n^2}{n}} \geq \frac{a_1 + a_2 + \cdots + a_n}{n}, \tag{5.11}$$

with equality if and only if $a_1 = a_2 = \cdots = a_n$.

4. (IMO shortlist) Let a_1, a_2, a_3, a_4 be positive reals. Prove that

$$\sum_{1 \leq i < j < k \leq 4} \frac{a_i^2 + a_j^2 + a_k^2}{a_i + a_j + a_k} \geq a_1 + a_2 + a_3 + a_4,$$

with equality if and only if $a_1 = a_2 = a_3 = a_4$.

5. Use Cauchy's inequality to give a third proof of (5.7).
6. (Leningrad) Given positive reals a, b, c and d, prove that

$$\frac{1}{a} + \frac{1}{b} + \frac{4}{c} + \frac{16}{d} \geq \frac{64}{a + b + c + d}.$$

7. (OIM) Let x, y and z be positive reals whose sum equals 3. Prove that

$$3\sqrt[3]{9} < \sqrt{2x + 3} + \sqrt{2y + 3} + \sqrt{2z + 3} \leq 3\sqrt{5}.$$

When does the right hand inequality become an equality?

8. Let $n > 2$ be an integer. Prove that

$$\sqrt{\binom{n}{1}} + \sqrt{\binom{n}{2}} + \cdots + \sqrt{\binom{n}{n}} < \sqrt{n(2^n - 1)}.$$

9. (Soviet Union—adapted) Given $x, y, z > 0$, apply Cauchy's inequality to prove that

$$\frac{x^2}{y^2} + \frac{y^2}{z^2} + \frac{z^2}{x^2} \geq \frac{y}{x} + \frac{z}{y} + \frac{x}{z},$$

with equality if and only if $x = y = z$.

10. (APMO) Let $a_1, a_2, \ldots, a_n$ and $b_1, b_2, \ldots, b_n$ be given positive reals, such that $a_1 + a_2 + \cdots + a_n = b_1 + b_2 + \cdots + b_n$. Show that

$$\sum_{k=1}^{n} \frac{a_k^2}{a_k + b_k} \geq \frac{1}{2} \sum_{k=1}^{n} a_k.$$

11. (TT) Let $a_1, a_2, \ldots, a_n$ be given positive reals. Prove that

$$\left(1 + \frac{a_1^2}{a_2}\right)\left(1 + \frac{a_2^2}{a_3}\right) \cdots \left(1 + \frac{a_n^2}{a_1}\right) \geq (1 + a_1)(1 + a_2)\ldots(1 + a_n).$$

12. (APMO) Let a, b and c be the lengths of the sides of a triangle. Show that

$$\sqrt{a + b - c} + \sqrt{b + c - a} + \sqrt{c + a - b} \leq \sqrt{a} + \sqrt{b} + \sqrt{c}$$

and explain when equality occurs.

13. Let $a_1, a_2, \ldots, a_n, b_1, b_2, \ldots, b_n$ and $c_1, c_2, \ldots, c_n$ be given positive real numbers. Show that

$$\frac{1}{n}\left(\sum_{k=1}^{n} a_k b_k c_k\right)^4 \leq \left(\sum_{k=1}^{n} a_k^4\right)\left(\sum_{k=1}^{n} b_k^4\right)\left(\sum_{k=1}^{n} c_k^4\right).$$

14. (IMO) Let a, b and c be positive reals such that $abc = 1$. Prove that

$$\frac{1}{a^3(b + c)} + \frac{1}{b^3(a + c)} + \frac{1}{c^3(a + b)} \geq \frac{3}{2}.$$

5.3 More on Inequalities

This section is devoted to the study of other important elementary inequalities. The first of them is generally attributed to the Bernoulli brothers,[5] being known as **Bernoulli's inequality**. In spite of its simplicity, we will see that it is quite useful in applications.

Proposition 5.17 (Bernoulli) *If n is a natural number and $x > -1$ is a real number, then*

$$(1 + x)^n \geq 1 + nx,$$

with equality holding for $n > 1$ if and only if $x = 0$.

---

[5] Jacob and Johann Bernoulli, Swiss mathematicians of the XVIII century.

Proof Let's make induction on n, the case $n = 1$ being immediate. Suppose, by induction hypothesis, that $(1 + x)^k \geq 1 + kx$; since $1 + x > 0$, we have

$$(1 + x)^{k+1} = (1 + x)(1 + x)^k \geq (1 + x)(1 + kx)$$
$$= 1 + (k + 1)x + kx^2 \geq 1 + (k + 1)x,$$

with equality if and only if $(1 + x)^k = 1 + kx$ and $kx^2 = 0$, i.e., if and only if $x = 0$.
□

Example 5.18 Given a natural number n and positive reals a and b, show that

$$\left(1 + \frac{a}{b}\right)^n + \left(1 + \frac{b}{a}\right)^n \geq 2^{n+1},$$

with equality if and only if $a = b$.

Proof Dividing both sides of the inequality by 2^n, we see that it suffices to prove that

$$\left(1 - \frac{1}{2} + \frac{a}{2b}\right)^n + \left(1 - \frac{1}{2} + \frac{b}{2a}\right)^n \geq 2.$$

Since $-\frac{1}{2} + \frac{a}{2b} > -1$ and $-\frac{1}{2} + \frac{b}{2a} > -1$, if we apply Bernoulli's inequality to both summands at the left hand side above and add the results, we get

$$\left(1 - \frac{1}{2} + \frac{a}{2b}\right)^n + \left(1 - \frac{1}{2} + \frac{b}{2a}\right)^n \geq 2 + n\left(\frac{a}{2b} + \frac{b}{2a} - 1\right).$$

Now, it suffices to apply the inequality between the arithmetic and geometric means to obtain

$$\frac{a}{2b} + \frac{b}{2a} - 1 \geq 2\sqrt{\frac{a}{2b} \cdot \frac{b}{2a}} - 1 = 0,$$

with equality if and only if $\frac{a}{2b} = \frac{b}{2a}$, i.e., if and only if $a = b$.
□

We continue by presenting an inequality known in the literature as **Chebyshev's inequality**.[6]

Theorem 5.19 (Chebyshev) *If $a_1, a_2, \ldots, a_n$ and $b_1, b_2, \ldots, b_n$ are real numbers such that*

$$a_1 \leq a_2 \leq \cdots \leq a_n \text{ and } b_1 \leq b_2 \leq \cdots \leq b_n,$$

---

[6] After Pafnuty Chebyshev, Russian mathematician of the XIX century.

then

$$\left(\frac{1}{n}\sum_{i=1}^{n}a_i\right)\left(\frac{1}{n}\sum_{i=1}^{n}b_i\right) \le \frac{1}{n}\sum_{i=1}^{n}a_ib_i,$$

with equality holding if and only if $a_1 = a_2 = \cdots = a_n$ *ou* $b_1 = b_2 = \cdots = b_n$.

Proof We have to show that

$$n\sum_{i=1}^{n}a_ib_i - \left(\sum_{i=1}^{n}a_i\right)\left(\sum_{i=1}^{n}b_i\right) \ge 0,$$

and for this it suffices to observe that the expression at the left hand side equals

$$\sum_{i,j=1}^{n}(a_i - a_j)(b_i - b_j), \tag{5.12}$$

which, in turn, is nonnegative (for, the a_i's and b_i's are equally ordered).

Now, if $a_1 = a_2 = \cdots = a_n$ or $b_1 = b_2 = \cdots = b_n$, it is immediate to check that equality holds in Chebyshev's inequality. Conversely, suppose that equality holds in such an inequality. Then, the argument of the previous paragraph assures that the expression in (5.12) must be equal to zero. Since $(a_i - a_j)(b_i - b_j) \ge 0$ for all indices $1 \le i, j \le n$, we must have $(a_i - a_j)(b_i - b_j) = 0$ for all $1 \le i, j \le n$. If there exists $1 \le k \le n$ such that $b_k < b_{k+1}$, then $b_1 \le \cdots \le b_k < b_{k+1} \le \cdots \le b_n$, and the condition $(a_i - a_{k+1})(b_i - b_{k+1}) = 0$ for every $1 \le i \le n$ gives $a_i = a_{k+1}$ for $1 \le i \le k$. Hence, $a_1 = a_2 = \cdots = a_k = a_{k+1}$. If we now start from $(a_i - a_k)(b_i - b_k) = 0$ for $k < i \le n$, we conclude in a similar way that $a_{k+1} = \cdots = a_n$. Therefore, all of the a_i's must be equal. $\square$

The following corollary collects an important consequence of Chebyshev's inequality.

Corollary 5.20 *If k is a natural number and $a_1, a_2, \ldots, a_n$ are positive reals, then*

$$\frac{a_1^k + a_2^k + \cdots + a_n^k}{n} \ge \left(\frac{a_1 + a_2 + \cdots + a_n}{n}\right)^k. \tag{5.13}$$

Moreover, if $k > 1$, then equality holds if and only if all of the a_i's are equal.

Proof Let's prove the inequality by induction on $k \ge 1$, noting that (5.13) is trivially true for $k = 1$ and all positive reals $a_1, \ldots, a_n$. By induction hypothesis, let $l > 1$ be a natural number such that (5.13) is true for $k = l - 1$ and all positive reals $a_1, a_2, \ldots, a_n$.

Since both sides of the inequality we wish to prove are invariant under permutations of the indices $1, 2, \ldots, n$, we can suppose, without any loss of generality, that $a_1 \leq a_2 \leq \cdots \leq a_n$. Then, $a_1^{l-1} \leq a_2^{l-1} \leq \cdots \leq a_n^{l-1}$, and it follows from Chebyshev' inequality that

$$\frac{1}{n} \sum_{i=1}^{n} a_i^l \geq \left(\frac{1}{n} \sum_{i=1}^{n} a_i \right) \left(\frac{1}{n} \sum_{i=1}^{n} a_i^{l-1} \right). \tag{5.14}$$

On the other hand, induction hypothesis gives

$$\frac{1}{n} \sum_{i=1}^{n} a_i^{l-1} \geq \left(\frac{1}{n} \sum_{i=1}^{n} a_i \right)^{l-1}, \tag{5.15}$$

and if we combine these two inequalities we get

$$\frac{1}{n} \sum_{i=1}^{n} a_i^l \geq \left(\frac{1}{n} \sum_{i=1}^{n} a_i \right) \left(\frac{1}{n} \sum_{i=1}^{n} a_i \right)^{l-1} = \left(\frac{1}{n} \sum_{i=1}^{n} a_i \right)^{l}.$$

Finally, let $l > 1$ and suppose we have equality in (5.13) when $k = l$. Then, the argument of the previous paragraph assures that we must have equality in (5.14) and (5.15). However, by the condition for equality in Chebyshev's inequality, the only way to have equality in (5.14) is having $a_1 = \cdots = a_n$. □

The two coming examples illustrate how one can apply Chebyshev's inequality.

Example 5.21 (Poland) Let $a_1, a_2, \ldots, a_n$ be positive reals with sum equal to s. Prove that

$$\frac{a_1}{s - a_1} + \frac{a_2}{s - a_2} + \cdots + \frac{a_n}{s - a_n} \geq \frac{n}{n - 1}.$$

Proof Suppose, without loss of generality, that $a_1 \leq a_2 \leq \cdots \leq a_n$. Then, $s - a_1 \geq s - a_2 \geq \cdots \geq s - a_n$ and, since $s - a_i > 0$ for every i, it follows that $\frac{1}{s-a_1} \leq \frac{1}{s-a_2} \leq \cdots \leq \frac{1}{s-a_n}$. Therefore, Chebyshev's inequality gives

$$\sum_{i=1}^{n} \frac{a_i}{s - a_i} = \sum_{i=1}^{n} \left(a_i \cdot \frac{1}{s - a_i} \right) \geq \frac{1}{n} \left(\sum_{i=1}^{n} a_i \right) \left(\sum_{i=1}^{n} \frac{1}{s - a_i} \right)$$

$$= \frac{s}{n} \left(\sum_{i=1}^{n} \frac{1}{s - a_i} \right). \tag{5.16}$$

On the other hand, it follows from Corollary 5.8 that

$$\left(\sum_{i=1}^{n} (s - a_i) \right) \left(\sum_{i=1}^{n} \frac{1}{s - a_i} \right) \geq n^2.$$

However, since $\sum_{i=1}^{n}(s - a_i) = (n-1)s$, this last inequality yields

$$\sum_{i=1}^{n}\frac{1}{s - a_i} \geq \frac{n^2}{(n-1)s}. \tag{5.17}$$

If we now combine (5.16) and (5.17), we arrive at the desired inequality.   □

Example 5.22 (Turkey) Let $n > 1$ be a natural number and $x_1, x_2, \ldots, x_n$ be positive reals such that $\sum_{i=1}^{n} x_i^2 = 1$. Find the least possible value of

$$\sum_{i=1}^{n}\frac{x_i^5}{x_1 + \cdots + \widehat{x}_i + \cdots + x_n},$$

where the hat over x_i at the denominator of the i-th summand indicates that it contains all of $x_1, x_2, \ldots, x_n$, except x_i.

Proof Let s denote the sum of the x_i's and suppose, without loss of generality, that $x_1 \leq x_2 \leq \cdots \leq x_n$. Then, $x_1^2 \leq x_2^2 \leq \cdots \leq x_n^2$ and $\frac{1}{s-x_1} \leq \frac{1}{s-x_2} \leq \cdots \leq \frac{1}{s-x_n}$.

The expression to be minimized can be written as

$$S = \sum_{i=1}^{n}\frac{1}{s - x_i} \cdot x_i^5.$$

Hence, if we apply Chebyshev's inequality twice, together with the inequality between the arithmetic and quadratic means (cf. Problem 3, page 127) and Corollary 5.8, we get

$$S \geq \frac{1}{n}\left(\sum_{i=1}^{n}\frac{1}{s - x_i}\right)\left(\sum_{i=1}^{n}x_i^5\right)$$

$$\geq \frac{1}{n^2}\left(\sum_{i=1}^{n}\frac{1}{s - x_i}\right)\left(\sum_{i=1}^{n}x_i\right)\left(\sum_{i=1}^{n}x_i^4\right)$$

$$\geq \frac{1}{n^3}\left(\sum_{i=1}^{n}\frac{1}{s - x_i}\right)\left(\sum_{i=1}^{n}x_i\right)\left(\sum_{i=1}^{n}x_i^2\right)^2$$

$$\geq \frac{s}{n^3}\left(\sum_{i=1}^{n}\frac{1}{s - x_i}\right) \geq \frac{s}{n^3} \cdot n^2\left(\sum_{i=1}^{n}(s - x_i)\right)^{-1}$$

$$= \frac{s}{n} \cdot \frac{1}{n(n-1)s} = \frac{1}{n^2(n-1)}.$$

In order to guarantee that $\frac{1}{n^2(n-1)}$ is the least possible value, we ought to show that it is attained. To this end, it is enough to see that, in view of the constraint $\sum_{i=1}^{n} x_i^2 = 1$, the condition for equality in Chebyshev' inequality gives $x_1 = x_2 = \cdots = x_n = \frac{1}{\sqrt{n}}$ if equality holds. However, computing with these values for x_1, ..., x_n, we see that all of the inequalities above become equalities, so that $\frac{1}{n^2(n-1)}$ is, indeed, the least possible value. $\qquad\square$

Next, we present an inequality known as the **rearrangement inequality**. We recall that a sequence $(x_1, x_2, \ldots, x_n)$ is a **permutation** or a **rearrangement** of $(a_1, a_2, \ldots, a_n)$ if these two sequences differ only by the order of their terms.[7]

Proposition 5.23 *Let $a_1 < a_2 < \cdots < a_n$ be given positive real numbers. If $(x_1, x_2, \ldots, x_n)$ is any permutation of $(a_1, a_2, \ldots, a_n)$, then*

$$\sum_{i=1}^{n-1} a_i a_{n-i} \leq \sum_{i=1}^{n-1} a_i x_i \leq \sum_{i=1}^{n-1} a_i^2,$$

with equality in the left (resp. right) inequality above if and only if $x_i = a_{n-i}$ (resp. $x_i = a_i$), for $1 \leq i \leq n$.

Proof Let us show how to maximize the sum $a_1 x_1 + a_2 x_2 + \cdots + a_n x_n$. (The argument to minimize it is completely analogous.)

Since the number of permutations[8] $(x_1, x_2, \ldots, x_n)$ of $(a_1, a_2, \ldots, a_n)$ is finite, there is at least one of them which maximizes the sum $a_1 x_1 + a_2 x_2 + \cdots + a_n x_n$.

If $(b_1, b_2, \ldots, b_n)$ is such a permutation, we want to show that $b_i = a_i$ for $1 \leq i \leq n$. To this end, it suffices to show that $b_1 < b_2 < \cdots < b_n$. By contradiction, suppose that there exist indices $1 \leq i < j \leq n$ for which $b_i > b_j$. Define a permutation $(b_1', b_2', \ldots, b_n')$ of the a_i's by setting

$$b_k' = \begin{cases} b_k, & \text{se} \quad k \neq i, j \\ b_i, & \text{se} \quad k = j \\ b_j, & \text{se} \quad k = i \end{cases}.$$

Then,

$$\sum_{i=1}^{n} a_i b_i' - \sum_{i=1}^{n} a_i b_i = (a_i b_i' + a_j b_j') - (a_i b_i + a_j b_j)$$

$$= (a_i b_j + a_j b_i) - (a_i b_i + a_j b_j)$$

$$= (a_i - a_j)(b_j - b_i) > 0.$$

---

[7] In the more precise language of functions (cf. Chap. 6), we say that $(x_1, x_2, \ldots, x_n)$ is a permutation of $(a_1, a_2, \ldots, a_n)$ if there is a *bijection* $\varphi : \{1, \ldots, n\} \rightarrow \{1, \ldots, n\}$, so that $x_i = a_{\varphi(i)}$, for $1 \leq i \leq n$.

[8] Actually, it is easy to show that there are exactly $n!$ such permutations. For a proof, see [5], or provide one yourself, by making an induction argument.

This is the same as

$$a_1 b_1' + a_2 b_2' + \cdots + a_n b_n' > a_1 b_1 + a_2 b_2 + \cdots + a_n b_n,$$

which, in turn, contradicts the fact that $(b_1, b_2, \ldots, b_n)$ is a permutation of $(a_1, a_2, \ldots, a_n)$ that maximizes the sum $a_1 x_1 + a_2 x_2 + \cdots + a_n x_n$. Therefore, $b_1 < b_2 < \cdots < b_n$. □

With essentially the same argument as above, one can easily extend the rearrangement inequality to the case in which $a_1 \leq a_2 \leq \cdots \leq a_n$. In this case, if $(x_1, x_2, \ldots, x_n)$ is any permutation of $(a_1, a_2, \ldots, a_n)$, then

$$\sum_{i=1}^{n-1} a_i a_{n-i} \leq \sum_{i=1}^{n-1} a_i x_i \leq \sum_{i=1}^{n-1} a_i^2,$$

with equality in the left (resp. right) inequality above *if* (and no more *if, and only if*) $x_i = a_{n-i}$ (resp. $x_i = a_i$), for $1 \leq i \leq n$. The coming example explores this more general form of the rearrangement inequality.

Example 5.24 Given positive reals a, b and c, show that:

(a) $a^3 + b^3 + c^3 \geq a^2 b + b^2 c + c^2 a$.
(b) $\frac{a+b+c}{abc} \leq \frac{1}{a^2} + \frac{1}{b^2} + \frac{1}{c^2}$.

Proof

(a) Suppose, without loss of generality, that $a \leq b \leq c$. (Other orderings of a, b and c would give rise to analogous arguments.) A direct application of the rearrangement inequality gives

$$a^3 + b^3 + c^3 = a^2 \cdot a + b^2 \cdot b + c^2 \cdot c \geq a^2 \cdot b + b^2 \cdot c + c^2 \cdot a.$$

(b) The symmetry of both sides with respect to a, b and c allows us to suppose again that $a \leq b \leq c$. The inequality to be proved is equivalent to

$$a^2 bc + ab^2 c + abc^2 \leq (ab)^2 + (bc)^2 + (ca)^2.$$

In order to get this, let's first observe that the condition $0 < a \leq b \leq c$ implies $ab \leq ac \leq bc$. Therefore, upon applying the general form of the rearrangement inequality, we obtain

$$a^2 bc + ab^2 c + abc^2 = ab \cdot ac + ab \cdot bc + ac \cdot bc$$
$$\leq (ab)^2 + (bc)^2 + (ca)^2.$$

□

A very useful idea in certain types of problems involving inequalities is to try to use arguments similar to the one of the proof of the rearrangement inequality. Let's see an example along these lines.

Example 5.25 (Taiwan) Let $n > 2$ be an integer. Compute the greatest possible value of the expression

$$\sum_{1 \leq i < j \leq n} x_i x_j (x_i + x_j),$$

when $(x_1, x_2, \ldots, x_n)$ varies over all sequences of nonnegative reals such that $x_1 + x_2 + \cdots + x_n = 1$.

Solution Let

$$E_n(x_1, \ldots, x_n) = \sum_{1 \leq i < j \leq n} x_i x_j (x_i + x_j)$$

and observe that the symmetry of the right hand side allows us to suppose that $x_1 \geq \cdots \geq x_n \geq 0$.

Let's first of all show that

$$E_n(x_1, \ldots, x_{n-2}, x_{n-1} + x_n, 0) \geq E_n(x_1, \ldots, x_n). \tag{5.18}$$

To this end, set $y_j = x_j$ for $1 \leq j < n - 1$, $y_{n-1} = x_{n-1} + x_n$ and $y_n = 0$, and note that $y_1 + y_2 + \cdots + y_n = 1$. For the sake of notation, let ΔE_n denote the difference

$$\Delta E_n = E_n(y_1, \ldots, y_n) - E_n(x_1, \ldots, x_n).$$

Then, we get

$$\Delta E_n = \sum_{1 \leq i < j \leq n-2} y_i y_j (y_i + y_j) + \sum_{i=1}^{n-2} y_i y_{n-1} (y_i + y_{n-1})$$

$$+ \sum_{i=1}^{n-1} y_i y_n (y_i + y_n) - \sum_{1 \leq i < j \leq n-2} x_i x_j (x_i + x_j)$$

$$- \sum_{1 \leq i < j \leq n-2} x_i x_{n-1} (x_i + x_{n-1}) - \sum_{1 \leq i < j \leq n-1} x_i x_n (x_i + x_n).$$

Taking the definition of the y_i's into account, we obtain

$$\Delta E_n = \sum_{i=1}^{n-2} x_i(x_{n-1} + x_n)(x_i + x_{n-1} + x_n)$$

$$- \sum_{i=1}^{n-2} x_i x_{n-1}(x_i + x_{n-1}) - \sum_{i=1}^{n-1} x_i x_n(x_i + x_n)$$

$$= \sum_{i=1}^{n-2} [x_i(x_{n-1} + x_n)(x_i + x_{n-1} + x_n) - x_i x_{n-1}(x_i + x_{n-1})$$

$$- x_i x_n(x_i + x_n)] - x_{n-1} x_n(x_{n-1} + x_n)$$

$$= \sum_{i=1}^{n-2} 2 x_i x_{n-1} x_n - x_{n-1} x_n(x_{n-1} + x_n)$$

$$= x_{n-1} x_n(2x_1 + \cdots + 2x_{n-2} - x_{n-2} - x_n).$$

Since $x_1 \geq x_2 \geq \cdots \geq x_n \geq 0$ and $n \geq 3$, it perspires that

$$2x_1 + \cdots + 2x_{n-2} - x_{n-1} - x_n \geq 2x_1 - x_{n-1} - x_n$$

$$= x_1 - x_{n-1} + x_1 - x_n \geq 0,$$

which, in turn, establishes (5.18).

Thus, in order to maximize $E_n(x_1, \ldots, x_n)$, we can restrict our attention to the sequences $(x_1, x_2, \ldots, x_n)$ of nonnegative reals for which $x_1 + \cdots + x_{n-1} + x_n = 1$ and $x_n = 0$. In this case, it is immediate to see that

$$E_n(x_1, \ldots, x_{n-1}, 0) = E_{n-1}(x_1, \ldots, x_{n-1}),$$

with $x_1, \ldots, x_{n-1}$ nonnegative reals for which $x_1 + \cdots + x_{n-1} = 1$. Then, we can repeat the same argument and, doing this several times, we conclude that we can suppose $x_i = 0$ for $i > 2$. Therefore, it is sufficient to maximize

$$E_2(x_1, x_2) = x_1 x_2(x_1 + x_2),$$

for nonnegative x_1 and x_2 satisfying $x_1 + x_2 = 1$. This, in turn, is immediate:

$$x_1 x_2(x_1 + x_2) = x_1 x_2 \leq \left(\frac{x_1 + x_2}{2} \right)^2 = \frac{1}{4}.$$

Finally, we have

$$E_n(x_1, \ldots, x_n) \leq E_2\left(\frac{1}{2}, \frac{1}{2} \right) = \frac{1}{4}.$$

$\square$

The last inequality we wish to consider is due to the XIX century Norwegian mathematician N. H. Abel. For this reason, it is known as **Abel's inequality**.

Theorem 5.26 (Abel) *Let $n > 1$ be a natural number and $a_1, a_2, \ldots, a_n, b_1, b_2, \ldots, b_n$ be given real numbers, with $a_1 \geq a_2 \geq \cdots \geq a_n \geq 0$. If M and m respectively denote the maximum and minimum elements of the set of sums $\{b_1, b_1 + b_2, \ldots, b_1 + b_2 + \cdots + b_n\}$, then*

$$ma_1 \leq a_1 b_1 + a_2 b_2 + \cdots + a_n b_n \leq Ma_1.$$

Proof Let's prove the right hand side inequality, the proof of the left hand side one being totally analogous.

Make $s_0 = 0$ and $s_i = b_1 + \cdots + b_i$ for $1 \leq i \leq n$. Then,

$$\sum_{i=1}^{n} a_i b_i = \sum_{i=1}^{n} a_i(s_i - s_{i-1}) = \sum_{i=1}^{n} a_i s_i - \sum_{i=0}^{n-1} a_{i+1} s_i$$

$$= \sum_{i=1}^{n-1} (a_i - a_{i+1}) s_i + a_n s_n$$

$$\leq \sum_{i=1}^{n-1} M(a_i - a_{i+1}) + Ma_n = Ma_1.$$

$\square$

For future reference, and in the notations of the proof of Abel's theorem, the identity

$$\sum_{i=1}^{n} a_i b_i = \sum_{i=1}^{n-1} (a_i - a_{i+1}) s_i + a_n s_n \tag{5.19}$$

is usually referred to as **Abel's identity**. As we shall see here and later (see Problem 18, page 243), it is almost as useful as the inequality itself.

We finish this section by presenting a beautiful application of Abel's inequality. In the course of the proof, we will use the fact that every set with k elements has exactly 2^k distinct subsets. For a proof of this last fact, we refer the reader to Problem 12, page 96, or to Chap. 1 of [5].

Example 5.27 (Romania) Do the following items:

(a) Let $n > 1$ be an integer and $x_1, \ldots, x_n, y_1, \ldots, y_n$ positive reals such that $x_1 y_1 < x_2 y_2 < \cdots < x_n y_n$ and, for $1 \leq k \leq n$, $x_1 + \cdots + x_k \geq y_1 + \cdots + y_k$. Prove that

$$\frac{1}{x_1} + \frac{1}{x_2} + \cdots + \frac{1}{x_n} \leq \frac{1}{y_1} + \frac{1}{y_2} + \cdots + \frac{1}{y_n}.$$

(b) Let $A = \{a_1, a_2, \ldots, a_n\}$ be a set of positive integers with the following property: the sums of the elements of any two nonempty subsets of A are always distinct. Prove that

$$\frac{1}{a_1} + \frac{1}{a_2} + \cdots + \frac{1}{a_n} < 2.$$

Proof

(a) First of all, observe that

$$\sum_{i=1}^{n} \frac{1}{y_i} - \sum_{i=1}^{n} \frac{1}{x_i} = \sum_{i=1}^{n} \frac{x_i - y_i}{x_i y_i}. \tag{5.20}$$

On the other hand, the condition $x_1 + \cdots + x_k \geq y_1 + \cdots + y_k$ for $1 \leq k \leq n$ can be rewritten as

$$(x_1 - y_1) + \cdots + (x_k - y_k) \geq 0$$

for $1 \leq k \leq n$. Thus, making $a_i = \frac{1}{x_i y_i}$ and $b_i = x_i - y_i$ for $1 \leq i \leq n$, we have $a_1 > a_2 > \cdots > a_n > 0$, $b_1 + \cdots + b_i \geq 0$ and $\frac{x_i - y_i}{x_i y_i} = a_i b_i$, for $1 \leq i \leq n$. Applying Abel's inequality to (5.20), we obtain

$$\sum_{i=1}^{n} \frac{1}{y_i} - \sum_{i=1}^{n} \frac{1}{x_i} \geq a_n \cdot \min\{b_1 + \cdots + b_i; \ 1 \leq i \leq n\} \geq 0.$$

(b) Suppose, without loss of generality, that $a_1 < a_2 < \cdots < a_n$, and let $B_k = \{a_1, \ldots, a_k\}$ for $1 \leq k \leq n$. The hypothesis on the set A assures that all of the $2^k - 1$ nonempty subsets of B_k have distinct sums of elements. However, since each of these sums is a natural number and $a_1 + \cdots + a_k$ is the greatest of them, we conclude that

$$a_1 + \cdots + a_k \geq 2^k - 1.$$

Now, observing that $2^k - 1 = 2^0 + 2^1 + \cdots + 2^{k-1}$, we have

$$a_1 + a_2 + \cdots + a_k \geq 2^0 + 2^1 + \cdots + 2^{k-1}$$

for $1 \leq k \leq n$. On the other hand, it is obvious that

$$2^0 a_1 < 2^1 a_2 < \cdots < 2^{n-1} a_n,$$

so that the inequality of item (a) gives

$$\frac{1}{a_1} + \frac{1}{a_2} + \cdots + \frac{1}{a_n} \leq \frac{1}{2^0} + \frac{1}{2^1} + \cdots + \frac{1}{2^{n-1}} < 2.$$

$\square$

For another proof of item (b) of the previous example, see Chap. 3 of [5].

Problems: Section 5.3

1. Given $n \in \mathbb{N}$, prove that $\left(1 + \frac{1}{n}\right)^n < \left(1 + \frac{1}{n+1}\right)^{n+1}$.

2. (USA) Given naturals m and n, let $a = \frac{m^{m+1} + n^{n+1}}{m^m + n^n}$. Prove that

$$a^m + a^n \geq m^m + n^n.$$

3. Let a, b and c be positive reals. Prove that

$$\frac{1}{a} + \frac{1}{b} + \frac{1}{c} \leq \frac{a^8 + b^8 + c^8}{a^3 b^3 c^3}.$$

4. (OIM) Find all real positive solutions of the system of equations

$$\begin{cases} x_1 + x_2 + \cdots + x_{1994} = 1994 \\ x_1^4 + x_2^4 + \cdots + x_{1994}^4 = x_1^3 + x_2^3 + \cdots + x_{1994}^3 \end{cases}.$$

5. (IMO shortlist) Let a_1, a_2, a_3, a_4 be positive reals. Prove that

$$\sum_{1 \leq i < j < k \leq 4} \frac{a_i^3 + a_j^3 + a_k^3}{a_i + a_j + a_k} \geq a_1^2 + a_2^2 + a_3^2 + a_4^2,$$

with equality if and only if $a_1 = a_2 = a_3 = a_4$.

6. Let $n > 1$ be an integer and $a_1, a_2, \ldots, a_n, b_1, b_2, \ldots, b_n$ be given real numbers, with $a_1 \leq a_2 \leq \cdots \leq a_n$ and $b_1 \leq b_2 \leq \cdots \leq b_n$. If $\lambda_1 \leq \lambda_2 \leq \cdots \leq \lambda_n$ are positive reals whose sum is equal to 1, prove that

$$\left(\sum_{i=1}^n \lambda_i a_i\right)\left(\sum_{i=1}^n \lambda_i b_i\right) \leq \sum_{i=1}^n \lambda_i a_i b_i$$

and give necessary and sufficient conditions for the equality. To which particular case there corresponds Chebyshev's inequality?

7. (IMO shortlist) Let a, b, c and d be nonnegative reals for which $ab + bc + cd + da = 1$. Prove that

$$\frac{a^3}{b+c+d} + \frac{b^3}{a+c+d} + \frac{c^3}{a+b+d} + \frac{d^3}{a+b+c} \geq \frac{1}{3}.$$

8. For positive reals a, b, c and $n \in \mathbb{N}$, prove that

$$\frac{a^n}{b+c} + \frac{b^n}{a+c} + \frac{c^n}{a+b} \geq \frac{a^{n-1} + b^{n-1} + c^{n-1}}{2}.$$

9. (IMO shortlist) Let x, y and z be positive reals such that $xyz = 1$. Prove that

$$\frac{x^3}{(1+y)(1+z)} + \frac{y^3}{(1+x)(1+y)} + \frac{z^3}{(1+x)(1+y)} \geq \frac{3}{4}.$$

10. (India) Let $n > 1$ be an integer and $x_1, x_2, \ldots, x_n$ be given positive reals whose sum is equal to 1. Prove that

$$\sum_{i=1}^{n} \frac{x_i}{\sqrt{1-x_i}} \geq \sqrt{\frac{n}{n-1}} \geq \frac{1}{\sqrt{n-1}} \sum_{i=1}^{n} \sqrt{x_i}.$$

11. (Slovenia) Given $2n$ positive reals $a_1, a_2, \ldots, a_{2n}$, how should we arrange them in pairs, such that the sum of the n products of the numbers of each pair is maximal?

12. (IMO) Let $(a_k)_{k \geq 1}$ be a sequence of pairwise distinct positive integers. Prove that, for every $n \in \mathbb{N}$, we have

$$\sum_{k=1}^{n} \frac{a_k}{k^2} \geq \sum_{k=1}^{n} \frac{1}{k}.$$

13. Do the following items:

    (a) If $x < y$ are given positive reals and $a = \frac{x+y}{2}$, prove that $a(x + y - a) \geq xy$.
    (b) Use item (a) to furnish another proof for the inequality between the arithmetic and geometric means.

14. (TT) Let $a_1, a_2, \ldots, a_n$ be given positive reals. Prove that

$$\left(1 + \frac{a_1^2}{a_2}\right)\left(1 + \frac{a_2^2}{a_3}\right) \cdots \left(1 + \frac{a_n^2}{a_1}\right) \geq \prod_{k=1}^{n}(1 + a_k).$$

15. (Taiwan) Let $n \geq 3$ be an integer and $x_1, x_2, \ldots, x_n$ be nonnegative reals whose sum equals 1. Prove that

$$x_1^2 x_2 + x_2^2 x_3 + x_3^2 x_4 + \cdots + x_n^2 x_1 \leq \frac{4}{27}.$$

16. If a, b, c and d are nonnegative reals such that $a \leq 1$, $a+b \leq 5$, $a+b+c \leq 14$ and $a+b+c+d \leq 30$, use Abel's inequality to prove that

$$\sqrt{a} + \sqrt{b} + \sqrt{c} + \sqrt{d} \leq 10.$$

17. Let $a_1, \ldots, a_n$ and $b_1, \ldots, b_n$ be real numbers such that $a_1 \geq a_2 \geq \cdots \geq a_n > 0$ and $b_1 \geq a_1$, $b_1 b_2 \geq a_1 a_2$, $\ldots$, $b_1 b_2 \ldots b_n \geq a_1 a_2 \ldots a_n$. Show that

$$b_1 + b_2 + \cdots + b_n \geq a_1 + a_2 + \cdots + a_n.$$

Chapter 6
The Concept of Function

With the algebraic background of the previous chapters at our disposal, we now begin the study of real functions of a single variable. After presenting the basic concepts of domain, codomain and image, some relevant examples of functions are given. Then, we introduce the notions of monotonicity and extremal values, discussing various examples that, in spite of their elementary character, will reveal themselves to be very useful. The chapter continues with the study of the operations of composition and inversion of functions, and culminates with the orchestration of the whole of it to a first study of implicitly defined functions. We end the chapter by discussing, in its last two sections, graphs of elementary functions.

6.1 Definitions and Examples

Let X and Y be two given nonempty sets. Informally, a **function** f from X to Y is a *rule* that associates to each $x \in X$ a unique $y \in Y$. It is sometimes useful to *visualize* a function $f : X \to Y$ in a more concrete way, by means of diagrams as that of Fig. 6.1, where each arrow indicates which element $y \in Y$ is associated to a given $x \in X$.

We write $f : X \to Y$ to denote that f is a function from X to Y. In this case, the element $y \in Y$ associated to $x \in X$ via f is denoted by $y = f(x)$, and is called the **image** of $x \in X$ by f. In the example of Fig. 6.1, we have $X = \{1, 2, 3\}$, $Y = \{a, b, c, d\}$ and $f(1) = a, f(2) = a, f(3) = c$. Thus, a is the image of 1 and 2 by f, and c is the image of 3 by f.

The discussion of the previous paragraph makes it clear that the definition of function allows, in the corresponding diagram, that one or more elements of Y *do not receive arrows*, or that one or more elements of Y *receive more than one arrow* (observe that both these possibilities are present in Fig. 6.1).

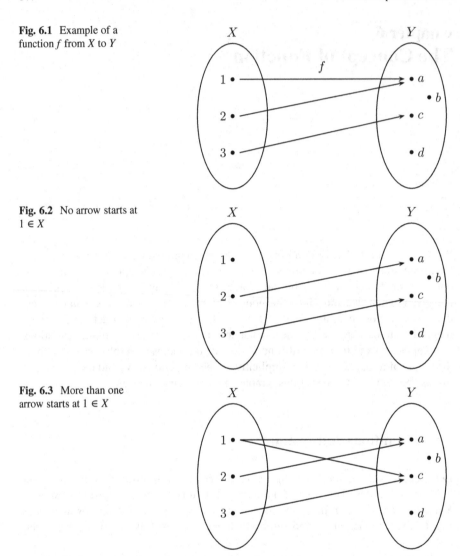

Fig. 6.1 Example of a function f from X to Y

Fig. 6.2 No arrow starts at $1 \in X$

Fig. 6.3 More than one arrow starts at $1 \in X$.

Note, however, that the diagrams in Fig. 6.2 and 6.3 do not correspond to functions. The situation of Fig. 6.2 is forbidden because there is no arrow departing from $1 \in X$. The situation of Fig. 6.3 is forbidden because more than one arrows departs from $1 \in X$.

The next three definitions isolate some quite useful types of functions.

Definition 6.1 Let X and Y be given nonempty sets. For a fixed element $c \in Y$, the **constant function** c from X to Y is the function $f : X \to Y$ such that $f(x) = c$ for every $x \in X$.

Thus, in the extreme case of the constant function equal to c, defined as above, every $x \in X$ is associated to a single $y \in Y$, namely, $y = c$. Nevertheless, the conditions required by the "definition" of function are fully satisfied, i.e., *every $x \in X$ is associated to a unique $y \in Y$*.

Definition 6.2 Let X be a given nonempty set. The **identity function** of X is the function $\mathrm{Id}_X : X \to X$, such that $\mathrm{Id}_X(x) = x$, for every $x \in X$.

As in the previous definition, it is immediate to see that the conditions required by the "definition" of function are satisfied, so that Id_X is indeed a function from X to itself.

For the next definition, given an arbitrary $n \in \mathbb{N}$, we let I_n be the set whose elements are the first n natural numbers, i.e., $I_1 = \{1\}, I_2 = \{1,2\}, I_3 = \{1,2,3\}$ and, more generally,

$$I_n = \{k \in \mathbb{N};\ 1 \le k \le n\}.$$

Definition 6.3 An **infinite sequence** of real numbers is a function $f : \mathbb{N} \to \mathbb{R}$. A **finite sequence** of real numbers is a function $f : I_n \to \mathbb{R}$, for some $n \in \mathbb{N}$.

As was anticipated in Chapter 3, given a sequence $f : \mathbb{N} \to \mathbb{R}$ (resp. $f : I_n \to \mathbb{R}$) and $k \in \mathbb{N}$ (resp. $k \in I_n$), one uses to write a_k, instead of $f(k)$, to denote the image of k by f. In this case, one also writes $(a_k)_{k \ge 1}$ (resp. $(a_k)_{1 \le k \le n}$) to denote the sequence as a whole, and says that a_k is its **k –th term**.

From a mathematically rigorous viewpoint, a function is a particular type of *relation*, according to Definition 6.4 below. In order to state it properly, we begin by recalling two simple facts from elementary set theory[1].

Given nonempty sets X and Y, and elements $x \in X$ and $y \in Y$, the **ordered pair** (x, y) is defined by

$$(x, y) = \{\{x\}, \{x, y\}\}.$$

From this, it is immediate to prove that, if $x, x' \in X$ and $y, y' \in Y$, then

$$(x, y) = (x', y') \Leftrightarrow x = x' \text{ and } y = y'.$$

The **cartesian product** of X and Y (in this order), denoted $X \times Y$, is the set

$$X \times Y = \{(x, y);\ x \in X \text{ and } y \in Y\}.$$

Definition 6.4 Given nonempty sets X and Y, a **relation** from X to Y (or between X and Y, in this order) is a subset $\mathcal{R}$ of the cartesian product $X \times Y$, i.e., $\mathcal{R}$ is a set of ordered pairs (x, y), with $x \in X$ and $y \in Y$. If $\mathcal{R}$ is a relation from X to X, we simply say that $\mathcal{R}$ is a *relation on X*.

---

[1]For a more thorough account of set theory, we refer the reader to the classic [15].

Example 6.5 Let $X = \{1, 2, 3\}$ and $Y = \{2, 3, 4, 5\}$. The set

$$\mathcal{R} = \{(x, y) \in X \times Y; \ x \geq y\}$$

is a relation from X to Y, given by $\mathcal{R} = \{(2, 2), (3, 2), (3, 3)\}$. Indeed, these are the only ordered pairs (x, y), with $x \in \{1, 2, 3\}$, $y \in \{2, 3, 4, 5\}$ and such that $x \geq y$.

The example above illustrates an obvious procedure we can use to define a specific relation $\mathcal{R}$ between given nonempty sets X and Y (in this order): it suffices to declare, somehow, a subset of the cartesian product $X \times Y$; those ordered pairs of $X \times Y$ that satisfy the given prescription will be precisely the elements of $\mathcal{R}$.

If $\mathcal{R}$ is a relation from X to Y, then $\mathcal{R} \subset X \times Y$, by definition. Conversely, once we choose an ordered pair $(x, y) \in X \times Y$, then either $(x, y) \in \mathcal{R}$ or $(x, y) \notin \mathcal{R}$. In the former case, we say that x and y are *related* by $\mathcal{R}$, and write $x \mathcal{R} y$; in the latter case, we say that x and y are *not related* by $\mathcal{R}$, and write $x \cancel{\mathcal{R}} y$. Thus, in symbols,

$$x \mathcal{R} y \Leftrightarrow (x, y) \in \mathcal{R}. \tag{6.1}$$

Therefore, with respect to the relation of Example 6.5, we have $3 \mathcal{R} 2$ but $2 \cancel{\mathcal{R}} 3$, since $2 \geq 3$ is false.

Among all types of relations we can consider, one of the most important ones is given by the coming definition.

Definition 6.6 Given nonempty sets X and Y, a relation $\mathcal{R}$ from X to Y is a **function** if the following condition is satisfied:

$$\forall \, x \in X, \ \exists \text{ a unique } y \in Y; \ x \mathcal{R} y.$$

As in the beginning of this section, given nonempty sets X and Y, it is customary to use smallcase latin letters like f, g, h etc to denote functions from X to Y. Also as before, we write $f : X \to Y$ if f is such a function, and write $f(x) = y$ if the ordered pair $(x, y) \in X \times Y$ belongs to f, i.e., if it is such that $x f y$. Notice that such a notation makes sense, for the definition of function assures that, if (x, y_1) and (x, y_2) are ordered pairs in $X \times Y$ such that $x f y_1$ and $x f y_2$, then $y_1 = y_2$. On the other hand, a moment's thought shows that the formal definition of function given above is simply the correct way of validating the informal definition given at the beginning of this section.

We will usually work with functions $f : X \to Y$ such that $X, Y \subset \mathbb{R}$. In these cases, we will generally indicate which element $f(x) \in Y$ is associated to a generic element $x \in X$ by means of a *formula* in x that ought to be seen as a rule to get $f(x)$ out of x. For instance, we can say "*consider the function $f : \mathbb{R} \to \mathbb{R}$ given by $f(x) = x^2$*" to mean that the function f associates, to each $x \in \mathbb{R}$, its square x^2. Observe that the defining requisites of a function are fulfilled, for, to each $x \in \mathbb{R}$, we have associated a single $f(x) \in \mathbb{R}$, namely, x^2. In particular, yet with respect to this example, we have $f(\sqrt{2}) = (\sqrt{2})^2 = 2, f(3) = 3^2 = 9$ etc.

When $X, Y \subset \mathbb{R}$ and $f : X \to Y$ is a function such that the element $f(x) \in Y$ associated to $x \in X$ is given by a *formula* in x, we will sometimes denote such a correspondence by writing

$$f : X \longrightarrow Y \\ x \longmapsto f(x) \, .$$

This way, the function of the previous paragraph, that associates to each $x \in \mathbb{R}$ its square x^2, could have been denoted in the following manner:

$$f : \mathbb{R} \longrightarrow \mathbb{R} \\ x \longmapsto x^2 \, .$$

Let's see another example.

Example 6.7 Consider the function $f : \mathbb{Q} \to \mathbb{R}$, given by

$$f(x) = \begin{cases} \sqrt{x^2 + 1}, & \text{if } x \leq 0 \\ x + 1, & \text{if } x > 0 \end{cases} .$$

We surely have defined a function, for the expressions that define $f(x)$ have sense in $\mathbb{R}$ and, although we must apply different formulas, according to the rational x satisfies $x \leq 0$ or $x > 0$, each rational x has a single, well defined image $f(x)$. Thus, for example, $f(-1) = \sqrt{(-1)^2 + 1} = \sqrt{2}$ (since $-1 \leq 0$), but $f(2) = 2 + 1 = 3$ (since $2 > 0$).

Observe that we could have defined $f(x)$ by writing

$$f(x) = \begin{cases} \sqrt{x^2 + 1}, & \text{if } x \leq 0 \\ x + 1, & \text{if } x \geq 0 \end{cases} .$$

If that were the case, then, albeit conditions $x \leq 0$ and $x \geq 0$ cover all rationals, they wouldn't be mutually exclusive: $x = 0$ would satisfy both. However, the formulas that should be applied in one case or the other would give the same result for $x = 0$ (since $\sqrt{0^2 + 1} = 0 + 1$), thus avoiding any possibility of inconsistence.

We sometimes say that a function f like that of the previous example is **defined by parts**. This expression alludes to the fact that there is a formula to compute $f(x)$ when $x \in (-\infty, 0]$, and another one to compute it when $x \in (0, +\infty)$ (or $x \in [0, +\infty)$, as one may wish).

When dealing with a function $f : X \to Y$, it's frequently useful to refer to the sets X and Y as the **domain** and **codomain** of the function, respectively; in this context, we will write $X = \text{Dom}\,(f)$. For instance, for the function f of Example 6.7, the domain and codomain are, respectively, $\mathbb{Q}$ and $\mathbb{R}$.

We will often work with functions $f : X \to \mathbb{R}$, with $X \subset \mathbb{R}$. In cases like these, we shall say that f is a **real function** (alluding to the fact that it assumes real values – i.e., that its codomain is $\mathbb{R}$) of a **real variable** (now alluding to the fact that a generic element x of the domain X of f – the *variable* of the function – is a real number).

In cases like those of the previous paragraph, if $f(x)$ is defined by a formula on x, then, unless explicitly stated otherwise, we will stick to the usage of taking X as the *largest possible domain*. In other words, in such cases we will take X to be the largest possible subset of $\mathbb{R}$ in which the mathematical operations that define the expression $f(x)$ have sense. We shall, then, say that X is the **maximal domain of definition**, or simply the **maximal domain** of f.

Example 6.8 Find the maximal domain of definition $X \subset \mathbb{R}$ of the function $f : X \to \mathbb{R}$, given by $f(x) = \frac{1}{\sqrt{x(x-1)}}$.

Solution For $\frac{1}{\sqrt{x(x-1)}}$ to be a real number, it's necessary and sufficient that $x(x-1)$ be positive. Thus,

$$X = \left\{ x \in \mathbb{R}; \ \frac{1}{\sqrt{x(x-1)}} \in \mathbb{R} \right\}$$
$$= \{ x \in \mathbb{R}; \ x(x-1) > 0 \}.$$

A product of two real numbers is positive only if both factors have the same sign. Therefore, we must have $x > 0$ and $x - 1 > 0$, or else $x < 0$ and $x - 1 < 0$. Hence, we must have $x > 1$ or $x < 0$, so that

$$X = (-\infty, 0) \cup (1, +\infty).$$
□

In the context of real functions of a real variable, there are standard ways of building new functions out of other known ones: we just have to use the arithmetic operations of their codomain $\mathbb{R}$. More precisely, given a nonempty set $X \subset \mathbb{R}$, a real number c and functions $f, g : X \to \mathbb{R}$ (with the same domain!), we define the functions $f + g, f \cdot g, c \cdot f : X \to \mathbb{R}$ by setting

$$(f + g)(x) = f(x) + g(x),$$
$$(f \cdot g)(x) = f(x) \cdot g(x),$$
$$(c \cdot f)(x) = c \cdot f(x),$$

for every $x \in X$.

Some remarks are in order. First of all, notice that the addition signs in the first equality above have distinct meanings: in the left hand side, the $+$ sign is used in the very *definition* of the function $f + g$, while in the right hand side $f(x) + g(x)$ stands for the usual addition of the real numbers $f(x)$ and $g(x)$. Analogous remarks are valid for the multiplication signs used in the definitions of the functions $f \cdot g$ and $c \cdot f$. Secondly, as is usual with real numbers, we shall generally omit multiplication signs, writing fg and cf instead of $f \cdot g$ and $c \cdot f$. This usage will not be source of any kind of confusion.

It is evident that $f + g, fg$ and cf are indeed functions from X to $\mathbb{R}$. On the other hand, by analogy with real numbers, $f + g$ and fg are called the **sum** and the **product** of f and g, respectively. Also, note that the product cf of the real number c by the function f can be seen as a particular case of the product of two functions: taking $g : X \to \mathbb{R}$ to be the function identically equal to c, we have $fg = cf$; yet for such a g, we shall denote $f + g$ simply by $f + c$, so that

$$(f + c)(x) = f(x) + c,$$

for every $x \in X$.

Example 6.9 Let f and g be functions from $\mathbb{R}$ to $\mathbb{R}$ given by $f(x) = \frac{x}{x^2+1}$ and $g(x) = -x + 3$, for every $x \in \mathbb{R}$. Then,

$$(f + g)(x) = f(x) + g(x) = \frac{x}{x^2 + 1} + (-x + 3)$$

$$= \frac{x + (x^2 + 1)(-x + 3)}{x^2 + 1} = \frac{-x^3 + 3x^2 + 3}{x^2 + 1},$$

$$(fg)(x) = f(x)g(x) = \frac{x}{x^2 + 1} \cdot (-x + 3) = \frac{-x^2 + 3x}{x^2 + 1}$$

and

$$(\sqrt{3}f)(x) = \sqrt{3}f(x) = \sqrt{3} \cdot \frac{x}{x^2 + 1} = \frac{x\sqrt{3}}{x^2 + 1}.$$

We leave to the reader the task of verifying that the operations of *addition* and *multiplication* of functions, defined as above, satisfy properties analogous to those satisfied by the corresponding operations with real numbers. More precisely, for functions $f, g, h : X \to \mathbb{R}$ and real numbers a, b, we have:

- *Commutativity*: $f + g = g + f$; $fg = gf$.
- *Associativity*: $f + (g + h) = (f + g) + h$; $f(gh) = (fg)h$; $a(bf) = (ab)f$.
- *Distributivity*: $f(g + h) = fg + fh$; $(a + b)f = af + bf$.

Finally, note that the associativity property of the operations of addition and multiplication of two functions allows us to define, in entirely analogous ways, the sum and the product of an arbitrary finite number of functions from X to $\mathbb{R}$. For example, given functions $f, g, h : X \to \mathbb{R}$, we define $f + g + h : X \to \mathbb{R}$ as either $(f + g) + h$ or $f + (g + h)$, since these are equal functions; analogously, $fgh : X \to \mathbb{R}$ can be defined as either $(fg)h$ or $f(gh)$.

For an additional extension of the discussion of operations on functions, we refer the reader to Problem 8.

Turning our attention to the codomain of a general function $f : X \to Y$, it's important for the reader to realize that Y *doesn't generally coincide* with the set of the images of the elements of X (which, anyhow, is a subset of Y). Let's illustrate this point using again the function f of Example 6.7. We have already observed that its codomain is the set $\mathbb{R}$ of real numbers. On the other hand, it's certain that its subset $\{f(x); x \in \mathbb{Q}\}$ doesn't contain real numbers less than 1. Indeed, since $x^2 + 1 \geq 1$ for any real x, we have

$$f(x) = \sqrt{x^2 + 1} \geq \sqrt{1} = 1,$$

whenever $x < 0$ is rational; on the other hand, for a rational $x > 0$, we have $f(x) = x + 1 > 1$. In any case,

$$\{f(x); x \in \mathbb{Q}\} \subset [1, +\infty),$$

which is a proper subset of the codomain $\mathbb{R}$ of f.

More generally, given a function $f : X \to Y$, the **image** of f is the set $\mathrm{Im}(f)$, whose elements are the images $f(x) \in Y$ of the elements $x \in X$:

$$\mathrm{Im}(f) = \{f(x) \in Y; x \in X\}.$$

In particular, we always have $\mathrm{Im}(f) \subset Y$.

In the example just discussed, we showed that the image of a function can be a proper subset of its codomain. Nevertheless, we didn't get an explicit description of the image of the function under consideration. As further developments will show, in specific situations this can be a somewhat hard task. For the time being, let's consider an illustrative example.

Example 6.10 Find an explicit description of the image of the function $f : \mathbb{Q}\setminus\{0\} \to \mathbb{Q}$ given by $f(r) = \frac{1}{b}$, if the nonzero rational r is written as $r = \frac{a}{b}$, with $a \in \mathbb{Z}, b \in \mathbb{N}$ and $\gcd(a, b) = 1$[2].

Solution The function f is *well defined*, in the sense that its definition is not ambiguous. Indeed, it's a standard fact that every nonzero rational number admits a unique representation as the quotient of two relatively prime integers, the denominator being a natural number; for example, $-\frac{4}{6} = \frac{-2}{3}$, so that $f\left(-\frac{4}{6}\right) = \frac{1}{3}$. On the other hand, in the notations of the statement of the example, since $b \in \mathbb{N}$, we immediately see that

$$\mathrm{Im}(f) \subset \left\{\frac{1}{b}; b \in \mathbb{N}\right\} = \left\{1, \frac{1}{2}, \frac{1}{3}, \frac{1}{4}, \dots\right\}.$$

Moreover, it is clear that all elements of the last set above do belong to the image of f, for $f\left(\frac{1}{b}\right) = \frac{1}{b}$, for every $b \in \mathbb{N}$. Hence, we conclude that

---

[2]For further details on the gcd of two nonzero integers, read again the introduction to Chapter 1 or, alternatively, read Section 6.2 of [5].

$$\text{Im}(f) = \left\{1, \frac{1}{2}, \frac{1}{3}, \frac{1}{4}, \ldots\right\}.$$

□

Unfortunately, there exists no algorithm that allows us to explicitly describe the image of an arbitrarily given function. Nevertheless, in subsequent chapters we will solve this problem for several important classes of functions.

We finish this section discussing the important concept of equality of functions. With respect to the function f of Example 6.7, it makes no sense to consider $f(\sqrt{2})$, since $\sqrt{2} \notin \mathbb{Q}$ and the domain of f is $\mathbb{Q}$. What we could do would be to consider, instead of f, the function $g : \mathbb{R} \to \mathbb{R}$, given by

$$g(x) = \begin{cases} \sqrt{x^2 + 1}, & \text{if } x \le 0 \\ x + 1, & \text{if } x > 0 \end{cases}.$$

Although the formulas that define $f(x)$ and $g(x)$ are the same, for f they can be applied only to $x \in \mathbb{Q}$, while for g they can be applied to every real x. Therefore, it makes no sense to think about f and g as being *equal* functions, just denoted in two different ways.

In the positive direction, we have the following

Definition 6.11 Functions $f : X \to Y$ and $g : W \to Z$ are **equal** if $X = W$, $Y = Z$ and $f(x) = g(x)$, for every $x \in X$.

If $f : X \to Y$ and $g : W \to Z$ are equal functions, we write $f = g$. We also stress that, according to the above definition, the equality of f and g doesn't merely reduce to $f(x) = g(x)$; it also means the equality $X = W$ of the domains of f and g, as well as the equality $Y = Z$ of the codomains of these two functions. If functions f and g as above are not equal, we will write $f \ne g$ and say that f and g are *different* or *distinct* functions.

Problems: Section 6.1

1. Find the maximal domain of definition of the function f, such that $f(x) = \frac{\sqrt{x-1}}{\sqrt{3-x}}$.
2. Find the maximal domain of definition of the function f, such that $f(x) = \sqrt{\frac{1}{2} - \sqrt{\frac{3}{2} - \sqrt{3 - \sqrt{x}}}}$.
3. Let $f : \mathbb{Q}_+^* \to \mathbb{Q}_+$ be the function defined by $f\left(\frac{a}{b}\right) = \frac{|a^2 - b^2|}{a^2 + b^2}$, if $a, b \in \mathbb{N}$ are relatively prime.

   (a) Compute $f(1), f(10)$ and $f\left(\frac{24}{36}\right)$.
   (b) Among the rationals $\frac{55}{73}, \frac{32}{257}$ and $\frac{101}{89}$, which do belong to the image of f?

4. Consider the function $f : \mathbb{R} \to \mathbb{R}$ given by $f(x) = x^3 - 2x^2 + 5x$, for every $x \in \mathbb{R}$. Prove that $f(x)$ has the same sign of x, for every real number $x \neq 0$.

5. Function $f : \mathbb{R} \to \mathbb{R}$ is such that $f(1) = 2, f(\sqrt{2}) = 4$ and $f(x+y) = f(x)f(y)$, for every $x, y \in \mathbb{R}$. Compute the value of $f(3 + \sqrt{2})$.

6. Let $f : \mathbb{R} \to \mathbb{R}$ be a function such that $f(x + y) = f(x) + f(y)$, for every reals x and y. If $(a_k)_{k \geq 1}$ is an AP of common difference r, prove that the sequence $(f(a_k))_{k \geq 1}$ is an AP of common difference $f(r)$.

7. Let $f : \mathbb{R} \to \mathbb{R}$ be a function such that $f(x + y) = f(x)f(y)$, for every reals x and y. If $(a_k)_{k \geq 1}$ is an AP of common difference r and such that $f(a_1) \neq 0$, prove that the sequence $(f(a_k))_{k \geq 1}$ is a GP of common ratio $f(r)$.

8. * Given a nonempty set $X \subset \mathbb{R}$ and functions $f, g : X \to \mathbb{R}$, extend the discussion in the text, furnishing adequate definitions to the **difference** $f - g$ and the **quotient** $\frac{f}{g}$ of the functions f and g.

9. * The **integer part** of a real number x, denoted $\lfloor x \rfloor$, is defined to be the greatest integer which is less than or equal to x. For instance, $\lfloor \pi \rfloor = 3, \lfloor -\frac{3}{2} \rfloor = -2$ and $\lfloor 1 \rfloor = 1$. Find the image of the **integer part function**

$$
\begin{aligned}
\lfloor \ \rfloor : \mathbb{R} &\longrightarrow \mathbb{R} \\
x &\longmapsto \lfloor x \rfloor
\end{aligned}, \tag{6.2}
$$

that associates to each $x \in \mathbb{R}$ its integer part $\lfloor x \rfloor$.

10. * The **fractional part** of a real number x, denoted $\{x\}$, is defined as $\{x\} = x - \lfloor x \rfloor$, where $\lfloor x \rfloor$ stands for the integer part of x (see the previous problem). For example, $\{\pi\} = \pi - 3, \{-\frac{3}{2}\} = -\frac{3}{2} - (-2) = \frac{1}{2}$ and $\{1\} = 1 - 1 = 0$. Find the image of the **fractional part function**

$$
\begin{aligned}
\{ \ \} : \mathbb{R} &\longrightarrow \mathbb{R} \\
x &\longmapsto \{x\}
\end{aligned}, \tag{6.3}
$$

that associates to each $x \in \mathbb{R}$ its fractional part $\{x\}$.

11. (TT) Prove that the n-th natural number which is not a perfect square is equal to $\lfloor n + \sqrt{n} + \frac{1}{2} \rfloor$, where $\lfloor \cdot \rfloor$ is defined as in Problem 9.

12. * Let $f : \mathbb{Q} \to \mathbb{Q}$ be a function such that $f(x+y) = f(x) + f(y)$, for all $x, y \in \mathbb{Q}$. Prove the following identities, for all $x, y \in \mathbb{Q}$ and $m, n \in \mathbb{Z}$, with $n \neq 0$:

(a) $f(0) = 0$ and $f(-x) = -f(x)$.
(b) $f(x - y) = f(x) - f(y)$.
(c) $f(mx) = mf(x)$.
(d) $f\left(\frac{1}{n}\right) = \frac{f(1)}{n}$.
(e) $f\left(\frac{m}{n}\right) = \frac{m}{n}f(1)$.

6.2 Monotonicity, Extrema and Image

We start this section concentrating ourselves in the problem of finding the image of a given function. To this end, recall that the *image* of a function $f : X \to Y$ is the set

$$\mathrm{Im}(f) = \{f(x) \in Y; \, x \in X\}$$
$$= \{y \in Y; \, y = f(x) \text{ for some } x \in X\}.$$

This second way of declaring $\mathrm{Im}(f)$ is particularly useful in case f is a real function of a real variable, i.e., if $f : X \to \mathbb{R}$, with $X \subset \mathbb{R}$. Indeed, for such an f, if the values $f(x)$ are given by a formula on $x \in X$, we can look at the problem of finding the image of f as that of finding the $y \in \mathbb{R}$ for which the equation $f(x) = y$ has at least one solution $x \in X$. Let's see some examples.

Example 6.12 An **affine function** is a function $f : \mathbb{R} \to \mathbb{R}$ such that $f(x) = ax + b$, for every real x, where a and b are given real numbers, with $a \neq 0$. A **linear function** is an affine function f as above, such that $b = 0$.

According to the previous paragraph, the image of an affine function f as above can be found by searching for the $y \in \mathbb{R}$ for which the equation $ax + b = y$ has at least one solution $x \in \mathbb{R}$. Since $a \neq 0$, this equation always admits the solution $x = \frac{y-b}{a}$. Therefore, we conclude that every $y \in \mathbb{R}$ belongs to the image of f and, hence, $\mathrm{Im}\,(f) = \mathbb{R}$.

Example 6.13 The **function of inverse proportionality** is the function $f : \mathbb{R} \setminus \{0\} \to \mathbb{R} \setminus \{0\}$ given by $f(x) = \frac{1}{x}$, for every $x \in \mathbb{R} \setminus \{0\}$.

In order to find its image, it suffices to find all $y \in \mathbb{R}$ for which there exists a real number $x \neq 0$ (i.e., x belonging to the domain of f), such that $f(x) = y$, i.e., such that $\frac{1}{x} = y$. If $y = 0$, it is clear that this equation doesn't admit solutions; on the other hand, if $y \neq 0$, this same equation admits the solution $x = \frac{1}{y} \neq 0$. Hence, $\mathrm{Im}\,(f) = \mathbb{R} \setminus \{0\}$.

It is time we define one of the most important classes of elementary real functions of a real variable.

Definition 6.14 A **quadratic**, or **second degree function** is a function $f : \mathbb{R} \to \mathbb{R}$ such that $f(x) = ax^2 + bx + c$ for every real x, where a, b and c are given real numbers, with $a \neq 0$. The **discriminant** Δ of f is the discriminant of the associated second degree trinomial $ax^2 + bx + c$, so that $\Delta = b^2 - 4ac$.

The problem of finding the image of a quadratic function is sufficiently important to be collected in the following

Proposition 6.15 *With respect to the quadratic function $f(x) = ax^2 + bx + c$, we have that:*

(a) *If $a > 0$, then $\mathrm{Im}\,(f) = \left[-\frac{\Delta}{4a}, +\infty\right)$.*
(b) *If $a < 0$, then $\mathrm{Im}\,(f) = \left(-\infty, -\frac{\Delta}{4a}\right]$.*

Moreover, in any of the above cases,

$$f(x) = -\frac{\Delta}{4a} \Leftrightarrow x = -\frac{b}{2a}.$$

Proof Following the general procedure described at the beginning of this section, it suffices to find all $y \in \mathbb{R}$ for which the equation $ax^2 + bx + c = y$, i.e., the second degree equation $ax^2 + bx + (c - y) = 0$, has at least one real solution. As we already know from Section 2.3, a necessary and sufficient condition for the existence of such (a) root(s) is that this last equation has nonnegative discriminant, i.e., that $b^2 - 4a(c - y) \geq 0$. Since we agreed to let $b^2 - 4ac = \Delta$, the y's we're looking for are precisely the solutions of the first degree inequality

$$\Delta + 4ay \geq 0.$$

Now, we separately consider the cases $a > 0$ and $a < 0$. If $a > 0$, then

$$4ay + \Delta \geq 0 \Leftrightarrow y \geq -\frac{\Delta}{4a},$$

and it follows that

$$\mathrm{Im}\,(f) = \left\{ y \in \mathbb{R};\ y \geq -\frac{\Delta}{4a} \right\} = \left[-\frac{\Delta}{4a}, +\infty \right);$$

if $a < 0$, then

$$4ay + \Delta \geq 0 \Leftrightarrow y \leq -\frac{\Delta}{4a}$$

and, hence,

$$\mathrm{Im}\,(f) = \left\{ y \in \mathbb{R};\ y \leq -\frac{\Delta}{4a} \right\} = \left(-\infty, -\frac{\Delta}{4a} \right].$$

To what was left to do, notice that, for $y \in \mathrm{Im}\,(f)$, the solutions of the equation $ax^2 + bx + c = y\ (\Leftrightarrow ax^2 + bx + (c - y) = 0)$ are

$$x = \frac{-b \pm \sqrt{b^2 - 4a(c - y)}}{2a} = \frac{-b \pm \sqrt{\Delta + 4ay}}{2a}. \tag{6.4}$$

Therefore, equation $f(x) = y$ admits a single solution if and only if $\Delta + 4ay = 0$, or, which is the same, if and only if $y = -\frac{\Delta}{4a}$; this being the case, we have from (6.4) that $x = -\frac{b}{2a}$. $\qquad\square$

To what comes next, we make the convention of saying that the quadratic function $f(x) = ax^2 + bx + c$ has *constant sign* if $f(x) \geq 0$ for every $x \in \mathbb{R}$, or $f(x) \leq 0$ for every $x \in \mathbb{R}$.

Corollary 6.16 *The quadratic function* $f(x) = ax^2 + bx + c$ *has constant sign if and only if* $\Delta \leq 0$. *In this case, we have* $af(x) \geq 0$ *for every* $x \in \mathbb{R}$. *In other words:*

(a) If $\Delta \leq 0$ *and* $a > 0$, *then* $f(x) \geq 0$ *for every* $x \in \mathbb{R}$.
(b) If $\Delta \leq 0$ *and* $a < 0$, *then* $f(x) \leq 0$ *for every* $x \in \mathbb{R}$.

Proof Let's look at the case $a > 0$, the case $a < 0$ being totally analogous. If $\Delta \leq 0$, it follows from the previous proposition that

$$f(x) \geq -\frac{\Delta}{4a} \geq 0, \ \forall \, x \in \mathbb{R}.$$

Conversely, suppose that $a > 0$ and that f has constant sign. Again from the previous proposition, the image of f contains positive numbers, so that we must have $f(x) \geq 0$, for every $x \in \mathbb{R}$. In particular,

$$-\frac{\Delta}{4a} = f\left(-\frac{b}{2a}\right) \geq 0.$$

Therefore, $\Delta \leq 0$. $\square$

Remark 6.17 A simple modification of the argument presented in the proof of the previous corollary allows us to conclude that

i. If $\Delta < 0$ and $a > 0$, then $f(x) > 0$ for all $x \in \mathbb{R}$.
ii. If $\Delta < 0$ and $a < 0$, then $f(x) < 0$ for all $x \in \mathbb{R}$.

From now on, we shall use these results whenever needed, without further comments.

The previous corollary can also be used to give a much simpler proof of Cauchy's inequality (cf. Theorem 5.13).

Example 6.18 Let $n > 1$ be an integer and $a_1, a_2, \ldots, a_n, b_1, b_2, \ldots, b_n$ be real numbers, such that at least one of the a_i's and at least one of the b_i's is nonzero. Consider the quadratic function

$$f(x) = (a_1x - b_1)^2 + (a_2x - b_2)^2 + \cdots + (a_nx - b_n)^2$$
$$= Ax^2 - 2Bx + C,$$

where $A = a_1^2 + a_2^2 + \cdots + a_n^2 > 0$ (for, at least one of the a_i's is nonzero), $B = a_1b_1 + a_2b_2 + \cdots + a_nb_n$ and $C = b_1^2 + b_2^2 + \cdots + b_n^2$.

Since $f(x)$ is a sum of squares, we have $f(x) \geq 0$ for every $x \in \mathbb{R}$. On the other hand, since $A > 0$, Corollary 6.16 gives $\Delta = 4(B^2 - AC) \leq 0$. Hence, $B^2 \leq AC$, or, which is the same, $|B| \leq \sqrt{A}\sqrt{C}$. Substituting the values of A, B and C, we get Cauchy's inequality.

According to the above reasoning, equality in Cauchy's inequality is equivalent, for the function f, to $\Delta = 0$. In turn, this is equivalent to the existence of a single $\alpha \in \mathbb{R}$ such that $f(\alpha) = 0$. However, since $f(\alpha)$ is a sum of squares, the only way we can have $f(\alpha) = 0$ is if each one of these squares vanishes, i.e., if

$$a_1\alpha - b_1 = a_2\alpha - b_2 = \cdots = a_n\alpha - b_n = 0.$$

Finally, since at least one of the b_i's is nonzero, we have $\alpha \neq 0$ and, writing $\lambda = \frac{1}{\alpha}$, we get

$$a_1 = \lambda b_1, \ a_2 = \lambda b_2, \ \ldots, \ a_n = \lambda b_n$$

as a necessary and sufficient condition for equality.

In order to continue in our study of function, we need a piece of terminology.

Definition 6.19 Let $I \subset \mathbb{R}$ be an interval. A function $f : I \to \mathbb{R}$ is said to be:

(a) **increasing**, if $f(x_1) < f(x_2)$, for all $x_1 < x_2$ in I.
(b) **decreasing**, if $f(x_1) > f(x_2)$, for all $x_1 < x_2$ in I.
(c) **nondecreasing**, if $f(x_1) \leq f(x_2)$, for all $x_1 < x_2$ in I.
(d) **nonincreasing**, if $f(x_1) \geq f(x_2)$, for all $x_1 < x_2$ in I.

Moreover, in any of the cases above, we say that the function f is **monotonic** in I[3].

Regarding the above definition, an interesting (and, as we shall see, important) problem is the one of finding the *monotonicity intervals* of a function $f : I \to \mathbb{R}$, where $I \subset \mathbb{R}$ is an interval. By that we mean to find the intervals $J \subset I$ such that f is increasing (resp. decreasing, nondecreasing or nonincreasing) in J. Here we shall see some elementary examples, postponing a more general analysis to Chapter 9.

Example 6.20 The affine function $f : \mathbb{R} \to \mathbb{R}$, given by $f(x) = ax + b$, is increasing if $a > 0$ and decreasing if $a < 0$.

Let's verify this claim in the case $a > 0$, the analysis of the case $a < 0$ being totally analogous. Letting $x_1 < x_2$ be two real numbers, it follows from $a > 0$ that

$$f(x_2) - f(x_1) = (ax_2 + b) - (ax_1 + b) = a(x_2 - x_1) > 0.$$

Therefore, f is increasing.

---

[3] In the notations of this definition, it is worth observing that, for some authors, a function f satisfying the condition of item (a) (resp. of item (b), (c) or (d)) is said to be *strictly increasing* (resp. *strictly decreasing, increasing* or *decreasing*).

Example 6.21 The function $f : [0, +\infty) \to \mathbb{R}$, given by $f(x) = \frac{x^2}{x+2}$, is increasing in the whole interval $[0, +\infty)$. To check this, take real numbers $0 \le a < b$. Then,

$$f(b) - f(a) = \frac{b^2}{b+2} - \frac{a^2}{a+2}$$

$$= \frac{1}{(a+2)(b+2)}[b^2(a+2) - a^2(b+2)],$$

and, since $(a+2)(b+2) > 0$, it suffices to show that $b^2(a+2) - a^2(b+2) > 0$. To this end, start by writing

$$b^2(a+2) - a^2(b+2) = b^2a - a^2b + 2(b^2 - a^2)$$

$$= ab(b-a) + 2(b-a)(b+a)$$

$$= (b-a)[ab + 2(b+a)].$$

Now, since $0 \le a < b$, both factors in the last product above are positive, so that $b^2(a+2) - a^2(b+2) > 0$.

Example 6.22 Function $f : \mathbb{R} \to \mathbb{R}$, given by $f(x) = x^3 + 2x$, for every $x \in \mathbb{R}$, is increasing. Indeed, for any real numbers $a < b$, we have

$$f(b) - f(a) = b^3 - a^3 + 2b - 2a$$

$$= (b-a)(b^2 + ba + a^2) + 2(b-a)$$

$$= (b-a)(b^2 + ab + a^2 + 2).$$

Since $b - a > 0$, it suffices to show that $a^2 + ab + b^2 + 2 > 0$; in order to do this, one possibility is to use the inequality between the arithmetic and geometric means for two numbers:

$$a^2 + b^2 + ab + 2 \ge 2|ab| + ab + 2 \ge |ab| + 2 > 0,$$

where, in the next to last inequality, we used the fact that $|\alpha| + \alpha \ge 0$, for every $\alpha \in \mathbb{R}$.

The coming proposition solves, for quadratic functions, the problem of finding the monotonicity intervals.

Proposition 6.23 *Let $a, b, c \in \mathbb{R}$, with $a \ne 0$, and $f(x) = ax^2 + bx + c$ for every $x \in \mathbb{R}$.*

(a) *If $a > 0$, then f is decreasing in $\left(-\infty, -\frac{b}{2a}\right]$ and increasing in $\left[-\frac{b}{2a}, +\infty\right)$.*

(b) *If $a < 0$, then f is increasing in $\left(-\infty, -\frac{b}{2a}\right]$ and decreasing in $\left[-\frac{b}{2a}, +\infty\right)$.*

Proof Let's do the proof of item (a), the proof of item (b) being totally analogous. For $x_2 > x_1 \geq -\frac{b}{2a}$, we have

$$f(x_2) - f(x_1) = a(x_2^2 - x_1^2) + b(x_2 - x_1)$$

$$= a(x_2 - x_1)\left(x_2 + x_1 + \frac{b}{a}\right) > 0,$$

since $x_2 > x_1 \geq -\frac{b}{2a}$ gives $x_2 - x_1 > 0$, as well as $x_2 + x_1 + \frac{b}{a} > 0$.   □

The next definition is, in a certain sense, complementary to Definition 6.19.

Definition 6.24 Let $I \subset \mathbb{R}$ be an interval and $f : I \to \mathbb{R}$ be a given function. We say that $y_0 \in \mathbb{R}$ is the **minimum value** of f in I if the two following conditions are satisfied:

(a) $\text{Im}(f) \subset [y_0, +\infty)$.
(b) $y_0 \in \text{Im}(f)$.

In this case, the real numbers $x_0 \in I$ such that $f(x_0) = y_0$ are called the **minimum points** of the function f.

Similarly, we define what one means by the **maximum value** and the **maximum points** of a function $f : I \to \mathbb{R}$ ($I \subset \mathbb{R}$ being an interval). The maximum and minimum points of a given function (provided they exist) are collectively called its **extreme points**; accordingly, the values the function takes at those points are its **extreme values**.

In Section 9.6, we shall see how to search extreme points for *differentiable* functions, i.e., functions possessing derivatives. For the time being, we shall content ourselves in analysing some elementary examples, the first of which being an immediate consequence of Proposition 6.15.

Proposition 6.25 *With respect to the quadratic function $f(x) = ax^2 + bx + c$, if $a > 0$ (resp. $a < 0$), then $-\frac{b}{2a}$ is the only minimum (resp. maximum) point of f. Moreover, the minimum (maximum) value of f is $-\frac{\Delta}{4a}$.*

The proposition above has several interesting applications, two of which are collected below, for the sake of illustrating its use. For the necessary geometric background, we refer the reader to [4].

Example 6.26 There is given a semicircle of 1cm of radius. A rectangle is so situated that one of its sides lies on the diameter of the semicircle, whereas its other two vertices lie on the semicircle itself. Compute the largest possible value for the area of the rectangle.

Solution Let Fig. 6.4 account for the described situation, so that AB is the diameter of the semicircle, O is its center and $PQRS$ is the given rectangle, with $PQ \subset AB$.

Setting $\overline{OQ} = x$ and $\overline{QR} = y$, the area of $PQRS$ equals $2xy$. On the other hand, applying Pitagoras' theorem to triangle OQR, we get $x^2 + y^2 = 1$ and, hence,

Fig. 6.4 Maximizing the
area of rectangle $PQRS$

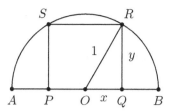

$$2xy = 2x\sqrt{1-x^2} = 2\sqrt{x^2(1-x^2)} = 2\sqrt{x^2-x^4}.$$

Now, making the substitution $z = x^2$, it follows from the last expression above for the area that it suffices to maximize the quadratic function $f(z) = z - z^2$, subjected to the condition $0 < z < 1$ (for, $x < \overline{OR} = 1$). By Proposition 6.25, such a function (without any further restrictions) admits $z = \frac{1}{2}$ as its only maximum point. Moreover, since $\frac{1}{2} \in (0, 1)$, it follows that the desired maximum value is $f\left(\frac{1}{2}\right) = \frac{1}{4}$. Therefore, the maximum value for the area is $2\sqrt{\frac{1}{4}} = 1$. $\qquad\Box$

Example 6.27 Given a triangle ABC in the plane, show that its baricenter is the only point P in the plane of ABC for which the sum $\overline{AP}^2 + \overline{BP}^2 + \overline{CP}^2$ attains its minimum possible value.

Proof Fix a cartesian system of coordinates in the plane, with respect to which $A(x_1, y_1)$, $B(x_2, y_2)$ and $C(x_3, y_3)$. If $P(x, y)$, then the formula for the distance between two points in the plane furnishes

$$\overline{AP}^2 + \overline{BP}^2 + \overline{CP}^2 = f(x) + g(y),$$

where

$$f(x) = \sum_{i=1}^{3}(x - x_i)^2 = 3x^2 - 2(x_1 + x_2 + x_3)x + (x_1^2 + x_2^2 + x_3^2)$$

and, analogously, $g(y) = 3y^2 - 2(y_1 + y_2 + y_3)y + (y_1^2 + y_2^2 + y_3^2)$.

Since x and y are independent variables, in order to minimize $\overline{AP}^2 + \overline{BP}^2 + \overline{CP}^2$ it suffices to minimize the quadratic functions f and g. To this end, invoking Proposition 6.25 we conclude that f and g attain their minimum values only at the points $x = \frac{1}{3}(x_1 + x_2 + x_3)$ and $y = \frac{1}{3}(y_1 + y_2 + y_3)$, respectively. However, it is a well known fact (cf. Chapter 6 of [4], for instance) that these are precisely the coordinates of the baricenter of triangle ABC. $\qquad\Box$

To address the problem of finding the maximum and/or minimum values of a given function, another elementary strategy which is sometimes useful is to resort to inequalities. In what follows we shall see two examples along these lines.

Example 6.28 Let $f : [0, +\infty) \to \mathbb{R}$ be the function given by $f(x) = \frac{x^2+1}{x+1}$, for every $x \in \mathbb{R}$. What is the minimum value of f? Does f attain a maximum value?

Solution First of all, note that

$$f(x) = \frac{x^2+1}{x+1} = \frac{x^2-1+2}{x+1}$$

$$= x - 1 + \frac{2}{x+1} = (x+1) + \frac{2}{x+1} - 2.$$

Therefore, applying (5.2) (with $a = x + 1$ and $b = \frac{2}{x+1}$), we get

$$(x+1) + \frac{2}{x+1} \geq 2\sqrt{(x+1) \cdot \frac{2}{x+1}} = 2\sqrt{2},$$

with equality if and only if $x + 1 = \frac{2}{x+1}$, i.e., if and only if $x^2 + 2x - 1 = 0$. Since $x \geq 0$, we conclude that equality takes place in the above inequality if and only if $x = \sqrt{2} - 1$. Thus, for $x \geq 0$ we have

$$f(x) = (x+1) + \frac{2}{x+1} - 2 \geq 2\sqrt{2} - 2,$$

so that $2\sqrt{2} - 2$ is the minimum possible value for f, and it is attained only at $x = \sqrt{2} - 1$.

To what is left, just notice that, for $n \in \mathbb{N}$, we have $f(n) = n - 1 + \frac{2}{n+1} \geq n - 1$. Therefore, f does not attain a maximum value. $\qquad\square$

Example 6.29 Find the maximum value and the maximum point(s) of the function $f : [0, +\infty) \to \mathbb{R}$, given by $f(x) = \frac{\sqrt{x}}{x^2+16}$.

Solution It follows from the inequality between the arithmetic and geometric means that

$$x^2 + 16 = x^2 + \frac{16}{3} + \frac{16}{3} + \frac{16}{3}$$

$$\geq 4\sqrt[4]{x^2 \cdot \frac{16}{3} \cdot \frac{16}{3} \cdot \frac{16}{3}} = \frac{32}{\sqrt[4]{27}} \cdot \sqrt{x},$$

so that

$$f(x) = \frac{\sqrt{x}}{x^2 + 16} \leq \frac{\sqrt[4]{27}}{32}.$$

Equality holds if and only if $x^2 = \frac{16}{3}$, which is the same as $x = \frac{4}{\sqrt{3}}$ (since $x \geq 0$).

Thus, $\frac{\sqrt[4]{27}}{32}$ is the maximum value of f and it is attained only at $x = \frac{4}{\sqrt{3}}$. $\qquad\square$

Problems: Section 6.2

1. Find the image of the function $f : \mathbb{R} \to \mathbb{R}$, given by $f(x) = \frac{1}{x^2+1}$, for every $x \in \mathbb{R}$.

2. * Find the image of the function $f : \mathbb{R}^* \to \mathbb{R}$, given by $f(x) = x + \frac{1}{x}$, for every $x \in \mathbb{R}^*$.

3. Let $I \subset \mathbb{R}$ be an interval, let $a \in I$ and $f : I \to \mathbb{R}$ be a given function. If f is increasing (resp. decreasing) in $(-\infty, a] \cap I$ and decreasing (resp. increasing) in $[a, +\infty) \cap I$, prove that a is the only maximum (resp. minimum) point of f in I.

4. * Let $X \subset \mathbb{R}$ be a nonempty set, $f : X \to \mathbb{R}$ be a given function and $c \in \mathbb{R}$. Relate the images of the functions f and $f + c$. More precisely, if $Y = \text{Im}(f)$, prove that $\text{Im}(f + c) = Y + c$, where $Y + c$ denotes the set

$$Y + c = \{y + c;\ y \in Y\}.$$

5. * Let $X \subset \mathbb{R}$ be a nonempty set, $f : X \to \mathbb{R}$ be a given function and $c \in \mathbb{R}^*$. Relate the images of the functions f and cf. More precisely, if $Y = \text{Im}(f)$, prove that $\text{Im}(cf) = cY$, where cY denotes the set

$$cY = \{cy;\ y \in Y\}.$$

6. Motivated by the *canonical form* of the second degree trinomial $ax^2 + bx + c$, from now on we shall say that

$$f(x) = a\left\{\left(x + \frac{b}{2a}\right)^2 - \frac{\Delta}{4a^2}\right\} \tag{6.5}$$

is the **canonical form** of the quadratic function $f(x) = ax^2 + bx + c$. Use this canonical form to give another proof of Proposition 6.15.

7. Let $f(x) = ax^2 + bx + c$ be a quadratic function for which $\Delta > 0$, and $x_1 < x_2$ be the roots of $f(x) = 0$. Prove the following items:

   (a) If $a > 0$, then $f(x) < 0 \Leftrightarrow x \in (x_1, x_2)$.
   (b) If $a < 0$, then $f(x) < 0 \Leftrightarrow x \notin [x_1, x_2]$.

8. Let $f(x) = ax^2 + bx + c$ be a quadratic function. If there exists a real number x_0 for which $af(x_0) < 0$, prove that $\Delta > 0$ and $x_0 \in (x_1, x_2)$, where $x_1 < x_2$ are the roots of the equation $f(x) = 0$.

9. Among all rectangles of a given perimeter, prove that the one of largest possible area is the square.

10. The cross section of a tunnel has the shape of a semicircle of radius 5m. The tunnel has two lanes of traffic, which go in opposite directions and are separated one from the other by a narrow median. The trucks of a transportation firm are to cross the tunnel to take goods from one city to another. If the trucks are 18m long, what should be their widths and heigths, so that they can carry the largest possible load volume per travel?

11. Compute the maximum value of the function $f : \mathbb{R} \to \mathbb{R}$ given by $f(x) = \frac{5x-1}{x^2+1}$, for every $x \in \mathbb{R}$.

12. Let $\alpha_1 < \alpha_2 < \cdots < \alpha_n$ be given reals, and $f : \mathbb{R} \to \mathbb{R}$ be given by

$$f(x) = |x - \alpha_1| + |x - \alpha_2| + \cdots + |x - \alpha_n|,$$

for every $x \in \mathbb{R}$. Prove that f attains a minimum value and compute it in terms of $\alpha_1, \alpha_2, \ldots, \alpha_n$.

13. In each of the following items, use the inequality between the arithmetic and geometric means to compute the maximum value of the given function:

(a) $f : \mathbb{R} \to \mathbb{R}$ given by $f(x) = \frac{x}{2x^2+3}$, for every $x \in \mathbb{R}$.

(b) $f : \mathbb{R} \to \mathbb{R}$ given by $f(x) = \frac{2x^2+5x+2}{x^2+1}$, for every $x \in \mathbb{R}$.

(c) $f : [0, 1] \to \mathbb{R}$ given by $f(x) = x(1 - x^3)$, for every $x \in [0, 1]$.

14. In each of the following items, use the inequality between the arithmetic and geometric means to compute the minimum value of the given function:

(a) $f : (0, +\infty) \to \mathbb{R}$ given by $f(x) = \frac{(x+10)(x+2)}{x+1}$, for every $x \in (0, +\infty)$.

(b) $f : (0, +\infty) \to \mathbb{R}$ given by $f(x) = x^2 + \frac{a}{x}$, for every $x \in (0, +\infty)$, where a is a positive real constant.

(c) $f : (0, +\infty) \to \mathbb{R}$ given by $f(x) = \frac{x^2}{x^3+a}$, for every $x \in (0, +\infty)$.

(d) $f : (0, +\infty) \to \mathbb{R}$ given by $f(x) = 6x + \frac{24}{x^2}$, for every $x \in (0, +\infty)$.

15. (Romania) Let $x, y \in \mathbb{R}$ be such that $x^2 - xy + y^2 \le 2$. Show that $x^4 + y^4 \le 8$ and explain when the equality holds.

16. (TT) Find all reals x, y, z and t such that

$$\begin{cases} y = x^3 + 2x \\ z = y^3 + 2y \\ t = z^3 + 2z \\ x = t^3 + 2t \end{cases}.$$

For the next problem, the reader might find it helpful to recall the discussion on the gcd of two nonzero integers, at the beginning of Chapter 1. alternatively, look at Chapter 6 of [5].

17. (OCM) Let $f : \mathbb{N} \to \mathbb{N}$ be a function such that $f(mn) = f(m) + f(n)$ whenever m and n are relatively prime. A natural number m is said to be a *strangulation point* of f if $n < m \Rightarrow f(n) < f(m)$ and $n > m \Rightarrow f(n) > f(m)$. If f has infinitely many strangulation points, show that it is an increasing function.

6.3 Composition of Functions

Functions $f : X \to Y$ and $g : Y \to Z$ give well defined rules for, departing from $x \in X$ via f, get $y = f(x) \in Y$ and then, via g, get $z = g(y) \in Z$. It thus seems reasonable that we may form a new function that allows us to go directly from X to Z. This is indeed the case, and the corresponding function is called the *composite* of f and g, according to the following

Definition 6.30 Given functions $f : X \to Y$ and $g : Y \to Z$, the **composite function** of f and g (in this order) is the function $g \circ f : X \to Z$, defined for $x \in X$ by

$$(g \circ f)(x) = g(f(x)).$$

Roughly speaking, the above definition means that, in order to find the image of $x \in X$ by $g \circ f$, it suffices to find the image of $f(x) \in Y$ by g. On the other hand, it is easy to verify that $g \circ f$, as defined above, is indeed a function. Also, observe that to form the composite of f and g, it is necessary that the domain of g equals the codomain of f. Let's see some examples.

Example 6.31 Let X and Y be nonempty sets and $f : X \to Y$ an arbitrary function. If $\mathrm{Id}_X : X \to X$ and $\mathrm{Id}_Y : Y \to Y$ are the identity functions on X and Y, respectively, then

$$f \circ \mathrm{Id}_X = f \quad \text{and} \quad \mathrm{Id}_Y \circ f = f.$$

Let's check the equality $f \circ \mathrm{Id}_X = f$, the other one being totally analogous. To this end, it suffices to note that $f \circ \mathrm{Id}_X$ is a function from X to Y such that, for evey $x \in X$,

$$(f \circ \mathrm{Id}_X)(x) = f(\mathrm{Id}_X(x)) = f(x).$$

Example 6.32 Let $f, g : \mathbb{R} \to \mathbb{R}$ be the functions given by $f(x) = x^2$ and $g(x) = \frac{1}{x^2+1}$, for every $x \in \mathbb{R}$. Then, $g \circ f$ and $f \circ g$ are functions from $\mathbb{R}$ to $\mathbb{R}$, with

$$(g \circ f)(x) = g(f(x)) = \frac{1}{(f(x))^2 + 1} = \frac{1}{(x^2)^2 + 1} = \frac{1}{x^4 + 1}$$

and

$$(f \circ g)(x) = f(g(x)) = (g(x))^2 = \left(\frac{1}{x^2 + 1}\right)^2 = \frac{1}{x^4 + 2x^2 + 1}.$$

The preceding example show an interesting phenomenon: it must happen that $g \circ f \neq f \circ g$. More precisely, it may happen that we can form $g \circ f$ but cannot form $f \circ g$ (or vice-versa); it suffices to have, for instance, $f : X \to Y$ and $g : Y \to Z$, with $X \neq Z$. Nevertheless, even if we can form both $g \circ f$ and $f \circ g$, it may well be the case that $g \circ f \neq f \circ g$.

Example 6.33 Let $f, g : (0, +\infty) \to (0, +\infty)$ be functions given by

$$f(x) = \frac{x^2 + 1}{3x^2} \quad \text{and} \quad (f \circ g)(x) = \frac{x + 2}{3},$$

for every $x \in (0, +\infty)$. Find an expression for $g(x)$ in terms of x.

Solution The definition of composite function gives

$$\frac{x + 2}{3} = (f \circ g)(x) = f(g(x)) = \frac{g(x)^2 + 1}{3g(x)^2},$$

so that $\frac{g(x)^2 + 1}{3g(x)^2} = \frac{x+2}{3}$ or, which is the same,

$$3g(x)^2 + 3 = 3(x + 2)g(x)^2.$$

Looking at this expression as a first degree equation in $g(x)^2$, we obtain $g(x)^2 = \frac{1}{x+1}$ and, hence, $g(x) = \pm\frac{1}{\sqrt{x+1}}$, for each $x > 0$. However, since g has positive image, we must have $g(x) = \frac{1}{\sqrt{x+1}}$, for every $x > 0$.   □

The operation of composition of functions, albeit not commutative, is associative, as the next proposition teaches us.

Proposition 6.34 *Given functions $f : X \to Y$, $g : Y \to Z$ and $h : Z \to W$, we have*

$$h \circ (g \circ f) = (h \circ g) \circ f.$$

Proof First of all, both $h \circ (g \circ f)$ and $(h \circ g) \circ f$ are functions from X to W. Hence, we only need to check that they associate, to each $x \in A$, a single element of W. To see this, just note that

$$(h \circ (g \circ f))(x) = h((g \circ f)(x)) = h((g(f(x)))$$
$$= (h \circ g)(f(x)) = ((h \circ g) \circ f)(x).$$

□

The previous proposition is quite important, for, it guarantees that if functions f, g and h (in this order) can be composed, then the composite function can be safely denoted by $h \circ g \circ f$, and we need not worry about which composition to perform first. As an immediate generalization of this remark, suppose functions f, g, h and l (in this order) can be composed. Then, the previous proposition gives

$$l \circ (h \circ (g \circ f)) = (l \circ h) \circ (g \circ f) = ((l \circ h) \circ g) \circ f = \cdots,$$

so that the order in which we insert parentheses does not alter the composition.

It is not difficult to see that the situation described in the last paragraph remains true for the composition (whenever possible) of any finite number of functions. In particular, given a nonempty set X, a function $f : X \to X$ and a natural number n, this last fact allows us to unambiguously define the n−th *composite function* $f^{(n)} : X \to X$, by setting

$$f^{(n)} = \underbrace{f \circ f \circ \cdots \circ f}_{n}.$$

Example 6.35 Let $f : \mathbb{R} \setminus \{-1, 1\} \to \mathbb{R} \setminus \{-1, 1\}$ be the function given by $f(x) = \frac{1-x}{1+x}$, for every $x \neq \pm 1$. For each $n \in \mathbb{N}$, find the expression that defines the n−th composite function $f^{(n)}$.

Solution Firstly, note that $f^{(n)} : \mathbb{R} \setminus \{-1, 1\} \to \mathbb{R} \setminus \{-1, 1\}$. Now, since

$$f^{(2)}(x) = (f \circ f)(x) = f(f(x)) = \frac{1 - f(x)}{1 + f(x)} = \frac{1 - \frac{1-x}{1+x}}{1 + \frac{1-x}{1+x}} = x,$$

we have $f^{(2)} = \mathrm{Id}_X$, the identity function of $X = \mathbb{R} \setminus \{-1, 1\}$. This gives us

$$f^{(3)} = f \circ f^{(2)} = f \circ \mathrm{Id}_X = f \text{ and } f^{(4)} = f \circ f^{(3)} = f \circ f = \mathrm{Id}_X.$$

In general, suppose (by induction hypothesis) that we have already proved that $f^{(2k-1)} = f$ and $f^{(2k)} = \mathrm{Id}_X$, for some integer $k \geq 1$. Then,

$$f^{(2k+1)} = f \circ f^{(2k)} = f \circ \mathrm{Id}_X = f$$

and

$$f^{(2k+2)} = f \circ f^{(2k+1)} = f \circ f = \mathrm{Id}_X.$$

Therefore, we conclude that $f^{(n)} = f$ whenever n is odd, and $f^{(n)} = \mathrm{Id}_X$ whenever n is even. $\square$

Given a function $f : X \to Y$, we've already seen some examples that illustrate the fact that the image of f is not necessarily equal to its codomain Y. On the other hand, we can also have two distinct elements of X with the same image via f. For an example, consider the quadratic function $f(x) = x^2$, with $x \in \mathbb{R}$; for every $x \in \mathbb{R}$, we have $f(x) = x^2 = (-x)^2 = f(-x)$. We attach special names to functions whose images coincide with their codomains, or which associate distinct images to distinct elements of their domains. This is set in the coming

Definition 6.36 A function $f : X \to Y$ is said to be:

(a) **Injective**, or **one-to-one**, or an **injection**, if, for every $y \in Y$, there exists at most one $x \in X$ such that $f(x) = y$.

(b) **Surjective**, or **onto**, or a **surjection**, if, for every $y \in Y$, there exists at least one $x \in X$ such that $y = f(x)$. In other words, this is the same as saying that the image of f is all of Y.

(c) **Bijective**, or **one-to-one onto**, or a **bijection**, if it is simultaneously injective and surjective.

An efficient way of verifying whether a function $f : X \to Y$ is injective or not is to verify whether the implication

$$f(x_1) = f(x_2) \Rightarrow x_1 = x_2 \qquad\qquad (6.6)$$

is true for all $x_1, x_2 \in X$. Accordingly, in order to prove that f is surjective, one must be capable of, for each $y \in Y$, guarantee the existence of at least one solution $x \in X$ for the equation $f(x) = y$. Let's see some examples.

Example 6.37 If $X \subset \mathbb{R}$ is a nonempty set and $f : X \to X$ is a function such that $f(f(x)) = x$ for every $x \in X$, then f is a bijection.

Proof Let x_1 and x_2 be elements of X for which $f(x_1) = f(x_2)$. According to (6.6), in order to show that f is injective it suffices to prove that $x_1 = x_2$. To this end, observe that $f(x_1) = f(x_2)$ implies $f(f(x_1)) = f(f(x_2))$ and, then (by using the given hypothesis), $x_1 = x_2$.

The surjectivity of f also follows from the hypothesis: for a fixed $y \in X$, taking $x = f(y) \in X$ we get $f(x) = f(f(y)) = y$, so that $y \in \text{Im}(f)$. $\square$

As a particular case of the previous example, the function of inverse proportionality (cf. Example 6.13) is a bijection from $\mathbb{R} \setminus \{0\}$ into itself.

Example 6.38 Let $f : [0, 1] \to [0, 1]$ be a surjective function, such that $|f(x_1) - f(x_2)| \leq |x_1 - x_2|$ for all $x_1, x_2 \in [0, 1]$. Prove that there are only two possibilities: either $f(x) = x$ for every $x \in [0, 1]$, or $f(x) = 1 - x$ for every $x \in [0, 1]$.

Proof Let $a, b \in [0, 1]$ be chosen in such a way that $f(a) = 0$ and $f(b) = 1$ (that it is possible to make such a choice follows from the surjectivity of f). Then, the given hypothesis allows us to write

$$1 = |1 - 0| = |f(b) - f(a)| \leq |b - a| \leq 1,$$

so that $|b - a| = 1$. However, the only $a, b \in [0, 1]$ such that $|b - a| = 1$ are $a = 0$ and $b = 1$, or vice-versa. Suppose that $a = 0$ and $b = 1$ (the other case can be haldled similarly), and take any $c \in (0, 1)$. Triangle inequality (5.10), together with the hypothesis on f, give

$$
\begin{aligned}
1 &= |f(1) - f(0)| \\
&\leq |f(1) - f(c)| + |f(c) - f(0)| \\
&\leq |1 - c| + |c - 0| \\
&= (1 - c) + c = 1.
\end{aligned}
$$

Hence, we must have $|f(c) - f(0)| = |c - 0|$ and, since $c, f(c) \geq 0$, we get $f(c) = c$. Finally, this is true for every $c \in [0, 1]$, so that $f(x) = x$ for every $x \in [0, 1]$. $\quad\square$

The coming proposition teaches us the way injective, surjective and bijective functions behave with respect to composition.

Proposition 6.39 *Let $f : X \to Y$ and $g : Y \to Z$ be given functions. Then:*

(a) $g \circ f$ injective $\Rightarrow f$ injective, but the converse is not necessarily true.
(b) $g \circ f$ surjective $\Rightarrow g$ surjective, but the converse is not necessarily true.
(c) g and f injective $\Rightarrow g \circ f$ injective.
(d) g and f surjective $\Rightarrow g \circ f$ surjective.
(e) g and f bijective $\Rightarrow g \circ f$ bijective.

Proof

(a) For $x_1, x_2 \in X$, we have

$$
\begin{aligned}
f(x_1) = f(x_2) &\Rightarrow g(f(x_1)) = g(f(x_2)) \\
&\Rightarrow (g \circ f)(x_1) = (g \circ f)(x_2) \\
&\Rightarrow x_1 = x_2,
\end{aligned}
$$

where, in the last passage, we used the fact that $g \circ f$ is injective. We now have to give an example in which f is injective but $g \circ f$ is not. To this end, take $X = Y = Z = \mathbb{R}$, $f(x) = x$ and $g(x) = x^2$.

(b) Choosing $z \in Z$ arbitrarily, the surjectivity of $g \circ f$ guarantees the existence of at least one $x \in X$ such that $z = (g \circ f)(x)$. Then, $z = g(f(x))$, so that g is also surjective. For the second part, take $X = Y = Z = \mathbb{R}$, $g(x) = x$ and $f(x) = x^2$; then, g is surjective, but $g \circ f$ is not.

(c) Let $x_1, x_2 \in X$. By using the injectivity of g and, then, that of f, we obtain

$$
\begin{aligned}
(g \circ f)(x_1) = (g \circ f)(x_2) &\Rightarrow g(f(x_1)) = g(f(x_2)) \\
&\Rightarrow f(x_1) = f(x_2) \\
&\Rightarrow x_1 = x_2,
\end{aligned}
$$

Hence, $g \circ f$ is also injective.

(d) Let $z \in Z$ be arbitrarily choosen. The surjectivity of g guarantees the existence of $y \in Y$ such that $z = g(y)$. On the other hand, the surjectivity of f assures the existence of $x \in X$ for which $f(x) = y$. Then, we have

$$(g \circ f)(x) = g(f(x)) = g(y) = z,$$

so that $g \circ f$ is surjective, too.

(e) It follows from items (c) and (d) that

$$g \text{ and } f \text{ bijective} \Rightarrow g \text{ and } f \text{ injective and surjective}$$

$$\Rightarrow g \circ f \text{ injective and surjective}$$

$$\Rightarrow g \circ f \text{ bijective.}$$

$\square$

Let's revisit Example 6.37 in light of the previous proposition.

Example 6.40 Let X be a nonempty set. If $f : X \to X$ is a function such that $f \circ f = \mathrm{Id}_X$, then f is a bijection.

Proof Indeed, since the identity function $\mathrm{Id}_X : X \to X$ is a bijection, it follows from items (a) and (b) of the previous proposition that f is injective and surjective, hence, bijective. $\square$

The material of this section also allows us to study the important concept of *countably infinite* sets, according to the following

Definition 6.41 An infinite set A is said to be **countable** if there exists an injective function $f : A \to \mathbb{N}$. In this case, we shall also say that A is a **countably infinite** set.

If A is countably infinite and $B \subset A$ is infinite, then B is also countably infinite. Indeed, by composing an injective function $f : A \to \mathbb{N}$ with the inclusion $\iota : B \to A$ (that sends each $x \in B$ to itself), we get the injective function $f \circ \iota : B \to \mathbb{N}$. In particular, every infinite subset of $\mathbb{N}$ is countable.

The set $\mathbb{Z}$ of integers is also countably infinite, for one can easily check that the function $f : \mathbb{Z} \to \mathbb{N}$ given by

$$f(x) = \begin{cases} -2x, & \text{if } x < 0 \\ 2x + 1, & \text{if } x \geq 0 \end{cases}$$

is a bijection. Perhaps a little more surprising is the following

Example 6.42 The cartesian product $\mathbb{N} \times \mathbb{N}$ is a countably infinite set.

Proof It suffices to define $f : \mathbb{N} \times \mathbb{N} \to \mathbb{N}$ by setting

$$f(m, n) = 2^{m-1}(2n - 1), \quad \forall \; m, n \in \mathbb{N}.$$

A straightforward argument on odd and even numbers shows that f is injective. (Actually, the Fundamental Theorem of Arithmetic – cf. introduction to Chapter 1 or Chapter 6 of [5] – assures that f is bijective.) ☐

With a little more effort (see Problem 21), we can prove that $\mathbb{Q}$ is countably infinite. However, as will be seen in Section 7.4 (cf. Example 7.46), $\mathbb{R}$ is **uncountable**, i.e., not countable. In turn, this implies (cf. Problem 22) that $\mathbb{I} = \mathbb{R} \setminus \mathbb{Q}$ is also uncountable. For yet another example of an uncountable set, see Problem 23.

The following lemma shows that the elements of every countably infinite set can be written as terms of a sequence.

Lemma 6.43 *If A is a countably infinite set, then there exists a bijection $f : \mathbb{N} \to A$. In particular, letting $a_n = f(n)$, we get $A = \{a_1, a_2, a_3, \dots\}$.*

Proof By definition, there exists an injective function $g : A \to \mathbb{N}$. Then, g induces a bijection (which we will also denote by g) from A to $B = \mathrm{Im}(g) \subset \mathbb{N}$. If we construct a bijection $h : \mathbb{N} \to B$, then $f = g \circ h : \mathbb{N} \to A$ will also be a bijection.

To what is left to do, we start by letting $b_1 = \min B$ and setting $h(1) = b_1$. Since B is infinite, we can let $b_2 = \min(B \setminus \{b_1\})$ and set $h(2) = b_2$. By the same token, let $b_3 = \min(B \setminus \{b_1, b_2\})$ and set $h(3) = b_3$. Continuing this way, we define a function $h : \mathbb{N} \to B$ such that, letting $h(k) = b_k$, we have $b_1 = \min B$ and

$$b_k = \min \left(B \setminus \{b_1, \dots, b_{k-1}\} \right)$$

for every natural $k > 1$. In particular, $b_1 < b_2 < b_3 < \cdots$ and h is injective.

If h was not surjective, there would exist $b \in B$ such that $b \neq b_1, b_2, b_3, \dots$. If $b_k < b$ for every $k \in \mathbb{N}$, then $B \subset \{1, 2, \dots, b\}$, a contradiction to the fact that B is infinite. Hence, there would exist a natural $m > 1$ such that $b_{m-1} < b < b_m$. However, since

$$h(m) = \min \left(B \setminus \{h(1), \dots, h(m-1)\} \right)$$
$$= \min \left(B \setminus \{b_1, \dots, b_{m-1}\} \right),$$

we should have defined $h(m)$ to be b, instead of b_m. Since this is a contradiction, we conclude that h is indeed surjective. ☐

The concept and properties of countable sets give rise to interesting results, one of which we collect in the following

Example 6.44 Show that it is possible to partition the set of natural numbers in two sets A and B satisfying the following conditions:

(a) Neither A nor B contains the terms of an infinite and nonconstant AP.
(b) For all distinct $x, y \in A$, we have $|x - y| \geq 2016$.

Proof An infinite and nonconstant AP of naturals is characterized by the ordered pair (a, r), where a is its first term and r is its common difference. Since $\mathbb{N} \times \mathbb{N}$

is countable, the same is true of the family $\mathcal{P}$ of infinite and nonconstant AP's of naturals. Hence, we can write $\mathcal{P} = \{s_1, s_2, s_3, \ldots\}$.

Now, let A_k be the set whose elements are the terms of s_k, and let A be defined in the following way: take $x_1 \in A_1$, $x_2 \in A_2$, ... such that $x_2 - x_1 \geq 2016$, $x_3 - x_2 \geq 2017$, $x_4 - x_3 \geq 2018$, ... (that this is possible follows from the fact that each A_k is an infinite set). Then, let $B = \mathbb{N} \setminus A$.

It is clear that A does not contain the terms of an infinite AP of common difference $r \in \mathbb{N}$, for the differences $x_j - x_{j-1}$ are eventually all greater than r. On the other hand, B does not contain the terms of an infinite and nonconstant AP either, for, if it did, we should have $B \supset A_k$ for some $k \in \mathbb{N}$; however, $x_k \in A_k \setminus B$. $\square$

Problems: Section 6.3

1. Let f and g be real functions of a real variable, given by $f(x) = x - \frac{7}{2}$ and $g(x) = x^2 - \frac{1}{4}$. Find the solution set of the inequality $|(g \circ f)(x)| > (g \circ f)(x)$.
2. Let f and g be real functions of a real variable, such that $f(x) = 2x + 7$ and $(f \circ g)(x) = x^2 - 2x + 3$, for every $x \in \mathbb{R}$. Find the expression that defines $g(x)$ in terms of x.
3. Let $f, g : \mathbb{R} \to \mathbb{R}$ be such that $g(x) = 2x - 3$ and $(f \circ g)(x) = 2x^2 - 4x + 1$. Find the expression that gives $f(x)$ in terms of x.
4. Let f and g be real functions of a real variable, given by $f(x) = ax + b$ and $g(x) = cx + d$ for every $x \in \mathbb{R}$, with $ac \neq 0$. Show that

$$f \circ g = g \circ f \Leftrightarrow (a-1)d = (c-1)b.$$

5. Let $f : \mathbb{R} \to \mathbb{R}$ be the function defined by

$$f(x) = \begin{cases} \frac{x+a}{x+b}, & \text{if } x \neq -b \\ -1, & \text{if } x = -b \end{cases}.$$

   If $f(f(x)) = x$ for every real x, compute the possible values of b.
6. * Let $I \subset \mathbb{R}$ be an interval and $f : I \to \mathbb{R}$ be an increasing or decreasing function. Prove that f is injective.
7. Let $I, J \subset \mathbb{R}$ be intervals and $f : I \to J$ and $g : J \to \mathbb{R}$ be given functions. If f and g are increasing (resp. if f is increasing and g is decreasing, or vice-versa), prove that $g \circ f$ is also increasing (resp. decreasing).
8. Let $f : \mathbb{R} \setminus \{0\} \to \mathbb{R}$ be a function such that

$$f(x)f\left(\frac{1}{x}\right) = 1 \quad \text{and} \quad f\left(x + \frac{1}{x}\right) = f(x) + f\left(\frac{1}{x}\right),$$

   for every real $x \neq 0$. If u and v are nonzero real numbers such that $u^2 + v^2 = 1$ and $f\left(\frac{1}{uv}\right) = 2$, compute the value of $f\left(\frac{u}{v}\right)$.

9. Given a function $f : X \to Y$, we define its *graph*[4] as the subset G_f of the cartesian product $X \times Y$ given by

$$G_f = \{(x, y) \in X \times Y; \; y = f(x)\}.$$

Let $F : X \to G_f$ be the function defined by $F(x) = (x, f(x))$, for every $x \in X$. Prove that F is a bijection.

10. * Let $\emptyset \neq X \subset \mathbb{R}$ be a union of intervals, which is symmetric with respect to $0 \in \mathbb{R}$. We say that a function $f : X \to \mathbb{R}$ is **even** (resp. **odd**) if $f(x) = f(-x)$ (resp. $f(x) = -f(-x)$), for every $x \in X$. If $\emptyset \neq X \subset \mathbb{R}$ is as above, prove that every function $f : X \to \mathbb{R}$ can be written, in a unique way, as a sum of an even and an odd function with domain X.

11. Let $f : \mathbb{R} \setminus \{0\} \to \mathbb{R}$ be a function such that $f\left(\frac{a}{b}\right) = f(a) - f(b)$, for every nonzero real numbers a and b. Prove that f is an even function.

12. Let $f : \mathbb{R} \to \mathbb{R}$ be an odd function. Decide whether the function $f \circ f$ is even, odd, or not even nor odd.

13. Let $g : \mathbb{R} \to \mathbb{R}$ be an odd function, such that $g(x) > 0$ whenever $x > 0$. Show that there exists a function $f : \mathbb{R} \to \mathbb{R}$ for which $g = f \circ f$.

14. Find all real values of k for which the image of the function $f : \mathbb{R} \setminus \{-1\} \to \mathbb{R}$, given by $f(x) = \frac{4x^2 + kx + k}{x + 1}$, equals $\mathbb{R} \setminus (-L, L)$, for some positive real L.

15. * A function $f : \mathbb{R} \to \mathbb{R}$ is **periodic** if there exists a smallest positive real number p, called the **period** of f, such that $f(x + p) = f(x)$ for every $x \in \mathbb{R}$. Given a periodic function $f : \mathbb{R} \to \mathbb{R}$, of period $p > 0$, do the following items:

    (a) Let $g : \mathbb{R} \to \mathbb{R}$ be also periodic of period p. If $f(x) = g(x)$ for every $x \in [0, p)$, prove that $f = g$.
    (b) Given $a \in \mathbb{R} \setminus \{0\}$, prove that the function $g : \mathbb{R} \to \mathbb{R}$, given by $g(x) = f(ax)$, is periodic of period $\frac{p}{|a|}$.

16. (Italy) Let $f : \mathbb{R} \to \mathbb{R}$ be a function such that $f(10 + x) = f(10 - x)$ and $f(20 + x) = -f(20 - x)$, for every real x. Prove that f is odd and find $p > 0$ such that $f(x + p) = f(x)$, for every $x \in \mathbb{R}$.

17. (IMO - adapted) Let $f : \mathbb{R} \to [0, 1]$ be a function such that, for a certain $a \in \mathbb{R}$, we have

$$f(x + a) = \frac{1}{2} + \sqrt{f(x) - f(x)^2},$$

for every $x \in \mathbb{R}$. Prove that f is periodic.

18. (Brazil) The function $f : \mathbb{Z} \to \mathbb{R}$ is such that $f(x) = x - 10$ for $x > 100$ and $f(x) = f(f(x + 11))$ for $x \leq 100$. Find the image of f.

---

[4] Graphs of real functions of a real variable will be one of the main objects of study along these notes, starting from Section 6.6.

19. (Hungary) Let $f : \mathbb{N} \to \mathbb{N}$ be a function satisfying the following conditions:

(a) $f(1) = 1$.
(b) $f(2n) = 2f(n) + 1$.
(c) $f(f(n)) = 4n + 1$.

Compute $f(1993)$.

20. Give an example of a surjective function $f : \mathbb{N} \to \mathbb{N}$ such that, for every $n \in \mathbb{N}$, the set $\{x \in \mathbb{N}; f(x) = n\}$ is infinite.

21. * The purpose of this problem is to show that $\mathbb{Q}$ is countably infinite. To this end, do the following items:

(a) If sets A and B are either finite or countably infinite sets, show that $A \times B$ is either finite or countably infinite.
(b) Construct a surjective function from $\mathbb{Z} \times \mathbb{N}$ into $\mathbb{Q}$.
(c) Conclude that there exists a surjection $f : \mathbb{N} \to \mathbb{Q}$ and, then, that $\mathbb{Q}$ is countable.

22. * The purpose of this problem is to show that $\mathbb{I} = \mathbb{R} \setminus \mathbb{Q}$ is uncountable, assuming that $\mathbb{R}$ itself is uncountable (this will be proved in Example 7.46). To this end, do the following items:

(a) Let $A_1, A_2, A_3, \ldots$ be a countably infinite collection of sets. Show that there exist $B_k \subset A_k$ such that $B_1, B_2, B_3, \ldots$ are pairwise disjoint and $\bigcup_{k \geq 1} A_k = \bigcup_{k \geq 1} B_k$.
(b) In the notations of (a), if each of $A_1, A_2, A_3, \ldots$ is finite or countably infinite, show that $\bigcup_{k \geq 1} A_k$ is also finite or countably infinite.
(c) Show that $\mathbb{I}$ is uncountable.

23. * Let $\mathcal{F}$ be the *family*[5] of infinite subsets A of $\mathbb{N}$ such that $\mathbb{N} \setminus A$ is also infinite, i.e.,

$$\mathcal{F} = \{A \subset \mathbb{N}; A \text{ and } \mathbb{N} \setminus A \text{ are infinite.}\}.$$

Show that $\mathcal{F}$ is an uncountable set.

6.4   Inversion of Functions

Among all functions $f : X \to Y$, the case of a bijection is, in a certain sense, the best possible. Indeed, in this case the elements of X and Y are in a *one-to-one onto correspondence*, so that to each element of X there corresponds a single element of Y via f, and vice-versa. When this happens, we can form a function $g : Y \to X$ by asking that

$$f(x) = y \Leftrightarrow g(y) = x.$$

---

[5]A **family** is a set whose elements are sets themselves.

At this point, a natural question is this: why cannot we use the same declaration above to define g when f is not bijective? From an intuitive point of view, if f is not surjective, then there exists an element y of Y which is not the image of any element of X via f; therefore, there is no natural way of using f and y to define $g(y)$. On the other hand, if f is not injective, then there exist distinct elements $x_1, x_2 \in X$ with the same image $y \in Y$ via f; if we were to try to define g by using f, there would also be no natural way of deciding which of x_1 and x_2 should be equal to $g(y)$.

Back to the case in which f is bijective, it is not difficult to see that g, defined as above, is indeed a function, moreover such that $(g \circ f)(x) = x$, for every $x \in X$, and $(f \circ g)(y) = y$, for every $y \in Y$. Yet in another way, we have $g \circ f = \mathrm{Id}_X$ and $f \circ g = \mathrm{Id}_Y$. Conversely, if $f : X \to Y$ and $g : Y \to X$ are functions such that $g \circ f = \mathrm{Id}_X$ and $f \circ g = \mathrm{Id}_Y$, then Proposition 6.39 guarantees that f must be a bijection, and Problem 1 assures that g is the *only* function satisfying these compositions.

We summarize the above discussion in the following

Definition 6.45 Let $f : X \to Y$ be a given bijection. The **inverse function** of f is the function $g : Y \to X$ such that, for every $y \in Y$,

$$g(y) = x \Leftrightarrow y = f(x).$$

From now on, whenever there is no danger of confusion, we shall denote the inverse (function) of a bijection $f : X \to Y$ by $f^{-1} : Y \to X$. Notice that the *exponent* -1 in the notation of inverse function has no arithmetic meaning; it simply calls the reader's attention to the fact that f^{-1} *does the opposite of f*, i.e., applies Y into X instead of X into Y, and *reverses the arrows* of the correspondences made by f.

Now, it naturally emerges the question of how to *effectively* compute the inverse of a given bijection. Although such a computation is generally more complicated than that of composite functions, in the case of real functions of a real variable $f : X \to Y$ we can reason in the following simple way: since $f^{-1}(y) = x \Leftrightarrow f(x) = y$, in order to find $f^{-1}(y) = x$ it suffices to solve, for $x \in X$, the equation $f(x) = y$. If we find a single solution $x \in X$ for every $y \in Y$, then f will be a bijection and x (computed in terms of y) will give the sought for expression for $f^{-1}(y)$. Let's implement such a reasoning in some relevant examples.

Example 6.46 Let a and b be given reals, with $a \neq 0$, and let $f : \mathbb{R} \to \mathbb{R}$ be the affine function given by $f(x) = ax + b$, for every $x \in \mathbb{R}$. Show that f is a bijection and find the expression of f^{-1}.

Proof Initially, note that

$$f(x) = y \Leftrightarrow ax + b = y \Leftrightarrow x = \frac{y - b}{a}.$$

Hence, f is a bijection and, by the definition of f^{-1}, this value of x is precisely equal to $f^{-1}(y)$, so that

$$f^{-1}(y) = \frac{y-b}{a}.$$

□

Example 6.47 If X is a nonempty set, then, analogously to the former example, the inverse function of the identity function Id_X is this function itself, so that we can write $(\mathrm{Id}_X)^{-1} = \mathrm{Id}_X$. Nevertheless, the inverse of a bijective function can be the function itself, even if it is not the identity function of some nonempty set; an example is furnished by the inverse proportionality function $f : \mathbb{R} \setminus \{0\} \to \mathbb{R} \setminus \{0\}$ (which we already know to be bijective), for which $f(x) = \frac{1}{x}$, for every $x \in \mathbb{R} \setminus \{0\}$. Indeed, since

$$f(x) = y \Leftrightarrow \frac{1}{x} = y \Leftrightarrow x = \frac{1}{y},$$

we conclude that

$$f^{-1}(y) = x = \frac{1}{y}$$

and, hence, $f^{-1} = f$.

Example 6.48 Let a, b and c be given real numbers, with $a > 0$. Propositions 6.15 and 6.23 guarantee that the quadratic function $f(x) = ax^2 + bx + c$, seen as a function

$$f : \left[-\frac{b}{2a}, +\infty\right) \to \left[-\frac{\Delta}{4a}, +\infty\right),$$

is a bijection. Compute the expression of its inverse.

Solution According with the discussion that preceded Example 6.46, in order to get the expressão of $f^{-1} : \left[-\frac{\Delta}{4a}, +\infty\right) \to \left[-\frac{b}{2a}, +\infty\right)$, it suffices to fix $y \in \left[-\frac{\Delta}{4a}, +\infty\right)$ and solve, for $x \in \left[-\frac{b}{2a}, +\infty\right)$, the equation $f(x) = y$, i.e., $ax^2 + bx + c - y = 0$. Since this is a second degree equation in y, the condition $x \geq -\frac{b}{2a}$ gives (remember that $a > 0$)

$$x = \frac{-b + \sqrt{b^2 - 4a(c-y)}}{2a} = \frac{-b + \sqrt{\Delta + 4ac}}{2a},$$

where $\Delta = b^2 - 4ac$ is the discriminant of f. Therefore,

$$f^{-1}(y) = \frac{-b + \sqrt{\Delta + 4ay}}{2a}.$$

□

Example 6.49 As a particular case of the previous example[6], the function $f :$ $[0, +\infty) \rightarrow [0, +\infty)$, given by $f(x) = x^2$ for every nonnegative real x, is a bijection having as its inverse the **square root function**

$$f^{-1} : [0, +\infty) \longrightarrow [0, +\infty)$$
$$x \longmapsto \sqrt{x}$$

We finish this section obtaining a useful formula relating composition and inversion of bijections.

Proposition 6.50 *If* $f : X \rightarrow Y$ *and* $g : Y \rightarrow Z$ *are bijections, then* $g \circ f : X \rightarrow Z$ *is also a bijection and*

$$(g \circ f)^{-1} = f^{-1} \circ g^{-1}.$$

Proof We already know, from item (e) of Proposition 6.39, that $g \circ f$ is bijective. On the other hand, since $(g \circ f)^{-1}$ and $f^{-1} \circ g^{-1}$ are both functions from Z to X, in order to check that $(g \circ f)^{-1} = f^{-1} \circ g^{-1}$ it suffices, by the uniqueness of the inverse (cf. Problem 1), to show that

$$(f^{-1} \circ g^{-1}) \circ (g \circ f) = \text{Id}_X \text{ and } (g \circ f) \circ (f^{-1} \circ g^{-1}) = \text{Id}_Z.$$

These verifications are quite straightforward and will be left for the reader as exercises. □

Problems: Section 6.4

1. * Let $f : X \rightarrow Y$ be a given function.

   (a) If $g : Y \rightarrow X$ is such that $g \circ f = \text{Id}_X$ and $f \circ g = \text{Id}_Y$, prove that f is a bijection.
   (b) Prove that there exists at most one function $g : Y \rightarrow X$ for which $g \circ f = \text{Id}_X$ and $f \circ g = \text{Id}_Y$.

2. * Complete the proof of Proposition 6.50, showing (in the notations of its statement) that

$$(f^{-1} \circ g^{-1}) \circ (g \circ f) = \text{Id}_X \text{ e } (g \circ f) \circ (f^{-1} \circ g^{-1}) = \text{Id}_Z.$$

---

[6]Here and in the previous example, we are relying on previous knowledge from the reader, namely, the existence of square roots of positive real numbers. Therefore, the present discussion will be rigorously justified only after Theorem 7.9, where we prove the existence of such roots. In this respect, the reader might also want to take a look at Example 8.36.

3. Let $f : [\frac{1}{2}, +\infty) \to [\frac{3}{4}, +\infty)$ be the function given by $f(x) = x^2 - x + 1$, for every $x \geq \frac{1}{2}$. Show that f is a bijection and obtain the defining expression of its inverse.

4. Let $f : \mathbb{R} \setminus \{2\} \to \mathbb{R} \setminus \{3\}$ be the function given by $f(x) = \frac{3x-5}{x-2}$. Show that f is a bijection and obtain the defining expression of its inverse.

5. Let $a, b, c, d \in \mathbb{R}^*$ and $f : \mathbb{R} \setminus \{-\frac{d}{c}\} \to \mathbb{R}$ be the function given by $f(x) = \frac{ax+b}{cx+d}$. Generalize the previous problem, showing that f defines a bijection from $\mathbb{R} \setminus \{-\frac{d}{c}\}$ onto $\mathbb{R} \setminus \{\frac{a}{c}\}$. Moreover, conclude that, if $d = -a$, then $f^{-1} = f$.

6. * Let n be a natural number and $f : [0, +\infty) \to [0, +\infty)$ be the function given by $f(x) = x^n$. Admitting that $\text{Im}(f) = [0, +\infty)$ (a fact to be proved in Example 8.36), show that f is an increasing bijection and get the expression for f^{-1}, thus defining the **n–th root function**.

7. Give an example of a bijection $f : \mathbb{R} \to \mathbb{R}$ for which both $f + f^{-1}$ and $f - f^{-1}$ are also bijections.

8. * Let $I, J \subset \mathbb{R}$ be intervals and $f : I \to J$ be a bijection. If f is increasing (resp. decreasing), prove that $f^{-1} : J \to I$ is also increasing (resp. decreasing).

9. Let $f : \mathbb{R} \to \mathbb{R}$ be a bijection and $g : \mathbb{R} \times \mathbb{R} \to \mathbb{R} \times \mathbb{R}$ be the function defined by $g(x, y) = (x^3, x - f(y))$. Prove that g is bijective and find the expression that defines its inverse in terms of f^{-1}.

10. (IMO) Let G be a (nonempty) set of affine functions possessing the folllowing properties:

    (a) If $f, g \in G$, then $f \circ g \in G$.
    (b) If $f \in G$, then $f^{-1} \in G$.
    (c) For every $f \in G$, there exists $x_f \in \mathbb{R}$ such that $f(x_f) = x_f$.

    Prove that there exists a real number x_0 such that $f(x_0) = x_0$ for every $f \in G$.

11. (France) Let $f : \mathbb{N} \to \mathbb{N}$ be a bijection. Prove that there exist natural numbers $a < b < c$ such that $f(a) + f(c) = 2f(b)$.

6.5   Defining Functions Implicitly

A function can be defined implicitly by a set of properties. For example, letting $g(x) = x + 1$ and $h(x) = x - 1$ for every $x \in \mathbb{R}$, the function $f : \mathbb{R} \to \mathbb{R}$ given by $f(x) = x^2$ satisfies

$$f(g(x)) = g(x)^2 \quad \text{and} \quad f(h(x)) = h(x)^2,$$

i.e., it is such that

$$f(x + 1) = (x + 1)^2 = x^2 + 2x + 1 \quad \text{and} \quad f(x + 1) = (x - 1)^2 = x^2 - 2x + 1;$$

therefore, for every $x \in \mathbb{R}$, we have

$$f(x + 1) - f(x - 1) = 4x.$$

Now, we can try to reverse the steps above, asking which functions $f : \mathbb{R} \to \mathbb{R}$ are such that

$$f(x + 1) - f(x - 1) = 4x, \ \forall \ x \in \mathbb{R}. \tag{6.7}$$

Clearly, $f(x) = x^2$ is not the only one, for, $f_c(x) = x^2 + c$ also satisfies (6.7), whatever the real constant c.

Now, let $f : \mathbb{R} \to \mathbb{R}$ be any real function satisfying (6.7), so that f is not *explicitly* given by its values. Since we merely know that f must satisfy that relation, we say that f is *implicitly defined*.

On the other hand, since we know that (6.7) must be satisfied, it allows us to find other relations f must also satisfy. For instance, if $g : \mathbb{R} \to \mathbb{R}$ is given by $g(x) = x^2$, then

$$f(g(x) + 1) - f(g(x) - 1) = 4g(x)$$

or, which is the same,

$$f(x^2 + 1) - f(x^2 - 1) = 4x^2, \ \forall \ x \in \mathbb{R}. \tag{6.8}$$

Thus, any function $f : \mathbb{R} \to \mathbb{R}$ satisfying (6.7) will also satisfy (6.8). Nevertheless, relation (6.8) may not be very useful to the task of finding all functions $f : \mathbb{R} \to \mathbb{R}$ that satisfy (6.7).

It is our purpose in this section to approach the problem of finding all functions implicitly defined by a certain (finite) set of relations. Since there is no general theory to accomplish this, we shall content ourselves in analysing a number of interesting examples. In turn, these examples will provide us with several useful techniques to be employed in such problems.

Let's start by looking at two simple examples.

Example 6.51 (Canada) Find all increasing functions $f : \mathbb{N} \to \mathbb{N}$, such that $f(2) = 2$ and $f(mn) = f(m)f(n)$, for every $m, n \in \mathbb{N}$.

Solution From $1 \le f(1) < f(2) = 2$ we get $f(1) = 1$; also, $f(4) = f(2)f(2) = 4$ and $f(8) = f(4)f(2) = 8$. Suppose, by induction hypothesis, that $f(2^k) = 2^k$ for a certain natural number k. Then,

$$f(2^{k+1}) = f(2^k)f(2) = 2^k \cdot 2 = 2^{k+1},$$

and it follows that $f(2^n) = 2^n$ for every nonnegative integer n.

Now, let n be a fixed natural number. Since f is increasing, we have

$$2^n = f(2^n) < f(2^n + 1) < \cdots < f(2^{n+1} - 1) < f(2^{n+1}) = 2^{n+1}.$$

However, $f(2^n + 1), f(2^n + 2), \ldots, f(2^{n+1} - 1)$ are natural numbers too, so that the only possibility is

$$f(2^n + 1) = 2^n + 1, \; f(2^n + 2) = 2^n + 2, \; \ldots, \; f(2^{n+1} - 1) = 2^{n+1} - 1.$$

Finally, since such a reasoning is valid for whatever natural n, we conclude that $f(m) = m$, for every $m \in \mathbb{N}$.   □

Example 6.52 (OIM) If $D = \mathbb{R} - \{-1, 0, 1\}$, find all functions $f : D \to \mathbb{R}$ such that, for every $x \in D$, we have

$$f(x)^2 f\left(\frac{1-x}{1+x}\right) = 64x.$$

Solution First of all, note that, since $x \neq 0$, we have $f(x)^2 f\left(\frac{1-x}{1+x}\right) \neq 0$ for every $x \in D$. In particular, $f(x) \neq 0$ for every $x \in D$. Now, let $g(x) = \frac{1-x}{1+x}$ for $x \in D$. The definition of D easily gives that $g(D) \subset D$, so that we can compose f and g. This way, for every $x \in D$, we have

$$f(g(x))^2 f\left(\frac{1-g(x)}{1+g(x)}\right) = 64g(x). \tag{6.9}$$

Substituting the defining formula for g in the above relation, we arrive at the equality

$$f\left(\frac{1-x}{1+x}\right) f\left(\frac{1-\frac{1-x}{1+x}}{1+\frac{1-x}{1+x}}\right) = 64\left(\frac{1-x}{1+x}\right)$$

or, which is the same,

$$f\left(\frac{1-x}{1+x}\right)^2 f(x) = 64\left(\frac{1-x}{1+x}\right),$$

for every $x \in D$. Squaring both sides of the relation given in the statement of the problem and dividing the result by the last relation above, we find

$$f(x)^3 = 64x^2 \left(\frac{1-x}{1+x}\right),$$

so that $f(x) = 4\sqrt[3]{x^2 \left(\frac{1-x}{1+x}\right)}$.

Up to this point we have only shown that, if there exists a function $f : D \to \mathbb{R}$ satisfying the stated relation, then it must be given by the last expression above. Hence, it is necessary that we verify that such an f does satisfy that relation, for every $x \in D$. Since such a verification is straightforward, we leave it to the reader.   □

Yet in respect to the previous example, a little more practice would have allowed us to get rid of formally defining g to, then, compose it with f in order to get (6.9). Instead, we could have just said

Substituting x by $\dfrac{1-x}{1+x}$ in the given relation, we obtain ...,

keeping in mind that this *substitution* is actually a composition of functions. From now on, whenever there is no danger of confusion, we shall adopt this language shortcut, which already appears in the next example. When reading it, try to identify the compositions that correspond to the employed substitutions.

Example 6.53 (Poland) Find all functions $f : \mathbb{R} \to \mathbb{R}$ such that, for every $x, y \in \mathbb{R}$, we have

$$(x - y)f(x + y) - (x + y)f(x - y) = 4xy(x^2 - y^2).$$

Solution Since this relation is valid for every $x, y \in \mathbb{R}$, it must be valid if we make $x = \frac{a+b}{2}$ and $y = \frac{a-b}{2}$, with $a, b \in \mathbb{R}$. Substituting these values for x and y, we get the relation

$$bf(a) - af(b) = (a^2 - b^2)ab,$$

which must be satisfied for every $a, b \in \mathbb{R}$. In particular, when $ab \neq 0$, dividing both sides of the last relation above by ab we get

$$\frac{f(a)}{a} - \frac{f(b)}{b} = a^2 - b^2,$$

for all $a, b \in \mathbb{R} \setminus \{0\}$. Hence, letting $g : \mathbb{R} \setminus \{0\} \to \mathbb{R}$ be the function given by $g(x) = \frac{f(x)}{x} - x^2$, it follows from the above that $g(a) = g(b)$, for all $a, b \in \mathbb{R} \setminus \{0\}$. In other words, g must be constant, so that there must exist a real number k such that $g(x) = k$, for every $x \in \mathbb{R} \setminus \{0\}$. But this is the same of saying that $f(x) = x^3 + kx$, for every $x \in \mathbb{R} \setminus \{0\}$.

On the other hand, making $x = y = 1$ in the relation given in the statement of the problem, we get $f(0) = 0$, for every function $f : \mathbb{R} \to \mathbb{R}$ that satisfies those conditions. Since $0^3 + k \cdot 0 = 0$, we conclude that every such function must be of the form $f(x) = x^3 + kx$, for every $x \in \mathbb{R}$ and some real constant k.

As in the previous example, we have to verify that every such function does satisfy the stated conditions. This amounts to straightforward algebrism, which will be left to the reader. $\square$

For the next example, we need the following

Definition 6.54 Let X be a nonempty set and $f : X \to X$ be a given function. An element $x_0 \in X$ is said to be a **fixed point** of f if $f(x_0) = x_0$.

If $I \subset \mathbb{R}$ is an interval, then a decreasing function $f : I \to I$ admits at most one fixed point. Indeed, if $x_1, x_2 \in I$ were fixed points of f, with $x_1 < x_2$, it would follow from the fact that f is decreasing that

$$x_1 = f(x_1) > f(x_2) = x_2,$$

a contradiction to $x_1 < x_2$.

Example 6.55 (Argentina) Let $f : \mathbb{R} \to \mathbb{R}$ be a decreasing function such that $f(x + f(x)) = x + f(x)$, for every real number x. Prove that $f(f(x)) = x$, for every real x.

Proof The hypotheses on f guarantee that $x + f(x)$ is a fixed point of f, for every $x \in \mathbb{R}$. On the other hand, according to the previous discussion, the decreasing character of f assures the existence of at most one fixed point for it. Therefore, there must exist $a \in \mathbb{R}$ such that $x + f(x) = a$, for every $x \in \mathbb{R}$, so that $f(x) = a - x$ for every $x \in \mathbb{R}$. Hence,

$$f(f(x)) = f(a - x) = a - (a - x) = x,$$

for every $x \in \mathbb{R}$.   □

The next example develops a body of ideas which will reveal themselves to be useful in a number of other situations involving implicitly defined functions. In particular, the first part of the presented reasoning solves Problem 12, page 152.

Example 6.56 Find all functions $f : \mathbb{R} \to \mathbb{R}$ such that $f(1) = 1$ and, for all $x, y \in \mathbb{R}$:

(a) $f(x + y) = f(x) + f(y)$.
(b) $f(xy) = f(x)f(y)$.

Solution Let f be a function satisfying the stated conditions. Making $x = y = 0$ in (a), we get

$$f(0) = f(0 + 0) = f(0) + f(0) = 2f(0),$$

so that $f(0) = 0$. Making $y = x$ in (a), we obtain

$$f(2x) = f(x + x) = f(x) + f(x) = 2f(x),$$

for every $x \in \mathbb{R}$. Now, making $y = 2x$ in (a), it comes that

$$f(3x) = f(x + 2x) = f(x) + f(2x) = f(x) + 2f(x) = 3f(x),$$

for every $x \in \mathbb{R}$. Repeating the above reasoning we easily conclude, by induction on $n \in \mathbb{N}$, that

$$f(nx) = nf(x), \ \forall \ n \in \mathbb{N}, \ x \in \mathbb{R} \tag{6.10}$$

In particular, making $x = 1$ in (6.10), we get $f(n) = n$, for every $n \in \mathbb{N}$. Letting $x = \frac{1}{n}$ in (6.10), it follows that

$$1 = f(1) = f\left(n \cdot \frac{1}{n}\right) = nf\left(\frac{1}{n}\right)$$

and, hence, $f\left(\frac{1}{n}\right) = \frac{1}{n}$. Finally, $x = \frac{1}{m}$ in (6.10), with $m \in \mathbb{N}$, furnishes

$$f\left(\frac{n}{m}\right) = f\left(n \cdot \frac{1}{m}\right) = nf\left(\frac{1}{m}\right) = n \cdot \frac{1}{m} = \frac{n}{m}.$$

In order to see what happens with negative rationals, let $y = -x$ in item (a), to get

$$0 = f(0) = f(x + (-x)) = f(x) + f(-x),$$

or, which is the same,

$$f(x) = -f(-x), \ \forall \ x \in \mathbb{R}. \tag{6.11}$$

In particular, if $x < 0$ is rational, it follows from (6.11), together with the fact that $-x$ is a positive rational, that $f(x) = -f(-x) = -(-x) = x$. Therefore, $f(x) = x$, for every $x \in \mathbb{Q}$.

Since $f(x) = x$ for every $x \in \mathbb{Q}$, we suspect that the identity function of $\mathbb{R}$ is the only one satisfying the stated conditions. To confirm that, we now turn our attention to item (b). First of all, let's show that, if $f(x) = 0$ for some $x \in \mathbb{R}$, then $x = 0$. Indeed, if $x \neq 0$, then, making $y = \frac{1}{x}$ in item (b), we would have

$$0 = f(x)f\left(\frac{1}{x}\right) = f\left(x \cdot \frac{1}{x}\right) = f(1) = 1,$$

which is a contradiction. Now, letting $y = x \neq 0$ in (b), we obtain

$$f(x^2) = f(x \cdot x) = f(x) \cdot f(x) = f(x)^2 > 0; \tag{6.12}$$

hence, if $x, y \in \mathbb{R}$, with $x < y$, and $a \neq 0$ is such that $y - x = a^2$, then, successively applying (a), (6.11) and (6.12), we get

$$f(y) - f(x) = f(y) + f(-x) = f(y - x) = f(a^2) = f(a)^2 > 0,$$

so that f is an increasing function.

Finally, suppose that there exists $a \in \mathbb{R}$ such that $f(a) < a$. We invoke the result of Problem 4, page 206, according to which there exists a rational number between any two real numbers. This allows us to choose a rational number r such that $f(a) < r < a$, and the increasing character of f gives

$$r = f(r) < f(a),$$

which is a contradiction. Analogously, we cannot have $f(a) > a$, so that the only possibility is $f(a) = a$. However, since $a \in \mathbb{R}$ was arbitrarily chosen, we conclude that $f(x) = x$ for every $x \in \mathbb{R}$.   □

Our last example shows that, for implicitly defined functions $f : \mathbb{N} \to \mathbb{N}$, elementary divisibility arguments are sometimes useful.

Example 6.57 (Lituania) Find all functions $f : \mathbb{N} \to \mathbb{N}$ such that, for all natural numbers m and n, we have

$$f(f(m) + f(n)) = m + n.$$

Solution Let's first prove that f is injective. To this end, let m and n be natural numbers such that $f(m) = f(n) = k$. Then, $f(2k) = f(f(n) + f(n)) = 2n$ and, analogously, $f(2k) = 2m$, so that $m = n$.

Now, let $k > 1$ be a natural number. From $(k - 1) + 2 = k + 1$, it follows that

$$f(f(k - 1) + f(2)) = k + 1 = f(f(k) + f(1)).$$

By the injectivity of f, we then get $f(k - 1) + f(2) = f(k) + f(1)$ or, which is the same, $f(k) - f(k - 1) = f(2) - f(1)$, for every $k > 1$. Writing this relation for $k = 2, 3, \ldots, n$ and adding the corresponding equalities, we arrive at

$$f(n) = (n - 1)(f(2) - f(1)) + f(1), \qquad (6.13)$$

for every natural $n > 1$. Letting $n = 2f(1)$ in the above relation, it comes that

$$2 = f(f(1) + f(1)) = f(2f(1)) = (2f(1) - 1)(f(2) - f(1)) + f(1)$$

and, hence,

$$f(2) - f(1) = \frac{2 - f(1)}{2f(1) - 1}.$$

Since $f(2) - f(1)$ is an integer, we conclude that $2f(1) - 1$ divides $2 - f(1)$. In particular, $2f(1) - 1 \leq |2 - f(1)|$, and from this inequality it is easy to conclude that the only possibility is $f(1) = 1$, so that $f(2) = 2$. It thus follows from (6.13) that $f(n) = n$, for every natural n.   □

Problems: Section 6.5

1. * Generalize the discussion of the paragraph that precedes Example 6.55, showing that, if $I \subset \mathbb{R}$ is an interval and $f, g : I \to \mathbb{R}$ are functions such that f is decreasing and g is increasing, then there exists at most one $x_0 \in I$ for which $f(x_0) = g(x_0)$.

2. Find all positive reals x for which $\sqrt{2 + \sqrt{x}} = \frac{8}{x}$.

3. Find all functions $f : \mathbb{Q} \to \mathbb{Q}$ such that

$$f\left(\frac{x+y}{2}\right) = \frac{f(x) + f(y)}{2},$$

for all $x, y \in \mathbb{Q}$.

4. (Austria) Find all functions $f : \mathbb{Z} \setminus \{0\} \to \mathbb{Q}$ such that, for all $x, y \in \mathbb{Z} \setminus \{0\}$ for which $x + y$ is a multiple of 3, we have

$$f\left(\frac{x+y}{3}\right) = \frac{f(x) + f(y)}{2}.$$

5. (Vietnam) Find all functions $f : \mathbb{R} \to \mathbb{R}$ such that

$$\frac{1}{2}f(xy) + \frac{1}{2}f(xz) - f(x)f(yz) \geq \frac{1}{4},$$

for all $x, y, z \in \mathbb{R}$.

6. (Spain) Find all increasing functions $f : \mathbb{N} \to \mathbb{N}$ such that $f(n + f(n)) = 2f(n)$, for every $n \in \mathbb{N}$.

7. (OIM shortlist) Find all functions $f : \mathbb{R} \to \mathbb{Z}$ which satisfy the following set of conditions:

   (a) $f(x + a) = f(x) + a$, for every $x \in \mathbb{R}$ and every $a \in \mathbb{Z}$.
   (b) $f(f(x)) = 0$, for $x \in [0, 1)$.

8. (Austrian-Polish) Prove that there doesn't exist a function $f : \mathbb{Z} \to \mathbb{Z}$ such that, for all $x, y \in \mathbb{Z}$, we have

$$f(x + f(y)) = f(x) - y.$$

9. (Romania) Find all functions $f : \mathbb{Z} \to \mathbb{Z}$ such that $f(0) = 1$ and

$$f(f(k)) + f(k) = 2k + 3,$$

for every $k \in \mathbb{Z}$.

10. (Romania) Let $k > 1$ be an odd integer and $A = \{x_1, x_2, \ldots, x_k\}$ be a set of k real numbers. Find all injective functions $f : A \to A$ such that

$$|f(x_1) - x_1| = |f(x_2) - x_2| = \cdots = |f(x_k) - x_k|.$$

11. (BMO) Let a be a given real number and $f : \mathbb{R} \to \mathbb{R}$ be a function such that $f(0) = \frac{1}{2}$ and

$$f(x+y) = f(x)f(a-y) + f(y)f(a-x),$$

for all $x, y \in \mathbb{R}$. Prove that f is constant.

12. Find all functions $f : \mathbb{Q} \to \mathbb{Q}_+^*$ such that $f(x+y) = f(x)f(y)$, for all $x, y \in \mathbb{Q}$.

13. Find all functions $f : [0, 1] \to [0, 1]$ such that $f(0) = 0, f(1) = 1$ and

$$f(x+y) + f(x-y) = 2f(x),$$

for all $x, y \in [0, 1]$ such that $x - y, x + y \in [0, 1]$.

14. (Lituania) Let $f : \mathbb{Z} \to \mathbb{Z}$ be such that $f(m^2 + f(n)) = f(m)^2 + n$, for all $m, n \in \mathbb{Z}$.

    (a) Prove that $f(0) = 0$ and $f(1) = 1$.
    (b) Find all such functions.

15. (OIM) Find all increasing functions $f : \mathbb{N} \to \mathbb{N}$ such that $f(yf(x)) = x^2 f(xy)$, for every $x, y \in \mathbb{N}$.

16. (IMO) Find all functions $f : [0, +\infty) \to \mathbb{R}$ satisfying the following set of conditions:

    (a) $f(xf(y))f(y) = f(x+y)$, for all $x, y \in [0, +\infty)$.
    (b) $f(2) = 0$ and $f(x) \neq 0$, for $0 \leq x < 2$.

17. (IMO) Let $S = \{x \in \mathbb{R}; \, x > -1\}$. Obtain all functions $f : S \to S$ satisfying the following conditions:

    (a) $f(x + f(y) + xf(y)) = y + f(x) + yf(x))$, for every $x, y \in S$.
    (b) $\frac{f(x)}{x}$ is increasing in each of the intervals $(-1, 0)$ and $(0, +\infty)$.

18. (IMO) Decide whether there exists a function $f : \mathbb{N} \to \mathbb{N}$ satisfying the following set of conditions:

    (a) $f(1) = 2$.
    (b) $f(n) < f(n + 1)$, for every $n \in \mathbb{N}$.
    (c) $f(f(n)) = f(n) + n$, for every $n \in \mathbb{N}$.

19. (Iran) Find all functions $f : \mathbb{R} \to \mathbb{R}$ such that

$$f(f(x+y)) = f(x+y) + f(x)f(y) - xy,$$

for all $x, y \in \mathbb{R}$.

20. (Poland) Find all functions $f : \mathbb{Q}_+^* \to \mathbb{Q}_+^*$ satisfying, for every positive rational number x, the following set of conditions:

    (a) $f(x + 1) = f(x) + 1$.
    (b) $f(x^3) = f(x)^3$.

6.6 Graphs of Functions

Given a function $f : X \to Y$, we define its **graph** to be the subset G_f of the cartesian product $X \times Y$, defined by

$$G_f = \{(x, y) \in X \times Y;\ y = f(x)\}. \tag{6.14}$$

If $f : X \to \mathbb{R}$ is a real function of a real variable, so that $X \subset \mathbb{R}$ is a finite union of intervals (possibly $X = \mathbb{R}$), then the graph of f has considerable geometric importance, for

$$G_f \subset X \times Y \subset \mathbb{R} \times \mathbb{R},$$

and this last set can be identified with the euclidean plane, furnished with a fixed cartesian coordinate system[7].

Our purpose in these last two sections of the chapter is to examine some elementary examples and study some simple properties of graphs of functions $f : X \to \mathbb{R}$, when $X \subset \mathbb{R}$ is a finite union of intervals. We postpone the discussion of the main properties of graphs of continuous and differentiable functions to chapters 8 and 9.

In all that follows, we fix a cartesian coordinate system is in the plane.

The first point worth observing is that not every subset of the cartesian plane can be the graph of a function. Indeed, suppose $f : X \to \mathbb{R}$ is a real function of a real variable, X being a finite union of intervals. If $(x_0, y_0) \in G_f$, then, by the very definition of graph, $x_0 \in X$ and $y_0 = f(x_0)$. On the other hand, for a fixed $x_0 \in X$, if $A_1(x_0, y_1)$ and $A_2(x_0, y_2)$ are points on the graph of f, then, again from the definition, we have

$$y_1 = f(x_0) = y_2,$$

so that $A_1 = A_2$. In short, for a given $x_0 \in X$, the vertical line $x = x_0$ of the cartesian coordinate system intersects the graph of f if and only if $x_0 \in X$; moreover, in this case such a line intersects the graph in *exactly one* point. Thus, the subset C depicted in Fig. 6.5 as a continuous curve doesn't represent the graph of any function $f : [-3, 3] \to \mathbb{R}$, since every vertical line contained in the gray strip intersects C in more than one point.

The graph of a real function of a real variable gives a quite simple geometric interpretation for the image of the function. To discuss it, let's consider the cartesian plane of Fig. 6.6, in which the graph of a function $f : [a, b] \to \mathbb{R}$ is drawn and its points of intersection with the horizontal line $y = y_0$ are marked. (There are three such points in Fig. 6.6, whose abscissas are denoted by α, β and γ.) Let (x_0, y_0) be

---

[7]We refer the reader to Chapter 6 of [4] for an adequate presentation of cartesian coordinate systems.

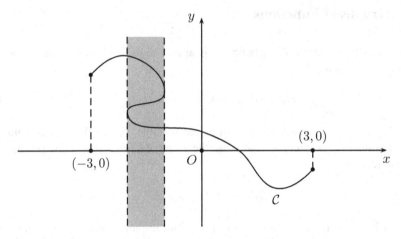

Fig. 6.5 A subset of the cartesian plane which is not the graph of a function

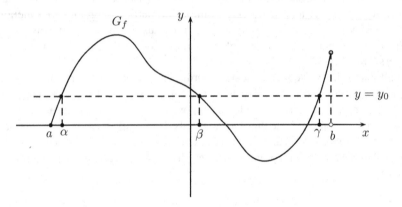

Fig. 6.6 Image × graph

an intersection point of the line and the graph. By pertaining to the graph of f, the
point (x_0, y_0) must be such that $x_0 \in [a, b)$ and $f(x_0) = y_0$, so that $x_0 = \alpha$, β or γ.
Conversely, let be given a point (x_0, y_0) of the cartesian plane, with $x_0 \in [a, b)$. It is
clear that (x_0, y_0) belongs to the horizontal line $y = y_0$; moreover, if x_0 is a solution
of the equation $f(x) = y_0$, i.e., if $f(x_0) = y_0$, then $x_0 = \alpha$, β or γ and we also
have $(x_0, y_0) \in G_f$. Therefore, the horizontal line $y = y_0$ intersects the graph of f
exactly when y_0 belongs to the image of f. The reasoning for an arbitrary function
$f : X \to \mathbb{R}$, with $X \subset \mathbb{R}$, is entirely analogous and allow us to conclude that

The image of f is precisely the set of $y_0 \in \mathbb{R}$ for which the horizontal line

$y = y_0$ intersects the graph of f.

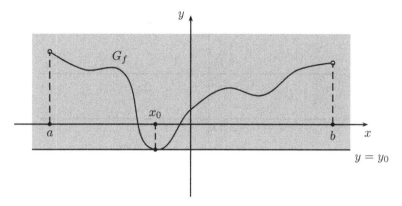

Fig. 6.7 Minimum point of $f : I \rightarrow \mathbb{R}$

Now, if $I \subset \mathbb{R}$ is an interval, then the *monotonicity* of a function $f : I \rightarrow \mathbb{R}$ also says a lot about the behavior of its graph. For instance, if we suppose that f increases (resp. decreases) in I, we conclude that, as long as the x increases in I, the values $f(x)$ increase (resp. decrease) in $\mathbb{R}$, so that the graph of f *rises* (resp. *falls*).

On the other hand, if $y_0 \in \mathbb{R}$ is the minimum value of $f : I \rightarrow \mathbb{R}$ and $x_0 \in I$ is a minimum point of f (cf. Definition 6.24), then, for every $x \in I$, the point $(x, f(x))$ *is above or coincides* with the point (x, y_0) (cf. Fig. 6.7). Yet in another way, the graph of f is entirely contained in the closed upper halfplane determined by the horizontal line $y = y_0$, touching it at the point (x_0, y_0).

Notice that the concepts of maximum value and maximum point of a function $f : I \rightarrow \mathbb{R}$ admit geometric interpretations analogous to those discussed above.

In the rest of this section we shall examine some important examples of graphs of functions.

Example 6.58 Let $f : \mathbb{R} \rightarrow \mathbb{R}$ be a constant function, with $f(x) = c$ for every $x \in \mathbb{R}$. The graph of f is the set

$$G_f = \{(x, y); \ x \in \mathbb{R} \text{ and } y = c\} = \{(x, c); \ x \in \mathbb{R}\},$$

i.e., it is the horizontal line $y = c$, which crosses the vertical axis at the point $(0, c)$ (cf. Fig. 6.8, where we consider the case $c > 0$).

Example 6.59 Recall (cf. Definition 6.2) that the identity function $\mathrm{Id}_\mathbb{R} : \mathbb{R} \rightarrow \mathbb{R}$ is given by $\mathrm{Id}_\mathbb{R}(x) = x$, for every $x \in \mathbb{R}$. Therefore, its graph is the set

$$G_{\mathrm{Id}_\mathbb{R}} = \{(x, y); \ x \in \mathbb{R} \text{ and } y = x\} = \{(x, x); \ x \in \mathbb{R}\}.$$

It is an easy exercise in euclidean geometry to verify that the points of the cartesian plane of the form (x, x) or $(-x, x)$ are precisely those situated on the bisectors of the angles formed by the coordinate axes, those of the form (x, x) belonging to the first

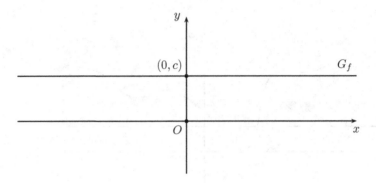

Fig. 6.8 Graph of the constant function $f(x) = c$, $\forall x \in \mathbb{R}$

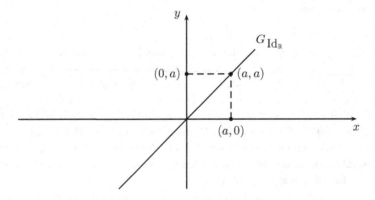

Fig. 6.9 Graph of the identity function $\mathrm{Id}_{\mathbb{R}}$

or third quadrants. Hence, the graph of the identity function $\mathrm{Id}_{\mathbb{R}}$ is the line shown in Fig. 6.9, and, from now on, will be called the *bisector of odd quadrants*. Observe that the set of points of the form $(x, -x)$, i.e., the *bisector of even quadrants*, is the graph of the function $f : \mathbb{R} \to \mathbb{R}$ given by $f(x) = -x$.

Example 6.60 The **modular function** is the function $f : \mathbb{R} \to \mathbb{R}$ given by $f(x) = |x|$. It follows immediately from the definition of modulus of a real number that

$$G_f = \{(x, |x|); \, x \in \mathbb{R}\}$$
$$= \{(x, |x|); \, x \in \mathbb{R}_+\} \cup \{(x, |x|); \, x \in \mathbb{R}_-\}$$
$$= \{(x, x); \, x \in \mathbb{R}_+\} \cup \{(x, -x); \, x \in \mathbb{R}_-\}.$$

Since the points $(x, -x)$ and (x, x) are symmetric with respect to the horizontal axis, the graph of the modular function is obtained by reflecting, along the horizontal axis, the portion of the graph of the function $\mathrm{Id}_{\mathbb{R}}$ situated in the third quadrant (cf. Fig. 6.10).

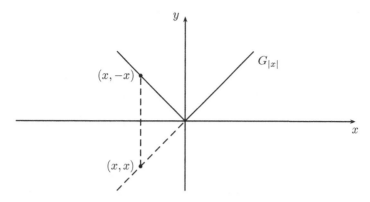

Fig. 6.10 Graph of the modular function $f(x) = |x|$

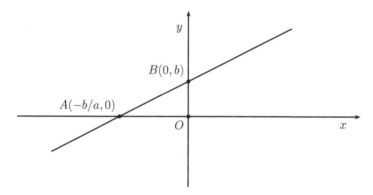

Fig. 6.11 Graph of the affine function $f(x) = ax + b$

Example 6.61 If $f(x) = ax + b$ is an affine function, then its graph is the subset of the cartesian plane given by

$$G_f = \{(x, y); x, y \in \mathbb{R} \text{ and } y = ax + b\}.$$

According to basic analytic geometry (cf. Chapter 6 of [4], for instance), the graph of f is the line of equation $y - ax - b = 0$, with slope a and passing through the points $A\left(-\frac{b}{a}, 0\right)$ and $B = (0, b)$. Fig. 6.11 depicts the graph of $f(x) = ax + b$ in the case $a, b > 0$.

For what comes next, recall (cf. Chapter 6 of [4], for instance) that, given in the plane a point F and a line d, with $F \notin d$, the **parabola** of **focus** F and **directrix** d (cf. Fig. 6.12) is the locus of the points P in the plane for which

$$\overline{PF} = \text{dist}(P, d),$$

Fig. 6.12 Parabola of focus
F and directrix d

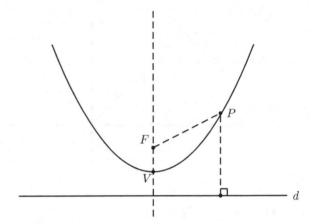

where dist(P, d) stands for the distance from P to d. The **axis** of the parabola is the line that passes through F and is perpendicular to d, while its **vertex** is the point V where it intersects its axis.

We now show that the graph of every quadratic function is a parabola. More precisely, we have the following result.

Theorem 6.62 *Given $a, b, c \in \mathbb{R}$, with $a \neq 0$, the graph of the quadratic function $f(x) = ax^2 + bx + c$ is the parabola whose axis is the line $x = -\frac{b}{2a}$ and whose vertex is $V\left(-\frac{b}{2a}, -\frac{\Delta}{4a}\right)$. Moreover, it "opens upwards" if $a > 0$, and "opens downwards" if $a < 0$.*

Proof Let's look for $x_0, y_0, k \in \mathbb{R}$ such that $y_0 \neq k$ and, letting $F(x_0, y_0)$ and d be the line $y = k$, we have

$$P \in G_f \iff \overline{PF} = \text{dist}(P, d).$$

To this end, set $P(x, y)$. We first notice that

$$P \in G_f \iff y = ax^2 + bx + c;$$

also, the formula for the distance between two points of the cartesian plane gives

$$\overline{PF} = \text{dist}(P, d) \iff \sqrt{(x - x_0)^2 + (y - y_0)^2} = |y - k|.$$

Therefore, we want to find x_0, y_0 and k such that

$$y = ax^2 + bx + c \iff (x - x_0)^2 + (y - y_0)^2 = (y - k)^2$$

$$\iff y = \frac{1}{2(y_0 - k)}x^2 - \frac{x_0}{y_0 - k}x + \frac{x_0^2 + y_0^2 - k^2}{2(y_0 - k)}.$$

Hence, it is natural to try to solve, with respect to x_0, y_0 and k, the system of equations

$$\frac{1}{2(y_0 - k)} = a, \quad -\frac{x_0}{y_0 - k} = b, \quad \frac{x_0^2}{2(y_0 - k)} + \frac{1}{2}(y_0 + k) = c.$$

This, in turn, is immediate: the first two equations readily give $x_0 = -\frac{b}{2a}$; then, substituting the first equation and the value of x_0 into the third equation, it comes that

$$y_0 + k = 2\left(c - x_0^2 \cdot \frac{1}{2(y_0 - k)}\right) = 2\left(c - \frac{b^2}{4a^2} \cdot a\right) = -\frac{\Delta}{2a};$$

finally, by solving the system of equations

$$y_0 - k = \frac{1}{2a}, \quad y_0 + k = -\frac{\Delta}{2a},$$

we get $y_0 = \frac{1 - \Delta}{4a}$ and $k = -\frac{1 + \Delta}{4a}$.

To what was left to prove, since the vertex of the parabola is the intersection of the line $x = -\frac{b}{2a}$ with the graph, its ordinate y equals

$$y = a\left(-\frac{b}{2a}\right)^2 + b\left(-\frac{b}{2a}\right) + c = -\frac{\Delta}{4a}.$$

$\square$

We finish this section with the coming proposition, which establishes an important relation between the graphs of a bijection and its inverse. Then, we use it to sketch the graphs of two important functions.

Proposition 6.63 *Let $I, J \subset \mathbb{R}$ be finite unions of interval. If $f : I \to J$ is a bijection, then the graphs of f and f^{-1} are symmetric with respect to the bisector of the odd quadrants of the cartesian plane.*

Proof Fix $a \in I$ and $b \in J$. From the definition of inverse function, we have

$$(a, b) \in G_f \Leftrightarrow b = f(a)$$

$$\Leftrightarrow a = f^{-1}(b)$$

$$\Leftrightarrow (b, a) \in G_{f^{-1}}.$$

However, since the points (a, b) and (b, a) are symmetric with respect to the line $y = x$, there is nothing left to do. $\square$

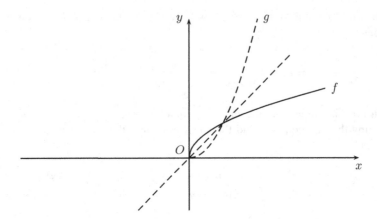

Fig. 6.13 Sketching the graph of $x \mapsto \sqrt[n]{x}$

Example 6.64 Sketch the graph of the square root function

$$f : [0, +\infty) \longrightarrow [0, +\infty)$$
$$x \longmapsto \sqrt{x} \quad .$$

Solution We saw at Example 6.49 that f is the inverse of $g : [0, +\infty) \to [0, +\infty)$, given by $g(x) = x^2$ for every $x \in [0, +\infty)$. Since we already know the graph of g, it follows from the previous proposition tha the graph of f is the symmetric of the graph of g with respect to the line $y = x$ (cf. Fig. 6.13). $\square$

Example 6.65 Recall that the inverse proportionality function is $f : \mathbb{R} \setminus \{0\} \to \mathbb{R} \setminus \{0\}$, such that $f(x) = \frac{1}{x}$, for every $x \in \mathbb{R} \setminus \{0\}$. With the previous discussion at our disposal, we can accurately sketch its graph. Indeed, f is clearly decreasing in $(0, +\infty)$; also, we already know from the Problem 10, page 171, that f is odd, so that (this time by the Problem 5, page 193) its graph is symmetric with respect to the origin of the cartesian plane; on the other hand, since f is its own inverse, the previous proposition assures that its graph is also symmetric with respect to the bisector of the odd quadrants; finally, we will see later that its graph is a *continuous* curve (i.e., one with no interruptions) which "*opens upwards*" in $(0, +\infty)$.

Since it is intuitively clear (and will be formalized later) that $f(x) = \frac{1}{x}$ becomes more and more close to zero as long as x increases without bound, we arrive at the sketch shown in Fig. 6.14. There, besides the above remarks, we plotted the auxiliary points $\left(n, \frac{1}{n}\right)$, for $1 \le n \le 4$ integer.

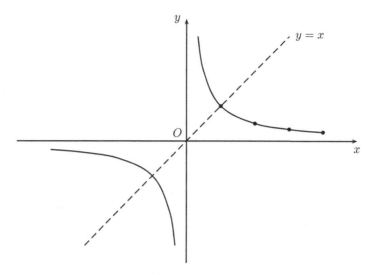

Fig. 6.14 Sketching the graph of $f(x) = \frac{1}{x}$

Problems: Section 6.6

1. If $f : [2, +\infty) \to [1, +\infty)$ is given by $f(x) = x^2 - 4x + 5$, show that f is a bijection and find the intersection points of the graphs of f and f^{-1}.

   For the next problem, given a nonempty set X and a function $f : X \to \mathbb{R}$, we say that f is **bounded** if there exists $M > 0$ such that $|f(x)| \le M$, for every $x \in X$.

2. * Let $I \subset \mathbb{R}$ be an interval and $f : I \to \mathbb{R}$ be a given function. If f is bounded, prove that its graph is contained in a horizontal strip of the cartesian plane bounded by two parallel lines.

3. * If $I \subset \mathbb{R}$ is an interval and $f : I \to I$ is a given function, show that the fixed points of f are the abscissas of the intersection points of the graph of f with the bisector of the odd quadrants.

4. * Let $I \subset \mathbb{R}$ be an interval and $f, g : I \to \mathbb{R}$ be given functions. Explain how to analytically identify the intersection points of the graphs of f and g.

   For the next problem, we think the reader will find it useful to read again the statement of Problem 10, page 171.

5. * Let $I \subset \mathbb{R}$ be a union of intervals, symmetric with respect to $0 \in \mathbb{R}$, and $f : I \to \mathbb{R}$ be a given function.

   (a) If f is even, prove that G_f is symmetric with respect to the vertical axis.
   (b) If f is odd, prove that G_f is symmetric with respect to the origin.

6. In each of the following items, sketch the graphs of the given real functions of a real variable in a single cartesian system:

(a) $f_1(x) = x^2, f_2(x) = x^4$ and $f_3(x) = x^3$.
(b) $f_1(x) = x, f_2(x) = x^3$ and $f_3(x) = x^5$.

7. Sketch the graph of the function $f : \mathbb{R} \to \mathbb{R}$ such that $f(x) = \sqrt[3]{x}$, for every $x \in \mathbb{R}$.

8. Sketch the graph of the integer part function $\lfloor \cdot \rfloor : \mathbb{R} \to \mathbb{R}$ (cf. Problema 9, página 152).

   For the next problem, we think the reader may find it useful to read the statement of Problem 15, page 171 again.

9. * Do the following items:

   (a) If $f : \mathbb{R} \to \mathbb{R}$ is periodic of period $p > 0$, explain how to draw the graph of f by knowing the portion of it in the interval $0 \leq x < p$.
   (b) Use item (a) to sketch the graph of the fractional part function $\{\ \} : \mathbb{R} \to \mathbb{R}$ (cf. Problem 10, page 152).

10. * Let $f : \mathbb{R} \to \mathbb{R}$ be a given function and $a \neq 0$ be a given real number. Prove that the graph of:

    (a) $g(x) = f(x + a)$ is obtained by translating the graph of f of $-a$ units in the direction of the horizontal axis.
    (b) $g(x) = f(x) + a$ is obtained by translating the graph of f of a units in the direction of the vertical axis.
    (c) $g(x) = -f(x)$ is obtained by reflecting the graph of f along the horizontal axis.
    (d) $g(x) = f(-x)$ is obtained by reflecting the graph of f along the vertical axis.
    (e) $g(x) = af(x)$ is obtained by vertically *stretching* (resp. *compressing*) the graph of f by a factor a, if $a > 1$ (resp. $0 < a < 1$).
    (f) $g(x) = f(ax)$ is obtained by horizontally *stretching* (resp. *compressing*) the graph of f by a factor a, if $a > 1$ (resp. $0 < a < 1$).

11. * Use the results of the previous problem to sketch the graph of the function $f : \mathbb{R} \setminus \{2\} \to \mathbb{R}$ given by $f(x) = \frac{x}{2-x}$, for every $x \in \mathbb{R} \setminus \{2\}$.

12. Let I be an interval and $f : I \to \mathbb{R}$ be a given function. Which relation does exist between the graphs of f and of the function $g : I \to \mathbb{R}$, given by $g(x) = |f(x)|$? Apply your conclusion, together with the result of Problem 10, to sketch the graphs of the functions listed below:

    (a) $g(x) = \frac{1}{|x+1|}$, for every $x \in \mathbb{R} \setminus \{-1\}$.
    (b) $g(x) = |x^2 - 4|$, for every $x \in \mathbb{R}$.
    (c) $g(x) = |x^2 - |x + 2| + 2|$, for every $x \in \mathbb{R}$.
    (d) $g(x) = 1 - \frac{1}{(x-2)^2}$, for every real $x \neq 2$.

13. Prove that the graph of the inverse proportionality function is obtained by a trigonometric (i.e., counterclockwise) rotation, of $\frac{\pi}{4}$ radians, of the hyperbola[8] of equation $x^2 - y^2 = 2$.

---

[8]For the necessary background on the equation of hyperbolas, we refer the reader to Chapter 6 of [4], for instance.

6.7 Trigonometric Functions

The **sine function** is the function $\sin : \mathbb{R} \to \mathbb{R}$, that associates to each $x \in \mathbb{R}$ the sine of an arc of x radians[9]:

$$\sin : \mathbb{R} \longrightarrow \mathbb{R}$$
$$x \longmapsto \sin x$$

Analogously, we define the **cosine function** by

$$\cos : \mathbb{R} \longrightarrow \mathbb{R}$$
$$x \longmapsto \cos x$$

where $\cos x$ stands for the cosine of an arc of x radians.

Some basic properties of the sine and cosine functions are collected in the following proposition, for which the reader may find it useful to recall the statements of Problem 15, page 171, and Problem 10, page 171.

Proposition 6.66 *Sine and cosine functions have image* $[-1, 1]$ *and are periodic of period* 2π. *Furthermore, sine function is odd, whereas cosine function is even.*

Proof Immediate from basic Trigonometry.   □

According to the above proposition and the discussion contained in Problem 15, page 171, in order to get an accurate sketch of the graph of the sine function, it suffices to make it in the interval $[-\pi, \pi]$, then *copying* this portion of the graph to each interval of the form $[-\pi + 2k\pi, \pi + 2k\pi]$, where $k \in \mathbb{Z}$.

On the other hand, since the sine function is odd, in order to get its graph in the interval $[-\pi, \pi]$, it suffices to sketch it in the interval $[0, \pi]$; once this is done, then, *reflecting* it across the origin of the cartesian plane, we obtain (from item (c) of Problem 10, page 171) the graph in the interval $[-\pi, \pi]$.

We shall prove in Section 8.1 (cf. Example 8.8) that the graph of the sine function is a *continuous curve*, i.e., a curve with interruptions. On the other hand, in Section 9.7 we shall show (cf. Example 9.66) that such a graph "*opens downwards*" in the interval $[0, \pi]$. For the time being, by assuming the validity of these facts, we can finally sketch the graph of the sine function.

Example 6.67 Gathering together the above information on the behavior of the sine function in the inteval $[0, \pi]$ (image, periodicity, continuity and concavity), together with the fact that it is increasing in $\left[0, \frac{\pi}{2}\right]$, decreasing in $\left[\frac{\pi}{2}, \pi\right]$ and such that $\sin(\pi - x) = \sin x$ for every $x \in \mathbb{R}$, we conclude that, in order to get a reasonably accurate sketch of its graph in that interval, we just need to compute the values of $\sin x$ for some values $x \in \left[0, \frac{\pi}{2}\right]$. This is done in the table below:

---

[9]For the necessary background on Trigonometry for this section, we refer the reader to Chapter 7 of [4].

Fig. 6.15 Graphs of the sine
and cosine functions

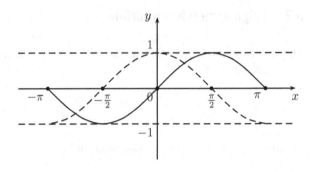

0	$\frac{\pi}{6}$	$\frac{\pi}{4}$	$\frac{\pi}{3}$	$\frac{\pi}{2}$
0	$1/2$	$\sqrt{2}/2$	$\sqrt{3}/2$	1

With all this at hand, we immediately get Fig. 6.15, first on the interval $[0, \pi]$,
then on the interval $[-\pi, \pi]$ and finally on $\mathbb{R}$.

We now recall one more piece of basic Trigonometry, which assures that

$$\cos x = \sin\left(x + \frac{\pi}{2}\right)$$

for every $x \in \mathbb{R}$. Hence, item (a) of Problem 10, page 194, guarantees that, once the
graph of the sine function is drawn, we can get a corresponding sketch for the graph
of the cosine function by *translating* the graph of the sine function of $-\frac{\pi}{2}$ in the
direction of the horizontal axis. In Fig. 6.15, the graph of the cosine function in the
interval $[-\pi, \pi]$ is given by the dashed curve contained in the strip of the cartesian
plane bounded by the (also dashed) lines $y = -1$ and $y = 1$.

Let us now look at a relevant application of the Ptolemy's formulae for the sine
and cosine of the sum and difference of two arcs, which tells us how to proceed to
study a function given as a *linear combination* of the sine and cosine functions.

Example 6.68 Given positive integers a and b, let $f : \mathbb{R} \to \mathbb{R}$ be the function such
that

$$f(x) = a \cos x + b \sin x,$$

for every $x \in \mathbb{R}$. Writing

$$a \cos x + b \sin x = \sqrt{a^2 + b^2} \left(\frac{a}{\sqrt{a^2 + b^2}} \cos x + \frac{b}{\sqrt{a^2 + b^2}} \sin x \right),$$

we notice that

$$\left(\frac{a}{\sqrt{a^2 + b^2}} \right)^2 + \left(\frac{b}{\sqrt{a^2 + b^2}} \right)^2 = 1.$$

Fig. 6.16 Defining angle α

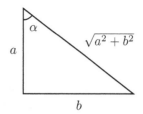

Therefore, point $P\left(\frac{a}{\sqrt{a^2+b^2}}, \frac{b}{\sqrt{a^2+b^2}}\right)$ belongs to the portion of the trigonometric circle situated in the first quadrant, so that there exists a real number $\alpha \in (0, \frac{\pi}{2})$ for which

$$\cos \alpha = \frac{a}{\sqrt{a^2+b^2}} \quad \text{and} \quad \sin \alpha = \frac{b}{\sqrt{a^2+b^2}}$$

(cf. Fig. 6.16). Now, it follows from the formula for the cosine of a difference that

$$\begin{aligned} f(x) &= a \cos x + b \sin x \\ &= \sqrt{a^2 + b^2}(\cos \alpha \cos x + \sin \alpha \sin x) \qquad (6.15) \\ &= \sqrt{a^2 + b^2} \cos(x - \alpha). \end{aligned}$$

In particular, since $|\cos(x - \alpha)| \le 1$, we get

$$|f(x)| = \sqrt{a^2 + b^2}|\cos(x - \alpha)| \le \sqrt{a^2 + b^2},$$

and it's not difficult to use (6.15) to prove that the image of f is precisely the interval $[-c, c]$, where $c = \sqrt{a^2 + b^2}$ (in this respect, see Problem 3).

We close this section with an elementary study of the **tangent function**, i.e., the function that associates, to each real x in its domain, the real number $\tan x = \frac{\sin x}{\cos x}$. Since

$$\cos x = 0 \Leftrightarrow x = \frac{\pi}{2} + k\pi, \ \exists \, k \in \mathbb{Z},$$

the (maximal) domain of definition of the tangent function is the set

$$D = \mathbb{R} \setminus \left\{\frac{\pi}{2} + k\pi; \, k \in \mathbb{Z}\right\},$$

so that the function we wish to study is

$$\begin{aligned} \tan : D &\longrightarrow \mathbb{R} \\ x &\longmapsto \tan x \end{aligned}.$$

For $x \in D$, we have

$$\tan(x + \pi) = \frac{\sin(x + \pi)}{\cos(x + \pi)} = \frac{-\sin x}{-\cos x} = \tan x,$$

and it's immediate to verify that there doesn't exist a real number $0 < p < \pi$ such that $\tan(x + p) = \tan x$ for every $x \in D$. Therefore, the tangent function is periodic of period π. Furthermore, since D is a subset of $\mathbb{R}$ which is symmetric with respect to 0 and

$$\tan(-x) = \frac{\sin(-x)}{\cos(-x)} = \frac{-\sin x}{\cos x} = -\tan x$$

for every $x \in D$, we conclude that the tangent function is odd.

From what was collected above, in order to sketch the graph of the tangent function it suffices to do it on the interval $\left[0, \frac{\pi}{2}\right)$. Indeed, the odd character of the tangent function assures that its graph on the interval $\left(-\frac{\pi}{2}, 0\right]$ is obtained by reflecting, around the origin of the cartesian system, the portion of it on the interval $\left[0, \frac{\pi}{2}\right)$. On the other hand, since we have sketched the desired graph on the interval $\left(-\frac{\pi}{2}, \frac{\pi}{2}\right)$, the periodicity of the tangent function will allows us to sketch the graph of it on all intervals of the form $\left(-\frac{\pi}{2} + k\pi, \frac{\pi}{2} + k\pi\right)$, with $k \in \mathbb{Z}$: it suffices to translate the graph on the interval $\left(-\frac{\pi}{2}, \frac{\pi}{2}\right)$ of $k\pi$ units in the direction of the horizontal axis, for every $k \in \mathbb{Z}$.

In Section 8.1 (cf. Problem 5, page 254), we shall prove that the graph of the tangent function, restricted to the interval $\left(-\frac{\pi}{2}, \frac{\pi}{2}\right)$, is a *continuous* curve, i.e., a curve without interruptions. On the other hand, in Section 9.7 (cf. Example 9.66) we shall establish the fact that the graph "*opens upwards*" on the interval $\left[0, \frac{\pi}{2}\right)$. For the time being, assuming the validity of these statements and in view of the discussion on the previous paragraphs, we can sketch the graph of the tangent function by tabulating some values of it on the interval $\left(0, \frac{\pi}{2}\right)$. The result is approximately that of Fig. 6.17.

Problems: Section 6.7

1. * The purpose of this problem is to introduce the **inverse trigonometric functions**. To this end, do the following items:

   (a) Show that the restriction of the sine function to the interval $\left[-\frac{\pi}{2}, \frac{\pi}{2}\right]$ defines an increasing bijection (which we also denote by sin, whenever there is no danger of confusion) sin $: \left[-\frac{\pi}{2}, \frac{\pi}{2}\right] \to [-1, 1]$; its inverse is the **arcsine**, denoted arcsin $: [-1, 1] \to \left[-\frac{\pi}{2}, \frac{\pi}{2}\right]$, which is also increasing. Then, compute arcsin $\frac{1}{2}$, arcsin 1 and arcsin(-1).

Fig. 6.17 Graph of the
tangent function

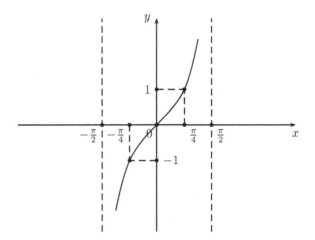

(b) Show that the restriction of the cosine function to the interval $[0, \pi]$ defines
a decreasing bijection (which we also denote by cos, whenever there is no
danger of confusion) $\cos : [0, \pi] \to [-1, 1]$; its inverse is the **arc-cosine**,
denoted $\arccos : [-1, 1] \to [0, \pi]$, which is also decreasing. Then, compute
$\arccos \frac{1}{2}$, $\arcsin 1$ and $\arcsin(-1)$.

(c) Show that the restriction of the tangent function to the interval $\left(-\frac{\pi}{2}, \frac{\pi}{2}\right)$
defines an increasing bijection (which we also denote by tan, whenever
there is no danger of confusion) $\tan : \left(-\frac{\pi}{2}, \frac{\pi}{2}\right) \to \mathbb{R}$; its inverse is the **arc-
tangent** function, denoted $\arctan : \mathbb{R} \to \left(-\frac{\pi}{2}, \frac{\pi}{2}\right)$, which is also increasing.
Then, compute $\arctan 1$, $\arctan \sqrt{3}$ e $\arctan \frac{1}{\sqrt{3}}$.

For item (a) of the next problem, recall that the **cotangent** of $x \neq k\pi$ $(k \in \mathbb{Z})$
is the real number $\cot x$, defined by $\cot x = \frac{\cos x}{\sin x}$. In particular, $\cot x = \frac{1}{\tan x}$.

2. In each of the items below, sketch the graph of the given function:

(a) $f : \mathbb{R} \setminus \{k\pi; k \in \mathbb{Z}\} \to \mathbb{R}$, given by $f(x) = \cot x$.
(b) $f : [-1, 1] \to \left[-\frac{\pi}{2}, \frac{\pi}{2}\right]$, given by $f(x) = \arcsin x$.
(c) $f : [-1, 1] \to [0, \pi]$, given by $f(x) = \arccos x$.
(d) $f : \mathbb{R} \to \left[-\frac{\pi}{2}, \frac{\pi}{2}\right]$, given by $f(x) = \arctan x$.

3. * Let a and b given real numbers, not both zero, and $f : \mathbb{R} \to \mathbb{R}$ be the function
given by

$$f(x) = a \cos x + b \sin x.$$

(a) Find the image set of f.
(b) Prove that f is periodic of period 2π.
(c) Sketch the graphs of f and of the sine function in a single cartesian
coordinate system.

4. * Let a and b be given real numbers, at least one of which is nonzero, and $f : \mathbb{R} \to \mathbb{R}$ be the function given by

$$f(x) = a\cos(\lambda x) + b\sin(\lambda x),$$

where λ is a nonzero real number. Show that f is periodic and compute its period.

5. Let $f : \mathbb{R} \to \mathbb{R}$ be the function given by $f(x) = 2\sin x + \cos(2x)$.

   (a) Compute the maximum and minimum values of f, as well as the real numbers x for which f attains these values.
   (b) Show that f is periodic, of period 2π.
   (c) Sketch the graph of f.

6. (Canada) Compute the number of real solutions of the equation $\sin x = \frac{x}{10}$.

7. Find the maximum value attained by the function $f : [-1, 1] \to \mathbb{R}$, given by $f(x) = 3x + 4\sqrt{5 - x^2}$.

8. (New Zealand) Let α be a given irrational number. Prove that the function $f : \mathbb{R} \to \mathbb{R}$, defined by

$$f(x) = \cos x + \cos(\alpha x),$$

for every $x \in \mathbb{R}$, is not periodic.

9. Find all integer values of n for which the function $f : \mathbb{R} \to \mathbb{R}$, given for $x \in \mathbb{R}$ by

$$f(x) = \cos(nx)\sin\left(\frac{5x}{n}\right),$$

is periodic of period 3π.

10. (Canada) Prove that the function $f : \mathbb{R} \to \mathbb{R}$ given by $f(x) = \sin(x^2)$ is not periodic.

Chapter 7
More on Real Numbers

This chapter proceeds with the study of real numbers by presenting the notion of convergence for (infinite) sequences and series of real numbers. Among other applications, we shall introduce one of the two most important numbers of Mathematics,[1] the number e. We shall also present a famous result of Kronecker on dense subsets of the real line, which will find several interesting applications, here and in the coming chapters.

7.1 Supremum and Infimum

We begin this chapter by examining the completeness of $\mathbb{R}$ from another viewpoint, for which we need to introduce some preliminary concepts.

A nonempty subset $X \subset \mathbb{R}$ is **bounded from above** if there exists a real number M such that

$$X \subset (-\infty, M].$$

In this case, we also say that M is an **upper bound** for X. Similarly, a nonempty set $X \subset \mathbb{R}$ is **bounded from below** if there exists a real number m such that

$$X \subset [m, +\infty).$$

In this case, we say that m is a **lower bound** for X. Finally, a nonempty set $X \subset \mathbb{R}$ is **bounded** if X is simultaneously bounded from above and from below.

---

[1] As the reader probably suspects, the other one is the number π, which is defined as the numerical value of the area of a circle of radius 1—see [4], for instance.

© Springer International Publishing AG 2017
A. Caminha Muniz Neto, *An Excursion through Elementary Mathematics, Volume I*,
Problem Books in Mathematics, DOI 10.1007/978-3-319-53871-6_7

Yet in another way, a nonempty set $X \subset \mathbb{R}$ is bounded (resp. bounded from above, bounded from below) if there exists a positive real number a such that

$$-a \leq x \leq a \text{ (resp. } x \leq a, \ x \geq -a\text{)}, \ \forall x \in X.$$

On the other hand, a nonempty set $X \subset \mathbb{R}$ which is not bounded (resp. not bounded from above, not bounded from below) is said to be **unbounded** (resp. **unbounded from above, unbounded from below**). In this case, given any positive real number a, one can always find an element $x \in X$ such that

$$x \notin [-a, a] \text{ (resp. } x > a, \ x < -a\text{)}.$$

Examples 7.1

(a) The set $X = \{1, \frac{1}{2}, \frac{1}{3}, \frac{1}{4}, \ldots\}$ is bounded. Indeed, 0 is a lower bound and 1 is an upper bound for X.
(b) Bounded intervals (cf. Definition 1.10 and subsequent discussion) are bounded sets, in the sense of the above discussion.

Example 7.2 If a nonempty set $X \subset \mathbb{R}$ is bounded from above, then the subset Y of $\mathbb{R}$ given by $Y = \{-x; \ x \in X\}$ is bounded from below, and conversely. Indeed, a real number a is an upper bound for X if and only if $-a$ is a lower bound for Y.

In spite of its apparent obviousness, we shall list our next example as an *axiom* of the natural number system, known as its **archimedian property**, after the Greek mathematician Archimedes of Syracuse.[2] Some important consequences of it are collected in Problems 2, 3 and 4, page 206.

Axiom 7.3 *The set $\mathbb{N}$ of natural numbers is unbounded from above.*

Continuing with the development of the theory, fix a bounded from above nonempty set $X \subset \mathbb{R}$. If $M \in \mathbb{R}$ is an upper bound for X, then $X \subset (-\infty, M]$. Nevertheless, it may happen that there exists $M' < M$ which still is an upper bound for X, i.e., is such that $X \subset (-\infty, M']$. Indeed, the condition $X \subset (-\infty, M]$ doesn't guarantee that, for $M' < M$, we have $X \cap (M', M] \neq \emptyset$; if it happens that $X \cap (M', M] = \emptyset$, then we will have $X \subset (-\infty, M']$, and M' will be another upper bound for X, which is less than M.

On the other hand, given $x \in X$, it is obvious that no $M' < x$ is an upper bound for X, since $x \in X \setminus (-\infty, M']$, i.e., $X \not\subset (-\infty, M']$. To put in another way, the set of upper bounds of a nonempty, bounded from above set $X \subset \mathbb{R}$ is bounded from below.

---

[2]Archimedes, who lived in the III century b.C., was the greatest scientist of his time. Among many important contributions to Mathematics and Physics, his seminal ideas on how to compute the area under a parabolic segment anticipated, in some 2000 years, the development of the Integral Calculus by Newton and Leibniz.

As we shall see later (cf. Problem 9, page 242), the completeness of $\mathbb{R}$, as postulated in Sect. 1.3, is a consequence of the following statement, which sharpens the discussion of the two previous paragraphs and will be taken as an axiom.

Axiom 7.4 *If a nonempty set $X \subset \mathbb{R}$ is bounded from above, then there exists a real number M satisfying the two following properties:*

(a) M is an upper bound for X.
(b) If $M' < M$, then M' is not an upper bound for X.

In the hypotheses and notations of the previous axiom, a moment's thought shows that there exists at most one real number M satisfying the properties there stated. Indeed, if two distinct real numbers M_1 and M_2 did the job, suppose, without loss of generality, that $M_1 < M_2$. Then, on the one hand, item (a) would guarantee that M_1 and M_2 are upper bounds for X; on the other, since $M_1 < M_2$, item (b) would guarantee that M_1 is not an upper bound for X.

Yet in the hypotheses and notations of the axiom, the discussion of the last paragraph allows us to say that M is *the* **least upper bound** or **supremum** of X. We denote $M = \sup(X)$ or $M = \text{lub}(X)$.

Examples 7.5

(a) The set $X = \{1, \frac{1}{2}, \frac{1}{3}, \frac{1}{4}, \ldots\}$ has 1 as an upper bound. On the other hand, since $1 \in X$, no real number less that 1 can be an upper bound of X, so that $\sup(X) = 1$.

(b) If $X = (1, 2)$ (an open interval), then 2, but no real number less than 2, is an upper bound for X. Indeed, if $1 < a < 2$, then the number $\frac{1+a}{2}$ is also greater than 1 and less than 2, so that $\frac{1+a}{2} \in X$. Now, since $a < \frac{1+a}{2} \in X$, the number a cannot be an upper bound for X. Therefore, $\sup(X) = 2$, and observe that $2 \notin X$.

If $\emptyset \neq Y \subset \mathbb{R}$ is bounded below, then one can prove, as an easy consequence of Theorem 7.4 (cf. Problem 6), that Y admits a **greatest lower bound** m, which is also said to be an **infimum** of Y. As in the case of nonempty, bounded above sets, one can easily prove that a nonempty, bounded below set Y has a unique greatest lower bound m; hence, we denote $m = \inf(Y)$ or $m = \text{glb}(Y)$.

For future use, the coming results collect some useful properties of the notions of sup and inf.

Proposition 7.6 *Let $X \subset \mathbb{R}$ be a nonempty, bounded above set, and $M = \sup(X)$. If $n \in \mathbb{N}$, then there exists $x \in X$ such that*

$$M - \frac{1}{n} < x \leq M.$$

Proof Since M is the least upper bound of X and $M - \frac{1}{n} < M$, the real number $M - \frac{1}{n}$ is no longer an upper bound of X. Hence, there exists $x \in X$ such that $x > M - \frac{1}{n}$. However, since $X \subset (-\infty, M]$, we must also have $x \leq M$. $\qquad\square$

Problem 10 states an analogous result for the greatest lower bound of nonempty, bounded below sets.

Proposition 7.7 *Let $X, Y \subset \mathbb{R}$ be nonempty sets. If $x \leq y$ for all $x \in X$ and $y \in Y$, then X is bounded above, Y is bounded below and*

$$\sup(X) \leq \inf(Y).$$

Proof Fix $y \in Y$ arbitrarily. Since $x \leq y$ for all $x \in X$, the number y is an upper bound for X. Hence, X is bounded above and, letting $M = \sup(X)$ (the *least* upper bound of X), we have $M \leq y$.

Since $y \in Y$ was arbitrarily chosen, the reasoning of the previous paragraph shows that, *for all $y \in Y$*, we have $M \leq y$. Therefore, M is a lower bound for Y, so that Y is bounded below and, letting $m = \inf Y$ (the *greatest* lower bound of Y), we have $M \leq m$. $\qquad\qquad\square$

The next result gives a sufficient condition for equality to happen in the previous proposition.

Proposition 7.8 *Let $X, Y \subset \mathbb{R}$ be nonempty sets, such that X is bounded above, Y is bounded below and $\sup X \leq \inf Y$. If, for every $n \in \mathbb{N}$, there exist $x_n \in X$ and $y_n \in Y$ satisfying $y_n - x_n < \frac{1}{n}$, then $\sup(X) = \inf(Y)$.*

Proof Let $M = \sup(X)$, $m = \inf(Y)$ and suppose that $M < m$. Since $x \leq M < m \leq y$ for all $x \in X$ and $y \in Y$, we would have $y - x \geq m - M$, for all $x \in X$ and $y \in Y$.

On the other hand, by choosing a natural number $n > \frac{1}{m-M}$ (which is possible, thanks to Axiom 7.3), our hypotheses would guarantee the existence of real numbers $x_n \in X$ and $y_n \in Y$ such that

$$y_n - x_n < \frac{1}{n} < m - M,$$

which is a contradiction!

Finally, since the assumption that $\sup(X) < \inf(Y)$ leads us to a contradiction, the only possibility is that $\sup(X) = \inf(Y)$. $\qquad\qquad\square$

We finish this section by presenting a nontrivial (and important) application of the concept of supremum of a bounded above nonempty set. To set the stage, we recall that, in the previous chapters, we several times relied upon the existence of square roots of positive real numbers. The next result establishes existence in a more general setting.

Theorem 7.9 *Given a positive real number x and a natural number $n > 1$, there exists a unique positive real number y such that $y^n = x$.*

Proof Let's consider just the case $x > 1$, referring the reader to Problem 1 for the case $0 < x < 1$ and observing that the case $x = 1$ is trivial. If

$$X = \{a \in \mathbb{R}; \ a \geq 0 \text{ and } a^n < x\},$$

then X is nonempty, since $0 \in X$. Also, X is bounded above, for, if $\alpha \geq x + 1$, then Problem 6, page 176, together with item (b) of Corollary 1.3, guarantees that $\alpha^n \geq (x+1)^n > x + 1 > x$; therefore, $\alpha \notin X$ and, thus, $X \subset (-\infty, x+1)$.

Being nonempty and bounded above, X has a least upper bound, say y. Since $1^n = 1 < x$, we have $1 \in X$ and, thus, $y \geq 1$. We shall show that $y^n = x$ in the following way:

(i) If $y^n < x$, we shall obtain a positive real number z such that $y^n < z^n < x$. Then, the first inequality will give $y < z$, whereas the second (with the aid of the result of Problem 6, page 176) $z \in X$. Therefore, we will get $y < z \in X$, thus contradicting the fact that y is an upper bound for X.

(ii) If $y^n > x$, we shall obtain a positive real number z for which $x < z^n < y^n$. Then, taking an arbitrary $a \in X$, it will follow from $a^n < x < z^n < y^n$ (again by the result of Problem 6, page 176) that $a < z < y$, so that z will be an upper bound for X which is less than y. This will contradict the fact that y is the least upper bound for X.

Once we have proved that $y^n < x$ and $y^n > x$ both lead to contradictions, the only possibility left will be $y^n = x$.

For the proof of (i), suppose $y^n < x$. If $z = y + \frac{1}{k}$, with $k \in \mathbb{N}$, then, once more by the result of Problem 6, page 176, we have $y^n < z^n$. We assert that $z^n < x$ for all sufficiently large k. Indeed, the binomial formula (cf. Theorem 4.20) gives

$$z^n = \left(y + \frac{1}{k}\right)^n = y^n + \sum_{j=1}^{n} \binom{n}{j} \frac{1}{k^j} y^{n-j};$$

however, since $\binom{n}{j} < 2^n$, $\frac{1}{k^j} < \frac{1}{k}$ and $y^{n-j} < y^n$ for $1 \leq j \leq n$, it follows that

$$z^n < y^n + \sum_{j=1}^{n} \frac{2^n}{k} y^n = y^n + \frac{n(2y)^n}{k};$$

therefore, we have $z^n < x$ provided $y^n + \frac{n(2y)^n}{k} < x$, i.e., $k > \frac{n(2y)^n}{x - y^n}$.

For the proof of (ii), suppose $y^n > x$. Letting $z = y - \frac{1}{k}$, with $k \in \mathbb{N}$, then, as in the previous paragraph anterior, we have $y^n > z^n$. We claim that $z^n > x$ for all sufficiently large k. To this end, arguing as above we have

$$z^n = \left(y - \frac{1}{k}\right)^n = y^n + \sum_{j=1}^{n} (-1)^j \binom{n}{j} \frac{1}{k^j} y^{n-j}$$

$$> y^n - \sum_{j=1}^{n} \binom{n}{j} \frac{1}{k^j} y^{n-j} > y^n - \frac{n(2y)^n}{k};$$

hence, $z^n > x$ whenever $y^n - \frac{n(2y)^n}{k} > x$, i.e., $k > \frac{n(2y)^n}{y^n - x}$. $\qquad\square$

For another, conceptually simpler, proof of the above theorem, we refer the reader to Example 8.36.

Problems: Section 7.1

1. * Yet in respect to roots of positive real numbers, do the following items:

    (a) If $0 < x < 1$ and $n > 1$ is natural, show that the real number $\frac{1}{\sqrt[n]{1/x}}$ (whose existence is guaranteed by Theorem 7.9, since $\frac{1}{x} > 1$) is the n-th root of x.

    (b) Given $a, b > 0$ and $m, n > 1$ naturals, show that $\sqrt[n]{ab} = \sqrt[n]{a}\sqrt[n]{b}$, $\sqrt[n]{\frac{a}{b}} = \frac{\sqrt[n]{a}}{\sqrt[n]{b}}$ and $\sqrt[mn]{a} = \sqrt[m]{\sqrt[n]{a}}$.

    (c) Given $0 < a < b$ and $n > 1$ natural, show that $\sqrt[n]{a} < \sqrt[n]{b}$.

    (d) Given $a > 0$ and $m > n > 1$ naturals, show that $\sqrt[m]{a} > \sqrt[n]{a}$ if $a > 1$, and $\sqrt[m]{a} < \sqrt[n]{a}$ if $0 < a < 1$.

    (e) Let $x < 0$ be a real number and $n \in \mathbb{N}$ be odd. If $y = -\sqrt[n]{-x}$, show that $y^n = x$ (the real number y is also called the **n-th root** of x) and extend the properties of items (b)–(d) to this case.

2. * Use the archimedian property of the set of natural numbers to prove the following items:

    (a) If $a \in \mathbb{R}$ is such that $0 \le a < \frac{1}{n}$ for every $n \in \mathbb{N}$, then $a = 0$.

    (b) If $a, b, c \in \mathbb{R}$, with $a > 0$, then there exists $n \in \mathbb{N}$ for which $an + b > c$.

3. * Let a and b be given rationals, with $a < b$. Prove that:

    (a) $a < \frac{a+b}{2} < b$ and $a < a + \frac{b-a}{\sqrt{2}} < b$.

    (b) The interval (a, b) contains infinitely many rational numbers and infinitely many irrational numbers.

4. * The purpose of this problem is to generalize the result of the previous one, showing that between any two given real numbers there is always a rational number and an irrational number (thanks to these properties, we say that $\mathbb{Q}$ and $\mathbb{R} \setminus \mathbb{Q}$ are **dense** in $\mathbb{R}$). To this end, let a and b be given real numbers, with $a < b$.

    (a) Show that it suffices to consider the case $a \ge 0$.

    (b) Prove that there exists $n \in \mathbb{N}$ such that $0 < \frac{1}{n} < b - a$ and $0 < \frac{\sqrt{2}}{n} < b - a$.

    (c) Letting $a \ge 0$ e $n \in \mathbb{N}$ be chosen as in (b), show that one of the numbers $\frac{1}{n}, \frac{2}{n}, \frac{3}{n}, \dots$ and one of the numbers $\frac{\sqrt{2}}{n}, \frac{2\sqrt{2}}{n}, \frac{3\sqrt{2}}{n}, \dots$ belong to the interval (a, b).

5. * A rational number $r \in [0, 1]$ is said to be **dyadic** if there exist $k, n \in \mathbb{Z}$ such that $0 \le n \le 2^k$ and $r = \frac{n}{2^k}$. Prove that the set of dyadic rationals is dense in $[0, 1]$, i.e., that for every $a \in [0, 1]$ and $\epsilon > 0$, there exists a dyadic rational in the interval $(a - \epsilon, a + \epsilon)$.

6. * If $Y \subset \mathbb{R}$ is nonempty and bounded below, prove that it has a greatest lower bound.

7. Let $X = \{x \in \mathbb{Q};\ 0 < x < 1\}$ and $Y = \{y \in \mathbb{R} \setminus \mathbb{Q};\ 0 < y < 1\}$. Prove that

$$\inf(X) = \inf(Y) = 0 \quad \text{and} \quad \sup(X) = \sup(Y) = 1.$$

8. If $X = \{|\sqrt{a} - \sqrt{b}|; \, a, b \in \mathbb{N}$ and $a \neq b\}$, compute $\inf(X)$.

9. Let

$$C_1 = (0, 1)$$
$$C_2 = (0, 1) \setminus (1/3, 2/3) = (0, 1/3] \cup [2/3, 1)$$
$$C_3 = ((0, 1/3] \setminus (1/9, 2/9)) \cup ([2/3, 1) \setminus (7/9, 8/9))$$
$$= (0, 1/9] \cup [2/9, 1/3] \cup [2/3, 7/9] \cup [8/9, 1).$$

More generally, for each $n \in \mathbb{N}$, obtain C_{n+1} from C_n, by erasing the open middle third of each of the intervals that form C_n. If $C = \bigcup_{n \geq 1} C_n$, show that $\inf(C) = 0$ and $\sup(C) = 1$.

10. * Let $Y \subset \mathbb{R}$ be a nonempty, bounded below set, with $m = \inf Y$. If $n \in \mathbb{N}$, prove that there exists $y \in Y$ such that

$$m \leq y < m + \frac{1}{n}.$$

11. Let $X, Y \subset \mathbb{R}$ be nonempty sets, such that X is bounded above, Y is bounded below and $\sup(X) = \inf(Y) = \alpha$. If $\alpha \notin X \cup Y$, prove that there exist elements $x_n \in X$ and $y_n \in Y$ for which $y_n - x_n < \frac{1}{n}$, for every $n \in \mathbb{N}$.

12. * Let $X \subset \mathbb{R}$ be a nonempty, bounded above set. Given $c \in \mathbb{R}$, let $cX = \{cx; \, x \in X\}$. Prove that:

(a) If $c > 0$, then $\sup(cX) = c \sup(X)$.
(b) If $c < 0$, then cX is bounded below and $\inf(cX) = c \sup(X)$.

Then, if X is bounded below, establish properties analogous to the ones listed above, relating $\inf X$ to the sup or the inf of cX, according to whether $c < 0$ or $c > 0$.

13. * Let $X, Y \subset \mathbb{R}$ be nonempty sets and $X + Y = \{x + y; \, x \in X$ and $y \in Y\}$.

(a) If X and Y are bounded above, prove that $X + Y$ is bounded above and $\sup(X + Y) = \sup(X) + \sup(Y)$.
(b) If X and Y are bounded below, prove that $X + Y$ is bounded below and $\inf(X + Y) = \inf(X) + \inf(Y)$.

14. * Let $X, Y \subset [0, +\infty)$ be nonempty, bounded above sets. If $XY = \{xy; \, x \in X$ and $y \in Y\}$, prove that XY is bounded above and such that $\sup(XY) = \sup(X) \cdot \sup(Y)$ and $\inf(XY) = \inf(X) \cdot \inf(Y)$.

15. (Hungary) Let $(R_n)_{n \geq 1}$ be a sequence of pairwise distinct rectangles in the cartesian plane, each of which having all vertices with integer coordinates and two sides along the axes. Prove that one can find two of them such that one contains the other.

16. (IMO) Let $f, g : \mathbb{R} \to \mathbb{R}$ be functions satisfying, for all $x, y \in \mathbb{R}$, the relation

$$f(x + y) + f(x - y) = 2f(x)g(y).$$

If f is not identically zero and $|f(x)| \leq 1$ for every $x \in \mathbb{R}$, prove that $|g(x)| \leq 1$, for every $x \in \mathbb{R}$.

7.2   Limits of Sequences

Given a sequence $(a_n)_{n\geq 1}$ of real numbers, we are interested in recognizing whether or not their terms are approaching a certain real number l, as n increases. For instance, if $a_n = \frac{1}{n}$, it is reasonable to say that the numbers a_n become closer and closer to 0 as n increases, since the result of the division of 1 by n is increasingly smaller as n increases. This naive point of view is formalized as follows.

Definition 7.10 A sequence $(a_n)_{n\geq 1}$ of real numbers **converges** to a real number l if, given an error $\epsilon > 0$ for the value of l, there exists an index $n_0 \in \mathbb{N}$ such that $|a_n - l| < \epsilon$ for every $n > n_0$.

Alternatively, if the sequence $(a_n)_{n\geq 1}$ converges to l, we say that it is **convergent** and that l is *a* **limit** of the sequence, which we denote by writing

$$a_n \xrightarrow{n} l \text{ or } \lim_{n\to+\infty} a_n = l.$$

Finally, a sequence which is not convergent is said to be **divergent**.

In general, if we diminish the error $\epsilon > 0$ and the condition "$|a_n - l| < \epsilon$ for every $n > n_0$" is to continue holding, then the natural number n_0 in the definition of convergent sequence tends to increase. In other words, in general n_0 depends on $\epsilon > 0$. Anyhow, what is important to assure the convergence of the sequence $(a_n)_{n\geq 1}$ is that, for an arbitrarily given error $\epsilon > 0$, we are capable of finding $n_0 \in \mathbb{N}$ such that

$$n > n_0 \Rightarrow |a_n - l| < \epsilon.$$

For the reader to get used to the important concept of convergent sequence, we collect below some elementary examples of convergent and divergent sequences.

Examples 7.11

(a) If $a_n = \frac{1}{n}$, then $a_n \xrightarrow{n} 0$: indeed, for a given $\epsilon > 0$, we have $|a_n - 0| < \epsilon$ provided $n > \frac{1}{\epsilon}$; thus, once we have chosen $n_0 \in \mathbb{N}$ such that $n_0 > \frac{1}{\epsilon}$, we will have $|a_n - 0| < \epsilon$ whenever $n > n_0$.

(b) If $a_n = (-1)^n$, then $(a_n)_{n\geq 1}$ is divergent: indeed, since the terms of the sequence are alternately equal to 1 and -1, it is impossible for them to (collectively) become closer to a single real number l (formalize this intuition).

(c) If $a_n = 1 + \frac{(-1)^n}{n}$, then $a_n \xrightarrow{n} 1$: this is so because $|a_n - 1| = \frac{1}{n}$, so that $|a_n - 1| < \epsilon$ for $n > \frac{1}{\epsilon}$.

(d) If $(a_n)_{n\geq 1}$ is a constant sequence, with $a_n = c$ for every $n \geq 1$, then $a_n \to c$.

Example 7.12 If $a_n = q^n$, with $0 < |q| < 1$, then $a_n \xrightarrow{n} 0$.

Proof Since $\frac{1}{|q|} > 1$, we can write $\frac{1}{|q|} = 1 + \alpha$, with $\alpha > 0$. Therefore, taking the first two terms in the binomial expansion formula, we get

$$\frac{1}{|q|^n} = (1 + \alpha)^n \geq 1 + n\alpha$$

and, hence,

$$|a_n - 0| = |q|^n \le \frac{1}{1 + n\alpha}.$$

Thus, if we wish that $|a_n - 0| < \epsilon$, it suffices to impose $\frac{1}{1+n\alpha} < \epsilon$ or, equivalently, $n > \frac{1}{\alpha}\left(\frac{1}{\epsilon} - 1\right)$. $\square$

Example 7.13 The sequence $(a_n)_{n \ge 1}$, given for $n \ge 1$ by $a_n = \sqrt{n+1} - \sqrt{n}$, converges to 0.

Proof Note that $a_n = \frac{1}{\sqrt{n+1}+\sqrt{n}}$. Thus, given $\epsilon > 0$ and taking $n_0 \in \mathbb{N}$ such that $n_0 > \frac{1}{4\epsilon^2}$, we have

$$n > n_0 \Rightarrow \sqrt{n+1} + \sqrt{n} > \sqrt{n_0 + 1} + \sqrt{n_0} > 2\sqrt{n_0} > \frac{1}{\epsilon}.$$

Therefore,

$$n > n_0 \Rightarrow |a_n - 0| = \frac{1}{\sqrt{n+1} + \sqrt{n}} < \epsilon.$$

$\square$

The notion of convergent sequence doesn't make it clear whether the correspondent limit is unique. Yet in another way, in principle it could happen that a certain sequence converges to more than one limit. The coming result shows that this is not so.

Proposition 7.14 *If the sequence $(a_n)_{n \ge 1}$ converges, then its limit is unique.*

Proof Let l_1 and l_2 be distinct real numbers, and suppose that the given sequence simultaneously converges to l_1 and l_2. Toking $\epsilon = \frac{1}{2}|l_1 - l_2| > 0$, the definition of convergence guarantees the existence of $n_1, n_2 \in \mathbb{N}$ such that

$$n > n_1 \Rightarrow |a_n - l_1| < \epsilon \text{ and } n > n_2 \Rightarrow |a_n - l_2| < \epsilon.$$

Therefore, triangle inequality gives

$$n > \max\{n_1, n_2\} \Rightarrow |l_1 - l_2| \le |a_n - l_1| + |a_n - l_2| < 2\epsilon = |l_1 - l_2|,$$

which is an absurd. $\square$

Thanks to the previous result, from now on we speak of *the* limit of a convergent sequence. In this respect, the next proposition collects two basic, albeit very important, properties of limits of convergent sequences. In order to state it properly, we define a **subsequence** of a sequence $(a_n)_{n \ge 1}$ as the restriction of the given sequence to an infinite subset $\mathbb{N}_1 = \{n_1 < n_2 < n_3 < \cdots\}$ of $\mathbb{N}$; in this case, we denote it by $(a_{n_k})_{k \in \mathbb{N}}$. Since the function $j \mapsto n_j$ from $\mathbb{N}_1$ to $\mathbb{N}$ is a bijection, every subsequence can actually be seen as a sequence.

Proposition 7.15 *Let $(a_n)_{n\geq 1}$ be a convergent sequence, with $\lim_{n\to+\infty} a_n = l$.*
Then:

(a) *If $a_n \geq a$ (resp. $a_n \leq a$), for every $n \geq 1$, then $l \geq a$ (resp. $l \leq a$).*
(b) *Every subsequence $(a_{n_k})_{k\geq 1}$ of $(a_n)_{n\geq 1}$ also converges to l.*

Proof

(a) Suppose that $a_n \geq a$ for every $n \geq 1$, and let's show that $l \geq a$ (the other case is completely analogous). By contradiction, if $l < a$, take $\epsilon = a - l > 0$. The definition of convergence guarantees the existence of an index $n_0 \in \mathbb{N}$ such that $n > n_0 \Rightarrow |a_n - l| < \epsilon$; in particular, given $n > n_0$, we have

$$a_n < l + \epsilon = l + (a - l) = a,$$

which is an absurd.
(b) Let $\epsilon > 0$ be given. Since $a_n \xrightarrow{n} l$, there exists a natural number n_0 such that $|a_n - l| < \epsilon$ for $n > n_0$. Since $n_1 < n_2 < n_3 < \cdots$, there exists an index n_i such that $n_j > n_0$ for $j \geq i$; hence, for all such j, we have $|a_{n_j} - l| < \epsilon$, which is the same as saying that $a_{n_k} \xrightarrow{k} l$. $\qquad\square$

In words, item (b) of the previous proposition says that, if the terms of a certain sequence come closer and closer to l as their indices increase, then the same is true for the terms of every subsequence of the given sequence. Item (b) of the previous proposition also has the following immediate corollary, which gives us a sufficient (and quite useful) condition for the divergence of a sequence. sequência.

Corollary 7.16 *If two subsequences of a given sequence converge to distinct limits, then the original sequence is divergent.*

Up to now, except for some very simple examples we haven't seen how one could find out the limit of a convergent sequence. In order to remedy this situation, we need to understand how to perform simple *arithmetic operations* with convergent sequences. We turn to this next, starting with an auxiliary result which is important in its own.

We say that a sequence $(a_n)_{n\geq 1}$ is **bounded** (resp. **bounded from above**, **bounded from below**) if the set $\{a_1, a_2, \ldots\}$ is bounded (resp. bounded from above, bounded from below), in the sense of the previous section.

Lemma 7.17 *Every convergent sequence is bounded.*

Proof If $(a_n)_{n\geq 1}$ is a convergent sequence with limit l, then there exists $n_0 \in \mathbb{N}$ such that

$$n > n_0 \Rightarrow |a_n - l| < 1.$$

This, together with the triangle inequality, gives

$$n > n_0 \Rightarrow |a_n| \leq |a_n - l| + |l| < 1 + |l|.$$

Finally, letting $L = \max\{1 + |a|, |a_1|, |a_2|, \dots, |a_{n_0-1}|\}$, we get $|a_n| < L$ for every $n \in \mathbb{N}$, so that the sequence is bounded. $\qquad\qquad\qquad\qquad\qquad\qquad\qquad\square$

Proposition 7.18 *Let $(a_n)_{n\geq 1}$ and $(b_n)_{n\geq 1}$ be convergent sequences, and c be any real number.*

(a) *If $a_n \xrightarrow{n} a$, then $ca_n \xrightarrow{n} ca$.*

(b) *If $a_n \xrightarrow{n} a$ and $b_n \xrightarrow{n} b$, then $a_n \pm b_n \xrightarrow{n} a \pm b$ and $a_n b_n \xrightarrow{n} ab$.*

(c) *If $a_n \xrightarrow{n} 0$ and $(b_n)_{n\geq 1}$ is bounded, then $a_n b_n \xrightarrow{n} 0$.*

(d) *If $a_n \xrightarrow{n} a$ and $b_n \xrightarrow{n} b$, with $b, b_n \neq 0$ for every $n \geq 1$, then $\frac{a_n}{b_n} \xrightarrow{n} \frac{a}{b}$.*

Proof

(a) If $c = 0$, then $ca_n = ca = 0$, and there is nothing to do. Suppose, then, that $c \neq 0$, and let $\epsilon > 0$ be given. Since $a_n \xrightarrow{n} a$, there exists $n_0 \in \mathbb{N}$ such that $n > n_0 \Rightarrow |a_n - a| < \frac{\epsilon}{|c|}$. Hence,

$$n > n_0 \Rightarrow |ca_n - ca| = |c||a_n - a| < |c| \cdot \frac{\epsilon}{|c|} = \epsilon.$$

(b) For the first part, let's prove that $a_n + b_n \xrightarrow{n} a + b$ (to prove that $a_n - b_n \to a - b$ is completely analogous). Given $\epsilon > 0$, the convergences $a_n \xrightarrow{n} a$ and $b_n \xrightarrow{n} b$ assure the existence of $n_1, n_2 \in \mathbb{N}$ such that

$$n > n_1 \Rightarrow |a_n - a| < \frac{\epsilon}{2} \text{ and } n > n_2 \Rightarrow |b_n - b| < \frac{\epsilon}{2}.$$

Therefore, taking $n > \max\{n_1, n_2\}$, we get

$$|(a_n + b_n) - (a + b)| \leq |a_n - a| + |b_n - b| < \frac{\epsilon}{2} + \frac{\epsilon}{2} = \epsilon.$$

For the second part, let $L > 0$ be such that $|b_n| < L$ for every $n \in \mathbb{N}$. Given $\epsilon > 0$, take $n_0 \in \mathbb{N}$ for which

$$n > n_0 \Rightarrow |a_n - a| < \frac{\epsilon}{2L} \text{ and } |b_n - b| < \frac{\epsilon}{2|a| + 1}.$$

Then

$$|a_n b_n - ab| = |a_n b_n - ab_n + ab_n - ab| \leq |a_n - a||b_n| + |a||b_n - b|$$

$$< \frac{\epsilon}{2L} \cdot L + |a| \cdot \frac{\epsilon}{2|a| + 1} < \frac{\epsilon}{2} + \frac{\epsilon}{2} = \epsilon.$$

(c) Let $L > 0$ be such that $|b_n| < L$ for every $n \geq 1$. Given $\epsilon > 0$, let's take $n_0 \in \mathbb{N}$ such that

$$n > n_0 \Rightarrow |a_n| < \frac{\epsilon}{L}.$$

Then,

$$n > n_0 \Rightarrow |a_n b_n - 0| = |a_n||b_n| < \frac{\epsilon}{L} \cdot L = \epsilon.$$

(d) By the second part of item (b), it suffices to show that $\frac{1}{b_n} \xrightarrow{n} \frac{1}{b}$. To this end, start by observing that

$$\left| \frac{1}{b_n} - \frac{1}{b} \right| = \frac{1}{|b|} \cdot \frac{|b_n - b|}{|b_n|} \leq \frac{1}{|b|} \cdot \frac{|b_n - b|}{|b| - |b_n - b|},$$

where we used the triangle inequality in the last passage above. Now, given $\epsilon > 0$, choose $n_0 \in \mathbb{N}$ such that

$$n > n_0 \Rightarrow |b_n - b| < \frac{\epsilon}{2}, \frac{|b|}{2}.$$

Then, for $n > n_0$, we have

$$\left| \frac{1}{b_n} - \frac{1}{b} \right| \leq \frac{1}{|b|} \cdot \frac{|b_n - b|}{|b| - |b|/2} = 2|b_n - b| < \epsilon.$$

□

Example 7.19 Let a be a positive real number. If $(a_n)_{n \geq 1}$ is given by $a_n = \sqrt[n]{a}$ for every $n \geq 1$, then $a_n \xrightarrow{n} 1$.

Proof If $a > 1$, then $a_n > 1$. Write $a_n = 1 + b_n$, so that $b_n > 0$. Since

$$a = a_n^n = (1 + b_n)^n \geq 1 + \binom{n}{1} b_n = 1 + n b_n,$$

we get $0 < b_n < \frac{a-1}{n}$. Hence, the squeezing principle (cf. Problem 6) guarantees that $b_n \xrightarrow{n} 0$, and item (b) of Proposition 7.18 gives $a_n = 1 + b_n \xrightarrow{n} 1$.

If $0 < a < 1$, let $a_n' = \frac{1}{a_n} = \sqrt[n]{\frac{1}{a}}$, so that $a_n' \xrightarrow{n} 1$ by the first part. Then, item (d) of Proposition 7.18 gives that $a_n \xrightarrow{n} 1$.

□

Example 7.20 The sequence $(a_n)_{n \geq 1}$, given by $a_n = \sqrt[n]{n}$ for every $n \geq 1$, converges to 1.

Proof As in the previous example, write $a_n = 1 + b_n$ for $n \geq 2$. Since $b_n > 0$, we have

$$n = a_n^n = (1 + b_n)^n \geq 1 + \binom{n}{1} b_n + \binom{n}{2} b_n^2 > \frac{n(n-1)}{2} \cdot b_n^2,$$

so that

$$0 < b_n^2 < \frac{2}{n-1}.$$

Hence, once more from the squeezing principle, the above inequality gives $b_n \xrightarrow{n} 0$ and, thus, $a_n = 1 + b_n \xrightarrow{n} 1$.   □

For what comes next, recall that a sequence $(a_n)_{n\geq 1}$ of real numbers is just a function $f : \mathbb{N} \to \mathbb{R}$, for which we write $a_n = f(n)$. Hence, it is natural to say that $(a_n)_{n\geq 1}$ is **monotonic** *increasing* (resp. *decreasing, nondecreasing, nonincreasing*) provided $a_n < a_{n+1}$ (resp. $a_n > a_{n+1}, a_n \leq a_{n+1}, a_n \geq a_{n+1}$), for every $n \geq 1$.

The most important result on limits of sequences is the theorem below, which is known in the mathematical literature as **Bolzano-Weierstrass theorem**.[3]

Theorem 7.21 (Bolzano-Weierstrass) *Every monotonic bounded sequence is convergent.*

Proof Suppose that $(a_n)_{n\geq 1}$ is a nondecreasing bounded sequence (the other cases can be dealt with similarly), i.e., that

$$a_1 \leq a_2 \leq a_3 \leq \cdots < M,$$

for some $M > 0$. Then, M is an upper bound for the set $A = \{a_1, a_2, a_3, \ldots\}$, so that A has a sup, say $\sup A = l$. We claim that $a_n \xrightarrow{n} l$. Indeed, let $\epsilon > 0$ be given; since $l - \epsilon$ is no longer an upper bound for A, some element of it is greater than $l - \epsilon$, say, $a_{n_0} > l - \epsilon$. Therefore, since $a_{n_0} \leq a_{n_0+1} \leq a_{n_0+2} \leq \cdots$, we conclude that $a_n > l - \epsilon$ for every $n \geq n_0$. Thus, for $n \geq n_0$, we have

$$l - \epsilon < a_n \leq l < l + \epsilon,$$

as we wished to show.   □

The previous theorem, together with the definition of convergence, assures that if a bounded sequence is monotonic from a certain term on, then it will be convergent. We explore this comment by revisiting the last two examples.

Example 7.22 Let a be a positive real number. If $(a_n)_{n\geq 1}$ is given by $a_n = \sqrt[n]{a}$ for every $n \geq 1$, then $a_n \xrightarrow{n} 1$.

Proof Assume $a > 1$ (the case $0 < a < 1$ can be dealt with as in the Example 7.19). Then, $a_1 > a_2 > a_3 > \cdots > 1$, and the Bolzano-Weierstrass theorem guarantees the existence of $l = \lim_{n\to+\infty} a_n \geq 1$. Item (a) of Proposition 7.15 gives $l \geq 1$, and item (b) guarantees that every subsequence of $(a_n)_{n\geq 1}$ also converges to l.

---

[3] After Bernard Bolzano and Karl Weierstrass, German mathematicians of the XIX century.

Therefore, $a_{k(k+1)} \xrightarrow{\,k\,} l$. Now, since $a_{k(k+1)} = \frac{\sqrt[k]{a}}{\sqrt[k+1]{a}}$, it follows from item (d) of Proposition 7.18 that

$$a_{k(k+1)} = \frac{\sqrt[k]{a}}{\sqrt[k+1]{a}} \xrightarrow{\,k\,} \frac{l}{l} = 1.$$

Therefore, $l = 1$.   □

Example 7.23 The sequence $(a_n)_{n \geq 1}$, given by $a_n = \sqrt[n]{n}$ for every $n \geq 1$, converges to 1.

Proof The initial terms of the sequence are $\sqrt{2}$, $\sqrt[3]{3}$, $\sqrt[4]{4}$, ..., and it is easy to directly show that $\sqrt{2} < \sqrt[3]{3}$ and $\sqrt[3]{3} > \sqrt[4]{4} > \sqrt[5]{5}$. Since $2^n \geq n^2$ for $n \geq 4$ (by induction, for instance), we get $a_2 \geq a_n > 1$ for $n \geq 4$, so that the sequence is bounded; hence, if we show that it is indeed decreasing from the third term on, its convergence will follow from Bolzano-Weierstrass theorem, with limit $l \geq 1$.

For what is left to do, for an integer $n > 2$ we have

$$\sqrt[n]{n} > \sqrt[n+1]{n+1} \Leftrightarrow n^{n+1} > (n+1)^n \Leftrightarrow n > \left(1 + \frac{1}{n}\right)^n.$$

Let's prove the last inequality above. For $n = 3$, it's immediate to check it numerically; for $n > 3$, it suffices to show that $\left(1 + \frac{1}{n}\right)^n < 3$. To this end, notice that

$$\left(1 + \frac{1}{n}\right)^n = 1 + \binom{n}{1}\frac{1}{n} + \binom{n}{2}\frac{1}{n^2} + \cdots + \binom{n}{n}\frac{1}{n^n}$$

and

$$\binom{n}{k}\frac{1}{n^k} = \frac{n!}{k!(n-k)!n^k} = \frac{1}{k!} \cdot \frac{n(n-1)\ldots(n-k+1)}{n^k} < \frac{1}{k!} \leq \frac{1}{2^{k-1}}.$$

Therefore,

$$\left(1 + \frac{1}{n}\right)^n = 1 + \binom{n}{1}\frac{1}{n} + \binom{n}{2}\frac{1}{n^2} + \cdots + \binom{n}{n}\frac{1}{n^n}$$

$$< 1 + 1 + \frac{1}{2} + \frac{1}{2^2} + \cdots + \frac{1}{2^{n-1}}$$

$$= 3 - \frac{1}{2^{n-1}} < 3.$$

In order to finish, we need to show that $l = 1$. To this end, first observe that the subsequence $a_{2k} = \sqrt[2k]{2k}$ also converges to l. On the other hand, $\sqrt[2k]{2k} = \sqrt[2k]{2} \cdot \sqrt{\sqrt[k]{k}}$, with $\sqrt[2k]{2} \xrightarrow{k} 1$. Now, it follows from Problem 2 that $\sqrt{\sqrt[k]{k}} \xrightarrow{k} \sqrt{l}$. Therefore, by applying item (b) of Proposition 7.18, we get

$$l = \lim_{k \to +\infty} \sqrt[2k]{2k} = \lim_{k \to +\infty} \sqrt[2k]{2} \cdot \lim_{k \to +\infty} \sqrt{\sqrt[k]{k}} = \sqrt{l},$$

$\square$

Sometimes, we have to show that a given sequence has at least a convergent subsequence (even if, as a whole, the sequence does not converge). In this sense, Theorem 7.25 below, also due to Weierstrass, provides a sufficient condition for the existence of such a subsequence. Before we state and prove it, we need to discuss an important auxiliary result, known as the **lemma of nested intervals**. In what follows if $I = [a, b]$, with $a, b \in \mathbb{R}$, we shall let $|I| = b - a$.

Lemma 7.24 *For* $n \in \mathbb{N}$, *let* $I_n = [a_n, b_n]$. *If* $I_1 \supset I_2 \supset I_3 \supset \dots$ *and* $\lim_{n \to +\infty} |I_n| = 0$, *then there exists a unique* $l \in \mathbb{R}$ *such that* $\bigcap_{n \geq 1} I_n = \{l\}$.

Proof First of all, note that the intersection of the intervals I_n, if not empty, has a single element; indeed, if there existed reals $a < b$ in such an intersection, we would have $[a, b] \subset \bigcap_{n \geq 1} I_n$; in particular, $[a, b] \subset I_n$ and, hence, $|I_n| \geq b - a$ for every $n \in \mathbb{N}$, thus contradicting the fact that $\lim_{n \to +\infty} |I_n| = 0$.

Secondly, the inclusions $I_1 \supset I_2 \supset I_3 \supset \dots$ give $a_1 \leq a_2 \leq a_3 \leq \dots \leq \dots \leq b_3 \leq b_2 \leq b_1$, and the Bolzano-Weierstrass theorem assures the existence of $l = \lim_{n \to +\infty} a_n$. We claim that $l \in \bigcap_{n \geq 1} I_n$.

Since $l = \sup\{a_n; n \in \mathbb{N}\}$, it follows that $a_n \leq l$ for every $n \in \mathbb{N}$. On the other hand, for a fixed $m \in \mathbb{N}$, we have $a_n \leq b_m$ for every $n \in \mathbb{N}$, and item (a) of Proposition 7.15 gives $l = \lim_{n \to +\infty} a_n \leq b_m$. However, since m was chosen arbitrarily, we have $l \leq b_m$ for every $m \in \mathbb{N}$. It thus follows that $l \in [a_m, b_m] = I_m$ for every $m \in \mathbb{N}$, as we wished to show. $\square$

We are now in position to prove Weierstrass theorem.

Theorem 7.25 (Weierstrass) *Every bounded sequence admits a convergent subsequence.*

Proof Let $(a_n)_{n \geq 1}$ be a given bounded sequence, and $I_0 = [a, b]$ be a closed and bounded interval containing all of its terms. One (possibly both) of the intervals $\left[a, \frac{a+b}{2}\right]$ and $\left[\frac{a+b}{2}, b\right]$, call it I_1, also contains infinitely many terms of the sequence. Do the same with I_1, obtaining a closed and bounded interval $I_2 \subset I_1$ such that $|I_2| = \frac{1}{2}|I_1|$ and I_2 contains infinitely many terms of the sequence $(a_n)_{n \geq 1}$. Proceeding inductively, we construct a nested sequence $I_1 \supset I_2 \supset I_3 \supset \dots$ of closed and bounded intervals, such that $|I_{k+1}| = \frac{1}{2}|I_k|$ and I_k contains infinitely many terms of the sequence $(a_n)_{n \geq 1}$, for every $k \geq 1$. Therefore, by the lemma of nested intervals, there exists $c \in \mathbb{R}$ such that $\bigcap_{k \geq 1} I_k = \{c\}$.

Now, choose $n_1 \in \mathbb{N}$ such that $a_{n_1} \in I_1$; then, after having chosen $n_j \in \mathbb{N}$ such that $a_{n_j} \in I_j$, choose $n_{j+1} \in \mathbb{N}$ such that $n_{j+1} > n_j$ and $a_{n_{j+1}} \in I_{j+1}$ (this is possible by the way the I_j's were defined). This way, we inductively construct a subsequence $(a_{n_k})_{k\geq 1}$ of $(a_n)_{n\geq 1}$, such that $a_{n_k} \in I_k$ for every $k \geq 1$. Since $|I_k| = \frac{1}{2^k}|I_0|$ and $a_{n_k}, c \in I_k$, we conclude that $|a_{n_k} - c| \leq \frac{1}{2^k}|I_0|$, for every $k \geq 1$; since $\frac{1}{2^k}|I_0| \xrightarrow{k} 0$, we conclude that $a_{n_k} \xrightarrow{k} c$. $\square$

The concept of convergent sequence gives a precise meaning to the geometric intuition that the terms of the given sequence come closer and closer to a certain real number (the limit of the sequence), as long as their indices increase. However, it is also reasonable to expect that, if the terms of a given sequence come close together, then the sequence should also converge. This is indeed the case and, in order to establish it, we start with the following

Definition 7.26 A sequence $(a_n)_{n\geq 1}$ is said to be a **Cauchy sequence** if, for every $\epsilon > 0$, there exists $n_0 \in \mathbb{N}$ such that

$$m, n > n_0 \Rightarrow |a_m - a_n| < \epsilon.$$

The fundamental result concerning Cauchy sequences is the content of the following

Theorem 7.27 *A sequence $(a_n)_{n\geq 1}$ is convergent if and only if is Cauchy.*

Proof Let $(a_n)_{n\geq 1}$ be a convergent sequence, with limit l. Given $\epsilon > 0$, the definition of convergence guarantees the existence of $n_0 \in \mathbb{N}$ such that

$$n > n_0 \Rightarrow |a_n - l| < \frac{\epsilon}{2}.$$

Hence, given naturals $m, n > n_0$, the triangle inequality gives

$$|a_m - a_n| \leq |a_m - l| + |a_n - l| < \frac{\epsilon}{2} + \frac{\epsilon}{2} = \epsilon,$$

and the sequence is Cauchy.

Conversely, let $(a_n)_{n\geq 1}$ be a Cauchy sequence. Then, there exists $n_0 \in \mathbb{N}$ such that $|a_m - a_n| < 1$ for $m, n > n_0$. In particular, $|a_m - a_{n_0+1}| < 1$ for every $m > n_0$, and the sequence has all of its terms contained in the set

$$\{a_1, a_2, \ldots, a_{n_0}\} \cup (a_{n_0+1} - 1, a_{n_0+1} + 1),$$

so that it is bounded. Hence, by the theorem of Bolzano-Weierstrass, the sequence $(a_n)_{n\geq 1}$ has a convergent subsequence, say, $a_{n_k} \xrightarrow{k} l$. Let us prove that, actually, $a_n \xrightarrow{n} l$.

Given $\epsilon > 0$, there exists $N_0 \in \mathbb{N}$ such that

$$n_k > N_0 \Rightarrow |a_{n_k} - l| < \frac{\epsilon}{2}.$$

On the other hand, since the sequence is Cauchy, there exists $N_1 \in \mathbb{N}$ such that

$$m, n > N_1 \Rightarrow |a_m - a_n| < \frac{\epsilon}{2}.$$

Letting $M = \max\{N_1, N_2\}$ and fixing $n_k > M$, we have, for $n > M$ and with the aid of the triangle inequality,

$$|a_n - l| \leq |a_n - a_{n_k}| + |a_{n_k} - l| < \frac{\epsilon}{2} + \frac{\epsilon}{2} = \epsilon,$$

as we wished to show. □

The coming example collects an interesting application of the above result.

Example 7.28 Let $(a_n)_{n \geq 1}$ be a sequence of real numbers such that

$$|a_{n+2} - a_{n+1}| \leq c|a_{n+1} - a_n|$$

for every $n \in \mathbb{N}$, where $0 < c < 1$ is a real constant. Show that this sequence is convergent.

Proof By Theorem 7.27, it suffices to show that $(a_n)_{n \geq 1}$ is a Cauchy sequence. To this end, iterating the inequality in the statement we get, for every $k \in \mathbb{N}$,

$$|a_{k+1} - a_k| \leq c^{k-1}|a_2 - a_1|.$$

Let n and p be given natural numbers. The above inequality, together with the triangle inequality, gives

$$|a_{n+p} - a_n| \leq \sum_{k=n}^{n+p-1} |a_{k+1} - a_k| \leq \sum_{k=n}^{n+p-1} c^{k-1}|a_2 - a_1|$$

$$= \left(\frac{c^{n-1} - c^{n+p}}{1 - c} \right) |a_2 - a_1|$$

$$< \frac{1}{1 - c} \cdot |a_2 - a_1|c^{n-1}.$$

We now note that, by Example 7.12, the last expression above tends to 0 when $n \to +\infty$. Therefore, given $\epsilon > 0$, there exists $n_0 \in \mathbb{N}$ such that $\frac{1}{1-c} \cdot |a_2 - a_1|c^{n-1} < \epsilon$ for every $n > n_0$. Hence, for $n > n_0$ and $p \in \mathbb{N}$, we get $|a_{n+p} - a_n| < \epsilon$, so that $(a_n)_{n \geq 1}$ is, indeed, a Cauchy sequence. □

Problems: Section 7.2

1. * Let $(a_n)_{n\geq 1}$ and $(b_n)_{n\geq 1}$ be convergent sequences of real numbers, with $\lim_{n\to+\infty} a_n = a$ and $\lim_{n\to+\infty} b_n = b$. Generalize item (a) of Proposition 7.15, showing that if $a_n \leq b_n$ for every $n \geq 1$, then $a \leq b$.

2. * Let $(a_n)_{n\geq 1}$ be a sequence of positive real numbers converging to $a > 0$. Show that $\sqrt[n]{a_n} \xrightarrow{n} \sqrt{a}$.

3. * Given $a \in \mathbb{R}$ such that $|a| > 1$, show that $\frac{a^n}{n!} \to 0$ when $n \to +\infty$.

4. Generalize the result of Example 7.12, showing that, given $k \in \mathbb{N}$ and $a \in \mathbb{R}$, with $|a| > 1$, we have $\frac{n^k}{a^n} \xrightarrow{n} 0$.

5. Let $(a_n)_{n\geq 1}$ and $(b_n)_{n\geq 1}$ be sequences of real numbers and, for each $n \in \mathbb{N}$, let $t_n \in [0, 1]$ be given. Denote by $(c_n)_{n\geq 1}$ the sequence defined by

$$c_n = (1 - t_n)a_n + t_n b_n,$$

for every $n \in \mathbb{N}$. If $a_n, b_n \xrightarrow{n} l$, show that $c_n \xrightarrow{n} l$.

6. * Prove the **squeezing theorem**: let $(a_n)_{n\geq 1}$, $(b_n)_{n\geq 1}$ and $(c_n)_{n\geq 1}$ be sequences of real numbers such that $a_n \leq b_n \leq c_n$, for every $n \in \mathbb{N}$. If $a_n, c_n \xrightarrow{n} l$, for some $l \in \mathbb{R}$, show that $b_n \xrightarrow{n} l$.

7. Compute the following limits:

   (a) $\lim_{n\to+\infty} \frac{n\sqrt{n}}{n^2+1}$.

   (b) $\lim_{n\to+\infty}(\sqrt{n^2 + an + b} - n)$, with $a, b \in \mathbb{R}$.

   (c) $\lim_{n\to+\infty} \sqrt[n]{1 + q^n}$, where $0 < q < 1$ is a real number.

   (d) $\lim_{n\to+\infty} \sqrt[n]{a^n + b^n}$, with a and b positive reals such that $a > b$.

8. * This problem extends the concept of limit of sequences to consider *infinite limits*. We say that a sequence $(a_n)_{n\geq 1}$ of real numbers converges to $+\infty$ (resp. $-\infty$) if, given $M > 0$, there exists $n_0 \in \mathbb{N}$ such that $n > n_0 \Rightarrow a_n > M$ (resp. $a_n < -M$). In this case, we denote $\lim_{n\to+\infty} a_n = +\infty$ (resp. $\lim_{n\to+\infty} a_n = -\infty$), or simply $a_n \xrightarrow{n} +\infty$ (resp. $a_n \xrightarrow{n} -\infty$). With respect to this concept, and given sequences $(a_n)_{n\geq 1}$ and $(b_n)_{n\geq 1}$ of real numbers, do the following items:

   (a) If $a_n \xrightarrow{n} \pm\infty$ and $(b_n)_{n\geq 1}$ is bounded, then $a_n + b_n \xrightarrow{n} \pm\infty$.

   (b) If $a_n \xrightarrow{n} \pm\infty$ and $b_n \geq c > 0$ (resp. $b_n \leq c < 0$) for every $n \geq 1$, then $a_n b_n \xrightarrow{n} \pm\infty$ (resp. $a_n b_n \xrightarrow{n} \mp\infty$).

   (c) If $b_n \xrightarrow{n} +\infty$ and there exists $c > 0$ such that $a_n \geq cb_n$ (resp. $a_n \leq -cb_n$) for every $n \geq 1$, then $a_n \xrightarrow{n} \pm\infty$.

9. Let q be a real number and $(a_n)_{n\geq 1}$ be the sequence defined by $a_n = q^n$. If $q > 1$, show that $a_n \xrightarrow{n} +\infty$. If $q < -1$, show that $(a_n)_{n\geq 1}$ does not converge to either $+\infty$ or $-\infty$.

10. * Given positive reals a and q, with $q < 1$, show that $a^n q^{2^n} \xrightarrow{n} 0$.

11. (IMO shortlist) Let $(a_n)_{n \geq 1}$ be a sequence of real numbers such that, for every $m, n \in \mathbb{N}$, we have

$$|a_m - a_n| \leq \frac{2mn}{m^2 + n^2}.$$

Show that the sequence is constant.

12. Let $(a_n)_{n \geq 1}$ be the sequence defined by $a_1 = 1$, $a_2 = \sqrt{1 + 1}$, $a_3 = \sqrt{1 + \sqrt{2}}$,

$a_4 = \sqrt{1 + \sqrt{1 + \sqrt{2}}}$, $a_5 = \sqrt{1 + \sqrt{1 + \sqrt{1 + \sqrt{2}}}}$, ... Show that $(a_n)_{n \geq 1}$
is convergent and compute its limit.

13. (Austrian-Polish) Let $(a_n)_{n \geq 1}$ be a sequence of positive reals, such that

$$a_{k+2} = \sqrt{a_{k+1}} + \sqrt{a_k},$$

for every $k \geq 1$. Prove that the sequence converges and compute its limit.

14. Let $n > 1$ be a given integer and $t_0, t_1, \ldots, t_n$ be given real numbers, such that $t_0 + t_1 + \cdots + t_n = 0$. Prove that the sequence $(a_k)_{k \geq 1}$, defined by

$$a_k = t_0 \sqrt{k} + t_1 \sqrt{k + 1} + \cdots + t_n \sqrt{k + n}$$

converges to 0.

15. (Romania) The sequence $(x_n)_{n \geq 1}$ is such that $\sqrt{x_{n+1} + 2} \leq x_n \leq 2$, for every $n \geq 1$. Find all possible values of x_{1986}.

16. (Leningrad) Let $(a_n)_{n \geq 1}$ be a sequence of real numbers such that

$$|a_m + a_n - a_{m+n}| \leq \frac{1}{m + n},$$

for all $m, n \in \mathbb{N}$. Prove that the sequence is an AP.

17. (Bulgaria) For each $n \in \mathbb{N}$, let

$$a_n = \frac{n + 1}{2^{n+1}} \left(\frac{2^1}{1} + \frac{2^2}{2} + \cdots + \frac{2^n}{n} \right).$$

Prove that the sequence $(a_n)_{n \geq 2}$ is decreasing and convergent, and compute its limit.

18. (Romania) Let k be a fixed natural number and $(a_n)_{n \geq 1}$ be the sequence defined by

$$a_n = \sqrt{k + \sqrt{k + \cdots + \sqrt{k}}},$$

with exactly n square roots.

(a) Show that $(a_n)_{n\geq 1}$ is convergent.

(b) Show that, if k is odd, then the limit of the sequence is an irrational number.

(c) Find all natural values of k for which sequence converges to an integer.

19. For each positive real a, let the sequence $(a_n)_{n\geq 1}$ be defined by $a_1 = 1$ and

$$a_{k+1} = \frac{1}{2}\left(a_k + \frac{a}{a_k}\right),$$

for every integer $k \geq 1$. Prove that the sequence converges to $\sqrt{a}$.

20. (TT) The set of natural numbers is partitioned into m disjoint, infinite and nonconstant arithmetic progressions, of common ratios $d_1, d_2, \ldots, d_m$. Prove that

$$\frac{1}{d_1} + \frac{1}{d_2} + \cdots + \frac{1}{d_m} = 1.$$

21. (OIMU) Let c and α be positive real constants[4] and Q be a square in the plane. Prove that there doesn't exist a surjection $f : [0, 1] \rightarrow Q$ for which

$$d(f(x), f(y)) \leq c|x - y|^{\alpha+1/2}$$

for all $0 \leq x, y \leq 1$, where we let $d(A, B) = \overline{AB}$ denote the euclidean distance between the points A and B in the plane.

22. (Turkey) Let $(a_n)_{n\geq 1}$ be a sequence of integers such that $0 < a_{n+1} - a_n < \sqrt{a_n}$, for every natural n. Given real numbers x and y, with $0 \leq x < y \leq 1$, prove that there exist natural numbers m and n such that

$$x < \frac{a_m}{a_n} < y.$$

23. (IMO shortlist) Let $(a_n)_{n\geq 1}$ be a sequence of positive reals. Show that

$$1 + a_n > a_{n-1}\sqrt[n]{2}$$

for infinitely many values of n.

---

[4]Powers of a positive basis with real exponents will be defined in Sect. 10.7. For the time being, you may assume that α is a positive rational, if you will.

7.3 Kronecker's Lemma

In this section, we apply some of the ideas exposed so far in this chapter to study the important concept of *dense* set, as well as to present an interesting geometric application of it. We start by recalling the following definition.

Definition 7.29 Given an interval I, we say that a subset X of I is **dense** (in I) if, for every $a \in I$ and $\epsilon > 0$, it happens that

$$X \cap (a - \epsilon, a + \epsilon) \neq \emptyset.$$

Intuitively, the density of X in I means that X is spreaded all over I. Problem 4, page 206, shows that both $\mathbb{Q}$ and $\mathbb{R} \setminus \mathbb{Q}$ are dense in $\mathbb{R}$, whereas Problem 5, page 206, shows that the set of *dyadic rationals*, i.e., rational numbers of the form $\frac{n}{2^k}$, where $n, k \in \mathbb{Z}_+$ are such that $0 \leq n \leq 2^k$, is dense in $[0, 1]$.

A quite useful result on the density (in $\mathbb{R}$) of certain of its subsets is the content of Theorem 7.31 and Corollary 7.32, which are collectively known in the mathematical literature as **Kronecker's lemma**.[5] The proofs we present, albeit not being the simplest ones, have the advantage of deriving from a circle of ideas which are interesting in themselves. First of all, we need yet another definition.

Definition 7.30 A nonempty subset G of $\mathbb{R}$ is said to be an **additive subgroup** of $\mathbb{R}$ if, for all $x, y \in G$, we have $x - y \in G$.

Evidently, $\{0\}$, $\mathbb{Z}$, $\mathbb{Q}$ and $\mathbb{R}$ itself are additive subgroups of $\mathbb{R}$. For a less obvious example, given real numbers $x_1, \ldots, x_k$, it is immediate to verify (see Problem 1) that the set

$$G_{x_1,\ldots,x_k} = \{a_1 x_1 + \cdots + a_k x_k;\ a_1, \ldots, a_k \in \mathbb{Z}\} \tag{7.1}$$

is also an additive subgroup of $\mathbb{R}$.

Now, let G be an arbitrary additive subgroup of $\mathbb{R}$ and take $x \in G$. By the above definition, we have $0 = x - x \in G$. Thus, for $x, y \in G$, we also have $-y = 0 - y \in G$ and, hence, $x + y = x - (-y) \in G$; therefore, G is closed under the operation of addition.

Hence, if $\alpha \in G$, then $2\alpha = \alpha + \alpha \in G$; moreover, if $k\alpha \in G$, for some $k \in \mathbb{N}$, then $(k + 1)\alpha = k\alpha + \alpha \in G$, so that $m\alpha \in G$, for every $m \in \mathbb{N}$. Since $0\alpha = 0 \in G$ and $(-k)\alpha = -k\alpha \in G$ for every $k \in \mathbb{N}$, it follows that

$$G_\alpha = \{m\alpha;\ m \in \mathbb{Z}\} \subset G. \tag{7.2}$$

The coming result collects two central facts on additive subgroups of $\mathbb{R}$.

---

[5] After Leopold Kronecker, German mathematician of the XIX century.

Theorem 7.31 (Kronecker) *Let $G \neq \{0\}$ be an additive subgroup of $\mathbb{R}$, and let $G_+^* = G \cap \mathbb{R}_+^*$.*

(a) *If $\inf(G_+^*) = 0$, then G is dense in $\mathbb{R}$.*
(b) *If $\inf(G_+^*) = \alpha > 0$, then $\alpha \in G$ and $G = G_\alpha$.*

Proof

(a) Suppose $\inf(G_+^*) = 0$, let $a \in \mathbb{R}$ and $\epsilon > 0$ be given. We have to show that $G \cap (a - \epsilon, a + \epsilon) \neq \emptyset$. Since $x \in G \Leftrightarrow -x \in G$, it suffices to analyse the case $a \geq 0$. If $a - \epsilon < 0$, we have $0 \in G \cap (a - \epsilon, a + \epsilon)$ and there is nothing to do. Suppose, then, that $a - \epsilon \geq 0$.

The hypothesis $\inf(G_+^*) = 0$ guarantees the existence of $x \in G_+^*$ such that $x < 2\epsilon$. Letting m be the greatest nonnegative integer such that $mx \leq a - \epsilon$, we claim that $(m + 1)x \in G \cap (a - \epsilon, a + \epsilon)$. Indeed, if $(m + 1)x \geq a + \epsilon$, we would have

$$mx \leq a - \epsilon < a + \epsilon \leq (m + 1)x,$$

so that

$$x = (m + 1)x - mx \geq (a + \epsilon) - (a - \epsilon) = 2\epsilon,$$

thus contradicting the choice of x. Hence, $(m + 1)x \in (a - \epsilon, a + \epsilon) \cap G$.

(b) Suppose $\inf(G_+^*) = \alpha > 0$. We initially claim that $\alpha \in G_+^*$. By the sake of arriving at a contradiction, suppose that $\alpha \notin G_+^*$. Then, the definition of infimum of a set would assure the existence of elements $\beta, \gamma \in G_+^*$ such that $\alpha < \beta < \gamma < 2\alpha$. However, since G is an additive subgroup of $\mathbb{R}$, it would follow from here that $\gamma - \beta \in G_+^*$, with

$$0 < \gamma - \beta < 2\alpha - \alpha = \alpha.$$

This contradicts the fact that $\alpha = \inf(G_+^*)$, thus showing that $\alpha \in G_+^*$.

Now, take any $x \in G_+^*$ and let $q = \left\lfloor \frac{x}{\alpha} \right\rfloor$ and $r = x - \alpha \left\lfloor \frac{x}{\alpha} \right\rfloor$, so that $q \in \mathbb{Z}_+$, $0 \leq r < \alpha$ and $x = q\alpha + r$. If $r > 0$, then the fact that G is an additive subgroup of $\mathbb{R}$ would imply $r = x - q\alpha \in G_+^*$, with $0 < r < \alpha$. Since this contradicts the fact that $\alpha = \inf(G_+^*)$, we conclude that $r = 0$ and, hence, $x = q\alpha \in G_\alpha$.

Therefore,

$$G_+^* \subset \{n\alpha;\ n \in \mathbb{N}\}$$

and, since the opposite inclusion was already established in (7.2), we actually have $G_+^* = \{n\alpha;\ n \in \mathbb{N}\}$. Finally, since $G = G_+^* \cup \{0\} \cup G_-^*$, where $G_-^* = \{-x;\ x \in G_+^*\}$, it is immediate to see that

$$G = \{m\alpha;\ m \in \mathbb{Z}\} = G_\alpha.$$

$\square$

In the coming corollary, we stick to the notation set forth in (7.1).

Corollary 7.32 (Kronecker) *If α is an irrational number, then the additive subgroup $G_{1,\alpha} = \{m + n\alpha;\ m, n \in \mathbb{Z}\}$ of $\mathbb{R}$ is dense in $\mathbb{R}$.*

Proof By the sake of simplicity of notation, let $G = G_{1,\alpha}$. By the previous theorem, in order to prove that G is dense in $\mathbb{R}$, it suffices to prove that $\inf(G_+^*) = 0$.

If this was not the case, then, once more from Kronecker's theorem, there would exist a positive real number β such that $\inf(G_+^*) = \beta > 0$ and $G = G_\beta$. Since both α and $1 + \alpha$ belong to G, there would exist distinct, nonzero integers m and n for which

$$\alpha = n\beta \text{ and } 1 + \alpha = m\beta.$$

Now, since α is irrational, the first equality above would give $n \neq 0$ and $\beta = \frac{\alpha}{n} \notin \mathbb{Q}$. On the other hand, we would also have

$$(m - n)\beta = (1 + \alpha) - \alpha = 1,$$

so that $\beta = \frac{1}{m-n} \in \mathbb{Q}$.

We have then reached a contradiction, which came from the supposition that G is not dense in $\mathbb{R}$. Therefore, G is indeed dense in $\mathbb{R}$, as we wished to show. $\qquad\square$

Our next corollary refines the conclusion of the previous one.

Corollary 7.33 *If α is an irrational number, then the following sets are dense in $\mathbb{R}$:*

(a) $A = \{m + n\alpha;\ m, n \in \mathbb{Z} \text{ and } m < 0 < n\}$.
(b) $B = \{m + n\alpha;\ m, n \in \mathbb{Z} \text{ and } n < 0 < m\}$.

Proof Without any loss of generality, we can suppose that $\alpha > 0$. Let's prove item (a), the proof of item (b) being totally analogous. Given $a \in \mathbb{R}$ and $\epsilon > 0$, we want to establish the existence of $x \in A$ such that $a - \epsilon < x < a + \epsilon$. Suppose that $a - \epsilon \geq 0$ (the remaining cases are entirely analogous), and let $\delta = \min\{\alpha, 2\epsilon\} > 0$.

We first claim that there exist $m, n \in \mathbb{Z}$ such that $m < 0 < n$ and $m + n\alpha \in A \cap (0, \delta)$. By contradiction, suppose that

$$m + n\alpha \in A \cap (0, \delta) \Rightarrow n \leq 0.$$

Choose (by the former corollary) $x_0 = m_0 + n_0\alpha \in A \cap (0, \delta)$, with the greatest possible $n_0 \leq 0$. Since Kronecker's lemma guarantees that $A \cap (0, x_0)$ is infinite, we can take $x_1 = m_1 + n_1\alpha \in A \cap (0, x_0)$, with $n_1 < n_0$. Then,

$$x_0 - x_1 = (m_0 - m_1) + (n_0 - n_1)\alpha \in A \cap (0, x_0) \subset A \cap (0, \delta),$$

which is a contradiction, for, $n_0 - n_1 > 0$. Hence, we can choose $m + n\alpha \in A \cap (0, \delta)$, with $n > 0$. This being said, if it were $m \geq 0$, we would have $m + n\alpha \geq \alpha \geq \delta$, which is another contradiction. Therefore, $m < 0$ and, thus, $A \cap (0, \delta) \neq \emptyset$.

Now, take $x \in A \cap (0, \delta)$ and consider all numbers of the form kx, with $k \in \mathbb{Z}_+$. Letting k_0 be the greatest nonnegative integer for which $k_0 x \leq a - \epsilon$, we claim that $(k_0 + 1)x \in (a - \epsilon, a + \epsilon)$. Indeed, if it were $(k_0 + 1)x \geq a + \epsilon$, we would have

$$k_0 x \leq a - \epsilon < a + \epsilon \leq (k_0 + 1)x$$

and, hence,

$$\delta > x = (k_0 + 1)x - k_0 x \geq (a + \epsilon) - (a - \epsilon) = 2\epsilon.$$

This contradicts the choice of δ.   □

The discussion of the following example uses a few simple facts on plane Euclidean Geometry, for which we refer the reader to [4].

Example 7.34 (Brazil) Let Π be an euclidean plane and $f : \Pi \to \Pi$ be a function such that

$$d(P, Q) = 1 \Rightarrow d(f(P), f(Q)) = 1,$$

for all $P, Q \in \Pi$. Prove that f is an **isometry** of Π, i.e., prove that, for all $P, Q \in \Pi$, one has $d(P, Q) = d(f(P), f(Q))$, where $d(X, Y) = \overline{XY}$ stands for the euclidean distance between the points X and Y.

Proof For $P \in \Pi$, we let $f(P)$ be systematically denoted by P', so that $d(P, Q) = \overline{PQ}$ and $d(f(P), f(Q)) = \overline{P'Q'}$. Firstly, let's show that f must preserve segments of length $\sqrt{3}$.

Claim 1

$$\overline{PQ} = \sqrt{3} \Rightarrow \overline{P'Q'} = \sqrt{3}.$$

Indeed, given points P and Q in the plane for which $\overline{PQ} = \sqrt{3}$, let's construct points R and S such that both QRS and PRS are equilateral triangles whose side lengths are equal to 1 (cf. Fig. 7.1).

Fig. 7.1 $\overline{PQ} = \sqrt{3} \Rightarrow$
$\overline{P'Q'} = \sqrt{3}$

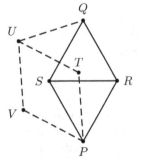

Fig. 7.2 $\overline{PQ} = 2 \Rightarrow$
$\overline{P'Q'} = 2$

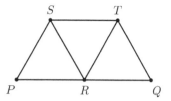

Let us counterclockwise rotate rhombus *PRQS* with center at *P*, until we get a rhombus *PTUV* such that $\overline{QU} = 1$.

Observe that the images P', R' and S' of P, R and S form an equilateral triangle of side lengths equal to 1. Since $\overline{Q'R'} = \overline{Q'S'} = 1$, it follows that $Q' = P'$ or $P'R'Q'S'$ is a rhombus congruent to *PRQS* (so that $\overline{P'Q'} = \sqrt{3}$). In order to discard the first possibility, it suffices to note that, if it were $P' = Q'$, then T', U' and V' would all be points on a circle centered at $P' = Q'$, while being vertices of an equilateral triangle of side lengths equal to 1, which is an absurd.

Claim 2 For every positive integer *n*, we have

$$\overline{PQ} = n \Rightarrow \overline{P'Q'} = n.$$

It suffices to establish the case $n = 2$, the general case being totally analogous. Let P and Q be such that $\overline{PQ} = 2$, and let R be the midpoint of PQ, such that $\overline{PR} = \overline{RQ} = 1$ (cf. Fig. 7.2). Let's consider points S and T such that PRS, RST and QRT are equilateral triangles of side lengths equal to 1, all situated on one of the half-planes determined by line $\overleftrightarrow{PQ}$. Using claim 1 twice, it's immediate that $\overline{P'Q'} = 2$.

Analogously, we can prove that

$$\overline{PQ} = n\sqrt{3} \Rightarrow \overline{P'Q'} = n\sqrt{3}.$$

Claim 3 $\overline{P'Q'} \geq \overline{PQ}$, for all points P and Q in the plane.

In order to prove this claim, let $\overline{PQ} = l$, such that l is neither a natural number nor a real number of the form $n\sqrt{3}$, for some $n \in \mathbb{N}$. By Corollary 7.33 (see, also, Problem 3), we can take sequences $(m_k)_{k \geq 1}$ e $(n_k)_{k \geq 1}$ of integers satisfying the following conditions:

  i. $m_k < 0 < n_k$, for every $k \geq 1$;
 ii. $\lim_{k \to +\infty} (m_k + n_k\sqrt{3}) = l$;
iii. $\max\{0, l - 1\} < m_k + n_k\sqrt{3} < l$, for every $k \geq 1$.

Let's first show that there exists a triangle of side lengths l, $-m_k$ and $n_k\sqrt{3}$. To this end, since $m_k + n_k\sqrt{3} < l$, we have $l + (-m_k) > n_k\sqrt{3}$; also, $l + (m_k + n_k\sqrt{3}) > 0$, which implies $l + n_k\sqrt{3} > -m_k$; finally, from $l - 1 < m_k + n_k\sqrt{3}$ we get

$$n_k\sqrt{3} + (-m_k) > n_k\sqrt{3} + m_k + 1 > l.$$

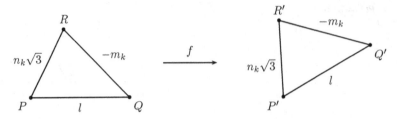

Fig. 7.3 $\overline{P'Q'} \geq \overline{PQ}$

Fig. 7.4 $\overline{PQ} = l \Rightarrow$
$\overline{P'Q'} = l$

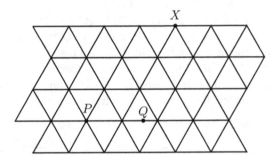

Therefore, triangle inequality assures the existence of a point $R \in \Pi \setminus \overleftrightarrow{PQ}$ such that $\overline{PR} = n_k\sqrt{3}$ and $\overline{RQ} = -m_k$; since we already have $\overline{PQ} = l$, there is nothing left to do.

It follows from what we did above (cf. Fig. 7.3) that $\overline{P'Q'} + \overline{R'Q'} \geq \overline{P'R'}$ or, which is the same, $\overline{P'Q'} \geq \overline{P'R'} - \overline{R'Q'}$. However, since $\overline{P'R'} = n_k\sqrt{3}$ and $\overline{R'Q'} = -m_k$, we get $\overline{P'Q'} \geq n_k\sqrt{3} + m_k$. On the other hand, since $n_k\sqrt{3} + m_k \xrightarrow{k} l$, it comes that $\overline{P'Q'} \geq l = \overline{PQ}$.

Let's consider again points P and Q in the plane, with $\overline{PQ} = l$. Tesselate the plane with equilateral triangles of side lengths all equal to 1, such that one of these triangles has one of its vertices at P and one of its sides on line $\overleftrightarrow{PQ}$ (cf. Fig. 7.4).

By what we did above, the images by f of the vertices of such a triangulation form the vertices of an analogous triangulation. On the other hand, if X is an arbitrary vertex of the original triangulation (see Fig. 7.4 once more), then $\overline{X'Q'} \geq \overline{XQ}$, for every point Q of the plane. Geometrically, this means that Q' doesn't belong to the interior of the disk centered at X' and having radius $\overline{XQ}$. However, since this is true for every vertex of the triangulation, we must necessarily have $\overline{P'Q'} = l$. □

Apart from the following problems, other interesting applications of Kronecker's lemma will appear in the context of continuity in Problem 14, page 264, as well as in Sect. 10.8 (cf. Examples 10.59 and 10.60), when we have at our disposal the concepts and elementary properties of logarithms.

Problems: Section 7.3

1. * Given real numbers $x_1, \ldots, x_k$, verify that the set

$$G_{x_1,\ldots,x_k} = \{a_1 x_1 + \cdots + a_k x_k;\ a_1, \ldots, a_k \in \mathbb{Z}\}$$

   is indeed an additive subgroup of $\mathbb{R}$.

   For the next problem, the reader might want to recall the definitions of *integer part* and *fractional part* of a real number, given in Problems 9 and 10, page 152.

2. The purpose of this problem is to give a direct proof of Corollary 7.32. To this end, given $\alpha \in \mathbb{R} \setminus \mathbb{Q}$, $a \in \mathbb{R}$ and $\epsilon > 0$, start by choosing $p \in \mathbb{N}$ such that $\frac{1}{p} < 2\epsilon$ and do the following items:

   (a) Show that at least two of the numbers $\{\alpha\}, \{2\alpha\}, \ldots, \{(p+1)\alpha\}$ belong to a single interval of the form $\left[\frac{k}{p}, \frac{k+1}{p}\right)$, for some integer k satisfying $0 \le k < p$.

   (b) Use the result of (a) to show that there exist $m', n' \in \mathbb{Z}$ such that $0 < m' + n'\alpha < \frac{1}{p}$.

   (c) Use the result of (b) to show that there exists $r \in \mathbb{Z}$ such that $r(m' + n'\alpha) \in (a - \epsilon, a + \epsilon)$.

3. * Given $\alpha, l \in \mathbb{R}$, with α irrational, show that there exist sequences $(m_k)_{k \ge 1}$ and $(n_k)_{k \ge 1}$ of integers satisfying the following conditions:

   (a) $m_k < 0 < n_k$, for every $k \ge 1$;
   (b) $\lim_{k \to +\infty}(m_k + n_k\alpha) = l$.

   For the next problem, we assume from the reader some familiarity with the basics of plane analytic geometry and vector algebra in the plane. We refer to Chaps. 6 and 8 of [4] for the necessary background.

4. A subset X of an euclidean plane Π is said to be *dense* in π provided X intersects every disk of Π. Now, let O be a fixed point in Π. A nonempty subset X of Π is said to be an *additive subgroup* of Π with respect to O provided the following condition is satisfied: for every $A, B \in X$, if $\overrightarrow{OA} - \overrightarrow{OB} = \overrightarrow{OC}$, then $C \in X$.

   (a) If $A_1, \ldots, A_n \in \Pi$ and

$$X_{A_1,\ldots,A_n} = \left\{ \sum_{k=1}^{n} m_k \overrightarrow{OA_k};\ m_k \in \mathbb{Z},\ \forall\ 1 \le k \le n \right\},$$

   show that $X_{A_1,\ldots,A_n}$ is an additive subgroup of Π with respect to O, which is not dense in it.

   (b) Choose A and B in Π such that $O \notin \overleftrightarrow{AB}$. If $\alpha \in \mathbb{R} \setminus \mathbb{Q}$ and

$$Y_{A,B} = \{m\overrightarrow{OA} + (n + p\alpha)\overrightarrow{OB};\ m, n, p \in \mathbb{Z}\},$$

show that $Y_{A,B}$ is an additive subgroup of Π with respect to O, which is not dense in it.

(c) Give an example of an additive subgroup X of Π with respect to O, different from Π itself and dense in it.

(d) Let X be an additive subgroup of Π with respect to O. Do the following items:

    i. Given a line r through O, look at it as a real line, with O representing 0. Prove that $X \cap r$ is either the empty set or an additive subgroup of r.

    ii. If there exist distinct lines r and s through O such that $X \cap r, X \cap s \neq \emptyset$, prove that X is dense in Π.

    iii. If X is not dense in Π and is not contained in a single line, prove that there exists a direction d in Π such that X is contained in the union of a family of equally spaced lines, all parallel to d.

Although the next problem does not really use Kronecker's lemma, this is the best place to put it.

5. Let α and β be positive irrationals such that $\frac{1}{\alpha} + \frac{1}{\beta} = 1$. Our purpose is to show that the sets

$$\{\lfloor k\alpha \rfloor;\ k \in \mathbb{N}\} \quad \text{and} \quad \{\lfloor k\beta \rfloor;\ k \in \mathbb{N}\}$$

form a partition of the natural numbers. To this end, do the following items:

(a) Show that if $\alpha > 1$ is irrational and $n \in \mathbb{N}$, then n is a term of the sequence $(\lfloor k\alpha \rfloor)_{k \geq 1}$ if and only if $\left\{\frac{n}{\alpha}\right\} > 1 - \frac{1}{\alpha}$.

(b) Given $n \in \mathbb{N}$, show that $\left\{\frac{n}{\alpha}\right\} + \left\{\frac{n}{\beta}\right\} = 1$.

(c) Conclude that either $\left\{\frac{n}{\alpha}\right\} > 1 - \frac{1}{\alpha}$ or $\left\{\frac{n}{\beta}\right\} > 1 - \frac{1}{\beta}$, but not both.

(d) Finish the proof.

In the notations above, one says that $(a_k)_{k \geq 1}$ and $(b_k)_{k \geq 1}$, given by $a_k = \lfloor k\alpha \rfloor$ and $b_k = \lfloor k\beta \rfloor$, are the **Beatty sequences**[6] corresponding to α and β.

7.4   Series of Real Numbers

Let $(a_n)_{n \geq 1}$ be a sequence of real numbers. By the **series**

$$\sum_{n=1}^{+\infty} a_n,$$

---

[6] After Samuel Beatty, Canadian mathematician of the XX century.

or simply $\sum_{n\geq 1} a_n$, we mean the sequence $(s_n)_{n\geq 1}$, where $s_n = a_1 + a_2 + \cdots + a_n$ for $n \geq 1$. The real number s_n is called the **n-th partial sum** of the series $\sum_{n\geq 1} a_n$, and we say that such a series **converges** to $s \in \mathbb{R}$ if the sequence $(s_n)_{n\geq 1}$ of its partial sums converges to s. In this case, we say that s is the **sum** of the series and write

$$\sum_{n\geq 1} a_n = s. \qquad (7.3)$$

In other words, whenever we write $\sum_{k\geq 1} a_k = s$, we will be saying that the terms of the sequence of finite sums $s_n = a_1 + a_2 + \cdots + a_n$ come closer and closer to the real number s, as long as $n \to +\infty$. It is in this sense that equality (7.3) must be thought of, as a limit.

We shall sometimes have a sequence $(a_n)_{n\geq 0}$ of reals, in which case the corresponding series will be denoted by $\sum_{n\geq 0} a_n$. We leave to the reader the (immediate) task of adapting the former and coming discussions to such a situation.

Our main interest in this section is to find out efficient criteria to decide whether a given series converges or not. If it doesn't converge, we shall say that it is a **divergent** series. Let's see two simple examples of divergent series.

Example 7.35 The series $\sum_{k\geq 1} k$ and $\sum_{k\geq 1}(-1)^k$ diverge.

Proof The first series diverges, for its n-th partial sum is $s_n = 1 + 2 + \cdots + n = \frac{n(n+1)}{2}$, so that $(s_n)_{n\geq 1}$ is a divergent sequence. In the second case, the n-th partial sum s_n of the given series is such that $s_n = 0$ if n is even and $s_n = -1$ if n is odd, so that $(s_n)_{n\geq 1}$ is also a divergent sequence.   □

Given a series $\sum_{n\geq 1} a_n$, we refer to a generic term a_n as the **general term** of the series. The following proposition gives a *necessary* condition on the general term of a series, if it is to converge.

Proposition 7.36 *If the series* $\sum_{k\geq 1} a_k$ *converges, then* $a_k \xrightarrow{n} 0$.

Proof Given $\epsilon > 0$, we want to prove that there exists $n_0 \in \mathbb{N}$ such that $n > n_0 \Rightarrow |a_n| < \epsilon$. To this end, let $l = \sum_{k\geq 1} a_k$. By the definition of convergence for series, there exists $n_0 \in \mathbb{N}$ such that

$$n \geq n_0 \Rightarrow |(a_1 + a_2 + \cdots + a_n) - l| < \frac{\epsilon}{2}.$$

Therefore, it follows from the triangle inequality that, for $n > n_0$, we have

$$|a_n| \leq |(a_1 + a_2 + \cdots + a_n) - l| + |l - (a_1 + a_2 + \cdots + a_{n-1})|$$
$$\leq \frac{\epsilon}{2} + \frac{\epsilon}{2} = \epsilon.$$

□

The converse to the above proposition is not true, namely, there are divergent series $\sum_{k\geq 1} a_k$ for which $a_k \xrightarrow{n} 0$. The classical example is that of the **harmonic series**, i.e., the series $\sum_{k\geq 1} \frac{1}{k}$, whose divergence is established in the coming example and will find further use in these notes.

Example 7.37 Given $n \in \mathbb{N}$, let m be the only natural number such that $2^m \leq n < 2^{m+1}$. Then,

$$\sum_{k=1}^{n} \frac{1}{k} \geq \frac{m}{2} + 1. \tag{7.4}$$

In particular, the harmonic series diverges.

Proof Note that, for every integer $k > 1$,

$$\frac{1}{2^{k-1}+1} + \frac{1}{2^{k-1}+2} + \cdots + \frac{1}{2^k} > \underbrace{\frac{1}{2^k} + \frac{1}{2^k} + \cdots + \frac{1}{2^k}}_{2^{k-1} \text{ times}} = \frac{1}{2}.$$

Hence,

$$1 + \frac{1}{2} + \sum_{j=3}^{n} \frac{1}{j} \geq 1 + \frac{1}{2} + \sum_{j=3}^{2^m} \frac{1}{j}$$

$$= 1 + \frac{1}{2} + \sum_{k=2}^{m} \left(\frac{1}{2^{k-1}+1} + \cdots + \frac{1}{2^k} \right)$$

$$> 1 + \frac{1}{2} + \sum_{k=2}^{m} \frac{1}{2} = 1 + \frac{m}{2}.$$

$\square$

In what comes next, we shall show that, for every rational $r > 1$, the series $\sum_{k\geq 1} \frac{1}{k^r}$ converges. To this end, we need to examine the convergence of a **geometric series**, i.e., of a series of the form

$$\sum_{k\geq 1} q^{k-1},$$

for a certain nonzero real number q. In this sense, we have the following important result.

Proposition 7.38 *Given $q \in \mathbb{R} \setminus \{0\}$, the geometric series $\sum_{k\geq 1} q^{k-1}$ converges if and only if $0 < |q| < 1$. Moreover, if this is so, then*

$$\sum_{k\geq 1} q^{k-1} = \frac{1}{1-q}.$$

Proof If $|q| \geq 1$, the geometric series diverges, since its general term q^{k-1} doesn't converge to 0. Suppose, then, that $0 < |q| < 1$, and let $s_n = 1 + q + \cdots + q^{n-1}$ be the n-th partial sum of the series. By the formula for the sum of the terms of a finite GP, we have

$$s_n = \frac{1-q^n}{1-q} = \frac{1}{1-q} - \frac{q^n}{1-q}.$$

Hence, in order to show that the series converges to $\frac{1}{1-q}$, it suffices to show that $q^n \xrightarrow{n} 0$. But this was done in Example 7.12. $\quad\square$

We can now discuss the promised example.

Example 7.39 If $r > 1$ is rational, then the series $\sum_{k \geq 1} \frac{1}{k^r}$ converges.[7]

Proof By the Bolzano-Weierstrass theorem, it suffices to show that the sequence $(s_n)_{n \geq 1}$ of the partial sums $s_n = \sum_{k=1}^{n} \frac{1}{k^r}$ is bounded. To this end, given $n \in \mathbb{N}$, take $m \in \mathbb{N}$ such that $2^m > n$. Then,

$$s_n \leq 1 + \left(\frac{1}{2^r} + \frac{1}{3^r} \right) + \cdots + \left(\frac{1}{(2^{m-1})^r} + \cdots + \frac{1}{(2^m - 1)^r} \right)$$

$$< 1 + 2 \cdot \frac{1}{2^r} + 4 \cdot \frac{1}{4^r} + \cdots + 2^{m-1} \cdot \frac{1}{2^{(m-1)r}}$$

$$< 1 + \frac{1}{2^{r-1}} + \frac{1}{4^{r-1}} + \cdots + \frac{1}{2^{(m-1)(r-1)}}$$

$$< \sum_{k \geq 0} \frac{1}{2^{(r-1)k}}.$$

However, since $r > 1$, we have $0 < \frac{1}{2^{r-1}} < 1$, and it follows from the previous proposition that

$$\sum_{k \geq 0} \frac{1}{2^{(r-1)k}} = \frac{2^{r-1}}{2^{r-1} - 1}.$$

Therefore, we conclude that $0 < s_n < \frac{2^{r-1}}{2^{r-1}-1}$ for every $n \in \mathbb{N}$, so that the sequence $(s_n)_{n \geq 1}$ is, indeed, bounded. $\quad\square$

Remark 7.40 For the sake of curiosity, we inform the reader that $\sum_{k \geq 1} \frac{1}{k^r} = \zeta(r)$, where $\zeta : (1, +\infty) \to \mathbb{R}$ stands for the famous **Riemann's zeta function**.[8]

---

[7] Had powers k^r, with $r > 0$ real, been defined (this will be done in Sect. 10.7), the reasoning presented in the proof would work equally well to show that the series $\sum_{k \geq 1} \frac{1}{k^r}$ converges.

[8] After Bernhard Riemann, German mathematician of the XIX century. For more on Riemann, see the footnote at page 354.

An elementary computation of $\zeta(2) = \frac{\pi^2}{6}$ will be hinted to at Problem 12, page 470. We also refer to Chap. 9 of [5]. Note, however, that for an *odd* natural number $m > 1$ the computation of the exact numerical value of $\zeta(m)$ is an open problem in Mathematics.

We finish our initial list of examples of series with an additional application of Bolzano-Weierstrass theorem to the convergence of series. In the coming example, we introduce one of the most important constants of Mathematics, the number e, which will play a major role in Sect. 10.7.

Example 7.41 The series $\sum_{k \geq 0} \frac{1}{k!}$ converges to an irrational number e, such that $2 < e < 3$. In symbols,

$$e = \sum_{k \geq 0} \frac{1}{k!}. \tag{7.5}$$

Proof Let $(s_n)_{n \geq 0}$ be the sequence of the partial sums of the given series, i.e.,

$$s_n = 1 + \frac{1}{1!} + \frac{1}{2!} + \cdots + \frac{1}{n!}.$$

For this sequence, we clearly have $1 = s_0 < s_1 < s_2 < \cdots$; on the other hand, since $k! > 2^{k-1}$ for every integer $k > 2$, we have, for an integer $n \geq 4$,

$$s_n = \sum_{k=0}^{n} \frac{1}{k!} < 1 + 1 + \frac{1}{2} + \frac{1}{6} + \sum_{k=4}^{n} \frac{1}{2^{k-1}} < \frac{8}{3} + \sum_{k \geq 4} \frac{1}{2^{k-1}} = \frac{35}{12},$$

where we used the formula for the sum of a geometric series in the last equality above. Hence, the sequence $(s_n)_{n \geq 0}$ is increasing and bounded, thus, convergent. Now, item (b) of Proposition 7.15, together with $s_2 = 2$ and $s_n < \frac{35}{12}$ for every integer $n \geq 4$, gives $2 < e < 3$ (for an approximation of e with five correct decimal places, see Problem 8).

We shall now show that e is irrational.[9] To this end, observe initially that, for natural numbers $1 < n < m$, we have

$$s_m - s_n = \sum_{k=n+1}^{m} \frac{1}{k!} < \sum_{k \geq n+1} \frac{1}{k!}$$

$$= \frac{1}{(n+1)!} \left(1 + \frac{1}{n+2} + \frac{1}{(n+2)(n+3)} + \cdots \right)$$

$$< \frac{1}{(n+1)!} \sum_{k \geq 0} \frac{1}{(n+2)^k} = \frac{1}{(n+1)!} \cdot \frac{n+2}{n+1}, \tag{7.6}$$

---

[9]For the reader's knowledge, we observe that the number e is, indeed, *transcendent*, i.e., it is not the root of a polynomial with rational coefficients. A proof of this fact is beyond the scope of these notes, and can be found in [11].

where, once more, we used the formula for the sum of a geometric series in the last equality above.

Therefore, yet for natural numbers $1 < n < m$, it follows from the above computations that

$$s_m = s_n + \sum_{k=n+1}^{m} \frac{1}{k!} < s_n + \frac{1}{(n+1)!} \cdot \frac{n+2}{n+1},$$

and item (a) of Proposition 7.15 gives

$$e = \lim_{m \to +\infty} s_m \leq s_n + \frac{1}{(n+1)!} \cdot \frac{n+2}{n+1}.$$

Thus,

$$s_n < e \leq s_n + \frac{1}{(n+1)!} \cdot \frac{n+2}{n+1}.$$

Multiplying this last inequality by $(n-1)!$ and noticing that $s_n = s_{n-1} + \frac{1}{n!}$, we conclude that

$$(n-1)!s_{n-1} + \frac{1}{n} < (n-1)!e \leq (n-1)!s_{n-1} + \frac{1}{n} + \frac{n+2}{n(n+1)^2},$$

Writing $t_n = n!s_n \in \mathbb{N}$ and observing that

$$\frac{1}{n} + \frac{n+2}{n(n+1)^2} = \frac{1}{n} + \frac{1}{(n+1)^2} + \frac{2}{n(n+1)^2}$$

$$\leq \frac{1}{3} + \frac{1}{4^2} + \frac{2}{3 \cdot 4^2} = \frac{21}{48} < 1,$$

for every integer $n > 2$, we finally arrive at the estimate

$$t_{n-1} < (n-1)!e < t_{n-1} + 1,$$

which is valid for every integer $n > 2$.

Now, suppose that $e = \frac{p}{q}$, with $p, q \in \mathbb{N}$. Making $n = q + 1 > 2$ in the above inequalities, we would get

$$t_q < (q-1)!p < t_q + 1,$$

with $t_q \in \mathbb{N}$. This is a contradiction. □

As an alternative to (7.5), we have the following result, which will be quite useful in Sect. 10.7.

Theorem 7.42 $e = \lim_{n \to +\infty} \left(1 + \frac{1}{n}\right)^n$.

Proof Let $a_n = \left(1 + \frac{1}{n}\right)^n$. Arguing as in the proof of Example 7.23, we get

$$a_n = 2 + \sum_{k=2}^{n} \binom{n}{k} \frac{1}{n^k}$$

$$= 2 + \sum_{k=2}^{n} \frac{1}{k!} \cdot \frac{n(n-1)(n-2)\ldots(n-k+1)}{n^k}$$

$$= 2 + \sum_{k=2}^{n} \frac{1}{k!} \left(1 - \frac{1}{n}\right)\left(1 - \frac{2}{n}\right)\ldots\left(1 - \frac{k-1}{n}\right)$$

$$< 2 + \sum_{k=2}^{n} \frac{1}{k!} = \sum_{k=0}^{n} \frac{1}{k!} < \sum_{k \geq 0} \frac{1}{k!} = e.$$

The above computations also give

$$a_n = 2 + \sum_{k=2}^{n} \frac{1}{k!} \left(1 - \frac{1}{n}\right)\left(1 - \frac{2}{n}\right)\ldots\left(1 - \frac{k-1}{n}\right)$$

$$< 2 + \sum_{k=2}^{n+1} \frac{1}{k!} \left(1 - \frac{1}{n+1}\right)\left(1 - \frac{2}{n+1}\right)\ldots\left(1 - \frac{k-1}{n+1}\right) = a_{n+1}.$$

Therefore, $(a_n)_{n \geq 1}$ is monotone increasing and bounded above, and hence there exists $l = \lim_{n \to +\infty} a_n$. In particular, since $a_n < e$ for every $n \in \mathbb{N}$, item (b) of Proposition 7.15 gives $l \leq e$.

Also from the computations above, given natural numbers $n > m \geq 2$ we can write

$$a_n = 2 + \sum_{k=2}^{n} \frac{1}{k!} \left(1 - \frac{1}{n}\right)\left(1 - \frac{2}{n}\right)\ldots\left(1 - \frac{k-1}{n}\right)$$

$$> 2 + \sum_{k=2}^{m} \frac{1}{k!} \left(1 - \frac{1}{n}\right)\left(1 - \frac{2}{n}\right)\ldots\left(1 - \frac{k-1}{n}\right)$$

$$> 2 + \sum_{k=2}^{m} \frac{1}{k!} \left(1 - \frac{1}{n}\right)\left(1 - \frac{2}{n}\right)\ldots\left(1 - \frac{m-1}{n}\right)$$

$$> \left(1 - \frac{1}{n}\right)\left(1 - \frac{2}{n}\right)\ldots\left(1 - \frac{m-1}{n}\right)\left(2 + \sum_{k=2}^{m} \frac{1}{k!}\right).$$

Therefore, it follows from Problem 1, page 218, we have

$$l = \lim_{n \to +\infty} a_n \geq \lim_{n \to +\infty} \left(1 - \frac{1}{n}\right)\left(1 - \frac{2}{n}\right)\cdots\left(1 - \frac{m-1}{n}\right)\left(2 + \sum_{k=2}^{m} \frac{1}{k!}\right)$$

$$= 2 + \sum_{k=2}^{m} \frac{1}{k!} = \sum_{k=0}^{m} \frac{1}{k!}.$$

However, since $m \in \mathbb{N}$ was arbitrarily chosen, we conclude that $l \geq \sum_{k=0}^{m} \frac{1}{k!}$ for every $m \in \mathbb{N}$. Then, letting $m \to +\infty$ and invoking again item (b) of Proposition 7.15, we finally obtain

$$l \geq \lim_{m \to +\infty} \sum_{k=0}^{m} \frac{1}{k!} = \sum_{k \geq 0} \frac{1}{k!} = e.$$

□

Back to the general development of the theory, the next result is the analogue, for series, of Proposition 7.18, and teaches us how to operate with convergent series.

Proposition 7.43 *If $\sum_{k \geq 1} a_k$ and $\sum_{k \geq 1} b_k$ are convergent series and c is a real number, then:*

(a) *$\sum_{k \geq 1} ca_k$ converges and $\sum_{k \geq 1} ca_k = c \sum_{k \geq 1} a_k$.*
(b) *$\sum_{k \geq 1}(a_k + b_k)$ converges and $\sum_{k \geq 1}(a_k + b_k) = \sum_{k \geq 1} a_k + \sum_{k \geq 1} b_k$.*

Proof

(a) If s_n denotes the n-th partial sum of the series $\sum_{k \geq 1} a_k$, then the n-th partial sum of the series $\sum_{k \geq 1} ca_k$ equals cs_n. Hence, according to item (a) of Proposition 7.18, $\sum_{k \geq 1} ca_k$ converges, and

$$\sum_{k \geq 1} ca_k = \lim_{n \to +\infty} cs_n = c \lim_{n \to +\infty} s_n = c \sum_{k \geq 1} a_k.$$

(b) If s_n and t_n are the n-th partial sums of the series $\sum_{k \geq 1} a_k$ and $\sum_{k \geq 1} b_k$, respectively, then the n-th partial sum of the series $\sum_{k \geq 1}(a_k + b_k)$ equals $s_n + t_n$. Therefore, item (b) of Proposition 7.18 assures the convergence of this last series, with

$$\sum_{k \geq 1}(a_k + b_k) = \lim_{n \to +\infty}(s_n + t_n) = \lim_{n \to +\infty} s_n + \lim_{n \to +\infty} t_n = \sum_{k \geq 1} a_k + \sum_{k \geq 1} b_k.$$

□

A quick analysis of the arguments presented in Examples 7.39 and 7.41 gives the following more general result, known as the **comparison test** for the convergence of series.

Proposition 7.44 *Let $(a_k)_{k\geq 1}$ and $(b_k)_{k\geq 1}$ be sequences of positive real numbers, such that $a_k \leq b_k$ for every $k \geq 1$. If the series $\sum_{k\geq 1} b_k$ converges, then so does the series $\sum_{k\geq 1} a_k$. Moreover,*

$$\sum_{k\geq 1} a_k \leq \sum_{k\geq 1} b_k.$$

Proof Letting $s_n = \sum_{k=1}^{n} a_k$ and $t_n = \sum_{k=1}^{n} b_k$, it follows from $0 < a_k \leq b_k$ that $0 < s_n \leq t_n$, for every $n \in \mathbb{N}$. Since the sequence $(t_n)_{n\geq 1}$ converges, it is bounded. Hence, the sequence $(s_n)_{n\geq 1}$ is monotonic and bounded, thus convergent, by Bolzano-Weierstrass theorem. To what is left to prove, it suffices to make $n \to +\infty$ in the inequality $s_n \leq t_n$ and apply the result of Problem 1, page 218.   □

Example 7.45 Is there a sequence $(a_k)_{k\geq 1}$ of positive real numbers such that both series $\sum_{k\geq 1} a_k$ and $\sum_{k\geq 1} \frac{1}{k^2 a_k}$ converge?

Solution Suppose there is such a series. Then, item (b) of Proposition 7.43, together with the inequality between the arithmetic and geometric means, would give us

$$\sum_{k\geq 1} a_k + \sum_{k\geq 1} \frac{1}{k^2 a_k} = \sum_{k\geq 1} \left(a_k + \frac{1}{k^2 a_k} \right) \geq \sum_{k\geq 1} 2\sqrt{a_k \cdot \frac{1}{k^2 a_k}} = \sum_{k\geq 1} \frac{2}{k}.$$

Therefore, by the comparison test for series, the harmonic series would be convergent, which is an absurd.   □

The coming example uses the theory of series to give a proof of the uncountability of $\mathbb{R}$.

Example 7.46 Problem 23, 172, shows that the family $\mathcal{F}$ of infinite subsets of $\mathbb{N}$ is uncountable. On the other hand, if $A = \{m_1 < m_2 < m_3 < \cdots\}$ is such a set, then the comparison test, together with the convergence of the geometric series $\sum_{j\geq 1} \frac{1}{2^j}$, guarantees the convergence of the series $\sum_{k\geq 1} \frac{1}{2^{m_k}}$.

Let $B = \{n_1 < n_2 < n_3 < \cdots\}$ be another infinite subset of $\mathbb{N}$. If we show that

$$\sum_{k\geq 1} \frac{1}{2^{m_k}} \neq \sum_{k\geq 1} \frac{1}{2^{n_k}},$$

then the correspondence $A \mapsto \sum_{k\geq 1} \frac{1}{2^{m_k}}$ defines an injection from $\mathcal{F}$ into $\mathbb{R}$, and this guarantees that $\mathbb{R}$ is uncountable. (Otherwise, by composing such a function with a bijection from $\mathbb{R}$ to $\mathbb{N}$, we would get an injection from $\mathcal{F}$ to $\mathbb{N}$, thus contradicting the uncountability of $\mathcal{F}$.)

What is left to do is quite similar to the proof of Example 4.12. Indeed, suppose we had

$$\sum_{k \geq 1} \frac{1}{2^{m_k}} = \sum_{k \geq 1} \frac{1}{2^{n_k}}.$$

Then,

$$\frac{1}{2^{m_1}} < \sum_{k \geq 1} \frac{1}{2^{m_k}} = \sum_{k \geq 1} \frac{1}{2^{n_k}} \leq \sum_{j \geq n_1} \frac{1}{2^j} = \frac{1}{2^{n_1 - 1}},$$

so that $m_1 \geq n_1$. By reversing the roles of the two series, we analogously conclude that $m_1 \leq n_1$, so that $m_1 = n_1$. Thus,

$$\sum_{k \geq 2} \frac{1}{2^{m_k}} = \sum_{k \geq 2} \frac{1}{2^{n_k}},$$

and a similar reasoning gives $m_2 = n_2$. Finally, by continuing this way, we get $m_k = n_k$ for every $k \geq 1$, so that $A = B$.

Back to the development of the theory, for a series $\sum_{k \geq 1} a_k$ with infinitely many positive and negative terms the results obtained so far say nothing about its convergence. We remedy this situation from now on, starting from the following

Definition 7.47 A series $\sum_{k \geq 1} a_k$ is said to be **absolutely convergent** provided the series $\sum_{k \geq 1} |a_k|$ converges.

The usefulness of the concept of absolutely convergent series stems from the coming proposition, as well as the subsequent example.

Proposition 7.48 *Every absolutely convergent series is convergent.*

Proof Let $\sum_{k \geq 1} a_k$ be an absolutely convergent series and, for each integer $n \geq 1$, let $s_n = a_1 + a_2 + \cdots + a_n$ and $t_n = |a_1| + |a_2| + \cdots + |a_n|$ be the n-th partial sums of the series $\sum_{k \geq 1} a_k$ and $\sum_{k \geq 1} |a_k|$. Given integers $m > n \geq 1$, we have

$$|s_m - s_n| = |a_{n+1} + a_{n+2} + \cdots + a_m|$$
$$\leq |a_{n+1}| + |a_{n+2}| + \cdots + |a_m|$$
$$= t_m - t_n.$$

Since $(t_n)_{n \geq 1}$ converges, it is a Cauchy sequence; hence, given $\epsilon > 0$, there exists $n_0 \in \mathbb{N}$ such that $m > n > n_0 \Rightarrow |t_m - t_n| < \epsilon$. With these ϵ and n_0, it follows from the above inequality that

$$m > n > n_0 \Rightarrow |s_m - s_n| \leq t_m - t_n < \epsilon,$$

and $(s_n)_{n\geq 1}$ is also a Cauchy sequence. Therefore, Theorem 7.27 guarantees the convergence of the sequence $(s_n)_{n\geq 1}$, as we wished to show.   □

The converse to the previous proposition is not valid, namely, there are convergent series which are not absolutely convergent. The classical example is a direct application of the coming result, which is known in the mathematical literature as the **Leibniz criterion**[10] for the convergence of **alternate series**.

Proposition 7.49 (Leibniz) *If* $(a_n)_{n\geq 1}$ *is a nonincreasing sequence of positive reals such that* $a_n \to 0$*, then the alternate series* $\sum_{k\geq 1}(-1)^{k-1}a_k$ *converges.*

Proof For each $n \in \mathbb{N}$, let $s_n = a_1 + a_2 + \cdots + a_n$. Condition $a_1 \geq a_2 \geq a_3 \geq \cdots > 0$ easily gives

$$s_1 \geq s_3 \geq s_5 \geq \cdots \geq s_6 \geq s_4 \geq s_2. \tag{7.7}$$

On the other hand, for each $m \in \mathbb{N}$ we have

$$|s_{2m-1} - s_{2m}| = a_{2m} \to 0,$$

which clearly guarantees, in conjunction with (7.7), that $(s_n)_{n\geq 1}$ is a Cauchy sequence. Therefore, $(s_n)_{n\geq 1}$ is convergent, as desired.   □

Example 7.50 The alternate series $\sum_{k\geq 1}\frac{(-1)^{k-1}}{k}$ converges, by a simple application of the former proposition (we shall compute its value in Problem 8, page 484). Nevertheless, the series formed by the absolute values of its terms (the harmonic series) diverges.

We now discuss quite a useful criterion for the convergence of series of nonzero real numbers. It is based on the *asymptotic behavior*[11] of the quotient of neighboring terms of the series, and is known as the **ratio test**.

Proposition 7.51 *Let* $(a_n)_{n\geq 1}$ *be a sequence of nonzero real numbers, such that* $\left|\frac{a_{n+1}}{a_n}\right| \to l$. *If* $l < 1$*, then the series* $\sum_{k\geq 1}a_k$ *is absolutely convergent; if* $l > 1$*, then the series* $\sum_{k\geq 1}a_k$ *is divergent.*

Proof Let's prove that the series $\sum_{k\geq 1}a_k$ is absolutely convergent if $l < 1$ (the proof of its divergence in case $l > 1$ is completely analogous).

---

[10]After Gottfried Wilhelm Leibniz, German mathematician and philosopher of the XVII century. Together with Sir Isaac Newton, Leibniz is considered to be one of the creators of the Differential and Integral Calculus. Up to this day, we still use some of the notations invented by Leibniz more than 300 years ago.

[11]In this context, this expression refers to the behavior of some expression that depends on $n \in \mathbb{N}$, when $n \to +\infty$.

Letting $l < 1$, we can take a real number q such that $l < q < 1$. The convergence $\frac{|a_{n+1}|}{|a_n|} \xrightarrow{n} l$ assures the existence of $n_0 \in \mathbb{N}$ such that

$$n \geq n_0 \Rightarrow \frac{|a_{n+1}|}{|a_n|} \leq q.$$

Hence, for $n \geq n_0$, we have

$$|a_n| = |a_{n_0}| \prod_{k=n_0}^{n-1} \frac{|a_{k+1}|}{|a_k|} \leq |a_{n_0}| q^{n-n_0}.$$

Thus, for $n \geq n_0$, the terms of the series $\sum_{k \geq 1} |a_k|$ are majorized by those of the series $\sum_{k \geq 1} |a_{n_0}| q^{n-n_0}$, which converges, by Propositions 7.43 and 7.38. Therefore, it follows from the comparison test that $\sum_{k \geq 1} |a_k|$ converges, which is the same as saying that $\sum_{k \geq 1} a_k$ is absolutely convergent. ☐

In the notations of the former proposition, we observe that, if $l = 1$, then the series $\sum_{k \geq 1} a_k$ may converge or diverge. Indeed, for $a_n = \frac{1}{n}$ we have

$$\frac{a_{k+1}}{a_k} = \frac{k}{k+1} \xrightarrow{k} 1,$$

albeit the series $\sum_{k \geq 1} \frac{1}{k}$ diverges; on the other hand, for $a_n = \frac{1}{n^2}$ we have

$$\frac{a_{k+1}}{a_k} = \frac{k^2}{(k+1)^2} \xrightarrow{k} 1,$$

while the series $\sum_{k \geq 1} \frac{1}{k^2}$ converges. On the positive side of things, we present the following

Example 7.52 Given a natural number m and a real number $q > 1$, explain whether the series $\sum_{k \geq 1} (-1)^{k-1} \frac{k^m}{q^k}$ converges or diverges.

Solution Letting $a_n = (-1)^{n-1} \frac{n^m}{q^n}$ for $n \geq 1$, we have

$$\left| \frac{a_{n+1}}{a_n} \right| = \frac{(n+1)^m}{q^{n+1}} \cdot \frac{q^n}{n^m} = \left(\frac{n+1}{n} \right)^m \cdot \frac{1}{q} \xrightarrow{n} \frac{1}{q} < 1.$$

Therefore, by the ration test, the given series is absolutely convergent, hence, convergent. ☐

We close this section by discussing the product of two absolutely convergent series.

Theorem 7.53 *Let $\sum_{i \geq 1} a_i$ and $\sum_{j \geq 1} b_j$ be absolutely convergent series. If*

$$c_k = \sum_{i+j=k} a_i b_j = \sum_{i=1}^{k-1} a_i b_{k-i} \tag{7.8}$$

for $k \geq 1$, then $\sum_{k \geq 1} c_k$ is absolutely convergent and such that

$$\sum_{k \geq 1} c_k = \left(\sum_{i \geq 1} a_i\right)\left(\sum_{j \geq 1} b_j\right).$$

Proof It suffices to show that, given $\epsilon > 0$, there exists $n_0 \in \mathbb{N}$ such that, for $n > n_0$, we have

$$\left|\sum_{k=1}^{2n} c_k - \left(\sum_{i \geq 1} a_i\right)\left(\sum_{j \geq 1} b_j\right)\right| < \epsilon \quad \text{and} \quad \left|\sum_{k=1}^{2n-1} c_k - \left(\sum_{i \geq 1} a_i\right)\left(\sum_{j \geq 1} b_j\right)\right| < \epsilon.$$

Let's guarantee the existence of $n_0 \in \mathbb{N}$ for which the first inequality above is true (the analysis of the validity of the second inequality is entirely analogous).

Given $n \in \mathbb{N}$, it follows from triangle inequality that

$$\left|\sum_{k=1}^{2n} c_k - \left(\sum_{i \geq 1} a_i\right)\left(\sum_{j \geq 1} b_j\right)\right| \leq \left|\sum_{k=1}^{2n}\left(\sum_{i+j=k} a_i b_j\right) - \left(\sum_{i=1}^{n} a_i\right)\left(\sum_{j=1}^{n} b_j\right)\right|$$

$$+ \left|\left(\sum_{i=1}^{n} a_i\right)\left(\sum_{j=1}^{n} b_j\right) - \left(\sum_{i \geq 1} a_i\right)\left(\sum_{j=1}^{n} b_j\right)\right|$$

$$+ \left|\left(\sum_{i \geq 1} a_i\right)\left(\sum_{j=1}^{n} b_j\right) - \left(\sum_{i \geq 1} a_i\right)\left(\sum_{j \geq 1} b_j\right)\right|.$$

Let A, B and C denote the first, second and third summands of the right hand side above, respectively, so that

$$B = \left|\sum_{i > n} a_i\right| \cdot \left|\sum_{j=1}^{n} b_j\right| \quad \text{and} \quad C = \left|\sum_{i \geq 1} a_i\right| \cdot \left|\sum_{j > n} b_j\right|.$$

The sequence $\left(\sum_{j=1}^{n} b_j\right)_{n \geq 1}$, being convergent, is bounded; therefore, there exists $M > 0$ such that $\left|\sum_{j=1}^{n} b_j\right| < M$, for every $n \geq 1$. On the other hand, since the series $\sum_{i \geq 1} a_i$ and $\sum_{j \geq 1} b_j$ converge, we have

$$\sum_{i > n} a_i = \sum_{i \geq 1} a_i - \sum_{i=1}^{n} a_i \xrightarrow{n} \sum_{i \geq 1} a_i - \sum_{i \geq 1} a_i = 0$$

and, analogously, $\sum_{j > n} b_j \xrightarrow{n} 0$. In order to estimate C, we can suppose, without loss of generality, that $\sum_{i \geq 1} a_i \neq 0$. Then, we can choose $n_1, n_2 \in \mathbb{N}$ such that

$$n > n_1 \Rightarrow \left|\sum_{i > n} a_i\right| < \frac{\epsilon}{3M} \quad \text{and} \quad n > n_2 \Rightarrow \left|\sum_{j > n} b_j\right| < \frac{\epsilon}{3\left|\sum_{i \geq 1} a_i\right|};$$

hence, for $n > \max\{n_1, n_2\}$, we have

$$B < \frac{\epsilon}{3M} \cdot M = \frac{\epsilon}{3} \quad \text{and} \quad C < \left| \sum_{i \geq 1} a_i \right| \cdot \frac{\epsilon}{3 \left| \sum_{i \geq 1} a_i \right|} = \frac{\epsilon}{3}.$$

In what concerns A, notice firstly that

$$A = \left| \sum_{k=1}^{2n} \left(\sum_{i+j=k} a_i b_j \right) - \left(\sum_{i=1}^{n} a_i \right) \left(\sum_{j=1}^{n} b_j \right) \right| = \left| \sum_{\substack{\max\{i,j\}>n \\ i+j \leq 2n}} a_i b_j \right|$$

$$\leq \sum_{\substack{\max\{i,j\}>n \\ i+j \leq 2n}} |a_i b_j| \leq \sum_{i=n+1}^{2n} \sum_{j \leq n} |a_i||b_j| + \sum_{j=n+1}^{2n} \sum_{i \leq n} |a_i||b_j|$$

$$\leq \left(\sum_{i=n+1}^{2n} |a_i| \right) \left(\sum_{j \geq 1} |b_j| \right) + \left(\sum_{j=n+1}^{2n} |b_j| \right) \left(\sum_{i \geq 1} |a_i| \right).$$

To estimate A, we can suppose that $\sum_{i \geq 1} |a_i| \neq 0$ and $\sum_{j \geq 1} |b_j| \neq 0$. Now, since the series $\sum_{i \geq 1} |a_i|$ and $\sum_{j \geq 1} |b_j|$ converge, the sequences $\left(\sum_{i=1}^{n} |a_i| \right)_{n \geq 1}$ and $\left(\sum_{j=1}^{n} |b_j| \right)_{n \geq 1}$ are Cauchy; therefore, there exist $n_3, n_4 \in \mathbb{N}$ such that

$$n > n_3 \Rightarrow \sum_{i=n+1}^{2n} |a_i| < \frac{\epsilon}{6 \sum_{j \geq 1} |b_j|} \quad \text{and} \quad n > n_4 \Rightarrow \sum_{j=n+1}^{2n} |b_j| < \frac{\epsilon}{6 \sum_{i \geq 1} |a_i|}.$$

Then, for $n > \max\{n_3, n_4\}$, we get

$$A \leq \frac{\epsilon}{6 \sum_{j \geq 1} |b_j|} \cdot \sum_{j \geq 1} |b_j| + \frac{\epsilon}{6 \sum_{i \geq 1} |a_i|} \cdot \sum_{i \geq 1} |a_i| = \frac{\epsilon}{3}.$$

Finally, by letting $n_0 = \max\{n_1, n_2, n_3, n_4\}$ and taking $n > n_0$, all of the previous estimates are valid, so that

$$\left| \sum_{k=1}^{2n} c_k - \left(\sum_{i \geq 1} a_i \right) \left(\sum_{j \geq 1} b_j \right) \right| \leq A + B + C < 3 \cdot \frac{\epsilon}{3} = \epsilon.$$

$\square$

Problems: Section 7.4

1. Let $(a_n)_{n\geq 1}$ be a sequence of positive real numbers defined by $a_1 = \frac{1}{2}$ and $a_{n+1} = a_n^2 + a_n$, for every $n \in \mathbb{N}$. Prove that $\sum_{k\geq 1} \frac{1}{a_k+1}$ converges and show that

$$\sum_{k\geq 1} \frac{1}{a_k + 1} = 2.$$

2. Sequence $(a_n)_{n\geq 1}$ is a nonconstant AP of nonzero real numbers. Prove that $\sum_{k\geq 1} \frac{1}{a_k a_{k+1}}$ converges and compute its sum.

3. Given a real number $a > 1$, prove that the series $\sum_{k\geq 1} \frac{2k-1}{a^k}$ converges and compute its sum.

4. Decide whether the series $\sum_{k\geq 1} \frac{1}{\sqrt{k+\sqrt{k^2-1}}}$ converges or diverges.

5. Prove that the series $\sum_{k>1000} \frac{1}{\sqrt{k^3-1000k^2}}$ converges.

6. Let $(a_n)_{n\geq 1}$ be an infinite, nonconstant AP of positive terms. Prove that:

   (a) $\sum_{k\geq 1} \frac{1}{a_k}$ diverges.

   (b) $\sum_{k\geq 1} \frac{1}{a_{2^k}}$ converges.

7. (NMC) Let A be a finite set of naturals, all of the form $2^a 3^b 5^c$, for some nonnegative integers a, b and c. Prove that

$$\sum_{x\in A} \frac{1}{x} < 4.$$

8. * The purpose of this problem is to show that $e \cong 2.71828$, with five correct decimal places. To this end, do the following items:

   (a) For every integer $n > 10$, show that

$$\frac{1}{10!} + \frac{1}{11!} + \cdots + \frac{1}{n!} < \frac{1}{10!} \left(1 + \frac{1}{11} + \frac{1}{11^2} + \cdots + \frac{1}{11^{n-10}} \right).$$

   (b) Use item (a), together with the fact that $10! > 2 \cdot 10^6$, to show that $0 < e - \sum_{k=1}^{10} \frac{1}{k!} < 10^{-6}$.

   (c) Conclude from (b) that 2.71828 approximates e with five correct decimal places.

9. * Given a sequence $(a_1, a_2, a_3, \ldots)$ of digits, prove that there exists a single $x \in \mathbb{R}$ such that, for a given $n \in \mathbb{N}$, we have

$$0 \leq x - \left(\frac{a_1}{10} + \frac{a_2}{10^2} + \cdots + \frac{a_k}{10^k} \right) \leq \frac{1}{10^n},$$

for every natural number $k \geq n$. In such a case (and as the reader is certainly used to), we write $x = 0.a_1a_2a_3\ldots$ and say that $0.a_1a_2a_3\ldots$ is the **decimal representation** of x.

10. Show that every real number $x \in (0, 1)$ admits a unique decimal expansion of the form $x = 0.a_1a_2a_3\ldots$, with $a_n \neq 0$ for infinitely many values of n. Then, use this fact to construct a surjective function $f : [0, 1] \rightarrow [0, 1] \times [0, 1]$.

11. Let $(a_n)_{n\geq 1}$ be a sequence of real numbers such that $\sum_{k\geq 1} a_k^2$ converges. Prove that, for every rational $\alpha > \frac{1}{2}$, $\sum_{k\geq 1} \frac{a_k}{k^\alpha}$ also converges.

12. Let $(a_n)_{n\geq 1}$ be a sequence of positive real numbers, such that the series $\sum_{k\geq 1} a_k$ converges. Prove that $\sum_{k\geq 1} \sqrt{a_k a_{k+1}}$ also converges.

13. Let $(F_n)_{n\geq 1}$ be the Fibonacci sequence, i.e., the sequence defined by $F_1 = 1$, $F_2 = 1$ and $F_{k+2} = F_{k+1} + F_k$, for every integer $k \geq 1$. Show that the series $\sum_{k\geq 1} \frac{1}{F_k}$ converges.

14. Let $(a_n)_{n\geq 1}$ be a sequence of positive reals such that $\sqrt[n]{a_n} \rightarrow l$.

    (a) If $l < 1$, show that the series $\sum_{k\geq 1} a_k$ converges.
    (b) If $l > 1$, show that the series $\sum_{k\geq 1} a_k$ diverges.
    (c) If $l = 1$, give examples showing that the series $\sum_{k\geq 1} a_k$ may converge or diverge.

    The convergence criterion given by the case $l < 1$ is known as the **root test**.

15. Let $\sum_{k\geq 1} a_k$ be an absolutely convergent series, with $\sum_{k\geq 1} a_k = 0$. Show that

$$\sum_{n\geq 1} \left(\frac{a_1}{(n-1)^2} + \frac{a_2}{(n-2)^2} + \cdots + \frac{a_{n-1}}{1^2} \right)$$

converges and compute the corresponding sum.

16. Prove that $e^{-1} = \sum_{k\geq 0} \frac{(-1)^k}{k!}$.

17. Let $\sum_{k\geq 1} a_k$ be an absolutely convergent series and $\varphi : \mathbb{N} \rightarrow \mathbb{N}$ be a bijection.

    (a) If $a_n \geq 0$ for every $n \in \mathbb{N}$, prove that $\sum_{k\geq 1} a_k = \sum_{k\geq 1} a_{\varphi(k)}$.
    (b) Write $a_n = a_n^+ - a_n^-$, where $a_n^+ = \max\{a_n, 0\}$ and $a_n^- = -\min\{a_n, 0\}$, for every $n \in \mathbb{N}$. Prove that $\sum_{k\geq 1} a_k^+$ and $\sum_{k\geq 1} a_k^-$ are both convergent.
    (c) Conclude that, in general, $\sum_{k\geq 1} a_k = \sum_{k\geq 1} a_{\varphi(k)}$.

    In view of item (c) above, we say that an absolutely convergent series is **commutatively convergent**.

    The convergence criterion for series stated in the next problem is due to N. Abel, and is known in the mathematical literature as **Abel's convergence test** or **Abel's convergence criterion**.

18. Let $(a_n)_{n\geq 1}$ and $(b_n)_{n\geq 1}$ be two sequences of real numbers satisfying the following conditions:

    (a) The sequence $(s_n)_{n\geq 1}$, defined by $s_n = a_1 + \cdots + a_n$ for every $n \in \mathbb{N}$, is bounded.

(b) $b_1 \geq b_2 \geq b_3 \geq \cdots > 0$ and $b_n \xrightarrow{n} 0$.

Prove that the series $\sum_{k \geq 1} a_k b_k$ converges.

19. Show that Abel's criterion implies Leibniz criterion.
20. Do the following items:

   (a) Given $a, h \in \mathbb{R}$, with $h \neq 2l\pi$ for every $l \in \mathbb{Z}$, show that

   $$\sum_{j=0}^{k} \sin(a + jh) = \frac{\sin\left(a + \frac{(k-1)h}{2}\right) \sin \frac{(k+1)h}{2}}{\sin \frac{h}{2}}.$$

   (b) Use Abel's criterion to show that $\sum_{k \geq 1} \frac{\sin k}{k}$ converges.

21. In a cartesian coordinate system centered at O, let $(A_n)_{n \geq 1}$ be the sequence of points such that $A_1 = (1, 0)$ and:

   (i) Triangle $OA_n A_{n+1}$ is rectangle at A_n and such that $\overline{A_n A_{n+1}} = 1$.
   (ii) Triangles $OA_{n+1}A_{n+2}$ and $OA_n A_{n+1}$ have disjoint interiors, for every $n \geq 1$.

   Prove that, when $n \to +\infty$, half-line $\overrightarrow{OA_n}$ revolves infinitely many times around O.

22. Let T_n be a right triangle whose side lengths are $4n^2$, $4n^4 - 1$ and $4n^4 + 1$, with $n \in \mathbb{N}$. Let α_n be the measure, in radians, of its internal angle opposite to the side of length $4n^2$. Show that

   $$\sum_{k \geq 1} \alpha_k = \frac{\pi}{2}.$$

Chapter 8
Continuous Functions

In this chapter we formalize the concept of continuous function, intuitively thought of as that of a function whose graph is a *curve without interruptions*. As a result of our discussion, we shall present sufficient criteria for a function to be continuous and, among other important results, shall show that every continuous function possesses the *intermediate value property*. Also, several interesting examples are scattered throughout the chapter.

8.1 The Concept of Continuity

Let's initially consider the function $f : \mathbb{R} \to \mathbb{R}$ given by $f(x) = \{x\}$ (the fractional part function), whose graph is sketched in Fig. 8.1. After a quick look at it, we would certainly feel comfortable in saying that such a graph is *discontinuous*, for, it presents several (actually, infinitely many) *jumps*. Also, note that this apparently doesn't happen with the graph of the function $g : \mathbb{R} \to \mathbb{R}$ given by $g(x) = x^2$, which should surely be called *continuous*.

It does emerge the question of how to find a reasonable criterion to identify the existence or absence of *jumps* in graphs, thus discerning between the two possibilities above. In order to develop some intuition on how to do it, let's restrict the domain of the function $x \mapsto \{x\}$ to the interval $[\frac{3}{4}, \frac{3}{2}]$. If we denote this new function still by f, then we easily conclude that

$$\mathrm{Im}(f) = \left[0, \frac{1}{2}\right] \cup \left[\frac{3}{4}, 1\right);$$

in particular, there exist values of y in the closed interval bounded by $f(\frac{3}{4})$ and $f(\frac{3}{2})$, for example $y = \frac{5}{8}$, such that no $x \in [\frac{3}{4}, \frac{3}{2}]$ satisfies $f(x) = y$.

© Springer International Publishing AG 2017
A. Caminha Muniz Neto, *An Excursion through Elementary Mathematics, Volume I*,
Problem Books in Mathematics, DOI 10.1007/978-3-319-53871-6_8

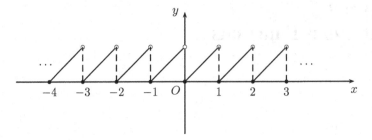

Fig. 8.1 Graph of $x \mapsto \{x\}$

On the other hand, it is easy to verify that this doesn't happen with the function g of the first paragraph. More precisely, let's fix an interval $[a, b] \subset \mathbb{R}$ and let $g([a, b])$ denote the image of the restriction of g to $[a, b]$, i.e.,

$$g([a, b]) = \{g(x) \in \mathbb{R}; \ x \in [a, b]\};$$

then, by setting $c = \max\{-a, b\}$, it's easy to show that

$$g([a, b]) = \begin{cases} \left[a^2, b^2\right], & \text{if } 0 \leq a < b \\ \left[0, c^2\right], & \text{if } a < 0 < b \ . \\ \left[b^2, a^2\right], & \text{if } a < b \leq 0 \end{cases}$$

In particular, for every y in the closed interval bounded by $g(a)$ and $g(b)$, there exists $x \in [a, b]$ ($x = \pm\sqrt{y}$, according to the case at hand), such that $g(x) = y$.

We are now in position to generalize the previous discussion with the following definition. To this end, in all that follows, unless we explicitly state otherwise, I denotes an interval and X a union of intervals of the real line.

Definition 8.1 A function $f : X \to \mathbb{R}$ has the **intermediate value property** (cf. Fig. 8.2) if, for every interval $[a, b] \subset X$ and every y_0 situated in the closed interval bounded by $f(a)$ and $f(b)$, there exists $x_0 \in [a, b]$ such that $y = f(x)$.

The previous discussion assures that the function $f(x) = \{x\}$, $x \in \mathbb{R}$, does not have the intermediate value property, while the function $g(x) = x^2$, $x \in \mathbb{R}$, has such a property. Therefore, it compels us to say that, if a function $f : X \to \mathbb{R}$ has the intermediate value property, then its graph should be *continuous*, i.e., with no *jumps*. Nevertheless, the reader can easily verify that the function $f : [0, 2] \to \mathbb{R}$, given by

$$f(x) = \begin{cases} x, & \text{if } 0 \leq x \leq 1 \\ x - \frac{1}{2}, & \text{if } 1 < x \leq 2 \end{cases} \ ,$$

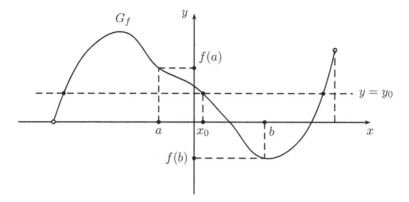

Fig. 8.2 The intermediate value property

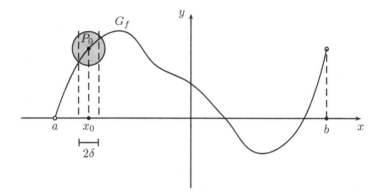

Fig. 8.3 Continuity from the correct viewpoint

does satisfy the intermediate value property, albeit its graph presents a *jump* at $x = 1$. Thus, the intermediate value property is not the correct way to formulate the concept of continuous function.

In order to adequately define the concept of continuous function, let's analyse the whole situation from another point of view. Let $f : (a, b) \to \mathbb{R}$ be a given function, $x_0 \in (a, b)$ be a fixed real number and $P_0(x_0, f(x_0)) \in G_f$. Also, let $x \in (a, b) \setminus \{x_0\}$ and $P(x, f(x)) \in G_f$. For the graph of f to be called *continuous* at x_0, our geometric intuition says that, the closer x is of x_0, the closer P should be of P_0.

More precisely, this *closedness* means (cf. Fig. 8.3) that, for a fixed, arbitrarily chosen *error* $r > 0$ for the position of the point P_0 (i.e., given an arbitrary disk $D(P_0; r)$, centered at P_0 and with radius r), we should have $P \in D$ whenever the abscissa x of P approximates x_0 within a sufficiently small error, say less than a certain $\delta > 0$ (one reads *delta*). In symbols, given an arbitrary $r > 0$, there must exist $\delta > 0$ such that

$$|x - x_0| < \delta \Rightarrow \overline{PP_0} < r. \tag{8.1}$$

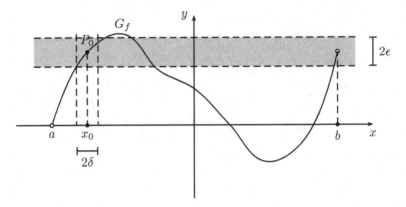

Fig. 8.4 Elaborating the correct definition of continuous function

Since every disk in the cartesian plane contains arbitrarily small rectangles with sides parallel to the axis, it is easy to show (cf. Problem 4) that the validity of the condition (8.1) is equivalent to the following alternative geometric description (cf. Fig. 8.4): given an arbitrary error $\epsilon > 0$ (one reads *epsilon*) for the *value of* f at x_0, i.e., given an arbitrary *horizontal strip*

$$\mathbb{R} \times (f(x_0) - \epsilon, f(x_0) + \epsilon)$$

of the cartesian plane, symmetric with respect to the point $P_0(x_0, f(x_0))$ (the gray strip in Fig. 8.4)), there must exist another error $\delta > 0$ such that, for every $x \in X$ satisfying $|x - x_0| < \delta$, the point $P(x, f(x))$ belongs to the gray strip. This being said, we can finally state the formal definition of continuity of a function (at a point).

Definition 8.2 A function $f : X \to \mathbb{R}$ is **continuous at the point** $x_0 \in X$ if the following condition is satisfied: given $\epsilon > 0$, there exists $\delta > 0$ such that

$$x \in X, \ |x - x_0| < \delta \Rightarrow |f(x) - f(x_0)| < \epsilon. \tag{8.2}$$

The function f is said to be **continuous** if it is continuous at every point $x_0 \in X$.

Example 8.3 Every constant function is continuous.

Proof Let c be a given real number and $f : \mathbb{R} \to \mathbb{R}$ be the function constant and equal to c. To show that f is continuous at $x_0 \in \mathbb{R}$, the definition of continuity asks that, given $\epsilon > 0$, we find $\delta > 0$ such that, for $x \in \mathbb{R}$, the validity of inequality $|x - x_0| < \delta$ implies that of $|f(x) - f(x_0)| < \epsilon$. However, since $|f(x) - f(x_0)| = |c - c| = 0$, inequality $|f(x) - f(x_0)| < \epsilon$ is always true, independently of any restriction on $|x - x_0|$. Yet in another way, taking δ equal to *any* positive real number, we will always have that $|x - x_0| < \delta \Rightarrow |f(x) - f(x_0)| < \epsilon$, for, the inequality $|f(x) - f(x_0)| < \epsilon$ cannot be false. □

In order to establish the continuity of other, less simple functions, we need to elaborate a little more the former definition, and we do this next.

Suppose we want to establish the continuity of $f : X \to \mathbb{R}$ at $x_0 \in X$. According to the given definition, we should assume that an error $\epsilon > 0$ for the value of $f(x_0)$ is given and, then, be capable of finding another error $\delta > 0$ for x_0 that turns true implication (8.2). In general, the following strategy is a good one:

(i) Starting from $x \in X$ subjected to an error $|x - x_0| < \delta$, we estimate *by excess* the error $|f(x) - f(x_0)|$ in terms of δ, obtaining an inequality of the type $|f(x) - f(x_0)| < E(\delta)$, where E represents a certain function of δ.

(ii) Then, we *impose* that such an error $E(\delta)$ doesn't surpass the desired error ϵ, thus finding the appropriate values of δ. Usually, this second step reduces to solving, for $\delta > 0$, inequality $E(\delta) \leq \epsilon$.

Once we execute the two steps above, if $\delta > 0$ satisfies $E(\delta) \leq \epsilon$, we will clearly have that

$$x \in X \text{ and } |x - x_0| < \delta \Rightarrow |f(x) - f(x_0)| < E(\delta) \leq \epsilon,$$

as we wished to show.

As a last remark, the coming examples will make it clear that, along the execution of items (i) and (ii), the function E will be generally implied.

Example 8.4 The modular function (cf. Example 6.60) is continuous.

Proof Given $x_0 \in \mathbb{R}$, if $|x - x_0| < \delta$, then triangle inequality gives

$$|f(x) - f(x_0)| = ||x| - |x_0|| \leq |x - x_0| < \delta.$$

Hence, if $\delta \leq \epsilon$, it will follow from the above that $|f(x) - f(x_0)| < \epsilon$ whenever $|x - x_0| < \delta$. □

Example 8.5 Let a and b be real numbers, with $a \neq 0$. If $f : \mathbb{R} \to \mathbb{R}$ is given by $f(x) = ax + b$, then f is continuous.

Proof Again, for a given $x_0 \in \mathbb{R}$, se $|x - x_0| < \delta$, we have

$$|f(x) - f(x_0)| = |(ax + b) - (ax_0 + b)| = |a||x - x_0| < |a|\delta.$$

Hence, if $|a|\delta \leq \epsilon$ (or, equivalently, $\delta < \frac{\epsilon}{|a|}$), we have $|f(x) - f(x_0)| < \epsilon$ whenever $|x - x_0| < \delta$. □

Example 8.6 The square root function $f : [0, +\infty) \to \mathbb{R}$, such that $f(x) = \sqrt{x}$ for $x \geq 0$, is continuous.

Proof If $x_0 = 0$ and $0 \leq x < \delta$, then $|f(x) - f(x_0)| = \sqrt{x} < \sqrt{\delta}$. Since $\sqrt{\delta} \leq \epsilon \Leftrightarrow \delta \leq \epsilon^2$, for $0 < \delta \leq \epsilon^2$ we have $|f(x) - f(x_0)| < \epsilon$ if $|x - 0| < \delta$.

Suppose, now, that $x_0 > 0$. Then, for $|x - x_0| < \delta$, we have

$$|f(x) - f(x_0)| = |\sqrt{x} - \sqrt{x_0}| = \frac{|x - x_0|}{\sqrt{x} + \sqrt{x_0}} \leq \frac{1}{\sqrt{x_0}}|x - x_0| < \frac{\delta}{\sqrt{x_0}}.$$

Fig. 8.5 $|\sin x| \leq |x|$

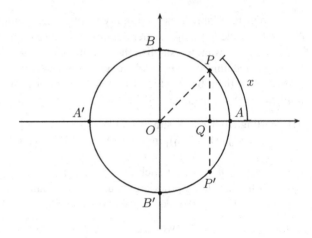

Note that $\frac{\delta}{\sqrt{x_0}} \leq \epsilon \Leftrightarrow \delta < \sqrt{x_0}\epsilon$. Therefore, taking $0 < \delta \leq \sqrt{x_0}\epsilon$, we conclude that $|x - x_0| < \delta \Rightarrow |f(x) - f(x_0)| < \epsilon$, as wished.   □

In what comes next, we establish the continuity of the sine and cosine functions. To this end, we shall need the following auxiliary result.

Lemma 8.7 *For every $x \in \mathbb{R}$, we have $|\sin x| \leq |x|$.*

Proof Firstly, let's show that $\sin x \leq x$ whenever $0 \leq x \leq \frac{\pi}{2}$. This inequality is obvious for $x = 0$; for $0 < x \leq \frac{\pi}{2}$, mark (cf. Fig. 8.5) point P in the first quadrant of the unit circle such that $\ell(\overset{\frown}{AP}) = x$.

Letting P' be the symmetric of P with respect to $\overleftrightarrow{AA'}$ and Q the intersection of $P'P$ and $A'A$, we have $\overline{P'P} = 2\,\overline{QP}$ and $\ell(\overset{\frown}{P'P}) = 2\ell(\overset{\frown}{AP})$. Since the length of every arc of a circle is greater than that of the corresponding chord, we get

$$2 \sin x = \overline{PP'} < \ell(\overset{\frown}{PP'}) = 2\ell(\overset{\frown}{AP}) = 2x.$$

Now, since $\sin(-x) = -\sin x$, it's immediate that $|\sin x| \leq |x|$ for $|x| \leq \frac{\pi}{2}$. Finally, for $|x| > \frac{\pi}{2}$, we have

$$|\sin x| \leq 1 < \frac{\pi}{2} < |x|.$$

□

Example 8.8 Sine and cosine functions are continuous.

Proof Fix $x_0 \in \mathbb{R}$. If $|x - x_0| < \delta$, it follows from the sum-to-product identities, together with the previous lemma, that

$$|\cos x - \cos x_0| = 2 \left|\sin\left(\frac{x - x_0}{2}\right)\right| \left|\sin\left(\frac{x + x_0}{2}\right)\right|$$

$$\leq 2 \left|\sin\left(\frac{x - x_0}{2}\right)\right| \leq 2 \left|\frac{x - x_0}{2}\right|$$

$$= |x - x_0| < \delta.$$

Hence, if $\delta \leq \epsilon$, then $|\cos x - \cos x_0| < \epsilon$ whenever $|x - x_0| < \delta$, as wished.

Finally, the reasoning for the sine function is completely analogous. $\square$

Later (cf. Proposition 8.19), we shall prove that if I is an interval and $f, g : I \to \mathbb{R}$ are continuous at $x_0 \in I$, then the functions $f \pm g, f \cdot g : I \to \mathbb{R}$ are also continuous at x_0. On the other hand, we shall show in Proposition 8.21 that, if g doesn't vanish in I, then the function $\frac{f}{g} : I \to \mathbb{R}$ is also continuous at x_0. For the time being, we assume these facts without proof, using them to present some additional examples of continuous functions.

Example 8.9 Given $a \in \mathbb{R}$ and $k \in \mathbb{N}$, the function $x \mapsto ax^k$ (from $\mathbb{R}$ to itself) can be seen as the product of a constant function (equal to a) and k copies of the identity function, $x \mapsto x$. Therefore, Examples 8.3 and 8.5, together with the discussion at the last paragraph (extended to the product of k continuous functions), assure that the function $x \mapsto ax^k$ is also continuous.

Applying once more the discussion at the paragraph that precedes this example (this time extended to the sum of a finite number of continuous functions), we conclude that every **polynomial function**, i.e., every function $f : \mathbb{R} \to \mathbb{R}$ of the form

$$f(x) = a_n x^n + a_{n-1} x^{n-1} + \cdots + a_1 x + a_0,$$

for some $n \in \mathbb{N}$ and $a_0, a_1, \ldots, a_n \in \mathbb{R}$, is continuous too.

Example 8.10 Given a polynomial function $f : \mathbb{R} \to \mathbb{R}$ (see the preceding example), we say that $x_0 \in \mathbb{R}$ is a **root** of f if $f(x_0) = 0$. It is possible to prove (cf. Chap. 15 of [5], for instance), that every polynomial function has a finite (possibly zero) number of real roots. Hence, if f and g are polynomial functions and $\mathcal{R}_g = \{x_1 < x_2 < \cdots < x_k\}$ is the set of real roots of g, the complement

$$\mathbb{R} \setminus \mathcal{R}_g = (-\infty, x_1) \cup (x_1, x_2) \cup \ldots \cup (x_{k-1}, x_k) \cup (x_k, +\infty)$$

is a finite union of open intervals. This way, we get the (well defined) **rational function**

$$\frac{f}{g} : \mathbb{R} \setminus \mathcal{R}_g \to \mathbb{R}.$$

The previous example, together with the discussion at the paragraph that precedes it, assures the continuity of rational functions, wherever they are defined.

Example 8.11 Function $f : \mathbb{R} \to \mathbb{R}$, given by

$$f(x) = \frac{x^4 \sin x - (x^2 + 1) \cos x}{x^8 + 2 + \cos x},$$

is continuous. Indeed, f is the quotient of the functions $g, h : \mathbb{R} \to \mathbb{R}$, such that $g(x) = x^4 \sin x - (x^2 + 1) \cos x$ and $h(x) = x^8 + 2 + \cos x$. The previous examples and discussions guarantee the continuity of g and h, and function h doesn't vanish, for $x^8 \geq 0$ and $2 + \cos x > 0$.

We close this section by examining the continuity of a composition. In this sense, the coming result is known as the **chain rule** for continuous functions.

Proposition 8.12 *If $X, Y \subset \mathbb{R}$ are unions of intervals and $f : X \to Y$, $g : Y \to \mathbb{R}$ are continuous functions, then $g \circ f : X \to \mathbb{R}$ is also continuous.*

Proof Let $x_0 \in X$ and $y_0 = f(x_0)$. Given $\epsilon > 0$, the continuity of g guarantees the existence of $\delta > 0$ such that

$$y \in Y, \ |y - y_0| < \delta \Rightarrow |g(y) - g(y_0)| < \epsilon. \tag{8.3}$$

On the other hand, the continuity of f assures the existence of $\delta' > 0$ such that

$$x \in X, \ |x - x_0| < \delta' \Rightarrow |\ \underbrace{f(x)}_{y} \ - \ \underbrace{f(x_0)}_{y_0}\ | < \delta.$$

Therefore, it follows from the above relations (with $y = f(x)$ in (8.3)) that

$$x \in X, \ |x - x_0| < \delta' \Rightarrow |f(x) - f(x_0)| < \delta \Rightarrow |g(f(x)) - g(f(x_0))| < \epsilon,$$

thus establishing the continuity of $g \circ f$ at x_0. $\square$

The two coming examples show typical applications of the chain rule for continuous functions.

Example 8.13 Function $f : \mathbb{R} \to \mathbb{R}$, given by $f(x) = \sin(x^2)$, is continuous. Indeed, if $g, h : \mathbb{R} \to \mathbb{R}$ are given by $g(x) = \sin x$ and $h(x) = x^2$, then g and h are continuous and $f = g \circ h$; hence, the chain rule assures the continuity of f.

Example 8.14 Function $f : \mathbb{R} \to \mathbb{R}$, given by

$$f(x) = \sqrt{\frac{x^2 |\sin(x^2)| + 1}{x^4 + x^2 + 2 + \sqrt{|\sin x|}}},$$

is continuous. Indeed, since f is the composition of the square root function (which is continuous) with the function $g : \mathbb{R} \to \mathbb{R}$ given by $g(x) = \frac{x^2 |\sin(x^2)| + 1}{x^4 + x^2 + 2 + \sqrt{|\sin x|}}$, the chain rule assures that it suffices to establish the continuity of g. Now, g is the quotient of the functions $x \mapsto x^2 |\sin(x^2)| + 1$ and $x \mapsto x^4 + x^2 + 2 + \sqrt{|\sin x|}$, which are continuous by the previous examples, together with the chain rule. Finally, observe that both the numerator and denominator of $g(x)$ are positive.

Let us finish this section with a result on the set of points of discontinuity of monotonic functions.

Proposition 8.15 *If $I \subset \mathbb{R}$ be an interval and $f : I \to \mathbb{R}$ is an nondecreasing function. then the set of points of discontinuity of f is finite or countably infinite.*

Proof Let

$$D = \{x_0 \in I; x_0 \text{ is not an endpoint of } I \text{ and } f \text{ is discontinuous at } x_0\}.$$

For each $x_0 \in D$, let $m_{x_0} = \sup\{f(x); x \in I \text{ and } x < x_0\}$ and $M_{x_0} = \inf\{f(x); x \in I \text{ and } x > x_0\}$. Problem 13 guarantees that $m_{x_0} < M_{x_0}$. Now, Problem 4, page 206 guarantees that we can choose $r_{x_0} \in \mathbb{Q} \cap (m_{x_0}, M_{x_0})$. On the other hand, if $x_0 < y_0$ are both in D, the fact that f is nondecreasing gives

$$M_{x_0} \le f\left(\frac{x_0 + y_0}{2}\right) \le m_{y_0},$$

so that the intervals (m_{x_0}, M_{x_0}) and (m_{y_0}, M_{y_0}) are disjoint.

Hence, the correspondence $x_0 \mapsto r_{x_0}$ defines an injective function from D into $\mathbb{Q}$, and it follows from Problem 21, page 172, that D is finite or countably infinite.   □

In the notations of the statement of the previous proposition, we call the reader's attention to the fact that Problem 14 will guarantee that every countably infinite subset of I is the set of points of discontinuity of a nonincreasing function.

Problems: Section 8.1

1. Prove that:

    (a) The function $f : \mathbb{R} \setminus \{0\} \to \mathbb{R}$, given by $f(x) = \frac{1}{x}$, has the intermediate value property and is continuous, but the function $f : \mathbb{R} \to \mathbb{R}$ given by

    $$f(x) = \begin{cases} \frac{1}{x}, & \text{if } x \ne 0 \\ 0, & \text{if } x = 0 \end{cases}$$

    doesn't have such property.
    (b) The function $f : \mathbb{R} \to \mathbb{R}$, given by

    $$f(x) = \begin{cases} x^2 + 1, & \text{if } x \ge 0 \\ -x, & \text{if } x < 0 \end{cases}$$

    doesn't have the intermediate value property.

2. Let $I \subset \mathbb{R}$ be an interval and $f : I \to \mathbb{R}$ be a continuous function. For a fixed $x_0 \in I$, let $g : I \to \mathbb{R}$ be the function given by

$$g(x) = \begin{cases} c, & \text{if } x = x_0 \\ f(x), & \text{if } x \neq x_0 \end{cases}.$$

Show that g is continuous if and only if $c = f(0)$, i.e., if and only if $g = f$.

3. In each of the following items, find out whether there exists a real value of c that turns the function $f : \mathbb{R} \to \mathbb{R}$ into a continuous one. Justify your answers.

(a) $f(x) = \begin{cases} 3x - 2, & \text{if } x < 0 \\ c, & \text{if } x = 0 \\ -2, & \text{if } x > 0 \end{cases}.$

(c) $f(x) = \begin{cases} 3x - 2, & \text{if } x < 0 \\ c, & \text{if } x = 0 \\ x^2, & \text{if } x > 0 \end{cases}.$

(b) $f(x) = \begin{cases} x \cos x, & \text{if } x \neq 0 \\ c, & \text{if } x = 0 \end{cases}.$

(d) $f(x) = \begin{cases} \frac{1}{x}, & \text{if } x \neq 0 \\ c, & \text{if } x = 0 \end{cases}.$

4. * Prove that every function continuous in the sense of relation (8.1) is continuous in the sense of Definition 8.2, and vice-versa.

5. * Let $D = \mathbb{R} \setminus \{\pi/2 + k\pi; \, k \in \mathbb{Z}\}$. Use the discussion at the paragraph that precedes Example 8.9 to establish the continuity of the tangent function,

$$\tan : D \longrightarrow \mathbb{R}$$
$$x \longmapsto \tan x.$$

6. If $f : X \to \mathbb{R}$ is a continuous function, explain why the function $|f| : X \to \mathbb{R}$ is also continuous.

7. Use the results of this section to establish the continuity of the function $f :$ $[-1, +\infty) \to \mathbb{R}$, given by $f(x) = \frac{\cos\sqrt{x+1}}{x^3 + 2}$.

8. Let $n > 1$ be a natural number and $f : [0, +\infty) \to \mathbb{R}$ be the n-th root function, $f(x) = \sqrt[n]{x}$ for $x \geq 0$.

   (a) Show that f is continuous at $x_0 = 0$.
   (b) If $x_0 > 0$ and $x \geq 0$, show that

$$|f(x) - f(x_0)| = \frac{|x - x_0|}{(\sqrt[n]{x})^{n-1} + (\sqrt[n]{x})^{n-2}\sqrt[n]{x_0} + \cdots + (\sqrt[n]{x_0})^{n-1}}$$

$$< \frac{1}{(\sqrt[n]{x_0})^{n-1}} \cdot |x - x_0|.$$

   (c) Use the result of (b) to conclude that f is continuous at x_0.

9. Use the results of this section to establish the continuity of the function $f : \mathbb{R} \to \mathbb{R}$, given by $f(x) = \sqrt[3]{\frac{x^4 - 2x^3 + 1}{x^2 + 1}}$.

10. Justify the continuity of the function $f : \mathbb{R} \to \mathbb{R}$, given by

$$f(x) = \begin{cases} 0, & \text{if } x = 0 \\ x \sin \frac{1}{x}, & \text{if } x \neq 0 \end{cases}.$$

11. * The **Dirichlet function**[1] is the function $f : [0, 1] \to \mathbb{R}$ such that

$$f(x) = \begin{cases} 0 \text{ if } x \notin \mathbb{Q} \\ 1 \text{ if } x \in \mathbb{Q} \end{cases}.$$

Prove that f is discontinuous at every $x \in [0, 1]$.
12. Let $f : [0, 1] \to \mathbb{R}$ be the function given by

$$f(x) = \begin{cases} 0, & \text{if } x \notin \mathbb{Q} \\ \frac{1}{n}, & \text{if } x = \frac{m}{n} \text{ with } m \in \mathbb{Z}_+, n \in \mathbb{N} \text{ and } \gcd(m, n) = 1 \end{cases}.$$

Prove that f is discontinuous at every rational number and continuous at every irrational number[2] of the interval $[0, 1]$.
13. Let $I \subset \mathbb{R}$ be an interval and $f : I \to \mathbb{R}$ be a nondecreasing function. If $x_0 \in I$ is not an endpoint of I, then

$$\sup\{f(x); x \in I \text{ and } x < x_0\} \leq \inf\{f(x); x \in I \text{ and } x > x_0\},$$

with equality if and only if f is continuous at x_0.
14. Let $D = \{x_1, x_2, x_3, \ldots\}$ be an arbitrary countably infinite set of reals. Let $f : \mathbb{R} \to \mathbb{R}$ be given by

$$f(x) = \sum_{\substack{n \in \mathbb{N}; \\ x_n < x}} \frac{1}{2^n},$$

where the sum is interpreted as being equal to 0 if $\{n \in \mathbb{N}; x_n < x\} = \emptyset$. Prove that:

(a) f is well defined and nondecreasing.
(b) The set of points of discontinuity of f is D.

---

[1] After Gustav Lejeune Dirichlet, German mathematician of the XIX century. Dirichlet made several important contributions to Mathematics, notably to Analysis and Number Theory. Even though the most famous of these is perhaps Dirichlet's theorem for primes in arithmetic progressions (about which we shall have more to say on [5]), Dirichlet made very important contributions to the theory of *Fourier Series*, which we will tangentially touch upon in Chap. 11.
[2] It is possible to prove that, given an interval I, there doesn't exist a function $f : I \to \mathbb{R}$ continuous at (every point of) $I \cap \mathbb{Q}$ and discontinuous at $I \cap (\mathbb{R} \setminus \mathbb{Q})$. A proof can be found in [1].

8.2   Sequential Continuity

In this section we relate the concepts of limit of a convergent sequence and continuity of a function. The main result is the coming one, which will be used several times in all that follows and assures that continuous functions are exactly those that preserve convergent sequences.

In all that follows, I denotes an interval of the real line.

Theorem 8.16 *A function $f : I \rightarrow \mathbb{R}$ is continuous if and only if the following condition is satisfied: for every $a \in I$ and every sequence $(a_n)_{n\geq 1}$ of elements of I, we have*

$$\lim_{n\to+\infty} a_n = a \Rightarrow \lim_{n\to+\infty} f(a_n) = f(a).$$

Proof Let's first assume that f is continuous. Then, given $a \in I$ and $\epsilon > 0$, there exists $\delta > 0$ such that

$$x \in I, \ |x - a| < \delta \Rightarrow |f(x) - f(a)| < \epsilon.$$

Now, let $(a_n)_{n\geq 1}$ be a sequence of elements of I converging to a. Then, there exists $n_0 \in \mathbb{N}$ such that

$$n > n_0 \Rightarrow |a_n - a| < \delta.$$

Taking together the two conditions above, we get

$$n > n_0 \Rightarrow |a_n - a| < \delta \Rightarrow |f(a_n) - f(a)| < \epsilon,$$

which is the same as saying that $\lim_{n\to+\infty} f(a_n) = f(a)$.

Conversely, suppose that f is not continuous at $a \in I$. Then, the definition of continuity guarantees the existence of $\epsilon > 0$ such that, for every $\delta > 0$, we have $|f(x) - f(a)| \geq \epsilon$ for some $x_\delta \in I$ satisfying $|x_\delta - a| < \delta$. In particular, taking $\delta = \frac{1}{n}$, with $n \in \mathbb{N}$, and writing $a_n = x_{1/n}$, we get

$$|a_n - a| < \frac{1}{n} \quad \text{and} \quad |f(a_n) - f(a)| \geq \epsilon.$$

Then, the above conditions assure that the sequence $(a_n)_{n\geq 1}$ thus constructed converges to a, whereas the sequence $(f(a_n))_{n\geq 1}$ doesn't converge to $f(a)$. $\square$

Let us illustrate the importance of the previous result with two examples, the first of which revisits Example 7.19 and shows that the idea used in the proof of Example 7.23 is quite natural.

Example 8.17 If $a > 0$ and $a_n = \sqrt[n]{a}$, then $a_n \xrightarrow{n} 1$.

Proof We make the proof in the case $a \geq 1$, the proof in the case $0 < a < 1$ being completely analogous. If $a \geq 1$, then $a_n \geq a_{n+1} \geq 1$, for every $n \in \mathbb{N}$. Hence, the sequence $(a_n)_{n \geq 1}$ is monotonic and bounded, thus convergent.

Setting $l = \lim_{n \to +\infty} a_n$, it follows from item (a) of Proposition 7.15 that $l \geq 1$. On the other hand, the continuity of the square root function (according to Example 8.6) and the previous result guarantee that

$$a_n \xrightarrow{n} l \Rightarrow \sqrt{a_n} \xrightarrow{n} \sqrt{l} \Rightarrow a_{2n} \xrightarrow{n} \sqrt{l},$$

where in the last implication we used the fact that $\sqrt{a_n} = a_{2n}$.

Since every subsequence of a convergent sequence converges to the same limit of the whole sequence, we have $a_{2n} \xrightarrow{n} l$. Therefore, the uniqueness of limits of convergent sequences gives $l = \sqrt{l}$, so that $l = 1$. □

Example 8.18 (Sweden) Find all continuous functions $f : \mathbb{R} \to \mathbb{R}$ such that $f(x) + f(x^2) = 0$, for every $x \in \mathbb{R}$.

Solution If f is any function satisfying the stated conditions, we shall show that f is identically zero.

Making $x = 0$ and $x = 1$, we get $f(0) = 0$ and $f(1) = 0$. Now, for $x > 0$ and applying the stated condition several times, we get

$$f(x) = -f(\sqrt{x}) = f(\sqrt[4]{x}) = -f(\sqrt[8]{x}) = \cdots .$$

In particular, $|f(\sqrt[2^n]{x})| = |f(x)|$ for every $n \in \mathbb{N}$.

Letting $a_n = \sqrt[2^n]{x}$, we already know that $a_n \xrightarrow{n} 1$. On the other hand, since $|f|$ is also continuous, Theorem 8.16 guarantees that

$$|f(a_n)| \xrightarrow{n} |f(1)| = 0.$$

However, since $|f(a_n)| = |f(x)|$ for every natural n, we also have $|f(a_n)| \xrightarrow{n} |f(x)|$. Therefore, the uniqueness of the limit of a convergent sequence gives $|f(x)| = 0$.

Finally, for $x < 0$ we have $x^2 > 0$. Hence, by what we've done above, $f(x^2) = 0$, so that $f(x) = f(x) + f(x^2) = 0$. □

Theorem 8.16 allows us to easily prove the claims made at the paragraph that precedes Example 8.9. We start by establishing the continuity of the sum and product of two continuous functions.

Proposition 8.19 *If $f, g : I \to \mathbb{R}$ are continuous at $x_0 \in I$, then $f \pm g, f \cdot g : I \to \mathbb{R}$ are also continuous at x_0.*

Proof Let $(a_n)_{n \geq 1}$ be a sequence in I, such that $\lim_{n \to +\infty} a_n = x_0$. According to Theorem 8.16, the continuity of f and g at x_0 gives

$$\lim_{n \to +\infty} f(a_n) = f(x_0) \quad \text{and} \quad \lim_{n \to +\infty} g(a_n) = g(x_0).$$

It thus follows from item (b) of Proposition 7.18 that

$$\lim_{n \to +\infty} (f \pm g)(a_n) = \lim_{n \to +\infty} (f(a_n) \pm g(a_n))$$
$$= \lim_{n \to +\infty} f(a_n) \pm \lim_{n \to +\infty} g(a_n)$$
$$= f(x_0) \pm g(x_0) = (f \pm g)(x_0).$$

Hence, once more from Theorem 8.16, we conclude that $f \pm g$ are continuous at x_0.

The reasoning for $f \cdot g$ uses item (c) of Proposition 7.18 and is totally similar to what we did above, so that it will be left to the reader. □

Before we look at the continuity of the quotient of two continuous functions, we need an auxiliary result, known as the **sign-preserving lemma** for continuous functions and which is important in itself.

Lemma 8.20 *Let $I \subset \mathbb{R}$ be an interval and $f : I \to \mathbb{R}$ be a continuous function. If $x_0 \in I$ is such that $f(x_0) > 0$ (resp. $f(x_0) < 0$), then there exists $\delta > 0$ such that*

$$x \in I, \ |x - x_0| < \delta \Rightarrow f(x) > \frac{f(x_0)}{2} \ (resp. \ f(x) < -\frac{f(x_0)}{2}).$$

In particular, f is still positive (resp. negative) in $I \cap (x_0 - \delta, x_0 + \delta)$.

Proof Let's do the proof in the case $f(x_0) > 0$, the proof in the other case being totally analogous.

The definition of continuity guarantees that, for $\epsilon = \frac{f(x_0)}{2} > 0$, there exists $\delta > 0$ such that

$$x \in I, \ |x - x_0| < \delta \Rightarrow |f(x) - f(x_0)| < \epsilon = \frac{f(x_0)}{2}.$$

On the other hand, this inequality implies

$$f(x) - f(x_0) > -\frac{f(x_0)}{2}$$

or, which is the same, $f(x) > \frac{f(x_0)}{2}$, for every $x \in I \cap (x_0 - \delta, x_0 + \delta)$. □

If $g : I \to \mathbb{R}$ is continuous at $x_0 \in I$ and such that $g(x_0) \neq 0$, the sign-preserving lemma assures the existence of $\delta > 0$ such that g doesn't vanish in the open interval $J = I \cap (x_0 - \delta, x_0 + \delta)$. Hence, by restricting g to J, if necessary, we can suppose, without loss of generality, that g doesn't vanish in I.

Proposition 8.21 *Let $f, g : I \to \mathbb{R}$ be continuous at $x_0 \in I$. If g doesn't vanish in I, then $\frac{f}{g} : I \to \mathbb{R}$ is continuous at x_0.*

Proof As in the proof of the previous proposition, if $(a_n)_{n \geq 1}$ is a sequence in I, such that $\lim_{n \to +\infty} a_n = x_0$, then

$$\lim_{n \to +\infty} f(a_n) = f(x_0) \quad \text{and} \quad \lim_{n \to +\infty} g(a_n) = g(x_0).$$

Now, item (e) of Proposition 7.18 gives

$$\lim_{n \to +\infty} \frac{f}{g}(a_n) = \lim_{n \to +\infty} \frac{f(a_n)}{g(a_n)} = \frac{\lim_{n \to +\infty} f(a_n)}{\lim_{n \to +\infty} g(a_n)} = \frac{f(x_0)}{g(x_0)} = \frac{f}{g}(x_0).$$

Invoking Theorem 8.16 once more, we conclude that $\frac{f}{g}$ is continuous at x_0. $\qquad\Box$

In what comes next, we introduce a *stronger* notion of continuity, and show that it is equivalent to the usual one for functions defined in intervals $I = [a, b] \subset \mathbb{R}$.

Definition 8.22 A function $f : I \to \mathbb{R}$ is **uniformly continuous** if the following condition is satisfied: for every $\epsilon > 0$, there exists $\delta > 0$ such that

$$x, y \in I, \ |x - y| < \delta \Rightarrow |f(x) - f(y)| < \epsilon. \tag{8.4}$$

In words, the difference between the definitions of continuous and uniformly continuous functions is that, for a uniformly continuous function and given $\epsilon > 0$, the number $\delta > 0$ whose existence is assured by the definition depends only on the chosen error $\epsilon > 0$, being the same *for all* $x, y \in I$; on the other hand, in the definition of continuity of a function at a point x_0 of its domain, such a δ depends on the error $\epsilon > 0$ as well as on x_0.

It is clear from the former definition that every uniformly continuous function is, in particular, continuous: it suffices to make $y = x_0$, where $x_0 \in I$ is arbitrarily chosen. On the other hand, Problems 4 and 6 bring examples of continuous functions which are not uniformly continuous.

We now pause to give an example of a uniformly continuous function which will be useful for later purposes (cf. Example 11.22). To this end, for $x \in \mathbb{R}$ we let $d(x)$ denote the distance of x to the nearest integer, so that

$$d(x) = \begin{cases} \{x\}, & \text{if } 0 \le \{x\} \le \frac{1}{2} \\ 1 - \{x\}, & \text{if } \frac{1}{2} \le \{x\} < 1 \end{cases}.$$

Example 8.23 In the above notations, let $a > 0$ be given. If $x, y \in \mathbb{R}$ are such that $|x - y| < \frac{1}{a}$, then $|d(ax) - d(ay)| \le a|x - y|$. In particular, $x \mapsto d(ax)$ is uniformly continuous.

Proof If $|x - y| < \frac{1}{a}$, then $|ax - ay| < 1$. In this case, the integers closest to ax and ay, if not equal, are consecutive. Assume this is so, and let n and $n + 1$ be the integers closest to ax and ay, respectively. Then,

$$n \le ax < n + \frac{1}{2} < ay < n + 1,$$

so that

$$\left| d(ax) - d(ay) \right| = \left| (n + 1 - ay) - (ax - n) \right| = \left| 2n + 1 - ay - ax \right|$$

$$= \left| \left(n + \frac{1}{2} - ay \right) + \left(n + \frac{1}{2} - ax \right) \right|$$

$$\leq \left| n + \frac{1}{2} - ay \right| + \left| n + \frac{1}{2} - ax \right|$$

$$= \left| ay - ax \right| = a|x - y|.$$

For the last part, given $\epsilon > 0$ take $\delta = \min\{\frac{1}{a}, \frac{\epsilon}{a}\} > 0$. Then,

$$|x - y| < \delta \Rightarrow |x - y| < \frac{1}{a} \Rightarrow \left| d(ax) - d(ay) \right| \leq a|x - y| < a \cdot \frac{\epsilon}{a} = \epsilon.$$

$\square$

The coming result proves that, for a function defined on a closed and bounded interval, the concepts of continuity and uniform continuity coincide.

Theorem 8.24 *Every continuous function $f : [a, b] \rightarrow \mathbb{R}$ is uniformly continuous.*

Proof By contradiction, suppose that f is continuous but (8.4) is not valid. Then, there exists $\epsilon > 0$ such that, for every $\delta > 0$, we can find $x, y \in [a, b]$ satisfying

$$|x - y| < \delta \quad \text{and} \quad |f(x) - f(y)| \geq \epsilon.$$

In particular, choosing $\delta = \frac{1}{n}$, with $n \in \mathbb{N}$, we conclude that there exist elements $x_n, y_n \in [a, b]$ such that

$$|x_n - y_n| < \frac{1}{n} \quad \text{and} \quad |f(x_n) - f(y_n)| \geq \epsilon. \tag{8.5}$$

Since $(x_n)_{n \geq 1}$ is a sequence in $[a, b]$, Weierstrass Theorem 7.25 guarantees the existence of a subsequence $(x_{n_k})_{k \geq 1}$ of $(x_n)_{n \geq 1}$, converging to some $x_0 \in [a, b]$. Then, it follows from triangle inequality that

$$|y_{n_k} - x_0| \leq |y_{n_k} - x_{n_k}| + |x_{n_k} - x_0| < \frac{1}{n_k} + |x_{n_k} - x_0| \xrightarrow{k} 0.$$

Hence, $(y_{n_k})_{k \geq 1}$ also converges to x_0, and Theorem 8.16, together with the continuity of $x \mapsto |x|$, gives

$$\lim_{k \rightarrow +\infty} |f(x_{n_k}) - f(y_{n_k})| = |f(x_0) - f(x_0)| = 0.$$

On the other hand, as a particular case of (8.5), we have $|f(x_{n_k}) - f(y_{n_k})| \geq \epsilon$, and such an inequality obviously contradicts the former limit.   □

The previous theorem admits the following important consequence.

Corollary 8.25 *Every continuous function $f : [a, b] \to \mathbb{R}$ is bounded.*

Proof Since such an f is uniformly continuous, given $\epsilon = 1$ there exists $\delta > 0$ such that

$$x, y \in [a, b], \ |x - y| < \delta \Rightarrow |f(x) - f(y)| < 1. \tag{8.6}$$

Now, choose real numbers $a = x_0 < x_1 < \cdots < x_k = b$ satisfying $x_i - x_{i-1} < \delta$ for $1 \leq i \leq k$, and let

$$M = \max\{|f(x_1)|, \ldots, |f(x_k)|\}. \tag{8.7}$$

For $x \in [a, b]$, there exists $1 \leq j \leq k$ such that $x \in [x_{j-1}, x_j]$; in particular, $|x - x_{j-1}| \leq |x_j - x_{j-1}| < \delta$. Then, the triangle inequality, together with (8.6) and (8.7), gives

$$|f(x)| \leq |f(x) - f(x_j)| + |f(x_j)| < 1 + |f(x_j)| \leq 1 + M.$$

Finally, since $x \in [a, b]$ as in the last paragraph was arbitrarily chosen, we get $|f(x)| \leq M + 1$ for every $x \in [a, b]$, so that f is bounded.   □

The last property of continuous functions we shall study in this section is also due to Weierstrass, and states that every continuous function $f : [a, b] \to \mathbb{R}$ assumes extreme values in the interval $[a, b]$.

Theorem 8.26 (Weierstrass) *If $f : [a, b] \to \mathbb{R}$ is continuous, then there exist $x_m, x_M \in [a, b]$ such that*

$$f(x_m) = \min\{f(x); \ x \in [a, b]\} \ and \ f(x_M) = \max\{f(x); \ x \in [a, b]\}.$$

Proof Let's prove the existence of x_m, that of x_M being totally analogous. Since f is bounded, it makes sense to let

$$m = \inf\{f(x); \ x \in [a, b]\}.$$

The definition of greatest lower bound assures (with the aid of Problem 10, page 207) the existence of a sequence $(x_n)_{n \geq 1}$ in $[a, b]$ such that $f(x_n) \to m$. Since every bounded sequence possesses a convergent subsequence, we can take a subsequence $(x_{n_k})_{k \geq 1}$ of $(x_n)_{n \geq 1}$, converging to a real number $x_0 \in [a, b]$. Then, Theorem 8.16 gives

$$\lim_{k \to +\infty} f(x_{n_k}) = f(x_0).$$

On the other hand, since the sequence $(f(x_{n_k}))_{k \geq 1}$ is a subsequence of $(f(x_n))_{n \geq 1}$ (which, in turn, converges to m), we conclude from Proposition 7.15, that $(f(x_{n_k}))_{k \geq 1}$ also converges to m. Therefore, the uniqueness of the limit of sequences gives $f(x_0) = m$. $\square$

Problems: Section 8.2

1. Find all continuous[3] functions $f : \mathbb{R} \to \mathbb{R}$ such that, for all $x, y \in \mathbb{R}$, we have $f(x + y) = f(x) + f(y)$.
2. (Romania) Let $f : \mathbb{R} \to \mathbb{R}$ be a surjective function satisfying the following property: for every divergent sequence $(a_n)_{n \geq 1}$, the sequence $(f(a_n))_{n \geq 1}$ is also divergent. Prove that f is bijective and that f^{-1} is continuous.
3. * Given an interval $I \subset \mathbb{R}$, we say that a function $f : I \to \mathbb{R}$ is **lipschitzian**[4] if there exists a constant $c > 0$ (called the **Lipschitz constant** of f) such that

$$|f(x) - f(y)| \leq c|x - y|,$$

   for all $x, y \in I$. For example, we saw in Example 8.8 that

$$| \sin x - \sin y | \leq |x - y| \quad \text{and} \quad | \cos x - \cos y | \leq |x - y|$$

   for all $x, y \in \mathbb{R}$, so that both sine and cosine functions are lipschitzian, with Lipschitz constant equal to 1. Prove that every lipschitzian function $f : I \to \mathbb{R}$ is uniformly continuous.
4. For an integer $n > 1$, show that the function $f : \mathbb{R} \to \mathbb{R}$, given by $f(x) = x^n$, is not uniformly continuous.
5. Let $I \subset \mathbb{R}$ be an interval and $f : I \to \mathbb{R}$ be a continuous function. Suppose that there exist $\epsilon > 0$ and sequences $(a_n)_{n \geq 1}$ and $(b_n)_{n \geq 1}$ in I, such that $|a_n - b_n| \xrightarrow{n} 0$, but $|f(a_n) - f(b_n)| \geq \epsilon$ for every $n \geq 1$. Show that f is not uniformly continuous.
6. Show that the function $f : (0, +\infty) \to \mathbb{R}$, given by $f(x) = \sin \frac{1}{x}$, is not uniformly continuous.
7. The purpose of this problem is to present another proof of Theorem 8.26. To this end, recall that Corollary 8.25 assures the boundedness of a continuous function $f : [a, b] \to \mathbb{R}$, so that there exist

$$m = \inf\{f(x); x \in [a, b]\} \quad \text{and} \quad M = \sup\{f(x); x \in [a, b]\}.$$

---

[3]It can be shown that the hypothesis of continuity is essential here. More precisely, there exists infinitely many functions $f : \mathbb{R} \to \mathbb{R}$ such that $f(x + y) = f(x) + f(y)$ for all $x, y \in \mathbb{R}$ but f is not linear. A proof of this fact is beyond the scope of these notes and can be found in [13].

[4]After Rudolf Lipschitz, German mathematician of the XIX century.

Now, do the following items:

(a) If $m \notin \text{Im}(f)$ and $g : [a,b] \to \mathbb{R}$ is the function given by $g(x) = \frac{1}{f(x)-m}$, for every $x \in [a,b]$, show that there exist $c > 0$ such that $g(x) \leq c$, for every $x \in [a,b]$.

(b) Yet under the hypothesis of item (a), conclude that $f(x) \geq m + \frac{1}{c}$ for every $x \in [a,b]$ and arrive at a contradiction.

(c) Argue in an analogous way to show that $M \in \text{Im}(f)$.

8. Given real numbers $a < b$, give an example of a continuous function $f : (a,b) \to \mathbb{R}$, which is unbounded from above and from below.

9. If $f : (a,b) \to \mathbb{R}$ is a continuous function and $P(x_0, y_0)$ is a point not belonging to the graph of f, we define the **distance** from P to G_f, denoted $d(P; G_f)$, by

$$d(P; G_f) = \inf\{\overline{A'P};\ A' \in G_f\}.$$

If $|y_0 - f(x_0)| < \min\{|x_0 - a|, |x_0 - b|\}$, prove that there exists $A \in G_f$ such that $d(P; G_f) = \overline{AP}$.

10. * Let $f : \mathbb{R} \to \mathbb{R}$ be the polynomial function given by

$$f(x) = a_n x^n + a_{n-1} x^{n-1} + \cdots + a_1 x + a_0$$

for $x \in \mathbb{R}$, where $a_0, a_1, \ldots, a_n \in \mathbb{R}$ and $a_n \neq 0$. If n is even and $a_n > 0$, prove that:

(a) $\frac{f(x)-a_0}{x^n} \geq a_n - \frac{|a_{n-1}|}{|x|} - \cdots - \frac{|a_1|}{|x|^{n-1}}$ if $x \neq 0$.

(b) $\frac{f(x)-a_0}{x^n} \geq a_n - \frac{1}{|x|}\sum_{j=1}^{n-1} |a_j|$ if $|x| \geq 1$.

(c) There exists $A > 0$ such that $f(x) > a_0$ for $|x| > A$.

(d) There exists $x_0 \in [-A, A]$ such that $f(x_0) = \min\{f(x); x \in \mathbb{R}\}$.

11. * The purpose of this problem is to prove the famous **Banach fixed point theorem**[5] in the real line. To this end, let $0 < c < 1$ and $f : \mathbb{R} \to \mathbb{R}$ be a function such that

$$|f(x) - f(y)| \leq c|x - y|, \quad \forall\ x, y \in \mathbb{R}. \tag{8.8}$$

Do the following items:

(a) Choose an arbitrary $x_0 \in \mathbb{R}$ and let $(x_n)_{n \geq 1}$ be such that $x_k = f(x_{k-1})$, for every $k \in \mathbb{N}$. Prove that $|x_{k+1} - x_k| \leq c|x_k - x_{k-1}|$ for every $k \in \mathbb{N}$.

(b) Use the result of Example 7.28 to conclude that $(x_n)_{n \geq 1}$ is convergent.

(c) If $\alpha = \lim_{n \to +\infty} x_n$, use the continuity of f (which is guaranteed by Problem 3) to show that α is the unique fixed point of f.

---

[5] After Stefan Banach, Polish mathematician of the XX century.

12. Give an example of a continuous function $f : \mathbb{R} \to \mathbb{R}$, without fixed points and such that $|f(x) - f(y)| < |x - y|$, for all $x, y \in \mathbb{R}$.
13. * Let $f : [0, 1] \to \mathbb{R}$ be a continuous function such that $f(r) \geq 0$ for every dyadic rational (cf. Problem 5, page 206) $r \in [0, 1]$. Prove that $f(x) \geq 0$, for every $x \in [0, 1]$.
14. (Berkeley) Let α be a given irrational number. Find all continuous functions $f : \mathbb{R} \to \mathbb{R}$ such that

$$f(x) = f(x + 1) = f(x + \alpha),$$

for every $x \in \mathbb{R}$.

8.3   The Intermediate Value Theorem

In this section, we show that a continuous function defined on an interval does satisfy the intermediate value property. Then, we present several interesting applications of this fact. We start by analysing the following special case, known as **Bolzano's theorem**.

Theorem 8.27 (Bolzano) *Let* $f : [a, b] \to \mathbb{R}$ *be a continuous function. If* $f(a)f(b) < 0$, *then there exists* $c \in (a, b)$ *such that* $f(c) = 0$.

Proof Suppose, without loss of generality, that $f(a) < 0 < f(b)$, and let

$$A = \{x \in [a, b]; f \text{ is negative on } [a, x]\}.$$

Since $a \in A$ by hypothesis, we have $A \neq \emptyset$. On the other hand, A is bounded (for, $A \subset [a, b]$), so that we can take $c = \sup A$. We shall show that $f(c) = 0$.

We initially assert that $c > a$. Indeed, since $f(a) < 0$, Lemma 8.20 guarantees the existence of $0 < \delta < b - a$ such that $f(x) < 0$ for $x \in [a, a + \delta)$; therefore, $c \geq a + \delta$.

Now, suppose $f(c) < 0$. Then, $c < b$ (for, $f(b) > 0$) and, once more from the sign-preserving lemma, there would exist $0 < \delta < b - c$ such that f is negative in $(c - \delta, c + \delta)$. However, since $c = \sup A$, we could take $d \in (c - \delta, c) \cap A$, so that $f < 0$ in $[a, d]$. Hence, we would have $f < 0$ in $[a, d] \cup (c - \delta, c + \delta) = [a, c + \delta)$, which is in contradiction to the fact that $c = \sup A$.

Finally, suppose $f(c) > 0$. Then (by invoking the sign-preserving lemma yet another time), there would exist $\delta > 0$ such that f is positive in $(c - \delta, c + \delta) \cap [a, b]$; in particular, $A \cap (c - \delta, c] = \emptyset$, and we would have $\sup A \leq c - \delta$, which is also a contradiction.

Therefore, the only left possibility is that $f(c) = 0$.   □

Example 8.28 If $f : \mathbb{R} \to \mathbb{R}$ is a polynomial function of the form

$$f(x) = a_n x^n + a_{n-1} x^{n-1} + \cdots + a_1 x + a_0,$$

with $a_0, a_1, \ldots, a_n \in \mathbb{R}$, $a_n \neq 0$ and n odd, then the image of f is $\mathbb{R}$. In particular, f has at least one real root.

Proof Given $d \in \mathbb{R}$, let $g(x) = f(x) - d$. Then, $g : \mathbb{R} \to \mathbb{R}$ is a polynomial function satisfying the same hypotheses as f and, for $c \in \mathbb{R}$, we have $f(c) = d$ if and only if $g(c) = 0$. This reasoning reduces the example to proving the existence of $c \in \mathbb{R}$ such that $f(c) = 0$. To this end, we shall use Bolzano's theorem.

Without loss of generality, let $a_n > 0$. For $x \neq 0$, several applications of the triangle inequality give

$$
\begin{aligned}
\frac{f(x)}{x^n} &= a_n + \frac{a_{n-1}}{x} + \cdots + \frac{a_1}{x^{n-1}} + \frac{a_0}{x^n} \\
&\geq a_n - \left| \frac{a_{n-1}}{x} + \cdots + \frac{a_1}{x^{n-1}} + \frac{a_0}{x^n} \right| \\
&\geq a_n - \left| \frac{a_{n-1}}{x} \right| - \cdots - \left| \frac{a_1}{x^{n-1}} \right| - \left| \frac{a_0}{x^n} \right| \\
&= a_n - \frac{|a_{n-1}|}{|x|} - \cdots - \frac{|a_1|}{|x|^{n-1}} - \frac{|a_0|}{|x|^n}.
\end{aligned}
$$

If $|x| \geq 1$, then $|x| \leq |x|^2 \leq \cdots \leq |x|^n$, so that

$$\frac{f(x)}{x^n} \geq a_n - \frac{1}{|x|} \sum_{j=0}^{n-1} |a_j|;$$

in turn, this last expression is positive for $|x| > \frac{1}{a_n} \sum_{j=0}^{n-1} |a_j|$.

The argument at the previous paragraph shows that $\frac{f(x)}{x^n} > 0$ at $x = \pm A$, whenever $A > \max \left\{ 1, \frac{1}{a_n} \sum_{j=0}^{n-1} |a_j| \right\}$. However, since n is odd, it follows that $f(-A) < 0 < f(A)$, and Bolzano's theorem assures the existence of $c \in [-A, A]$ such that $f(c) = 0$. □

Our next result is a refinement of Bolzano's theorem, which shows that every continuous function defined on an interval satisfies the intermediate value property. For this reason, this result is known as the **intermediate value theorem** (we abbreviate **IVT**).

Theorem 8.29 (IVT) *Let $f, g : [a, b] \to \mathbb{R}$ be continuous functions. If $f(a) < g(a)$ and $f(b) > g(b)$ (or vice-versa), then there exists $c \in (a, b)$ such that $f(c) = g(c)$. In particular, if a real number d belongs to the interval with endpoints $f(a)$ and $f(b)$, then there exists $c \in [a, b]$ such that $f(c) = d$.*

Proof For the first claim, note that $h = f - g$ is continuous and such that $h(a)h(b) = (f(a) - g(a))(f(b) - g(b)) < 0$. Hence, Bolzano's theorem gives $c \in (a, b)$ for which $h(c) = 0$, i.e., such that $f(c) = g(c)$.

The particular case of the second claim is obtained by taking g to be the constant function, equal to d. $\square$

As a first application of the IVT, let us show that the image of a continuous function defined on an interval is also an interval.

Corollary 8.30 *If $I \subset \mathbb{R}$ is an interval and $f : I \to \mathbb{R}$ is a continuous function, then the image of f is also an interval. If $I = [a, b]$, then there exist real numbers $c \leq d$ such that* $\mathrm{Im}(f) = [c, d]$.

Proof Let's first consider the case $I = [a, b]$. By Weierstrass Theorem 8.26, there exist $x_m, x_M \in [a, b]$ such that f attains its minimum and maximum values at x_m and x_M, respectively. Letting $f(x_m) = c$ and $f(x_M) = d$, we have $\mathrm{Im}(f) \subset [c, d]$. On the other hand, for a fixed $y \in [c, d]$, the IVT gives a real x, belonging to the interval of endpoints x_m and x_M and such that $f(x) = y$. In particular, $\mathrm{Im}(f) \supset [c, d]$.

For the general case, note that every interval I can be written as $I = \bigcup_{n \geq 1}[a_n, b_n]$, with $[a_1, b_1] \subset [a_2, b_2] \subset \ldots$ (for instance, if $I = (a, b)$, we could take $a_n = a + \frac{1}{n}$ and $b_n = b - \frac{1}{n}$, provided $n > \frac{1}{2(b-a)}$—the remaining cases can be dealt with in analogous ways). Now, letting $[c_n, d_n]$ be the image of the interval $[a_n, b_n]$ by f, it's easy to show that $[c_1, d_1] \subset [c_2, d_2] \subset \ldots$ and, hence, that $\bigcup_{n \geq 1}[c_n, d_n]$ is an interval J. On the other hand, it's also easy to see that, since $I = \bigcup_{n \geq 1}[a_n, b_n]$, we have

$$\mathrm{Im}(f) = \bigcup_{n \geq 1}[c_n, d_n] = J.$$

$\square$

In what follows, we discuss some other interesting applications of the IVT.

Example 8.31 Let $f : [0, 1] \to [0, 1]$ be a continuous function. Prove that there exists a real number $0 \leq c \leq 1$ such that $f(c) = c$ (i.e., prove that f has at least one fixed point).

Proof If $f(0) = 0$ or $f(1) = 1$, there's nothing left to do; otherwise, suppose $f(0) > 0$ and $f(1) < 1$. Letting $g : [0, 1] \to \mathbb{R}$ be the function given by $g(x) = x$, we then have $f(a) > g(a)$ and $f(b) < g(b)$. Hence, the IVT gives $0 < c < 1$ satisfying $f(c) = g(c)$ or, which is the same, $f(c) = c$. $\square$

Example 8.32 (Bulgaria) Let $m, n \geq 1$ be given integers. For $x > 0$, compute the number of real solutions of the equation

$$\frac{1}{x} + \frac{1}{x^2} + \frac{1}{x^3} + \cdots + \frac{1}{x^m} = x + \sqrt{x} + \sqrt[3]{x} + \cdots + \sqrt[n]{x}.$$

Solution We claim that the given equation has just one real solution. To this end, note first that, for $x > 0$ and a natural k, the function $x \mapsto \frac{1}{x^k}$ is decreasing, whereas the function $x \mapsto \sqrt[k]{x}$ is increasing. Now, since a finite sum of increasing functions on the same domain is increasing and a finite sum of decreasing functions on the same domain is decreasing (prove these facts!), we conclude that the functions $f, g : (0, +\infty) \to \mathbb{R}$ given by

$$f(x) = \frac{1}{x} + \frac{1}{x^2} + \frac{1}{x^3} + \cdots + \frac{1}{x^m} \quad \text{and} \quad g(x) = x + \sqrt{x} + \sqrt[3]{x} + \cdots + \sqrt[n]{x}$$

are decreasing and increasing, respectively. On the other hand, since the solutions of the given equation correspond to the positive values of x for which $f(x) = g(x)$, Problem 1, page 183, guarantees that such equation has at most one real solution.

In order to show that a real solution indeed exists, let's use the IVT, initially observing that both f and g are continuous functions. We then consider two cases separately:

- for $x > 1$, we have

$$\frac{1}{x^m} < \frac{1}{x^{m-1}} < \cdots < \frac{1}{x^2} < \frac{1}{x} \quad \text{and} \quad x > \sqrt{x} > \cdots > {}^{n-1}\!\sqrt{x} > \sqrt[n]{x},$$

so that $f(x) < \frac{m}{x}$ and $g(x) > n\sqrt[n]{x}$. In particular, $f(x) < g(x)$ if $\frac{m}{x} < n\sqrt[n]{x}$, i.e., if $x > \left(\frac{m}{n}\right)^{\frac{n}{n+1}}$. Hence, $f(x) < g(x)$ for $x > \max\left\{1, \left(\frac{m}{n}\right)^{\frac{n}{n+1}}\right\}$.

- for $0 < x < 1$, we have

$$\frac{1}{x^m} > \frac{1}{x^{m-1}} > \cdots > \frac{1}{x^2} > \frac{1}{x} \quad \text{and} \quad x < \sqrt{x} < \cdots < {}^{n-1}\!\sqrt{x} < \sqrt[n]{x},$$

so that $f(x) > \frac{m}{x}$ and $g(x) < n\sqrt[n]{x}$. In particular, $f(x) > g(x)$ if $\frac{m}{x} > n\sqrt[n]{x}$, i.e., if $x < \left(\frac{m}{n}\right)^{\frac{n}{n+1}}$. Hence, $f(x) > g(x)$ for $0 < x < \min\left\{1, \left(\frac{m}{n}\right)^{\frac{n}{n+1}}\right\}$.

Finally, the previous discussion assures that we can take real numbers $0 < a < 1 < b$ satisfying $f(a) > g(a)$ and $f(b) < g(b)$, and the IVT provides $c \in (a, b)$ such that $f(c) = g(c)$.   □

Example 8.33 (Romania) There exists a continuous function $f : \mathbb{R} \to \mathbb{R}$ such that

$$f(x) \in \mathbb{Q} \Leftrightarrow f(x + 1) \notin \mathbb{Q}?$$

Solution Suppose such an f exists, and let $g : \mathbb{R} \to \mathbb{R}$ be given by $g(x) = f(x + 1) - f(x)$. Then, g is continuous (by the chain rule for continuous functions) and, by the stated conditions, transforms every real number into an irrational number. However, since every nondegenerate interval contains irrational numbers (see Problem 4, page 206), the only way of not contradicting the IVT is that g is constant. Thus, there exists an irrational number α such that $f(x + 1) - f(x) = \alpha$ for every $x \in \mathbb{R}$. Therefore,

$$f(x + 2) - f(x) = f(x + 2) - f(x + 1) + f(x + 1) - f(x) = 2\alpha, \tag{8.9}$$

also for every $x \in \mathbb{R}$.

We now assert that there exists $x_0 \in \mathbb{R}$ such that $f(x_0) \in \mathbb{Q}$. Indeed, take any real number a; if $f(a)$ is not rational, it follows from the hypotheses that $f(a+1)$ is rational, and it suffices to take $x_0 = a$ or $x_0 = a+1$.

Finally, for x_0 as in the former paragraph, the hypotheses on f guarantee that

$$f(x_0) \in \mathbb{Q} \Rightarrow f(x_0 + 1) \notin \mathbb{Q} \Rightarrow f(x_0 + 2) \in \mathbb{Q}.$$

Hence, $f(x_0 + 2) - f(x_0) \in \mathbb{Q}$, which contradicts (8.9) and finishes the proof. □

For the coming example, recall that a function $f : \mathbb{R} \to \mathbb{R}$ is *even* if $f(-x) = f(x)$, for every real x.

Example 8.34 Let $f : \mathbb{R} \to \mathbb{R}$ be a continuous function such that $f(f(x)) = x^2 + 1$, for every $x \in \mathbb{R}$. Prove that f is even.

Proof First of all, note that for every $x \in \mathbb{R}$ we have

$$f(f(x)) = x^2 + 1 \Rightarrow f(f(f(x))) = f(x^2 + 1)$$
$$\Rightarrow f(x)^2 + 1 = f(x^2 + 1)$$
$$\Rightarrow f(x)^2 + 1 = f(-x)^2 + 1$$
$$\Rightarrow f(x) = \pm f(-x)$$

Now, if $f(\alpha) = f(\beta) = 0$, then

$$\alpha^2 + 1 = f(f(\alpha)) = f(0) = f(f(\beta)) = \beta^2 + 1,$$

so that $\alpha = \pm \beta$ and f has at most two zeros. Now, there are three possibilities:

- $f(x) \neq 0$, for every real x: by the IVT, f has a constant sign in $\mathbb{R}$. However, since $f(x) = \pm f(-x)$ for every real x, it follows that $f(x) = f(-x)$ for every real x, and f is even.
- f has a single zero, say at $x = \alpha$: since $f(-\alpha) = \pm f(\alpha) = 0$, we must have $\alpha = -\alpha$, so that $\alpha = 0$. On the other hand,

$$f(f(0)) = 0^2 + 1 = 1 \Rightarrow f(0) \neq 0$$

which is a contradiction.
- f has exactly two zeros, at $x = \alpha$ and $x = -\alpha$, for some $\alpha > 0$: again by the IVT, f has a constant sign in $(-\alpha, \alpha)$. Let $g(x) = f(x) - x$. If $f > 0$ in $(-\alpha, \alpha)$, then

$$g(0) = f(0) - 0 > 0 \quad \text{and} \quad g(\alpha) = f(\alpha) - \alpha = -\alpha < 0,$$

so that there exists $0 < c < \alpha$ for which $g(c) = 0$. If $f < 0$ in $(-\alpha, \alpha)$, then

$$g(0) = f(0) - 0 < 0 \quad \text{and} \quad g(-\alpha) = f(-\alpha) - (-\alpha) = \alpha > 0,$$

so that there exists $-\alpha < c < 0$ for which $g(c) = 0$. In any case, f admits a fixed point c. However,

$$f(c) = c \Rightarrow c = f(f(c)) = c^2 + 1 \Rightarrow c^2 - c + 1 = 0,$$

which is impossible.   □

We close this section by using the IVT to prove the continuity of the inverse of a continuous function defined on an interval.

Theorem 8.35 *Let $I \subset \mathbb{R}$ be an interval and $f : I \to \mathbb{R}$ be a continuous function. Then, f is injective if and only if f is increasing or decreasing. Moreover, in this case:*

(a) The image J of f is an interval of the same type (i.e., open, half-open or closed) as I.
(b) $f^{-1} : J \to I$ is continuous.

Proof If f is not injective, then f is neither increasing nor decreasing. Conversely, if f is neither increasing nor decreasing, then there exist $a < b < c$ in I such that $f(a) \leq f(b) \geq f(c)$ or $f(a) \geq f(b) \leq f(c)$. Suppose that $f(a) \leq f(b) \geq f(c)$ (the other case is entirely analogous) and choose $d \in \mathbb{R}$ such that

$$\max\{f(a), f(c)\} \leq d \leq f(b).$$

The IVT assures the existence of $x_0 \in (a, b)$ and $x_1 \in (b, c)$ (therefore, $x_0, x_1 \in I$) such that $f(x_0) = d$ and $f(x_1) = d$. In particular, $f(x_0) = f(x_1)$ and f is not injective.

(a) Corollary 8.30 showed that J is an interval. Since f is injective, we can suppose that f is increasing (again, the case of a decreasing f is analogous). Let $I = (a, b)$, with $a, b \in \mathbb{R}$ (the remaining cases can also be dealt with in analogous ways).

   If $\operatorname{Im}(f) = (c, d]$ or $(-\infty, d]$, with $c, d \in \mathbb{R}$, take $x_0 \in (a, b)$ such that $f(x_0) = d$. Since f is increasing, for $x \in (x_0, b)$ we have $f(x) > f(x_0) = d$, thus contradicting the fact that $f(x_0) \in \operatorname{Im}(f)$. Analogously, we show that $\operatorname{Im}(f) \neq [c, d), [c, +\infty), [c, d]$, with $c, d \in \mathbb{R}$. Therefore, $\operatorname{Im}(f)$ is also an open interval.

(b) Let's look at the case of an increasing f, with $I = (a, b)$ and $\operatorname{Im}(f) = (c, d)$ (once more, the analysis of the other ones is quite similar). We start by observing (according to Problem 8, page 176), that f^{-1} is also increasing. Now, for a fixed $y_0 \in (c, d)$, let $x_0 = f^{-1}(y_0)$. Given $\epsilon > 0$, we want $\delta > 0$ such that

$$y \in (c, d), \ |y - y_0| < \delta \Rightarrow |f^{-1}(y) - f^{-1}(y_0)| < \epsilon. \tag{8.10}$$

To this end, let $\delta_0 = \min\{y_0 - c, d - y_0\}$ and take $0 < \delta < \delta_0$ (so that the condition $|y - y_0| < \delta$ suffices to guarantee that $y \in (c, d)$). Letting $x = f^{-1}(y)$, we have $y = f(x)$ and can rewrite (8.10) in the following way: we want $0 < \delta < \delta_0$ such that

$$|f(x) - f(x_0)| < \delta \Rightarrow |x - x_0| < \epsilon.$$

Notice that we can pick $\epsilon > 0$ so small that $x_0 \pm \epsilon \in (a, b)$ (otherwise, change the originally given $\epsilon > 0$ by a smaller one, so that this additional condition is satisfied). Recalling that f is increasing, take

$$0 < \delta < \min\{\delta_0, f(x_0 + \epsilon) - f(x_0), f(x_0) - f(x_0 - \epsilon)\}.$$

Then,

$$f(x) - f(x_0) < \delta \Rightarrow f(x) - f(x_0) < f(x_0 + \epsilon) - f(x_0)$$
$$\Rightarrow f(x) < f(x_0 + \epsilon) \Rightarrow x < x_0 + \epsilon$$

and, analogously,

$$f(x) - f(x_0) > -\delta \Rightarrow x > x_0 - \epsilon.$$

In any case, the choice of δ gives

$$|f(x) - f(x_0)| < \delta \Rightarrow -\delta < f(x) - f(x_0) < \delta$$
$$\Rightarrow -\epsilon < x - x_0 < \epsilon$$
$$\Rightarrow |x - x_0| < \epsilon,$$

as we wished to show.   □

As a first application of the previous result, we shall now give an alternative proof for Theorem 7.9 and Problem 8, page 254.

Example 8.36 Given $n \in \mathbb{N}$, the **n-th power function** is $f : \mathbb{R} \to \mathbb{R}$ given by $f(x) = x^n$. If n is even, write also f to denote its restriction to $[0 + \infty)$. Since f is continuous, increasing and $f(k) = k^n > k$ for every $k \in \mathbb{N}$, the IVT gives $\text{Im}(f) = [0, +\infty)$. Hence, for every $y \geq 0$ there exists a single $x \geq 0$ such that $x^n = y$, and we denote by $x = \sqrt[n]{y}$. The previous theorem now guarantees the continuity of the function

$$f^{-1} : [0, +\infty) \longrightarrow [0, +\infty)$$
$$y \longmapsto \sqrt[n]{y} \quad .$$

Analogously, if n is odd, then f is continuous, increasing, $\text{Im}(f) = \mathbb{R}$ and (letting $\sqrt[n]{y}$ be as above)

$$f^{-1} : \mathbb{R} \longrightarrow \mathbb{R}$$
$$y \longmapsto \sqrt[n]{y}$$

is continuous. In both cases, f^{-1} is called the **n-th root function**.

The next example establishes the continuity of the inverse trigonometric functions, introduced in Problem 1, page 198.

Example 8.37 The restriction of the sine function to the interval $[-\frac{\pi}{2}, \frac{\pi}{2}]$, which we shall also denote by $\sin : [-\frac{\pi}{2}, \frac{\pi}{2}] \to [-1, 1]$, is a continuous and increasing bijection. The arc-sine function is its inverse $\arcsin : [-1, 1] \to [-\frac{\pi}{2}, \frac{\pi}{2}]$, so that, for $x \in [-1, 1]$ and $y \in [-\frac{\pi}{2}, \frac{\pi}{2}]$,

$$\arcsin x = y \Leftrightarrow \sin y = x.$$

Thanks to Theorem 8.35, the arc-sine function os increasing and continuous.

Similarly, the inverse of $\cos : [0, \pi] \to [-1, 1]$ is the arc-cosine function $\arccos : [-1, 1] \to [0, \pi]$, which is continuous (again by Theorem 8.35) and decreasing. Note also that, for $x \in [-1, 1]$ and $y \in [0, \pi]$, we have

$$\arccos x = y \Leftrightarrow \cos y = x.$$

Finally, the arc-tangent function $\arctan : \mathbb{R} \to \left(-\frac{\pi}{2}, \frac{\pi}{2}\right)$ is the inverse of the restriction $\tan : \left(-\frac{\pi}{2}, \frac{\pi}{2}\right) \to \mathbb{R}$ of the tangent function, so that it is also continuous and increasing. Also, for $x \in \mathbb{R}$ and $y \in \left(-\frac{\pi}{2}, \frac{\pi}{2}\right)$, we have

$$\arctan x = y \Leftrightarrow \tan y = x.$$

Problems: Section 8.3

1. The continuous function $f : (-\infty, 1) \cup (1, +\infty) \to \mathbb{R}$, given by

$$f(x) = \begin{cases} x, & \text{if } x \in (-\infty, 1) \\ x^2 + 1, & \text{if } x \in (1, +\infty) \end{cases},$$

   satisfies $f(0) < \frac{3}{2} < f(2)$ but there doesn't exist any c in the domain of f such that $f(c) = \frac{3}{2}$. Why does this example do not contradict the IVT?
2. Compute the number of real solutions of each of the following equations:

   (a) $|x| + 1 = x^4$.

(b) $\cos x = x^2$.

(c) $\sin x = \frac{x}{4}$.

3. Let $f : \mathbb{R} \to \mathbb{R}$ be given by $f(x) = x^3 \sin x$, for every $x \in \mathbb{R}$. Show that its graph intersects every non vertical real line an infinite number of times.

4. Let $n > 1$ be an integer. Write down an explicit expression for a continuous function $f : [0, 1] \to [0, 1]$ with exactly n fixed points.

5. * Let I be an interval and X be a nonempty subset of I having the following properties:

  (i) For every $x_0 \in X$, there exists $\delta > 0$ such that $I \cap (x_0 - \delta, x_0 + \delta) \subset X$.

  (ii) If $(a_n)_{n \geq 1}$ is a sequence of points of X and $l \in I$ is such that $a_n \to l$, then $l \in X$.

  Show that $X = I$.

6. (Bulgaria) Let $n > 1$ be an integer and $a_1, a_2, \ldots, a_n$ be given positive real numbers. Prove that the equation

$$\sqrt{1 + a_1 x} + \sqrt{1 + a_2 x} + \cdots + \sqrt{1 + a_n x} = nx$$

has exactly one positive real solution.

7. Let $x_1, x_2, \ldots, x_n$ be real numbers in the interval $[0, 1]$. Prove that there exists $x \in [0, 1]$ such that

$$|x - x_1| + |x - x_2| + \cdots + |x - x_n| = \frac{n}{2}.$$

8. (Leningrad) Let $f : \mathbb{R} \to \mathbb{R}$ be a continuous function satisfying, for every $x \in \mathbb{R}$, the relation $f(x)f(x + 2) + f(x + 1) = 0$. If $f(0) \geq 0$, show that there exist infinitely many values of x for which $f(x) = 0$.

9. (Leningrad) Let $f : \mathbb{R} \to \mathbb{R}$ be a continuous function such that $f(x)f(f(x)) = 1$, for every real x. If $f(1000) = 999$, compute $f(500)$.

10. (Australia) Find all real values of a for which, for every continuous function $f : [0, 1] \to \mathbb{R}$ satisfying the condition $f(0) = f(1)$, there exists $x_0 \in [0, 1 - a]$ such that $f(x_0) = f(x_0 + a)$.

11. Let $f : \mathbb{R} \to \mathbb{R}$ be such that $f(x + 1)f(f(x) + 1) + 1 = 0$, for every real x. Prove that f is not continuous.

12. Let $f : [0, 1] \to [0, 1]$ be a continuous function, such that $(f \circ f)(x) = x$ for every $x \in [0, 1]$.

  (a) Show that f is an increasing or decreasing bijection.

  (b) If f is increasing, show that $f(x) = x$ for every $x \in [0, 1]$.

  (c) Show that there are infinitely many possibilities for f, if it is decreasing.

13. (Leningrad) The continuous functions $f, g : [0, 1] \to [0, 1]$ are such that $f \circ g = g \circ f$. If f is nondecreasing, prove that there exists $0 \leq a \leq 1$ such that $f(a) = g(a) = a$.

14. (Crux) Let $f : \mathbb{R} \to \mathbb{R}$ be a continuous function which assumes positive and negative values. Given a natural number $k > 2$, prove that there exists a nonconstant AP $(a_1, a_2, \ldots, a_k)$ such that

$$f(a_1) + f(a_2) + \cdots + f(a_k) = 0.$$

15. (TT) Prove that, for each natural number n, the graph of any continuous increasing function $f : [0, 1] \to [0, 1]$ can be covered by n closed rectangles, each of which having area $\frac{1}{n^2}$ and sides parallel to the coordinate axis.

16. (Leningrad) Let $f : \mathbb{R} \to \mathbb{R}$ be a continuous function such that, for every real x, we have $f(x + f(x)) = f(x)$. Prove that f is constant.

17. (Belarus) Find all functions $f, g, h : \mathbb{R} \to \mathbb{R}$ such that, for every $x, y \in \mathbb{R}$, we have

$$f(x + y^3) + g(x^3 + y) = h(xy).$$

Chapter 9
Limits and Derivatives

In this chapter we study derivatives of functions and some of their applications. Along the way, we show how to use derivatives to solve the problem of finding the monotonicity (resp. concavity) intervals of a differentiable (resp. twice differentiable) function, as well as to build an accurate sketch of the graph of such functions. In turn, the analysis of such problems will motivate several interesting applications of the concept of derivative to problems of maxima and minima.

9.1 Some Heuristics I

As a motivation for what is to come, in this section we heuristically discuss two problems which gave birth to the notion of *derivative* of a function.

Let's start by trying to define the *tangent line* to the graph of a function $f : (a, b) \to \mathbb{R}$ at a point $A(x_0, f(x_0))$ on it, so that $x_0 \in (a, b)$. Taking $x_1 \in (a, b) \setminus \{x_0\}$ and letting $B_1(x_1, f(x_1))$, we say that $\overleftrightarrow{AB_1}$ is a *secant* to the graph of f and passing through A. In Fig. 9.1, we consider secants $\overleftrightarrow{AB_1}$ and $\overleftrightarrow{AB_2}$ to the graph of f.

A little geometric intuition makes it plausible to guess that a generic secant line $\overleftrightarrow{AB}$ should come closer and closer to the tangent line to the graph of f at A, as long as B approaches A along the graph (or, which is the same, as long as x approaches x_0 in (a, b)).

If $x_1 \in (a, b) \setminus \{x_0\}$ and $B(x_1, f(x_1))$, elementary analytic geometry (cf. Chap. 6 of [4], for instance) shows that $\overleftrightarrow{AB}$ has equation

$$y - f(x_0) = \left(\frac{f(x_1) - f(x_0)}{x_1 - x_0} \right) (x - x_0); \tag{9.1}$$

© Springer International Publishing AG 2017

A. Caminha Muniz Neto, *An Excursion through Elementary Mathematics, Volume I*, Problem Books in Mathematics, DOI 10.1007/978-3-319-53871-6_9

Fig. 9.1 Secants $\overleftrightarrow{AB_1}$ and
$\overleftrightarrow{AB_2}$ to the graph of f

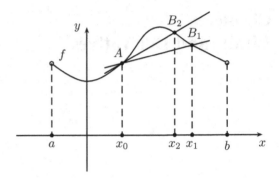

on the other hand, if we can define the line tangent to the graph of f at A, and it is
not vertical, then its equation must be of the form

$$y - f(x_0) = m(x - x_0), \qquad (9.2)$$

for some $m \in \mathbb{R}$.

Comparing (9.1) and (9.2), and taking into account the discussion at the
previous paragraph, we are led to the conclusion that the quotients $\frac{f(x_1) - f(x_0)}{x_1 - x_0}$ should
approximate m better and better, provided x_1 approximates x_0 better and better. Thus,
we conclude that the line tangent to the graph of f at A is that of Eq. (9.2), where m
is the *limit value* of the quotients $\frac{f(x_1) - f(x_0)}{x_1 - x_0}$, when x_1 comes closer and closer to x_0.

This *limit value* of the quotients $\frac{f(x_1) - f(x_0)}{x_1 - x_0}$ when x_1 approaches x_0 is the *derivative*
of the function f at x_0, which we denote $f'(x_0)$ or, in classical notation, $\frac{df}{dx}(x_0)$.
Writing x in place of x_1, we summarize the previous discussion by "*defining*"

$$\frac{df}{dx}(x_0) = \lim_{x \to x_0} \frac{f(x) - f(x_0)}{x - x_0},$$

where the notation $\lim\limits_{x \to x_0}$ means that the closer x is to x_0, the closer $\frac{f(x) - f(x_0)}{x - x_0}$ is
to $\frac{df}{dx}(x_0)$. Hence, for small values of $|x - x_0|$, we expect to have

$$\frac{df}{dx}(x_0) \cong \frac{f(x) - f(x_0)}{x - x_0}.$$

Classically, the differences $f(x) - f(x_0)$ and $x - x_0$ were respectively denoted by
$\Delta f\big|_{x_0}^{x}$ and $\Delta x\big|_{x_0}^{x}$, and were called the *variations* of f and of the independent variable
in the interval $[x_0, x]$. Accordingly, $\frac{\Delta f}{\Delta x}\big|_{x_0}^{x}$ is the *rate of change* of f (with respect to
the independent variable) in the interval $[x_0, x]$. Therefore, for small $|x - x_0|$, we have

$$\frac{df}{dx}(x_0) \cong \frac{\Delta f}{\Delta x}\bigg|_{x_0}^{x}.$$

Now, suppose we are in a situation where x represents time and f represents some quantity that evolves with time. Writing t in place of x and t_0 in place of x_0, it follows from the above that

$$\frac{df}{dt}(t_0) = \lim_{t \to t_0} \left. \frac{\Delta f}{\Delta t} \right|_{t_0}^{t},$$

so that $\frac{df}{dt}(t_0)$ can naturally be called the *instantaneous rate of change* of f with respect to t, at instant t_0.

As a particular situation of that of the previous paragraph, if $f(t)$ measures the *displacement* of a point object along a real line at instant t, then $\frac{df}{dt}(t_0)$ is the *instantaneous velocity* of the point object at instant t_0, and is the starting point of the discussion of Kinematics at most Physics books.

The circle of ideas we developed up to this point, and much more, bears the name of *Differential Calculus* and is considered to be the creation of two of the most brilliant minds mankind ever produced, Newton and Leibniz.[1]

The rest of this chapter is devoted to a thorough development of the concepts, results and applications of the Differential Calculus.

9.2 Limits of Functions

Let be given a continuous function $f : (a, b) \to \mathbb{R}$ and $x_0 \in (a, b)$. Recalling our intuitive discussion so far, we conclude that if there exists a reasonable notion of tangent line to the graph of f at the point $A(x_0, f(x_0))$, and if such a line is not vertical, then its equation must have the form (9.2), where

$$m = \lim_{x \to x_0} \frac{f(x) - f(x_0)}{x - x_0}. \tag{9.3}$$

Here, the expression at the right hand side above intuitively represents the *limit value* of the quotients $\frac{f(x)-f(x_0)}{x-x_0}$ as x *goes to* (i.e., approaches) x_0, provided such a "*limit*" exists in a proper sense (to be precised in a while).

Since our discussion rests upon rather naive grounds, we start to fix this now by setting some preliminaries and giving a rigorous definition of the notion of limit of a function at a point.

Definition 9.1 Given $a \in \mathbb{R}$, a **neighborhood** of a is an interval of the form $(a - r, a + r)$, where r is a positive real number. In this case, we say that r is the **radius** of the neighborhood $(a - r, a + r)$ and that each $x \in (a - r, a + r)$ approximates a with error less than r.

---

[1]Gottfried Wilhem Leibniz, German mathematician and philosopher of the XVII century. Together with Sir Isaac Newton, Leibniz is considered to be one of the creators of the Differential and Integral Calculus. Some of the notations used in Calculus up to this day go back to Leibniz, having survived the ruthless test of time.

Let $I \subset \mathbb{R}$ be an interval, let $x_0 \in I$ and $f : I \setminus \{x_0\} \to \mathbb{R}$ be a given function. For a given real number L, our first task is to formulate a precise definition for the claim that $f(x)$ can be taken as close to L as we wish, provided we take $x \in I$ sufficiently close to (but different from) x_0.

A little reflection allows us to conclude that a reasonable formulation of this concept is obtained if we ask that, to each given neighborhood of L, there exists a neighborhood of x_0 that is applied to the former one by f.

Since a neighborhood of x_0 is completely determined by its radius r, and

$$x \in (x_0 - r, x_0 + r) \setminus \{x_0\} \Leftrightarrow 0 < |x - x_0| < r,$$

we can summarize the previous discussion in the following definition (see Fig. 9.2).

Definition 9.2 Let $I \subset \mathbb{R}$ be an interval, let $x_0 \in I$ and $f : I \setminus \{x_0\} \to \mathbb{R}$ be a given function. We say that f has **limit** L when x **goes to** x_0, and write

$$\lim_{x \to x_0} f(x) = L, \tag{9.4}$$

if, to each given real number $\epsilon > 0$, there corresponds a real number $\delta > 0$ such that

$$x \in I, \ 0 < |x - x_0| < \delta \Rightarrow |f(x) - L| < \epsilon. \tag{9.5}$$

In words, (9.5) happens if, to each arbitrarily given error $\epsilon > 0$ for L, there corresponds an error $\delta > 0$ for x_0 such that, if $x \in I \setminus \{x_0\}$ approximates x_0 in I with error less that δ, then $f(x)$ approximates L with error less than ϵ.

Geometrically, we want that, to each $x \in I$ sufficiently close to (but different from) x_0, the point of the graph of f with abscissa x belongs to the gray strip of Fig. 9.2.

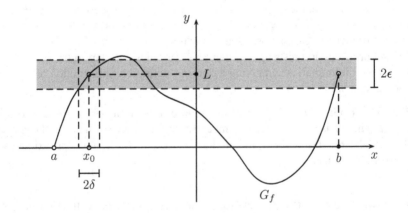

Fig. 9.2 Limit of a function at a point

At this point (and as it happened in our study of continuity), it is worth stressing that, to justify the validity of a specific limit by applying the definition given above is the same as to play the following *cat and mouse game*: to each arbitrarily given error $\epsilon > 0$ for the limit candidate L, we ought to be able to find an error $\delta > 0$ for x_0 (which, in general, will depend both on ϵ as on x_0 itself) so that the validity of the condition $0 < |x - x_0| < \delta$ for some $x \in I$ implies $|f(x) - L| < \epsilon$.

Having gone through the study of continuity, the reader has surely acquired some familiarity with the dynamics of finding which $\delta > 0$ are adequate to some given $\epsilon > 0$. In spite of this, and as a remembrance, we shall carefully work out two examples below. In all that follows, and whenever there is no danger of confusion, we shall frequently omit explicit references to the domain or codomain of the functions f involved, thus focusing on the expressions that define $f(x)$ in terms of the independent variable x. Whenever this is so, we shall implicitly assume that the domain of f is the maximal one (cf. Sect. 6.1), whereas its codomain is the set $\mathbb{R}$ of real numbers.

Examples 9.3

(a) $\lim_{x \to 2}(-2x + 7) = 3$: let $\epsilon > 0$ be given. Departing from $x \in \mathbb{R}$ such that $0 < |x - 2| < \delta$, we have

$$|(-2x + 7) - 3| = |-2x + 4| = 2|x - 2| < 2\delta.$$

Therefore, choosing $\delta > 0$ so that $2\delta \leq \epsilon$, we conclude that

$$x \in \mathbb{R}, \ 0 < |x - 2| < \delta \Rightarrow |(-2x + 7) - 3| < 2\delta \leq \epsilon,$$

as wished.

(b) $\lim_{x \to 3} x^2 = 9$: again, let $\epsilon > 0$ be given. Starting with $x \in \mathbb{R}$ such that $0 < |x - 3| < \delta$, we have

$$|x^2 - 9| = |x - 3||x + 3| < \delta|x - 3 + 6|$$
$$\leq \delta(|x - 3| + 6) < \delta(\delta + 6),$$

where we used the triangle inequality in the next to last passage above. Hence, if it is possible to choose $\delta > 0$ so that $\delta(\delta + 6) \leq \epsilon$, we will have

$$x \in \mathbb{R}, \ 0 < |x - 3| < \delta \Rightarrow |x^2 - 9| < \delta(\delta + 6) \leq \epsilon,$$

as wished. It now suffices to show that it is possible to choose such a $\delta > 0$, and to this end we just have to solve the inequality $\delta(\delta + 6) \leq \epsilon$. By doing so, we get

$$0 < \delta \leq \sqrt{\epsilon + 9} - 3.$$

Looking back at the previous examples, we conclude that if we have a function $f : I \setminus \{x_0\} \to \mathbb{R}$ and want to prove that $\lim_{x \to x_0} f(x) = L$, then, as in Sect. 8.1, the key to find out which $\delta > 0$ works for a given $\epsilon > 0$ is to reason in the following way: starting with an $x \in I$ such that $0 < |x - x_0| < \delta$, we estimate *from above* the error $|f(x) - L|$ in terms of δ, thus getting an inequality of the form $|f(x) - L| < E(\delta)$. Here, E represents a certain function of δ; in item (a) we got $E(\delta) = 2\delta$, whereas in item (b) we found $E(\delta) = \delta(\delta + 6)$. Then, we impose that such an error $E(\delta)$ does not surpass ϵ, thus finding the appropriate values of δ (usually, this second step reduces to solving, for $\delta > 0$, the inequality $E(\delta) \leq \epsilon$). Finally, if $\delta > 0$ satisfies $E(\delta) \leq \epsilon$, then we clearly have that

$$x \in I, \ 0 < |x - x_0| < \delta \Rightarrow |f(x) - L| < E(\delta) \leq \epsilon,$$

as wished.

Notice that the only difference between the discussion above and that of Sect. 8.1 lies in the fact that, here, we have to start by assuming $|x - x_0| > 0$, i.e., $x \neq x_0$. On the other hand, and as the coming proposition explains, such a difference was not relevant to the previous examples because both functions $x \mapsto -2x + 7$ and $x \mapsto x^2$ are continuous.

Proposition 9.4 *Let $I \subset \mathbb{R}$ be an interval and $f : I \to \mathbb{R}$ be a given function. For $x_0 \in I$, we have f continuous at x_0 if and only if*

$$\lim_{x \to x_0} f(x) = f(x_0). \tag{9.6}$$

Proof First of all, let f be continuous at x_0. Definition 8.2 guarantees that, given $\epsilon > 0$, there exists $\delta > 0$ such that

$$x \in I, \ |x - x_0| < \delta \Rightarrow |f(x) - f(x_0)| < \epsilon.$$

In particular, if $x \in I$ and $0 < |x - x_0| < \delta$, we still have $|f(x) - f(x_0)| < \epsilon$, so that (9.5) is satisfied. Therefore, $\lim_{x \to x_0} f(x) = f(x_0)$.

Conversely, let $\lim_{x \to x_0} f(x) = f(x_0)$. According to the definition of limit, given $\epsilon > 0$, there exists $\delta > 0$ such that

$$x \in I, \ 0 < |x - x_0| < \delta \Rightarrow |f(x) - f(x_0)| < \epsilon.$$

However, since the condition $|f(x) - f(x_0)| < \epsilon$ is trivially satisfied for $x = x_0$, we can certainly write the implication above as

$$x \in I, \ |x - x_0| < \delta \Rightarrow |f(x) - f(x_0)| < \epsilon.$$

Hence, f is continuous at x_0. $\square$

Example 9.5 Given $f : \mathbb{R} \setminus \{\pm 1\} \to \mathbb{R}$ such that $f(x) = \frac{x^3 - 3x^2 + 2}{x^2 - 1}$ for every $x \neq \pm 1$, we want to compute $\lim_{x \to 1} f(x)$. To this end, we initially observe that 1 is a root of both the numerator and the denominator of the defining expression for $f(x)$, with

$x^3 - 3x^2 + 2 = (x-1)(x^2 - 2x - 2)$ and $x^2 - 1 = (x-1)(x+1)$. Therefore, on $\mathbb{R} \setminus \{\pm 1\}$ we have

$$f(x) = \frac{(x-1)(x^2 - 2x - 2)}{(x-1)(x+1)} = \frac{x^2 - 2x - 2}{x+1}.$$

Now, since the function $g : \mathbb{R} \setminus \{-1\} \to \mathbb{R}$ given by $g(x) = \frac{x^2 - 2x - 2}{x+1}$ is continuous, the previous proposition gives

$$\lim_{x \to 1} f(x) = \lim_{x \to 1} g(x) = g(1) = -\frac{3}{2}.$$

A rather important fact on limits of functions is that if $\lim_{x \to x_0} f(x) = L$, with $L \neq 0$, then there exists a neighborhood of x_0 such that f has the same sign of L in all points of its domain which belong to this neighborhood (with the possible exception of x_0 itself). This is essentially the content of the coming result, which extends Lemma 8.20 to limits of functions and for this reason is also known as the **sign-preserving lemma**.

Lemma 9.6 *Let $I \subset \mathbb{R}$ be an interval, $x_0 \in I$ and $f : I \setminus \{x_0\} \to \mathbb{R}$ be a given function. If $\lim_{x \to x_0} f(x) = L$, with $L \neq 0$, then there exists $\delta > 0$ such that*

$$x \in I \quad and \quad 0 < |x - x_0| < \delta \Rightarrow \begin{cases} L/2 < f(x) < 3L/2, & if\ L > 0 \\ -3L/2 < f(x) < -L/2, & if\ L < 0 \end{cases}.$$

Proof Suppose $L > 0$ (the other case is completely analogous). By the definition of limit, given $\epsilon = \frac{L}{2} > 0$ there exists $\delta > 0$ such that

$$x \in I,\ 0 < |x - x_0| < \delta \Rightarrow |f(x) - L| < \frac{L}{2}.$$

Hence, for each such x we have $-\frac{L}{2} \leq f(x) - L < \frac{L}{2}$, which is the same as $\frac{L}{2} < f(x) < \frac{3L}{2}$. $\qquad\square$

We now collect some arithmetic properties on limits that pretty much simplify their computation. For item (c) of the coming proposition, it is worth observing the following: given an interval $I \subset \mathbb{R}$, $x_0 \in I$ and $g : I \setminus \{x_0\} \to \mathbb{R}$ such that $\lim_{x \to x_0} g(x) = M \neq 0$, Lemma 9.6 guarantees the existence of $r > 0$ such that g doesn't vanish on $J \setminus \{x_0\}$, where $J = I \cap (x_0 - r, x_0 + r)$. Hence, when considering the function $\frac{1}{g}$ such that $\frac{1}{g}(x) = \frac{1}{g(x)}$, we will always implicitly assume that its domain is restricted to $J \setminus \{x_0\}$.

Proposition 9.7 *Let $I \subset \mathbb{R}$ be an interval, let $x_0 \in I$ and $f, g : I \setminus \{x_0\} \to \mathbb{R}$ be given functions. If $\lim_{x \to x_0} f(x) = L$ and $\lim_{x \to x_0} g(x) = M$, then:*

(a) $\lim_{x \to x_0} (f \pm g)(x) = L \pm M$.

(b) $\lim_{x \to x_0} (f \cdot g) = L \cdot M$.

(c) $\lim_{x \to x_0} \left(\frac{f}{g} \right)(x) = \frac{L}{M}$, *provided $M \neq 0$.*

Proof In all cases below, we let $\epsilon > 0$ be given.

(a) Let's do the proof for $f + g$, the proof for $f - g$ being completely analogous. As was previously hinted, our strategy will be to try to estimate $|(f+g)(x)-(L+M)|$ by excess, in terms of $|f(x) - L|$ and $|g(x) - M|$. This is easily accomplished with the aid of the triangle inequality, which gives

$$|(f + g)(x) - (L + M)| = |(f(x) - L) + (g(x) - M)|$$
$$\leq |f(x) - L| + |g(x) - M|.$$

Thus, in order that $|(f + g)(x) - (L + M)| < \epsilon$ for $x \in I$ close to (but different from) x_0, it suffices for us to have $|f(x) - L| < \frac{\epsilon}{2}$ and $|g(x) - M| < \frac{\epsilon}{2}$.

Since $\frac{\epsilon}{2} > 0$ and $\lim_{x \to x_0} f(x) = L$, $\lim_{x \to x_0} g(x) = M$, the definition of limit guarantees the existence of $\delta_1, \delta_2 > 0$ such that

$$x \in I, \ 0 < |x - x_0| < \delta_1 \Rightarrow |f(x) - L| < \frac{\epsilon}{2}$$

and

$$x \in I, \ 0 < |x - x_0| < \delta_2 \Rightarrow |g(x) - M| < \frac{\epsilon}{2}.$$

Therefore, letting $\delta = \min\{\delta_1, \delta_2\} > 0$ and taking $x \in I$ such that $0 < |x - x_0| < \delta$, we get

$$|(f + g)(x) - (L + M)| \leq |f(x) - L| + |g(x) - M|$$
$$\leq \frac{\epsilon}{2} + \frac{\epsilon}{2} = \epsilon.$$

(b) Here again, we estimate $|(fg)(x) - LM|$ by excess in terms of $|f(x) - L|$ and $|g(x) - M|$. To this end, it follows from triangle inequality that

$$|(fg)(x) - LM| = |f(x)(g(x) - M) + (f(x) - L)M|$$
$$\leq |f(x)||g(x) - M| + |f(x) - L||M|$$
$$\leq (|f(x) - L| + |L|)|g(x) - M| + |f(x) - L||M|$$
$$= |f(x) - L||g(x) - M| + |L||g(x) - M| + |M||f(x) - L|.$$

Therefore, in order that $|(fg)(x) - LM| < \epsilon$ for $x \in I$ close to (but different from) x_0, it suffices for us to have $|f(x) - L||g(x) - M|$, $|L||g(x) - M|$ and $|M||f(x) - L|$ all less than $\frac{\epsilon}{3}$. In turn, these inequalities hold provided we have

$$|f(x) - L|, |g(x) - M| < \sqrt{\frac{\epsilon}{3}},$$

$$|g(x) - M| < \frac{\epsilon}{3(|L| + 1)} \quad \text{and} \quad |f(x) - L| < \frac{\epsilon}{3(|M| + 1)}.$$

In short, it suffices to have

$$|f(x) - L| < \min\left\{ \sqrt{\frac{\epsilon}{3}}, \frac{\epsilon}{3(|M| + 1)} \right\}$$

and

$$|g(x) - M| < \min\left\{ \sqrt{\frac{\epsilon}{3}}, \frac{\epsilon}{3(|L| + 1)} \right\}.$$

Now, letting $\epsilon_1 = \min\left\{ \sqrt{\frac{\epsilon}{3}}, \frac{\epsilon}{3(|M|+1)} \right\} > 0$ and $\epsilon_2 = \min\left\{ \sqrt{\frac{\epsilon}{3}}, \frac{\epsilon}{3(|L|+1)} \right\} > 0$, the definition of limit guarantees the existence of errors $\delta_1, \delta_2 > 0$ such that $|f(x) - L| < \epsilon_1$ for every $x \in I$ such that $0 < |x - x_0| < \delta_1$, and $|g(x) - M| < \epsilon_2$ for every $x \in I$ such that $0 < |x - x_0| < \delta_2$. Hence, setting $\delta = \min\{\delta_1, \delta_2\} > 0$, we conclude that $x \in I$ and $0 < |x - x_0| < \delta$ imply both $|f(x) - L| < \epsilon_1$ and $|g(x) - M| < \epsilon_2$, as needed.

(c) The sign-preserving lemma gives $\delta_0 > 0$ such that $|g(x)| > \frac{|M|}{2}$ for all $x \in I$ satisfying $0 < |x - x_0| < \delta_0$. Sticking to the notations set forth in the paragraph that immediately precedes this proposition, we look at $\frac{f}{g}$ as the function $\frac{f}{g} : J \setminus \{x_0\} \to \mathbb{R}$, where $J = I \cap (x_0 - \delta_0, x_0 + \delta_0)$.

Since $\frac{f}{g} = f \cdot \frac{1}{g}$, by item (b) it suffices to show that $\lim_{x \to x_0} \frac{1}{g(x)} = \frac{1}{M}$. To this end, and taking into account that $|g(x)| > \frac{|M|}{2}$ for $x \in J \setminus \{x_0\}$, we obtain (for such an x)

$$\left| \frac{1}{g(x)} - \frac{1}{M} \right| = \frac{|g(x) - M|}{|g(x)||M|} \leq \frac{2}{M^2}|g(x) - M|.$$

Therefore, in order that $\left| \frac{1}{g(x)} - \frac{1}{M} \right| < \epsilon$, we need to have $|g(x) - M| < \frac{M^2\epsilon}{2}$. To fulfill the last condition above, we choose (by invoking once more the definition of limit) $\delta_1 > 0$ for which

$$x \in J, \ 0 < |x - x_0| < \delta_1 \Rightarrow |g(x) - M| < \frac{M^2\epsilon}{2}.$$

Finally, letting $\delta = \min\{\delta_0, \delta_1\} > 0$ and taking $x \in J$ such that $0 < |x - x_0| < \delta$, we have $|g(x)| < \frac{|M|}{2}$ and $|g(x) - M| < \frac{M^2\epsilon}{2}$, as needed. $\qquad \square$

An easy induction allows us to extend the formulas of items (a) and (b) of the previous proposition to a finite number of functions. More specifically, if I is an interval, $x_0 \in I$ and $f_1, \ldots, f_n : I \setminus \{x_0\} \to \mathbb{R}$ are such that $\lim_{x \to x_0} f_j(x) = L_j$ for $1 \le j \le n$, then

$$\lim_{x \to x_0} (f_1 \pm f_2 \pm \cdots \pm f_n)(x) = L_1 \pm L_2 \pm \cdots \pm L_n \tag{9.7}$$

and

$$\lim_{x \to x_0} (f_1 f_2 \ldots f_n)(x) = L_1 L_2 \ldots L_n. \tag{9.8}$$

In particular, if $k \in \mathbb{N}$ and $f : I \setminus \{x_0\} \to \mathbb{R}$ is such that $\lim_{x \to x_0} f(x) = L$, then

$$\lim_{x \to x_0} f(x)^k = L^k.$$

From now on, we assume the validity of these remarks without further comments.

Our next result is usually referred to as the **squeezing theorem**. As we shall see right after its proof, it is quite useful for the actual computation of limits.

Proposition 9.8 *Let I be an interval, $x_0 \in I$ and $f, g, h : I \setminus \{x_0\} \to \mathbb{R}$ be such that $g(x)$ belongs to the interval of ends $f(x)$ and $h(x)$ for every $x \in I \setminus \{x_0\}$. If $\lim_{x \to x_0} f(x) = \lim_{x \to x_0} h(x) = L$, then $\lim_{x \to x_0} g(x)$ exists and is also equal to L.*

Proof Given $\epsilon > 0$, we want to find $\delta > 0$ for which the conditions $x \in I$ and $0 < |x - x_0| < \delta$ imply $|g(x) - L| < \epsilon$. To this end, if $f(x) \le g(x) \le h(x)$, then $f(x) - L \le g(x) - L \le h(x) - L$ and is easy to conclude that

$$|g(x) - L| \le \max\{|f(x) - L|, |h(x) - L|\}.$$

If $h(x) \le g(x) \le f(x)$, one reaches the inequality above in essentially the same way.

Now, the definition of limit gives $\delta_1, \delta_2 > 0$ such that

$$x \in I, \ 0 < |x - x_0| < \delta_1 \Rightarrow |f(x) - L| < \epsilon$$

and

$$x \in I, \ 0 < |x - x_0| < \delta_2 \Rightarrow |h(x) - L| < \epsilon.$$

Therefore, letting $\delta = \min\{\delta_1, \delta_2\} > 0$ and taking $x \in I$ such that $0 < |x - x_0| < \delta$, we get $|f(x) - L| < \epsilon$ and $|h(x) - L| < \epsilon$, so that

$$|g(x) - L| \le \max\{|f(x) - L|, |h(x) - L|\} < \epsilon.$$

$\square$

Concerning the computation of limits, the coming corollary is quite a useful tool. For its statement, recall that a function $f : I \to \mathbb{R}$ is *bounded* if there exists $M > 0$ such that

$$|f(x)| \le M, \ \forall \ x \in I.$$

Corollary 9.9 *Let I be an interval, $x_0 \in I$ and $f, g : I \setminus \{x_0\} \to \mathbb{R}$ be such that f is bounded and $\lim_{x \to x_0} g(x) = 0$. Then, $\lim_{x \to x_0} f(x)g(x) = 0$, even if $\lim_{x \to x_0} f(x)$ doesn't exist.*

Proof If $|f(x)| \le M$ for every $x \in I \setminus \{x_0\}$, then $0 \le |f(x)g(x)| \le M|g(x)|$ for every $x \in I \setminus \{x_0\}$. Since $\lim_{x \to x_0} |g(x)| = 0$ (check this!), it follows from the squeezing principle that $\lim_{x \to x_0} |f(x)g(x)| = 0$. Therefore, $\lim_{x \to x_0} f(x)g(x) = 0$. $\qquad\square$

Example 9.10 If $f : \mathbb{R} \setminus \{0\} \to \mathbb{R}$ is given by $f(x) = \sin \frac{1}{x}$, then $|f(x)| \le 1$ for every $x \in \mathbb{R} \setminus \{0\}$. Since $\lim_{x \to 0} x = 0$, it follows from the previous proposition that $\lim_{x \to 0} x \cdot \sin \frac{1}{x} = 0$. Note that this result agrees with Problem 10, page 255, and (9.6).

The squeezing theorem also allows us to compute the **fundamental trigonometric limit**, which will reveal itself to be of crucial importance in the next section.

Lemma 9.11 $\lim_{x \to 0} \frac{\sin x}{x} = 1$.

Proof Since we wish to compute a limit, we can restrict ourselves to the interval $|x| < \frac{\pi}{2}$. Suppose first that $x > 0$. Letting $\ell(\overset{\frown}{AB}) = x$ be the length of the arc $\overset{\frown}{AB}$ (cf. Fig. 9.3), we have

$$\sin x = \overline{BD} < \overline{AB} < \ell(\overset{\frown}{AB}) = x,$$

so that $\frac{\sin x}{x} < 1$. On the other hand, it is well known (cf. Chap. 5 of [5], for instance) that the area of the circular sector AOB equals $\pi \cdot \frac{x}{2\pi} = \frac{x}{2}$, whereas the area of the

Fig. 9.3 The fundamental trigonometric limit

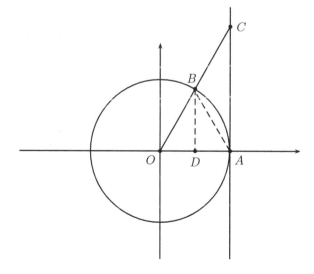

triangle AOC equals $\frac{1}{2}\overline{AC}$; hence, we have

$$x = \ell(\overset{\frown}{AB}) < \overline{AC} = \tan x,$$

so that $\cos x < \frac{\sin x}{x}$.

Combining the two inequalities deduced above, we obtain

$$\cos x < \frac{\sin x}{x} < 1 \qquad\qquad (9.9)$$

for $0 < x < \frac{\pi}{2}$. However, since both $x \mapsto \cos x$ and $x \mapsto \frac{\sin x}{x}$ are even functions, these inequalities still hold true for $-\frac{\pi}{2} < x < 0$. Then, it follows from (9.9), together with the squeezing theorem and the continuity of the cosine function, that $\lim_{x\to 0} \frac{\sin x}{x}$ exists and equals 1. $\qquad\square$

Remark 9.12 The attentive reader may object the former proof under the following argument: the very definitions of the sine and cosine functions, as given in elementary school, rely upon the fact that a circle of radius 1 has length 2π; in turn, as we shall see in Sect. 10.10, a thorough computation of length of a circle involves the use of the derivative of the sine function, so that we are in presence of a circular reasoning. Although this is indeed the case, we observe that we will fix this situation in Sect. 11.5, when we shall construct the sine and cosine functions without invoking any concepts or results of Euclidean Geometry. In particular, at that future time we will show, by purely analytical means, that $\lim_{x\to 0} \frac{\sin x}{x} = 1$ and $\sin' x = \cos x$. For the time being, we suggest that the reader takes the proof of the previous lemma simply as a heuristic argument for justifying the validity of both these results.

Let be given an open interval I, a point $x_0 \in I$ and a function $f : I \setminus \{x_0\} \to \mathbb{R}$. In a way entirely analogous to what we have done so far, we can define for f the concept of **one-sided** or **lateral** limits at x_0. More precisely:

(i) If $I \cap (-\infty, x_0) \neq \emptyset$, we say that L is the **left-handed limit** of f as x goes to x_0, and write $\lim_{x\to x_0-} f(x) = L$, if, for each given $\epsilon > 0$, there exists $\delta > 0$ such that

$$x \in I, \ x_0 - \delta < x < x_0 \Rightarrow |f(x) - L| < \epsilon. \qquad\qquad (9.10)$$

(ii) If $I \cap (x_0, +\infty) \neq \emptyset$, we say that L is the **right-handed limit** of f as x goes to x_0, and write $\lim_{x\to x_0+} f(x) = L$, if, for each given $\epsilon > 0$, there exists $\delta > 0$ such that

$$x \in I, \ x_0 < x < x_0 + \delta \Rightarrow |f(x) - L| < \epsilon. \qquad\qquad (9.11)$$

The idea behind the notation $x \to x_0-$ (resp. $x \to x_0+$) should be clear to the reader: we write $x \to x_0-$ (resp. $x \to x_0+$) in left-handed (resp. right-handed)

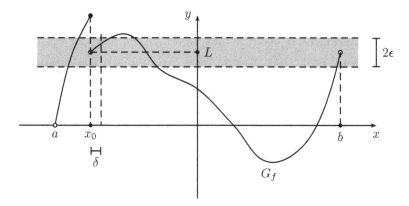

Fig. 9.4 The right-handed limit $\lim_{x \to x_0+} f(x)$

limits, for x approaches x_0 *from the left* (resp. *from the right*), i.e., through values smaller (resp. larger) than x_0 itself. In yet another way, when $x \to x_0-$ (resp. $x \to x_0+$), we have $x = x_0 - $ (something positive) (resp. $x = x_0 + $ (something positive)).

Figure 9.4 gives a geometric interpretation for the notion of right-handed limit. Note that, for $0 < |x - x_0| < \delta$, the point $(x, f(x))$ only belongs to the gray strip if $x_0 < x < x_0 + \delta$. We left to the reader the task of drawing a figure that provides the analogous geometric interpretation for left-handed limits.

Yet with respect to one-sided limits, it is immediate to state and prove for them results analogous to those of Proposition 9.7 (cf. Problem 4). We also stress that, for a function $f : (a, x_0) \to \mathbb{R}$, the notions of limite and left-handed limit of f at x_0 coincide (for, there is simply no way of approaching x_0 from the right, while belonging to the domain of f). Thus, in such a case we shall generally write $\lim_{x \to x_0} f(x) = L$ instead of $\lim_{x \to x_0-} f(x) = L$. Obviously, for a given function $f : (x_0, b) \to \mathbb{R}$, analogous remarks are true for limits and right-handed limits at x_0. In this respect, see also Problem 5.

We close this section by defining the notions of *infinite limits* and *limits at infinity*, and sketching some elementary (though useful) results on them.

Definition 9.13 Let be given an interval I, a point $x_0 \in I$ and a function $f : I \setminus \{x_0\} \to \mathbb{R}$. We write

$$\lim_{x \to x_0} f(x) = +\infty$$

if, for any given $M > 0$, there exists $\delta > 0$ such that

$$x \in I, \ 0 < |x - x_0| < \delta \Rightarrow f(x) > M.$$

We leave to the reader the task of stating definitions analogous to the one above for the concepts of $\lim_{x \to x_0} f(x) = -\infty$ and $\lim_{x \to x_0\pm} f(x) = \pm\infty$.

The next result establishes two simple, yet useful, arithmetic rules for infinite limits. Since its proof resembles that of Proposition 9.7, we also leave it as an exercise for the reader.

Proposition 9.14 *Let be given an interval I, a point $x_0 \in I$ and functions f, g : $I \setminus \{x_0\} \to \mathbb{R}$.*

(a) If $\lim_{x \to x_0} f(x) = L > 0$ and $\lim_{x \to x_0} g(x) = \pm\infty$, then $\lim_{x \to x_0} f(x)g(x) = \pm\infty$.
(b) If $\lim_{x \to x_0} f(x) = L < 0$ and $\lim_{x \to x_0} g(x) = \pm\infty$, then $\lim_{x \to x_0} f(x)g(x) = \mp\infty$.

The prototype of infinite limits are those collected in the coming example, whose proof is also an easy exercise for the reader.

Example 9.15

(a) If $f : (x_0, +\infty) \to \mathbb{R}$ is given by $f(x) = \frac{1}{x - x_0}$, then $\lim_{x \to x_0} f(x) = +\infty$.
(b) If $f : (-\infty, x_0) \to \mathbb{R}$ is given by $f(x) = \frac{1}{x - x_0}$, then $\lim_{x \to x_0} f(x) = -\infty$.

Let's now turn our attention to **limits at infinity**.

Definition 9.16 Given $f : (a, +\infty) \to \mathbb{R}$ and $L \in \mathbb{R}$, we write $\lim_{x \to +\infty} f(x) = L$ if, for a given $\epsilon > 0$, there exists $A > a$ such that

$$x > A \Rightarrow |f(x) - L| < \epsilon.$$

As before, we leave to the reader the task of stating definitions analogous to the above one for $\lim_{x \to -\infty} f(x) = L$ and $\lim_{x \to \pm\infty} f(x) = \pm\infty$. On the other hand, note that Propositions 9.7, 9.8 and 9.14 remain true if we change $I \setminus \{x_0\}$ by $(a, +\infty)$ (resp. $(-\infty, b)$) and x_0 by $+\infty$ (resp. $-\infty$); moreover, the proofs of the corresponding results in these cases are essentially the same.

Let us see an interesting (and useful) example of application of Proposition 9.14.

Example 9.17 Let $n \in \mathbb{N}$ and $f : \mathbb{R} \to \mathbb{R}$ be the polynomial function given by

$$f(x) = a_n x^n + a_{n-1} x^{n-1} + \cdots + a_1 x + a_0,$$

with $a_0, a_1, \ldots, a_n \in \mathbb{R}$ and $a_n > 0$. Prove that:

(a) If n is even, then $\lim_{|x| \to +\infty} f(x) = +\infty$.
(b) If n is odd, then $\lim_{x \to \pm\infty} f(x) = \pm\infty$.

Proof Since $\lim_{|x| \to +\infty} x^n = +\infty$ if n is even, $\lim_{x \to \pm\infty} x^n = \pm\infty$ if n is odd and $\lim_{|x| \to +\infty} \left(a_n + \frac{a_{n-1}}{x} + \cdots + \frac{a_1}{x^{n-1}} + \frac{a_0}{x^n} \right) = a_n$ (verify these claims!), it suffices to write

$$f(x) = x^n \left(a_n + \frac{a_{n-1}}{x} + \cdots + \frac{a_1}{x^{n-1}} + \frac{a_0}{x^n} \right)$$

and, then, apply Proposition 9.14 (with $\pm\infty$ in place of x_0). □

One of the most important applications of infinite limits and limits at infinity is in identifying horizontal and vertical *asymptotes* of (the graph of) a function, according to the coming

Definition 9.18 If I is an interval, $x_0 \in I$ and $f : I \setminus \{x_0\} \to \mathbb{R}$ is a function such that $\lim_{x \to x_0} f(x) = \pm\infty$, then we say that line $x = x_0$ is a **vertical asymptote** of (the graph of) f. Analogously, if $f : (a, +\infty) \to \mathbb{R}$ (resp. $f : (-\infty, b) \to \mathbb{R}$) is such that $\lim_{x \to +\infty} f(x) = L$ (resp. $\lim_{x \to -\infty} f(x) = L$), then we say that line $y = L$ is a **horizontal asymptote** of (the graph of) f.

Examples 9.19

(a) Figure 6.17 sketches the graph of the tangent function in the interval $(-\frac{\pi}{2}, \frac{\pi}{2})$. In view of the limits $\lim_{x \to \frac{\pi}{2}} \sin x = 1$ and $\lim_{x \to \frac{\pi}{2}} \cos x = 0$, it follows from Proposition 9.14 that $\lim_{x \to \frac{\pi}{2}} \tan x = +\infty$. Analogously, $\lim_{x \to -\frac{\pi}{2}} \tan x = -\infty$, so that lines $x = \pm\frac{\pi}{2}$ are vertical asymptotes of this graph.

(b) Figure 9.5 sketches the graph of $f : \mathbb{R} \setminus \{-1\} \to \mathbb{R}$, given by $f(x) = \frac{2x}{x+1}$. This graph can be obtained from that of the inverse proportionality function (cf. Fig. 6.14) with the aid of Problem 10, page 194. Indeed, since

$$\frac{2x}{x+1} = \frac{2x + 2 - 2}{x+1} = 2 - \frac{2}{x+1}, \qquad (9.12)$$

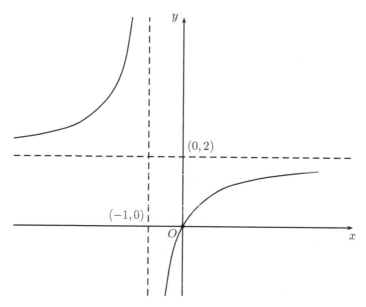

Fig. 9.5 Graph of $x \mapsto \frac{2x}{x+1}$

we can sketch the graph of f in the following way: firstly, we sketch the graph of $x \mapsto \frac{1}{x}$; then, we translate it one unit to the left, thus getting the graph of $x \mapsto \frac{1}{x+1}$; thirdly, we stretch this second graph in the vertical direction by factor 2, obtaining the graph of $x \mapsto \frac{2}{x+1}$; we then reflect the result along the horizontal axis, arriving at the graph of $x \mapsto -\frac{2}{x+1}$; we end by translating this last graph two units above, thus finally obtaining the graph of $x \mapsto \frac{2x}{x+1}$.

Since $\lim_{x \to -1\pm} 2x = \mp 2$ and $\lim_{x \to -1\pm} \frac{1}{x+1} = \pm\infty$, it follows from Proposition 9.14 that $\lim_{x \to -1\pm} \frac{2x}{x+1} = \mp\infty$, so that line $x = -1$ is indeed a vertical asymptote of the graph. On the other hand, it easily follows from (9.12) that $\lim_{x \to \pm\infty} \frac{2x}{x+1} = 2$, so that line $y = 2$ is a horizontal asymptote of the graph. Finally, note that these results are in perfect accordance with the geometric intuition given by the sketch of the graph.

Problems: Section 9.2

1. Let I be an interval and $x_0 \in I$. Given a function $f : I \setminus \{x_0\} \to \mathbb{R}$, prove that if $\lim_{x \to x_0} f(x)$ exists, then it is unique.
2. * Let I be an interval, $x_0 \in I$ and $f, g : I \setminus \{x_0\} \to \mathbb{R}$ be such that there exist $\lim_{x \to x_0} f(x) = L$ and $\lim_{x \to x_0} g(x) = M$. If $f(x) \geq g(x)$ for every $x \in I \setminus \{x_0\}$, prove that $L \geq M$.
3. * Establish the generalizations of items (a) and (b) of Proposition 9.7, as discussed right after the proof of it.
4. * Extend Proposition 9.7 to one-sided limits.
5. * Let $f : (a, b) \to \mathbb{R}$ be a given function, and $x_0 \in (a, b)$. Prove that $\lim_{x \to x_0} f(x)$ exists if and only if the one-sided limits $\lim_{x \to x_0+} f(x)$ and $\lim_{x \to x_0-} f(x)$ *exist and are equal*. In this case, show also that $\lim_{x \to x_0} f(x)$ equals this common value.
6. Prove Proposition 9.14 and its analogue for limits at infinity.
7. * Let $-\infty \leq a < b \leq +\infty$ and $f : (a, b) \to \mathbb{R}$ be a continuous and increasing (resp. decreasing) function. If $\lim_{x \to a+} f(x) = L$ and $\lim_{x \to b-} f(x) = M$, with $-\infty \leq L, M \leq +\infty$, prove that $\text{Im}(f) = (L, M)$.
8. Compute the following limits:
   (a) $\lim_{x \to 0} \frac{x \sin x}{1 - \cos x}$.
   (b) $\lim_{x \to 0} \frac{1 - \cos x}{x^2}$.
   (c) $\lim_{x \to 0} \frac{\sin(2x)}{\sin(3x)}$
   (d) $\lim_{x \to 1} \frac{\sin \frac{\pi}{2}(x-1)}{x-1}$.
   (e) $\lim_{x \to 1} \frac{\cos\left(\frac{\pi x}{2}\right)}{x-1}$.
   (f) $\lim_{x \to 0} \frac{1 - \sqrt[3]{\cos x}}{x^2}$.
9. If $f, g : (a, +\infty) \to \mathbb{R}$ are such that f is bounded and $\lim_{x \to +\infty} g(x) = 0$, prove that $\lim_{x \to +\infty} f(x)g(x) = 0$. Then, use this fact to compute the two limits below:
   (a) $\lim_{x \to +\infty} \frac{\sin x}{x}$.
   (b) $\lim_{x \to +\infty} \frac{\sin x}{x + \cos x}$.

10. * Let I be an interval, $x_0 \in I$ and $f : I \setminus \{x_0\} \to \mathbb{R}$ be such that $\lim_{x \to x_0} f(x) = L$. If $(a_n)_{n \geq 1}$ is a sequence in $I \setminus \{x_0\}$ satisfying $\lim_{n \to +\infty} a_n = x_0$, prove that $\lim_{n \to +\infty} f(a_n) = L$. Then, use this fact to conclude that $\lim_{x \to 0} \sin \frac{1}{x}$ does not exist.

11. * Given $f : (A, +\infty) \to \mathbb{R}$, we say that line $y = ax+b$ is an **oblique asymptote** (of the graph) of f provided $\lim_{x \to +\infty} (f(x) - (ax + b)) = 0$. If this is so, prove that

$$a = \lim_{x \to +\infty} \frac{f(x)}{x} \quad \text{and} \quad b = \lim_{x \to +\infty} (f(x) - ax).$$

Then, elaborate the concept of oblique asymptote $y = ax+b$ for $f : (-\infty, A) \to \mathbb{R}$ and show how to compute a and b in such a case.

12. In each of the following items, find out (cf. previous problem) the horizontal, vertical and oblique asymptotes of the given function:

(a) $f : \mathbb{R} \setminus \{0\} \to \mathbb{R}$ such that $f(x) = x + \frac{1}{x}$ for every real $x \neq 0$.

(b) $f : \mathbb{R} \setminus (-a, a) \to \mathbb{R}$ such that $f(x) = \frac{b}{a}\sqrt{x^2 - a^2}$ for $|x| \geq a$. (Notice— cf. Chap. 6 of [4], for instance—that the graph of f is the portion of the hyperbola $\frac{x^2}{a^2} - \frac{y^2}{b^2} = 1$ situated in the upper semiplane of the cartesian plane.)

(c) $f : (0, +\infty) \to \mathbb{R}$ such that $f(x) = x^2 \sin \frac{1}{x}$ for $x > 0$.

13. Compute $\lim_{x \to +\infty} \left(\sqrt{x + \sqrt{x + \sqrt{x}}} - \sqrt{x} \right)$.

14. Let $f : (0, +\infty) \to (0, +\infty)$ be an increasing function. If $\lim_{x \to +\infty} \frac{f(2x)}{f(x)} = 1$, prove that $\lim_{x \to +\infty} \frac{f(nx)}{f(x)} = 1$ for every $n \in \mathbb{N}$.

15. Let $(a_n)_{n \geq 1}$ be the sequence defined by $a_1 = 2\sqrt{2}$ and, for an integer $n > 1$,

$$a_n = 2^n \sqrt{2 - \sqrt{2 + \sqrt{2 + \cdots + \sqrt{2}}}},$$ with n square root signs. In this respect, do the following items:

(a) If $b_n = \sqrt{2 + \sqrt{2 + \cdots + \sqrt{2}}}$ for $n > 1$, with $n-1$ square root signs, show that $b_n^2 = 2 + b_{n-1}$ and conclude that $b_n < 2$ for every $n > 1$.

(b) Write $b_n = 2 \cos \theta_n$, with $0 < \theta_n < \frac{\pi}{2}$, to conclude that $\theta_n = \frac{1}{2}\theta_{n-1}$ for every $n > 2$.

(c) Show that $a_n = 2^{n+1} \sin \frac{\theta_n}{2} = 2^{n+1} \sin \frac{\theta_1}{2^n}$. Then, use this fact to show that $\lim_{n \to +\infty} a_n = \pi$.

16. (IMO) Find all functions $f : (0, +\infty) \to (0, +\infty)$ satisfying the following conditions:

(a) $f(xf(y)) = yf(x)$ for all $x, y > 0$.

(b) $\lim_{x \to +\infty} f(x) = 0$.

9.3   Basic Properties of Derivatives

With the concept of limit at our disposal, we return to the discussion that lead to (9.3).

Definition 9.20 Let $I \subset \mathbb{R}$ be an open interval and $f : I \to \mathbb{R}$ be a given function. For a fixed $x_0 \in I$, we say that f is **differentiable** at x_0 if the limit

$$\lim_{x \to x_0} \frac{f(x) - f(x_0)}{x - x_0}$$

does exist. In this case, this limit is called the **derivative** of f at x_0, and is denoted $f'(x_0)$.

In the settings of the previous definition, we observe that the existence of $f'(x_0)$ is equivalent to the existence of the limit

$$\lim_{h \to 0} \frac{f(x_0 + h) - f(x_0)}{h}.$$

Indeed, on the one hand, letting $x_0 + h = x$ we have $h = x - x_0$, and obviously $h \to 0 \Leftrightarrow x \to x_0$. On the other hand, when we write $f(x_0 + h)$ we are tacitly assuming that h is so small that $x_0 + h \in I$; however, since we are computing a limit and I is open, such a supposition doesn't impose any restriction upon the definition of derivative.

In short, f being differentiable at $x_0 \in I$, we have

$$f'(x_0) = \lim_{x \to x_0} \frac{f(x) - f(x_0)}{x - x_0} = \lim_{h \to 0} \frac{f(x_0 + h) - f(x_0)}{h}. \qquad (9.13)$$

From now on, whenever convenient we refer to any one of the fractions above as the **Newton's quotient** of f at x_0.

The coming examples compute derivatives of some simple functions.

Example 9.21

(a) If $f : \mathbb{R} \to \mathbb{R}$ is a constant function, then f is differentiable and $f'(x_0) = 0$ for every $x_0 \in \mathbb{R}$.
(b) If $n \in \mathbb{N}$ and $f : \mathbb{R} \to \mathbb{R}$ is such that $f(x) = x^n$ for every $x \in \mathbb{R}$, then f is differentiable, with $f'(x_0) = n x_0^{n-1}$ for every $x_0 \in \mathbb{R}$.
(c) If $n \in \mathbb{Z}$ is negative and $f : \mathbb{R} \setminus \{0\} \to \mathbb{R}$ is given by $f(x) = x^n$, then f is differentiable and $f'(x_0) = n x_0^{n-1}$ for every $x_0 \in \mathbb{R} \setminus \{0\}$.

Proof (a) Let's apply the first equality in (9.13): letting $f(x) = c$ for every $x \in \mathbb{R}$, we get

$$f'(x_0) = \lim_{x \to x_0} \frac{f(x) - f(x_0)}{x - x_0} = \lim_{x \to x_0} \frac{c - c}{x - x_0} = \lim_{x \to x_0} 0 = 0.$$

(b) Let's apply the second equality in (9.13): by Newton's binomial formula, we get

$$\frac{f(x_0 + h) - f(x_0)}{h} = \frac{1}{h}((x_0 + h)^n - x_0^n) = \frac{1}{h}\left(\sum_{k=0}^{n}\binom{n}{k}x_0^{n-k}h^k - x_0^n\right)$$

$$= \sum_{k=1}^{n}\binom{n}{k}x_0^{n-k}h^{k-1} = nx_0^{n-1} + \sum_{k=2}^{n}\binom{n}{k}x_0^{n-k}h^{k-1}.$$

Then, successively applying (9.7) and (9.6), we obtain

$$f'(x_0) = \lim_{h\to 0}\left(nx_0^{n-1} + \sum_{k=2}^{n}\binom{n}{k}x_0^{n-k}h^{k-1}\right)$$

$$= nx_0^{n-1} + \sum_{k=2}^{n}\lim_{h\to 0}\left(\binom{n}{k}x_0^{n-k}h^{k-1}\right) = nx_0^{n-1}.$$

(c) Let $n = -m$, with $m \in \mathbb{N}$. Then,

$$\frac{f(x) - f(x_0)}{x - x_0} = \frac{1}{x - x_0}\left(\frac{1}{x^m} - \frac{1}{x_0^m}\right) = -\frac{1}{x^m x_0^m}\left(\frac{x^m - x_0^m}{x - x_0}\right).$$

Hence, Proposition 9.7 and (9.6) furnish

$$f'(x_0) = -\lim_{x\to x_0}\frac{1}{x^m x_0^m}\left(\frac{x^m - x_0^m}{x - x_0}\right) = -\lim_{x\to x_0}\frac{1}{x^m x_0^m} \cdot \lim_{x\to x_0}\left(\frac{x^m - x_0^m}{x - x_0}\right)$$

$$= -\frac{1}{x_0^{2m}} \cdot \lim_{x\to x_0}\left(\frac{x^m - x_0^m}{x - x_0}\right).$$

Now, applying the result of item (b) (with m in place of n), we get

$$f'(x_0) = -\frac{1}{x_0^{2m}} \cdot mx_0^{m-1} = -mx_0^{-m-1} = nx_0^{n-1}.$$

$\square$

Example 9.22 The sine and cosine functions are differentiable, with $\sin' x_0 = \cos x_0$ and $\cos' x_0 = -\sin x_0$ for every $x_0 \in \mathbb{R}$.

Proof Let's do the proof for the sine function, the proof for the cosine function being completely analogous.

It follows from the product formulas of Trigonometry that

$$\frac{\sin(x_0 + h) - \sin x_0}{h} = \frac{\sin \frac{h}{2}}{\frac{h}{2}} \cdot \cos\left(x_0 + \frac{h}{2}\right).$$

Since the cosine function is continuous, Proposition 9.7, (9.6) and the fundamental trigonometric limit (cf. Lemma 9.11) give

$$\sin' x_0 = \lim_{h \to 0} \frac{\sin \frac{h}{2}}{\frac{h}{2}} \cdot \cos\left(x_0 + \frac{h}{2}\right)$$

$$= \lim_{h \to 0} \frac{\sin \frac{h}{2}}{\frac{h}{2}} \cdot \lim_{h \to 0} \cos\left(x_0 + \frac{h}{2}\right)$$

$$= \lim_{h \to 0} \frac{\sin h}{h} \cdot \cos x_0 = \cos x_0.$$

$\square$

Remarks 9.23 Before we proceed with the development of the theory, let's pause to make two useful remarks.

1. Since derivatives are limits, given a function $f : [a, b) \to \mathbb{R}$ we can consider the **right-handed derivative** of f at a, i.e., the one-sided limit

$$f'_+(a) = \lim_{x \to a+} \frac{f(x) - f(a)}{x - a},$$

provided it exists. Accordingly, for $f : (a, b] \to \mathbb{R}$ we can consider the **left-handed derivative** at b, i.e., the limit

$$f'_-(b) = \lim_{x \to b-} \frac{f(x) - f(b)}{x - b}$$

(also as before, provided such a limit exists). Moreover, whenever we say that a function $f : [a, b] \to \mathbb{R}$ is **differentiable**, we shall implicitly assume that its derivatives at $x = a$ and $x = b$ are one-sided ones, and shall write simply $f'(a)$ and $f'(b)$ (instead of $f'_+(a)$ and $f'_-(b)$) to denote them. From now on, such conventions will be in force, without further comments.

2. As in Sect. 9.1, we shall sometimes denote the derivative of a function f at a point x_0 by using Leibniz's classical notation $\frac{df}{dx}(x_0)$, so that

$$\frac{df}{dx}(x_0) = \lim_{x \to x_0} \frac{f(x) - f(x_0)}{x - x_0}.$$

The definition of derivative, together with the heuristic discussion of Sect. 9.1, allows us to present another important

Fig. 9.6 Tangent to the graph of $x \mapsto \frac{1}{x}$

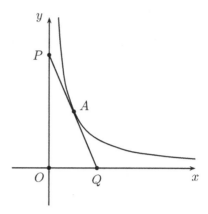

Definition 9.24 Let I be an interval and $f : I \to \mathbb{R}$ be a function differentiable at $x_0 \in I$. The **tangent line** to the graph of f at the point $(x_0, f(x_0))$ is the line that passes through this point and has slope $f'(x_0)$.

If $f : I \to \mathbb{R}$ is differentiable at x_0, it follows from the definition above and elementary analytic geometry (cf. Chap. 6 of [4], for instance) that the tangent line to its graph at $(x_0, f(x_0))$ has equation

$$y - f(x_0) = f'(x_0)(x - x_0). \tag{9.14}$$

Example 9.25 Let $\mathcal{H}$ be the portion of the curve $xy = 1$ contained in the first quadrant of a cartesian plane,[2] and P and Q be points along the axes, such that line $\overleftrightarrow{PQ}$ is tangent to $\mathcal{H}$ at the point A (cf. Fig. 9.6). Prove that:

(a) $\overline{AP} = \overline{AQ}$.
(b) If O is the origin of the cartesian plane, then the area of triangle POQ doesn't depend on the position of A along $\mathcal{H}$.

Proof It is clear that $\mathcal{H}$ coincides with the graph of $f : (0, +\infty) \to \mathbb{R}$, given by $f(x) = \frac{1}{x}$. If $x_0 > 0$ and A is the point $(x_0, f(x_0)) = (x_0, \frac{1}{x_0})$, then the line tangent to G_f at A has equation $y - \frac{1}{x_0} = f'(x_0)(x - x_0)$. Example 9.21 gives $f'(x_0) = -\frac{1}{x_0^2}$, so that the equation of the tangent line can be written as $y - \frac{1}{x_0} = -\frac{1}{x_0^2}(x - x_0)$ or, which is the same, $x_0^2 y + x = 2x_0$.

Making $x = 0$ and then $y = 0$ in such an equation, we get $P\left(0, \frac{2}{x_0}\right)$ and $Q(2x_0, 0)$, or vice-versa. In any case, items (a) and (b) follow immediately:

---

[2]Such a portion is a branch of an equilateral hyperbola (cf. Chap. 6 of [4], for instance), but we shall not use this fact.

(a) It's a well know fact (again, see Chap. 6 of [4], for instance) that the midpoint of PQ has abscissa and ordinate respectively equal to $\frac{1}{2} \cdot \frac{2}{x_0} = \frac{1}{x_0}$ and $\frac{1}{2} \cdot 2x_0 = x_0$. Hence, such a point coincides with A.

(b) Since POQ is a right triangle, we have

$$A(POQ) = \frac{1}{2}\overline{OP} \cdot \overline{OQ} = \frac{1}{2} \cdot \frac{2}{x_0} \cdot 2x_0 = 2.$$

$\square$

Continuing with the analysis of the concept of derivative, let's show that differentiability is *stronger* than continuity.

Proposition 9.26 *If a function $f : I \rightarrow \mathbb{R}$ is differentiable at $x_0 \in I$, then f is continuous at x_0.*

Proof Note first that, for $x \in I \setminus \{x_0\}$, we have

$$f(x) = f(x_0) + \left(\frac{f(x) - f(x_0)}{x - x_0}\right)(x - x_0).$$

On the other hand, since we are assuming the existence of $\lim_{x \to x_0} \frac{f(x)-f(x_0)}{x-x_0} = f'(x_0)$, it follows from Proposition 9.7 that

$$\lim_{x \to x_0} f(x) = \lim_{x \to x_0}\left(f(x_0) + \left(\frac{f(x)-f(x_0)}{x-x_0}\right)(x-x_0)\right)$$

$$= f(x_0) + \lim_{x \to x_0}\left(\frac{f(x)-f(x_0)}{x-x_0}\right)\lim_{x \to x_0}(x-x_0)$$

$$= f(x_0) + f'(x_0) \cdot 0 = f(x_0).$$

Therefore, Proposition 9.4 assures the continuity of f at x_0. $\square$

The coming example shows that there exist continuous functions which are not differentiable in an arbitrary finite set of points of their domains.[3]

Example 9.27 Given distinct $a_1, a_2, \ldots, a_n \in \mathbb{R}$, the function $f : \mathbb{R} \rightarrow \mathbb{R}$ such that $f(x) = \sum_{j=1}^{n}|x - a_j|$ is continuous, though not differentiable at any of $a_1, a_2, \ldots, a_n$.

Proof The given function is obviously continuous. For what is left to do note that, for a fixed $1 \leq k \leq n$, we have

$$f(x) = |x - a_k| + \sum_{\substack{1 \leq j \leq n \\ j \neq k}}|x - a_j|,$$

---

[3]Later, in Chap. 11, we shall see an example of a continuous function $f : \mathbb{R} \rightarrow \mathbb{R}$ that is not differentiable at any point.

and in an appropriate neighborhood of a_k each summand $|x - a_j|$, with $j \neq k$, equals $x - a_j$ or $a_j - x$. Therefore, it suffices to show that $x \mapsto |x - a_k|$ is not differentiable at a_k.

For the sake of notation, write $f(x) = |x - a|$, with $a \in \mathbb{R}$. Then, for $x \neq a$, we have

$$\frac{f(x) - f(a)}{x - a} = \frac{|x - a|}{x - a} = \begin{cases} 1, & \text{if } x > a \\ -1, & \text{if } x < a \end{cases}.$$

Hence, $\lim_{x \to a+} \frac{f(x) - f(a)}{x - a} = 1$ and $\lim_{x \to a-} \frac{f(x) - f(a)}{x - a} = -1$, so that Problem 5, page 290, assures that $\lim_{x \to a} \frac{f(x) - f(a)}{x - a}$ does not exist. □

If $I \subset \mathbb{R}$ is an interval and $f : I \to \mathbb{R}$ is differentiable at each $x_0 \in I$, we shall simply say that f is *differentiable function*. In this case, the **derivative function** $f' : I \to \mathbb{R}$ of f associates to $x \in I$ the derivative $f'(x)$ of f at x.

If I and J are intervals and $f : I \to J$ is a differentiable bijection, the last proposition guarantees that f is continuous. Moreover, Theorem 8.35 assures that $f^{-1} : J \to I$ is also continuous. Our next result explains when f^{-1} is differentiable.

Theorem 9.28 *Let I and J be intervals and $f : I \to J$ be a differentiable bijection. For $x_0 \in I$ and $y_0 = f(x_0) \in J$, we have $f^{-1} : J \to I$ differentiable at y_0 if and only if $f'(x_0) \neq 0$. Moreover, if this is so, then*

$$(f^{-1})'(y_0) = \frac{1}{f'(x_0)}. \tag{9.15}$$

Proof For h sufficiently close to (but different from) 0, let $x = x_0 + h$, $l = f(x_0 + h) - f(x_0)$ and $y = y_0 + l$. Then, $y = f(x_0) + l = f(x_0 + h)$, so that

$$h = (x_0 + h) - x_0 = f^{-1}(y_0 + l) - f^{-1}(y_0).$$

Hence, the continuity of f and f^{-1} guarantee that $h \to 0 \Leftrightarrow l \to 0$. Also, notice that

$$\left(\frac{f(x_0 + h) - f(x_0)}{h} \right) \left(\frac{f^{-1}(y_0 + l) - f^{-1}(y_0)}{l} \right) = \frac{l}{h} \cdot \frac{h}{l} = 1.$$

Now, suppose that f^{-1} is differentiable at y_0. Making $h \to 0$ (which, as we have just seen, is equivalent to making $l \to 0$), it follows from item (b) of Proposition 9.7 and the computations above that

$$1 = \lim_{h \to 0} \left[\left(\frac{f(x_0 + h) - f(x_0)}{h} \right) \left(\frac{f^{-1}(y_0 + l) - f^{-1}(y_0)}{l} \right) \right]$$

$$= \lim_{h \to 0} \left(\frac{f(x_0 + h) - f(x_0)}{h} \right) \lim_{l \to 0} \left(\frac{f^{-1}(y_0 + l) - f^{-1}(y_0)}{l} \right)$$

$$= f'(x_0)(f^{-1})'(y_0).$$

Therefore, $f'(x_0) \neq 0$ and $(f^{-1})'(y_0) = \frac{1}{f'(x_0)}$.

Conversely, suppose that $f'(x_0) \neq 0$. Since

$$\frac{f^{-1}(y_0 + l) - f^{-1}(y_0)}{l} = \frac{1}{(f(x_0 + h) - f(x_0))/h},$$

item (c) of Proposition 9.7, together with the fact that $l \to 0 \Leftrightarrow h \to 0$, furnishes

$$\lim_{l \to 0} \left(\frac{f^{-1}(y_0 + l) - f^{-1}(y_0)}{l} \right) = \lim_{l \to 0} \left(\frac{1}{(f(x_0 + h) - f(x_0))/h} \right)$$

$$= \lim_{h \to 0} \left(\frac{1}{(f(x_0 + h) - f(x_0))/h} \right) = \frac{1}{f'(x_0)}.$$

Therefore, f^{-1} is differentiable at y_0. □

Example 9.29 Let $n > 1$ be an integer and $f : [0, +\infty) \to [0, +\infty)$ be the n-th root function, i.e., $f(x) = \sqrt[n]{x} = x^{1/n}$. Then, f is not differentiable at $x = 0$ but is differentiable at every $x > 0$, with $f'(x) = \frac{1}{n} \cdot x^{1/n-1}$.

Proof Recall that $f = g^{-1}$, where $g : [0, +\infty) \to [0, +\infty)$ is such that $g(y) = y^n$. Hence, it follows from the previous result that f is differentiable at $x = g(y)$ if and only if $g'(y) \neq 0$. However, since $g'(y) = ny^{n-1}$, we have f differentiable at $x = g(y)$ if and only if $y \neq 0$ or, which is the same, $x \neq 0$. In this last case, recalling that $x = g(y) = y^n \Leftrightarrow y = \sqrt[n]{x}$, (9.15) gives

$$f'(x) = \frac{1}{g'(y)} = \frac{1}{ny^{n-1}} = \frac{1}{n(\sqrt[n]{x})^{n-1}} = \frac{1}{n} \cdot x^{1/n-1}.$$

□

For an important generalization of the previous example, see Problem 4, page 310.

We close this section by presenting the concept of higher order derivatives. To this end, let be given an interval I and a function $f : I \to \mathbb{R}$.

We say that f is **twice differentiable** at $x_0 \in I$ if there exists a neighborhood $(x_0 - r, x_0 + r)$ of x_0 such that f is differentiable in $I \cap (x_0 - r, x_0 + r)$ and $f' : I \cap (x_0 - r, x_0 + r) \to \mathbb{R}$ is differentiable at $x_0 \in I$. In this case, we also say that $(f')'(x_0)$ is the **second derivative** of f at x_0, and write $f''(x_0) = (f')'(x_0)$. If f is twice differentiable at each $x_0 \in I$, we shall simply say that f is twice differentiable in I, and let $f'' : I \to \mathbb{R}$ denote its **second derivative function**.

More generally, let $k \in \mathbb{N}$ and suppose we have already defined what is meant by the k-th derivative $f^{(k)}(x_0)$ of f at $x_0 \in I$. If f is k times differentiable in I (i.e., if $f^{(k)}(x_0)$ exists for every $x_0 \in I$), we let $f^{(k)} : I \to \mathbb{R}$ denote the k-th derivative[4] function of f. If f is k times differentiable in a neighborhood $(x_0 - r, x_0 + r)$ of x_0 and $f^{(k)} : I \cap (x_0 - r, x_0 + r) \to \mathbb{R}$ is differentiable at x_0, then we say that f is $k + 1$ times differentiable at x_0, and write

$$f^{(k+1)}(x_0) = (f^{(k)})'(x_0)$$

to denote its $(k + 1)$-th derivative at x_0.

If $f : I \to \mathbb{R}$ is k times differentiable in I and $f^{(k)} : I \to \mathbb{R}$ is continuous, we shall say that f is **k times continuously differentiable** (or simply **continuously differentiable**, if $k = 1$) in I. Finally, if f is k times differentiable in I for every $k \in \mathbb{N}$, then we say that f is **infinitely differentiable** in I.

Example 9.30 If $n \in \mathbb{Z} \setminus \{0\}$ and $f : \mathbb{R} \to \mathbb{R}$ is given by $f(x) = x^n$, it easily follows from item (b) of Example 9.21 and Problem 1 that f is infinitely differentiable in $\mathbb{R} \setminus \{0\}$ (in $\mathbb{R}$, if $n \geq 0$), with

$$f^{(k)}(x) = n(n - 1) \ldots (n - k + 1)x^{n-k}$$

for every $k \in \mathbb{N}$. In particular, if $n \geq 0$ and $k > n$, then $f^{(k)}$ vanishes identically.

Example 9.31 Functions $\sin, \cos : \mathbb{R} \to \mathbb{R}$ are infinitely differentiable, with

$$\sin^{(4k)} = \sin, \ \sin^{(4k+1)} = \cos, \ \sin^{(4k+2)} = -\sin, \ \sin^{(4k+3)} = -\cos$$

and

$$\cos^{(4k)} = \cos, \ \cos^{(4k+1)} = -\sin, \ \cos^{(4k+2)} = -\cos, \ \cos^{(4k+3)} = \sin,$$

for every $k \in \mathbb{N}$.

Problems: Section 9.3

1. * Let $I \subset \mathbb{R}$ be an interval and $f : I \to \mathbb{R}$ be differentiable at $x_0 \in I$. For a fixed $c \in \mathbb{R}$, prove that:

   (a) Function $g : I \to \mathbb{R}$ given by $g(x) = cf(x)$ is differentiable at x_0, with $g'(x_0) = cf'(x_0)$.
   (b) Function $h : I \to \mathbb{R}$ given by $h(x) = f(x) + c$ is differentiable at x_0, with $h'(x_0) = f'(x_0)$.

---

[4]Since $f^{(k)}$ is also used to denote the composition of $f : I \to I$ with itself, k times, we will rely on the context to clear any danger of confusion.

The next problem establishes a simple particular case of the *chain rule* (cf. Theorem 9.19).

2. * Let $f : (a, b) \to \mathbb{R}$ be differentiable at $x_0 \in (a, b)$. For $c > 0$ (resp. $c < 0$), let $g : \left(\frac{a}{c}, \frac{b}{c}\right) \to \mathbb{R}$ (resp. $g : \left(\frac{b}{c}, \frac{a}{c}\right) \to \mathbb{R}$) be the function given by $g(x) = f(cx)$. Show that g is differentiable at $\frac{x_0}{c}$, with $g'\left(\frac{x_0}{c}\right) = cf'(x_0)$.

3. * Let I be an interval and $k \in \mathbb{N}$. If $f, g : I \to \mathbb{R}$ are k times differentiable in $x_0 \in I$, prove that $f + g$ and fg are k times differentiable at x_0.

4. * In Example 8.37, we defined arcsin : $[-1, 1] \to [-\frac{\pi}{2}, \frac{\pi}{2}]$ and arccos : $[-1, 1] \to [0, \pi]$ as the inverses of sin : $[-\frac{\pi}{2}, \frac{\pi}{2}] \to [-1, 1]$ and cos : $[0, \pi] \to [-1, 1]$, respectively. Use Theorem 9.28 to prove that the restrictions of arcsin and arccos to the open interval $(-1, 1)$ are differentiable, with

$$\text{arcsin}' x = \frac{1}{\sqrt{1 - x^2}} \quad \text{and} \quad \text{arccos}' x = -\frac{1}{\sqrt{1 - x^2}},$$

para todo $x \in (-1, 1)$.

5. Compute the indicated derivatives (in each case, assume that the domain of the corresponding function is the maximal one):

(a) $f'(1)$; $f(x) = \frac{1}{4x^2}$.

(b) $f'(2)$; $f(x) = \frac{\sqrt{x}}{5}$.

(c) $f'(2)$; $f(x) = \sqrt{3x}$.

(d) $f'(1)$; $f(x) = \sqrt[3]{2x}$.

(e) $f'(\pi)$; $f(x) = \sin(3x)$.

(f) $f'\left(\frac{\pi}{2}\right)$; $f(x) = 4\cos\left(\frac{x}{3}\right)$.

(g) $f'\left(\frac{1}{4}\right)$; $f(x) = \arcsin(2x)$.

(h) $f'\left(\frac{1}{6}\right)$; $f(x) = 2\arccos(3x)$.

6. Let $f, g : (a, b) \to \mathbb{R}$ be differentiable functions whose graphs pass through a given point A. We say that these graphs are **tangent** to each other at A if their tangent lines at A coincide. In this respect, do the following items:

(a) Show that, for $m, n \in \mathbb{N}$, the graphs of $f(x) = x^n$ and $g(x) = x^m$ are tangent to each other at $(0, 0)$, but are tangent at $(1, 1)$ only if $m = n$.

(b) Let $f : (a, b) \to (a, b)$ be a differentiable bijection such that f^{-1} is also differentiable. If the graphs of f and f^{-1} are tangent to each other at $A(x_0, y_0)$, compute the possible values of $f'(x_0)f'(y_0)$.

7. Let $I \subset \mathbb{R}$ be an interval and $f : I \to \mathbb{R}$ be continuous at $x_0 \in I$. If the function $x \mapsto xf(x)$ $(x \in I)$ is differentiable at x_0, prove that f is also differentiable at x_0.

8. * Let I be an open interval, $x_0 \in I$ and $f : I \to \mathbb{R}$ be differentiable at x_0, with $f'(x_0) = L$. Given sequences $(a_n)_{n \geq 1}$ and $(b_n)_{n \geq 1}$ in I, with $a_n < x_0 < b_n$ for every $n \geq 1$, show that

$$\lim_{n \to +\infty} \frac{f(b_n) - f(a_n)}{b_n - a_n} = L.$$

9. Given a differentiable function $f : \mathbb{R} \to \mathbb{R}$, we say that a real number x_0 is a **root** of f provided $f(x_0) = 0$. **Newton's method**[5] for the computation of numerical approximations of the roots of f guarantees that, under certain conditions and for an appropriate choice of $\alpha \in \mathbb{R}$, the sequence $(a_n)_{n \geq 1}$ defined by $a_1 = \alpha$ and $a_{n+1} = a_n - \frac{f(a_n)}{f'(a_n)}$ for $n \geq 1$ (of course assuming $f'(a_n) \neq 0$ for every $n \geq 1$) converges to a root of f. In this respect, do the following items:

(a) Show that a_{n+1} is the point at which the tangent line to the graph of f at $(a_n, f(a_n))$ intersects the horizontal axis.

(b) Given $a > 0$, explain how Newton's method make it natural the definition of the sequence $(a_n)_{n \geq 1}$ of Problem 19, page 220.

10. Given $R > 0$, let $f : (-R, R) \to \mathbb{R}$ be given by $f(x) = \sqrt{R^2 - x^2}$. Prove that:

(a) The graph of f is the upper semicircle of radius R, centered at the origin of the cartesian plane and with diameter along the horizontal axis.

(b) For $x_0 \in (-R, R)$, we have $f'(x_0) = -\frac{x_0}{\sqrt{R^2 - x_0^2}}$.

(c) Given $x_0 \in (-R, R)$ and $A(x_0, f(x_0))$, the tangent line to the graph of f, obtained according to Definition 9.24, coincides with the straight line that passes through A and is perpendicular to the radius OA.

11. Let I be an open interval and $f : I \to \mathbb{R}$ be a differentiable function. Given $a, b \in \mathbb{R}$, suppose that there exists a straight line r, tangent to the graph of f and passing through the point (a, b). If (x_0, y_0) is the point at which r is tangent to the graph of f, show that

$$\begin{cases} y_0 = f(x_0) \\ y_0 - b = f'(x_0)(x_0 - a) \end{cases}.$$

Then, let r and r' be the straight lines that pass through the point $(1, -1)$ and are tangent to the parabola $y = \frac{x^2}{4}$; use the system above to find the points at which r e r' touch the parabola.

For the next problem, the reader may find it helpful to read the proof of Theorem 6.62 again, as well as to review the elementary facts on conics (cf. Chap. 6 of [4], for instance).

12. Let $\mathcal{P}$ be the parabola with focus F and directrix d. If P is a point on d and $A, B \in \mathcal{P}$ are such that $\overleftrightarrow{AP}$ and $\overleftrightarrow{BP}$ are tangent to $\mathcal{P}$, show that $F \in AB$.

For the next problem, the reader may find it convenient to have at his/her disposal an extension of the definition of $\lim_{x \to x_0} f(x)$ to the case of a function $f : X \to \mathbb{R}$, where $X \subset \mathbb{R}$ is a nonempty set and $x_0 \in \mathbb{R}$ is the limit of a sequence of distinct elements of X (for instance—and this will essentially be the case of our interest –, we can have $X = \mathbb{Q}$ and $x_0 = \sqrt{2}$). Fortunately, such an extension is formally identical to the one we have already met: we say

---

[5]We shall have more to say on Newton's method in Problem 7, page 455.

that $\lim_{x \to x_0} f(x)$ exists and is equal to L if, for every $\epsilon > 0$, there exists $\delta > 0$ (depending on ϵ) such that

$$x \in X \text{ and } 0 < |x - x_0| < \delta \Rightarrow |f(x) - L| < \epsilon.$$

13. (OIMU) Let $f : (0, +\infty) \to \mathbb{R}$ be given by

$$f(x) = \begin{cases} 0, & \text{if } x \notin \mathbb{Q} \\ \frac{1}{q^3}, & \text{if } x = \frac{p}{q} \text{ with } p, q \in \mathbb{N}; \ \gcd(p, q) = 1 \end{cases}.$$

If $k \in \mathbb{N}$ is not a perfect square, show that f is differentiable at $x = \sqrt{k}$.

9.4 Computing Derivatives

This section is devoted to the presentation of the usual *differentiation rules*, i.e., formulas that show how to compute derivatives of certain functions constructed in terms of two given differentiable functions.

For item (c) of the coming proposition, we recall that if $g : I \to \mathbb{R}$ is differentiable at $x_0 \in I$, then g is continuous at x_0; therefore, if $g(x_0) \neq 0$, then there exists an interval $J \subset I$, containing x_0 and such that $g \neq 0$ in J. Thus, we think of $\frac{f}{g}$ as defined in such a J.

Proposition 9.32 *If $f, g : I \to \mathbb{R}$ are differentiable at $x_0 \in I$, then:*

(a) $f \pm g$ is differentiable at x_0, with $(f \pm g)'(x_0) = f'(x_0) \pm g'(x_0)$.
(b) fg is differentiable at x_0, with $(fg)'(x_0) = f'(x_0)g(x_0) + f(x_0)g'(x_0)$.
(c) If $g(x_0) \neq 0$, then $\frac{f}{g}$ is differentiable at x_0, with

$$\left(\frac{f}{g}\right)'(x_0) = \frac{f'(x_0)g(x_0) - f(x_0)g'(x_0)}{g(x_0)^2}.$$

Proof

(a) Let us prove that $f + g$ is differentiable at x_0, with $(f + g)'(x_0) = f'(x_0) + g'(x_0)$ (the claims relative to $f - g$ can be proved in quite analogous ways). Since

$$\frac{(f + g)(x) - (f + g)(x_0)}{x - x_0} = \frac{f(x) - f(x_0)}{x - x_0} + \frac{g(x) - g(x_0)}{x - x_0}$$

for $x \in I \setminus \{x_0\}$, the arithmetic properties on limits of functions guarantee that, if $f'(x_0)$ and $g'(x_0)$ exist, then $(f + g)'(x_0)$ also does and

$$(f+g)'(x_0) = \lim_{x \to x_0} \frac{(f+g)(x) - (f+g)(x_0)}{x - x_0}$$

$$= \lim_{x \to x_0} \left(\frac{f(x) - f(x_0)}{x - x_0} + \frac{g(x) - g(x_0)}{x - x_0} \right)$$

$$= \lim_{x \to x_0} \frac{f(x) - f(x_0)}{x - x_0} + \lim_{x \to x_0} \frac{g(x) - g(x_0)}{x - x_0}$$

$$= f'(x_0) + g'(x_0).$$

(b) Firstly, note that

$$\frac{(fg)(x) - (fg)(x_0)}{x - x_0} = \left(\frac{f(x) - f(x_0)}{x - x_0} \right) g(x) + \left(\frac{g(x) - g(x_0)}{x - x_0} \right) f(x_0). \quad (9.16)$$

Now, since g is differentiable at x_0, it follows from Proposition 9.26 that g is continuous at x_0, so that

$$\lim_{x \to x_0} g(x) = g(x_0).$$

Therefore, letting $x \to x_0$ in (9.16) and applying the arithmetic properties on limits of functions, we get

$$\lim_{x \to x_0} \frac{(fg)(x) - (fg)(x_0)}{x - x_0} = \lim_{x \to x_0} \frac{f(x) - f(x_0)}{x - x_0} \cdot \lim_{x \to x_0} g(x)$$

$$+ f(x_0) \cdot \lim_{x \to x_0} \frac{g(x) - g(x_0)}{x - x_0}$$

$$= f'(x_0)g(x_0) + f(x_0)g'(x_0).$$

(c) Let us first consider the case in which f is constant and equal to 1, and let us show that $\left(\frac{1}{g} \right)'(x_0) = -\frac{g'(x_0)}{g(x_0)^2}$. Since $g(x) \neq 0$ for $x \in J \subset I$, we have for $x \in J \setminus \{x_0\}$ that

$$\frac{1}{x - x_0} \left(\frac{1}{g(x)} - \frac{1}{g(x_0)} \right) = - \left(\frac{g(x) - g(x_0)}{x - x_0} \right) \cdot \frac{1}{g(x)g(x_0)}.$$

From the equality above and invoking the differentiability (and continuity) of g at x_0, we get

$$\lim_{x \to x_0} \frac{1}{x - x_0} \left(\frac{1}{g(x)} - \frac{1}{g(x_0)} \right) = -g'(x_0) \cdot \frac{1}{g(x_0)^2}.$$

For the general case, we use the particular case above, together with the result of (b):

$$\left(\frac{f}{g}\right)'(x_0) = \left(f \cdot \frac{1}{g}\right)'(x_0) = f'(x_0)\left(\frac{1}{g}\right)(x_0) + f(x_0)\left(\frac{1}{g}\right)'(x_0)$$

$$= \frac{f'(x_0)}{g(x_0)} - f(x_0) \cdot \frac{g'(x_0)}{g(x_0)^2} = \frac{f'(x_0)g(x_0) - f(x_0)g'(x_0)}{g(x_0)^2}.$$

□

Example 9.33 Let $f : \mathbb{R} \to \mathbb{R}$ be the polynomial function given by

$$f(x) = a_n x^n + a_{n-1} x^{n-1} + \cdots + a_1 x + a_0,$$

with $a_0, a_1, \ldots, a_n \in \mathbb{R}$ and $a_n \neq 0$. Applying Example 9.21 and item (a) of the previous proposition several times, we conclude that f is differentiable, with

$$f'(x) = n a_n x^{n-1} + (n-1) a_{n-1} x^{n-2} + \cdots + 2 a_2 x + a_1$$

for every $x \in \mathbb{R}$. Then, $f' : \mathbb{R} \to \mathbb{R}$ is also polynomial, and an easy inductive argument shows that f is indeed infinitely differentiable, with $f^{(n+1)}$ vanishing identically.

For the next example, recall from Trigonometry that the **secant function** sec : $\mathbb{R} \setminus \{\frac{\pi}{2} + k\pi; k \in \mathbb{Z}\} \to \mathbb{R}$ is given by $\sec x = \frac{1}{\cos x}$, for every real x in its domain.

Example 9.34 If $D = \{\frac{\pi}{2} + k\pi; \ k \in \mathbb{Z}\}$, then the tangent function tan : $\mathbb{R} \setminus D \to \mathbb{R}$ is differentiable, with $\tan' x = \sec^2 x$ for every $x \in \mathbb{R} \setminus D$. Indeed, item (c) of the previous proposition, together with the formulae deduced in Example 9.22, furnish

$$\tan' x = \frac{\sin' x \cos x - \sin x \cos' x}{\cos^2 x} = \frac{\cos^2 x + \sin^2 x}{\cos^2 x} = \frac{1}{\cos^2 x} = \sec^2 x.$$

We now show that the arc-tangent function (cf. Example 8.37) is differentiable and compute its derivative.

Example 9.35 Function arctan : $\mathbb{R} \to \left(-\frac{\pi}{2}, \frac{\pi}{2}\right)$ is differentiable, with

$$\arctan' y = \frac{1}{1 + y^2}$$

for every $y \in \mathbb{R}$.

Proof First of all, observe that $\tan' x = \sec^2 x \neq 0$ for every $x \in \left(-\frac{\pi}{2}, \frac{\pi}{2}\right)$. Hence, it follows from Theorem 9.28 that arctan is differentiable. Moreover, if $y \in \mathbb{R}$ and $x = \arctan y$, then $y = \tan x$ and, thanks to (9.15) and the result of the former example, we get

$$\arctan' y = \frac{1}{\tan' x} = \frac{1}{\sec^2 x} = \frac{1}{1 + \tan^2 x} = \frac{1}{1 + y^2}.$$

□

The most important of all differentiation rules is the **chain rule**. Loosely speaking, it establishes the differentiability of the composite of two differentiable functions, and shows how to compute such a derivative.

More precisely, let I and J be intervals and $g : I \to J$ and $f : J \to I$ be given functions, with g differentiable at $x_0 \in I$ and f differentiable at $y_0 = g(x_0) \in J$. Suppose that, in some neighborhood $I \cap (x_0 - r, x_0 + r)$ of x_0, equation $g(x) = y_0$ has x_0 as its only root. Then, for $x \in I \cap (x_0 - r, x_0 + r)$, we can safely write

$$\frac{(f \circ g)(x) - (f \circ g)(x_0)}{x - x_0} = \frac{f(g(x)) - f(g(x_0))}{g(x) - g(x_0)} \cdot \frac{g(x) - g(x_0)}{x - x_0}.$$

Since g is continuous at x_0, we get $\lim_{x \to x_0} g(x) = g(x_0)$, so that

$$\lim_{x \to x_0} \frac{(f \circ g)(x) - (f \circ g)(x_0)}{x - x_0} = \lim_{x \to x_0} \frac{f(g(x)) - f(g(x_0))}{g(x) - g(x_0)} \cdot \lim_{x \to x_0} \frac{g(x) - g(x_0)}{x - x_0}$$

$$= \lim_{y \to y_0} \frac{f(y) - f(y_0)}{y - y_0} \cdot g'(x_0) = f'(y_0) g'(x_0)$$

$$= f'(g(x_0)) g'(x_0).$$

As we shall see in Theorem 9.19, this formula continues to hold in the general case, albeit the proof is much more involving, for, $g(x) = y_0$ can have infinitely many solutions in any neighborhood of x_0 (for instance, think of g as the function f of Problem 11, and solve the equation $g(x) = 0$). Hence, we need to develop some preliminaries.

Let $I \subset \mathbb{R}$ be an open interval and $f : I \to \mathbb{R}$ be differentiable at $x_0 \in I$. Set

$$I - x_0 = \{h \in \mathbb{R}; x_0 + h \in I\},$$

so that $I - x_0$ is an open interval containing 0. Define $r : I - x_0 \to \mathbb{R}$ by letting

$$r(h) = f(x_0 + h) - f(x_0) - f'(x_0)h.$$

Since f is continuous at x_0, we have

$$\lim_{h \to 0} r(h) = \lim_{h \to 0} (f(x_0 + h) - f(x_0) - f'(x_0)h) = 0 = r(0),$$

so that r is continuous at 0. Also, the definition of $f'(x_0)$ gives

$$\lim_{h \to 0} \frac{r(h)}{h} = \lim_{h \to 0} \left(\frac{f(x_0 + h) - f(x_0)}{h} - f'(x_0) \right) = 0.$$

(This last computation, together with $r(0) = 0$, actually shows that r is differentiable at 0 with $r'(0) = 0$. Nevertheless, we won't use this fact here.)

Conversely, we have the following auxiliary result.

Lemma 9.36 *Let $I \subset \mathbb{R}$ be an open interval, $f : I \to \mathbb{R}$ be a given function and $x_0 \in I$. If there exists a real number L such that the function $r : I - x_0 \to \mathbb{R}$, given by $r(h) = f(x_0 + h) - f(x_0) - Lh$, satisfies the condition $\lim_{h \to 0} \frac{r(h)}{h} = 0$, then f is differentiable at x_0, with $f'(x_0) = L$.*

Proof Since $\frac{r(h)}{h} = \frac{f(x_0+h)-f(x_0)}{h} - L$, the condition $\lim_{h \to 0} \frac{r(h)}{h} = 0$ is equivalent to

$$\lim_{h \to 0} \frac{f(x_0 + h) - f(x_0)}{h} = L.$$

$\square$

Let I be an open interval and $f : I \to \mathbb{R}$ be a function continuous at $x_0 \in I$. For every $L \in \mathbb{R}$, the graph of the affine function $h \mapsto f(x_0) + Lh$ passes through $(x_0, f(x_0))$. This being said, it is frequently useful to rephrase the previous discussion by saying that f is differentiable at x_0 if and only if f admits a *best affine approximation in a neighborhood of x_0*, in the sense of Lemma 9.36.

In other words, letting $f : I \to \mathbb{R}$ be differentiable at $x_0 \in I$, we have

$$f(x_0 + h) = f(x_0) + f'(x_0)h + r(h), \qquad (9.17)$$

with $r : I - x_0 \to \mathbb{R}$ such that $\lim_{h \to 0} \frac{r(h)}{h} = 0$. On the other hand, if $h \mapsto f(x_0) + Lh$ is an affine approximation of f in a neighborhood of x_0, such that setting $r(h) = f(x_0 + h) - f(x_0) - Lh$ we have $\lim_{h \to 0} \frac{r(h)}{h} = 0$, then f is differentiable at x_0 and $L = f'(x_0)$.

We shall sometimes refer to (9.17), valid for all $h \in I - x_0$, as **Taylor's formula of order 1** for f in a neighborhood of x_0. Notice that its real content is captured by the fact that $\lim_{h \to 0} \frac{r(h)}{h} = 0$ if f is differentiable at x_0.

For what comes next, it's more useful for us to write (9.17) as

$$f(x_0 + h) = f(x_0) + f'(x_0)h + R(h)h, \qquad (9.18)$$

with $R : I - x_0 \to \mathbb{R}$ given by

$$R(h) = \begin{cases} \frac{r(h)}{h}, & \text{if } h \neq 0 \\ 0, & \text{if } h = 0 \end{cases}.$$

Notice that the continuity of r implies that of R at every $h \in I - x_0$ different from 0, whereas condition $\lim_{h \to 0} \frac{r(h)}{h} = 0$ assures that R is continuous also at 0.

Conversely, if f can be written as

$$f(x_0 + h) = f(x_0) + Lh + R(h)h$$

in $I - x_0$, with R being continuous in $I - x_0$ and such that $R(0) = 0$, then a slight modification of the proof of Lemma 9.36 gives that f is differentiable at x_0, with $f'(x_0) = L$.

We can finally state and prove the chain rule in the general case.

Theorem 9.37 (Chain Rule) *Let I and J be open intervals and $g : I \to J$ and $f : J \to \mathbb{R}$ given functions. If g is differentiable at $x_0 \in I$ and f is differentiable at $g(x_0) \in J$, then $f \circ g : I \to \mathbb{R}$ is differentiable at x_0, with*

$$(f \circ g)'(x_0) = f'(g(x_0))g'(x_0). \tag{9.19}$$

Proof Let $y_0 = g(x_0)$. According to (9.18), the differentiability of f at y_0 allows us to write

$$f(y_0 + t) = f(y_0) + f'(y_0)t + s(t)t \tag{9.20}$$

for every $t \in J - y_0$, with $s : J - y_0 \to \mathbb{R}$ continuous and such that $s(0) = 0$. Analogously, the differentiability of g at x_0 furnishes, for $h \in I - x_0$,

$$g(x_0 + h) = g(x_0) + g'(x_0)h + r(h)h, \tag{9.21}$$

with $r : I - x_0 \to \mathbb{R}$ continuous and such that $r(0) = 0$.

Now, for $h \in I - x_0$, it follows from (9.21) that

$$
\begin{aligned}
(f \circ g)(x_0 + h) - (f \circ g)(x_0) &= f(g(x_0 + h)) - f(g(x_0)) \\
&= f(g(x_0) + g'(x_0)h + r(h)h) - f(y_0) \\
&= f(y_0 + t(h)) - f(y_0),
\end{aligned}
$$

where $t(h) = g'(x_0)h + r(h)h$. Hence, for $h \in I - x_0$, it follows from this and (9.20) that

$$
\begin{aligned}
(f \circ g)(x_0 + h) - (f \circ g)(x_0) &= f(y_0 + t(h)) - f(y_0) \\
&= f'(y_0)t(h) + s(t(h))t(h) \\
&= f'(y_0)(g'(x_0)h + r(h)h) + s(t(h))t(h) \\
&= f'(g(x_0))g'(x_0)h + f'(y_0)r(h)h + s(t(h))t(h) \\
&= f'(g(x_0))g'(x_0)h + f'(y_0)r(h)h \\
&\quad + s(t(h))(g'(x_0) + r(h))h \\
&= (f \circ g)(x_0) = f'(g(x_0))g'(x_0)h + R(h)h,
\end{aligned}
$$

where $R(h) = f'(y_0)r(h) + s(t(h))(g'(x_0) + r(h))$ for $h \in I - x_0$.

Finally, the discussion at the next to last paragraph before the statement of the chain rule guarantees that, in order to show that $f \circ g$ is differentiable at x_0 and that (9.19) is true, it suffices to prove that R is continuous at $I - x_0$, with $R(0) = 0$. But this is straightforward from the chain rule for continuous functions (cf. Proposition 8.12), together with the fact that r, s and t are continuous, with $r(0) = s(0) = t(0) = 0$.   □

Example 9.38 The chain rule allows us to compute the derivative of the cosine function from that of the sine function. Indeed, since $\cos x = (\sin \circ g)(x)$, where $g : \mathbb{R} \to \mathbb{R}$ is given by $g(x) = \frac{\pi}{2} - x$, it follows from the chain rule that

$$\cos' x = (\sin \circ g)'(x) = \sin' g(x) \cdot g'(x) = \cos g(x) \cdot (-1)$$

$$= -\cos\left(\frac{\pi}{2} - x\right) = -\sin x.$$

Example 9.39 Let $R > 0$ and $f : (-R, R) \to \mathbb{R}$ be given by $f(x) = \sqrt{R^2 - x^2}$. Define $g : (0, +\infty) \to \mathbb{R}$ and $h : (-R, R) \to (0, +\infty)$ by letting $g(x) = \sqrt{x}$ and $h(x) = R^2 - x^2$, so that $f = g \circ h$. Since $g'(y) = \frac{1}{2\sqrt{y}}$ for $y > 0$, the chain rule gives, for $x \in (-R, R)$,

$$f'(x) = (g \circ h)'(x) = g'(h(x))h'(x) = \frac{1}{2\sqrt{h(x)}} \cdot (-2x) = -\frac{x}{\sqrt{R^2 - x^2}}.$$

A faster way of using the chain rule to compute the derivative of f at $x \in (-R, R)$ is this: first write $f(x)^2 + x^2 = R^2$, then differentiate both sides with respect to x (and with the aid of the chain rule) to get $2f(x)f'(x) + 2x = 0$; hence, $f'(x) = -\frac{x}{f(x)} = -\frac{x}{\sqrt{R^2 - x^2}}$.

Example 9.40 We can use the chain rule, together with the formula for differentiation of a product, to establish the differentiation formula for a quotient. To this end, let $f, g : I \to \mathbb{R}$ be differentiable in I, with $g(x) \neq 0$ for every $x \in I$. Letting $\tau : \mathbb{R} \setminus \{0\} \to \mathbb{R}$ denote the inverse proportionality function, so that $\tau(x) = x^{-1}$ for every $x \neq 0$, it follows from item (c) of Example 9.21 that τ is differentiable, with $\tau'(x) = -x^{-2}$ for every $x \in \mathbb{R} \setminus \{0\}$. Now, notice that $\frac{f(x)}{g(x)} = f(x) \cdot (\tau \circ g)(x)$ for every $x \in I$, so that the formula for differentiation of a product and the chain rule give

$$\left(\frac{f}{g}\right)'(x) = f'(x) \cdot (\tau \circ g)(x) + f(x) \cdot (\tau \circ g)'(x)$$

$$= f'(x) \cdot \frac{1}{g(x)} + f(x) \cdot \tau'(g(x))g'(x)$$

$$= \frac{f'(x)}{g(x)} + f(x)(-g(x)^{-2})g'(x)$$

$$= \frac{f'(x)g(x) - f(x)g'(x)}{g(x)^2}.$$

The coming example is considerably more sophisticated than the previous ones.

Example 9.41 Prove that $f : \mathbb{R} \to \mathbb{R}$, given by $f(x) = \sin(x^2)$, is not periodic.

Proof First of all, note that if $f : \mathbb{R} \to \mathbb{R}$ is continuously differentiable and periodic, then its derivative $f' : \mathbb{R} \to \mathbb{R}$, besides being continuous, is also periodic. Indeed, if $f(x) = f(x + p)$ for some real $p > 0$ and every $x \in \mathbb{R}$, then the chain rule gives $f'(x) = f'(x + p)$ for every x. Therefore,

$$\text{Im}(f') = \text{Im}(f'_{|[0,p]}),$$

and the continuity of f', together with Corollary 8.30, guarantees that f' is a bounded function.

Now, letting f be the given function and $g : \mathbb{R} \to \mathbb{R}$ be given by $g(x) = x^2$, we have $f(x) = (\sin \circ g)(x)$, so that the chain rule furnishes

$$f'(x) = \sin' g(x) \cdot g'(x) = 2x \cos(x^2).$$

Hence, if $f(x) = \sin(x^2)$ were periodic, the discussion at the preceding paragraph would guarantee that f' would be bounded. Since this is obviously false, we have reached a contradiction. □

The following corollary of the chain rule will be of interest later.

Corollary 9.42 *Let I and J be open intervals and $g : I \to J$ and $f : J \to \mathbb{R}$ given functions. If g is n times differentiable at $x_0 \in I$ and f is n times differentiable at $g(x_0) \in J$, then $f \circ g : I \to \mathbb{R}$ is n times differentiable at x_0.*

Proof We make induction on n, relying on the chain rule for the case $n = 1$. Assume the corollary to be true when $n = k$, and let f and g be $k + 1$ times differentiable at x_0.

In a neighborhood of x_0, we have $(f \circ g)' = (f' \circ g)g'$, with f' and g' being k times differentiable at that point. The induction hypothesis guarantees that $f' \circ g$ is k times differentiable at x_0, from where Problem 3, page 300 (applied to $f' \circ g$ and g') assures that $(f \circ g)' = (f' \circ g)g'$ is also k times differentiable at x_0. Therefore, $f \circ g$ is $k + 1$ times differentiable at x_0. □

Problems: Section 9.4

1. Compute the derivatives of the given functions, assuming in each case that the function and its derivative are defined in their maximal domains:

   (a) $f(x) = x^3 - 2x^2 + 7x + 1$.
   (b) $f(x) = x\sqrt[3]{x} + x^4$.

   (c) $f(x) = \frac{1}{x} + \frac{1}{x^2}$
   (d) $f(x) = \cos x + \tan x$.

(e) $f(x) = x^3 \sin x$.

(f) $f(x) = 7\sqrt{x}(x^2 + 4)$.

(g) $f(x) = \frac{x^2-3x}{2x^3+1}$.

(h) $f(x) = \frac{\sin x}{1+\cos x}$.

2. Let I be an interval and $f : I \to (0, +\infty)$ be differentiable at $x_0 \in I$. If $g : I \to \mathbb{R}$ is given by $g(x) = \sqrt[n]{f(x)}$, show that g is differentiable at x_0, with $g'(x_0) = \frac{1}{n} f(x_0)^{\frac{1}{n}-1} f'(x_0)$.

3. Compute the derivatives of the given functions, assuming in each case that the function and its derivative are defined in their maximal domains:

(a) $f(x) = (\cos x)^2$.

(b) $f(x) = \cos(x^2)$.

(c) $f(x) = \sqrt{1 - x\sqrt{1-x}}$.

(d) $f(x) = \sqrt{1 - x\cos(x^2)}$.

(e) $f(x) = \arcsin(1 - x^2)$.

(f) $f(x) = \frac{x\arccos(x^2)}{x^2+1}$.

(g) $f(x) = \frac{\operatorname{tg}\sqrt{x}}{1+x^3}$.

(h) $f(x) = \sin(\sin(\cos x))$.

4. Let r be a nonzero rational number and $f : (0, +\infty) \to \mathbb{R}$ be given by $f(x) = x^r$. Prove that f is differentiable, with $f'(x) = rx^{r-1}$ for every $x > 0$.

For the next problem we recall a few facts on roots of polynomial functions. Given a polynomial function f with real coefficients and degree $n > 1$, and a real root α of f, the **division algorithm for polynomials** (cf. Chap. 14 of [5]) guarantees the existence of an integer $1 \leq k \leq n$ and a polynomial g of degree $n - k$ such that

$$f(x) = (x - \alpha)^k g(x), \tag{9.22}$$

with $g(\alpha) \neq 0$. In this case, we say that k is the **multiplicity** of α as a root of f.

5. Let $f : \mathbb{R} \to \mathbb{R}$ be a nonconstant polynomial function. Show that:

(a) If α is a root of f and $f(x) = (x - \alpha)^k g(x)$, with $g(\alpha) \neq 0$, then, for every $x \in \mathbb{R}$ such that $f(x) \neq 0$, we have

$$\frac{f'(x)}{f(x)} = \frac{k}{x - \alpha} + \frac{g'(x)}{g(x)}.$$

(b) If $f(x) = a(x - \alpha_1)(x - \alpha_2) \ldots (x - \alpha_n)$, with $a, \alpha_1, \ldots, \alpha_n \in \mathbb{R}$ and $a \neq 0$, then, for $x \neq \alpha_1, \ldots, \alpha_n$, we have

$$\frac{f'(x)}{f(x)} = \sum_{j=1}^{n} \frac{1}{x - \alpha_j}.$$

6. If $f : \mathbb{R} \setminus \{3\} \to \mathbb{R}$ is given by $f(x) = \frac{2x+1}{x-3}$, compute $f^{(5)}(0)$.

7. Let $f : [0, +\infty) \to \mathbb{R}$ be given by

$$f(x) = \sqrt{x + \sqrt{x + \cdots + \sqrt{x + \sqrt{x + 1}}}},$$

with ten square root signs. Compute $f'(0)$.

8. For $k \in \mathbb{N}$, show that $\arctan^{(k)}(x) = \frac{p_k(x)}{(1+x^2)^k}$, where p_k is a polynomial of degree $k - 1$ and leading coefficient $(-1)^{k-1}k!$.

9. Find the common tangents to the parabolas of equations $y = x^2$ and $y = 2 + (x - 3)^2$.

For the next problem, the reader may find it helpful to review the elementary material on conics, at the level of Chap. 6 of [4], for instance.

10. Let $\mathcal{E}$ be the ellipse of equation $\frac{x^2}{a^2} + \frac{y^2}{b^2} = 1$ and $\mathcal{H}$ the hyperbola of equation $\frac{x^2}{a'^2} - \frac{y^2}{b'^2} = 1$. We say that $\mathcal{E}$ and $\mathcal{H}$ are **confocal** if they have the same foci. In this respect, show that:

   (a) $a^2 - b^2 = a'^2 + b^2$.
   (b) $\mathcal{E}$ and $\mathcal{H}$ intersect in four distinct points; moreover, if (x_0, y_0) is one of them, then

$$x_0^2 \left(\frac{1}{a^2 b'^2} + \frac{1}{a'^2 b^2} \right) = \frac{1}{b^2} + \frac{1}{b'^2} \text{ and } y_0^2 \left(\frac{1}{a^2 b'^2} + \frac{1}{a'^2 b^2} \right) = \frac{1}{a'^2} - \frac{1}{a^2}.$$

   (c) If P is one of the intersection points of item (b), then the straight lines tangent to $\mathcal{E}$ and $\mathcal{H}$ at P are perpendicular.

11. Do the following items:

   (a) If $f : \mathbb{R} \to \mathbb{R}$ is given by

$$f(x) = \begin{cases} x^2 \sin \frac{1}{x}, & \text{if } x \neq 0 \\ 0, & \text{if } x = 0 \end{cases},$$

   show that it is differentiable at every real number and that f' is not continuous at 0.
   (b) For real numbers $x_1 < x_2 < \ldots < x_n$, give an example of a differentiable function $g : \mathbb{R} \to \mathbb{R}$ such that g' is discontinuous at $x_1, x_2, \ldots, x_n$.

12. Prove that $f : (0, +\infty) \to \mathbb{R}$, given by $f(x) = \sin \sqrt{x}$, is not periodic.

13. (Putnam) Let $a_1, a_2, \ldots, a_n$ be given reals and $f : \mathbb{R} \to \mathbb{R}$ be such that $f(x) = a_1 \sin x + a_2 \sin(2x) + \cdots + a_n \sin(nx)$ for every $x \in \mathbb{R}$. If $|f(x)| \leq |\sin x|$ for every $x \in \mathbb{R}$, prove that $|a_1 + 2a_2 + \cdots + na_n| \leq 1$.

14. (OBMU) Given nonzero real numbers $a_1, a_2, \ldots, a_n$, show that the function $f : \mathbb{R} \to \mathbb{R}$, such that

$$f(x) = \sum_{j=1}^{n} a_j \cos(jx)$$

for every $x \in \mathbb{R}$, has period 2π.

15. * Prove the following version of **l'Hôpital's rule**[6]: let $f, g : [a, b) \to \mathbb{R}$ be two
    functions differentiable at a and such that $f(a) = g(a) = 0$. If $g \neq 0$ in (a, b)
    and $g'(a) \neq 0$, show that

$$\lim_{x \to a} \frac{f(x)}{g(x)} = \frac{f'(a)}{g'(a)}.$$

9.5 Rôlle's Theorem and Applications

Let $f : [a, b] \to \mathbb{R}$ be continuous in $[a, b]$, differentiable in (a, b) and such that
$f(a) = f(b) = 0$. In addition, suppose that f is nonzero in at least one point of the
interval (a, b). Arguing heuristically, translate the horizontal axis parallel to itself,
until it becomes tangent to the graph of f at a point $(c, f(c))$ (cf. Fig. 9.7). Then, on
the one hand, such a tangent line must have slope $f'(c)$; on the other, being parallel
to the horizontal axis, its slope must be equal to 0, so that $f'(c) = 0$.

The first result of this section, which is known in the literature as **Rôlle's
theorem**,[7] puts the discussion of the previous paragraph in solid grounds.

Lemma 9.43 (Rôlle) *Let $f : [a, b] \to \mathbb{R}$ be continuous in $[a, b]$ and differentiable
in (a, b). If $f(a) = f(b) = 0$, then there exists $c \in (a, b)$ such that $f'(c) = 0$.*

Proof Let $c, d \in [a, b]$ be the points in which f attains its minimum and maximum
values, respectively. If $c, d \in \{a, b\}$, then for every $x \in [a, b]$ we have $0 = f(c) \leq
f(x) \leq f(d) = 0$. Therefore, f vanishes identically in $[a, b]$ and there is nothing left
to do.

Else, suppose that $c \in (a, b)$ (the case of $d \in (a, b)$ can be dealt with
analogously). Since $f(x) \geq f(c)$ for every $x \in [a, b]$, we conclude that

Fig. 9.7 Rôlle's theorem

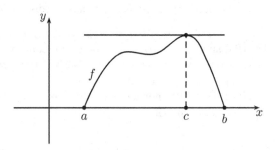

---

[6]After Guillaume F. Antoine, Marquis de l'Hôpital, French mathematician of the XVII century. A
more refined version of l'Hôpital's rule will be presented at Proposition 9.50.

[7]After Michel Rôlle, French mathematician of the XVII century.

$$x \in (c, b] \Rightarrow \frac{f(x) - f(c)}{x - c} \geq 0 \text{ and } x \in [a, c) \Rightarrow \frac{f(x) - f(c)}{x - c} \leq 0.$$

Hence, it follows from the results of Problems 2 and 5, page 290, that

$$f'(c) = \lim_{x \to c+} \frac{f(x) - f(c)}{x - c} \geq 0 \text{ and } f'(c) = \lim_{x \to c-} \frac{f(x) - f(c)}{x - c} \leq 0.$$

Thus, $f'(c) = 0$. □

The next example shows a typical application of Rôlle's theorem.

Example 9.44 Given $a, b, c \in \mathbb{R}$, show that the equation $4ax^3 + 3bx^2 + 2cx = a + b + c$ has at least one real root between 0 and 1.

Proof First of all, note that a standard application of the IVT is not conclusive, for, if $f(x) = 4ax^3 + 3bx^2 + 2cx - (a + b + c)$, then $f(0) = -(a + b + c)$ and $f(1) = 3a + 2b + c$, so that $f(0)$ and $f(1)$ can have equal signs (for instance, for $a = 1$ and $b = c = -\frac{2}{3}$).

On the other hand, if $f(x) = ax^4 + bx^3 + cx^2 - (a + b + c)x$, then $f(0) = f(1) = 0$. Hence, Rôlle's theorem assures the existence of $x_0 \in (0, 1)$ such that $f'(x_0) = 0$. Now, note that this is the same as saying that x_0 is a root of the given equation. □

Let's (heuristically) analyse a situation more general than that of Rôlle's theorem. More precisely, let's consider again a function $f : [a, b] \to \mathbb{R}$, continuous in $[a, b]$ and differentiable in (a, b), but such that $f(a)$ and $f(b)$ can assume any values whatsoever. The slope of the secant line to the graph of f passing through $(a, f(a))$ and $(b, f(b))$ equals $\frac{f(b) - f(a)}{b - a}$. We translate it parallel to itself until it becomes tangent to the graph at a point $(c, f(c))$, with $c \in (a, b)$ (cf. Fig. 9.8). Then, on the one hand, the slope of this tangent line equals $f'(c)$; on the other, since it is parallel to the original secant line, its slope equals that of the secant, so that

$$f'(c) = \frac{f(b) - f(a)}{b - a}.$$

This result is the content of the **mean value theorem** (MVT, shortly) of Lagrange, and it is the content of the coming result.

Theorem 9.45 (Lagrange) *If $f : [a, b] \to \mathbb{R}$ is continuous in $[a, b]$ and differentiable in (a, b), then there exists $c \in (a, b)$ such that*

$$\frac{f(b) - f(a)}{b - a} = f'(c).$$

Proof Let $g : [a, b] \to \mathbb{R}$ be given by

$$g(x) = f(x) - \left(\left(\frac{b - x}{b - a} \right) f(a) + \left(\frac{x - a}{b - a} \right) f(b) \right),$$

Fig. 9.8 Lagrange's mean
value theorem

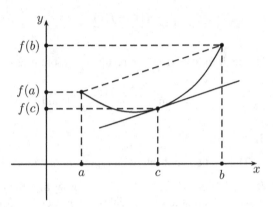

for every $x \in [a, b]$. (Note that the function of x between parentheses is precisely the
one that defines the secant to the graph of f passing through $(a, f(a))$ and $(b, f(b))$.)

Clearly, g is continuous in $[a, b]$ and differentiable in (a, b), with $g(a) = g(b) = 0$.
Hence, by Rôlle's theorem, there exists $c \in (a, b)$ such that $g'(c) = 0$. Now, an easy
computation gives

$$g'(x) = f'(x) - \left(\frac{-f(a)}{b-a} + \frac{f(b)}{b-a} \right),$$

so that

$$0 = g'(c) = f'(c) - \frac{f(b) - f(a)}{b-a}. \qquad \square$$

Example 9.46 For $x, y \in \left(-\frac{\pi}{2}, \frac{\pi}{2} \right)$, show that $|\tan x - \tan y| \geq |x - y|$.

Proof If $x = y$, there is nothing to do. Otherwise, it follows from the MVT of
Lagrange the existence of a real c between x and y, such that

$$\tan x - \tan y = (\tan c)(x - y) = \frac{1}{\cos^2 c}(x - y).$$

However, since $|\cos c| \leq 1$, we get

$$|\tan x - \tan y| = \frac{1}{\cos^2 c}|x - y| \geq |x - y|. \qquad \square$$

The following corollary sets an important consequence of Lagrange's MVT.

Corollary 9.47 *Let $I \subset \mathbb{R}$ be an interval and $f, g : I \to \mathbb{R}$ be two differentiable
functions. If $f' = g'$, then $f - g$ is constant.*

Proof Since $f' = g'$ if and only if $(f - g)' = 0$, it suffices to show that $f' = 0$ implies f constant.

Let $f' = 0$ and fix $a < b$ in I. Since f is differentiable in I, it is certainly continuous in $[a, b]$ and differentiable in (a, b). Therefore, Lagrange's MVT gives $c \in (a, b)$ such that $\frac{f(b)-f(a)}{b-a} = f'(c) = 0$. Hence, $f(b) = f(a)$, and since a and b were arbitrarily chosen in I, it follows that f is constant. $\qquad\square$

Example 9.48 If $f, g : \mathbb{R} \to \mathbb{R}$ are differentiable functions such that $f'(x) = g(x)$ and $g'(x) = -f(x)$ for every $x \in \mathbb{R}$, then $f(x) = f(0) \cos x + g(0) \sin x$ and $g(x) = g(0) \cos x - f(0) \sin x$, for every $x \in \mathbb{R}$.

Proof Letting $h(x) = f(x) \sin x + g(x) \cos x$ and $l(x) = f(x) \cos x - g(x) \sin x$, we have $h(0) = g(0), l(0) = f(0)$ and

$$h'(x) = f'(x) \sin x + f(x) \cos x + g'(x) \cos x - g(x) \sin x$$
$$= (f'(x) - g(x)) \sin x + (f(x) + g'(x)) \cos x = 0.$$

Analogously, $l'(x) = 0$ for every $x \in \mathbb{R}$, so that the previous corollary guarantees that h and l are constant functions. Hence,

$$\begin{cases} f(x) \sin x + g(x) \cos x = g(0) \\ f(x) \cos x - g(x) \sin x = f(0) \end{cases}.$$

The above inequalities can be seen as a linear system of two equations in the unknowns $f(x)$ and $g(x)$. We can easily solve it to get $f(x) = f(0) \cos x + g(0) \sin x$ and $g(x) = g(0) \cos x - f(0) \sin x$. $\qquad\square$

In what comes next, we discuss an important generalization of Lagrange's MVT to two given functions, known as **Cauchy's mean value theorem**. Note that (9.23) reduces to the MVT of Lagrange if $g(x) = x$.

Theorem 9.49 (Cauchy) *If $f, g : [a, b] \to \mathbb{R}$ are continuous in $[a, b]$ and differentiable in (a, b), then there exists $c \in (a, b)$ such that*

$$(f(b) - f(a))g'(c) = (g(b) - g(a))f'(c). \tag{9.23}$$

Proof Let $h : [a, b] \to \mathbb{R}$ be given by

$$h(x) = (f(b) - f(a))g(x) - (g(b) - g(a))f(x),$$

for every $x \in [a, b]$. It is immediate that h is continuous in $[a, b]$ and differentiable in (a, b), so that, by Lagrange's MVT, there exists $c \in (a, b)$ satisfying

$$h'(c) = \frac{h(b) - h(a)}{b - a}.$$

Now, since $h(a) = h(b) = f(b)g(a) - f(a)g(b)$, we have $h'(c) = 0$ and it suffices to observe that

$$h'(c) = (f(b) - f(a))g'(c) - (g(b) - g(a))f'(c).$$
□

In Calculus, it is often the case we have continuous functions $f, g : [a, b) \to \mathbb{R}$ such that $f(a) = g(a) = 0$, and we need to evaluate (if it exists) the limit

$$\lim_{x \to a} \frac{f(x)}{g(x)}.$$

In general, we refer to such a quotient as an *indeterminacy* of the form $\frac{0}{0}$ as $x = a$. Problem 15, page 311, taught us how to deal with it when f and g are differentiable at a, with $g'(a) \neq 0$. Cauchy's MVT allows us to get a more refined version of that result, which is also known in the literature as **l'Hôpital's rule**.

Proposition 9.50 (l'Hôpital's rule for $\frac{0}{0}$) *Let* $f, g : (a, b) \to \mathbb{R}$ *be differentiable functions such that* $\lim_{x \to a} f(x) = \lim_{x \to a} g(x) = 0$. *If* $g, g' \neq 0$ *in* (a, b), *then*

$$\lim_{x \to a} \frac{f'(x)}{g'(x)} = L \Rightarrow \lim_{x \to a} \frac{f(x)}{g(x)} = L.$$

Proof Start by continuously extending f and g to $[a, b)$, setting $f(a) = g(a) = 0$. For $x \in (a, b)$, it follows from Cauchy's MVT that

$$\frac{f(x)}{g(x)} = \frac{f(x) - f(a)}{g(x) - g(a)} = \frac{f'(c(x))}{g'(c(x))}, \qquad (9.24)$$

for some $c(x) \in (a, x)$; in particular, we clearly have $\lim_{x \to a} c(x) = a$.

Now, if $\lim_{x \to a} \frac{f'(x)}{g'(x)} = L$, a slight modification in the hint given to Problem 10, page 291, guarantees that $\lim_{x \to a} \frac{f'(c(x))}{g'(c(x))} = L$. Hence, it readily follows from (9.24) that $\lim_{x \to a} \frac{f(x)}{g(x)} = L$.
□

Example 9.51 We use l'Hôpital's rule for $\frac{0}{0}$ to compute $\lim_{x \to 0} \frac{1 - \cos x}{\sqrt{x}}$. (Note that $x \mapsto \sqrt{x}$ is not differentiable at $x = 0$, so that we cannot use the result of Problem 15, page 311). Indeed, both $f(x) = 1 - \cos x$ and $g(x) = \sqrt{x}$ (for $x > 0$) satisfy the hypotheses of the previous proposition and are such that

$$\lim_{x \to 0} \frac{f'(x)}{g'(x)} = \lim_{x \to 0} \frac{\sin x}{1/(2\sqrt{x})} = \lim_{x \to 0} 2\sqrt{x} \sin x = 0.$$

Therefore, $\lim_{x \to 0} \frac{f(x)}{g(x)} = 0$.

Problems: Section 9.5

1. Let $f : \mathbb{R} \to \mathbb{R}$ be a twice differentiable function, such that $f''(x) + f(x) = 0$ for every $x \in \mathbb{R}$. Prove that

$$f(x) = f(0) \cos x + f'(0) \sin x,$$

for every $x \in \mathbb{R}$.

2. Let $\lambda > 0$ and $f : \mathbb{R} \to \mathbb{R}$ be twice differentiable and such that $f''(x) + \lambda f(x) = 0$ for every $x \in \mathbb{R}$. Prove that

$$f(x) = f(0) \cos(\sqrt{\lambda} x) + \frac{f'(0)}{\sqrt{\lambda}} \sin(\sqrt{\lambda} x),$$

for every $x \in \mathbb{R}$.

3. * Let I be an interval and $f : I \to \mathbb{R}$ be a twice differentiable function, such that f'' is constant. Prove that f either vanishes identically or is a polynomial function of degree at most 2. More precisely, if $x_0 \in I$, show that

$$f(x) = f(x_0) + f'(x_0)(x - x_0) + \frac{f''(x_0)}{2}(x - x_0)^2.$$

4. (OIMU) Find all functions $f : \mathbb{R} \to \mathbb{R}$ such that $(f(x) - f(y))^2 \leq |x - y|^3$, for every $x, y \in \mathbb{R}$.

5. Let $I \subset \mathbb{R}$ be an interval and $c \in I$. Let $f : I \to \mathbb{R}$ be a function continuous in I and differentiable in $I \setminus \{c\}$. If $\lim_{x \to c} f'(x) = l$, prove that f is differentiable in c, with $f'(c) = l$.

6. (Putnam) Let $a_0, a_1, \ldots, a_n$ be real numbers such that $\frac{a_0}{1} + \frac{a_1}{2} + \cdots + \frac{a_n}{n+1} = 0$. Show that the polynomial function $a_0 + a_1 x + \cdots + a_n x^n$ has at least one real root.

7. Show that the equation $x^2 = x \sin x + \cos x$ has exactly two real roots.

8. Redo Problem 8, page 300, this time under the assumption that $f : I \to \mathbb{R}$ is continuously differentiable in I.

9. In each item below, compute the given limit:

   (a) $\lim_{x \to 0} \frac{1 - \cos x}{x^2}$.

   (b) $\lim_{x \to 0} \frac{1 - \sqrt[3]{\cos x}}{x^2}$.

   (c) $\lim_{x \to 0} \frac{\sqrt{\cos x} - \sqrt[3]{\cos x}}{x^2}$.

   (d) $\lim_{x \to -1} \frac{\sin^2(\pi x)}{(x^3 + 1)^2}$.

10. Let $f, g : (0, \frac{\pi}{2}) \to \mathbb{R}$ be given by $f(x) = x^{3/2} \sin \frac{1}{x}$ and $g(x) = x^{1/3} - \cos x + 1$. Compute, if it exists, $\lim_{x \to 0} \frac{f(x)}{g(x)}$.

11. If $f : (a, b) \to \mathbb{R}$ is twice differentiable, show that, for every $x_0 \in (a, b)$,

$$\lim_{h \to 0} \frac{f(x_0 + h) + f(x_0 - h) - 2f(x_0)}{h^2} = f''(x_0).$$

12. Let $f : [a, b) \to \mathbb{R}$ be twice continuously differentiable in $[a, b)$, with $f(a) = 0$ and $f''(a) \neq 0$.

   (a) Show that there exists $c \in (a, b)$ such that $f(x) - (x - a)f'(x) \neq 0$, for every $x \in (a, c)$.

   (b) Compute $\lim_{x \to a} \frac{(x-a)f(x)}{f(x)-(x-a)f'(x)}$ in terms of $f(a)$, $f'(a)$ and $f''(a)$ (item (a) assures that the denominator doesn't vanish in the interval (a, c)).

13. * Prove the following version of l'Hôpital's rule for indeterminacies $\frac{0}{0}$ at $+\infty$: let $a > 0$ and $f, g : (a, +\infty) \to \mathbb{R}$ be differentiable functions such that $g, g' \neq 0$ in $(a, +\infty)$. If $\lim_{x \to +\infty} f(x) = \lim_{x \to +\infty} g(x) = 0$, then

$$\lim_{x \to +\infty} \frac{f'(x)}{g'(x)} = L \Rightarrow \lim_{x \to +\infty} \frac{f(x)}{g(x)} = L.$$

14. Let $f : \mathbb{R} \to \mathbb{R}$ be a differentiable function such that $\lim_{x \to +\infty} f(x)$ and $\lim_{x \to +\infty} xf'(x)$ exist and are finite. Compute the possible values of this second limit.

15. Prove the following version of l'Hôpital's rule for indeterminacies $\frac{\pm\infty}{\pm\infty}$ at $a \in \mathbb{R}$: let $f, g : (a, b) \to \mathbb{R}$ be differentiable functions such that $g, g' \neq 0$ in (a, b) and $\lim_{x \to a} g(x) = +\infty$ (or $-\infty$). Then,

$$\lim_{x \to a} \frac{f'(x)}{g'(x)} = L \Rightarrow \lim_{x \to a} \frac{f(x)}{g(x)} = L.$$

16. The differentiable function $f : \mathbb{R} \to \mathbb{R}$ is such that $|f'(x)| \leq c < 1$ for every real x, where c is a positive real constant. Show that there exists only one $x_0 \in \mathbb{R}$ such that $f(x_0) = x_0$.

17. * Let $I \subset \mathbb{R}$ be an open interval, $x_0 \in I$ and $f : I \to \mathbb{R}$ a continuous function, which is differentiable in $I \setminus \{x_0\}$. If there exists $L = \lim_{x \to x_0} f'(x)$, prove that f is differentiable at x_0, with $f'(x_0) = L$.

18. (Romania) Prove that there does not exist a differentiable function $f : (0, +\infty) \to (0, +\infty)$ such that $f(x)^2 \geq f(x + y)(f(x) + y)$ for every $x, y \in (0, +\infty)$.

19. (OBMU) Find all differentiable functions $f : \mathbb{R} \to \mathbb{R}$ such that $f(0) = 0$ and $|f'(x)| \leq |f(x)|$ for every $x \in \mathbb{R}$.

9.6   The First Variation of a Function

We saw in Sect. 8.2 that every continuous function $f : [a, b] \to \mathbb{R}$ attains extreme (i.e., maximum and minimum) values in $[a, b]$. In this section, we shall discuss procedures that, among other things, allow us to effectively compute the corresponding extreme points of such an f, provided it is differentiable in (a, b).

In all that follows, I denotes an interval, whose **interior** is the interval obtained by excluding its endpoints, if they exist. Thus, if $I = [a, b]$, $[a, b)$, $(a, b]$ or (a, b), then the interior of I is the open interval (a, b); analogously, if $I = [a, +\infty)$ or $(a, +\infty)$, its interior is $(a, +\infty)$, and if $I = (-\infty, b]$ or $(-\infty, b)$, its interior is $(-\infty, b)$.

Broadening the situation of the first paragraph above, our purpose in this section is to solve the problem of finding extreme points and values of functions $f : I \to \mathbb{R}$ which are continuous in I and differentiable in the interior of I. To this end, we begin with the following

Definition 9.52 Given a function $f : I \to \mathbb{R}$, we say that $x_0 \in I$ is a **local maximum** (resp. **local minimum**) point for f if there exists $\delta > 0$ such that $f(x_0) \geq f(x)$ (resp. $f(x_0) \leq f(x)$), for every $x \in I \cap (x_0 - \delta, x_0 + \delta)$.

In Fig. 9.9, points x_0, x_0' and x_0'' are of local minimum for the function $f : (a, b) \to \mathbb{R}$ whose graph is depicted. Observe that, in a neighborhood of each of these points, the values of f are no less than those it assumes at x_0, x_0' and x_0'' (as indicated by the three dashed horizontal line segments). Also, x_0' is the only *global* minimum point of f (i.e., a point in which f attains the minimum possible value in (a, b)), whereas x_0 and x_0'' are not global minimum points. Finally, notice that f has also two maximum local points (identify them!), albeit none of which is a global maximum point.

Generically, local maximum or minimum points of $f : I \to \mathbb{R}$ are called its **extreme points**, or simply its **extrema**. If I is an open interval and $f : I \to \mathbb{R}$ is differentiable, we shall show below that its extrema (if any) can be found among the points where f' vanishes. Since these vanishing points will play quite an important role in subsequent discussions, it is worth to start by naming them.

Definition 9.53 Let $I \subset \mathbb{R}$ be an open interval and $f : I \to \mathbb{R}$ a differentiable function. We say that $x_0 \in I$ is a **critical point** of f if $f'(x_0) = 0$.

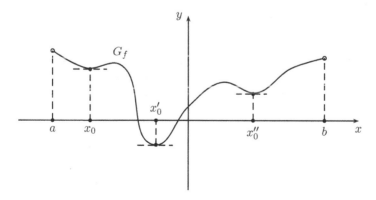

Fig. 9.9 Local minimum points of $f : I \to \mathbb{R}$

The notion of critical point has an obvious geometric interpretation: letting I be an open interval, it follows from (9.14) that the critical points of a differentiable function $f : I \to \mathbb{R}$ are precisely those points in which the tangent lines to its graph are horizontal.

We can now state and prove the following fundamental result, which is known as the **first derivative test** for extreme points. Although its proof is quite similar to that of Rôlle's theorem, we repeat it here for the sake of completeness.

Proposition 9.54 *If $I \subset \mathbb{R}$ is an open interval and $f : I \to \mathbb{R}$ is a differentiable function, then every extreme point of f is also critical.*

Proof Assume that $x_0 \in I$ is a local minimum point of f (the proof in the other case is completely analogous). Take $\delta > 0$ such that $(x_0 - \delta, x_0 + \delta) \subset I$ (since I is open, such a δ always exists) and

$$0 < |x - x_0| < \delta \Rightarrow f(x) - f(x_0) \geq 0.$$

For $x_0 < x < x_0 + \delta$ we have $\frac{f(x) - f(x_0)}{x - x_0} \geq 0$, so that

$$f'(x_0) = \lim_{x \to x_0+} \frac{f(x) - f(x_0)}{x - x_0} \geq 0.$$

An analogous reasoning with $x_0 - \delta < x < x_0$ gives us

$$f'(x_0) = \lim_{x \to x_0-} \frac{f(x) - f(x_0)}{x - x_0} \leq 0.$$

Therefore, $f'(x_0) = 0$.   □

As a corollary to the first derivative test, we have the following search criterion for extreme points.

Corollary 9.55 *Let $f : I \to \mathbb{R}$ be continuous in I and differentiable in the interior of I. If f attains a global (maximum or minimum) value in I, then the corresponding extreme point is either an endpoint of I or a critical point of f.*

Proof Suppose that f attains a global minimum value in I (the case of a global maximum value can be treated analogously). Let $x_0 \in I$ be such that $f(x_0) = \min\{f(x); x \in I\}$. If x_0 is an endpoint of I, there is nothing left to do. Otherwise, x_0 belongs to the interior of I; however, since the interior de I is an open interval, the first derivative test assures that x_0 is a critical point of f.   □

The coming example makes it clear that the above corollary solves the problem of finding the global extreme values of a continuous function $f : [a, b] \to \mathbb{R}$, differentiable in (a, b), as long as we know how to find the roots of the equation $f'(x) = 0$.

Example 9.56 Find the maximum value of $f : [0, 1] \to \mathbb{R}$, given by $f(x) = x - x^4$.

Solution Since f is continuous in $[0, 1]$, we know from Theorem 8.26 that f assumes a maximum value in $[0, 1]$. By the previous corollary, the corresponding maximum point is 0, 1 or a critical point of f. Now, since $f'(x) = 1 - 4x^3$, we conclude that $\frac{1}{\sqrt[3]{4}}$ is the only critical point of f, with $f(\frac{1}{\sqrt[3]{4}}) = \frac{3}{\sqrt[3]{4}}$. Finally, since $f(0) = f(1) = 0 < \frac{3}{\sqrt[3]{4}}$, it follows that the maximum value of f is $\frac{3}{\sqrt[3]{4}}$.   □

Our next result, a direct consequence of Lagrange's MVT, teaches us how to get the monotonicity intervals of a differentiable function. From now on, we shall refer to it as the study of the **first variation** of a differentiable function.[8]

Proposition 9.57 *If $f : I \to \mathbb{R}$ is continuous in I and differentiable in the interior of I, then:*

(a) $f' \geq 0$ *in the interior of I if and only if f is nondecreasing in I.*
(b) *If $f' > 0$ in the interior of I, then f is increasing in I, but the converse is not necessarily true.*
(c) $f' \leq 0$ *in the interior of I if and only if f is nonincreasing in I.*
(d) *If $f' < 0$ in the interior of I, then f is decreasing in I, but the converse is not necessarily true.*

Proof Let's prove only items (a) and (b), the proofs of items (c) and (d) being entirely analogous.

(a) Suppose first that f is nondecreasing in I. For a given x_0 in the interior of I, take $\delta > 0$ such that $(x_0 - \delta, x_0 + \delta) \subset I$. If $x_0 < x < x_0 + \delta$, then $x \in I$ and $f(x) - f(x_0) \geq 0$. Hence, $\frac{f(x)-f(x_0)}{x-x_0} \geq 0$ and, making $x \to x_0+$, we get

$$f'(x_0) = \lim_{x \to x_0+} \frac{f(x) - f(x_0)}{x - x_0} \geq 0.$$

Conversely, suppose $f' \geq 0$ in the interior of I. If $a, b \in I$ are such that $a < b$, the continuity of f in I guarantees its continuity in $[a, b]$. On the other hand, since (a, b) is contained in the interior of I, we have f differentiable in (a, b), so that Lagrange's MVT guarantees the existence of $c \in (a, b)$ for which

$$\frac{f(b) - f(a)}{b - a} = f'(c) \geq 0.$$

Therefore, $f(b) \geq f(a)$, and f is nondecreasing in I.
(b) Assume that $f' > 0$ in the interior of I, and let $a < b$ be two points of I. Again by Lagrange's MVT, there exists $c \in (a, b)$ for which

---

[8]This terminology alludes to the key role of the first derivative in this result, as well as to the fact that, classically, the first derivative of a (differentiable) function was called its *first variation*.

$$\frac{f(b) - f(a)}{b - a} = f'(c) > 0.$$

Hence, $f(b) > f(a)$ and f is increasing in I.

To see that the converse is not generally true, let $f : \mathbb{R} \to \mathbb{R}$ be given by $f(x) = x^3$. Then, f is increasing in the whole real line, albeit $f'(0) = 0$. □

The coming example shows how to use Proposition 9.57 to approach problems involving maxima and minima of differentiable functions.

Example 9.58 Let $f : [0, +\infty) \to \mathbb{R}$ be given by $f(x) = \frac{x^2+3}{x+1}$. Show that f attains a global minimum in $[0, +\infty)$ and compute this minimum value.

Solution If f is to attain a global minimum in $[0, +\infty)$, then Corollary 9.55 guarantees that the corresponding extreme point is either 0 or a critical point of f. Computing f', we get

$$f'(x) = \frac{2x(x+1) - (x^2+3) \cdot 1}{(x+1)^2} = \frac{x^2 + 2x - 3}{(x+1)^2},$$

so that $f'(x) = 0$ if and only if $x = 1$ (recall that we must have $x \geq 0$).

On the other hand, since $x^2 + 2x - 3 = (x+3)(x-1)$, we conclude that f' is negative in the interval $(0, 1)$ and positive in the interval $(1, +\infty)$. Hence, Proposition 9.57 says that f decreases in $[0, 1]$ and increases in $[1, +\infty)$, so that it really attains a global minimum, at $x = 1$. Therefore, the minimum value of f is $f(1) = 2$. □

We now apply the first variation of functions to give a third proof of the inequality between the arithmetic and geometric means.

Example 9.59 Given an integer $n > 1$ and positive reals $a_1, a_2, \ldots, a_n$ we have

$$\frac{a_1 + a_2 + \cdots + a_n}{n} \geq \sqrt[n]{a_1 a_2 \ldots a_n},$$

with equality if and only if $a_1 = a_2 = \cdots = a_n$.

Proof Let's make induction on $n > 1$, relying in the discussion that led to (5.2) for the case $n = 2$. Given an integer $k > 2$, suppose we have established the inequality for $k - 1$ arbitrary positive real numbers, with equality if and only if they are all equal.

Given k positive reals $a_1, \ldots, a_{k-1}, a_k$, let $f : (0, +\infty) \to \mathbb{R}$ be defined for $x > 0$ by

$$f(x) = a_1 + \cdots + a_{k-1} + x - k\sqrt[k]{a_1 \ldots a_{k-1}x}.$$

We shall show that $f(x) \geq 0$ for every $x > 0$, with equality if and only if $x = a_1$ and $a_1 = \cdots = a_{k-1}$; in particular, it will follow that $f(a_k) \geq 0$, with equality if and only if $a_1 = \cdots = a_{k-1} = a_k$, as wished. To this end, note that f is differentiable, with

$$f'(x) = 1 - \sqrt[k]{a_1 \dots a_{k-1}}\, x^{\frac{1}{k}-1}.$$

Hence, letting $x_0 = {}^{k-1}\!\sqrt{a_1 \dots a_{k-1}}$, we have $f' < 0$ in $(0, x_0)$, $f'(x_0) = 0$ and $f' > 0$ in $(x_0, +\infty)$, so that Proposition 9.57 assures f to decrease in $(0, x_0]$ and increase in $[x_0, +\infty)$. Therefore, f attains a global minimum (only) at $x = x_0$.

Finally, a straightforward substitution gives

$$f(x_0) = a_1 + \cdots + a_{k-1} - (k-1)\, {}^{k-1}\!\sqrt{a_1 \dots a_{k-1}}, \tag{9.25}$$

so that $f(x_0) \geq 0$ by induction hypothesis. Thus,

$$f(x) \geq f(x_0) \geq 0$$

for every $x > 0$, with equality if and only if $x = x_0$ and $f(x_0) = 0$. It now suffices to observe that, thanks to (9.25) and the induction hypothesis, we have $f(x_0) = 0$ if and only if $a_1 = \dots = a_{k-1}$. $\qquad\square$

We shall sometimes use the first variation in the form of the following

Corollary 9.60 *Let I be an open interval, $f : I \to \mathbb{R}$ be twice differentiable and $x_0 \in I$ be a critical point of f. If $f'' > 0$ (resp. $f'' < 0$) in I, then x_0 is the only global minimum (resp. maximum) point of f. In particular, x_0 is the only critical point of f.*

Proof Suppose $f'' > 0$ in I (the case $f'' < 0$ in I is totally analogous). Since $f'' = (f')'$, item (b) of Proposition 9.57 (applied to f', in place of f) guarantees that f' increases in I. However, since $f'(x_0) = 0$, it follows that $f'(x) < 0$ for $x < x_0$ and $f'(x) > 0$ for $x > x_0$ (in particular, x_0 is the only critical point of f). Then, items (b) and (d) of Proposition 9.57 (this time applied to f) guarantee that f decreases in $I \cap (-\infty, x_0]$ and increases in $I \cap [x_0, +\infty)$. Therefore, x_0 is the only global minimum point of f. $\qquad\square$

With the aid of the previous corollary, we can given an interesting application of the first variation to Euclidean Geometry.

Example 9.61 We are given an angle $\angle AOB$ such that $A\widehat{O}B < 90°$ (cf. Fig. 9.10), as well as a point P inside it. Show how to choose points $X \in \overrightarrow{OA}$ and $Y \in \overrightarrow{OB}$ such that X, P and Y are collinear and XY has the minimum possible length.

Fig. 9.10 Line segment $\overline{XY}$
of minimum length

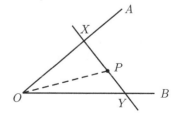

Proof In what follows, we shall rely on [4] for the necessary background in Euclidean Geometry.

Taking angles in radians, let $A\widehat{O}P = \alpha$, $B\widehat{O}P = \beta$ and $O\widehat{X}Y = \theta$, so that α and β are known and θ is variable (i.e., depends on the position of X along $\overrightarrow{OA}$). Since the sum of the internal angles of triangle OXY equals π radians, we have $0 < \theta < \pi - \alpha - \beta$ and $O\widehat{Y}X = \pi - \alpha - \beta - \theta$.

Applying the Sine Law (cf. Sect. 7.3 of [4]) to triangles XOP and YOP, we get

$$\frac{\overline{PX}}{\overline{OP}} = \frac{\sin\alpha}{\sin\theta} \quad\text{and}\quad \frac{\overline{PY}}{\overline{OP}} = \frac{\sin\beta}{\sin(\alpha + \beta + \theta)},$$

so that

$$\overline{XY} = \overline{PX} + \overline{PY} = \overline{OP}\left(\frac{\sin\alpha}{\sin\theta} + \frac{\sin\beta}{\sin(\alpha + \beta + \theta)}\right).$$

Hence, letting $f : (0, \pi - \alpha - \beta) \to \mathbb{R}$ be given by

$$f(\theta) = \frac{\sin\alpha}{\sin\theta} + \frac{\sin\beta}{\sin(\alpha + \beta + \theta)},$$

it suffices to show that there exists a single $\theta \in (0, \pi - \alpha - \beta)$ in which f attains its global minimum.

The first derivative test assures that, if f has a global minimum point, then it must be a critical one. On the other hand, an easy computation furnishes

$$f'(\theta) = -\frac{\sin\alpha\cos\theta}{\sin^2\theta} - \frac{\sin\beta\cos(\alpha + \beta + \theta)}{\sin^2(\alpha + \beta + \theta)}.$$

Substituting $\gamma = \pi - \alpha - \beta$, we easily get

$$\lim_{\theta\to 0} f'(\theta) = -\infty \quad\text{and}\quad \lim_{\theta\to\gamma} f'(\theta) = +\infty.$$

Therefore, the IVT (applied to f', which is clearly continuous on θ) guarantees the existence of $\theta_0 \in (0, \gamma)$ such that $f'(\theta_0) = 0$. Now,

$$f''(\theta) = \frac{\sin\alpha(1 + \cos^2\theta)}{\sin^3\theta} + \frac{\sin\beta(1 + \cos^2(\alpha + \beta + \theta))}{\sin^3(\alpha + \beta + \theta)} > 0,$$

since $\alpha, \beta, \theta, \alpha + \beta + \theta \in (0, \pi)$. Hence, Corollary 9.60 says that θ_0 is the only global minimum point of f. $\square$

We finish this section by applying Proposition 9.57 to prove the following result, known as **Darboux's theorem**.[9]

Theorem 9.62 (Darboux)   *If $I \subset \mathbb{R}$ is an interval and $f : I \to \mathbb{R}$ is a differentiable function, then the function $f' : I \to \mathbb{R}$, even if discontinuous, satisfies the mean value property.*

Proof Let $a < b$ in I and d be a real number between $f(a)$ and $f(b)$. We wish to establish the existence of $c \in [a, b]$ such that $f(c) = d$. To this end, we consider two cases separately:

(i) $f'(a) < 0 < f'(b)$ (or vice-versa): if f is not injective in $[a, b]$, then there exist $\alpha < \beta$ in $[a, b]$ such that $f(\alpha) = f(\beta)$. By Lagrange's MVT, there exists $c \in (\alpha, \beta)$ (hence, $c \in (a, b)$) such that $f'(c) = \frac{f(\beta)-f(\alpha)}{\beta-\alpha} = 0$. If f is injective in $[a, b]$, it follows from Theorem 8.35 that f is monotonic in $[a, b]$. Therefore, the previous proposition gives $f' \geq 0$ in $[a, b]$ or $f' \leq 0$ in $[a, b]$, which is not the case.

(ii) $f'(a) < d < f'(b)$ (or vice-versa): let $g : I \to \mathbb{R}$ be given by $g(x) = f(x) - dx$, so that g is differentiable, with $g'(x) = f'(x) - d$ for every $x \in I$. Hence, $g'(a) < 0 < g'(b)$ or vice-versa, and item (i) assures the existence of $c \in (a, b)$ such that $g'(c) = 0$. This is the same as $f'(c) = d$.   □

Problems: Section 9.6

1. Prove Proposition 6.25 applying the results of this section.
2. In each of the following items, find the maximum and minimum values (if these exist) of the given functions:

    (a) $f : [0, +\infty) \to \mathbb{R}$ given by $f(x) = \frac{\sqrt{x}}{x^2+16}$.
    (b) $f : \mathbb{R} \to \mathbb{R}$ given by $f(x) = \frac{x}{ax^2+b}$, where a and b are given positive real numbers.
    (c) $f : (0, +\infty) \to \mathbb{R}$ given by $f(x) = x^2 + \frac{a}{x}$, where a is a given positive real number.
    (d) $f : (0, +\infty) \to \mathbb{R}$ given by $f(x) = \frac{x^2}{x^3+a}$, where a is a given positive real number.

3. Let $f : \mathbb{R} \to \mathbb{R}$ be given by $f(x) = x^3 + ax^2 + bx + c$, where $a, b, c \in \mathbb{R}$ are real constants. Discuss the first variation of f in terms of a, b and c.
4. * Show that $x \geq \sin x \geq x - \frac{x^3}{3!}$ and $\cos x \geq 1 - \frac{x^2}{2}$, for every $x \geq 0$.
5. In a square $ABCD$ of diagonals AC and BD, the sides have length 2. Mark the midpoint P of side AB and, then, a point $Q \in AD$ and a point $R \in CD$ such that

---

[9] After Jean-Gaston Darboux, French mathematician of the XIX and XX centuries.

$P\widehat{Q}R = 90°$. Find the position of Q along side AD such that the area of triangle PQR is as large as possible.

6. Let $f : (a, b) \rightarrow \mathbb{R}$ be a differentiable function and $P(x_0, y_0)$ be a point not belonging to the graph of f. If there exists $A \in G_f$ such that

$$\overline{AP} = \min\{\overline{A'P}; \ A' \in G_f\},$$

we say that $\overline{AP}$ is the distance from P to G_f. In such a case, prove that $\overleftrightarrow{AP} \perp r$, where r denotes the tangent line to G_f at A.

7. Let $f : (a, b) \rightarrow \mathbb{R}$ and $g : (c, d) \rightarrow \mathbb{R}$ be differentiable functions whose graphs do not intersect each other. If there exist points $A(\alpha, f(\alpha))$ and $B(\beta, f(\beta))$ such that

$$\overline{AB} = \min\{\overline{A'B'}; \ A' \in G_f \text{ and } B' \in G_g\}, \qquad (9.26)$$

we say that $\overline{AB}$ is the distance between the graphs of f and g. In such a case, Show that $f'(\alpha) = g'(\beta)$.

Given differentiable functions $f : (a, b) \rightarrow \mathbb{R}$ and $g : (c, d) \rightarrow \mathbb{R}$ whose graphs do not intersect, the result of the previous problem assures that the problem of finding points $A \in G_f$ and $B \in G_g$ satisfying (9.26) is equivalent to that of minimizing $F : (a, b) \times (c, d) \rightarrow \mathbb{R}$ given by

$$F(x, y) = (x - y)^2 + (f(x) - g(y))^2,$$

subject to the *constraint* $f'(x) = g'(y)$. Nevertheless, the pair of functions $f, g : (0, +\infty) \rightarrow \mathbb{R}$ given by $f(x) = \frac{1}{x}$ and $g(x) = 0$ shows that it is not always the case that such a pair of points $A \in G_f$ and $B \in G_g$ actually exist. The next problem applies the above discussion to a case in the positive side of things.

8. Compute the distance between the graphs of functions $f, g : \mathbb{R} \rightarrow \mathbb{R}$ such that $f(x) = x^2$ e $g(x) = 1 - (x - 3)^2$, admitting the (geometrically plausible) fact that such a distance is attained by a certain pair of points $A \in G_f$ and $B \in G_g$.

9. In Fig. 9.11, we have $\overline{BC} = 2\overline{BH}$. Compute the largest possible value of the measure of angle $B\widehat{A}C$.

10. $ABCD$ is an isosceles trapezoid of bases AD and BC, with $\overline{AD} > \overline{BC}$, and legs AB and CD. If $\overline{BC} = a$ and $\overline{AB} = \overline{CD} = b$, compute the measure of the angles $B\widehat{A}D = A\widehat{D}C$ for the area of $ABCD$ to be as large as possible.

Fig. 9.11 Maximizing the measure of angle $B\widehat{A}C$

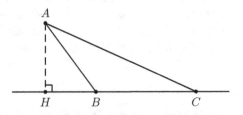

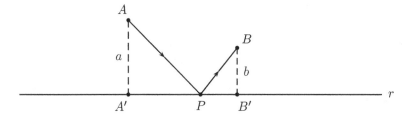

Fig. 9.12 Minimizing $\overline{AP} + \overline{BP}$

Fig. 9.13 Minimizing
$\overline{AP} \cdot \overline{BP}$

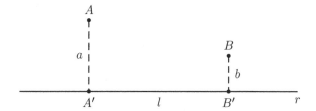

11. We are given an angular region $\angle AOB$, such that $A\widehat{O}B < 90°$ (cf. Fig. 9.10), and a point P inside it. Choose points $X \in \overrightarrow{OA}$ and $Y \in \overrightarrow{OB}$ in such a way that X, P are Y collinear. Show that $\overrightarrow{PX} \cdot \overline{PY}$ is minimum if and only if OXY is an isosceles triangle, with basis XY.

12. Let a line r and points A and B be given, as shown in Fig. 9.12. The distances from A and B to r are respectively equal to a and b, and c is the length of the line segment joining the feet of the perpendiculars dropped from A and B to r. Find the position of a point P along r such that the sum $\overline{AP} + \overline{BP}$ is as small as possible.

13. In Fig. 9.13, we have $\overleftrightarrow{AA'}, \overleftrightarrow{BB'} \perp r$, $\overline{AA'} = a$, $\overline{BB'} = b$ and $\overline{A'B'} = l$. If $l \le \sqrt{2(a^2 + b^2)}$, show that there exists a single point $P \in r$ such that $\overline{AP} \cdot \overline{BP}$ attains its minimum possible value. Show also that $P \in A'B'$.

14. Show that $\left(\frac{m+1}{n+1}\right)^{n+1} \ge \left(\frac{m}{n}\right)^n$, for every $m, n \in \mathbb{N}$.

15. For each real number $k > 1$, compute the minimum possible value of $x + y$, where x and y are real numbers satisfying $(x + \sqrt{1 + x^2})(y + \sqrt{1 + y^2}) = k$.

16. (Romania) Compute the minimum possible value of $x + y + \frac{2}{x+y} + \frac{1}{2xy}$, for distinct and positive real numbers x and y.

17. (BMO) In a triangle ABC, let $\sin^{23} \frac{\widehat{A}}{2} \cos^{48} \frac{\widehat{B}}{2} = \sin^{23} \frac{\widehat{B}}{2} \cos^{48} \frac{\widehat{A}}{2}$. Compute the ratio $\frac{\overline{AC}}{\overline{BC}}$.

18. (OIMU) Let $f : [0, 1] \to [0, 1]$ be continuous in $[0, 1]$, differentiable in $(0, 1)$ and such that $f(0) = 0$ and $f(1) = 1$. Prove that there exist distinct $a, b \in [0, 1]$ such that $f'(a)f'(b) = 1$.

19. (Leningrad) Let $f : \mathbb{R} \to \mathbb{R}$ be a polynomial function such that

$$f(x) - f'(x) - f''(x) + f'''(x) \ge 0$$

for every $x \in \mathbb{R}$. Prove that $f(x) \ge 0$ for every $x \in \mathbb{R}$.

9.7   The Second Variation of a Function

Along this section we continue to study how the behavior of the derivatives of a function influence the shape of its graph. As before, in all that follows $I \subset \mathbb{R}$ denotes an interval.

Definition 9.63 A function $f : I \to \mathbb{R}$ is said to be **convex** if, for all $a, b \in I$ and $t \in [0, 1]$, we have

$$f((1 - t)a + tb)) \leq (1 - t)f(a) + tf(b).$$

As t varies from 0 to 1, basic Analytic Geometry guarantees that the points of the form $(1 - t)(a, f(a)) + t(b, f(b))$ trace out the line segment secant with endpoints $(a, f(a))$ and $(b, f(b))$. Hence, $f : I \to \mathbb{R}$ is convex if and only if, for all $a < b$ in I, the portion of the graph of f between lines $x = a$ and $x = b$ does not intersect the open half-plane situated above the straight line passing through $(a, f(a))$ and $(b, f(b))$. Yet in another way, letting

$$\mathcal{R}_+(f) = \{(x, y) \in \mathbb{R}^2;\ x \in I \text{ and } y \geq f(x)\},$$

f is convex if and only if $\mathcal{R}_+(f)$ is a convex region of the plane.

The next definition refines the notion of convexity for functions (pay particular attention to the range of values of t, when compared to those in the former definition).

Definition 9.64 A function $f : I \to \mathbb{R}$ is **strictly convex** if, for all distinct $a, b \in I$ and every $t \in (0, 1)$, we have

$$f((1 - t)a + tb)) < (1 - t)f(a) + tf(b).$$

Analogously to the above, we say that $f : I \to \mathbb{R}$ is (**strictly**) **concave** if $-f$ is (strictly) convex. Hence, f is concave if and only if

$$f((1 - t)a + tb)) \geq (1 - t)f(a) + tf(b)$$

for all $a, b \in I$ and $t \in [0, 1]$. Accordingly, $f : I \to \mathbb{R}$ is **strictly concave** if and only if $f((1 - t)a + tb)) > (1 - t)f(a) + tf(b)$ for all distinct $a, b \in I$ and every $t \in (0, 1)$.

Properties of (strictly) concave functions are easily derived from those of (strictly) convex ones, just changing f by $-f$ when necessary. For this reason, in all that follows we shall restrict ourselves to the analysis of convex and strictly convex functions, leaving to the reader the task of adapting it to concave and strictly concave ones.

Definition 9.63 guarantees that every strictly convex function is convex. Nevertheless, the converse is not true in general, as the coming example shows.

Example 9.65 Every affine function is convex, albeit not strictly convex. Indeed, if $f : \mathbb{R} \to \mathbb{R}$ is given by $f(x) = Ax + B$, with $A, B \in \mathbb{R}$, it is immediate to verify that $f((1 - t)a + tb) = (1 - t)f(a) + tf(b)$, for all $a, b \in \mathbb{R}$ and every $t \in [0, 1]$.

In Corollary 9.70 we will show that, if I is an open interval, then a twice differentiable $f : I \to \mathbb{R}$ is convex if and only if $f''(x) \geq 0$ for every $x \in I$. We will also show that, if $f''(x) > 0$ for every $x \in I$, then f is strictly convex. By assuming the truth of these results for the time being, we now list some examples of convex and concave functions (note that when we sketched the graphs of these functions we *implicitly* assumed that they were indeed convex or concave, according to the case at hand).

Example 9.66

(a) The restriction of the inverse proportionality function to the set of positive reals is strictly convex. Indeed, letting $f(x) = \frac{1}{x}$ for $x > 0$, we have $f''(x) = \frac{2}{x^3} > 0$.
(b) Letting $n > 1$ be an integer, the function $f : (0, +\infty) \to \mathbb{R}$ given by $f(x) = x^n$ is also strictly convex, for, $f''(x) = n(n - 1)x^{n-2} > 0$.
(c) The sine function is strictly concave in the interval $(0, \pi)$, since $\sin'' x = -\sin x < 0$ in $(0, \pi)$.
(d) Since $\tan'' x = 2 \tan x \sec x > 0$ in $\left(0, \frac{\pi}{2}\right)$, the tangent function is strictly convex in this interval.

Remark 9.67 If $f : I \to \mathbb{R}$ is convex, then, letting $t = \frac{1}{2}$ in Definition 9.63, we get $f(\frac{a+b}{2}) \leq \frac{f(a)+f(b)}{2}$ for all $a, b \in I$. Conversely, if f is continuous and such that $f(\frac{a+b}{2}) \leq \frac{f(a)+f(b)}{2}$ for all $a, b \in I$, it is possible to show (see Problem 8) that f is convex. By the same token, it is also possible to show (see Problem 9) that a continuous $f : I \to \mathbb{R}$ is strictly convex if $f(\frac{a+b}{2}) < \frac{f(a)+f(b)}{2}$ for all distinct $a, b \in I$. In turn, these remarks allow us to use the elementary inequalities we had studied to establish the (strictly) convex or concave character of several common functions. In this respect, take a look at Problem 11.

Back to the development of the theory, let $f : I \to \mathbb{R}$ be a convex function. Then, given $a < b$ in I and $0 < t < 1$, we have

$$f((1 - t)a + tb) \leq (1 - t)f(a) + tf(b). \tag{9.27}$$

Letting $x = (1 - t)a + tb$, we have $t = \frac{x-a}{b-a}$ and $1 - t = \frac{b-x}{b-a}$. Therefore, the last inequality above can be written as

$$f(x) \leq \frac{b - x}{b - a} \cdot f(a) + \frac{x - a}{b - a} \cdot f(b).$$

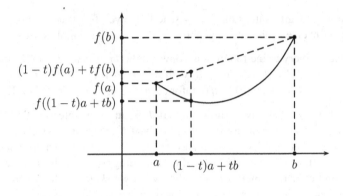

Fig. 9.14 Graph of a convex function

In turn, this gives

$$\frac{f(x) - f(a)}{x - a} \leq \frac{1}{x - a}\left(\frac{b - x}{b - a}\cdot f(a) + \frac{x - a}{b - a}\cdot f(b) - f(a)\right) = \frac{f(b) - f(a)}{b - a}$$

and, analogously, $\frac{f(b) - f(x)}{b - x} \geq \frac{f(b) - f(a)}{b - a}$. In short, letting f be convex in I, we have

$$\frac{f(x) - f(a)}{x - a} \leq \frac{f(b) - f(a)}{b - a} \leq \frac{f(b) - f(x)}{b - x} \tag{9.28}$$

for all $a < x < b$ in I.
  Conversely, if

$$\frac{f(x) - f(a)}{x - a} \leq \frac{f(b) - f(x)}{b - x}, \tag{9.29}$$

for all $a < x < b$ in I, then $(b - x)(f(x) - f(a)) \leq (x - a)(f(b) - f(x))$ or, which is the same,

$$(b - a)f(x) \leq (b - x)f(a) + (x - a)f(b).$$

Substituting $x = (1 - t)a + tb$ in this inequality, we immediately recover (9.27), so that f is convex.
  Notice that the quotients in (9.28) are the slopes of the secants to the graph of f and passing through the pairs of points $(a, f(a))$ and $(x, f(x))$, $(a, f(a))$ and $(b, f(b))$, $(x, f(x))$ and $(b, f(b))$, respectively. On the other hand, such slopes are the tangents of the trigonometric angles measured from the horizontal axis to those secants. Since the tangent function increases in each of the intervals $(0, \frac{\pi}{2})$ and $(\frac{\pi}{2}, \pi)$, the reader can easily use the comments above to grasp the "turning behavior" of those secants, as we walk along the horizontal axis in the positive direction (see Fig. 9.14).

In what follows, we use the previous discussion to show that a convex function, defined in an open interval, is continuous.

Proposition 9.68 *If I is an open interval and $f : I \to \mathbb{R}$ is convex, then f is continuous.*

Proof Fix $x_0 \in I$ and $a < x_0 < b$ also in I (this is possible due to the openness of I). For $x \in (x_0, b)$, the first inequality in (9.28), with x_0 in place of a, gives $\frac{f(x)-f(x_0)}{x-x_0} \leq \frac{f(b)-f(x_0)}{b-x_0}$; in turn, the second inequality in (9.28), with x_0 in place of x and x in place of b, gives $\frac{f(x)-f(x_0)}{x-x_0} \geq \frac{f(x_0)-f(a)}{x_0-a}$. By combining these two inequalities, we get

$$\left(\frac{f(x_0) - f(a)}{x_0 - a}\right)(x - x_0) \leq f(x) - f(x_0) \leq \left(\frac{f(b) - f(x_0)}{b - x_0}\right)(x - x_0).$$

Letting $x \to x_0+$, it follows from these inequalities, together with the squeezing theorem (cf. Proposition 9.8) that

$$\lim_{x \to x_0+} f(x) = f(x_0).$$

Arguing in an analogous way, we conclude that $\lim_{x \to x_0-} f(x) = f(x_0)$. Therefore, Problem 5, page 290, guarantees that $\lim_{x \to x_0} f(x) = f(x_0)$, and Proposition 9.4 shows that f is continuous at x_0. $\square$

As we have previously commented, we shall now show in a moment that the (strictly) convex character of a twice differentiable function is intimately related to the sign of its second derivative. This, in turn, will be a straightforward consequence of the coming result, in which we assume that the function under consideration is merely differentiable.

Theorem 9.69 *If I is an open interval and $f : I \to \mathbb{R}$ is differentiable, then:*

(a) f is convex in I if and only if f' is nondecreasing in I.
(b) f is strictly convex in I if and only if f' is increasing in I.

Proof Firstly, let f be convex in I. Given $a < b$ in I and $x \in (a, b)$, it follows from (9.28) that

$$\frac{f(x) - f(a)}{x - a} \leq \frac{f(b) - f(a)}{b - a} \leq \frac{f(b) - f(x)}{b - x}.$$

However, since f is differentiable in I, letting $x \to a+$ in the first inequality and $x \to b-$ in the second one, we obtain

$$f'(a) \le \frac{f(b) - f(a)}{b - a} \le f'(b).$$

Hence, f' is nondecreasing in I.

If f is strictly convex, we refine the above reasoning bay taking $c \in (a, b)$ and $x \in (a, c)$, $y \in (c, b)$. Then, (9.28) and the strict convexity of f give

$$\frac{f(x) - f(a)}{x - a} < \frac{f(c) - f(a)}{c - a} < \frac{f(b) - f(c)}{b - c} < \frac{f(b) - f(y)}{b - y}.$$

Now, letting $x \to a+$ and $y \to b-$, we get

$$f'(a) \le \frac{f(c) - f(a)}{c - a} < \frac{f(b) - f(c)}{b - c} \le f'(b),$$

so that f is increasing in I.

Conversely, suppose that f' is nondecreasing in I. To show that f is convex, we previously saw that it suffices to prove that

$$\frac{f(x) - f(a)}{x - a} \le \frac{f(b) - f(x)}{b - x}$$

for all $a < x < b$ in I. Since I is open and f is differentiable in I, we have f continuous in $[a, b]$ and differentiable in (a, b). Hence, Lagrange's MVT assures the existence of $\alpha \in (a, x)$ and $\beta \in (x, b)$ such that $\frac{f(x) - f(a)}{x - a} = f'(\alpha)$ and $\frac{f(b) - f(x)}{b - x} = f'(\beta)$. However, since $\alpha < \beta$ and f' is assumed to be nondecreasing, we have $f'(\alpha) \le f'(\beta)$. Thus,

$$\frac{f(x) - f(a)}{x - a} = f'(\alpha) \le f'(\beta) = \frac{f(b) - f(x)}{b - x},$$

as we wished to show.

In case f' is increasing in I, the above reasoning gives

$$\frac{f(x) - f(a)}{x - a} = f'(\alpha) < f'(\beta) = \frac{f(b) - f(x)}{b - x}.$$

Hence, we get a strict inequality in (9.29) for all $a < x < b$ in I, and f is strictly convex.   □

Corollary 9.70 *Let I be an open interval and $f : I \to \mathbb{R}$ be twice differentiable.*

(a) f is convex in I if and only if $f'' \ge 0$ in I.
(b) If $f'' > 0$ in I, then f is strictly convex in I.

Proof Item (a) follows immediately from item (a) of the previous result, together with item (a) of Proposition 9.57 (applied to f').

In what concerns (b), if $f'' > 0$ in I, then item (b) of Proposition 9.57 (applied once more to f') guarantees that f' increases in I. Therefore, item (b) of the previous theorem assures that f is strictly convex. $\square$

Let us see a simple application of this result.

Example 9.71 Let $f : \mathbb{R} \setminus \{-\sqrt{2}, \sqrt{2}\} \to \mathbb{R}$ be given by $f(x) = \frac{x}{2-x^2}$. Find the intervals of the real line in which f is strictly convex or strictly concave.

Solution Computing the first and second derivatives of f, we get $f'(x) = \frac{2+x^2}{(2-x^2)^2}$ and $f''(x) = \frac{2x}{(2-x^2)^3(6+x^2)}$. Hence,

$$f''(x) > 0 \Leftrightarrow \frac{x}{2-x^2} > 0 \Leftrightarrow \begin{cases} x > 0 \text{ and } 2 - x^2 > 0 \\ \text{or} \\ x < 0 \text{ and } 2 - x^2 < 0 \end{cases} \Leftrightarrow \begin{cases} 0 < x < \sqrt{2} \\ \text{or} \\ x < -\sqrt{2} \end{cases}.$$

We readily conclude that f is strictly convex in each of the intervals $(-\infty, -\sqrt{2})$ and $(0, \sqrt{2})$, and strictly concave in each of the intervals $(-\sqrt{2}, 0)$ and $(\sqrt{2}, +\infty)$. $\square$

The former corollary is usually referred to as the study of the **second variation** of a twice differentiable function. We now wish to look at the points in which the graph of f changes from being convex to being concave, or vice-versa. Before that, it is useful to have the following definition at our disposal.

Definition 9.72 Let I be an open interval and $f : I \to \mathbb{R}$ be a continuous function. A point $x_0 \in I$ is said to be an **inflection point** of f if there exists $\delta > 0$ such that f is convex in $I \cap (x_0 - \delta, x_0)$ and concave in $I \cap (x_0, x_0 + \delta)$, or vice-versa.

The coming corollary gives a necessary condition to be satisfied by the inflection points of twice differentiable functions.

Corollary 9.73 *Let I be an open interval and $f : I \to \mathbb{R}$ be a twice differentiable function. If $x_0 \in I$ is an inflection point of f, then $f''(x_0) = 0$.*

Proof Suppose f is convex in $(x_0 - \delta, x_0)$ and concave in $(x_0, x_0 + \delta)$, with $\delta > 0$ so small that $(x_0 - \delta, x_0 + \delta) \subset I$ (the other case can be dealt with in an analogous way). Then, Corollary 9.70 gives $f''(x) \geq 0$ in $(x_0 - \delta, x_0)$ and $f''(x) \leq 0$ in $(x_0, x_0 + \delta)$.

Suppose $f''(x_0) < 0$, and fix an arbitrary $a \in (x_0 - \delta, x_0)$, so that $f''(x_0) < 0 \leq f''(a)$. Since $f''(x_0) < \frac{f''(x_0)}{2} < f''(a)$, Darboux's Theorem 9.62 assures the existence of $b \in (a, x_0)$ such that $f''(b) = \frac{f''(x_0)}{2} < 0$. However, since $b \in (a, x_0) \Rightarrow b \in (x_0 - \delta, x_0)$, we should also have $f''(b) \geq 0$, which is a contradiction.

Analogously, we also reach a contradiction by supposing that $f''(x_0) > 0$. Hence, the only possibility is that $f''(x_0) = 0$. $\square$

We now state and prove **Jensen's inequality**,[10] a result that gives greater flexibility to the applications of convex functions to problems of maxima and minima.

Theorem 9.74 (Jensen) *Let $I \subset \mathbb{R}$ be an open interval and $f : I \to \mathbb{R}$ be a convex function. If $x_1, x_2, \ldots, x_n \in I$ and $t_1, t_2, \ldots, t_n \in (0, 1)$, with $t_1 + t_2 + \cdots + t_n = 1$, then $t_1 x_1 + t_2 x_2 + \cdots + t_n x_n \in I$ and*

$$f(t_1 x_1 + t_2 x_2 + \cdots + t_n x_n) \leq t_1 f(x_1) + t_2 f(x_2) + \cdots + t_n f(x_n). \tag{9.30}$$

Moreover, if f is strictly convex, then equality happens if and only if $x_1 = x_2 = \cdots = x_n$.

Proof Suppose f is strictly convex (the case of a merely convex f is totally analogous) and let's prove Jensen's inequality by induction on $n > 1$. Case $n = 2$ follows from the definition of strict convexity for f, since condition $t_1 + t_2 = 1$ is equivalent to $t_1 = t$ and $t_2 = 1 - t$.

Suppose that, for a certain $n > 1$ and all $x_1, x_2, \ldots, x_n \in I$ and $t_1, t_2, \ldots, t_n \in (0, 1)$, with $t_1 + t_2 + \cdots + t_n = 1$, we have $t_1 x_1 + t_2 x_2 + \cdots + t_n x_n \in I$ and

$$f(t_1 x_1 + t_2 x_2 + \cdots + t_n x_n) \leq t_1 f(x_1) + t_2 f(x_2) + \cdots + t_n f(x_n),$$

with equality if and only if $x_1 = x_2 = \cdots = x_n$. Let's consider $x_1, x_2, \ldots, x_n$, $x_{n+1} \in I$ and $t_1, t_2, \ldots, t_n, t_{n+1} \in (0, 1)$ such that $t_1 + t_2 + \cdots + t_n + t_{n+1} = 1$. Define

$$y = \frac{t_1 x_1 + t_2 x_2 + \cdots + t_n x_n}{1 - t_{n+1}} = s_1 x_1 + s_2 x_2 + \cdots + s_n x_n,$$

with $s_j = \frac{t_j}{1 - t_{n+1}}$ for $1 \leq j \leq n$. Then, $s_1, s_2, \ldots, s_n > 0$ and

$$\sum_{j=1}^{n} s_j = \frac{1}{1 - t_{n+1}} \sum_{j=1}^{n} t_j = \frac{1}{1 - t_{n+1}} \cdot (1 - t_{n+1}) = 1,$$

so that $s_j \in (0, 1)$ for $1 \leq j \leq n$. Hence, induction hypothesis gives $y \in I$ and, thus,

$$t_1 x_1 + t_2 x_2 + \cdots + t_n x_n + t_{n+1} x_{n+1} = (1 - t_{n+1})y + t_{n+1} x_{n+1} \in I.$$

Now, by invoking the strict convexity of f we obtain

$$f(t_1 x_1 + t_2 x_2 + \cdots + t_{n+1} x_{n+1}) = f((1 - t_{n+1})y + t_{n+1} x_{n+1})$$
$$\leq (1 - t_{n+1}) f(y) + t_{n+1} f(x_{n+1}),$$

---

[10] After Johan Jensen, Danish engineer and mathematician of the XIX and XX centuries.

with equality if and only if $y = x_{n+1}$. On the other hand, also by induction hypothesis we have

$$f(y) = f(s_1x_1 + s_2x_2 + \cdots + s_nx_n) \leq s_1f(x_1) + s_2f(x_2) + \cdots + s_nf(x_n)$$

$$= \frac{1}{1 - t_{n+1}}(t_1f(x_1) + t_2f(x_2) + \cdots + t_nf(x_n)),$$

with equality if and only if $x_1 = x_2 = \cdots = x_n$.

Taking the two inequalities above together, we conclude that

$$f(t_1x_1 + t_2x_2 + \cdots + t_{n+1}x_{n+1}) \leq (1 - t_{n+1})f(y) + t_{n+1}f(x_{n+1})$$

$$\leq (t_1f(x_1) + \cdots + t_nf(x_n)) + t_{n+1}f(x_{n+1}),$$

with equality happening if and only if $y = x_{n+1}$ and $x_1 = x_2 = \cdots = x_n$. Finally, it's immediate to verify that these conditions are equivalent to $x_1 = x_2 = \cdots = x_n = x_{n+1}$. $\square$

Most often, we shall apply Jensen's inequality in the form of the coming corollary, which is stated for the reader's convenience.

Corollary 9.75 (Jensen) *If $I \subset \mathbb{R}$ is an open interval and $f : I \to \mathbb{R}$ is a convex function, then, for $x_1, x_2, \ldots, x_n \in I$, we have*

$$f\left(\frac{x_1 + x_2 + \cdots + x_n}{n}\right) \leq \frac{f(x_1) + f(x_2) + \cdots + f(x_n)}{n}. \tag{9.31}$$

Moreover, if f is strictly convex, then equality happens if and only if $x_1 = x_2 = \cdots = x_n$.

Proof Let $t_1 = t_2 = \cdots = t_n = \frac{1}{n}$ in the previous result. $\square$

We leave to the reader the task of stating Jensen's inequality in the case of a (strictly) concave function f. Presently, we concentrate ourselves in discussing some applications of it.

Example 9.76 Given $n > 1$ positive reals $a_1, a_2, \ldots, a_n$, prove that

$$(a_1 + a_2 + \cdots + a_n)\left(\frac{1}{a_1} + \frac{1}{a_2} + \cdots + \frac{1}{a_n}\right) \geq n^2,$$

with equality if and only if $a_1 = a_2 = \cdots = a_n$.

Proof Example 9.66 showed that the inverse proportionality function $f : (0, +\infty) \to \mathbb{R}$, given by $f(x) = \frac{1}{x}$, is strictly convex. Hence, it follows from (9.31) that

$$\frac{n}{a_1 + a_2 + \cdots + a_n} = f\left(\frac{a_1 + a_2 + \cdots + a_n}{n}\right)$$

$$\leq \frac{f(a_1) + f(a_2) + \cdots + f(a_n)}{n}$$

$$= \frac{1/a_1 + 1/a_2 + \cdots + 1/a_n}{n}$$

with equality if and only if $a_1 = a_2 = \cdots = a_n$. Now, it suffices to observe that inequality

$$\frac{n}{a_1 + a_2 + \cdots + a_n} \leq \frac{1/a_1 + 1/a_2 + \cdots + 1/a_n}{n}$$

is precisely what we wished to get. $\square$

Example 9.77 (BMO) Let $n > 1$ and $a_1, \ldots, a_n$ be positive reals with sum equal to 1. For each $1 \leq i \leq n$, let $b_i = a_1 + \cdots + a_{i-1} + a_{i+1} + \cdots + a_n$. Prove that

$$\frac{a_1}{1 + b_1} + \frac{a_2}{1 + b_2} + \cdots + \frac{a_n}{1 + b_n} \geq \frac{n}{2n - 1},$$

with equality if and only if $a_1 = a_2 = \cdots = a_n = \frac{1}{n}$.

Proof Substituting $b_i = 1 - a_i$ for $1 \leq i \leq n$, it suffices to prove that

$$\frac{a_1}{2 - a_1} + \frac{a_2}{2 - a_2} + \cdots + \frac{a_n}{2 - a_n} \geq \frac{n}{2n - 1}.$$

To this end, we claim first that $f : (-\infty, 2) \to \mathbb{R}$, given by $f(x) = \frac{x}{2-x}$, is strictly convex. Indeed, one readily computes $f''(x) = \frac{4}{(2-x)^3} > 0$.

Hence, by applying Jensen's inequality we get

$$\sum_{i=1}^{n} f(a_i) \geq nf\left(\frac{1}{n}\sum_{i=1}^{n} a_i\right) = nf\left(\frac{1}{n}\right) = \frac{2}{2n - 1},$$

with equality if and only if $a_1 = a_2 = \cdots = a_n$, i.e., if and only if $a_1 = a_2 = \cdots = a_n = \frac{1}{n}$. $\square$

Our next example applies Jensen's inequality to solve an interesting geometric problem.

Example 9.78 Let Γ be a semicircle of radius R and diameter $A_0 A_1$. For each integer $n > 2$, show that there exists a single convex n-gon $A_0 A_1 A_2 \ldots A_{n-1}$ satisfying the following conditions:

(a) $A_2, \ldots, A_{n-1} \in \Gamma$.
(b) The area of $A_0 A_1 A_2 \ldots A_{n-1}$ is as large as possible.

Fig. 9.15 A convex n-gon inscribed in a semicircle

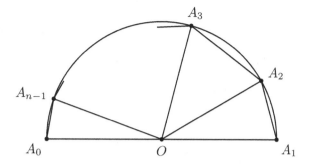

Proof Figure 9.15 depicts a convex n-gon $A_0 A_1 A_2 \ldots A_{n-1}$ inscribed in Γ. Let $A_i \widehat{O} A_{i+1} = \alpha_i$ for $1 \leq i \leq n-1$, where O is the center of Γ (and with the convention that $A_n = A_0$). Hence, $\alpha_1 + \alpha_2 + \cdots + \alpha_{n-1} = \pi$. Applying the sine formula for the area of a triangle (cf. Sect. 7.3 of [4]), we get

$$A(A_0 A_1 \ldots A_{n-1}) = \sum_{i=1}^{n-1} A(A_i O A_{i+1}) = \sum_{i=1}^{n-1} \frac{1}{2} R^2 \sin A_i \widehat{O} A_{i+1}$$

$$= \frac{1}{2} R^2 \sum_{i=1}^{n-1} \sin \alpha_i.$$

Now, since the sine function is strictly concave in the interval $[0, \pi]$, it follows from (9.31) that

$$\sum_{i=1}^{n-1} \sin \alpha_i \leq (n-1) \sin \left(\frac{1}{n-1} \sum_{i=1}^{n-1} \alpha_i \right) = (n-1) \sin \frac{\pi}{n-1},$$

with equality if and only if $\alpha_1 = \cdots = \alpha_{n-1} = \frac{\pi}{n-1}$. Therefore, there is indeed only one convex n-gon satisfying the prescribed conditions.   □

Problems: Section 9.7

1. Let $I, J \subset \mathbb{R}$ be intervals, and $g : I \to J$ and $f : J \to \mathbb{R}$ be two strictly convex functions. If f is increasing, prove that $f \circ g : I \to \mathbb{R}$ is also strictly convex.
2. Let $I, J \subset \mathbb{R}$ be intervals and $f : I \to J$ be a continuous bijection. If f is strictly convex, prove that $f^{-1} : J \to I$ is either strictly convex or strictly concave in J.
3. * Prove that the sum of a finite number of strictly convex (resp. strictly concave) functions on a common domain is also a strictly convex (resp. strictly concave).
4. Let $f : (-\infty, 1) \to \mathbb{R}$ be the function given by $f(x) = \frac{x}{\sqrt{1-x}}$. Find the intervals in which f is convex (resp. concave).

5. Let $a, b, c \in \mathbb{R}$ and $f : \mathbb{R} \to \mathbb{R}$ be given by $f(x) = x^3 + ax^2 + bx + c$. Find the inflection points of f, as well as the intervals along which it is convex or concave.

6. Let $f : (0, +\infty) \to \mathbb{R}$ be given by $f(x) = x^2 \sin \frac{1}{x}$. Show that f has infinitely many inflection points.

7. Let $f : (0, +\infty) \to \mathbb{R}$ be a twice differentiable, nondecreasing and convex function. If $g : (0, +\infty) \to \mathbb{R}$ is given by $g(x) = xf(x)$, show that g is also convex.

8. Let $f : I \to \mathbb{R}$ be a continuous function such that $f(\frac{x+y}{2}) \leq \frac{f(x)+f(y)}{2}$ for all $x, y \in I$ (resp. $f(\frac{x+y}{2}) \geq \frac{f(x)+f(y)}{2}$ for all $x, y \in I$). Prove that f is convex (resp. concave).

9. Let $f : I \to \mathbb{R}$ be continuous and such that $f(\frac{a+b}{2}) < \frac{f(a)+f(b)}{2}$ (resp. $f(\frac{a+b}{2}) > \frac{f(a)+f(b)}{2}$), for all distinct $a, b \in I$. Prove that f is strictly convex (resp. concave).

10. Let $f : (0, +\infty) \to \mathbb{R}$ be a continuous, increasing (resp. nondecreasing) and convex (resp. strictly convex) function. Prove that $g : (0, +\infty) \to \mathbb{R}$, given by $g(x) = xf(x)$, is also strictly convex.

11. Use the results of Problems 8 and 9 to establish the strictly concave or convex character of each of the functions of Example 9.66.

12. Let I be an open interval and $f : I \to \mathbb{R}$ be a twice differentiable convex function. Show that, for every $x_0 \in I$, the graph of f is not below its tangent line at x_0. More precisely, show that, for all $x, x_0 \in I$, we have

$$f(x) \geq f(x_0) + f'(x_0)(x - x_0).$$

13. (Romania) Let $I \subset \mathbb{R}$ be an interval and $f : I \to \mathbb{R}$ be strictly convex and increasing. Prove that the sequence $(f(n))_{n \geq 1}$ does not contain an infinite arithmetic progression.

14. Let $k > 1$ be an integer and $a_1, a_2, \ldots, a_n$ be positive reals. Prove that

$$\frac{a_1^k + a_2^k + \cdots + a_n^k}{n} \geq \left(\frac{a_1 + a_2 + \cdots + a_n}{n} \right)^k,$$

with equality if and only if $a_1 = a_2 = \cdots = a_n$.

15. Let Γ be the circle of center O and radius R, $n > 2$ be a given integer and $A_1 A_2 \ldots A_n$ be a convex n-gon inscribed in Γ.

   (a) Suppose that O does not belong to the interior of $A_1 A_2 \ldots A_n$. Use a geometric argument to show that there are convex n-gons inscribed in Γ and with area strictly greater than that of $A_1 A_2 \ldots A_n$.

   (b) Use Jensen's inequality to show that, among all convex n-gons inscribed in Γ, the regular ones are those of maximal area.

16. Let be given a circle Γ, with center O and radius r, and an integer $n > 2$. Among all convex n-gons $A_1 A_2 \ldots A_n$ circumscribed to Γ, prove that the regular ones are those of minimum perimeter.

17. Let ABC be an acute triangle of sides $\overline{AB} = c$, $\overline{AC} = b$ and $\overline{BC} = a$. If R is its circumradius (i.e., the radius of the circle circumscribed to ABC), show that

$$a + b + c \leq 3R\sqrt{3},$$

with equality if and only if ABC is equilateral.

18. (Romania—adapted) Let ABC be an equilateral triangle of height h, let P be an interior point and x, y and z be the distances from P to the sides of ABC.

(a) Prove that $x + y + z = h$.
(b) Find the least possible value of $\frac{h-x}{h+x} + \frac{h-y}{h+y} + \frac{h-z}{h+z}$.

19. (Turkey) Let $n > 1$ be an integer and $x_1, x_2, \ldots, x_n$ be positive reals with sum equal to 1. Prove that

$$\sum_{i=1}^{n} \frac{x_i}{\sqrt{1 - x_i}} \geq \sqrt{\frac{n}{n-1}} \geq \frac{\sqrt{x_1} + \cdots + \sqrt{x_n}}{\sqrt{n-1}},$$

with equality in any of the above inequalities if and only if $x_1 = \cdots = x_n = \frac{1}{n}$.

9.8 Sketching Graphs

Let be given an interval I and a continuous function $f : I \to \mathbb{R}$, which is twice differentiable in the interior of I. *In principle*, the theory developed so far allows us to draw a reasonably accurate sketch of the graph of f. Indeed:

(i) The results collected in Problem 10, page 194, allows us to possibly reduce the task of sketching the graph of f to doing so in an appropriate interval $J \subset I$.
(ii) The first derivative test guarantees that the extreme points of f are either the endpoints of I or the solutions of equation $f'(x) = 0$.
(iii) The study of the first variation of f assures that the intervals along which f increases (resp. decreases) are the solution sets of the inequality $f'(x) > 0$ (resp. $f'(x) < 0$).
(iv) The inflection points of f are found among the solutions of equation $f''(x) = 0$.
(v) The study of the second variation of f guarantees that the intervals along which f is strictly convex (resp. strictly concave) are the solution sets of the inequality $f''(x) > 0$ (resp. $f''(x) < 0$).
(vi) The tangent lines to the graph of f in its inflection points help us in sketching the graph in a neighborhood of each of those points.
(vii) Definition 9.18 and Problem 11, page 291, teach us how to find (if any) the asymptotes to the graph of f.

(viii) The computation of the limits $\lim_{x \to \pm\infty} f(x)$ (in case they make sense) allows us to better understand the behavior of the graph of f for large values of $|x|$.

(ix) Plotting some other points along the graph can be quite helpful in sketching it.

In this section, we illustrate the application of the program just delineated to sketching the graphs of some simple functions.

Example 9.79 Let $D = \mathbb{R} \setminus \{\frac{\pi}{2} + k\pi; \ k \in \mathbb{Z}\}$. Sketch the graph of the secant function $\sec : D \to \mathbb{R}$.

Solution

(i) First of all, note that it suffices to sketch the graph of sec in $(-\frac{\pi}{2}, \frac{\pi}{2}) \cup (\frac{\pi}{2}, \frac{3\pi}{2})$, for, since the cosine function is periodic of period 2π, so is the secant function. On the other hand, since $\cos(x + \pi) = -\cos x$ for every $x \in \mathbb{R}$, we have $\sec(x + \pi) = -\sec x$ for every $x \in D$; hence, Problem 10, page 194, guarantees that the portion of the graph of sec in the interval $(\frac{\pi}{2}, \frac{3\pi}{2})$ is obtained from that in the interval $(-\frac{\pi}{2}, \frac{\pi}{2})$ by means of a reflection along the horizontal axis, followed by a translation of π units, parallel to that same axis.

In view of the above, from now on we restrict the analysis of the graph of the secant function to the interval $(-\frac{\pi}{2}, \frac{\pi}{2})$.

(ii) and (iii) Since

$$\sec' x = (\cos^{-1} x)' = -(\cos x)^{-2}(-\sin x) = \frac{\sin x}{\cos^2 x} = \tan x \sec x, \qquad (9.32)$$

we have $\sec' x = 0 \Leftrightarrow \sin x = 0 \Leftrightarrow x = 0$. (Indeed, $0 < \cos x \le 1 \Rightarrow \sec x \ge 1$, with equality if and only if $\cos x = 1$, i.e., if and only if $x = 0$. Hence, $x = 0$ is an extreme point—thus, a critical one—of sec.) On the other hand, sec increases $\Leftrightarrow \sec' x > 0 \Leftrightarrow \sin x > 0 \Leftrightarrow x \in (0, \frac{\pi}{2})$; therefore, sec decreases in $(-\frac{\pi}{2}, 0)$.

(iv) and (v) Firstly, Example 9.34 and (9.32) furnish

$$\sec'' x = \tan' x \sec x + \tan x \sec' x = \sec^3 x + \tan^2 x \sec x$$

$$= \sec x(\sec^2 x + \tan^2 x).$$

Therefore, $\sec'' \ne 0$, and sec has no inflection points. Also, since sec is strictly convex if and only if $\sec'' x > 0$, which in turn happens if and only if $\cos x > 0$, we conclude that sec is strictly convex in $(-\frac{\pi}{2}, \frac{\pi}{2})$.

(vii) Since sec is periodic, it graph does not contain horizontal or oblique asymptotes. On the other hand, we easily get $\lim_{x \to \frac{\pi}{2}^-} \sec x = \lim_{x \to -\frac{\pi}{2}^+} \sec x = +\infty$, so that the vertical lines $x = \pm\frac{\pi}{2}$ are asymptotes of the graph of sec.

(viii) Since sec is periodic and nonconstant, there does not exist $\lim_{x \to \pm\infty} f(x)$.

(ix) Computing $\cos x$ for $x = 0, \pm\frac{\pi}{12}, \pm\frac{\pi}{6}, \pm\frac{\pi}{4}, \pm\frac{\pi}{3}$ and $\pm\frac{11\pi}{12}$, we plot points $(0, 1)$, $\left(\pm\frac{\pi}{6}, \frac{2}{\sqrt{3}}\right)$, $\left(\pm\frac{\pi}{4}, \sqrt{2}\right)$ and $\left(\pm\frac{\pi}{3}, 2\right)$ on the graph of sec.

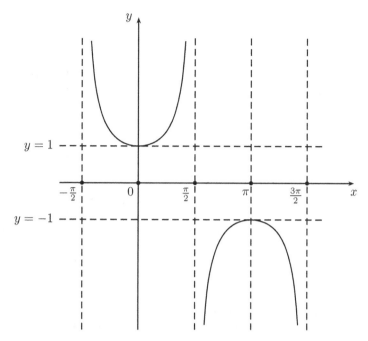

Fig. 9.16 Graph of sec in $\left(-\frac{\pi}{2}, \frac{\pi}{2}\right) \cup \left(\frac{\pi}{2}, \frac{3\pi}{2}\right)$

Gathering together the above information, we sketch the graph of sec in the interval $\left(-\frac{\pi}{2}, \frac{\pi}{2}\right) \cup \left(\frac{\pi}{2}, \frac{3\pi}{2}\right)$ as shown in Fig. 9.16.

□

Example 9.80 Sketch the graph of the polynomial function $f : \mathbb{R} \to \mathbb{R}$, given by $f(x) = 2x^3 + 4x^2 + 2x - 1$.

Solution

(ii) and (iii) Since $f'(x) = 6x^2 + 8x + 2$, the critical points of f are $x = -1$ and $x = -\frac{1}{3}$. Thus, $f'(x) < 0 \Leftrightarrow x \in (-1, -\frac{1}{3})$, and f decreases $\Leftrightarrow f'(x) < 0 \Leftrightarrow x \in (-1, -\frac{1}{3})$. Therefore, f increases $\Leftrightarrow x \in (-\infty, -1) \cup (-\frac{1}{3}, +\infty)$. It follows that $x = -1$ (resp. $x = -\frac{1}{3}$) is a local maximum (resp. minimum) point of f.

(iv) and (v) Since $f''(x) = 12x + 8$ and f is strictly convex (resp. concave) in an interval I if and only if $f''(x) > 0$ (resp. $f''(x) < 0$) in I, we conclude that f is strictly convex (resp. concave) in $(-\frac{2}{3}, +\infty)$ (resp. $(-\infty, -\frac{2}{3})$). Finally, $x = -\frac{2}{3}$ is the only inflection point of the graph of f.

(vi) Notice that $f(-\frac{2}{3}) = -\frac{31}{27}$, and the tangent line to the graph of f in $x = -\frac{2}{3}$ has slope $f'(-\frac{2}{3}) = -\frac{2}{3}$. Letting $(0, c)$ be the point where such a tangent line intersects the vertical axis, we have $\frac{c-(-31/27)}{0-(-2/3)} = -\frac{2}{3}$, so that $c = -\frac{43}{27}$.

Fig. 9.17 Graph of the
polynomial function
$f(x) = 2x^3 + 4x^2 + 2x - 1$

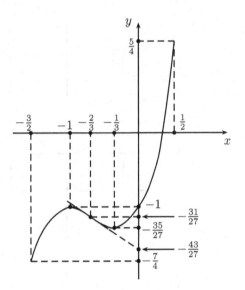

(vii) Since f is defined in the whole real line, its graph does not possess vertical
asymptotes. On the other hand, since $\lim_{x\to\pm\infty} \frac{f(x)}{x} = \lim_{x\to\pm\infty}(2x^2 + 4x +$
$2 - \frac{1}{x}) = +\infty$, Problem 11, page 291, guarantees that the graph of f does not
have oblique asymptotes.

(viii) It follows from Example 9.17 that $\lim_{x\to\pm\infty} f(x) = \pm\infty$.

(ix) Since $f(0) = -1$, it follows from (viii) and the IVT that f has a positive root
α. Computing $f(x)$ for $x = -\frac{3}{2}, -1, -\frac{1}{3}, 0, \frac{1}{4}, \frac{1}{3}, \frac{1}{2}$ and 1, we plot points
$(-\frac{3}{2}, -\frac{7}{4}), (-1, -1), (-\frac{1}{3}, -\frac{35}{27}), (0, -1), (\frac{1}{4}, -\frac{7}{32}), (\frac{1}{3}, \frac{1}{27}), (\frac{1}{2}, \frac{5}{4})$ and $(1, 7)$
on the graph of f. (Recall that, in item (vi), we had already get point $(-\frac{2}{3}, -\frac{31}{27})$
on the graph.) In particular, we conclude that $\alpha \in (\frac{1}{4}, \frac{1}{3})$.

Finally, as in the previous example, we gather the above information to sketch
the graph of f in Fig. 9.17.

□

Example 9.81 Sketch the graph of $f : \mathbb{R} \setminus \{-1\} \to \mathbb{R}$, given by $f(x) = x^2 + \frac{1}{x+1}$.

Solution

(ii) and (iii) Since $f'(x) = 2x - \frac{1}{(x+1)^2} = \frac{2x^3+4x^2+2x-1}{(x+1)^2}$, we have $f'(x) = 0 \Leftrightarrow$
$2x^3 + 4x^2 + 2x - 1 = 0$. Hence, the only critical point of f is the only positive
root α of the polynomial function of the previous example, so that $\alpha \in (\frac{1}{4}, \frac{1}{3})$.
Then, $f'(x) > 0 \Leftrightarrow 2x^3 + 4x^2 + 2x - 1 > 0 \Leftrightarrow x > \alpha$, and we conclude that
f increases in $(\alpha, +\infty)$. Note that α is a minimum local point of f.

(iv) and (v) Since $f''(x) = 2 + \frac{2}{(x+1)^3}$, we have f strictly convex (resp. concave)
in an interval I if and only if $\frac{1}{x+1} > -1$ (resp. $\frac{1}{x+1} < -1$) in I. Therefore, f is
strictly convex in each of the intervals $(-\infty, -2)$ and $(-1, +\infty)$, and strictly
concave in $(-2, -1)$. Moreover, its only inflection point is $x = -2$.

(vi) Note that $f(-2) = 3$, and the tangent line to the graph of f at $x = -2$ has slope $f'(-2) = -5$. Letting $(0, c)$ be the point in which such a tangent line intersects the vertical axis, we have $\frac{c-3}{0-(-2)} = -5$, so that $c = -7$.

(vii) Since $x = -1$ is the only point for which $f(x)$ is not defined, the vertical line $x = -1$ is the only candidate for vertical asymptote of the graph. This is indeed the case, for, $\lim_{x \to -1-} f(x) = -\infty$ and $\lim_{x \to -1+} f(x) = +\infty$. On the other hand, since $\lim_{x \to \pm\infty} \frac{f(x)}{x} = \lim_{x \to \pm\infty} \left(x + \frac{1}{x(x+1)} \right) = \pm\infty$, Problem 11, page 291, assures that the graph has no oblique asymptotes.

(viii) It follows from Example 9.17 that $\lim_{x \to \pm\infty} f(x) = +\infty$.

(ix) Computing $f(x)$ for $x = -\frac{5}{2}, -\frac{3}{2}, -\frac{5}{4}, -\frac{3}{4}, -\frac{1}{2}, -\frac{1}{4}, 0, \frac{1}{4}, \frac{1}{2}, 1$ and 2, we plot points $(-\frac{5}{2}, \frac{67}{12})$, $(-\frac{3}{2}, -\frac{7}{4})$, $(-\frac{5}{4}, -\frac{39}{16})$, $(-\frac{3}{4}, \frac{73}{16})$, $(-\frac{1}{2}, \frac{9}{4})$, $(-\frac{1}{4}, \frac{67}{48})$, $(0, -1)$, $(\frac{1}{4}, -\frac{7}{32})$, $(\frac{1}{2}, \frac{5}{4})$, $(1, \frac{3}{2})$ and $(2, \frac{13}{3})$ on the graph of f. (Recall that, in item (vi), we had already get point $(-2, 3)$ on the graph.)

Once more, we collect all of the above to sketch the graph of f, as shown in Fig. 9.18. Note that, for the purpose of a better qualitative picture, we adopted different scales on the horizontal and vertical axes.

□

Fig. 9.18 Graph of $f(x) = x^2 + \frac{1}{x+1}$

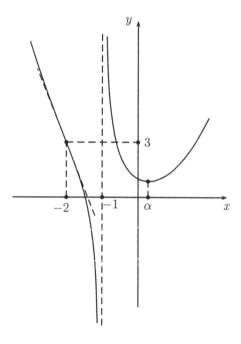

Problems: Section 9.8

1. Sketch the graph of the **cosecant** function, $\csc : \mathbb{R} \setminus \{k\pi; \ k \in \mathbb{Z}\} \to \mathbb{R}$, such that $\csc x = \frac{1}{\sin x}$.

2. In each of the items below, sketch the graph of the given function (defined in its maximal domain). For each such function (and whenever pertinent), compute explicitly or estimate the critical points, monotonicity intervals, inflection points, tangent lines at inflection points, intervals of convexity or concavity, asymptotes and behavior at infinity:

(a) $f(x) = x^4 + 2x^2 - 2x + 1$.

(b) $f(x) = x + \frac{1}{x}$.

(c) $f(x) = \frac{1}{x^2+1}$.

(d) $f(x) = \frac{1}{x^3+1}$.

(e) $f(x) = \frac{x}{2} + \frac{1}{x^2+1}$.

(f) $f(x) = \frac{x}{x^2+1}$.

(g) $f(x) = \frac{x}{x^2-3x+2}$.

(h) $f(x) = \frac{x}{(x^2+1)(x^2-1)}$.

3. * Let $f : (a, b) \to \mathbb{R}$ be twice differentiable at $x_0 \in (a, b)$, with $f''(x_0) \neq 0$. Since f is in particular differentiable at x_0, the tangent line r to the graph at $A(x_0, f(x_0))$, is not vertical. We define the **osculating circle** to the graph at A as the circle Γ tangent to r at A that *better approximates the shape of the graph* in a neighborhood of x_0, in the following sense: taking $c \in (a, x_0)$ and $d \in (x_0, b)$ such that one of the arcs of Γ situated in the strip defined by the vertical lines $x = c$ and $x = d$ is the graph of a function $g : (c, d) \to \mathbb{R}$, we have $g(x_0) = f(x_0)$, $g'(x_0) = f'(x_0)$ and $g''(x_0) = f''(x_0)$. In this respect, do the following items:

(a) Let $O(\alpha, \beta)$ be the center of Γ. If $f'(x_0) = 0$, show that $\alpha = x_0$; if $f'(x_0) \neq 0$, show that $\frac{\beta - f(x_0)}{\alpha - x_0} = -\frac{1}{f'(x_0)}$.

(b) Show that the equation of Γ is $(x - \alpha)^2 + (y - \beta)^2 = R^2$, where $R = \sqrt{(x_0 - \alpha)^2 + (f(x_0) - \beta)^2}$ is its radius.

(c) For $x \in (c, d)$, substitute $y = g(x)$ in (b) and differentiate the relation thus obtained to get $g'(x_0) = \frac{x_0 - \alpha}{\beta - g(x_0)} = f'(x_0)$ and $g''(x_0) = \frac{1 + f'(x_0)^2}{\beta - f(x_0)}$.

(d) Conclude that $\beta = f(x_0) + \frac{1 + f'(x_0)^2}{f''(x_0)}$ and $\alpha = x_0 - f'(x_0)\left(\frac{1+f'(x_0)^2}{f''(x_0)}\right)$. From this, and with respect to the line r, show that O is located in the upper (resp. lower) half-plane, provided f is strictly convex (resp. strictly concave) in a neighborhood of x_0.

(e) Show that $R = \frac{(1+f'(x_0)^2)^{3/2}}{|f''(x_0)|}$. (In particular, observe that the osculating circle at A is uniquely determined by the conditions given at the statement of the problem.)

(f) Show that the geometric vector $\overrightarrow{u} = (1, f'(x_0))$, of origin A, is parallel to the line r. Show also that, upon applying to $\overrightarrow{u}$ a counterclockwise rotation with center A and angle $90°$, we get a vector of the form $\lambda \overrightarrow{AO}$, with $\lambda \in \mathbb{R}^*$.

The radius R of Γ is the **curvature radius** of the graph of f at x_0 (or A); the **curvature** of the graph at x_0 is the real number

Fig. 9.19 Osculating circle
to the graph of $x \mapsto \frac{1}{x}$ at A

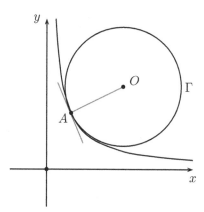

$$k = \frac{\sigma}{R} = \frac{\sigma f''(x_0)}{(1 + f'(x_0)^2)^{3/2}},$$

where $\sigma = \frac{\lambda}{|\lambda|} \in \{\pm 1\}$ is the sign of λ. (Observe that the bigger R, the closer $|k|$ is to 0. This fact reflects gives geometric intuition to the term *curvature*.)

(g) Let $R > 0$ and $f : (-R, R) \to \mathbb{R}$ be given by $f(x) = \sqrt{R^2 - x^2}$ (so that the graph of f is a semicircle of radius R). Show that the curvature radius of the graph of f is constant and equal to R.

As an illustration, Fig. 9.19 brings the osculating circle to the graph of $f(x) = \frac{1}{x}$, $x > 0$, at a point A.

4. The purpose of this problem is to sketch the graph of $f : \mathbb{R} \to \mathbb{R}$, given by

$$f(x) = \begin{cases} 0, & \text{se } x = 0 \\ x \sin \frac{1}{x}, & \text{se } x \neq 0 \end{cases}.$$

To this end, do the following items:

(a) Show that it suffices to consider the case $x \geq 0$.
(b) Show that the set of critical points of f in $(0, +\infty)$ is the set of reals x such that $x = \frac{1}{y}$, where $y > 0$ satisfies the equation $\sin y - y \cos y = 0$.
(c) Show that the solutions $y > 0$ to this equation form a sequence $(y_n)_{n \geq 1}$, such that $n\pi < y_n < n\pi + \frac{\pi}{2}$ and the sequence $(n\pi + \frac{\pi}{2} - y_n)_{n \geq 1}$ decreases to 0 when $n \to +\infty$.
(d) If $x_n = \frac{1}{y_n}$ for $n \geq 1$, show that x_{2k-1} is a local minimum point and x_{2k} is a local maximum point for f, for every $k \geq 1$.
(e) Show that $|f(x_1)| > |f(x_2)| > |f(x_3)| > \cdots \to 0$.
(f) f increases in $[x_1, +\infty)$, with $\lim_{x \to +\infty} f(x) = 1$.
(g) Mostre que f is strictly convex in $(\frac{1}{2k\pi}, \frac{1}{(2k-1)\pi})$ and strictly concave in $(\frac{1}{(2k+1)\pi}, \frac{1}{2k\pi})$, for every integer $k \geq 1$.
(h) Sketch the graph of f.

Chapter 10
Riemann's Integral

This chapter completes the task of establishing the fundamentals of Calculus, this time studying the operation of integration on functions. As we shall see in the next section, in its most simple form this reduces to the computation of areas under the graphs of nonnegative continuous functions $f : [a, b] \to \mathbb{R}$, suggesting that geometric intuition will play a strong role throughout. Among other byproducts of the coming discussions, we prove the irrationality of π and study the properties of two of the most ubiquitous functions of Mathematics, the exponential and logarithmic functions with base e.

10.1 Some Heuristics II

Our story begins in the III century B.C., with the great Archimedes, who considered (and solved) the problem of computing the area under an arc of parabola, by using the *method of exhaustion*.

In modern notation and in a slightly more general situation, the heuristics behind such a method is the following: given a nonnegative function $f : [a, b] \to \mathbb{R}$ (cf. Fig. 10.1), one wants to compute the area of the region $\mathcal{R}$ of the cartesian plane, situated under the graph of f and above the horizontal axis, so that

$$\mathcal{R} = \{(x, y) \in \mathbb{R}^2; \ a \leq x \leq b \text{ and } 0 \leq y \leq f(x)\}.$$

To this end, one starts by dividing the interval $[a, b]$ into k equal intervals, with the aid of the points $a = t_0 < t_1 < \cdots < t_k = b$ such that $t_i - t_{i-1} = \frac{b-a}{k}$ for $1 \leq i \leq k$. Then (cf. Fig. 10.2), one considers the portion of $\mathcal{R}$ contained in the vertical strip bounded by the straight lines $x = t_{i-1}$ and $x = t_i$, and approximates its area, from below, by that of the greatest rectangle contained therein and having the $[t_{i-1}, t_i]$ as one of its sides; accordingly, one approximates the area of that portion of

© Springer International Publishing AG 2017
A. Caminha Muniz Neto, *An Excursion through Elementary Mathematics, Volume I*,
Problem Books in Mathematics, DOI 10.1007/978-3-319-53871-6_10

Fig. 10.1 The region $\mathcal{R}$
under the graph of f

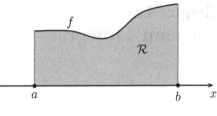

Fig. 10.2 Approximating the
area of $\mathcal{R}$ from above and
from below

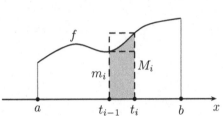

$\mathcal{R}$, from above, by that of the smallest rectangle containing it and having the interval
$[t_{i-1}, t_i]$ as one of its sides.

One now approximates the whole area of $\mathcal{R}$ from below and from above,
computing, in each case, the sums of the areas of the k rectangles of the previous
paragraph. Assuming that f attains its minimum and maximum values in each
interval $[t_{i-1}, t_i]$ (which is always the case if f is continuous), and letting m_i and
M_i respectively denote such values (so that m_i and M_i are the lengths of the heights
of the rectangles one is considering), one obtains the sums

$$\underline{A}(f;k) := \sum_{i=1}^{k} m_i(t_i - t_{i-1}) = \left(\frac{b-a}{k}\right) \sum_{i=1}^{k} m_i \qquad (10.1)$$

and

$$\overline{A}(f;k) := \sum_{i=1}^{k} M_i(t_i - t_{i-1}) = \left(\frac{b-a}{k}\right) \sum_{i=1}^{k} M_i. \qquad (10.2)$$

Letting $A(\mathcal{R})$ denote the area of $\mathcal{R}$, one then gets

$$\underline{A}(f;k) \leq A(\mathcal{R}) \leq \overline{A}(f;k),$$

and hopes that $\underline{A}(f;k)$ and $\overline{A}(f;k)$ do approximate $A(\mathcal{R})$ better and better as $k \to
+\infty$.

In order to illustrate the difficulties involved, let's consider the more particular
situation of the function $f : [0, b] \to \mathbb{R}$ given by $f(x) = x^n$, where n is a given
natural number. As before, letting $k \in \mathbb{N}$ and $0 = t_0 < t_1 < \cdots < t_k = b$ be
the partition of $[0, b]$ such that $t_i - t_{i-1} = \frac{b}{k}$ for $1 \leq i \leq k$, we have $t_i = \frac{bi}{k}$ for
$0 \leq i \leq k$. Also, since $x \mapsto x^n$ is increasing, it follows that $m_i = f(t_{i-1}) = t_{i-1}^n$ and

$M_i = f(t_i) = t_i^n$. Thus,

$$\overline{A}(f;k) = \frac{b}{k} \cdot \sum_{i=1}^{k} t_i^n = \frac{b}{k} \cdot \sum_{i=1}^{k} \left(\frac{bi}{k}\right)^n = \left(\frac{b}{k}\right)^{n+1} \sum_{i=1}^{k} i^n \qquad (10.3)$$

and, analogously,

$$\underline{A}(f;k) = \left(\frac{b}{k}\right)^{n+1} \sum_{i=1}^{k} (i-1)^n. \qquad (10.4)$$

It easily follows from these computations that

$$\overline{A}(f;k) - \underline{A}(f;k) = \frac{b^{n+1}}{k}; \qquad (10.5)$$

hence, $\underline{A}(f;k) - \overline{A}(f;k) \to 0$ as $k \to +\infty$. Therefore, at least in this case, we conclude that $\underline{A}(f;k)$ and $\overline{A}(f;k)$ approximate $A(\mathcal{R})$ better and better, as the number k of rectangles increases.

To compute the actual value of $A(\mathcal{R})$, we use the result of Problem 21, page 86, which guarantees that

$$\sum_{i=1}^{k} (i-1)^n < \frac{(k-1)^{n+1}}{n+1} < \sum_{i=1}^{k} i^n.$$

Combining these inequalities with (10.3) and (10.4), we get

$$\underline{A}(f;k) < \frac{b^{n+1}}{n+1} \left(1 - \frac{1}{k}\right)^{n+1} < \overline{A}(f;k). \qquad (10.6)$$

However, since both $A(\mathcal{R})$ and $\frac{b^{n+1}}{n+1}\left(1 - \frac{1}{k}\right)^{n+1}$ belong to the closed interval $\left[\overline{A}(f;k), \underline{A}(f;k)\right]$, relation (10.5) gives

$$\left| A(\mathcal{R}) - \frac{b^{n+1}}{n+1} \left(1 - \frac{1}{k}\right)^{n+1} \right| \leq \overline{A}(f;k) - \underline{A}(f;k) < \frac{b^{n+1}}{k}.$$

Making $k \to +\infty$, we then get

$$A(\mathcal{R}) = \lim_{k \to +\infty} \frac{b^{n+1}}{n+1} \left(1 - \frac{1}{k}\right)^{n+1} = \frac{b^{n+1}}{n+1}.$$

Fig. 10.3 Approximating the
area under the graph of f

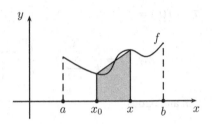

Let's now return (again in a modern setting) to the general problem of the computation of the area of the region under the graph of a given nonnegative function $f : [a, b] \rightarrow \mathbb{R}$ (cf. Fig. 10.3).

For each real number $x \in [a, b]$, let's denote by $A(x)$ the area of the region of the plane situated under the graph of f, above the horizontal axis and between the vertical lines of abscissas a and x (with the convention that $A(a) = 0$). For a fixed $x_0 \in [a, b)$ and taking $x_0 < x \leq b$, we have

$$A(x) - A(x_0) = A([x_0, x]), \tag{10.7}$$

where $A([x_0, x])$ denotes the area of the portion of the region just described, situated between the vertical lines of abscissas x_0 and x.

For small values of $x - x_0$ (when compared to the values of f along the interval $[x_0, x]$), it is reasonable to assume that a good approximation for the value of $A([x_0, x])$ is the area of the trapezoid with base lengths $f(x_0)$ and $f(x)$ and having the line segment $[x_0, x]$ as one of its legs (the gray trapezoid of Fig. 10.3). Since the altitude of such a trapezoid equals $x - x_0$, it follows from (10.7) that

$$A(x) - A(x_0) = A([x_0, x]) \cong \left(\frac{f(x_0) + f(x)}{2} \right) (x - x_0)$$

for small values of $x - x_0$. This way, we conclude that

$$\frac{A(x) - A(x_0)}{x - x_0} \cong \frac{f(x_0) + f(x)}{2} \tag{10.8}$$

for small values of $x - x_0$, and it is also reasonable to assume that, the closest x is to x_0, the better is such an approximation.

Therefore, if f is a continuous function, (10.8) suggests that the *area function* $A : [a, b] \rightarrow \mathbb{R}$ is differentiable, with

$$A'(x_0) = \lim_{x \to x_0} \frac{A(x) - A(x_0)}{x - x_0} = \lim_{x \to x_0} \left(\frac{f(x_0) + f(x)}{2} \right) = f(x_0), \tag{10.9}$$

In words, the heuristic reasoning above suggests that we can recover the continuous function $f : [a, b] \rightarrow \mathbb{R}$ out of the corresponding area function $A : [a, b] \rightarrow \mathbb{R}$. In the case of $f(x) = x^n$, for instance, we got $A(x) = \frac{x^{n+1}}{n+1}$, so that $A'(x) = x^n$.

In modern language, given a nonnegative continuous function $f : [a, b] \to \mathbb{R}$, we say that the area function $A : [a, b] \to \mathbb{R}$ is an *indefinite integral* for f, and write

$$A(x) = \int_a^x f(t)dt. \tag{10.10}$$

Here, the symbol $\int_a^x$ recalls the fact that the area under the graph of f, from a to x, was computed with the aid of sums that approximated the desired area better and better ($\int$ is a stylized capital S); in turn, the symbol $f(t)dt$ recalls the fact that the summands of those sums had the form $m_i(t_i - t_{i-1})$, where m_i denoted the smallest value of f along the interval $[t_{i-1}, t_i]$ (hence, $m_i = f(t)$ for some $t \in [t_{i-1}, t_i]$) and $t_i - t_{i-1} = \Delta t_i$ was the i-th difference along the interval $[a, b]$ (Leibniz systematically used the Greek letter Δ to denote finite differences, and such a notation survived to this day).

Taking (10.9) and (10.10) together, we get

$$\frac{d}{dx} \int_a^x f(t)dt = f(x), \tag{10.11}$$

and this is essentially the content of the *Fundamental Theorem of Calculus*, which suggests that the operations of *integration* (i.e., of computing integrals) and derivation are somewhat inverses of each another.

The rest of this chapter puts all of the above in solid grounds, developing several interesting applications along the way.

10.2 The Concept of Integral

This section introduces the concept of integral for *bounded* functions $f : [a, b] \to \mathbb{R}$ and establishes the *integrability* of two important classes of such functions, namely, the continuous and the monotone ones. To this end, we begin by fixing some notations.

A **partition** of an interval $[a, b]$ is the choice of a finite subset

$$P = \{a = x_0 < x_1 < x_2 < \cdots < x_k = b\} \tag{10.12}$$

of $[a, b]$. Given such a partition and a *bounded* (but not necessarily nonnegative) function $f : [a, b] \to \mathbb{R}$, we shall systematically denote

$$m_j = \inf\{f(x); \, x_{j-1} \le x \le x_j\} := \inf_{[x_{j-1}, x_j]} f$$

and

$$M_j = \sup\{f(x); \, x_{j-1} \le x \le x_j\} := \sup_{[x_{j-1}, x_j]} f.$$

Note that both m_j and M_j are well defined, thanks to the boundedness of f.

The analogues of the lower and upper approximations for the area of the region under the graph of f, discussed in the previous section, are the **lower sum** and the **upper sum** of f with respect to P, which are respectively defined by

$$s(f;P) = \sum_{j=1}^{k} m_j(x_j - x_{j-1}) \text{ and } S(f;P) = \sum_{j=1}^{k} M_j(x_j - x_{j-1}).$$

For a fixed partition P, we certainly have $m_j \leq M_j$ for every j, so that $s(f;P) \leq S(f;P)$. Hence, also in accordance with the discussion undertaken in the previous section, we would like to declare a bounded function $f : [a,b] \to \mathbb{R}$ as being *integrable* if

$$\sup\{s(f;P); P \text{ is a partition of } [a,b]\} = \inf\{S(f;P); P \text{ is a partition of } [a,b]\}.$$

However, some care is needed, for, the inequality $s(f;P) \leq S(f;P)$ (which is valid when we take the *same* partition P at both sides) *does not necessarily* imply a relation between the supremum and infimum above. This is due to the fact that there is no relation between the generic partition P that declares the set at the left hand side above and the one that declares the set at the right hand side (the choice of the same letter to represent them was just a matter of notational convenience).

Thus, before we proceed towards a coherent definition of integrable function and integral, we ought to compare $s(f;P)$ and $S(f;Q)$, for two generic partitions P and Q of $[a,b]$. We do that in the coming lemma.

Lemma 10.1 *Let $f : [a,b] \to \mathbb{R}$ be a bounded function. Given partitions P and Q of $[a,b]$ such that $P \subset Q$, we have*

$$s(f;P) \leq s(f;Q) \text{ and } S(f;Q) \leq S(f;P).$$

Proof Let P be as in (10.12). We initially consider the case in which $Q = P \cup \{x'\}$, with $x' \neq x_0, x_1, \ldots, x_k$, and let $i \in \{1, 2, \ldots, k\}$ be such that $x_{i-1} < x' < x_i$. Then,

$$S(f;Q) = \sum_{j=1}^{i-1} M_j(x_j - x_{j-1}) + (\sup_{[x_{i-1},x']} f)(x' - x_{i-1})$$

$$+ \left(\sup_{[x',x_i]} f\right)(x_i - x') + \sum_{j=i+1}^{k} M_j(x_j - x_{j-1}). \tag{10.13}$$

Now, since $[x_{i-1},x'], [x',x_i] \subset [x_{i-1},x_i]$, we get

$$\sup_{[x_{i-1},x']} f, \sup_{[x',x_i]} f \leq \sup_{[x_{i-1},x_i]} f = M_i,$$

so that

$$\left(\sup_{[x_{i-1}, x']} f\right)(x' - x_{i-1}) + \left(\sup_{[x', x_i]} f\right)(x_i - x') \le M_i(x' - x_{i-1}) + M_i(x_i - x')$$

$$= M_i(x_i - x_{i-1}).$$

Hence, it follows from (10.13) that

$$S(f; Q) \le \sum_{j=1}^{i-1} M_j(x_j - x_{j-1}) + M_i(x_i - x_{i-1}) + \sum_{j=i+1}^{k} M_j(x_j - x_{j-1})$$

$$= \sum_{j=1}^{k} M_j(x_j - x_{j-1}) = S(f; P).$$

We now consider an arbitrary partition Q containing P. Since Q is a finite set, we can pass from P to Q in a finite number of steps, by adjoining to P the points of $Q \setminus P$, one at a time. In doing so, we obtain partitions $P = P_1 \subset P_2 \subset \cdots \subset P_l = Q$ such that, for $2 \le i \le l$, we get P_i by adding a single point to P_{i-1}. Then, the first part of the proof gives

$$S(f; P) = S(f; P_1) \ge S(f; P_2) \ge \cdots \ge S(f; P_l) = S(f; Q).$$

Finally, the proof of the inequality involving lower sums is completely analogue and will be left to the reader (see Problem 1). □

As an immediate consequence of the previous lemma, given a bounded function $f : [a, b] \to \mathbb{R}$ and (any) partitions P and Q of $[a, b]$, we have

$$s(f; P) \le s(f; P \cup Q) \le S(f; P \cup Q) \le S(f; Q). \tag{10.14}$$

In particular, given a partition Q of $[a, b]$, the above inequalities guarantee that $S(f; Q)$ us an upper bound for the set $\{s(f; P); P \text{ is a partition of } [a, b]\}$ of the lower sums of f. Hence, Theorem 7.4 guarantees that such a set does have a supremum, which will be denoted $\sup_P s(f; P)$:

$$\sup_P s(f; P) = \sup\{s(f; P); P \text{ is a partition of } [a, b]\}.$$

On the other hand, since the supremum of a set bounded from above is the smallest of its upper bounds (one of which is $S(f; Q)$), we conclude that

$$\sup_P s(f; P) \le S(f; Q).$$

However, since the partition Q was arbitrarily chosen, we conclude that this last inequality is valid for every partition Q of $[a, b]$. To put in another way, the real number $\sup_P s(f; P)$ is a *lower* bound for $\{S(f; Q);\ Q$ is a partition of $[a, b]\}$, the set of the *upper* sums of f. Hence, it follows from Problem 6, page 206, that such a set has an infimum, which will be denoted by $\inf_Q S(f; Q)$:

$$\inf_Q S(f; Q) = \inf\{S(f; Q);\ Q \text{ is a partition of } [a, b]\}.$$

Finally, since the infimum of a set bounded from below is the largest of its lower bounds (one of which is $\sup_P s(f; P)$), it follows that

$$\sup_P s(f; P) \leq \inf_Q S(f; Q). \tag{10.15}$$

We are finally in position to present the central definition of this chapter.

Definition 10.2 A bounded function $f : [a, b] \to \mathbb{R}$ is **Riemann integrable**[1] if

$$\sup_P s(f; P) = \inf_P S(f; P).$$

As we have anticipated at the first paragraph of this section, continuous and monotone functions with domain $[a, b]$ are integrable. Nevertheless, with what we have at our disposal at this point, it is more convenient to start by presenting the classical example of a bounded *nonintegrable* function.

Example 10.3 Recall from Problem 11, page 255) that **Dirichlet's function** is $f : [0, 1] \to \mathbb{R}$ such that

$$f(x) = \begin{cases} 0 \text{ if } x \notin \mathbb{Q} \\ 1 \text{ if } x \in \mathbb{Q} \end{cases}.$$

Fixed a partition $P = \{0 = x_0 < x_1 < x_2 < \cdots < x_k = 1\}$ of $[0, 1]$, Problem 4, page 206, assures that all of the intervals $[x_{j-1}, x_j]$ contain rational and irrational numbers. Therefore, for $1 \leq j \leq k$ we have $m_j = 0$ and $M_j = 1$, so that

$$s(f; P) = \sum_{j=1}^{k} m_j(x_j - x_{j-1}) = 0$$

---

[1]When we look at large to Riemann's legacy to Geometry and Analysis, we easily come to the conclusion that they largely surpass the adequate formalization of the concept of integral that bears his name. (As the attentive reader has certainly noticed, this chapter is entitled *Riemann's Integral*!) Indeed, *Semi-Riemannian Geometry* gave Albert Einstein the adequate theoretical apparatus for the development of General Relativity. Other of Riemann's creations, known today as *Riemann's zeta function*, has shown to be an ubiquitous object in Number Theory.

and

$$S(f;P) = \sum_{j=1}^{k} M_j(x_j - x_{j-1}) = \sum_{j=1}^{k} (x_j - x_{j-1}) = 1.$$

Then,

$$\sup_{P} s(f;P) = 0 < 1 = \inf_{P} S(f;P),$$

and f is nonintegrable.

In order to discuss the integrability of functions in a more transparent way, it is worth to recast the definition of integrable function as in the coming result, which is known as **Cauchy's integrability criterion**.

Theorem 10.4 (Cauchy)   *A bounded function $f : [a,b] \to \mathbb{R}$ is integrable if and only if the following condition is satisfied: for every $\epsilon > 0$, there exists a partition P_ϵ of $[a,b]$ such that*

$$S(f;P_\epsilon) - s(f;P_\epsilon) < \epsilon. \tag{10.16}$$

Proof Firstly, suppose that f is integrable, so that $\sup_P s(f;P) = \inf_P S(f;P)$. Let this common value be denoted by I, and let $\epsilon > 0$ be given. Since $I = \sup_P s(f;P)$ is the least upper bound for the set of lower sums of f and $I - \frac{\epsilon}{2} < I$, there exists a partition P_1 of $[a,b]$ such that $I - \frac{\epsilon}{2} < s(f;P_1)$. Analogously, since $I = \inf_P S(f;P)$ is the greatest lower bound for the set of upper sums of f and $I + \frac{\epsilon}{2} > I$, there exists a partition P_2 of $[a,b]$ such that $S(f;P_2) < I + \frac{\epsilon}{2}$. Letting $P_\epsilon = P_1 \cup P_2$, it follows from (10.14) that

$$S(f;P_\epsilon) - s(f;P_\epsilon) \leq S(f;P_2) - s(f;P_1) < \left(I + \frac{\epsilon}{2}\right) - \left(I - \frac{\epsilon}{2}\right) = \epsilon.$$

Conversely, assume that the stated condition is satisfied. Then, given $\epsilon > 0$, and taking a partition P_ϵ as in the statement, it follows from (10.15) that

$$0 \leq \inf_{P} S(f;P) - \sup_{P} s(f;P) \leq S(f;P_\epsilon) - s(f;P_\epsilon) < \epsilon.$$

However, since $\epsilon > 0$ was arbitrarily chosen, it follows that $\inf_P S(f;P) - \sup_P s(f;P) = 0$, as wished. $\square$

We can finally establish the integrability of monotone and continuous functions.

Theorem 10.5   *Every monotone function $f : [a,b] \to \mathbb{R}$ is integrable.*

Proof Suppose f to be nondecreasing, the case of a nonincreasing function being totally analogous. Then, the image of contained in the interval $[f(a), f(b)]$, so that it is clearly bounded. Let $\epsilon > 0$ be given. By Cauchy's integrability criterion, in order to establish the integrability of f it suffices to find a partition $P = \{a = x_0 < x_1 < \cdots < x_k = b\}$ of $[a, b]$ such that condition (10.16) is satisfied for $P_\epsilon = P$.

Given a partition P as above, f being nondecreasing implies

$$m_j = \inf\{f(x); \ x_{j-1} \le x \le x_j\} = f(x_{j-1})$$

and, analogously, $M_j = f(x_j)$. Hence,

$$S(f; P) - s(f; P) = \sum_{j=1}^{k} M_j(x_j - x_{j-1}) - \sum_{j=1}^{k} m_j(x_j - x_{j-1})$$

$$= \sum_{j=1}^{k} (f(x_j) - f(x_{j-1}))(x_j - x_{j-1}).$$

Letting $\delta = \max\{x_j - x_{j-1}; \ 1 \le j \le k\}$, the last equality above and (3.15) give

$$S(f; P) - s(f; P) < \sum_{j=1}^{k} (f(x_j) - f(x_{j-1}))\delta = \delta(f(b) - f(a)).$$

Therefore, it suffices to choose P such that $\delta(f(b) - f(a)) < \epsilon$, which happens whenever $\delta < \frac{\epsilon}{f(b)-f(a)+1}$, for instance. $\square$

Theorem 10.6 *Every continuous function $f : [a, b] \to \mathbb{R}$ is integrable.*

Proof Corollary 8.25 guarantees that such an f is bounded. Hence, again by Cauchy's integrability criterion and given $\epsilon > 0$, in order to establish the integrability of f it suffices to find a partition $P = \{a = x_0 < x_1 < \cdots < x_k = b\}$ of $[a, b]$ such that (10.16) is satisfied for $P_\epsilon = P$.

As in the proof of the previous of the previous result,

$$S(f; P) - s(f; P) = \sum_{j=1}^{k} M_j(x_j - x_{j-1}) - \sum_{j=1}^{k} m_j(x_j - x_{j-1})$$

$$= \sum_{j=1}^{k} (M_j - m_j)(x_j - x_{j-1}).$$

(10.17)

Now, Theorem 8.24 assures that f is uniformly continuous. Therefore, according to Definition 8.22, given $\epsilon' > 0$ there exists $\delta > 0$ such that

$$x, y \in [a, b] \text{ and } |x - y| < \delta \Rightarrow |f(x) - f(y)| < \epsilon'.$$

Thus, starting with a partition P such that $x_j - x_{j-1} < \delta$ for $1 \leq j \leq k$, we conclude that

$$x_{j-1} \leq x, y \leq x_j \Rightarrow |x - y| < \delta \Rightarrow |f(x) - f(y)| < \epsilon'.$$

However, since $M_j = \sup\{f(x); x_{j-1} \leq x \leq x_j\}$ and $m_j = \inf\{f(y); x_{j-1} \leq y \leq x_j\}$, it easily follows from the above inequality that

$$x_j - x_{j-1} < \delta \Rightarrow M_j - m_j \leq \epsilon', \ \forall \ 1 \leq j \leq k. \tag{10.18}$$

Finally, letting P be a partition such that $x_j - x_{j-1} < \delta$ for $1 \leq j \leq k$, it follows from (10.17), (10.18) and (3.15) that

$$S(f; P) - s(f; P) = \sum_{j=1}^{k} (M_j - m_j)(x_j - x_{j-1})$$

$$\leq \sum_{j=1}^{k} \epsilon'(x_j - x_{j-1}) = \epsilon'(b - a).$$

Therefore, in order to get $S(f; P) - s(f; P) < \epsilon$ it suffices to start with $\epsilon' = \frac{\epsilon}{2(b-a)}$. $\square$

As anticipated in the end of the previous section, if $f : [a, b] \to \mathbb{R}$ is an integrable function, we let

$$\int_a^b f(x)dx \tag{10.19}$$

denote the integral of f on the interval $[a, b]$. Note also that it does not matter whether we denote the integral of f on $[a, b]$ as above or writing $\int_a^b f(t)dt$. Indeed, such a change in notation is equivalent to changing the notation for the independent variable of f, which certainly does not alter the value of the integral.

In spite of what we have done so far, unfortunately we do not have yet a general procedure for computing integrals of (integrable) specific functions $f : [a, b] \to \mathbb{R}$. We shall remedy this situation in Sect. 10.5. However, see Problems 2, 4, 5 and 10.

Problems: Section 10.2

1. * Complete the proof of Lemma 10.1 showing that, if $f : [a, b] \to \mathbb{R}$ is bounded and P and Q are partitions of $[a, b]$ such that $P \subset Q$, then $s(f; P) \leq s(f; Q)$.
2. * If $f : [a, b] \to \mathbb{R}$ is a constant function, say $f(x) = c$ for every $x \in [a, b]$, show that

$$\int_a^b f(x)dx = c(b - a).$$

3. Let $f : [a, b] \to \mathbb{R}$ be a monotone function and $P_k = \{a = x_0 < x_1 < \cdots < x_k = b\}$ be a **uniform** partition of $[a, b]$, i.e., such that $x_j - x_{j-1} = \frac{b-a}{k}$ for $1 \le j \le k$. Show that

$$\int_a^b f(x)dx = \lim_{k \to +\infty} s(f; P_k) = \lim_{k \to +\infty} S(f; P_k).$$

4. * Given a nonnegative function $f : [a, b] \to \mathbb{R}$, check that the approximations (10.1) and (10.2) for the area of the region $\mathcal{R}$ under the graph of f are particular cases of lower and upper sums of f. Then, use this fact, (10.6) and the result of the previous problem to show that, given $n \in \mathbb{N}$ and $b > 0$, we have

$$\int_0^b x^n dx = \frac{b^{n+1}}{n+1}.$$

5. In order to compute $\int_a^b \sqrt{x}\, dx$, do the following items:

   (a) Show that, for reals $u \ge v \ge 0$, one has

$$\frac{2}{3}(u\sqrt{u} - v\sqrt{v}) - \frac{1}{3}(u - v)(\sqrt{u} - \sqrt{v}) < \frac{1}{2}(\sqrt{u} + \sqrt{v})(u - v)$$

   and

$$\frac{1}{2}(\sqrt{u} + \sqrt{v})(u - v) < \frac{2}{3}(u\sqrt{u} - v\sqrt{v}).$$

   (b) If $P_k = \{a = x_0 < x_1 < \cdots < x_k = b\}$ is a uniform partition of $[a, b]$, use the result of (a) to show that

$$\frac{2}{3}(b\sqrt{b} - a\sqrt{a}) - \frac{(b - a)(\sqrt{b} - \sqrt{a})}{3k} < s(\sqrt{x}; P_k) + \frac{\sqrt{b} - \sqrt{a}}{2k}$$

   and

$$s(\sqrt{x}; P_k) + \frac{\sqrt{b} - \sqrt{a}}{2k} < \frac{2}{3}(b\sqrt{b} - a\sqrt{a}).$$

   (c) Conclude that $\int_a^b \sqrt{x}\, dx = \frac{2}{3}(b\sqrt{b} - a\sqrt{a})$.

6. Assume that $\int_0^1 \frac{1}{\sqrt{x+1}} dx = 2(\sqrt{2} - 1)$. (We shall compute this integral in item (d) of Problem 1, page 397.) If $a_n = \frac{1}{\sqrt{n}}\left(\frac{1}{\sqrt{1+n}} + \frac{1}{\sqrt{2+n}} + \cdots + \frac{1}{\sqrt{2n}} \right)$ for $n \in \mathbb{N}$, show that $(a_n)_{n \ge 1}$ converges to $2(\sqrt{2} - 1)$.

7. We shall compute $\int_0^1 \frac{1}{1+x^2}\, dx = \frac{\pi}{4}$ in Example 10.42. Use this fact to show that, if $a_n = n\left(\frac{1}{n^2+1^2} + \frac{1}{n^2+2^2} + \cdots + \frac{1}{n^2+n^2}\right)$, then the sequence $(a_n)_{n\geq1}$ converges to $\frac{\pi}{4}$.

8. Let $f : [a, b] \to \mathbb{R}$ be a bounded function and $\alpha > 0$ be given. If for each $k \in \mathbb{N}$ there exists a partition P_k of $[a, b]$ satisfying $S(f; P_k) - s(f; P_k) < \frac{\alpha}{k}$, show that f is integrable, with

$$\int_a^b f(x)dx = \lim_{k\to+\infty} s(f; P_k) = \lim_{k\to+\infty} S(f; P_k).$$

For the coming problem, the reader might find it convenient to read again the statement of Problem 3, page 262.

9. Let $f : [a, b] \to \mathbb{R}$ be a Lipschitz function, with Lipschitz constant c, and let $P_k = \{a = x_0 < x_1 < \cdots < x_k = b\}$ be a uniform partition of interval $[a, b]$.

   (a) Show that $S(f; P_k) - s(f; P_k) < \frac{c(b-a)^2}{k}$.
   (b) For $1 \leq j \leq k$, choose a real number $\xi_{kj} \in [x_{j-1}, x_j]$ and define the sum $\Sigma(f; P_k; \xi_k)$ by $\Sigma(f; P_k; \xi_k) = \sum_{j=1}^{k} f(\xi_{kj})(x_j - x_{j-1})$. Prove that

   $$\int_a^b f(x)dx = \lim_{k\to+\infty} \Sigma(f; P_k; \xi_k).$$

10. * Do the following items:

    (a) Given $a, h \in \mathbb{R}$, with $h \neq 2l\pi$ for every $l \in \mathbb{Z}$, prove that

    $$\sum_{j=0}^{k} \sin(a + jh) = \frac{\sin\left(a + \frac{(k-1)h}{2}\right)\sin\frac{(k+1)h}{2}}{\sin\frac{h}{2}}$$

    and

    $$\sum_{j=0}^{k} \cos(a + jh) = \frac{\cos\left(a + \frac{(k-1)h}{2}\right)\sin\frac{(k+1)h}{2}}{\sin\frac{h}{2}}$$

    (b) Show that $\int_a^b \sin x\, dx = \cos a - \cos b$ and $\int_a^b \cos x\, dx = \sin b - \sin a$.

11. Let $f : [0, 1] \to \mathbb{R}$ be given by

    $$f(x) = \begin{cases} 0, & \text{if } x = 0 \text{ or } x \notin \mathbb{Q} \\ \frac{1}{n}, & \text{if } x = \frac{m}{n} \text{ with } m, n \in \mathbb{N} \text{ and } \gcd(m, n) = 1 \end{cases}.$$

    Prove that f is integrable.

10.3   Riemann's Theorem and Some Remarks

In this section we state and prove an important theorem of Riemann on the characterization of Riemann integrable functions, as well as make some other important remarks on the theory of integration we are developing. In particular, we rigorously define the concept of area for the region under the graph of a nonnegative integrable function, which was informally discussed in the first section of the chapter.

I. RIEMANN'S THEOREM. Usually, the presentation of the concept of integral in Calculus courses makes use of the concept of *Riemann sums*. More precisely, given a bounded function $f : [a, b] \to \mathbb{R}$, one defines f as being (**Riemann**) **integrable** if there exists a real number I (the **Riemann integral** of f) satisfying the following condition: given $\epsilon > 0$, there exists $\delta > 0$ such that, for every partition $P = \{a = x_0 < x_1 < \cdots < x_k = b\}$ of $[a, b]$ and every choice of points $\xi_j \in [x_{j-1}, x_j]$, $1 \le j \le k$, one has

$$\max\{|x_j - x_{j-1}|; \ 1 \le j \le k\} < \delta \Rightarrow \left| \sum_{j=1}^{k} f(\xi_j)(x_j - x_{j-1}) - I \right| < \epsilon. \qquad (10.20)$$

In the above notations, we say that $\max\{|x_j - x_{j-1}|; \ 1 \le j \le k\}$ is the **norm** of the partition P and that $\sum_{j=1}^{k} f(\xi_j)(x_j - x_{j-1})$ is the **Riemann sum** of f with respect to P and to the chosen intermediate points $\xi = (\xi_j)_{1 \le j \le k}$. Letting

$$|P| = \max\{|x_j - x_{j-1}|; \ 1 \le j \le k\} \ \text{ and } \ \Sigma(f; P; \xi) = \sum_{j=1}^{k} f(\xi_j)(x_j - x_{j-1}),$$

we summarize (10.20) by saying that *I is the limit of the Riemann sums* $\Sigma(f; P; \xi)$, *when* $|P| \to 0$ *and for every choice* $\xi = (\xi_j)$ *of intermediate points relative to P.* Moreover, in this case we simply write

$$I = \lim_{|P| \to 0} \Sigma(f; P; \xi).$$

The following result shows that the above definition of integral coincides with the one we have adopted in the previous section. The proof of it can surely be omitted on a first reading.

Theorem 10.7 (Riemann) *A bounded function* $f : [a, b] \to \mathbb{R}$ *is integrable (in the sense of the previous section) if and only if the limit* $\lim_{|P| \to 0} \Sigma(f; P; \xi)$ *of its Riemann sums exists, does not depending on the chosen intermediate points of* ξ. *Moreover, in this case one has*

$$\int_a^b f(x)dx = \lim_{|P| \to 0} \Sigma(f; P; \xi).$$

Proof We first assume that f is integrable in $[a, b]$, with $\int_a^b f(x)dx = I$.

Given $\epsilon > 0$, we want to find $\delta > 0$ such that, for every partition P of $[a, b]$, it is the case that $|P| < \delta \Rightarrow |\Sigma(f; P; \xi) - I| < \epsilon$, for every choice ξ of intermediate points. To this end, start by choosing (by Cauchy's integrability criterion) a partition $P_0 = \{a = x_0 < x_1 < \ldots < x_k = b\}$ of $[a, b]$ such that $S(f; P_0) - s(f; P_0) < \frac{\epsilon}{4}$.

For a general partition $P = \{a = y_0 < y_1 < \ldots < y_l = b\}$ of $[a, b]$, there are two different kinds of intervals (y_{i-1}, y_i): those which contain at least one of the x_j's and those which do not.

If we choose $0 < \delta < \frac{1}{2}|P_0|$, then $|P| < \delta$ implies $y_i - y_{i-1} < \frac{1}{2}(x_j - x_{j-1})$ for all $1 \leq i \leq l, 1 \leq j \leq k$. Therefore, each interval (y_{i-1}, y_i) contains at most one of the x_j's and each interval (x_{j-1}, x_i) contains at least three of the y_i's. This way, we can write

$$\Sigma(f; P; \xi) = \Sigma'(f; P; \xi) + \Sigma''(f; P; \xi),$$

where $\Sigma'(f; P; \xi)$ (resp. $\Sigma''(f; P; \xi)$) denotes the collection of those summands of $\Sigma(f; P; \xi)$ corresponding to indices i such that $(y_{i-1}, y_i) \cap P_0 \neq \emptyset$ (resp. $(y_{i-1}, y_i) \cap P_0 = \emptyset$).

Let $\xi = (\xi_1, \ldots, \xi_l)$ be an arbitrary choice of intermediate points for P, and observe that

$$|\Sigma(f; P; \xi) - I| \leq |\Sigma'(f; P; \xi)| + |\Sigma''(f; P; \xi) - I|. \tag{10.21}$$

We now let $M = \sup_{[a,b]} |f|$ and estimate each of the summands at the right hand side above.

(i) Estimating $|\Sigma'(f; P; \xi)|$: the restriction imposed on δ forces $\Sigma'(f; P; \xi)$ to have at most $k-1$ summands (at most one for each of $x_1, \ldots, x_{k-1}$). Since $|f(\xi_i)| \leq M$ and $y_i - y_{i-1} < \delta$, we get

$$|\Sigma'(f; P; \xi)| = |\Sigma' f(\xi_i)(y_i - y_{i-1})| \leq \Sigma' |f(\xi_i)|(y_i - y_{i-1}) < M\delta(k - 1).$$

(ii) Estimating $|\Sigma''(f; P; \xi) - I|$: we start by noticing that

$$|\Sigma''(f; P; \xi) - I| \leq |\Sigma''(f; P; \xi) - s(f; P_0)| + |s(f; P_0) - I|$$

$$< |\Sigma''(f; P; \xi) - s(f; P_0)| + \frac{\epsilon}{4}, \tag{10.22}$$

so we are left to estimating $|\Sigma''(f; P; \xi) - s(f; P_0)|$. To this end, first recall that if $1 \leq i \leq l$ is one of the indices that fall into Σ'', then (y_{i-1}, y_i) does not contain any of the x_j's, but there is exactly one index $1 \leq j \leq k$ such that $[y_{i-1}, y_i] \subset [x_{j-1}, x_j]$. Hence, we can write

$$\Sigma''(f; P; \xi) = \Sigma_1''(f; P; \xi) + \cdots + \Sigma_k''(f; P; \xi),$$

where $\Sigma_j''(f; P; \xi)$ stands for those summands of $\Sigma''(f; P; \xi)$ corresponding to indices $1 \le i \le l$ such that $[y_{i-1}, y_i] \subset [x_{j-1}, x_j]$. Therefore, if $m_j = \inf_{[x_{j-1}, x_j]} f$, then the triangle inequality gives

$$|\Sigma''(f; P; \xi) - s(f; P_0)| \le \sum_{j=1}^{k} |\Sigma_j''(f; P; \xi) - m_j(x_j - x_{j-1})|.$$

In order to deal with this last sum, fix $1 \le j \le k$ and let $1 \le i \le l$ be such that $[y_{i-1}, y_i] \subset [x_{j-1}, x_j]$. Recall that our choice of δ forces $[x_{j-1}, x_j]$ to contain at least another point of P. Therefore, if $r_j \le i-1$ and $s_j \ge i$ are respectively the least and greatest indices such that $(y_{r_j}, y_{s_j}) \subset (x_{j-1}, x_j)$, then $s_j - r_j \ge 2$ and we have

$$y_{r_j-1} < x_{j-1} \le y_{r_j} < \ldots < y_{i-1} < y_i < \ldots < y_{s_j} \le x_j < y_{s_j+1}.$$

This way, and letting $S_j = |\Sigma_j''(f; P; \xi) - m_j(x_j - x_{j-1})|$ and $M_j = \sup_{[x_{j-1}, x_j]} f$, we estimate S_j by writing

$$S_j \le \left| \sum_{t=r_j+1}^{s_j} f(\xi_t)(y_t - y_{t-1}) - m_j(y_{r_j} - x_{j-1}) - \sum_{t=r_j+1}^{s_j} m_j(y_t - y_{t-1}) \right.$$

$$\left. - m_j(x_j - y_{s_j}) \right|$$

$$\le \sum_{t=r_j+1}^{s_j} |f(\xi_t) - m_j|(y_t - y_{t-1}) + |m_j|(y_{r_j} - x_{j-1}) + |m_j|(x_j - y_{s_j})$$

$$\le \sum_{t=r_j+1}^{s_j} (M_j - m_j)(y_t - y_{t-1}) + 2|m_j|\delta$$

$$\le (M_j - m_j)(y_{s_j} - y_{r_j-1}) + 2M\delta$$

$$\le (M_j - m_j)(x_j - x_{j-1}) + 2M\delta,$$

where we used the facts that $|f(\xi_t) - m_j| \le M_j - m_j$, $0 < y_{r_j} - x_{j-1}, x_j - y_{s_j} < \delta$ and $y_{s_j} - y_{r_j-1} \le x_j - x_{j-1}$.

Hence, we get from (10.22) that

$$|\Sigma''(f; P; \xi) - s(f; P_0)| \le \sum_{j=1}^{k} S_j \le \sum_{j=1}^{k} \left((M_j - m_j)(x_j - x_{j-1}) + 2M\delta \right)$$

$$= S(f; P_0) - s(f; P_0) + 2M\delta k < \frac{\epsilon}{4} + 2M\delta k.$$

Finally, gathering together (10.21), (10.22) and the estimates of items (i) and (ii), we arrive at

$$|\Sigma(f; P; \xi) - I| < \frac{\epsilon}{2} + M\delta(3k - 1).$$

Therefore, if we further ask that $0 < \delta < \frac{\epsilon}{2M(3k-1)}$, we get

$$|P| < \min\left\{\frac{1}{2}|P_0|, \frac{\epsilon}{2M(3k - 1)}\right\} \Rightarrow |\Sigma(f; P; \xi) - I| < \epsilon,$$

as we wished to show.

The proof of the converse is left to the reader as an exercise (see Problem 9). □

In spite of the fact that the introduction of the integral by means of Riemann sums was the actual way the theory historically evolved, the approach by means of lower and upper sums makes it technically much easier to deal with the material of the next section. Nevertheless, as we shall see right now, the characterization of the integral by Riemann's theorem allows us to deal with the heuristic discussion of the previous section (concerning area computations) in a mathematically precise way.

Given a bounded and nonnegative function $f : [a, b] \to \mathbb{R}$, let $\mathcal{R}$ denote the **region under the graph** of f (and above the horizontal axis and between the vertical lines $x = a$ and $x = b$):

$$\mathcal{R} = \{(x, y) \in \mathbb{R}^2; a \le x \le b \text{ and } 0 \le y \le f(x)\}.$$

Let $P_k = \{a = x_0 < x_1 < \cdots < x_k = b\}$ be a *uniform partition* of $[a, b]$, i.e., such that $x_j - x_{j-1} = \frac{b-a}{k}$ for $1 \le j \le k$.

If f is continuous, then Theorem 8.26 guarantees that, for $1 \le j \le k$, there exist $\xi_{kj}, \xi'_{kj} \in [x_{j-1}, x_j]$ such that

$$f(\xi_{kj}) = \min_{[x_{j-1}, x_j]} f \quad \text{and} \quad f(\xi'_{kj}) = \max_{[x_{j-1}, x_j]} f.$$

Hence, the lower and upper approximations $\underline{A}(f; k)$ and $\overline{A}(f; k)$ for the area of $\mathcal{R}$ (respectively defined by (10.1) and (10.2), page 348)) respectively coincide with the Riemann sums $\Sigma(f; P_k; \xi_k)$ and $\Sigma(f; P_k; \xi'_k)$ (here, note that $\xi_k = (\xi_{kj})$ and $\xi'_k = (\xi'_{kj})$. However, since $|P_k| = \frac{1}{k}$, it follows from Riemann's theorem that

$$\int_a^b f(x)dx = \lim_{k \to +\infty} \Sigma(f; P_k; \xi_k) = \lim_{k \to +\infty} \underline{A}(f; k)$$

and, analogously,

$$\int_a^b f(x)dx = \lim_{k \to +\infty} \overline{A}(f; k).$$

Thus, $\int_a^b f(x)dx$ is the only reasonable value for the area of $\mathcal{R}$.

On the other hand, if f is merely integrable (albeit also nonnegative), then, except for changing the minimum and maximum values of f along $[x_{j-1}, x_j]$ by its infimum and supremum along such an interval, $\underline{A}(f; k)$ and $\overline{A}(f; k)$ also reduce to the lower and upper sums $s(f; P_k)$ and $S(f; P_k)$, respectively.

We sum up the above discussion in the following

Definition 10.8 Let $f : [a, b] \to \mathbb{R}$ be a nonnegative integrable function. We define the area of the region $\mathcal{R}$ under its graph as

$$A(\mathcal{R}) = \int_a^b f(x)dx.$$

In the next section, the above definition will give heuristic arguments towards the validity of some of the *operational properties* of Riemann's integral.

II. LEBESGUE'S THEOREM AND INTEGRAL. In Theorem 10.6, we proved that every continuous function $f : [a, b] \to \mathbb{R}$ is integrable. We also saw, in Example 10.3, that the Dirichlet function (which, by Problem 11, page 255, is discontinuous at every point of the interval $[0, 1]$) is not integrable. On the other hand, we shall prove in the next section (cf. Proposition 10.19) that a *piecewise continuous* function $f : [a, b] \to \mathbb{R}$ (i.e., a function with a finite number of points of discontinuity $x_0 \in [a, b]$, such that for all of them the lateral limits $\lim_{x \to x_0 \pm} f(x)$ do exist) is still integrable. A more dramatic example is given by the function of Problem 11, page 359, which is integrable but has an infinite number of points of discontinuity.

The last paragraph suggests that there might be a more intimate relation between the integrability of a bounded function $f : [a, b] \to \mathbb{R}$ and the *size* of the set D_f of its points of discontinuity. This is indeed the case, and a precise formulation of such a connection requires the concept of a *set of measure zero*, which we now give.

Definition 10.9 We say that $X \subset \mathbb{R}$ is a set of (**Lebesgue**[2]) **measure zero**, or a **null set** if, given $\epsilon > 0$, there exist open intervals $I_1, I_2, I_3, \ldots$ such that

$$X \subset \bigcup_{j \geq 1} I_j \quad \text{and} \quad \sum_{j \geq 1} |I_j| < \epsilon,$$

where $|I_j|$ stands for the length of I_j.

The following result links the concept of null set to that of Riemann integrable functions.

Theorem 10.10 (Lebesgue) *A bounded function $f : [a, b] \to \mathbb{R}$ is Riemann integrable if and only if the set D_f of its points of discontinuity has measure zero.*

---

[2]After Henri Lebesgue, French mathematician of the XX century.

We shall not prove Lebesgue's theorem here, and refer the interested reader to [1, 9] or [20]. The reason is that, apart from the Problem 4 (which will be obtained by more elementary methods in the coming section) and Problem 5, page 468 (for which we present two proofs, one of which does not make use of Lebesgue's theorem), we will not use it anywhere else in the book. Nevertheless, it is important to realize that it gives a unified explanation for the status, with respect to integration, of all of the functions of the next to last paragraph. Indeed, since the set of points of discontinuity of a continuous (resp. piecewise continuous) function is empty (resp. finite), hence of measure zero, such functions are integrable. On the other hand, it is possible to prove that $\mathbb{Q}$ is a null set, while $[0, 1]$ is not (for the case of $\mathbb{Q}$, see Problem 2); this explains why the function of Problem 11, page 359 is integrable (for, its set of points of discontinuity is $\mathbb{Q} \cap [0, 1]$, which—being a subset of $\mathbb{Q}$—is certainly a null set), as well as why the Dirichlet function is not integrable (for, its set of points of discontinuity is the whole interval $[0, 1]$).

Let $\mathbb{I}$ denote the set of irrational numbers. It is possible to prove that the set $\mathbb{I} \cap [a, b]$, of the irrational numbers of the closed interval $[a, b]$, is not a null set. Hence, Lebesgue's theorem assures that a function $f : [a, b] \to \mathbb{R}$ whose set of points of discontinuity is $\mathbb{I} \cap [a, b]$ is not integrable. However, as we stressed before, such a function does not exist (a proof can be found in [1], for instance).

Lebesgue's contributions to integration theory go far beyond Theorem 10.10. At the dawn of the XX century, he introduced a much more refined and flexible concept of integral in his doctor's thesis. In order to get a glimpse of it, we shall first of all need the following

Definition 10.11 The **characteristic function** of a set $A \subset \mathbb{R}$ is the function $\mathcal{X}_A :$ $[a, b] \to \mathbb{R}$ such that

$$\mathcal{X}_A(x) = \begin{cases} 1, & \text{if } x \in A \\ 0, & \text{if } x \notin A \end{cases}.$$

Now, let $f : [a, b] \to [0, L]$ be a nonnegative bounded function. In the context of Riemann's integral, we start by choosing a partition $P = \{a = x_0 < x_1 < \cdots < x_k = b\}$ of the *domain* $[a, b]$ of f. Then, we define the **step functions** $f_{P,-}, f_{P,+} :$ $[a, b] \to \mathbb{R}$ by letting

$$f_{P,-} = \sum_{j=1}^{k} m_j \mathcal{X}_{[x_{j-1}, x_j]} \text{ and } f_{P,+} = \sum_{j=1}^{k} M_j \mathcal{X}_{[x_{j-1}, x_j]},$$

where $m_j = \inf_{[x_{j-1}, x_j]} f$ and $M_j = \sup_{[x_{j-1}, x_j]} f$. In the next section, we shall show (cf. Example 10.20) that such functions are Riemann integrable, with

$$\int_a^b f_{P,-}(x)dx = \sum_{j=1}^{k} m_j(x_j - x_{j-1}) \text{ and } \int_a^b f_{P,+}(x)dx = \sum_{j=1}^{k} M_j(x_j - x_{j-1}).$$

Thus, in accordance with the previous section, f is Riemann integrable if

$$\sup_P \int_a^b f_{P,-}(x)dx = \inf_P \int_a^b f_{P,+}(x)dx;$$

moreover, in this case the integral of f coincides with this common value.

On the contrary, in the realm of Lebesgue's integral, we start with a partition $Q = \{0 = y_0 < y_1 < \cdots < y_l = L\}$ of the codomain $[0, L]$ of f. Then, for $1 \le j \le k - 1$ we take the **inverse image** A_j of the interval $[y_{j-1}, y_j)$ by f, which is defined by

$$A_j = \{x \in [a, b]; f(x) \in [y_{j-1}, y_j)\};$$

we also consider the inverse image A_k of the interval $[y_{k-1}, y_k]$, such that

$$A_k = \{x \in [a, b]; f(x) \in [y_{k-1}, y_k]\}.$$

If these sets are not too complicated (in a sense that doesn't interest us at this moment), one can show that it is possible to associate to A_j a nonnegative real number $m(A_j)$, which we call the **Lebesgue measure** of A_j. Moreover, this is done in such a way that if A_j is an interval then $m(A_j)$ coincides with the length of A_j; in this case, we say that A_j is a (Lebesgue) **measurable set**.

Assuming that all of the A_j's are measurable, we consider the **simple function**

$$f_Q = \sum_{j=1}^{k} y_{j-1} \mathcal{X}_{A_j}$$

and define its Lebesgue integral, denoted $\int_{[a,b]} f_Q$, by

$$\int_{[a,b]} f_Q = \sum_{j=1}^{k} y_{j-1} m(A_j). \tag{10.23}$$

Finally, if for every partition Q of $[0, L]$ the resulting sets A_j are measurable, then we say that f is a **measurable function**; if this is so, we define the **Lebesgue integral** of f, denoted $\int_{[a,b]} f$, by

$$\int_{[a,b]} f = \sup_Q \left\{ \int f_Q; Q \text{ is a partition of } [0, L] \right\}. \tag{10.24}$$

Although we have "defined" the Lebesgue integral just for nonnegative bounded functions, it is possible to consider the notion of Lebesgue integral in the context of bounded measurable (but not necessarily nonnegative) functions $f : [a, b] \to \mathbb{R}$. (Actually, we could consider the even more general case of an *unbounded* measurable function, but this will not play a role here).

It is possible to show that the Lebesgue integral is more comprehensive than the Riemann integral, in two ways: on the one hand, every (bounded) Riemann integrable function $f : [a, b] \to \mathbb{R}$ is also Lebesgue integrable and the values of both integrals of f coincide; on the other, there exist bounded functions $f : [a, b] \to \mathbb{R}$ which are not Riemann integrable but are Lebesgue integrable (the Dirichlet function is one such—see Problem 5).

Moreover, as we shall comment on Sect. 11.2 (cf. Remark 11.15), Lebesgue integration also has at its disposal a number of *convergence results* whose proofs are inaccessible in the context of Riemann integration, but which reveal themselves to be central tools for the study of the deeper properties of *sequences and series of functions*. Actually, such results are one of the main reasons behind the fact that Lebesgue integration is more adequate than Riemann integration for many purposes in Mathematics and its applications. To name one relevant example, it is completely indispensable to the modern study of Partial Differential Equations and Differential Geometry.

If this is so, the reader might ask why not to study the Lebesgue integral in advance, relegating the Riemann integral to a museum. The reason is that, on the one hand, in spite of its greater flexibility, an adequate presentation of the concepts needed to develop Lebesgue's notion of integral is considerably more complicated than what we have done so far to the Riemann integral (in (10.24), the sets A_j may be very complicated); on the other hand, and as we shall see later in this chapter, Riemann's integral suffices to the discussion of several interesting problems.

For the interested reader, we suggest the references [20] or [27] for rather elementary introductions to the Lebesgue measure and integral.

Problems: Section 10.3

1. In item (c) of Problem 1, page 397, we shall show that $\int_0^1 x \sin(\pi x)dx = \frac{1}{\pi}$. Use this fact to show that, if $a_n = \sum_{k=1}^{n} \frac{k\pi}{n^2} \sin\left(\frac{k\pi}{n}\right)$, then $(a_n)_{n \geq 1}$ converges and $\lim_{n \to +\infty} a_n = 1$.
2. * Show that $\mathbb{Q}$ is a null set.
3. Prove that if $A_1, A_2, \ldots \subset \mathbb{R}$ are null sets, then so is $\bigcup_{j \geq 1} A_j$.
4. Use Lebesgue's theorem to prove that if $f, g : [a, b] \to \mathbb{R}$ are Riemann integrable, then $fg : [a, b] \to \mathbb{R}$ is also Riemann integrable.
5. Show that Dirichlet's function $f : [0, 1] \to \mathbb{R}$ (cf. Example 10.3) is Lebesgue integrable, with $\int_{[0,1]} f = 0$.
6. If $f : [a, b] \to [0, L]$ is a monotone function, we have showed in Theorem 10.5 that f is Riemann integrable. Show that f is also Lebesgue integrable, with $\int_a^b f(x)dx = \int_{[a,b]} f$.

7. (Berkeley) Let $f : [0, +\infty) \to [0, +\infty)$ be a continuous and increasing bijection. Show that

$$\int_0^a f(x)dx + \int_0^b f^{-1}(x)dx \geq ab,$$

for every positive reals a and b.

8. (Leningrad) Let $f, g : [0, 1] \to [0, 1]$ be continuous functions, with f nondecreasing. Prove that

$$\int_0^1 (f \circ g)(x)dx \leq \int_0^1 f(x)dx + \int_0^1 g(x)dx.$$

9. * Complete the proof of Riemann's theorem.

10.4 Operating with Integrable Functions

This section is devoted to the derivation of some useful *operational properties* for the Riemann integral. In spite of the fact that such properties will be extensively used along the rest of the book, their proofs can be omitted in a first reading, with essentially no loss of continuity.

Along all of this section, given bounded functions $f, g : [a, b] \to \mathbb{R}$ and a partition $P = \{a = x_0 < x_1 < \cdots < x_k = b\}$ of $[a, b]$, we shall denote the infimums of f and g in $[x_{j-1}, x_j]$ by $m_j(f)$ and $m_j(g)$, respectively; accordingly, $M_j(f)$ and $M_j(g)$ will denote the supremums of f and g in the same interval $[x_{j-1}, x_j]$, also respectively.

If $f, g : [a, b] \to \mathbb{R}$ are nonnegative integrable functions, with $f \leq g$, then the regions $\mathcal{R}_f$ e $\mathcal{R}_g$, respectively situated under the graphs of f and g, are such that $\mathcal{R}_f \subset \mathcal{R}_g$; we therefore expect that $A(\mathcal{R}_f) \leq A(\mathcal{R}_g)$. This is indeed the case, and we usually refer to it by saying that the Riemann integral is **monotonic**. The general case is as follows.

Proposition 10.12 *If $f, g : [a, b] \to \mathbb{R}$ are integrable functions such that $f \leq g$, then $\int_a^b f(x)dx \leq \int_a^b g(x)dx$.*

Proof Letting $P = \{a = x_0 < x_1 < \cdots < x_k = b\}$ denote a partition of $[a, b]$, it follows at once from $f \leq g$ that $M_j(f) \leq M_j(g)$ for $1 \leq j \leq k$. Hence,

$$S(f; P) = \sum_{j=1}^k M_j(f)(x_j - x_{j-1}) \leq \sum_{j=1}^k M_j(g)(x_j - x_{j-1}) = S(g; P).$$

However, since $\int_a^b f(x)dx \leq S(f; P)$, we conclude that

$$\int_a^b f(x)dx \leq S(g; P),$$

for every partition P of $[a, b]$. Thus, $\int_a^b f(x)dx$ is a lower bound for the set of the upper sums $S(g; P)$, so that

$$\int_a^b f(x)dx \leq \inf_P S(g; P) = \int_a^b g(x)dx.$$

$\square$

For what comes next, we consider again a nonnegative integrable function f : $[a, b] \to \mathbb{R}$ and a positive real number c. Problem 10, page 194, guarantees that the region $\mathcal{R}_{cf}$ under the graph of the function cf can be obtained from the region $\mathcal{R}_f$ under the graph of f by *vertically stretching* R_f by a factor c; we thus hope that $A(\mathcal{R}_{cf}) = cA(\mathcal{R}_f)$. On the other hand if $g : [a, b] \to \mathbb{R}$ is another integrable and nonnegative function, then, for each $x_0 \in [a, b]$, the segment of the vertical line $x = x_0$ contained in $\mathcal{R}_{f+g}$ (the region under the graph of $f + g$) has length equal to the sum of the lengths of the segments of such a line which are contained in $\mathcal{R}_f$ and $\mathcal{R}_g$; thus suggests that $\mathcal{R}_{f+g}$ can be obtained by *glueing* $\mathcal{R}_g$ *right above* $\mathcal{R}_f$ and, hence, that we should have $A(\mathcal{R}_{f+g}) = A(\mathcal{R}_f) + A(\mathcal{R}_g)$.

The next result shows that the two properties of the Riemann integral hinted by the heuristic reasonings above are actually true. From now on, we shall refer to these properties by saying that the Riemann integral is respectively **linear** and **additive**.

Proposition 10.13 *Let $f, g : [a, b] \to \mathbb{R}$ be integrable, and let $c \in \mathbb{R}$. Then:*

(a) $cf : [a, b] \to \mathbb{R}$ *is integrable, with* $\int_a^b cf(x)dx = c \int_a^b f(x)dx$.
(b) $f + g : [a, b] \to \mathbb{R}$ *is integrable, with* $\int_a^b (f(x)+g(x))dx = \int_a^b f(x)dx + \int_a^b g(x)dx$.

Proof (a) As in the proof of the previous proposition, let $P = \{a = x_0 < x_1 < \cdots < x_k = b\}$ be a partition of $[a, b]$.

If $c > 0$, it is immediate to verify (see Problem 12, page 207) that

$$m_j(cf) = \inf_{[x_{j-1}, x_j]} (cf) = c \cdot \inf_{[x_{j-1}, x_j]} f = c\, m_j(f)$$

and, analogously, $M_j(cf) = cM_j(f)$. Thus,

$$s(cf; P) = \sum_{j=1}^k m_j(cf)(x_j - x_{j-1}) = c \sum_{j=1}^k m_j(f)(x_j - x_{j-1}) = c\, s(f; P)$$

and, analogously, $S(cf; P) = c S(f; P)$. The integrability of f and the elementary properties of the concepts of supremum and infimum (see Problem 12, page 207, again) give us

$$\sup_P s(cf; P) = \sup_P (c\, s(f; P)) = c \sup_P s(f; P) = c \int_a^b f(x)dx$$

and

$$\inf_{P} S(cf; P) = \inf_{P}(c\, S(f; P)) = c\, \inf_{P} s(f; P) = c \int_{a}^{b} f(x)dx.$$

If $c < 0$, the deduction of the two relations above is entirely analogous; one just needs to observe that $m_j(cf) = c\, M_j(f)$, $M_j(cf) = c\, m_j(f)$, $s(cf; P) = c\, S(f; P)$ and $S(cf; P) = c\, s(f; P)$, so that

$$\sup_{P} s(cf; P) = \sup_{P}(c\, S(f; P)) = c\, \inf_{P} S(f; P) = c \int_{a}^{b} f(x)dx$$

and, in the same way, $\inf_P S(cf; P) = c \int_a^b f(x)dx$.

In any case, we have $\inf_P S(cf; P) = \sup_P s(cf; P) = c \int_a^b f(x)dx$. Therefore, cf is integrable, with

$$\int_{a}^{b} cf(x)dx = \inf_{P} S(cf; P) = c \int_{a}^{b} f(x)dx.$$

(b) Let P and Q be partitions of $[a, b]$ such that $P \cup Q = \{a = x_0 < x_1 < \cdots < x_k = b\}$. It readily follows from Problem 13, page 207, that

$$\sup_{[x_{j-1}, x_j]} (f + g) \leq \sup_{[x_{j-1}, x_j]} f + \sup_{[x_{j-1}, x_j]} g,$$

or (with respect to the partition $P \cup Q$) $M_j(f + g) \leq M_j(f) + M_j(g)$. Hence,

$$S(f + g; P \cup Q) = \sum_{j=1}^{k} M_j(f + g)(x_j - x_{j-1})$$

$$\leq \sum_{j=1}^{k} (M_j(f) + M_j(g))(x_j - x_{j-1})$$

$$= S(f; P \cup Q) + S(g; P \cup Q)$$

$$\leq S(f; P) + S(g; Q).$$

However, since $\inf_R S(f + g; R) \leq S(f + g; P \cup Q)$, the inequalities above give

$$\inf_{R} S(f + g; R) \leq S(f; P) + S(g; Q),$$

for all partitions P, Q and R of $[a, b]$.

In the last inequality above, taking the infimum over all partitions P and Q of $[a, b]$ and invoking Problem 13, page 207, again, we get

$$\inf_R S(f + g; R) \leq \inf\{S(f; P) + S(g; Q); \ P, Q \text{ partitions of } [a, b]\}$$

$$= \inf_P S(f; P) + \inf_Q S(g; Q)$$

$$= \int_a^b f(x)dx + \int_a^b g(x)dx. \tag{10.25}$$

Arguing as we have done up to this moment, we successively get $m_j(f + g) \geq m_j(f) + m_j(g)$, $s(f + g; P \cup Q) \geq s(f; P) + s(g; Q)$ and $\sup_R s(f + g; R) \geq s(f; P) + s(g; Q)$. Hence, by using Problem 13, page 207, yet another time, we obtain

$$\sup_R s(f + g; R) \geq \sup_P s(f; P) + \sup_Q s(g; Q)$$

$$= \int_a^b f(x)dx + \int_a^b g(x)dx. \tag{10.26}$$

Finally, since $\sup_R s(f + g; R) \leq \inf_R S(f + g; R)$, it follows from (10.25) and (10.26) that

$$\inf_R S(f + g; R) = \sup_R s(f + g; R) = \int_a^b f(x)dx + \int_a^b g(x)dx.$$

Thus, $f + g$ is integrable, with $\int_a^b (f(x) + g(x))dx = \int_a^b f(x)dx + \int_a^b g(x)dx$. $\square$

An obvious corollary of the previous proposition is that, if $f, g : [a, b] \to \mathbb{R}$ are integrable functions, then $f - g : [a, b] \to \mathbb{R}$ is also integrable, with

$$\int_a^b (f(x) - g(x))dx = \int_a^b f(x)dx - \int_a^b g(x)dx. \tag{10.27}$$

Indeed, item (a) guarantees the integrability of $-f$, whereas item (b) guarantees that of $f - g = f + (-g)$. Relation (10.27) now follows from the formulas of items (a) and (b) of the previous proposition:

$$\int_a^b (f(x) - g(x))dx = \int_a^b f(x)dx + \int_a^b (-g(x))dx$$

$$= \int_a^b f(x)dx + (-1)\int_a^b g(x)dx.$$

If $f : [a, b] \to \mathbb{R}$ is a continuous function, the chain rule assures that $|f| = |\cdot| \circ f : [a, b] \to \mathbb{R}$ is also continuous; in particular, $|f|$ is also integrable. In what follows, we establish the integrability of $|f|$ by supposing that f is only integrable. We also obtain a quite useful inequality relating the integrals of f and $|f|$, which is known as the **triangle inequality for integrals**.

Proposition 10.14 *If* $f : [a, b] \to \mathbb{R}$ *is an integrable function, then the function* $|f| : [a, b] \to \mathbb{R}$ *is also integrable, and the following inequality holds:*

$$\left| \int_a^b f(x)dx \right| \le \int_a^b |f(x)|dx. \tag{10.28}$$

Proof Let $f_+, f_- : [a, b] \to [0, +\infty)$ be given by

$$f_+(x) = \max\{f(x), 0\} \text{ and } f_-(x) = -\min\{f(x), 0\}.$$

It is immediate to verify that $f = f_+ - f_-$ and $|f| = f_+ + f_-$. Hence, if f_+ is integrable, then the discussion immediately subsequent to Proposition 10.13 guarantees that $f_- = f_+ - f$ is also integrable; however, this being the case, item (b) of that proposition assures that the same holds for $|f| = f_+ + f_-$.

Now, in order to prove (10.28), note first of all that the monotonicity of the integral, together with the fact that $f_- \ge 0$, gives $\int_a^b f_-(x)dx \ge 0$. In turn, by successively applying (10.27), this inequality and the formula of item (b) of the previous proposition we get

$$\int_a^b f(x)dx = \int_a^b f_+(x)dx - \int_a^b f_-(x)dx$$

$$\le \int_a^b f_+(x)dx + \int_a^b f_-(x)dx$$

$$= \int_a^b (f_+(x)dx + f_-(x))dx$$

$$= \int_a^b |f(x)|dx. \tag{10.29}$$

(For another proof of (10.28) we refer the reader to Problem 7.) Analogously,

$$-\int_a^b f(x)dx = \int_a^b (-f)(x)dx \le \int_a^b |(-f)(x)|dx = \int_a^b |f(x)|dx,$$

so that

$$-\int_a^b |f(x)|dx \le \int_a^b f(x)dx \le \int_a^b |f(x)|dx,$$

which is equivalent to (10.28).

We are left to showing the integrability of f_+. To this end, let $P = \{a = x_0 < x_1 < \cdots < x_k = b\}$ be a partition of $[a, b]$. If $f(x) \geq 0$ for some $x \in [x_{j-1}, x_j]$, then $M_j(f_+) = M_j(f)$; on the other hand, from $f_+ \geq f$ we get $m_j(f_+) \geq m_j(f)$, so that

$$M_j(f_+) - m_j(f_+) \leq M_j(f) - m_j(f).$$

If $f(x) < 0$ for all $x \in [x_{j-1}, x_j]$, then $f_+ = 0$ in $[x_{j-1}, x_j]$ and, then,

$$M_j(f_+) - m_j(f_+) = 0 \leq M_j(f) - m_j(f).$$

In any case, we have $M_j(f_+) - m_j(f_+) \leq M_j(f) - m_j(f)$, so that

$$\begin{aligned}
S(f_+; P) - s(f_+; P) &= \sum_{j=1}^{k}(M_j(f_+) - m_j(f_+))(x_j - x_{j-1}) \\
&\leq \sum_{j=1}^{k}(M_j(f) - m_j(f))(x_j - x_{j-1}) \\
&= S(f; P) - s(f; P).
\end{aligned}$$

Now, since f is integrable, Cauchy's criterion for integrability assures that, given $\epsilon > 0$, we can choose the partition P in such a way that $S(f; P) - s(f; P) < \epsilon$. However, this being the case, it follows from the last inequality above that $S(f_+; P) - s(f_+; P) < \epsilon$. Therefore, again by Cauchy's criterion, we conclude that f_+ is integrable. $\qquad\square$

Example 10.15 Given a function $f : [a, b] \to \mathbb{R}$, it may well happen that $|f|$ is integrable but f is not. The classical example is furnished by $f : [0, 1] \to \mathbb{R}$ such that

$$f(x) = \begin{cases} -1, & \text{if } x \notin \mathbb{Q} \\ 1, & \text{if } x \in \mathbb{Q} \end{cases}.$$

An argument analogous to that of Example 10.3 guarantees that f is not integrable. On the other hand, $|f|$ is constantly equal to 1, hence integrable.

The first and last inequalities in (10.29) make it clear the validity of the triangle inequality for integrals: the first inequality guarantees that $\int_a^b f(x)dx$ can be computed as the difference between the areas of the regions $\mathcal{R}_+$ and $\mathcal{R}_-$ of the cartesian plane, where $\mathcal{R}_+$ is under the graph of f and above the horizontal axis, while $\mathcal{R}_-$ is above the graph of f and below the horizontal axis; on the other hand, the last equality gives $\int_a^b |f(x)|dx$ as the sum of the areas of $\mathcal{R}_+$ and $\mathcal{R}_-$.

Our next result guarantees that the integral is also additive with respect to the domains $[a, c]$ and $[c, b]$ of two integrable functions. By heuristically reasoning with a nonnegative function $f : [a, b] \to \mathbb{R}$ whose restrictions to the intervals $[a, c]$ and $[c, b]$ are both integrable, we conclude that this is quite a plausible result. Indeed, letting $\mathcal{R}$, $\mathcal{R}_{|[a,c]}$ and $\mathcal{R}_{|[b,c]}$ denote the regions of the cartesian plane respectively under the graphs of f and of its restrictions to $[a,c]$ and $[c,b]$, we have $\mathcal{R} = \mathcal{R}_{|[a,c]} \cup \mathcal{R}_{|[b,c]}$, such that $\mathcal{R}_{|[a,c]}$ and $\mathcal{R}_{|[b,c]}$ have no interior points in common; therefore, one expects that $A(\mathcal{R}) = A(\mathcal{R}_{|[a,c]}) + A(\mathcal{R}_{|[b,c]})$.

Proposition 10.16 *Let a function $f : [a,b] \to \mathbb{R}$ and a real number $c \in (a,b)$ be given. If the restrictions of f to the intervals $[a,c]$ and $[c,b]$ are both integrable, then f is also integrable, with*

$$\int_a^b f(x)dx = \int_a^c f(x)dx + \int_c^b f(x)dx.$$

Proof Let $f_{|[a,c]}$ and $f_{|[c,b]}$ denote the restrictions of f to the intervals $[a,c]$ and $[c,b]$, respectively. Given $\epsilon > 0$, the integrabilities of $f_{|[a,c]}$ and $f_{|[c,b]}$ assure, by means of the Cauchy criterion, the existence of a partition P of $[a,c]$ and Q of $[c,b]$ such that

$$S(f_{|[a,c]}; P) - s(f_{|[a,c]}; P) < \frac{\epsilon}{2} \text{ and } S(f_{|[c,b]}; Q) - s(f_{|[c,b]}; Q) < \frac{\epsilon}{2}.$$

If $R = P \cup Q$, then R is a partition of $[a,b]$ and it is immediate that

$$S(f; R) = S(f_{|[a,c]}; P) + S(f_{|[c,b]}; Q) \text{ and } s(f; R) = s(f_{|[a,c]}; P) + s(f_{|[c,b]}; Q).$$

Hence,

$$S(f; R) - s(f; R) = S(f_{|[a,c]}; P) - s(f_{|[a,c]}; P) + S(f_{|[c,b]}; Q) - s(f_{|[c,b]}; Q) < \epsilon,$$

so that, by invoking Cauchy's criterion once more, we deduce the integrability of f.

For what is left to do, let $\Delta = \int_a^b f(x)dx - \int_a^c f(x)dx - \int_c^b f(x)dx$. In the notations of the above discussion, we have

$$\Delta \leq S(f; R) - \int_a^c f(x)dx - \int_c^b f(x)dx$$

$$= S(f_{|[a,c]}; P) + S(f_{|[c,b]}; Q) - \int_a^c f(x)dx - \int_c^b f(x)dx$$

$$\leq S(f_{|[a,c]}; P) + S(f_{|[c,b]}; Q) - s(f_{|[a,c]}; P) - s(f_{|[c,b]}; Q)$$

$$= S(f_{|[a,c]}; P) - s(f_{|[a,c]}; P) + S(f_{|[c,b]}; Q) - s(f_{|[c,b]}; Q) < \epsilon;$$

analogously,

$$\Delta \geq s(f; R) - \int_a^c f(x)dx - \int_c^b f(x)dx$$

$$\geq s(f_{|[a,c]}; P) + s(f_{|[c,b]}; Q) - S(f_{|[a,c]}; P) - S(f_{|[c,b]}; Q)$$

$$= -(S(f_{|[a,c]}; P) - s(f_{|[a,c]}; P)) - (S(f_{|[c,b]}; Q) - s(f_{|[c,b]}; Q)) > -\epsilon.$$

Therefore, $|\Delta| < \epsilon$. Finally, since $\epsilon > 0$ was chosen arbitrarily, we conclude that $\Delta = 0$. $\qquad\square$

The previous proposition shall let us present yet another class of examples of integrable functions. However, before we can do so, we need the following auxiliary result.

Lemma 10.17 *If $f, g : [a, b] \to \mathbb{R}$ are such that f is integrable and $f = g$ in (a, b), then g is integrable and $\int_a^b f(x)dx = \int_a^b g(x)dx$.*

Proof Let's consider the case in which $f = g$ in $[a, b]$, leaving the general case to the reader (cf. Problem 8).

If $P = \{a = x_0 < x_1 < \cdots < x_k = b\}$ is a partition of $[a, b]$, the coincidence of f and g in $[x_0, x_{k-1}]$ furnishes

$$|s(f; P) - s(g; P)| = \left| \sum_{j=1}^{k} \left(\inf_{[x_{j-1}, x_j]} f - \inf_{[x_{j-1}, x_j]} g \right)(x_j - x_{j-1}) \right|$$

$$= \left| \left(\inf_{[x_{k-1}, x_k]} f - \inf_{[x_{k-1}, x_k]} g \right)(x_k - x_{k-1}) \right|$$

$$\leq |f(b) - g(b)| \, (b - x_{k-1});$$

analogously,

$$|S(f; P) - S(g; P)| \leq |f(b) - g(b)|(b - x_{k-1}).$$

It follows from the computations above and the triangle inequality that

$$S(g; P) - s(g; P) = (S(g; P) - S(f; P)) + (S(f; P) - s(f; P))$$
$$+ (s(f; P) - s(g; P))$$
$$\leq |S(g; P) - S(f; P)| + (S(f; P) - s(f; P))$$
$$+ |s(f; P) - s(g; P)|$$
$$\leq 2|f(b) - g(b)|(b - x_{k-1}) + (S(f; P) - s(f; P)).$$

Now, given $\epsilon > 0$, the integrability of f guarantees, by means of the Cauchy criterion, the existence of a partition P such that $S(f; P) - s(f; P) < \frac{\epsilon}{2}$ and (refining P, if necessary) $b - x_{k-1} < \frac{\epsilon}{4(|f(b)-g(b)|+1)}$. With such a partition P, the computations above give

$$S(g; P) - s(g; P) \leq 2|f(b) - g(b)|(b - x_{k-1}) + (S(f; P) - s(f; P))$$
$$\leq 2|f(b) - g(b)| \cdot \frac{\epsilon}{4(|f(b) - g(b)| + 1)} + \frac{\epsilon}{2} < \epsilon.$$

Therefore, by resorting to the Cauchy criterion once more, we conclude that g is integrable.

For the equality of the integrals of f and g, note that (again by the computations above)

$$\int_a^b f(x)dx \leq S(f;P) = (S(f;P) - S(g;P)) + S(g;P)$$

$$\leq |S(f;P) - S(g;P)| + S(g;P)$$

$$\leq |f(b) - g(b)|(b - x_{k-1}) + S(g;P).$$

Hence,

$$\int_a^b f(x)dx - \int_a^b g(x)dx \leq |f(b) - g(b)|(b - x_{k-1}) + S(g;P) - \int_a^b g(x)dx.$$

Choosing the partition P in such a way that $b - x_{k-1} < \frac{\epsilon}{2(|f(b)-g(b)|+1)}$ and $S(g;P) - \int_a^b g(x)dx < \frac{\epsilon}{2}$ (this last choice being possible thanks to the Cauchy criterion), we get

$$\int_a^b f(x)dx - \int_a^b g(x)dx \leq |f(b) - g(b)|(b - x_{k-1}) + S(g;P) - \int_a^b g(x)dx$$

$$\leq |f(b) - g(b)| \cdot \frac{\epsilon}{2(|f(b) - g(b)| + 1)} + \frac{\epsilon}{2} < \epsilon.$$

However, since $\epsilon > 0$ was chosen arbitrarily, it follows from the above that

$$\int_a^b f(x)dx - \int_a^b g(x)dx \leq 0.$$

Finally, changing the roles of f and g in the reasoning above (which is perfectly valid, for we have already established the integrability of g), we get the opposite inequality between the integrals, thus showing that they are equal.   □

We now need the coming

Definition 10.18 A function $f : [a, b] \to \mathbb{R}$ is **piecewise continuous** if there exist real numbers $a = x_0 < x_1 < \cdots < x_k = b$ such that f is continuous in the interval (x_{j-1}, x_j), for $1 \leq j \leq k$, and the lateral limits $\lim_{x \to a+} f(x)$, $\lim_{x \to b-} f(x)$ and $\lim_{x \to x_j \pm} f(x)$ exist, for $1 \leq j < k$.

Figure 10.4 sketches the graph of a piecewise continuous function $f : [a, b] \to \mathbb{R}$, which is discontinuous in exactly three points.

Proposition 10.19 If $f : [a, b] \to \mathbb{R}$ is piecewise continuous, then f is integrable, with

$$\int_a^b f(x)dx = \sum_{j=1}^k \int_{x_{j-1}}^{x_j} f(x)dx. \qquad (10.30)$$

Fig. 10.4 A piecewise
continuous function
$f : [a, b] \to \mathbb{R}$

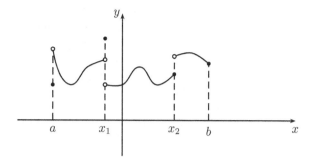

Proof If $f_j : [x_{j-1}, x_j] \to \mathbb{R}$ is such that $f_j = f$ in (x_{j-1}, x_j) and $f_j(x_{j-1}) = \lim_{x \to x_{j-1}+} f(x)$, $f_j(x_j) = \lim_{x \to x_j-} f(x)$, then f_j is continuous, thus integrable. Now, Lemma 10.17 assures that the restriction of f to the interval $[x_{j-1}, x_j]$ is also integrable, with $\int_{x_{j-1}}^{x_j} f(x)dx = \int_{x_{j-1}}^{x_j} f_j(x)dx$. Finally, by repeatedly applying Proposition 10.16, we conclude that f is integrable in $[a, b]$ and that (10.30) is valid. □

For the coming example, the reader may find it useful to review the concept of characteristic function of a set, in Definition 10.11.

Example 10.20 Let $a = x_0 < x_1 < \cdots < x_k = b$ be a partition of the interval $[a, b]$, and $f : [a, b] \to \mathbb{R}$ be defined by

$$f(x) = \sum_{j=1}^{k} c_j \mathcal{X}_{[x_{j-1}, x_j]},$$

with $c_j \in \mathbb{R}$ for $1 \le j \le k$. Since f is clearly piecewise continuous, the previous proposition guarantees its integrability, with

$$\int_a^b f(x)dx = \sum_{j=1}^{k} \int_{x_{j-1}}^{x_j} f(x)dx.$$

On the other hand, Problem 2, page 357, gives

$$\int_{x_{j-1}}^{x_j} f(x)dx = \int_{x_{j-1}}^{x_j} c_j dx = c_j(x_j - x_{j-1}).$$

Therefore, by combining both inequalities above, we get

$$\int_a^b f(x)dx = \sum_{j=1}^{k} c_j(x_j - x_{j-1}).$$

Our next result establishes the converse of Proposition 10.16 and will be of crucial importance for the actual computation of integrals, in the next section.

Proposition 10.21 *Let $f : [a, b] \to \mathbb{R}$ be integrable and $a < c < b$. Then, the restrictions of f to the intervals $[a, c]$ and $[c, b]$ (which, whenever no danger of confusion appears, will be denoted simply by f) are also integrable, with*

$$\int_a^b f(x)dx = \int_a^c f(x)dx + \int_c^b f(x)dx. \qquad (10.31)$$

Proof It suffices to prove the integrability of the restrictions of f to $[a, c]$ and $[c, b]$. Indeed, once we have done this, (10.31) will follow from Proposition 10.16.

For what is left to do, given $\epsilon > 0$, Cauchy's criterion assures the existence of a partition R of $[a, b]$ such that $S(f; R) - s(f; R) < \epsilon$. Letting $R' = R \cup \{c\}$, it follows from Lemma 10.1 that $S(f; R') - s(f; R') \leq S(f; R) - s(f; R) < \epsilon$; therefore, we can suppose from the very beginning that $c \in R$.

Let $f_{|[a,c]}$ denote the restriction of f to the interval $[a, c]$. If $R = \{a = x_0 < x_1 < \cdots < x_l = c < \cdots < x_k = b\}$ and $P = \{a = x_0 < x_1 < \cdots < x_l = c\}$, then P is a partition of $[a, c]$, such that

$$S(f_{|[a,c]}; P) - s(f_{|[a,c]}; P) = \sum_{j=1}^{l}(M_j(f) - m_j(f))(x_j - x_{j-1})$$

$$\leq \sum_{j=1}^{k}(M_j(f) - m_j(f))(x_j - x_{j-1})$$

$$= S(f; R) - s(f; R) < \epsilon.$$

Hence, again by Cauchy's criterion, $f_{|[a,c]}$ is integrable.

Analogously, the restriction of f to the interval $[c, b]$ is integrable.   □

The last result of this section shows that the product of two integrable functions is also integrable. Nevertheless, as Problem 10 shows, the value of the integral of the product of the functions cannot be computed from the values of the integrals of the factors. (However, see Problem 12.)

Proposition 10.22 *If $f, g : [a, b] \to \mathbb{R}$ are integrable functions, then $fg : [a, b] \to \mathbb{R}$ is also integrable.*

Proof Suppose we have established the integrability of the square of an integrable function. Then, $(f + g)^2, f^2$ and g^2 will be integrable, so that, since $fg = \frac{1}{2}((f + g)^2 - f^2 - g^2)$, several applications of Proposition 10.13, together with the remark right after it, guarantee the integrability of fg.

Let's then show that f^2 is integrable. To this end, note first that, since f is bounded, there exists a real constant c such that $f + c \geq 0$ in $[a, b]$. Looking at c as a constant function in $[a, b]$ and writing $f^2 = (f + c)^2 - 2cf - c^2$, we conclude (quite similarly as in the above reasoning) that f^2 is integrable if $(f + c)^2$ has this property. It then suffices to show that, if f is integrable and nonnegative, then f^2 is also integrable.

For what is left to do, assume f to be integrable and nonnegative, so that f^2 is surely bounded. Now, given a partition $P = \{a = x_0 < x_1 < \cdots < x_k = b\}$ of $[a, b]$, it follows from Problem 14, page 207, that

$$\sup_{[x_{j-1}, x_j]} f^2 = (\sup_{[x_{j-1}, x_j]} f)^2 = M_j^2$$

and, analogously, $\inf_{[x_{j-1}, x_j]} f^2 = m_j^2$. Hence,

$$S(f^2; P) - s(f^2; P) = \sum_{j=1}^{k} (M_j^2 - m_j^2)(x_j - x_{j-1})$$

$$= \sum_{j=1}^{k} (M_j + m_j)(M_j - m_j)(x_j - x_{j-1})$$

$$\leq 2 \sup_{[a,b]} f \cdot \sum_{j=1}^{k} (M_j - m_j)(x_j - x_{j-1})$$

$$= 2 \sup_{[a,b]} f \cdot (S(f; P) - s(f; P)).$$

Let $\epsilon > 0$ be given. By invoking Cauchy's criterion again, and thanks to the integrability of f, we can suppose that $S(f; P) - s(f; P) < \frac{\epsilon}{2(\sup_{[a,b]} f) + 1}$. Therefore, the above computations give

$$S(f^2; P) - s(f^2; P) < 2(\sup_{[a,b]} f) \cdot \frac{\epsilon}{2(\sup_{[a,b]} f) + 1} < \epsilon.$$

Hence, once more from Cauchy's criterion, f^2 is integrable. $\square$

Problems: Section 10.4

1. Give an example of a differentiable function $f : [a, b] \to \mathbb{R}$ such that f' is not bounded (hence, not integrable).

2. Let $f : [a, b] \to \mathbb{R}$ be continuous, with a finite number of zeros in the interval $[a, b]$, say $x_1 < x_2 < \cdots < x_k$. If

$$R_f^+ = \{(x, y) \in \mathbb{R}^2; a \leq x \leq b \quad \text{and} \quad 0 \leq y \leq f(x)\}$$

and

$$R_f^- = \{(x, y) \in \mathbb{R}^2;\ a \le x \le b \quad \text{and}\quad f(x) \le y \le 0\},$$

prove that

$$\int_a^b f(x)dx = A(R_f^+) - A(R_f^-).$$

For the next problem, given continuous functions $f, g : [a, b] \to \mathbb{R}$, such that $f(x) \le g(x)$ for every $x \in [a, b]$, let

$$R_{fg} = \{(x, y) \in \mathbb{R}^2;\ a \le x \le b \text{ and } f(x) \le y \le g(x)\}$$

be the portion of the cartesian plane situated *between the graphs* of f and g. Let the area of R_{fg} be defined by

$$A(R_{fg}) = \int_a^b (g(x) - f(x))dx.$$

3. Prove **Cavalieri's principle**: for $i = 1, 2$, let $f_i, g_i : [a, b] \to \mathbb{R}$ be continuous functions such that $f_i(x) \le g_i(x)$ for every $x \in [a, b]$. If, for every $x \in [a, b]$, the length of the line segment joining $(x, f_1(x))$ to $(x, g_1(x))$ equals that of the line segment joining $(x, f_2(x))$ to $(x, g_2(x))$, then $A(R_{f_1 g_1}) = A(R_{f_2 g_2})$.

4. Let $\lfloor \cdot \rfloor : \mathbb{R} \to \mathbb{R}$ be the integer part function (cf. Problem 9, page 152). Given $n \in \mathbb{N}$, compute $\int_0^n \lfloor x \rfloor dx$.

5. Let $\{\cdot\} : \mathbb{R} \to \mathbb{R}$ be the fractional part function (cf. Problem 10, page 152). Given $n \in \mathbb{N}$, do the following items:

   (a) Show that $\int_n^{n+1} \{x\}dx = \int_0^1 \{x\}dx$.
   (b) Compute $\int_0^n \{x\}dx$.

6. * Let $f, g : [a, b] \to \mathbb{R}$ be continuous functions such that $f(x) \le g(x)$ for every $x \in [a, b]$. If $\int_a^b f(x)dx = \int_a^b g(x)dx$, prove that $f = g$.

7. Let $f : [a, b] \to \mathbb{R}$ be an integrable function. Assuming that $|f|$ is also integrable, use the fact that $-|f(x)| \le f(x) \le |f(x)|$ for every $x \in [a, b]$ to deduce (10.28).

8. * Complete the proof of Lemma 10.17, by examining the case in which $f = g$ in (a, b).

9. Given $0 \le a < b$ and $n \in \mathbb{N}$, compute $\int_a^b x^n dx$. Then compute $\int_a^b f(x)dx$, where $f(x) = \sum_{j=0}^n a_j x^j$, with $a_0, a_1, \ldots, a_n \in \mathbb{R}$ and $a_n \ne 0$.

10. Give examples of continuous and nonnegative functions $f, g; [a, b] \to \mathbb{R}$, such that $\int_a^b f(x)dx > 0$ and $\int_a^b g(x)dx > 0$ but $\int_a^b f(x)g(x)dx = 0$.

11. * Let $f : [a, b] \to \mathbb{R}$ be an integrable function. For $0 < \epsilon < \frac{b-a}{2}$, show that

$$\lim_{\epsilon \to 0} \int_{a+\epsilon}^{b-\epsilon} f(x)dx = \int_a^b f(x)dx.$$

12. * Prove **Cauchy's inequality for integrals**: if $f, g : [a, b] \to \mathbb{R}$ are integrable
     functions, then

$$\left| \int_a^b f(x)g(x)dx \right| \leq \left(\int_a^b f(x)^2 dx \right)^{1/2} \left(\int_a^b g(x)^2 dx \right)^{1/2}.$$

Moreover, if f and g are continuous, show that equality happens if and only if
there exists $\lambda \in \mathbb{R}$ such that $f(x) = \lambda g(x)$ for every $x \in [a, b]$.

13. * If $f : [a, b] \to \mathbb{R}$ is a continuous and convex function, prove that

$$\frac{f(a) + f(b)}{2} \leq \frac{1}{b - a} \int_a^b f(x)dx.$$

10.5   The Fundamental Theorem of Calculus

The main purpose of this section is to turn (10.11) into a theorem. Along all that
follows, unless explicitly stated otherwise, we shall let I be an interval of the real
line and $f : I \to \mathbb{R}$ be integrable in each interval $[a, b] \subset I$.

For $c \in I$ and $[a, b] \subset I$, we set

$$\int_c^c f(x)dx = 0 \quad \text{and} \quad \int_b^a f(x)dx = - \int_a^b f(x)dx.$$

With these conventions at hand, and applying the second part of Proposition 10.21,
it is immediate to verify that

$$\int_a^b f(x)dx = \int_a^c f(x)dx + \int_c^b f(x)dx, \tag{10.32}$$

for every $a, b, c \in I$ (i.e., not only when $a < b$ and $c \in (a, b)$).

This being said, the following definition will be of paramount importance for
what is to come.

Definition 10.23 Let $I \subset \mathbb{R}$ be an interval and $f : I \to \mathbb{R}$ be integrable in each
interval $[a, b] \subset I$. For a fixed $c \in I$, the **indefinite integral** of f **based** at c is the
function $F : I \to \mathbb{R}$, given by

$$F(x) = \int_c^x f(t)dt.$$

If $f, F : I \to \mathbb{R}$ are as in the previous definition, (10.32) promptly gives

$$F(b) - F(a) = \int_c^b f(t)dt - \int_c^a f(t)dt = \int_a^b f(t)dt, \tag{10.33}$$

for all $a, b \in I$.

We can now state and prove the major result of this section, which is known in mathematical literature as the **fundamental theorem of Calculus (FTC)**.

Theorem 10.24 (FTC) *Let $I \subset \mathbb{R}$ be an interval, let $f : I \to \mathbb{R}$ be integrable in each interval $[a, b] \subset I$ and $F : I \to \mathbb{R}$ be the indefinite integral of f based at $c \in I$. If f is continuous at $x_0 \in I$, then F is differentiable at x_0, with $F'(x_0) = f(x_0)$.*

Proof For $x \in I \setminus \{x_0\}$ and successively applying (10.33), Problem 2, page 357, the additivity of the integral and the triangle inequality for integrals, we get

$$\left| \frac{F(x) - F(x_0)}{x - x_0} - f(x_0) \right| = \left| \frac{1}{x - x_0} \int_{x_0}^{x} f(t)dt - f(x_0) \right|$$

$$= \frac{1}{|x - x_0|} \left| \int_{x_0}^{x} (f(t) - f(x_0))dt \right|$$

$$\leq \frac{1}{|x - x_0|} \left| \int_{x_0}^{x} |f(t) - f(x_0)|dt \right|.$$

(In the last inequality above, the modulus outside the integral is due to the fact that, if $x < x_0$, then $\int_{x_0}^{x} |f(t) - f(x_0)|dt = - \int_{x}^{x_0} |f(t) - f(x_0)|dt$.)

The continuity of f at x_0 guarantees that, given $\epsilon > 0$, there exists $\delta > 0$ such that

$$t \in I, \ |t - x_0| < \delta \Rightarrow |f(t) - f(x_0)| < \epsilon.$$

Hence, for $0 < |x - x_0| < \delta$, we have $|t - x_0| < \delta$ for every t belonging to the interval with endpoints x_0 and x, so that $|f(t) - f(x_0)| < \epsilon$. It thus follows from the monotonicity of the integral, and once more from the result of Problem 2, page 357, that

$$\left| \int_{x_0}^{x} |f(t) - f(x_0)|dt \right| \leq \left| \int_{x_0}^{x} \epsilon \, dt \right| = \epsilon |x - x_0|.$$

Finally, the above computations assure that, for $x \in I$ such that $0 < |x - x_0| < \delta$, we have

$$\left| \frac{F(x) - F(x_0)}{x - x_0} - f(x_0) \right| \leq \frac{1}{|x - x_0|} \cdot \epsilon |x - x_0| = \epsilon.$$

Hence, F is differentiable at x_0, with $F'(x_0) = f(x_0)$. $\square$

The FTC is the ingredient that was missing for us to get an easy way of computing integrals of various elementary functions. However, before we can actually do that, we need one more piece of terminology.

Definition 10.25 Let $I \subset \mathbb{R}$ be an interval and $f : I \to \mathbb{R}$ be integrable in each interval $[a, b] \subset I$. A **primitive** for f in I is a differentiable function $F : I \to \mathbb{R}$ such that $F' = f$ in I.

In terms of the previous definition, the FTC guarantees that each indefinite integral of a continuous function $f : I \to \mathbb{R}$ is a primitive of f in I. Indeed, letting $F : I \to \mathbb{R}$ be the indefinite integral of F based at $c \in I$, it follows from the FTC and the continuity of f in I that $F'(x) = f(x)$ for every $x \in I$. Yet in another way, we have that:

$$f \text{ continuous} \implies \frac{d}{dx} \int_c^x f(t)dt = f(x). \tag{10.34}$$

The coming result assures that, even for functions which are merely integrable, there are no other possible primitives. In other words, it guarantees that, even if f is just an integrable function, there are no other possible solutions for (10.37).

Theorem 10.26 *Let $I \subset \mathbb{R}$ be an interval and $f : I \to \mathbb{R}$ be integrable in each interval $[a, b] \subset I$. If $F : I \to \mathbb{R}$ is a primitive for f, then, for a given $x_0 \in I$, we have*

$$F(x) = F(x_0) + \int_{x_0}^x f(t)dt \tag{10.35}$$

for every $x \in I$.

Proof Suppose $x > x_0$ (the case $x = x_0$ is trivial and the case $x < x_0$ can be dealt with in an analogous way, taking into account that $\int_{x_0}^x f(t)dt = -\int_x^{x_0} f(t)dt$).

Let $P = \{x_0 < x_1 < \cdots < x_k = x\}$ be a partition of $[x_0, x]$. Since F is differentiable in I, it is continuous in each interval $[a, b] \subset I$. Hence, Lagrange's MVT gives $\xi_j \in (x_{j-1}, x_j)$ such that

$$F(x_j) - F(x_{j-1}) = F'(\xi_j)(x_j - x_{j-1}) = f(\xi_j)(x_j - x_{j-1}),$$

for $1 \leq j \leq k$. Then, we get

$$\begin{aligned}
F(x) - F(x_0) &= \sum_{j=1}^k (F(x_j) - F(x_{j-1})) \\
&= \sum_{j=1}^k f(\xi_j)(x_j - x_{j-1}) \\
&= \Sigma(f; P; \xi).
\end{aligned}$$

Now recall that, according to Riemann's Theorem 10.7,

$$\int_{x_0}^x f(t)dt = \lim_{|P| \to 0} \Sigma(f; P; \xi),$$

for every choice $\xi = (\xi_j)$ of intermediate points relative to P. Therefore, in view of the computations in the previous paragraph, there is no other choice than to have (10.35). □

If $f : I \to \mathbb{R}$ has a primitive, then, thanks to the previous result, one uses to denote a generic primitive of f by writing $\int f(t)dt$. The reader must be careful not to making confusion with the notations $\int f(t)dt$ and $\int_a^b f(t)dt$: while the first one refers to a *differentiable function* from I to $\mathbb{R}$ whose derivative coincides with f in I, the second denotes a *real number*. Also, note that the theorem guarantees that if $\int f(t)dt$ is a primitive of f in I, then the primitives of f in I are the functions of the form

$$\int f(t)dt + C,$$

where C is a real constant.

With Theorem 10.26 at our disposal, the coming corollary assures that, if $f : [a, b] \to \mathbb{R}$ is an integrable function, then, in order to *effectively compute* $\int_a^b f(t)dt$, it suffices to find a primitive for f. From now on, given a continuous function $F : [a, b] \to \mathbb{R}$, we denote

$$F(x)\Big|_{x=a}^{x=b} = F(b) - F(a).$$

Corollary 10.27 *If $f : [a, b] \to \mathbb{R}$ is an integrable function and $F : [a, b] \to \mathbb{R}$ is a primitive for f, then*

$$\int_a^b f(t)dt = F(x)\Big|_{x=a}^{x=b}. \tag{10.36}$$

Proof Let $x_0 = a$ and $x = b$ in (10.35). □

As we anticipated before, the former corollary provides us with a general strategy for computing the actual value of $\int_a^b f(t)dt$: it suffices to get to visualize the *integrand function* f as the derivative of some function F, then applying (10.36) to obtain $\int_a^b f(t)dt = F(b) - F(a)$. That is why we usually look at (10.36) as

$$\int_a^b F'(t)dt = F(x)\Big|_{x=a}^{x=b}.$$

Let us see a couple of examples.

Example 10.28 Let $f : \mathbb{R} \to \mathbb{R}$ be such that $f(x) = \sum_{j=0}^n a_j x^j$ for every $x \in \mathbb{R}$, where $a_0, a_1, \ldots, a_n$ are given real numbers, with $a_n \neq 0$. Since $F(x) = \sum_{j=0}^n \frac{a_j}{j+1} x^{j+1}$ is clearly a primitive of f, given reals $a < b$ we have

$$\int_a^b f(t)dt = F(x)\Big|_{x=a}^{x=b} = \sum_{j=0}^n \frac{a_j}{j+1}(b^{j+1} - a^{j+1}).$$

Example 10.29 Compute $\int_0^\pi \sin x\,dx$, $\int_0^\pi \sin^2 x\,dx$ and $\int_0^\pi \cos^2 x\,dx$.

Solution Since $-\cos$ is a primitive for $\sin$, the previous corollary gives

$$\int_0^\pi \sin x\,dx = -\cos x\Big|_{x=0}^{x=\pi} = -\cos\pi + \cos 0 = 2.$$

With respect to the second integral, since $\sin^2 x = \frac{1}{2}(1 - \cos 2x)$ and $\frac{d}{dx}\sin 2x = 2\cos 2x$, we have (again by that result)

$$\int_0^\pi \sin^2 x\,dx = \frac{1}{2}\int_0^\pi (1 - \cos 2x)dx = \frac{1}{2}\left(x - \frac{1}{2}\sin 2x\right)\Big|_{x=0}^{x=\pi} = \frac{\pi}{2}.$$

Finally, since $\sin^2 x + \cos^2 x = 1$, we have

$$\int_0^\pi \cos^2 x\,dx = \int_0^\pi (1 - \sin^2 x)dx = \int_0^\pi 1\,dx - \int_0^\pi \sin^2 x\,dx$$

$$= x\Big|_{x=0}^{x=\pi} - \int_0^\pi \sin^2 x\,dx = \pi - \frac{\pi}{2} = \frac{\pi}{2}.$$

$\square$

Another way of rephrasing (10.34), which is quite important for the theory of ordinary differential equations[3] is collected in the following result.

Proposition 10.30 *Let $I \subset \mathbb{R}$ be an interval and $f : I \to \mathbb{R}$ be a continuous function. Given $x_0 \in I$ and $y_0 \in \mathbb{R}$, the **initial value problem***

$$\begin{cases} y' = f(x) \\ y(x_0) = y_0 \end{cases} \tag{10.37}$$

has as its only solution the function $F : I \to \mathbb{R}$ given by

$$F(x) = y_0 + \int_{x_0}^x f(t)dt. \tag{10.38}$$

Proof A solution of (10.37) is a differentiable function $F : I \to \mathbb{R}$ such that $F'(x) = f(x)$ for every $x \in I$ and $F(x_0) = y_0$. Since f is continuous, (10.34) guarantees that F, defined as in (10.38), is a solution of (10.37).

---

[3]In spite of being quite an interesting and important subject, we shall not have occasion to give a systematic development of the theory of ODE's in these notes. For the interested reader, we recommend the outstanding book of professor George Simmons, [23].

On the other hand, if $F_1, F_2 : I \to \mathbb{R}$ solve (10.37), then $F_1' = f = F_2'$. Therefore, Corollary 9.47 guarantees that $F_1 - F_2$ is constant. However, since $(F_1 - F_2)(x_0) = y_0 - y_0 = 0$, we conclude that $F_1 - F_2 = 0$, i.e., $F_1 = F_2$.   □

Most often, the task of explicitly finding a primitive of a given integrable function (if possible) is not immediate. That's why Calculus courses generally spend considerable time in developing specific techniques for the computation of integrals, which are generically referred to as *integration techniques*. We discuss the most relevant parts of this toolkit in the rest of this section and along the next one, starting with the formula for **integration by parts**.

Proposition 10.31 *If $f, g : [a, b] \to \mathbb{R}$ are differentiable functions with integrable derivatives, then*

$$\int_a^b f'(x)g(x)dx = f(x)g(x)\Big|_{x=a}^{x=b} - \int_a^b f(x)g'(x)dx. \tag{10.39}$$

Proof Firstly, note that the integrals in both sides of (10.39) are well defined. Indeed, f and g, being differentiable, are continuous; on the other hand, since f' and g' are integrable, Proposition 10.22 guarantees that $f'g$ and fg' are also integrable.

Now, since $(fg)' = f'g + fg'$, Corollary 10.27 gives

$$\int_a^b (f'(x)g(x) + f(x)g'(x))dx = \int_a^b (fg)'(x)dx = f(x)g(x)\Big|_{x=a}^{x=b}.$$

Finally, in order to obtain (10.39), it suffices to apply the additivity of the integral to the left hand side of the equality above.   □

Example 10.32 Compute $\int_0^\pi x \sin x \, dx$ and $\int_0^\pi x^2 \cos x \, dx$.

Solution For the first integral, letting $f(x) = -\cos x$ and $g(x) = x$ in (10.39), we get

$$\int_0^\pi x \sin x \, dx = \int_0^\pi x(-\cos' x)dx = -x \cos x \Big|_{x=0}^{x=\pi} - \int_0^\pi (-\cos x)dx$$

$$= \pi + \int_0^\pi \cos x \, dx = \pi + \sin x \Big|_{x=0}^{x=\pi} = \pi.$$

For the second, letting $f(x) = \sin x$ and $g(x) = x^2$ in (10.39), we obtain

$$\int_0^\pi x^2 \cos x \, dx = \int_0^\pi x^2 \sin' x \, dx = x^2 \sin x \Big|_{x=0}^{x=\pi} - \int_0^\pi 2x \sin x \, dx$$

$$= -2 \int_0^\pi x \sin x \, dx = -2\pi,$$

where we used the result of the first part in the last equality above.   □

Note that, in terms of primitives, the integration by parts formula gives

$$\int f'(x)g(x)dx = f(x)g(x) - \int f(x)g'(x)dx.$$

In words, a primitive for $f'g$ can be obtained by subtracting a primitive for fg' from fg.

The following particular case of integration by parts is frequently useful.

Corollary 10.33 *If $f : [a, b] \to \mathbb{R}$ is differentiable and f' is integrable, then*

$$\int_a^b f(x)dx = xf(x)\Big|_a^b - \int_a^b xf'(x)dx. \tag{10.40}$$

Proof Changing the roles of f and g in (10.39), we get

$$\int_a^b f(x)g'(x)dx = f(x)g(x)\Big|_a^b - \int_a^b f'(x)g(x)dx.$$

Now, it suffices to let $g(x) = x$ for every $x \in [a, b]$. □

Remark 10.34 The derivative of a differentiable function may not be integrable. For an example, see Problem 1, page 379.

Our coming example will find a prominent role in the proof of Theorem 10.56, in Sect. 10.7.

Example 10.35 Let $n \in \mathbb{Z}_+$ and $I_n = \int_0^{\pi/2}(\cos x)^n dx$. Show that:

(a) $nI_n = (n-1)I_{n-2}$ for every integer $n \geq 2$.
(b) $I_{2k} = \frac{(2k)!}{(2^k k!)^2} \cdot \frac{\pi}{2}$ for every integer $k \geq 0$, and $I_{2k-1} = \frac{(2^k k!)^2}{2k(2k)!}$ for every integer $k \geq 1$.

Proof (a) For $n \geq 2$, integration by parts gives

$$I_n = \int_0^{\pi/2}(\cos x)^{n-1}\cos x\, dx = \int_0^{\pi/2}(\cos x)^{n-1}\sin' x\, dx$$

$$= (\cos x)^{n-1}\sin x\Big|_0^{\pi/2} - \int_0^{\pi/2}(n-1)(\cos x)^{n-2}(-\sin x)\,\sin x\, dx$$

$$= (n-1)\int_0^{\pi/2}(\cos x)^{n-2}(1 - \cos^2 x)dx$$

$$= (n-1)\int_0^{\pi/2}(\cos x)^{n-2}dx - (n-1)\int_0^{\pi/2}\cos^n x\, dx$$

$$= (n-1)I_{n-2} - (n-1)I_n.$$

Hence, $nI_n = (n-1)I_{n-2}$.

(b) It's immediate that $I_0 = \int_0^{\pi/2} dx = x\Big|_{x=0}^{x=\frac{\pi}{2}} = \frac{\pi}{2}$. By induction hypothesis, suppose that $I_{2m} = \frac{(2m)!}{(2^m m!)^2} \cdot \frac{\pi}{2}$ for some integer $m \geq 0$. Letting $n = 2m + 2$ in the recurrence relation of item (a), we get

$$\begin{aligned}
I_{2m+2} &= \frac{2m+1}{2m+2} \cdot I_{2m} = \frac{2m+1}{2m+2} \cdot \frac{(2m)!}{(2^m m!)^2} \cdot \frac{\pi}{2} \\
&= \frac{(2m+2)(2m+1)}{(2(m+1))^2} \cdot \frac{(2m)!}{(2^m m!)^2} \cdot \frac{\pi}{2} \\
&= \frac{(2m+2)!}{(2^{m+1}(m+1)!)^2} \cdot \frac{\pi}{2}.
\end{aligned}$$

Thus, the first part of (c) is valid for every $k \geq 0$.

Finally, the proof of the formula for I_{2k-1} is completely analogous, and will be left as an exercise to the reader.

$\square$

Let's finish this section by using the FTC to present a proof of the irrationality of π. Our discussion somewhat follows part B.17 of another marvelous book of professor G. Simmons [22], and (as quoted there) is an elaboration of ideas of C. Hermite by I. Niven. We remark, however, that the irrationality of π was first established by the Swiss mathematician of the XVIII century J. H. Lambert.

Theorem 10.36 (Lambert) π *is irrational.*

Proof If π were rational, say $\pi = \frac{p}{q}$, with $p, q \in \mathbb{N}$, then we would have $\pi^2 = \frac{p^2}{q^2}$, with $p^2, q^2 \in \mathbb{N}$. Thus, suppose, for the sake of contradiction, that $\pi^2 = \frac{a}{b}$, with $a, b \in \mathbb{N}$.

Let's start by taking a twice continuously differentiable function $g : [0, 1] \to \mathbb{R}$, so that

$$\frac{d}{dx}(g'(x)\sin(\pi x) - \pi g(x)\cos(\pi x)) = (g''(x) + \pi^2 g(x))\sin(\pi x).$$

The FTC gives

$$\begin{aligned}
\pi(g(1) + g(0)) &= (g'(x)\sin(\pi x) - \pi g(x)\cos(\pi x))\Big|_{x=0}^{x=1} \\
&= \int_0^1 (g''(x) + \pi^2 g(x))\sin(\pi x)dx \qquad (10.41) \\
&= \int_0^1 (g''(x) + \pi^2 g(x))\sin(\pi x)dx.
\end{aligned}$$

Loosely speaking, Lambert's ingenious idea was to use the fact that π^2 is rational to try to choose g in such a way that, on the one hand, $g(0), g(1) \in \mathbb{Z}$, while, on the other,

$$\left| \int_0^1 (g''(x) + \pi^2 g(x)) \sin(\pi x) dx \right| < 1.$$

In light of (10.41), this would clearly give us an absurd.

Actually, departing from a continuous function f and using the assumption that $\pi^2 = \frac{a}{b}$, set

$$\begin{aligned} g(x) &= b^n (\pi^{2n} f(x) - \pi^{2n-2} f^{(2)}(x) + \pi^{2n-4} f^{(4)}(x) - \cdots) \\ &= a^n f(x) - b a^{n-1} f^{(2)}(x) + b^2 a^{n-2} f^{(4)}(x) - \cdots, \end{aligned} \tag{10.42}$$

where $n \in \mathbb{N}$.

The second equality above assures that

$$f^{(j)}(0), f^{(j)}(1) \in \mathbb{Z}, \ \forall j \in \mathbb{Z}_+ \Rightarrow g(0), g(1) \in \mathbb{Z}.$$

On the other hand, a straightforward computation using the first equality in (10.42) gives

$$g''(x) + \pi^2 g(x) = b^n \pi^{2n+2} f(x) = \frac{a^n}{b} f(x);$$

Hence, to reach the desired contradiction, f must be also such that

$$\frac{a^n}{b} \left| \int_0^1 f(x) \sin(\pi x) dx \right| < 1. \tag{10.43}$$

Since (from the triangle inequality for integrals)

$$\left| \int_0^1 f(x) \sin(\pi x) dx \right| \leq \int_0^1 |f(x) \sin(\pi x)| dx \leq \int_0^1 |f(x)| dx$$

but we have no hints on the size of a, in order to get (10.43) it is natural to choose f in such a way that

$$\max\{|f(x)|; \ 0 \leq x \leq 1\} \leq \frac{C}{n!},$$

for some positive constant C. Once we have done that, we will have

$$\frac{a^n}{b} \left| \int_0^1 f(x) \sin(\pi x) dx \right| \leq \frac{a^n}{b n!},$$

which, for a given $a > 0$, is less that 1 for a sufficiently large n.

We now have enough clues to search for f, and a moment's thought shows

$$f(x) = \frac{1}{n!}x^n(1-x)^n$$

to be a strong candidate. Indeed, for such an f we obviously have $|f(x)| \leq 1$ for $0 \leq x \leq 1$, so that it suffices to check that $f^{(j)}(0), f^{(j)}(1) \in \mathbb{Z}$ for every $j \in \mathbb{Z}_+$.

What is left to do is this is relatively easy. First of all, it follows from the binomial formula that

$$f(x) = \frac{1}{n!}\sum_{k=n}^{2n} a_k x^k,$$

with $a_k \in \mathbb{Z}$ for $n \leq k \leq 2n$. Therefore, if $0 \leq j \leq n$ or $j > 2n$, we obviously have $f^{(j)}(0) = 0$ and $f^{(j)}(1) = 0$. On the other hand, if $n < j \leq 2n$, then

$$
\begin{aligned}
f^{(j)}(x) &= \frac{1}{n!}\sum_{k=j}^{2n} k(k-1)\ldots(k-j+1)a_k x^{k-j} \\
&= \sum_{k=j}^{2n} \frac{j!}{n!} \cdot \frac{k!}{(k-j)!j!}a_k x^{k-j} \\
&= \sum_{k=j}^{2n} j(j-1)\ldots(n+1)\binom{k}{j}a_k x^{k-j},
\end{aligned}
$$

so that $f^{(j)}(0)$ and $f^{(j)}(1)$ also belong to $\mathbb{Z}$. $\square$

Problems: Section 10.5

1. In each of the items below, compute the given integral:

   (a) $\int_0^{2\pi}(1-\cos t)^2 dt$.
   (b) $\int_0^{2\pi}\sqrt{1-\cos t}\, dt$.
   (c) $\int_0^1 x\sin(\pi x)dx$.

2. In the two items below, compute the primitives:

   (a) $\int(\sin x)^3\cos x\, dx$.
   (b) $\int x^2 \sin x\, dx$.

3. Given $m, n \in \mathbb{N}$, show that

$$\int_0^{2\pi} \sin(mx)\sin(nx)dx = \int_0^{2\pi}\cos(mx)\cos(nx)dx = \begin{cases} 0, & \text{if } m \neq n \\ \pi, & \text{if } m = n \end{cases}.$$

4. In each of the following items, solve the indicated initial value problem in the largest possible interval $I \subset \mathbb{R}$:

(a) $\begin{cases} y' = \sqrt{x} \\ y(1) = 2 \end{cases}$.

(b) $\begin{cases} y' = \sec^2 x \\ y(\frac{\pi}{6}) = 1 \end{cases}$.

5. * Let $g, h : [c, d] \to \mathbb{R}$ be differentiable functions and $f : [a, b] \to \mathbb{R}$ be continuous. Show that

$$\frac{d}{dx} \int_{g(x)}^{h(x)} f(t)dt = f(h(x))h'(x) - f(g(x))g'(x).$$

6. * Let $f : \mathbb{R} \to \mathbb{R}$ be a continuous and periodic function, with period $p > 0$. Prove that $\int_a^{a+p} f(t)dt = \int_0^p f(t)dt$ for every $a \in \mathbb{R}$.

7. Let $f : [a, b] \to \mathbb{R}$ be the restriction of an affine function to the interval $[a, b]$. If f is nonnegative, the region R_f is either a right triangle or a right trapezoid. Show that, in each of these cases, the value for the area of R_f obtained with the aid of the FTC coincides with the one obtained by means of the ordinary formulas of Euclidean Geometry for the areas of these polygons (cf. [4], for instance).

8. Let $f : (a, b) \to \mathbb{R}$ be continuous, nonnegative and increasing (resp. decreasing). Prove that every primitive of f is strictly convex (resp. strictly concave).

9. Given $n \in \mathbb{N}$, compute $\sum_{k=0}^{n} \frac{1}{(k+1)(k+2)} \binom{n}{k}$ in terms of n.

10. Prove the **mean value theorem for integrals**: given a continuous function $f : [a, b] \to \mathbb{R}$, there exists $c \in [a, b]$ such that

$$\frac{1}{b-a} \int_a^b f(x)dx = f(c).$$

11. Let $f : [a, b] \to \mathbb{R}$ be a continuously differentiable function, with $f(a) = f(b) = 0$. Prove that

$$\left(\int_a^b f(x)^2 dx \right)^2 \leq 4 \left(\int_a^b x^2 f(x)^2 dx \right) \left(\int_a^b f'(x)^2 dx \right).$$

12. (Romania) Let $\mathcal{F}$ be the set of continuous functions $f : [0, \pi] \to \mathbb{R}$ such that

$$\int_0^\pi f(x) \sin x dx = \int_0^\pi f(x) \cos x dx = 1.$$

Compute $\inf_{f \in \mathcal{F}} \int_0^\pi f(x)^2 dx$.

13. * Let $n \in \mathbb{Z}_+$ and $I_n = \int_0^{\pi/2} (\cos x)^n dx$. The purpose of this problem is to show that $\lim_{n \to +\infty} \frac{I_{n+1}}{I_n} = 1$. To this end, do the following items:

(a) Show that $I_{n+1} \leq I_n$ for every integer $n \geq 0$.
(b) Use item (a) of Example 10.35 to conclude that each of the sequences $\left(\frac{I_{2k-1}}{I_{2k-2}}\right)_{k\geq 1}$ and $\left(\frac{I_{2k}}{I_{2k-1}}\right)_{k\geq 1}$ is nondecreasing.
(c) Conclude that both sequences of the previous item converge to limits ℓ_0 and ℓ_1, say, such that $\ell_0, \ell_1 > 0$.
(d) Show that $\frac{I_{n+1}}{I_n} = \frac{n}{n+1}\left(\frac{I_n}{I_{n-1}}\right)^{-1}$. Then, let $n = 2k \to +\infty$ to get $\ell_0\ell_1 = 1$.
(e) Conclude that $\ell_0 = \ell_1 = 1$ and, then, that $\lim_{n\to+\infty} \frac{I_{n+1}}{I_n} = 1$.

10.6   The Change of Variables Formula

Continuing with the development of the theory, we shall now discuss the **change of variables formula**. As the subsequent discussion will make it evident, such a formula will provide another useful integration technique.

Theorem 10.37 *Let $f : [a,b] \to \mathbb{R}$ be continuous. If $g : [c,d] \to [a,b]$ is differentiable and g' is integrable, then*

$$\int_{g(c)}^{g(d)} f(t)dt = \int_c^d f(g(s))g'(s)ds. \tag{10.44}$$

Proof In order to compute the integral at the left hand side, we start by taking a primitive $F : [a,b] \to \mathbb{R}$ for f, whose existence follows from the FTC. Then, thanks to Corollary 10.27, we get

$$\int_{g(c)}^{g(d)} f(t)dt = \int_{g(c)}^{g(d)} F'(t)dt = F(x)\Big|_{x=g(c)}^{x=g(d)}$$

$$= F(g(d)) - F(g(c)) = (F \circ g)(s)\Big|_{s=c}^{s=d}.$$

Now, since F and g are differentiable, the chain rule guarantees that the function $F \circ g : [c,d] \to \mathbb{R}$ is also differentiable, with

$$(F \circ g)'(s) = F'(g(s))g'(s) = f(g(s))g'(s) = (f \circ g)(s)g'(s).$$

On the other hand, since f and g are continuous, we have $f \circ g$ continuous, thus, integrable; however, since g' is integrable, it follows from Proposition 10.22 that $(f \circ g)g'$ integrable. Hence, applying Corollary 10.27 once more, we have

$$(F \circ g)(s)\Big|_{s=c}^{s=d} = \int_c^d (F \circ g)'(s)ds = \int_c^d f(g(s))g'(s)ds.$$

$\square$

Remark 10.38 For a variant of the previous result, applicable to the case of a piecewise continuous f, see Problem 6.

The corollary below isolates a simple, yet quite useful, particular case of the change of variables formula.

Corollary 10.39 *If $f : [a, b] \to \mathbb{R}$ is continuous and $\lambda \in \mathbb{R}^*$, then*

$$\int_a^b f(\lambda s)\, ds = \frac{1}{\lambda} \int_{\lambda a}^{\lambda b} f(t)\, dt. \tag{10.45}$$

Proof For $\lambda > 0$ it suffices to let $g : [a, b] \to [\lambda a, \lambda b]$ be given by $g(s) = \lambda s$ in Theorem 10.37. Indeed, with such a choice we get

$$\int_{\lambda a}^{\lambda b} f(t)\, dt = \int_{g(a)}^{g(b)} f(t)\, dt = \int_a^b f(g(s))g'(s)\, ds = \int_a^b f(\lambda s)\lambda\, ds.$$

For $\lambda < 0$, let $g : [a, b] \to [\lambda b, \lambda a]$ be given by $g(s) = \lambda s$. Then, arguing as in the previous case and recalling that $\int_b^a = -\int_a^b$ and $\int_{\lambda b}^{\lambda a} = -\int_{\lambda a}^{\lambda b}$, we obtain

$$\int_{\lambda a}^{\lambda b} f(t)\, dt = -\int_{\lambda b}^{\lambda a} f(t)\, dt = -\int_{g(b)}^{g(a)} f(t)\, dt$$

$$= -\int_b^a f(g(s))g'(s)\, ds = \int_a^b f(\lambda s)\lambda\, ds.$$

$\square$

We shall sometimes refer to (10.44) as the formula for **integration by substitution**. In what follows, we give some heuristics on such an expression.

Given a continuous function $f : [a, b] \to \mathbb{R}$, in order to compute $\int_a^b f(t)dt$ with the aid of the FTC we need to obtain a primitive for F. In this sense, and after a careful examination of the expression of $f(t)$, we may eventually notice that it will be quite simplified if we perform an adequate *substitution of variables*, changing t by an expression $g(s)$, depending on a new variable s.

However, since f is defined in the interval $[a, b]$, for such a substitution to make sense it is necessary that $g(s)$ belongs to the interval $[a, b]$ when s varies in some interval $[c, d]$. In general, the best way of identifying such an interval $[c, d]$ is solving inequalities $a \leq g(s) \leq b$, thus finding the appropriate values of c and d. Then, by *formally differentiating* the equality $t = g(s)$, we get $dt = g'(s)\, ds$, so that the original integral must be *corrected* by the factor $g'(s)$ before it can be transformed into a new integral.

In short, the execution of both passages described below give us the formula for **integration by substitution** in any particular case we encounter:

$$\int_a^b f(t)\, dt \xrightarrow[t=g(s)]{\text{substituting}} \int_?^? f(g(s))g'(s)\, ds \xrightarrow[a \leq g(s) \leq b]{\text{solving}} \int_c^d f(g(s))g'(s)\, ds.$$

Finally, we observe that for the last integral above to make sense, the function $s \mapsto f(g(s))g'(s)$ must be integrable. Hence (as we saw in the proof of the theorem) it is natural to suppose that g is differentiable and g' is integrable in $[c, d]$.

Let's give a concrete example, showing how the informal discussion of the two previous paragraphs simplifies the use of the change of variables formula in the actual computation of integrals.

Example 10.40 Compute $\int_{\sqrt{\pi}}^{2\sqrt{\pi}} t\sin(t^2)\, dt$.

Solution Performing the substitution $t = \sqrt{s}$, we get

$$\int_{\sqrt{\pi}}^{2\sqrt{\pi}} t\sin(t^2)\, dt = \int_{?}^{?} \sqrt{s}(\sin s)\frac{1}{2\sqrt{s}}ds = \frac{1}{2}\int_{?}^{?} \sin s\, ds,$$

where the correction factor $\frac{1}{2\sqrt{s}}$ was obtained by *formally differentiating* equality $t = \sqrt{s}$, to get $dt = \frac{1}{2\sqrt{s}}ds$.

In order to correct the integration interval $[\sqrt{\pi}, 2\sqrt{\pi}]$, we solve the inequalities $\sqrt{\pi} \leq \sqrt{s} \leq 2\sqrt{\pi}$, thus getting $\pi \leq s \leq 4\pi$. Hence,

$$\int_{\sqrt{\pi}}^{2\sqrt{\pi}} t\sin(t^2)\, dt = \frac{1}{2}\int_{\pi}^{4\pi} \sin s\, ds = \frac{1}{2}(-\cos s)\Big|_{s=\pi}^{s=4\pi} = -1,$$

where we used the FTC in the last equality.   □

Also with respect to the previous example (and in accordance to Theorem 10.37), note that all of the performed substitutions are valid. Indeed, what we really did was to use $g : [\pi, 4\pi] \rightarrow [\sqrt{\pi}, 2\sqrt{\pi}]$ such that $g(s) = \sqrt{s}$, and such a function is differentiable, with $g'(s) = \frac{1}{2\sqrt{s}}$ integrable in the interval $[\pi, 4\pi]$.

Example 10.41 Given $n \in \mathbb{Z}_+$, compute $\int_0^{\pi/2}(\sin x)^n dx$.

Solution In Example 10.35, we computed $I_n = \int_0^{\pi/2}(\cos x)^n dx$. Here, we shall use integration by substitution to show that $I_n = \int_0^{\pi/2}(\sin x)^n dx$ too. To this end, start by recalling that $\cos x = \sin(\frac{\pi}{2} - x)$. Therefore, letting $x = \frac{\pi}{2} - y$ (so that $dx = -dy$), we get

$$\int_0^{\pi/2}(\sin x)^n dx = \int_{?}^{?}(\sin(\pi/2 - y))^n(-1)dy = -\int_{?}^{?}(\cos y)^n dy.$$

The correct endpoints of the last integral can be easily found once we note that $0 \leq \frac{\pi}{2} - y \leq \frac{\pi}{2}$ if and only if $0 \leq y \leq \frac{\pi}{2}$. Nevertheless, one has to be careful, for, $x = 0 \Rightarrow y = \frac{\pi}{2}$ and $x = \frac{\pi}{2} \Rightarrow y = 0$. Hence, we get

$$\int_0^{\pi/2}(\sin x)^n dx = -\int_{\pi/2}^{0}(\cos y)^n dy = \int_0^{\pi/2}(\cos y)^n dy = I_n.$$

□

From a more formal viewpoint, the last substitution above reduces to using g : $[0, \frac{\pi}{2}] \to [0, \frac{\pi}{2}]$ such that $g(x) = \frac{\pi}{2} - x$. Since $g(0) = \frac{\pi}{2}, g(\frac{\pi}{2}) = 0$ and $g'(x) = -x$, we have

$$\int_0^{\pi/2} (\sin x)^n dx = \int_{g(\frac{\pi}{2})}^{g(0)} (\sin x)^n dx = \int_{\frac{\pi}{2}}^0 (\sin g(x))^n g'(x) dx$$

$$= - \int_{\frac{\pi}{2}}^0 (\sin(\pi/2 - x))^n dx = \int_0^{\frac{\pi}{2}} (\cos x)^n dx.$$

One of the most important usages of integration by substitution is related to **trigonometric substitution**. To get to the point, suppose we have a continuous function $f : [-1, 1] \to \mathbb{R}$ and wish to compute $\int_{-1}^1 f(x) dx$. Then, recalling that the restriction of the sine function to the interval $[-\frac{\pi}{2}, \frac{\pi}{2}]$ is differentiable, has integrable derivative and satisfies $\sin(-\frac{\pi}{2}) = -1$ and $\sin \frac{\pi}{2} = 1$, it follows from (10.44) (with $g(t) = \sin t$) that

$$\int_{-1}^1 f(x) dx = \int_{\sin(-\frac{\pi}{2})}^{\sin \frac{\pi}{2}} f(x) \, dx = \int_{-\frac{\pi}{2}}^{\frac{\pi}{2}} f(\sin t) \cos t \, dt.$$

Depending on the algebraic expression that defines $f(x)$ in terms of x, the last integral above may be much simpler to compute that the original one.

Although the discussion on the previous paragraph has a number of useful variations (depending on the more adequate trigonometric function to be used), we shall not try to list them all. Instead, we collect two relevant examples of trigonometric substitutions.

Example 10.42 Compute $\int_0^1 \frac{1}{1+x^2} dx$.

Proof Recalling, from Example 9.35, that $\frac{d}{dx} \arctan x = \frac{1}{1+x^2}$, we may use the FTC to get

$$\int_0^1 \frac{1}{1+x^2} dx = \arctan x \Big|_{x=0}^{x=1} = \arctan 1 = \frac{\pi}{4}.$$

Alternatively, by performing the trigonometric substitution $x = \tan t$ and noticing that $0 \leq x \leq 1 \Leftrightarrow 0 \leq t \leq \frac{\pi}{4}$, we get, from the formula for integration by substitution and Example 9.34,

$$\int_0^1 \frac{1}{1+x^2} dx = \int_0^{\frac{\pi}{4}} \frac{1}{1+\tan^2 t} \cdot \tan' t \, dt = \int_0^{\frac{\pi}{4}} \frac{1}{\sec^2 t} \cdot \sec^2 t \, dt$$

$$= \int_0^{\frac{\pi}{4}} 1 \, dt = t \Big|_{t=0}^{t=\frac{\pi}{4}} = \frac{\pi}{4}.$$

$\square$

Example 10.43 Let Γ be the circle with center O and radius R, and choose a cartesian coordinate system such that $O(0, 0)$. If $f : [-R, R] \to \mathbb{R}$ is given by $f(x) = \sqrt{R^2 - x^2}$, then the region R_f under the graph of f (and above the horizontal axis is the half-disk bounded by Γ and lying in the upper half-plane. Then,

$$A(R_f) = \int_{-R}^{R} f(x)dx = \int_{-R}^{R} \sqrt{R^2 - x^2}dx.$$

The trigonometric substitution $x = R \cos t$, with $0 \le t \le \pi$, gives

$$A(R_f) = \int_{\pi}^{0} \sqrt{R^2 - R^2 \cos^2 t}(-R \sin t)dt$$

$$= \int_{0}^{\pi} R^2 \sin^2 t\, dt = \frac{\pi R^2}{2},$$

where we used the result of Example 10.29 in the last equality.

A similar reasoning proves that the area of the half-disk bounded by Γ and lying in the lower half-plane also equals $\frac{\pi R^2}{2}$. Hence, the area of the whole disk equals $2 \cdot \frac{\pi R^2}{2} = \pi R^2$, as the reader surely predicted from earlier studies of Euclidean Geometry.

Remark 10.44 As was pointed out in Remark 9.12, in Chap. 11 the sine and cosine functions will be constructed without relying in any concepts or results of Euclidean Geometry. In turn, this will allow us to compute the derivatives of the sine and cosine functions also without reference to Euclidean Geometry. Once this has been done, the previous example will provide us with a genuine way of computing the area of a circle with radius R.

Back to the general discussion of integration by substitution, let intervals I and J and functions $g : I \to J$ and $f : J \to \mathbb{R}$ be given, such that f is continuous and g is continuously differentiable. It's immediate to verify that (10.44) furnishes the following equality of primitives:

$$\left(\int f(t)dt \right)(g(x)) = \int f(g(x))g'(x)dx + C, \tag{10.46}$$

where C is a real constant. We now examine a specific relevant case.

Example 10.45 Let I be an interval, $f : I \to (0, +\infty)$ be a continuously differentiable function and r be a nonzero rational number. Since

$$\frac{d}{dx}f(x)^r = rf(x)^{r-1}f'(x),$$

we see that $\frac{1}{r}f(x)^r$ is a primitive for the function $x \mapsto f(x)^{r-1}f'(x)$.

In other words, we have

$$\int f(x)^{r-1} f'(x) dx = \frac{1}{r} f(x)^r + C, \tag{10.47}$$

an equality that can be easily remembered from (10.46), with f in place of g and $x \mapsto x^{r-1}$ in place of f: on the one hand, the FTC gives us

$$\int^{f(x)} t^{r-1} dt = \frac{t^r}{r} \Big|^{f(x)} + C = \frac{1}{r} f(x)^r + C;$$

on the other, the variable substitution $t = f(x)$ gives

$$\int^{f(x)} t^{r-1} dt = \int f(x)^{r-1} f'(x) dx.$$

We conclude this section by observing that, apart from the problems collected below, we shall see some other relevant examples of integration by trigonometric substitution in the coming sections.

Problems: Section 10.6

1. In each of the items below, compute the given integral:

   (a) $\int_0^1 \frac{1}{\sqrt{x+1}} dx$.

   (b) $\int_0^{1/2} \frac{x}{\sqrt{1-x^2}} dx$.

   (c) $\int_1^{\sqrt{2}} \frac{x}{1+x^4} dx$.

2. In each of the items below, compute the given primitives, together with their maximal domains of definition:

   (a) $\int x^2 \sqrt{1 + x^3} dx$.

   (b) $\int x^5 \sqrt{1 + x^3} dx$.

   (c) $\int x^3 \sqrt{1 - x^2} dx$.

   (d) $\int \cos \sqrt{x} \, dx$.

3. Let $f : [-a, a] \to \mathbb{R}$ be a continuous function. Show that:

   $$\int_{-a}^a f(x) dx = \begin{cases} 2 \int_0^a f(x) dx, & \text{if } f \text{ is even} \\ 0, & \text{if } f \text{ is odd} \end{cases}.$$

4. Compute $\int_{-1}^1 \frac{\sin(\pi x)}{1+x^2} dx$ and $\int_{-\frac{1}{2}}^{\frac{1}{2}} \frac{\cos(\pi x)}{1+x^3} dx$.

For the coming example the reader might find it useful to review the basics on ellipses, in Chap. 6 of [4], for instance.

5. Prove that the area of an ellipse of principal axes of lengths $2a$ and $2b$ equals πab.

6. Prove the following version of Theorem 10.37: if $f : [a, b] \to \mathbb{R}$ is piecewise continuous and $g : [c, d] \to [a, b]$ is increasing (resp. decreasing) and differentiable, with integrable derivative, then

$$\int_{g(c)}^{g(d)} f(t)dt = \int_{c}^{d} f(g(s))g'(s)ds.$$

7. If $f : [a, b] \to [a, b]$ is an increasing and differentiable bijection, compute all possible values of $\int_a^b (f(x) + f^{-1}(x))dx$.

For the next problem, the reader may find it convenient to review the concept of reciprocal polynomial, in Problem 28, page 44.

8. Let p be a polynomial of degree n. Show that p is reciprocal if and only if

$$\int_1^x \frac{p(t)}{t^{n/2+1}}dt = \int_{1/x}^1 \frac{p(t)}{t^{n/2+1}}dt,$$

for every real $x > 1$.

9. Let $f : [1, +\infty) \to \mathbb{R}$ be a differentiable function such that $f(1) = 1$ and $f'(x) = \frac{1}{x^2 + f(x)^2}$, for every $x \geq 1$. Prove that $\lim_{x \to +\infty} f(x)$ does exist and is at most $1 + \frac{\pi}{4}$.

10. Let $I \subset \mathbb{R}$ be an interval and $g : I \to I$ be a continuously differentiable function of *finite order*, i.e., such that $g^{(m)} = \text{Id}_I$ for some $m \in \mathbb{N}$, where $g^{(m)} = g \circ \ldots \circ g$, the composition of g with itself m times. If $x \in I$ and $f : I \to \mathbb{R}$ is a continuous function *semi-invariant* over g, in the sense that $(f \circ g)g' = f$, compute the possible values of $\int_x^{g(x)} f(t)dt$.

10.7   Logarithms and Exponentials

This section brings a fairly standard presentation of the main properties of two of the most important functions of Mathematics, namely, the natural logarithm and the exponential functions, postponing to the next section some interesting and important applications of them.

Definition 10.46 The **natural logarithm function** $\log : (0, +\infty) \to \mathbb{R}$ is defined by

$$\log x = \int_1^x \frac{1}{t}dt.$$

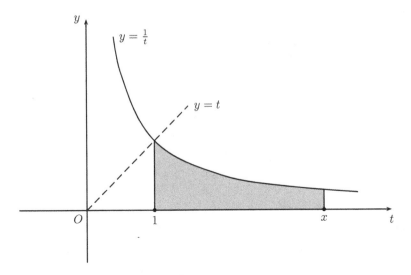

Fig. 10.5 Definition of $\log x$ for $x > 1$

It readily follows from the above definition that $\log 1 = 0$, $\log x > 0$ if $x > 1$ and $\log x < 0$ if $0 < x < 1$. Moreover, letting $f : (0, +\infty) \to \mathbb{R}$ denote the inverse proportionality function, so that $f(x) = \frac{1}{x}$ for every $x > 0$, then, in terms of the area of the region of the first quadrant situated under the graph of f (cf. Fig. 10.5), we have

$$\log x = \begin{cases} A(f_{|[1,x]}), & \text{if } x > 1 \\ 0, & \text{if } x = 1 \ . \\ -A(f_{|[x,1]}), & \text{if } x < 1 \end{cases}$$

The FTC assures that log is a differentiable function, with

$$\log' x = \frac{1}{x}.$$

Therefore, log is infinitely differentiable and, since $\log' > 0$, the study of the first variation of functions shows that log is increasing. Also, since

$$\log'' x = -\frac{1}{x^2} < 0,$$

it follows from Corollary 9.70 that log is strictly concave.

The coming result and their corollaries bring more important properties of the natural logarithm.

Proposition 10.47 *For $x, y > 0$, we have*

$$\log(xy) = \log x + \log y. \tag{10.48}$$

Proof For the first part, the additivity of the integral gives

$$\log xy = \int_1^{xy} \frac{1}{t} dt = \int_1^x \frac{1}{t} dt + \int_x^{xy} \frac{1}{t} dt = \log x + \int_x^{xy} \frac{1}{t} dt.$$

Now, if $g : [1,y] \to [x, xy]$ is given by $g(t) = xt$, the change of variables Theorem 10.37 furnishes

$$\int_x^{xy} \frac{1}{t} dt = \int_{g(1)}^{g(y)} \frac{1}{t} dt = \int_1^y \frac{1}{g(s)} \cdot g'(s) ds = \int_1^y \frac{1}{xs} \cdot x ds$$

$$= \int_1^y \frac{1}{s} \cdot ds = \log y.$$

$\square$

Corollary 10.48 *For $x > 0$ and $r \in \mathbb{Q}$, we have $\log x^r = r \log x$.*

Proof Applying (10.48) with $y = x$ we get $\log x^2 = 2 \log x$. If we assume that $\log x^k = k \log x$ for every $x > 0$ and some $k \in \mathbb{N}$, then letting $y = x^k$ in (10.48) we get

$$\log x^{k+1} = \log(x \cdot x^k) = \log x + \log x^k$$

$$= \log x + k \log x = (k + 1) \log x.$$

Therefore, it follows by induction that $\log x^n = n \log x$ for every $x > 0$ and every $n \in \mathbb{N}$.

We now apply this particular case with $x^{1/n}$ in place of x to obtain

$$\log x = \log(x^{1/n})^n = n \log x^{1/n}$$

or, which is the same, $\log x^{1/n} = \frac{1}{n} \log x$.

In view of the particular cases of the two previous paragraphs and letting $r = \frac{m}{n}$, with $m, n \in \mathbb{N}$, we have

$$\log x^r = \log x^{m/n} = \log(x^{1/n})^m = m \log x^{1/n} = m \cdot \frac{1}{n} \log x = r \log x.$$

In order to extend this last equality for negative rational exponents, start by letting $y = x^{-1}$ in (10.48) to get

$$0 = \log 1 = \log(x \cdot x^{-1}) = \log x + \log x^{-1},$$

so that $\log x^{-1} = -\log x$. Therefore, if r is a negative rational and $s = -r > 0$, we get

$$\log x^r = \log(x^s)^{-1} = -\log x^s = -(s \cdot \log x) = r \log x.$$

$\square$

Corollary 10.49 $\lim_{x \to 0+} \log x = -\infty$ and $\lim_{x \to +\infty} \log x = +\infty$. *In particular,* $\mathrm{Im}\,(\log) = \mathbb{R}$.

Proof Fix $a > 1$. Since $\log a > 0$, it follows from the previous corollary that

$$\log a^n = n \log a \xrightarrow{n} +\infty \quad \text{and} \quad \log a^{-n} = -n \log a \xrightarrow{n} -\infty.$$

On the other hand, since $0 < \frac{1}{a} < 1$, Example 7.12 furnishes

$$\lim_{n \to +\infty} a^{-n} = \lim_{n \to +\infty} (1/a)^n = 0 \quad \text{and} \quad \lim_{n \to +\infty} a^n = \lim_{n \to +\infty} \frac{1}{a^{-n}} = +\infty.$$

The limits above, together with the fact that log is an increasing function, guarantee that $\lim_{x \to 0+} \log x = -\infty$ and $\lim_{x \to +\infty} \log x = +\infty$. Hence, it follows from Problem 7, page 290, that $\mathrm{Im}\,(f) = \mathbb{R}$. □

We now relate the number e, defined in Example 7.41, to the natural logarithm function.

Theorem 10.50 $\log e = 1$.

Proof First of all, note (cf. Fig. 10.6, where we sketched the portion of the graph of $t \mapsto \frac{1}{t}$ from $t = 1$ to $t = 1 + \frac{1}{n}$) that

$$\log \left(1 + \frac{1}{n} \right) = \int_1^{1+\frac{1}{n}} \frac{1}{t}\,dt < \int_1^{1+\frac{1}{n}} dt = t \Big|_{t=1}^{t=1+\frac{1}{n}} = \left(1 + \frac{1}{n} \right) - 1 = \frac{1}{n}$$

and, analogously

$$\log \left(1 + \frac{1}{n} \right) = \int_1^{1+\frac{1}{n}} \frac{1}{t}\,dt > \int_1^{1+\frac{1}{n}} \frac{1}{1 + \frac{1}{n}}\,dt = \frac{n}{n+1} \cdot \frac{1}{n} = \frac{1}{n+1}.$$

Hence,

$$\frac{1}{n+1} < \log \left(1 + \frac{1}{n} \right) < \frac{1}{n}, \tag{10.49}$$

Fig. 10.6 Estimating $\log \left(1 + \frac{1}{n} \right)$

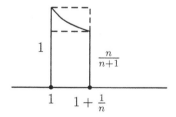

so that $\frac{n}{n+1} < n \log \left(1 + \frac{1}{n}\right) < 1$. The squeezing theorem (Problem 6, page 218) then gives $\lim_{n \to +\infty} n \log \left(1 + \frac{1}{n}\right) = 1$ or, which is the same,

$$\lim_{n \to +\infty} \log \left(1 + \frac{1}{n}\right)^n = 1.$$

Now, since (according to Theorem 7.42) $e = \lim_{n \to +\infty} \left(1 + \frac{1}{n}\right)^n$ and log is continuous, we have

$$\log e = \log \lim_{n \to +\infty} \left(1 + \frac{1}{n}\right)^n = \lim_{n \to +\infty} \log \left(1 + \frac{1}{n}\right)^n = 1.$$

$\square$

Gathering together the information obtained so far on the natural logarithm, we sketch its graph on Fig. 10.7.

The sketch above of the graph of the logarithm function suggests that it grows relatively slowly. The next result quantifies this suspicion.

Theorem 10.51 *For an integer $n > 1$, we have $\lim_{x \to +\infty} \frac{\log x}{\sqrt[n]{x}} = 0$.*

Proof Firstly, observe that $t > t^{1 - \frac{1}{2n}}$ for $t > 1$. Hence, $\frac{1}{t} < \frac{1}{t^{1 - \frac{1}{2n}}}$ for $t > 1$, so that, for $x > 1$,

$$\log x = \int_1^x \frac{1}{t} dt < \int_1^x \frac{1}{t^{1 - \frac{1}{2n}}} dt = \int_1^x t^{\frac{1}{2n} - 1} dt$$

$$= \frac{t^{\frac{1}{2n}}}{1/2n} \Big|_{t=1}^{t=x} = 2n(\sqrt[2n]{x} - 1) < 2n \sqrt[2n]{x}.$$

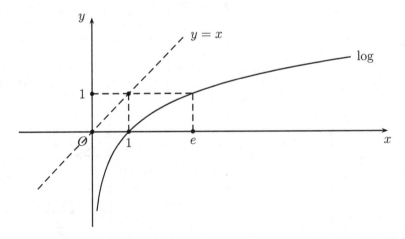

Fig. 10.7 Graph of log : $(0, +\infty) \to \mathbb{R}$

Therefore, for $x > 1$ we have

$$0 < \frac{\log x}{\sqrt[n]{x}} < 2n \cdot \frac{\sqrt[2n]{x}}{\sqrt[n]{x}} = \frac{2n}{\sqrt[2n]{x}}.$$

Now, since $\lim_{x \to +\infty} \frac{2n}{\sqrt[2n]{x}} = 0$, the squeezing theorem (cf. Problem 6, page 218) guarantees that $\frac{\log x}{\sqrt[n]{x}} \to 0$ when $x \to +\infty$. $\qquad\qquad\Box$

Since the natural logarithm function $\log : (0, +\infty) \to \mathbb{R}$ is a continuous and increasing bijection, we can consider its inverse

$$\exp : \mathbb{R} \to (0, +\infty), \qquad\qquad (10.50)$$

called the **exponential function**. Theorem 8.35 and Problem 8, page 176, guarantee that exp is also a continuous and increasing bijection. Moreover, for given $x, y \in \mathbb{R}$, with $x > 0$, we have

$$\log x = y \Leftrightarrow x = \exp(y);$$

in particular, it follows from $\log 1 = 0$ and $\log e = 1$ that $\exp(0) = 1$ and $\exp(1) = e$.

On the other hand, since $\log' x = \frac{1}{x} \neq 0$, Theorem 9.28 assures that exp is a differentiable function such that, for $x > 0$ and $y = \log x$,

$$\exp'(y) = \frac{1}{\log' x} = \frac{1}{1/x} = x = \exp(y).$$

In short,

$$\exp' = \exp,$$

which in particular shows that exp is infinitely differentiable. Also, since $\exp'' = \exp > 0$, it follows from Corollary 9.70 that exp is a strictly convex function.

The coming result translates, for the exponential function, the properties of the natural logarithm expressed in Proposition 10.47 and Corollary 10.48.

Proposition 10.52 *For $x, y \in \mathbb{R}$, we have:*

(a) $\exp(x + y) = \exp(x) \cdot \exp(y)$.
(b) $\exp(r) = e^r$, *if $r \in \mathbb{Q}$.*

Proof

(a) Letting $a = \exp(x)$ and $b = \exp(y)$, we have $a, b > 0$ and $\log a = x$, $\log b = y$. Hence,

$$x + y = \log a + \log b = \log(ab),$$

so that $\exp(x + y) = ab$. The above computations also give

$$\exp(x) \cdot \exp(y) = ab = \exp(x + y).$$

(b) Since log and exp are inverses of each other and $\log e = 1$, it follows from Proposition 10.47 that

$$\log \exp(r) = r = \log e^r.$$

Thus, the injectivity of log gives $\exp(r) = e^r$.

$\square$

Thanks to the item (b) of the former proposition, from now on we shall write

$$\exp(x) = e^x,$$

stressing that the right hand side *coincides with the usual meaning of* e^x when $x \in \mathbb{Q}$, and *defines* e^x when $x \notin \mathbb{Q}$. In this new notation, item (a) of the previous result is written as

$$e^{x+y} = e^x \cdot e^y,$$

for all $x, y \in \mathbb{R}$.

More generally, the exponential function allows us to rigorously define what is meant by the *power* a^x, for positive a (and real x). Indeed, in the notation just introduced, we set

$$a^x = e^{x \log a}, \tag{10.51}$$

so that

$$a^{x+y} = e^{(x+y) \log a} = e^{x \log a} \cdot e^{y \log a} = a^x \cdot a^y$$

and

$$\log a^x = \log(e^{x \log a}) = x \log a \tag{10.52}$$

for all $x, y \in \mathbb{R}$.

Notice that if $r \in \mathbb{Q}$, say $r = \frac{m}{n}$, with $m \in \mathbb{Z}$ and $n \in \mathbb{N}$, Corollary 10.48 gives

$$a^r = e^{\frac{m}{n} \cdot \log a} = e^{\log \sqrt[n]{a^m}} = \sqrt[n]{a^m};$$

in other words, for $x = r \in \mathbb{Q}$ definition (10.51) matches with the meaning we have attributed to a^r up to now. In particular, we have $a^0 = 1$ and $a^1 = a$.

If $a = 1$, it follows from (10.51) that $a^x = 1$ for every $x \in \mathbb{R}$. On the other hand, given a positive real $a \neq 1$, the **exponential function of basis a** is $f_a : \mathbb{R} \to (0, +\infty)$ such that $f_a(x) = a^x$.

Relation (10.51), together with the properties of the exponential function (of basis e) and the chain rule, guarantee that f_a is a differentiable bijection, with derivative

$$f_a'(x) = \exp^{x \log a} \cdot \log a = a^x \log a. \tag{10.53}$$

In particular, since $\log a > 0 \Leftrightarrow a > 1$, we get

$$f_a \text{ is increasing } \Leftrightarrow f_a' > 0 \Leftrightarrow a > 1;$$

accordingly, f_a is decreasing if and only if $0 < a < 1$.

The inverse of f_a is defined to be the function $\log_a : (0, +\infty) \to \mathbb{R}$, which is called the **base a logarithm function**, whose properties will be the object of Problem 1. Thus, the natural logarithm function is just the base e logarithm function.

An important variant of the above discussion is the following: given an interval I and functions $f : I \to (0, +\infty)$ and $g : I \to \mathbb{R}$, we define $f(x)^{g(x)}$ by letting

$$f(x)^{g(x)} = e^{g(x) \log f(x)}.$$

Observe that the requisite of $f(x)$ being positive gives sense to this definition. Also, the chain rule applied to the right hand side of the equality above assures that the function $x \in I \mapsto f(x)^{g(x)}$ is differentiable whenever f and g are so. In this respect, see Problem 9.

We can now state and prove a useful extension of Theorem 10.50.

Theorem 10.53 $e = \lim_{x \to +\infty} \left(1 + \frac{1}{x}\right)^x$.

Proof A reasoning analogous to that in the proof of Theorem 10.50 gives, for $x > 0$,

$$\frac{1}{x+1} < \log\left(1 + \frac{1}{x}\right) < \frac{1}{x}. \tag{10.54}$$

It thus follows from (10.52) that $\log\left(1 + \frac{1}{x}\right)^x = x \log(1 + \frac{1}{x})$.

Multiplying both inequalities above by $x > 0$, we get

$$\frac{x}{x+1} < \log\left(1 + \frac{1}{x}\right)^x < 1,$$

so that the squeezing theorem yields

$$\lim_{x \to +\infty} \log\left(1 + \frac{1}{x}\right)^x = 1.$$

Finally, since the exponential function is continuous, we can write

$$
e = \exp\left(\lim_{x \to +\infty} \log\left(1 + \frac{1}{x}\right)^x\right)
$$

$$
= \lim_{x \to +\infty} \exp\left(\log\left(1 + \frac{1}{x}\right)^x\right)
$$

$$
= \lim_{x \to +\infty} \left(1 + \frac{1}{x}\right)^x.
$$

$\square$

We now use Theorem 10.51 to show that the exponential function increases more rapidly than any polynomial when $x \to +\infty$.

Theorem 10.54 *If p is a polynomial, then $\lim_{x \to +\infty} \frac{p(x)}{e^x} = 0$.*

Proof Let $p(x) = a_m x^m + a_{m-1} x^{m-1} + \cdots + a_1 x + a_0$, with $a_m \neq 0$. Since

$$
\frac{p(x)}{e^x} = \sum_{k=0}^{m} a_k \cdot \frac{x^k}{e^x},
$$

it suffices to show that $\lim_{x \to +\infty} \frac{x^n}{e^x} = 0$ for every $n \in \mathbb{N}$. To this end, letting $y = e^x$ we have $x = \log y$ and, then,

$$
\frac{x^n}{e^x} = \frac{(\log y)^n}{y} = \left(\frac{\log y}{\sqrt[n]{y}}\right)^n.
$$

Finally, since $y \to +\infty$ when $x \to +\infty$, Theorem 10.51 gives

$$
\lim_{x \to +\infty} \frac{x^n}{e^x} = \lim_{y \to +\infty} \left(\frac{\log y}{\sqrt[n]{y}}\right)^n = \left(\lim_{y \to +\infty} \frac{\log y}{\sqrt[n]{y}}\right)^n = 0.
$$

$\square$

We close this section by deriving *asymptotic estimates* on the size of $\log n$ and $n!$. In other words, we estimate the sizes of $\log n$ and $n!$ for large values of n. Our first result is due to Euler.[4]

Theorem 10.55 (Euler)   *Given $n \in \mathbb{N}$, we have*

$$
\lim_{n \to +\infty} \left(\sum_{k=1}^{n} \frac{1}{k} - \log n\right) = \gamma, \tag{10.55}
$$

where γ is a positive real constant.

---

[4]The Swiss mathematician Leonhard Euler is one of the most prolific mathematicians of all times, even if we stick only to relevant Mathematics. His contributions impressively vary from Geometry to Combinatorics, passing through Number Theory, Calculus and Physics. In each of these areas of knowledge there is at least one, if not several, celebrated *Euler's theorem*.

Proof First of all, observe that

$$0 < \sum_{k=1}^{n} \frac{1}{k} - \log n = \sum_{k=1}^{n} \frac{1}{k} - \sum_{k=1}^{n-1} (\log(k+1) - \log k)$$

$$= \sum_{k=1}^{n-1} \left(\frac{1}{k} - \log \left(1 + \frac{1}{k} \right) \right) + \frac{1}{n}.$$

Now, since $\frac{1}{n} \to 0$ when $n \to +\infty$, it suffices to prove that the series

$$\sum_{k \geq 1} \left(\frac{1}{k} - \log \left(1 + \frac{1}{k} \right) \right) \tag{10.56}$$

converges to a positive sum γ.

Applying inequality (10.49) with k in place of n, we conclude that (10.56) is a series of positive terms, such that

$$\sum_{k=1}^{n-1} \left(\frac{1}{k} - \log \left(1 + \frac{1}{k} \right) \right) < \sum_{k=1}^{n-1} \left(\frac{1}{k} - \frac{1}{k+1} \right) = 1 - \frac{1}{n}.$$

Therefore, Proposition 7.44 guarantees that it does converge. □

The positive number γ defined by (10.55) is known in mathematical literature as the **Euler-Mascheroni constant**, and its value with five correct decimal places is $\gamma \cong 0,57721$. Up to this day it is unknown whether γ is rational or irrational, albeit there is a strong suspicion that it should be irrational.

Notice that (10.55) can be written as

$$\log n \cong \sum_{k=1}^{n} \frac{1}{k} - \gamma, \tag{10.57}$$

with error less than $\frac{1}{n}$ (cf. Problem 19).

In order to motivate what comes next, let $n \in \mathbb{N}$. The arithmetic-geometric mean inequality (cf. Theorem 5.7) gives

$$n! = 1 \cdot 2 \cdot \ldots \cdot n < \left(\frac{1+2+\cdots+n}{n} \right)^n = \left(\frac{n+1}{2} \right)^n;$$

thus, for sufficiently large n, we have

$$\frac{n!}{(n/2)^n} < \left(1 + \frac{1}{n} \right)^n \cong e.$$

On the other hand, since $n + 1 \leq 2(n-1) < 3(n-2) < \cdots$ for $n \geq 3$, we get

$$(n!)^2 = (1 \cdot n)(2 \cdot (n-1)) \ldots ((n-1) \cdot 2)(n \cdot 1)$$

$$\geq n^2(n+1)^{n-2} = \left(\frac{n}{n+1}\right)^2 (n+1)^n;$$

therefore, again for sufficiently large n,

$$\frac{n!}{n^{n/2}} \geq \frac{n}{n+1}\left(1 + \frac{1}{n}\right)^{n/2} \cong \sqrt{e}.$$

The coming theorem refines the above naive estimates, giving a sharp *asymptotic estimate* for $n!$. In order to properly state and prove it, it is now convenient to introduce some notations. Given sequences $(a_n)_{n\geq 1}$ and $(b_n)_{n\geq 1}$ of positive reals, we write $a_n \sim b_n$ and $a_n = o(b_n)$ as shorthands for the following possible asymptotic behaviors of $\frac{a_n}{b_n}$:

$$a_n \sim b_n \Leftrightarrow \lim_{n \to +\infty} \frac{a_n}{b_n} = 1 \quad \text{and} \quad a_n = o(b_n) \Leftrightarrow \lim_{n \to +\infty} \frac{a_n}{b_n} = 0.$$

In particular, $a_n = o(1)$ if $\lim_{n \to +\infty} a_n = 0$. Also, if $\alpha > 0$ and $\alpha a_n \sim b_n$, then

$$\lim_{n \to +\infty} \frac{b_n - \alpha a_n}{a_n} = 0,$$

so that $\frac{b_n - \alpha a_n}{a_n} = o(1)$; consequently,

$$b_n = a_n + (b_n - a_n) = \left(\alpha + \frac{b_n - \alpha a_n}{a_n}\right) a_n = (\alpha + o(1))a_n. \qquad (10.58)$$

We are finally in position to prove the following result, due to the Scottish mathematician of the XVIII century James Stirling and known in mathematical literature as **Stirling's formula**.

Theorem 10.56 (Stirling) $n! \sim \sqrt{2\pi n}\left(\frac{n}{e}\right)^n$.

Proof We have to show that $\lim_{n \to +\infty}\left(\frac{n^n e^{-n}\sqrt{n}}{n!}\right) = \frac{1}{\sqrt{2\pi}}$ or, which is the same, that

$$\lim_{n \to +\infty} \log\left(\frac{n^n e^{-n}\sqrt{2\pi n}}{n!}\right) = \log \frac{1}{\sqrt{2\pi}}.$$

Firstly, we shall show that the limit at the left hand side above actually exists. To this end, note that

$$\log\left(\frac{n^n e^{-n}\sqrt{n}}{n!}\right) = n \log n - n - \log \frac{n!}{\sqrt{n}} = (n \log n - n) - \frac{1}{2} \log \frac{(n!)^2}{n},$$

with

$$n \log n - n = \sum_{k=1}^{n-1} [((k+1)\log(k+1) - (k+1)) - (k \log k - k)]$$

$$= \sum_{k=1}^{n-1} \int_k^{k+1} \log x \, dx - 1$$

(we used the result of Problem 4 in the last equality) and

$$\frac{1}{2} \log \frac{(n!)^2}{n} = \frac{1}{2} (\log(1 \cdot 2) + \log(2 \cdot 3) + \cdots + \log((n-1)n))$$

$$= \sum_{k=1}^{n-1} \frac{\log k + \log(k+1)}{2}.$$

Hence,

$$\log \left(\frac{n^n e^{-n} \sqrt{n}}{n!} \right) = \sum_{k=1}^{n-1} \left(\int_k^{k+1} \log x \, dx - \frac{\log k + \log(k+1)}{2} \right) - 1. \qquad (10.59)$$

For $k \geq 1$, let $a_k = \int_k^{k+1} \log x \, dx - \frac{\log k + \log(k+1)}{2}$. Problem 13 (with $a = k$, $b = k+1$ and $f = \log$, which is concave) furnishes

$$a_k = \int_k^{k+1} \log x \, dx - \frac{\log k + \log(k+1)}{2} > 0.$$

On the other hand, by successively applying the results of Problems 4 and 20, we get

$$\log \left(k + \frac{1}{2} \right) - \int_k^{k+1} \log x \, dx = \log \left(k + \frac{1}{2} \right) - (k+1) \log(k+1) + k \log k + 1 > 0.$$

Therefore,

$$a_k < \log \left(k + \frac{1}{2} \right) - \frac{\log k + \log(k+1)}{2}$$

$$= \frac{1}{2} \log \frac{(k+\frac{1}{2})^2}{k(k+1)} = \frac{1}{2} \log \left(1 + \frac{1}{4k(k+1)} \right)$$

$$< \frac{1}{2} \cdot \frac{1}{4k(k+1)} = \frac{1}{8} \left(\frac{1}{k} - \frac{1}{k+1} \right),$$

where we used the result of Problem 5 in the last inequality above.

We conclude that $\sum_{k\geq 1} a_k$ is a series of positive terms, such that

$$\sum_{k=1}^{n-1} a_k < \sum_{k=1}^{n-1} \frac{1}{8} \left(\frac{1}{k} - \frac{1}{k+1} \right) = \frac{1}{8}.$$

Thus, the comparison test assures that $\sum_{k\geq 1} a_k$ does converge, and it follows from (10.59) that there exists

$$\lim_{n\to+\infty} \log \left(\frac{n^n e^{-n} \sqrt{n}}{n!} \right) = \lim_{n\to+\infty} \sum_{k=1}^{n-1} a_k - 1 = \sum_{k\geq 1} a_k - 1.$$

For the actual computation of the desired limit, let $c > 0$ be such that $\sum_{k\geq 1} a_k - 1 = \log c$. Since

$$\lim_{n\to+\infty} \log \left(\frac{n^n e^{-n} \sqrt{n}}{n!} \right) = \log c,$$

we have

$$c = e^{\log c} = \exp \left(\lim_{n\to+\infty} \log \left(\frac{n^n e^{-n} \sqrt{n}}{n!} \right) \right)$$

$$= \lim_{n\to+\infty} \exp \log \left(\frac{n^n e^{-n} \sqrt{n}}{n!} \right)$$

$$= \lim_{n\to+\infty} \left(\frac{n^n e^{-n} \sqrt{n}}{n!} \right).$$

It follows from (10.58) (with $a_n = n^n e^{-n} \sqrt{n}$, $b_n = n!$ and $\alpha = \frac{1}{c}$) that

$$n! = \left(\frac{1}{c} + o(1) \right) n^n e^{-n} \sqrt{n}.$$

Now recall that, according to Example 10.35, if $I_n = \int_0^{\pi/2} (\cos x)^n dx$ for $n \in \mathbb{N}$, then

$$I_{2k} = \frac{(2k)!}{(2^k k!)^2} \cdot \frac{\pi}{2} \quad \text{and} \quad I_{2k-1} = \frac{(2^k k!)^2}{2k(2k)!}$$

for every integer $k \geq 1$. Hence,

$$\frac{I_{2k}}{I_{2k-1}} = \frac{((2k)!)^2 2k}{2^{4k}(k!)^4} \cdot \frac{\pi}{2}$$

$$= \frac{\left[\left(\frac{1}{c} + o(1)\right)(2k)^{2k} e^{-2k} \sqrt{2k}\right]^2 2k}{2^{4k}\left[\left(\frac{1}{c} + o(1)\right) k^k e^{-k} \sqrt{k}\right]^4} \cdot \frac{\pi}{2}$$

$$= \frac{2\pi}{\left(\frac{1}{c} + o(1)\right)^2} \xrightarrow{k} 2\pi c^2.$$

However, Problem 13, page 391 showed that $\frac{I_{2k}}{I_{2k-1}} \xrightarrow{k} 1$. Therefore, the above computations give $2\pi c^2 = 1$ or, which is the same, $c = \frac{1}{\sqrt{2\pi}}$. □

Example 10.57 For $n \in \mathbb{N}$, the binomial number $\binom{2n}{n}$ is the largest of the $2n + 1$ coefficients of the binomial expansion of $(x + y)^{2n}$. This gives

$$(2n + 1)\binom{2n}{n} > \sum_{k=0}^{2n}\binom{2n}{k} = 2^{2n},$$

so that $\binom{2n}{n} > \frac{2^{2n}}{2n+1}$. Stirling's formula considerably improves this estimate, for, with the aid of it, we get

$$\binom{2n}{n} = \frac{(2n)!}{(n!)^2} \sim \frac{\sqrt{2\pi \cdot 2n}\left(\frac{2n}{e}\right)^{2n}}{\left(\sqrt{2\pi n}\right)^2\left(\frac{n}{e}\right)^{2n}} = \frac{2^{2n}}{\sqrt{\pi n}}.$$

Problems: Section 10.7

1. * Given $0 < a \neq 1$, prove that $\log_a$ is an infinitely differentiable bijection, with

$$\log_a x = \frac{\log x}{\log a} \quad \text{and} \quad \log'_a x = \frac{1}{x \log a}, \tag{10.60}$$

for all $0 < a, x \neq 1$. Moreover, given positive reals a, b, c, x and y, with $a, b, c \neq 1$, prove also that:

(a) $\log = \log_e$.
(b) $\log_a$ is increasing if $a > 1$ and decreasing if $0 < a < 1$.
(c) $\log_a(xy) = \log_a x + \log_a y$.
(d) $\log_a c = \log_b c \cdot \log_a b$.
(e) $\log_a b \cdot \log_b c \cdot \log_c a = 1$.

2. Which of e^π or π^e is bigger? Justify your answer!
3. Find all $a, b \in \mathbb{N}$ such that $a \neq b$ and $a^b = b^a$. Justify your answer!
4. * Compute $\int \log x \, dx = x \log x - x + C$.
5. * Show that $\log(x + 1) < x$ for every $x > 0$.
6. Use Corollary 10.48, together with the result of Problem 8, page 338, to give another proof of the fact that log is strictly concave.
7. Find all negative values of a such that $f : \mathbb{R} \to \mathbb{R}$, given by $f(x) = e^x + ax^3$, has a single inflection point.
8. * Let $\alpha \neq 0$ be a real number and $f : (0, +\infty) \to \mathbb{R}$ be the function given by $f(x) = x^\alpha$. Show that $f'(x) = \alpha x^{\alpha-1}$ for every $x > 0$.
9. Given an interval I and differentiable functions $f : I \to (0, +\infty)$ and $g : I \to \mathbb{R}$, compute the derivative of the function $h : I \to \mathbb{R}$ defined by $h(x) = f(x)^{g(x)}$, for every $x \in I$.
10. Given $a, b > 0$, let $f : [0, 1] \to \mathbb{R}$ be defined by $f(x) = x^a(1 - x)^b$. Show that f attains a maximum value and compute it.
11. The set of rationals is not closed with respect to powers; for instance, if $a = b = \frac{1}{2}$, then $a, b \in \mathbb{Q}$, albeit $a^b = \frac{1}{\sqrt{2}}$, an irrational number. If a and b are positive irrationals, is it always true that a^b is irrational? Justify your answer.
12. If a is a positive real, prove that $\lim_{x\to+\infty} \left(1 + \frac{a}{x}\right)^x = e^a$.
13. Find all positive values of a for which $\lim_{x\to+\infty} \left(\frac{x+a}{x-a}\right)^x = e$.
14. Let $f : \mathbb{R} \to \mathbb{R}$ be a differentiable function such that $f(0) = 0$ and $f'(x) > f(x)$ for every $x \in \mathbb{R}$. Show that $f(x) > 0$ for every $x > 0$.
15. * The purpose of this problem is to compute primitives for the function $t \mapsto \sec^3 t$, where sec denotes the restriction of the secant function to an interval of the form $\left(-\frac{\pi}{2} + k\pi, \frac{\pi}{2} + k\pi\right)$, with $k \in \mathbb{Z}$. To this end, do the following items:

(a) Show that $\sec' t = \tan t \cdot \sec t$ and $\frac{d}{dt} \log|\sec t - \tan t| = -\sec t$.
(b) Conclude that $\int \sec t \, dt = -\log|\sec t - \tan t| + C$.
(c) Write $\sec^2 t = 1 + \tan^2 t$ and integrate $\int \sec^3 t \, dt$ by parts to get

$$\int \sec^3 t \, dt = -\log|\sec t - \tan t| + \sec t \cdot \tan t - \int \sec^3 t \, dt$$

and, hence,

$$\int \sec^3 t \, dt = \frac{1}{2}(\sec t \cdot \tan t - \log|\sec t - \tan t|) + C.$$

16. * The **hyperbolic sine** and **hyperbolic cosine**, respectively denoted sinh, cosh : $\mathbb{R} \to \mathbb{R}$, are defined by

$$\sinh x = \frac{e^x - e^{-x}}{2} \quad \text{and} \quad \cosh x = \frac{e^x + e^{-x}}{2}.$$

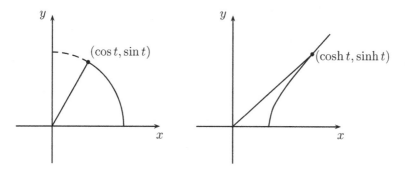

Fig. 10.8 Circular and hyperbolic sines and cosines

With respect to them, do the following items:

(a) Show that $\cosh^2 x - \sinh^2 x = 1$, $\sinh' x = \cosh x$ and $\cosh' x = \sinh x$.

(b) Show that $\cosh$ is an even and strictly convex function with image $[1, +\infty)$, such that $\cosh x = 1 \Leftrightarrow x = 0$.

(c) Show that $\sinh$ is an odd strictly increasing function with image $\mathbb{R}$. Show also that $\sinh$ is strictly convex in $(0, +\infty)$ and strictly concave in $(-\infty, 0)$, and that $x = 0$ is its only inflection point.

(d) Sketch the graphs of the hyperbolic sine and hyperbolic cosine in a single cartesian system of coordinates; in this same cartesian system, sketch also the graphs of $x \mapsto \frac{1}{2}e^x$ $(x > 0)$ and $x \mapsto \frac{1}{2}e^{-x}$, $x \mapsto -\frac{1}{2}e^{-x}$ $(x < 0)$.

(e) The graph of the hyperbolic cosine is known as the **catenary**.[5] Given $x_0 \in \mathbb{R}$, show that the curvature of the catenary at x_0 (cf. Problem 3, page 344) equals $k(x_0) = \frac{1}{\cosh^2 x_0}$.

The next problem explains the names *hyperbolic sine* and *hyperbolic cosine*. To put it in context, recall (cf. Fig. 10.8, left) that the area of the circular sector of radius 1, contained in the first quadrant of the cartesian plane, centered at the origin and bounded by the radii that join the origin to the points $(1, 0)$ and $(\cos t, \sin t)$ equals $\frac{t}{2}$.

17. For all $t \in \mathbb{R}$, observe that the point $(\cosh t, \sinh t)$ belongs to the *right branch* of the hyperbola $x^2 - y^2 = 1$, as well as to the straight line $y = (\tanh t)x$, where $\tanh t = \frac{\sinh t}{\cosh t}$. For a fixed $t > 0$, let A be the area of the bounded portion of the cartesian plane bounded by the horizontal axis, the right branch of $x^2 - y^2 = 1$ and the straight line $y = (\tanh t)x$ (cf. Fig. 10.8, right). Show that $A = \frac{t}{2}$.

18. Given $\lambda > 0$, show that $\lim_{x \to 0+} x^\lambda \log x = 0$.

---

[5]For the reader's curiosity, the *surface of revolution* generated by the rotation of the catenary along the horizontal axis is an example of a *minimal surface*, in this case a *catenoid*. Minimal surfaces have very interesting properties, some of which can be found in the excellent book of professor M. P. do Carmo [8].

19. * With respect to (10.57), show that

$$0 < \log n - \Big(\sum_{k=1}^{n} \frac{1}{k} - \gamma \Big) < \frac{1}{n}.$$

20. * If $f : [1, +\infty) \to \mathbb{R}$ is given by

$$f(x) = \log(x + 1/2) - (x + 1) \log(x + 1) + x \log x + 1,$$

   prove that f is positive in $[1, +\infty)$.
21. Show that $\lim_{n \to +\infty} \frac{n}{\sqrt[n]{n!}} = e$.
22. Let $0 < \alpha < 1$ be given and $k, n \in \mathbb{N}$ be such that $k > \alpha n$. Prove that $\binom{kn}{n} \sim \frac{((k+1)e)^n}{\sqrt{2\pi n}}$.

10.8 Miscellaneous

This section gathers some interesting applications of the natural logarithm and exponential function. We start by examining three examples which show that the fact that we now have logarithms and exponentials at our disposal gives greater flexibility to the material discussed so far.

Example 10.58 (Thailand) Find all functions $f : \mathbb{N} \to \mathbb{R}$ satisfying the following conditions:

(a) f is increasing.
(b) $f(1) = 1$ and $f(2) = 4$.
(c) $f(xy) = f(x)f(y)$ for all $x, y \in \mathbb{N}$.

Solution An easy induction gives $f(2^k) = 2^{2k}$, for every $k \in \mathbb{N}$.

Take an integer $a > 2$ and suppose, for the sake of contradiction, that $f(a) \le a^2 - 1$. Since $\mathbb{Q}$ is dense in $\mathbb{R}$, we can choose $m, n \in \mathbb{N}$ such that

$$\frac{\log 2^2}{\log a^2} < \frac{m}{n} < \frac{\log 2^2}{\log(a^2 - 1)},$$

so that $2^n < a^m$ and $(a^2 - 1)^m < 2^{2n}$. Hence, successive applications of (a) and (c) give us

$$f(2^n) = 2^{2n} > (a^2 - 1)^m \ge f(a)^m = f(a^m) > f(2^n),$$

which is a contradiction.

A similar reasoning shows that $f(a) \ge a^2 + 1$ also leads to a contradiction, so that the only left possibility is $f(a) = a^2$.   □

The next example shows how to use logarithms to guarantee the existence of powers of a fixed (and almost arbitrary) given basis beginning (from the left) by a prescribed set of algarisms. At this point, the reader might find it helpful to review the material of Sect. 7.3, specially Corollary 7.33, as well as the statement of Problem 1, page 411.

Example 10.59 Let $a > 1$ be an integer which is not a power of 10, and s be a finite sequence of algarisms, the first of which is nonzero. Prove that there exists a power of a whose decimal representation starts, from the left, with the sequence s of algarisms.

Proof We want to show that there exist $m, n \in \mathbb{N}$ for which

$$s \cdot 10^m \le a^n \le (s+1) \cdot 10^m.$$

Taking logarithms in base 10, this is the same as showing that there exist naturals m and n for which $\log_{10} s + m \le n \log_{10} a \le \log_{10}(s+1) + m$ or, which is the same,

$$\log_{10} s \le -m + n \log_{10} a \le \log_{10}(s+1).$$

If we show that $\log_{10} a$ is irrational, the proof will be finished thanks to Corollary 7.33 of Kronecker's lemma.

Suppose, by contradiction, that $\log_{10} a \in \mathbb{Q}$, say $\log_{10} a = \frac{p}{q}$, with $p, q \in \mathbb{N}$ relatively prime. Then $a^q = 10^p$, and the Fundamental Theorem of Arithmetic (cf. introduction to Chap. 1), allow us to write $a = 2^k 5^l b$, for some nonnegative integers k, l and some natural b relatively prime with 10. But these two facts give the equality $2^{kq} 5^{lq} b^q = 2^p 5^p$, from which we deduce that $b = 1$, $kq = p$ and $lq = p$. Thus, we conclude that $k = l$, which would give us $a = 2^k 5^k = 10^k$, a contradiction. □

We now refine the kind of argument of the previous example to establish the best possible converse of Problem 16, page 11.

Example 10.60 Prove that there exist infinitely many naturals n such that the decimal representations of 2^n and 5^n start (at left) with a 3.

Proof If 2^n and 5^n are to begin with a 3, it is necessary and sufficient that there exist naturals k and l such that $3 \cdot 10^k < 2^n < 4 \cdot 10^k$ and $3 \cdot 10^l < 5^n < 4 \cdot 10^l$ or, taking logarithms in base 10, that

$$\log_{10} 3 < -k + n \log_{10} 2 < \log_{10} 4 \quad \text{and} \quad \log_{10} 3 < -l + n \log_{10} 5 < \log_{10} 4.$$

In turn, this is equivalent to asking that the fractional parts[6] of both $n \log_{10} 2$ and $n \log_{10} 5$ belong to the interval $(\log 3, \log 4)$.

---

[6] Recall that the fractional part of $x \in \mathbb{R}$, denoted $\{x\}$, is defined as $\{x\} = x - \lfloor x \rfloor$, and belongs to the interval $[0, 1)$.

Since $n \log_{10} 2 + n \log_{10} 5 = n$ and $\{n\} = 0$, such fractional parts are at the same distance from $\frac{1}{2}$. Finally, since $\frac{1}{2} \in (\log_{10} 3, \log_{10} 4)$, we conclude that it suffices to show that there exist infinitely many naturals k and n such that

$$\log_{10} 3 < -k + n \log_{10} 2 < \log_{10} 4.$$

However, as we have shown in the solution to the previous example, $\log_{10} 2$ is irrational, so the existence of such k and n follows again from Corollary 7.33 of Kronecker's lemma.   □

We now turn to inequalities. Up to this moment, we have given or hinted at three different proofs of the inequality between the arithmetic and geometric means (one in Theorem 5.7, another in Problem 13, page 140, and a third one in Example 9.59). Jensen's inequality applied to the natural logarithm function provides a fourth one, which is particularly straightforward. However, before we present it, it is worth to point out the following generalization of (10.48), which can be easily established by induction: for $x_1, x_2, \ldots, x_n > 0$, we have

$$\log(x_1 x_2 \ldots x_n) = \log x_1 + \log x_2 + \cdots + \log x_n. \tag{10.61}$$

Example 10.61 Let $n \geq 2$ be an integer and $x_1, x_2, \ldots, x_n$ be positive reals. Since $\log : (0, +\infty) \to \mathbb{R}$ is strictly concave, the version of Corollary 9.75 for strictly concave functions gives us

$$\frac{\log x_1 + \log x_2 + \cdots + \log x_n}{n} \leq \log \left(\frac{x_1 + x_2 + \cdots + x_n}{n} \right).$$

If we successively apply (10.61) and Corollary 10.48 (with $x = x_1 x_2 \ldots x_n$ and $r = \frac{1}{n}$) to the left hand side, we get

$$\log \sqrt[n]{x_1 x_2 \ldots x_n} \leq \log \left(\frac{x_1 + x_2 + \cdots + x_n}{n} \right), \tag{10.62}$$

with equality if and only if $x_1 = x_2 = \cdots = x_n$. But since $\log$ is an increasing function, it follows from here that

$$\sqrt[n]{x_1 x_2 \ldots x_n} \leq \frac{x_1 + x_2 + \cdots + x_n}{n},$$

with equality if and only if $x_1 = x_2 = \cdots = x_n$.

Essentially the same idea as that of the proof of the previous example allows one to obtain the important *Young' inequality*.[7]

---

[7] After William H. Young, English mathematician of the XIX and XX centuries.

Proposition 10.62 (Young) *Let p and q be positive reals such that $\frac{1}{p} + \frac{1}{q} = 1$. Given $a, b > 0$, we have*

$$ab \leq \frac{a^p}{p} + \frac{b^q}{q}, \tag{10.63}$$

with equality if and only if $a^p = b^q$.

Proof Given positive reals x_1, x_2, t_1 and $t_2 > 0$ such that $t_1 + t_2 = 1$, strictly concave character of the natural logarithm function in $(0, +\infty)$, together with Jensen's inequality, gives

$$\log(t_1 x_1 + t_2 x_2) \geq t_1 \log x_1 + t_2 \log x_2,$$

with equality if and only if $x_1 = x_2$.

Letting $t_1 = \frac{1}{p}, t_2 = \frac{1}{q}, x_1 = a^p$ and $x_2 = b^q$, we get

$$\log\left(\frac{a^p}{p} + \frac{b^q}{q}\right) \geq \frac{1}{p} \log a^p + \frac{1}{q} \log b^q = \log ab,$$

with equality if and only if $a^p = b^q$.

Finally, since log is strictly increasing, it follows from the above inequality that

$$\frac{a^p}{p} + \frac{b^q}{q} \geq ab,$$

with equality if and only if $a^p = b^q$. □

The inequality of the coming result is a consequence of Young's inequality and is known as **Hölder's inequality**.[8] Note that, when $p = q = 2$, such an inequality reduces to Cauchy's inequality (cf. Theorem 5.13). Nevertheless, it is only after defining a^x for reals $a, x > 0$ that we can make sense of it.

Theorem 10.63 (Hölder) *Let $a_1, a_2, \ldots, a_n$ and $b_1, b_2, \ldots, b_n$ be given positive reals, and $p, q > 0$ be such that $\frac{1}{p} + \frac{1}{q} = 1$. Then,*

$$\sum_{i=1}^{n} a_i b_i \leq \left(\sum_{i=1}^{n} a_i^p\right)^{1/p} \left(\sum_{i=1}^{n} b_i^q\right)^{1/q},$$

with equality if and only if

$$\frac{a_1^p}{b_1^q} = \frac{a_2^p}{b_2^q} = \ldots = \frac{a_n^p}{b_n^q}.$$

---

[8] After Otto Hölder, German mathematician of the XIX and XX centuries.

Proof Letting $A = \left(\sum_{i=1}^{n} a_i^p \right)^{1/p}$ and $B = \left(\sum_{i=1}^{n} b_i^q \right)^{1/q}$, we have

$$\sum_{i=1}^{n} a_i b_i \le AB \Leftrightarrow \sum_{i=1}^{n} \frac{a_i}{A} \cdot \frac{b_i}{B} \le 1.$$

Now, letting $x_i = \frac{a_i}{A}$ and $y_i = \frac{b_i}{B}$, we have

$$\sum_{i=1}^{n} x_i^p = \frac{1}{A^p} \sum_{i=1}^{n} a_i^p = 1, \quad \sum_{i=1}^{n} y_i^q = \frac{1}{B^q} \sum_{i=1}^{n} b_i^q = 1$$

and want to prove that $\sum_{i=1}^{n} x_i y_i \le 1$. But this is straightforward from Young's inequality:

$$\sum_{i=1}^{n} x_i y_i \le \sum_{i=1}^{n} \left(\frac{x_i^p}{p} + \frac{y_i^q}{q} \right) = \frac{1}{p} \sum_{i=1}^{n} x_i^p + \frac{1}{q} \sum_{i=1}^{n} y_i^q = 1.$$

To have equality, we must have $x_i^p = y_i^q$ for $1 \le i \le n$ or, which is the same, $\frac{a_i^p}{A^p} = \frac{b_i^q}{B^q}$ for $1 \le i \le n$. Yet in another way, we must have

$$\frac{a_1^p}{b_1^q} = \frac{a_2^p}{b_2^q} = \dots = \frac{a_n^p}{b_n^q} = \frac{A^p}{B^q}.$$

Conversely, it's immediate to verify that if the above condition is satisfied then we do have equality. $\square$

A version of Hölder's inequality for integrals will be the object of Problem 8 of the next section.

We finish this section by discussing two more elaborate applications of Jensen's inequality, the first of which appeared as a problem at the 2001 IMO. We divide it into two examples.

Example 10.64 Sketch the graph of the function $f : \mathbb{R} \to \mathbb{R}$ given by

$$f(x) = \frac{1}{\sqrt{1 + 8e^x}}.$$

Solution Firstly, note that $0 < f(x) < 1$ for every $x \in \mathbb{R}$. Moreover, since $\lim_{x \to +\infty} e^x = +\infty$ and $\lim_{x \to -\infty} e^x = 0$, we have

$$\lim_{x \to +\infty} f(x) = 0 \quad \text{and} \quad \lim_{x \to -\infty} f(x) = 1.$$

It follows that the graph of f is entirely contained in the strip of the cartesian plane bounded by the horizontal lines $y = 0$ and $y = 1$, and that these lines are asymptotes to the graph. On the other hand, since $f(0) = \frac{1}{3}$, the graph intersects the vertical axis at the point $\left(0, \frac{1}{3}\right)$.

Now, an immediate computation with the aid of the chain rule gives

$$f'(x) = -\frac{4e^x}{(1 + 8e^x)^{3/2}} < 0,$$

for every $x \in \mathbb{R}$, so that f decreases along the whole real line. Also, computing f'' (again with the aid of the chain rule) we get

$$f''(x) = -\frac{4e^x}{(1 + 8e^x)^{5/2}}(1 - 4e^x).$$

Therefore, the sign of f'' coincides with that of $4e^x - 1$, so that f is strictly convex in $(-2\log 2, +\infty)$, strictly concave in $(-\infty, -2\log 2)$ and has $x_0 = -2\log 2$ as its only inflection point.

Finally, noticing that $-2\log 2 \cong -1.38, f(-2\log 2) = \frac{1}{\sqrt{3}} \cong 0.58$ and gathering together the above information, we get Fig. 10.9 as a reasonably accurate sketch for the graph of f.

□

We now come to the promised IMO problem.

Example 10.65 (IMO) Prove that, for every positive reals a, b and c, one has

$$\frac{a}{\sqrt{a^2 + bc}} + \frac{b}{\sqrt{b^2 + ac}} + \frac{c}{\sqrt{c^2 + ab}} \geq 1.$$

Proof Note, first of all, that the expression at the left hand side of the inequality equals

$$\frac{1}{\sqrt{1 + 8\frac{bc}{a^2}}} + \frac{1}{\sqrt{1 + 8\frac{ac}{b^2}}} + \frac{1}{\sqrt{1 + 8\frac{ab}{c^2}}}.$$

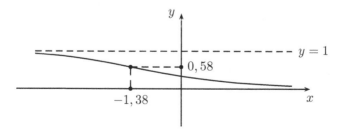

Fig. 10.9 Graph of the function $x \mapsto \frac{1}{\sqrt{1+8e^x}}$

Letting $\frac{bc}{a^2} = e^x$, $\frac{ac}{b^2} = e^y$ and $\frac{ab}{c^2} = e^z$ we have $x + y + z = 0$ and wish to prove that

$$f(x) + f(y) + f(z) \geq 1,$$

where $f : \mathbb{R} \to \mathbb{R}$ is the function of the previous example. To what is left to do, there are three possibilities:

(i) $x, y, z > -2 \log 2$: Jensen's inequality gives

$$f(x) + f(y) + f(z) \geq 3f\left(\frac{x+y+z}{3}\right) = 3f(0) = 1.$$

(ii) $x, y \leq -2 \log 2 < z$: then

$$f(x) + f(y) + f(z) \geq f(x) + f(y) \geq 2f(-2 \log 2) = \frac{2}{\sqrt{3}} > 1.$$

(iii) $x \leq -2 \log 2 < y, z$: again by Jensen's inequality, we get

$$f(x) + f(y) + f(z) \geq f(x) + 2f\left(\frac{y+z}{2}\right) = f(x) + 2f\left(-\frac{x}{2}\right).$$

If $g : (-\infty, -2 \log 2] \to \mathbb{R}$ is given by

$$g(x) = f(x) + 2f\left(-\frac{x}{2}\right),$$

observe that

$$\lim_{x \to -\infty} g(x) = \lim_{x \to -\infty} f(x) + 2 \lim_{x \to +\infty} f(x) = 1$$

and

$$g(-2 \log 2) = \frac{1}{\sqrt{3}} + \frac{2}{\sqrt{17}} > 1.$$

On the other hand,

$$g'(x) = f'(x) - f'\left(-\frac{x}{2}\right) = -\frac{4e^x}{(1+8e^x)^{3/2}} + \frac{4e^{-x/2}}{(1+8e^{-x/2})^{3/2}},$$

so that $g'(x) < 0$ if and only if $-2 \log 7 < x < -2 \log 2$. Hence,

$$x \in (-\infty, -2 \log 7] \Rightarrow g(x) > 1$$

and

$$x \in (-2\log 7, -2\log 2) \Rightarrow g(x) > g(-2\log 2) > 1.$$

□

The subsequent discussion derives an inequality which widely generalizes[9] that between the arithmetic and geometric means and is known as the **inequality of power means**. We start by defining what one means for such power means.

Definition 10.66 Given positive reals $a_1, a_2, \ldots, a_n$ and $\alpha \in \mathbb{R}$, we define the α-th **power mean** of $a_1, a_2, \ldots, a_n$ as the positive real number $M_\alpha = M_\alpha(a_1, \ldots, a_n)$, such that

$$M_\alpha = \begin{cases} \left(\dfrac{a_1^\alpha + a_2^\alpha + \cdots + a_n^\alpha}{n}\right)^{1/\alpha}, & \text{if } \alpha \neq 0 \\ \sqrt[n]{a_1 a_2 \ldots a_n}, & \text{if } \alpha = 0 \end{cases}.$$

Theorem 10.67 Let $a_1, a_2, \ldots, a_n$ be given positive reals. If $\alpha < \beta$ are any reals, then

$$\min_{1 \leq i \leq n} \{a_i\} \leq M_\alpha \leq M_\beta \leq \max_{1 \leq i \leq n} \{a_i\}, \tag{10.64}$$

with equality in any of the inequalities above if and only if all of the a_i's are equal.

Proof We can assume, without loss of generality, that not all of the a_i's are equal.

Let's momentarily assume that (10.64) is true (with strict inequalities) for all $0 < \alpha < \beta$. Given reals $x < y < 0$, we have $-x > -y > 0$; since $\frac{1}{a_1}, \ldots, \frac{1}{a_n}$ are also positive and not all equal, it follows from our assumptions that

$$\min_{1 \leq i \leq n} \left\{\frac{1}{a_i}\right\} < M_{-y}\left(\frac{1}{a_1}, \ldots, \frac{1}{a_n}\right) < M_{-x}\left(\frac{1}{a_1}, \ldots, \frac{1}{a_n}\right) < \max_{1 \leq i \leq n} \left\{\frac{1}{a_i}\right\}.$$

Now, an easy computation gives

$$M_{-t}\left(\frac{1}{a_1}, \ldots, \frac{1}{a_n}\right) = M_t(a_1, \ldots, a_n)^{-1}$$

for every $t \in \mathbb{R}$, so that the above inequalities are equivalent to

$$\left(\max_{1 \leq i \leq n} \{a_i\}\right)^{-1} < M_y(a_1, \ldots, a_n)^{-1} < M_x(a_1, \ldots, a_n)^{-1} < \left(\min_{1 \leq i \leq n} \{a_i\}\right)^{-1}.$$

---

[9]Another such remarkable generalization, due to Newton and McLaurin, is the object of Sect. 17.2 of [5].

or, which is the same,

$$\min_{1\leq i\leq n}\{a_i\} < M_x(a_1,\ldots,a_n) < M_y(a_1,\ldots,a_n) < \max_{1\leq i\leq n}\{a_i\}.$$

Hence, we conclude that it suffices to analyse the cases $0 < \alpha < \beta$ and $0 = \alpha < \beta$.

Also without loss of generality, assume that $a_1 = \max_{1\leq i\leq n}\{a_i\}$. Since not all of the a_i's are equal, we get

$$M_\beta(a_1,\ldots,a_n) = \left(\frac{a_1^\beta + a_2^\beta + \cdots + a_n^\beta}{n}\right)^{1/\beta} < \left(\frac{na_1^\beta}{n}\right)^{1/\beta} = a_1 = \max_{1\leq i\leq n}\{a_i\}$$

and, analogously, $\min_{1\leq i\leq n}\{a_i\} < M_\alpha$. Thus, we are left to showing that

$$0 < \alpha < \beta \Rightarrow M_0 < M_\alpha < M_\beta.$$

The first inequality above is equivalent to

$$\frac{a_1^\alpha + a_2^\alpha + \cdots + a_n^\alpha}{n} > \sqrt[n]{a_1^\alpha a_2^\alpha \cdots a_n^\alpha},$$

which, in turn, follows immediately from the inequality between the arithmetic and geometric means.

In order to prove the second inequality, letting $M_\beta = K$ it suffices to show that

$$\frac{1}{K}\left(\frac{a_1^\alpha + a_2^\alpha + \cdots + a_n^\alpha}{n}\right)^{1/\alpha} < 1$$

or, which is the same,

$$\left(\frac{(a_1/K)^\alpha + (a_2/K)^\alpha + \cdots + (a_n/K)^\alpha}{n}\right)^{1/\alpha} < 1.$$

Letting $b_i = (a_i/K)^\beta$, we conclude that not all of the b_i's are equal, but satisfy

$$\frac{b_1 + b_2 + \cdots + b_n}{n} = \frac{1}{K^\beta}\left(\frac{a_1^\beta + a_2^\beta + \cdots + a_n^\beta}{n}\right) = 1;$$

moreover, since $(a_i/K)^\alpha = b_i^{\alpha/\beta}$, we want to show that

$$\left(\frac{b_1^{\alpha/\beta} + b_2^{\alpha/\beta} + \cdots + b_n^{\alpha/\beta}}{n}\right)^{1/\alpha} < 1. \tag{10.65}$$

To what is left to do, let $f : (0, +\infty) \to \mathbb{R}$ be given by $f(x) = x^{\alpha/\beta}$. Since $0 < \alpha/\beta < 1$, two applications of the result of Problem 8, page 412, give

$$f''(x) = \frac{\alpha}{\beta}\left(\frac{\alpha}{\beta} - 1\right) x^{\alpha/\beta - 2} < 0,$$

so that f is strictly concave. It thus follows from Jensen's inequality that

$$\frac{f(b_1) + \cdots + f(b_n)}{n} < f\left(\frac{b_1 + \cdots + b_n}{n}\right) = f(1) = 1,$$

which is exactly (10.65). □

Given $a_1, a_2, \ldots, a_n > 0$, it is worth noticing that the function $f : \mathbb{R} \to \mathbb{R}$, defined for $\alpha \in \mathbb{R}$ by

$$f(\alpha) = M_\alpha(a_1, \ldots, a_n),$$

is continuous. Actually, f is obviously continuous at any $\alpha \neq 0$, so that it suffices to show that

$$\lim_{\alpha \to 0} f(\alpha) = f(0).$$

To this end, the continuity of $x \mapsto e^x$ and $x \mapsto \log x$ assures that we need only show that

$$\lim_{\alpha \to 0} \log f(\alpha) = \log f(0)$$

or, which is the same,

$$\lim_{\alpha \to 0} \frac{1}{\alpha} \log\left(\frac{a_1^\alpha + a_2^\alpha + \cdots + a_n^\alpha}{n}\right) = \frac{\log(a_1 a_2 \cdots a_n)}{n}. \qquad (10.66)$$

However, since

$$\lim_{\alpha \to 0} \log\left(\frac{a_1^\alpha + a_2^\alpha + \cdots + a_n^\alpha}{n}\right) = \log 1 = 0,$$

an application of l'Hôpital's rule to the left hand side of (10.66), together with (10.53), gives

$$\lim_{\alpha \to 0} \frac{1}{\alpha} \log \left(\frac{a_1^\alpha + a_2^\alpha + \cdots + a_n^\alpha}{n} \right) = \lim_{\alpha \to 0} \frac{d}{d\alpha} \log \left(\frac{a_1^\alpha + a_2^\alpha + \cdots + a_n^\alpha}{n} \right)$$

$$= \lim_{\alpha \to 0} \frac{(\log a_1) a_1^\alpha + \cdots + (\log a_n) a_n^\alpha}{a_1^\alpha + \cdots + a_n^\alpha}$$

$$= \frac{\log a_1 + \cdots + \log a_n}{n}$$

$$= \frac{\log(a_1 a_2 \ldots a_n)}{n}.$$

Problems: Section 10.8

1. Prove that equation $a^2 = 2^a$ has exactly three real roots, one of which is irrational.

2. (Israel) Find all real solutions of the system of equations

$$\begin{cases} x + \log(x + \sqrt{x^2 + 1}) = y \\ y + \log(y + \sqrt{y^2 + 1}) = z \\ z + \log(z + \sqrt{z^2 + 1}) = x \end{cases}.$$

3. (Croatia) Prove that there exists no polynomial function $p : \mathbb{R} \to \mathbb{R}$ such that $p(x) = \log x$, for every $x \in \mathbb{R}$.

4. Given $n \in \mathbb{N}$, show that the decimal representation of n has exactly $\lfloor \log_{10} n \rfloor + 1$ algarisms.

5. (Brazil) Let $m \in \mathbb{N}$ and $a_1, a_2, \ldots, a_m \in \{0, 1, 2, \ldots, 9\}$, with $a_1 \neq 0$. Given $k \in \mathbb{N}$, prove that there exists $n \in \mathbb{N}$ such that the decimal representation of n^k starts (from the left) with the sequence of algarismos $a_1 a_2 \ldots a_m$.

6. (Russia) Show that, for any reals $1 < a < b < c$, one has

$$\log_a (\log_a b) + \log_b (\log_b c) + \log_c (\log_c a) > 0.$$

7. (Brazil) Give an example of a function $f : [0, +\infty) \to \mathbb{R}$ such that $f(0) = 0$ and $f(2x + 1) = 3f(x) + 5$ for every $x \in \mathbb{R}$.

8. (South Korea) Find all functions $f : \mathbb{Z}_+ \to \mathbb{Z}_+$ satisfying the following conditions:

   (a) $2f(m^2 + n^2) = f(m)^2 + f(n)^2$, for all $m, n \in \mathbb{Z}_+$.
   (b) If $m, n \in \mathbb{Z}_+$, with $m \geq n$, then $f(m^2) \geq f(n^2)$.

9. (Romania) Given $a > 0$, show that

$$\lim_{n \to +\infty} n \int_0^1 \frac{x^n}{x^n + a} dx = \log \left(\frac{a + 1}{a} \right).$$

10. Let x, y and z be positive reals whose sum of squares equals 8. Prove that the sum of their cubes is greater than or equal to $16\sqrt{\frac{2}{3}}$.

11. Prove the following generalization of the inequality between the arithmetic and geometric means, known as the **weighted arithmetic-geometric means inequality**: given positive reals $a_1, a_2, \ldots, a_n$ and naturals $k_1, k_2, \ldots, k_n$ such that $\frac{1}{k_1} + \frac{1}{k_2} + \cdots + \frac{1}{k_n} = 1$, we have

$$\frac{a_1^{k_1}}{k_1} + \frac{a_2^{k_2}}{k_2} + \cdots + \frac{a_n^{k_n}}{k_n} \geq a_1 a_2 \ldots a_n,$$

with equality if and only if $a_1 = a_2 = \ldots = a_n$.

12. (USA) Given positive real numbers a, b and c, prove that

$$a^a b^b c^c \geq \left(\frac{a+b+c}{3}\right)^{a+b+c},$$

with equality if and only if $a = b = c$.

13. (Leningrad) Let $x_1, x_2, \ldots, x_n$ be nonnegative reals with sum equal to $\frac{1}{2}$. Show that

$$\prod_{j=1}^{n} \frac{1 - x_j}{1 + x_j} \leq \left(\frac{2n-1}{2n+1}\right)^n.$$

14. Let $0 < x_1, x_2, x_3, x_4 < \pi$ be real numbers such that $x_1 + x_2 + x_3 + x_4 = 2\pi$. Prove that

$$\prod_{i=1}^{4} \frac{\sin x_i}{x_i} \leq \frac{16}{\pi^4}.$$

15. (IMO shortlist) Let $a_1, a_2, \ldots, a_n$ be real numbers greater than or equal to 1. Prove that

$$\frac{1}{a_1 + 1} + \frac{1}{a_2 + 1} + \cdots + \frac{1}{a_n + 1} \geq \frac{n}{\sqrt[n]{a_1 a_2 \ldots a_n} + 1}.$$

16. (IMO shortlist) Let $a_1, a_2, \ldots, a_n$ be real numbers in the interval $\left[0, \frac{1}{2}\right]$, with $a_1 + a_2 + \cdots + a_n < 1$. Prove that

$$\frac{a_1 a_2 \ldots a_n [1 - (a_1 + a_2 + \cdots + a_n)]}{(a_1 + a_2 + \cdots + a_n)(1 - a_1)(1 - a_2) \ldots (1 - a_n)} \leq \frac{1}{n^{n+1}}.$$

17. Prove **Minkowski's inequality**[10]: given an integer $k > 1$ and positive reals a_1, $a_2, \ldots, a_n, b_1, b_2, \ldots, b_n$, we have

$$\sqrt[k]{\sum_{i=1}^{n}(a_i + b_i)^k} \leq \sqrt[k]{\sum_{i=1}^{n} a_i^k} + \sqrt[k]{\sum_{i=1}^{n} b_i^k},$$

with equality if and only if $\frac{a_1}{b_1} = \frac{a_2}{b_2} = \cdots = \frac{a_n}{b_n}$.

10.9   Improper Integration

The purpose of this section is to extend Riemann's integral to functions which are not necessarily bounded and are defined in arbitrary intervals.

If $I \subset \mathbb{R}$ is an interval, then I is of one of the forms $[A, B]$, (A, B), $[A, B)$, $(A, B]$, $(A, +\infty)$, $[A, +\infty)$, $(-\infty, A)$, $(-\infty, A]$ or $(-\infty, +\infty) = \mathbb{R}$, with $A, B \in \mathbb{R}$. In order to avoid the need of examining each one of these cases separately, and for general notational convenience, we:

(i) shall refer to the real numbers A, B, or to the symbols $-\infty$, $+\infty$ as the *endpoints* of I (albeit $-\infty$ and $+\infty$ are just *formal* symbols, and not real numbers), which will be generically denoted by α and β;

(ii) impose that $-\infty < A, B < +\infty$, for all $A, B \in \mathbb{R}$.

Thus, if I has endpoints α and β, with $\alpha < \beta$, then

$$(\alpha, \beta) \subset I \text{ and } I \setminus (\alpha, \beta) \subset \{\alpha, \beta\}.$$

For instance, if I has endpoints $\alpha = A$ and $\beta = +\infty$, we have $(A, +\infty) \subset I$ and $I \setminus (A, +\infty) \subset \{A, +\infty\}$, so that $I = (A, +\infty)$ or $[A, +\infty)$.

Let I be a fixed interval and $f : I \to \mathbb{R}$ be a given function, such that f is integrable in each interval $[a, b] \subset I$. (In particular, if $I = [a, b]$, then we assume that f is integrable in I.) This implies that f is bounded in each interval $[a, b] \subset I$, albeit it may well happen that f is not bounded in I. (Nevertheless, recall that if f is continuous in I, then, even if f is unbounded in I, Weierstrass' Theorem 8.26 assures that it is bounded in each interval $[a, b] \subset I$.)

Definition 10.68 Let $I \subset \mathbb{R}$ be an interval with endpoints α and β, with $\alpha < \beta$, and $f : I \to \mathbb{R}$ be integrable in each interval $[a, b] \subset I$. We say that f is **integrable**

---

[10]After Hermann Minkowski, German mathematician of the XIX and XX centuries. Minkowski made several seminal contributions to different branches of Mathematics. His *geometric method* on Algebraic Number Theory proved to be quite fruitful for the development of that theory. Also, Minkowski's four dimensional space is the correct setting for the geometric description of special relativity.

in I if, for some $x_0 \in (\alpha, \beta)$, the limits

$$\lim_{x \to \alpha} \int_x^{x_0} f(t) \, dt \quad \text{and} \quad \lim_{y \to \beta} \int_{x_0}^y f(t) \, dt \tag{10.67}$$

do exist. This being the case, we define the **improper integral** of f in I, denoted $\int_\alpha^\beta f(t) \, dt$, by setting

$$\int_\alpha^\beta f(t) \, dt = \lim_{x \to \alpha} \int_x^{x_0} f(t) \, dt + \lim_{y \to \beta} \int_{x_0}^y f(t) \, dt,$$

and say that the improper integral $\int_\alpha^\beta f(t) \, dt$ **converges**.

It is important to realize that the former definition does not depend on the chosen $x_0 \in I$. More precisely, suppose that $x_0 \in I$ is chosen so that the limits (10.67) do exist, and let x_0' be another point in I. Then,

$$\int_x^{x_0'} f(t) \, dt = \int_x^{x_0} f(t) \, dt + \int_{x_0}^{x_0'} f(t) \, dt,$$

so that the limit $\lim_{x \to \alpha} \int_x^{x_0'} f(t) \, dt$ also exists, with

$$\lim_{x \to \alpha} \int_x^{x_0'} f(t) \, dt = \lim_{x \to \alpha} \int_x^{x_0} f(t) \, dt + \int_{x_0}^{x_0'} f(t) \, dt.$$

Analogously, $\lim_{y \to \beta} \int_{x_0'}^y f(t) \, dt$ does exist, with

$$\lim_{y \to \beta} \int_{x_0'}^y f(t) \, dt = \int_{x_0'}^{x_0} f(t) \, dt + \lim_{y \to \beta} \int_{x_0}^y f(t) \, dt.$$

Therefore, since $\int_{x_0}^{x_0'} f(t) \, dt + \int_{x_0'}^{x_0} f(t) \, dt = 0$, we get

$$\lim_{x \to \alpha} \int_x^{x_0'} f(t) \, dt + \lim_{y \to \beta} \int_{x_0'}^y f(t) \, dt = \lim_{x \to \alpha} \int_x^{x_0} f(t) \, dt + \lim_{y \to \beta} \int_{x_0}^y f(t) \, dt.$$

On the other hand, in case $I = [a, b]$, we claim that the definition of integral given in Chap. 10 coincides with the present one. In order to check that, let $f : [a, b] \to \mathbb{R}$ be (bounded and) integrable in $[a, b]$, and fix $x_0 \in (a, b)$. Then f is integrable in $[a, x_0]$ and, for $x \in (a, x_0)$, we have

$$\int_x^{x_0} f(t) \, dt = \int_a^{x_0} f(t) \, dt - \int_b^x f(t) \, dt.$$

Letting M be an upper bound for $|f|$ in $[a, b]$, it follows from the triangle inequality for integrals and Proposition 10.12 that

$$\left| \int_a^x f(t)\, dt \right| \leq \int_a^x |f(t)|\, dt \leq M(x - a),$$

so that the squeezing principle gives $\lim_{x \to a} \int_a^x f(t)\, dt = 0$. Therefore,

$$\lim_{x \to a} \int_x^{x_0} f(t)\, dt = \int_a^{x_0} f(t)\, dt - \lim_{x \to a} \int_a^x f(t)\, dt = \int_a^{x_0} f(t)\, dt.$$

Since f is also integrable in $[x_0, b]$, a similar reasoning allows us to conclude that

$$\lim_{y \to b} \int_{x_0}^y f(t)\, dt = \int_{x_0}^b f(t)\, dt - \lim_{y \to b} \int_y^b f(t)\, dt = \int_{x_0}^b f(t)\, dt.$$

Finally, by invoking Proposition 10.21, we get

$$\lim_{x \to a} \int_x^{x_0} f(t)\, dt + \lim_{y \to b} \int_{x_0}^y f(t)\, dt = \int_a^{x_0} f(t)\, dt + \int_{x_0}^b f(t)\, dt = \int_a^b f(t)\, dt.$$

Thanks to the above discussion, we can restrict the study of improper integrals to the cases in which $I = (A, B]$, $[A, B)$, $(-\infty, B]$ or $[A, +\infty)$, with $A, B \in \mathbb{R}$. Along the rest of this section, we stick to the cases $I = [A, B)$ and $[A, +\infty)$, leaving to the reader the task of formulating the analogous results when $I = (A, B]$ or $(-\infty, B]$.

Let's start by looking at two examples.

Example 10.69 The improper integral $\int_0^{+\infty} e^{-t}\, dt$ does converge. More precisely, we have $\int_0^{+\infty} e^{-t}\, dt = 1$.

Proof For $x > 0$, it follows from the FTC that

$$\int_0^x e^{-t}\, dt = -e^{-t}\Big|_{t=0}^{t=x} = 1 - e^{-x}.$$

Now, Theorem 10.54 gives $\lim_{x \to +\infty} e^{-x} = 0$, so that $\int_0^{+\infty} e^{-t}\, dt = 1$. $\square$

Example 10.70 The improper integral $\int_1^{+\infty} t^\alpha\, dt$ converges if and only if $\alpha < -1$.

Proof For $x > 1$ and $\alpha \neq -1$, note that

$$\int_1^x t^\alpha\, dt = \frac{1}{\alpha + 1} t^{\alpha+1}\Big|_{t=1}^{t=x} = \frac{1}{\alpha + 1}(x^{\alpha+1} - 1). \tag{10.68}$$

Now, since $x^{\alpha+1} = e^{(\alpha+1)\log x}$, we have (again by Theorem 10.54) that

$$\lim_{x \to +\infty} x^{\alpha+1} = \begin{cases} 0, & \text{if } \alpha + 1 < 0 \\ +\infty, & \text{if } \alpha + 1 > 0 \end{cases}.$$

Hence, the right hand side of (10.68) has a (finite) limit when $x \to +\infty$ if and only if $\alpha + 1 < 0$, i.e., if and only if $\alpha < -1$.

Finally, if $\alpha = -1$, then

$$\int_1^x t^{-1}\, dt = \log t \Big|_{t=1}^{t=x} = \log x \to +\infty$$

when $x \to +\infty$. □

Also with respect to the previous example, since

$$\lim_{x \to +\infty} \frac{x^\alpha}{x^{-1}} = \lim_{x \to +\infty} x^{\alpha+1} = 0$$

if and only if $\alpha < -1$, we use to say that $\int_1^{+\infty} t^\alpha\, dt$ converges if and only if x^α decays to zero more rapidly than x^{-1} when $x \to +\infty$. See also Problem 3, which deals with the case $\int_0^1 t^\alpha dt$.

The coming theorem is the fundamental result on the convergence of improper integrals. Note that it is the analogue, for improper integrals, of Theorem 7.27. For its statement, as well as for remark just following its proof, the reader may find it convenient to read again Problem 8, page 218.

Theorem 10.71 *Let* $f : [A, +\infty) \to \mathbb{R}$ *be integrable in each interval* $[a, b] \subset [A, +\infty)$. *The following assertions are equivalent:*

(a) $\int_A^{+\infty} f(t)\, dt$ *converges.*

(b) *For every* $\epsilon > 0$, *there exists* $M > A$ *such that* $x_1, x_2 > M \Rightarrow \left| \int_{x_1}^{x_2} f(t)\, dt \right| < \epsilon.$

(c) *There exists* $L \in \mathbb{R}$ *such that, for every sequence* $(x_n)_{n \geq 1}$ *in* $[A, +\infty)$ *and satisfying* $\lim_{n \to +\infty} x_n = +\infty$, *one has* $\lim_{n \to +\infty} \int_A^{x_n} f(t)\, dt = L.$

Proof Let $F : [A, +\infty) \to \mathbb{R}$ be defined by $F(x) = \int_A^x f(t)\, dt$.

(a) $\Rightarrow$ (b): assume that $\int_A^{+\infty} f(t)\, dt$ converges, with $\int_A^{+\infty} f(t)\, dt = L$. By definition, we have $\lim_{x \to +\infty} F(x) = L$, so that, given $\epsilon > 0$, there exists $M > 0$ for which

$$x > M \Rightarrow |F(x) - L| < \frac{\epsilon}{2}.$$

Hence, for $x_1, x_2 > M$, triangle inequality gives

$$\left| \int_{x_1}^{x_2} f(t)\, dt \right| = |F(x_2) - F(x_1)|$$
$$\leq |F(x_2) - L| + |L - F(x_1)|$$
$$< \frac{\epsilon}{2} + \frac{\epsilon}{2} = \epsilon.$$

(b) $\Rightarrow$ (c): assume that condition (b) holds, and let $(x_n)_{n\geq1}$ be a sequence such that $x_n \geq A$ for every $n \geq 1$ and $\lim_{n\to+\infty} x_n = +\infty$. Given $\epsilon > 0$, take M as in the statement and $n_0 \in \mathbb{N}$ such that $n > n_0 \Rightarrow x_n > M$. Then, for $m, n > n_0$, we have $x_m, x_n > M$, so that

$$|F(x_m) - F(x_n)| = \left| \int_{x_n}^{x_m} f(t) \, dt \right| < \epsilon.$$

Therefore, $(F(x_n))_{n\geq1}$ is a Cauchy sequence, thus convergent, by Theorem 7.27.

If $L = \lim_{n\to+\infty} F(x_n)$, we claim that $\lim_{n\to+\infty} F(x'_n) = L$ for every sequence $(x'_n)_{n\geq1}$ satisfying $x'_n \geq A$ for $n \geq 1$ and $\lim_{n\to+\infty} x'_n = +\infty$. Indeed, the discussion in the previous paragraph (with $(x'_n)_{n\geq1}$ in place of $(x_n)_{n\geq1}$) assures the existence of $L' = \lim_{n\to+\infty} F(x'_n)$. However, if it was $L' \neq L$, then the sequence $(x''_n)_{n\geq1}$, given for $k \geq 1$ by $x''_{2k-1} = x_{2k-1}$ and $x''_{2k} = x'_{2k}$, would also converge to $+\infty$ and be such that

$$\lim_{k\to+\infty} F(x''_{2k-1}) = L \quad \text{and} \quad \lim_{k\to+\infty} F(x''_{2k}) = L'.$$

Since $L \neq L'$, the limit $\lim_{n\to+\infty} F(x''_n)$ would not exist, which contradicts the argument of the previous paragraph (this time with $(x''_n)_{n\geq1}$ in place of $(x_n)_{n\geq1}$).

(c) $\Rightarrow$ (a): suppose that there exists $L \in \mathbb{R}$ such that, for every sequence $(x_n)_{n\geq1}$ in $[A, +\infty)$ with $\lim_{n\to+\infty} x_n = +\infty$, we have $\lim_{n\to+\infty} F(x_n) = L$. If $\lim_{x\to+\infty} F(x)$ does not exist or is different from L, then there exist $\epsilon > 0$ and, for every $B > A$, a real number $x_B > B$ such that $|F(x_B) - L| \geq \epsilon$. Taking B successively equal to $A + 1, A + 2, A + 3, \ldots$, we would get a sequence $x_1, x_2, x_3, \ldots$ satisfying $x_n > A + n$ and $|F(x_n) - L| \geq \epsilon$ for every $n \geq 1$. Therefore, $\lim_{n\to+\infty} x_n = +\infty$ but $(F(x_n))_{n\geq1}$ does not converge to L, which is an absurd. $\square$

Remark 10.72 It is immediate to adapt the previous result to the case of a function $f : [A, B) \to \mathbb{R}$ which is integrable in each interval $[A, b]$ with $A < b < B$. More precisely, and as the reader can easily verify, the following assertions are equivalent:

(a) $\int_A^B f(t) \, dt$ converges.
(b) For every $\epsilon > 0$, there exists $\delta > 0$ such that $B - \delta < x_1, x_2 < B \Rightarrow$
$$\left| \int_{x_1}^{x_2} f(t) \, dt \right| < \epsilon.$$
(c) There exists $L \in \mathbb{R}$ such that, for every sequence $(x_n)_{n\geq1}$ in $[A, B)$ and satisfying $\lim_{n\to+\infty} x_n = B$, one has $\lim_{n\to+\infty} \int_A^{x_n} f(t) \, dt = L$.

We now establish an important consequence of the previous theorem, which is known in mathematical literature as the **comparison test** for improper integrals.

Proposition 10.73 *Let $f, g : [A, +\infty) \to \mathbb{R}$ be such that $|f(x)| \leq g(x)$ for every $x \in [A, +\infty)$. If f and g are integrable in every interval $[a, b] \subset [A, +\infty)$ and $\int_A^{+\infty} g(t) \, dt$ converges, then $\int_A^{+\infty} f(t) \, dt$ also converges and*

$$\left| \int_A^{+\infty} f(t) \, dt \right| \leq \int_A^{+\infty} g(t) \, dt.$$

Proof Firstly, notice that $g(x) \geq 0$ for every $x \geq A$, so that $\int_A^{+\infty} g(t)\, dt \geq 0$. Then, given $\epsilon > 0$, the convergence of $\int_A^{+\infty} g(t)\, dt$, together with Theorem 10.71, assures the existence of $M > A$ such that

$$x_1, x_2 > M \Rightarrow \int_{x_1}^{x_2} g(t)\, dt < \epsilon.$$

Now, the triangle inequality for integrals, together with Proposition 10.12, guarantees that, also for $x_1, x_2 > M$, we have

$$\left| \int_{x_1}^{x_2} f(t)\, dt \right| \leq \int_{x_1}^{x_2} |f(t)|\, dt \leq \int_{x_1}^{x_2} g(t)\, dt < \epsilon.$$

Then, once more from Theorem 10.71, we get the convergence of the integrals $\int_A^{+\infty} f(t)\, dt$ and $\int_A^{+\infty} |f(t)|\, dt$.

Finally, it follows from what we did above, together with the properties of limits of functions (cf. Sect. 9.2) that

$$\left| \int_A^{+\infty} f(t)\, dt \right| = \left| \lim_{x \to +\infty} \int_A^x f(t)\, dt \right| = \lim_{x \to +\infty} \left| \int_A^x f(t)\, dt \right|$$

$$\leq \lim_{x \to +\infty} \int_A^x |f(t)|\, dt \leq \lim_{x \to +\infty} \int_A^x g(t)\, dt$$

$$= \int_A^{+\infty} g(t)\, dt.$$

$\square$

Remark 10.74 As in the previous remark, the former proposition can be easily adapted to deal with the case $I = [A, B)$. More precisely, if $f, g : [A, B) \to \mathbb{R}$ are integrable in each interval $[A, b]$ with $A < b < B$, and such that $|f(x)| \leq g(x)$ for every $x \in [A, B)$, then the convergence of $\int_A^B g(t)\, dt$ implies that of $\int_A^B f(t)\, dt$; moreover, one also has

$$\left| \int_A^B f(t)\, dt \right| \leq \int_A^B g(t)\, dt.$$

Example 10.75 We shall use Proposition 10.73 to establish the convergence of **Dirichlet's integral**

$$\int_0^{+\infty} \frac{\sin t}{t}\, dt.$$

To this end, start by observing that the fundamental trigonometric limit (cf. Lemma 9.11) assures that the function $t \mapsto \frac{\sin t}{t}$ extends continuously to 0. Therefore, it suffices to establish the convergence of $\int_1^{+\infty} \frac{\operatorname{sen} t}{t} \, dt$.

To what is left, let $x > 1$. Integrating by parts, we get

$$\int_1^x \frac{\sin t}{t} \, dt = -\frac{\cos t}{t} \Big|_{t=1}^{t=x} - \int_1^x \frac{\cos t}{t^2} \, dt$$

$$= \cos 1 - \frac{\cos x}{x} - \int_1^x \frac{\cos t}{t^2} \, dt.$$

Since $|\cos x| \leq 1$ for every x, the computations above guarantee that $\int_1^{+\infty} \frac{\sin t}{t} \, dt$ converges if and only if $\int_1^{+\infty} \frac{\cos t}{t^2} \, dt$ converges. Hence, we have reduced our work to the convergence of this last improper integral.

Finally, it is enough to apply the comparison test, noticing that $\left| \frac{\cos x}{x^2} \right| \leq \frac{1}{x^2}$ and (according to Example 10.70) $\int_1^{+\infty} \frac{1}{t^2} \, dt$ does converge.

Problem 10, page 436, brings another proof of the convergence of Dirichlet's integral. On the other hand, it is possible to show (cf. [10, 20] or [27], for instance) that

$$\int_0^{+\infty} \frac{\sin t}{t} \, dt = \frac{\pi}{2}.$$

In turn, this computation is of fundamental importance to the theory of *Fourier series*.[11] Figure 10.10 sketches the graph of $x \mapsto \frac{\sin x}{x}$.

Back to the development of the theory, Proposition 10.73 assures that if $f : [A, +\infty) \to \mathbb{R}$ is integrable in each interval $[a, b] \subset [A, +\infty)$ and $\int_A^{+\infty} |f(t)| \, dt$ converges, then $\int_A^{+\infty} f(t) \, dt$ also converges and

$$\left| \int_A^{+\infty} f(t) \, dt \right| \leq \int_A^{+\infty} |f(t)| \, dt \tag{10.69}$$

(recall that $|f|$ is also integrable in each interval $[a, b] \subset [A, +\infty)$, by Proposition 10.14). In such a case, we say that $\int_A^{+\infty} f(t) \, dt$ is **absolutely convergent**. Not every convergent improper integral is absolutely convergent; actually, Problem 11, page 436, shows that $\int_0^{+\infty} \frac{\sin t}{t} \, dt$ is not absolutely convergent.

We finish this section by showing that the link between the theory of convergent series and convergent improper integrals is actually deeper. More precisely, the coming result establishes an equivalence between the convergence of the integral of a positive function and that of a certain series. For this reason, it is known as the **integral test** for absolute convergence of series.

---

[11] After Jean-Baptiste-Joseph Fourier, French engineer and mathematician of the XIX century. For more on Fourier see the footnote on page 469. A glimpse on the theory of Fourier Series is the content of Problems 11 to 17, page 468.

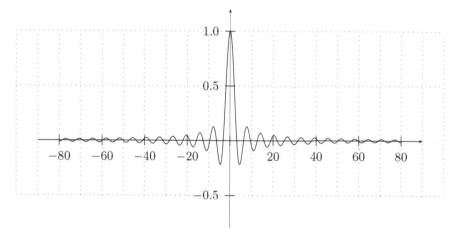

Fig. 10.10 Graph of $x \mapsto \frac{\sin x}{x}$

Theorem 10.76 *Let $n_0 \in \mathbb{N}$ and $f : [n_0, +\infty) \to \mathbb{R}$ be a decreasing function, integrable in each interval $[a, b] \subset [n_0, +\infty)$ and such that $\lim_{x \to +\infty} f(x) = 0$. Then*

$$\int_{n_0}^{+\infty} f(t) \, dt \text{ converges} \iff \sum_{k \geq n_0} f(k) \text{ converges.}$$

Proof Let $g, h : [n_0, +\infty) \to \mathbb{R}$ be the functions given, in the interval $[k, k+1)$ (for each $k \geq n_0$), by $g(x) = f(k+1)$ and $h(x) = f(k)$.

The conditions on f guarantee that $0 \leq g(x) \leq f(x) \leq h(x)$ for every $x \geq n_0$. Hence, the comparison test for improper integrals assures that

$$\int_{n_0}^{+\infty} f(t) \, dt \text{ converges} \Rightarrow \int_{n_0}^{+\infty} g(t) \, dt \text{ converges}$$

and

$$\int_{n_0}^{+\infty} h(t) \, dt \text{ converges} \Rightarrow \int_{n_0}^{+\infty} f(t) \, dt \text{ converges.}$$

On the other hand, for an integer $n > n_0$, we have

$$\int_{n_0}^{n} g(t) \, dt = \sum_{k=n_0+1}^{n} f(k) \text{ and } \int_{n_0}^{n} h(t) \, dt = \sum_{k=n_0}^{n-1} f(k).$$

Finally, let's apply equivalence (a) $\Leftrightarrow$ (c) of Theorem 10.71, with $x_k = k$ for every $k \geq n_0$:

(i) If $\int_{n_0}^{+\infty} f(t)\, dt$ converges, then $\int_{n_0}^{+\infty} g(t)\, dt$ converges and, hence

$$\int_{n_0}^{+\infty} g(t)\, dt = \lim_{n \to +\infty} \int_{n_0}^{n} g(t)\, dt = \lim_{n \to +\infty} \sum_{k=n_0+1}^{n} f(k) = \sum_{k \geq n_0+1} f(k).$$

Therefore, $\sum_{k \geq n_0} f(k)$ converges, with $\sum_{k \geq n_0} f(k) = f(n_0) + \sum_{k \geq n_0+1} f(k)$.

(ii) If $\sum_{k \geq n_0} f(k)$ converges, then

$$\int_{n_0}^{+\infty} h(t)\, dt = \lim_{n \to +\infty} \int_{n_0}^{n} h(t)\, dt = \lim_{n \to +\infty} \sum_{k=n_0}^{n-1} f(k) = \sum_{k \geq n_0} f(k),$$

so that $\int_{n_0}^{+\infty} h(t)\, dt$ converges. Hence, what we did before assures the convergence of $\int_{n_0}^{+\infty} f(t)\, dt$.

$\square$

Example 10.77 The integral test allows us to easily establish the divergence of the harmonic series, as well as the convergence of the series $\sum_{k \geq 1} \frac{1}{k^r}$, for every real number $r > 1$. For the harmonic series, letting $f : [1, +\infty) \to \mathbb{R}$ be given by $f(x) = \frac{1}{x}$, we have

$$\int_{1}^{+\infty} \frac{1}{t}\, dt = \lim_{x \to +\infty} \int_{1}^{x} \frac{1}{t}\, dt = \lim_{x \to +\infty} \log x = +\infty.$$

Analogously, taking $f : [1, +\infty) \to \mathbb{R}$ given by $f(x) = \frac{1}{x^r}$, with $r > 1$, we have

$$\int_{1}^{+\infty} \frac{1}{t^r}\, dt = \lim_{x \to +\infty} \int_{1}^{x} \frac{1}{t^r}\, dt = \lim_{x \to +\infty} \frac{1}{1-r} \left(\frac{1}{x^{r-1}} - 1 \right) = \frac{1}{r-1}.$$

Let's now see one more interesting example.

Example 10.78 Examine the convergence of the series $\sum_{k \geq 2} \frac{1}{k \log k}$.

Solution Let $f : [2, +\infty) \to \mathbb{R}$ be given by $f(x) = \frac{1}{x \log x}$, so that f is continuous, decreasing and such that $\lim_{x \to +\infty} f(x) = 0$. The integral test says that $\sum_{k \geq 2} \frac{1}{k \log k}$ converges if and only if $\int_{2}^{+\infty} f(t)\, dt$ also does.

To what is left, note that the substitution of variable $s = \log t$ (cf. Problem 2, page 435) gives

$$\int_2^{+\infty} f(t)\,dt = \int_2^{+\infty} \frac{1}{t\log t}\,dt = \int_{\log 2}^{+\infty} \frac{1}{s}\,ds$$

$$= \lim_{s\to+\infty}(\log s - \log 2) = +\infty.$$

Hence, $\int_2^{+\infty} f(t)\,dt$ diverges, so that $\sum_{k\geq 2}\frac{1}{k\log k}$ also diverges. □

Problems: Section 10.9

1. Examine the convergence of the following improper integrals:

   (a) $\int_{-\frac{\pi}{2}}^{\frac{\pi}{2}} \sec t\,dt.$

   (b) $\int_0^{+\infty} e^{-t}\sin t\,dt.$

   (c) $\int_0^1 \frac{1}{\log t}\,dt.$

   (d) $\int_{-\infty}^{+\infty} \frac{1}{1+t^2}\,dt.$

2. * Prove the following version of the **change of variables theorem for improper integrals**: let $f : [A, +\infty) \to \mathbb{R}$ be continuous and $g : [B, +\infty) \to [A, +\infty)$ be differentiable, with $\lim_{x\to+\infty} g(x) = +\infty$ and g' integrable in each interval $[a, b] \subset [B, +\infty)$. Then, $\int_{g(B)}^{+\infty} f(t)\,dt$ converges if and only if $\int_B^{+\infty} f(g(s))g'(s)\,ds$ converges. Moreover, if this is so, we also have

$$\int_{g(B)}^{+\infty} f(t)\,dt = \int_B^{+\infty} f(g(s))g'(s)\,ds.$$

3. * Show that $\int_0^1 t^\alpha\,dt$ converges if and only if $\alpha > -1$. Then, for $\alpha < 0$, conclude that $\int_0^1 t^\alpha\,dt$ converges if and only if x^α grows to $+\infty$ more slowly than x^{-1} as $x\to 0+$.

4. (Berkeley) Prove that both integrals $\int_0^\infty \cos(x^2)dx$ and $\int_0^\infty \sin(x^2)dx$ converge.

5. (Berkeley—adapted) Show that:

   (a) $\int_0^\pi \log(\sin x)dx$ and $\int_0^\pi \log|\cos x|dx$ converge.

   (b) $\int_0^\pi \log(\sin x)dx = \int_0^\pi \log|\cos x|dx = -\pi\log 2.$

6. Examine the convergence of the following series, where $\alpha > 0$ is a given real number:

   (a) $\sum_{k\geq 2}\frac{1}{(\log k)^\alpha}.$

   (b) $\sum_{k\geq 2}\frac{1}{k(\log k)^\alpha}.$

   (c) $\sum_{k\geq 2}\frac{1}{k(\log k)(\log\log k)^\alpha}.$

7. Show that $\int_0^{+\infty} \frac{\log t}{1+t^2} dt$ converges and compute its value.

   The next problem extends the results of Problem 12, page 381, and of Theorem 10.63.

8. * Let $I \subset \mathbb{R}$ be an interval with endpoints α and β, $f, g : I \to (0, +\infty)$ be integrable functions and $p, q > 0$ such that $\frac{1}{p} + \frac{1}{q} = 1$.

   (a) If $\int_\alpha^\beta f(t)^p \, dt = \int_\alpha^\beta g(t)^q \, dt = 1$, show that $\int_\alpha^\beta f(t)g(t) \, dt \le 1$.
   (b) Deduce **Hölder's inequality** for integrals:

   $$\int_\alpha^\beta f(t)g(t) \, dt \le \left(\int_\alpha^\beta f(t)^p \, dt \right)^{1/p} \left(\int_\alpha^\beta g(t)^q \, dt \right)^{1/q}.$$

9. In the proof of Theorem 10.55, we saw that in order to guarantee the existence of $\lim_{n \to +\infty} \left(\sum_{k=1}^n \frac{1}{k} - \log n \right)$ it suffices to assure the convergence of the series

   $$\sum_{k \ge 1} \left(\frac{1}{k} - \log(k+1) + \log k \right).$$

   To this end, let $f : [1, +\infty) \to \mathbb{R}$ be given by $f(x) = \frac{1}{x} - \log(x+1) + \log x$ for $x \ge 1$, and do the following items:

   (a) Show that f is decreasing, with $\lim_{x \to +\infty} f(x) = 0$.
   (b) Apply the integral test to establish the convergence of the series.

10. The purpose of this problem is to give another proof of the convergence of the improper integral $\int_0^{+\infty} \frac{\sin t}{t} \, dt$. To this end, do the following items:

    (a) Prove that it suffices to show the convergence of the series $\sum_{k \ge 1} \frac{\sin k}{k}$.
    (b) In turn, use Abel's criterion (cf. Problem 18, page 243) to show that this amounts to establishing the boundedness of the sequence $(s_n)_{n \ge 1}$, such that $s_n = \sum_{k=1}^n \sin k$ for $n \in \mathbb{N}$.
    (c) Apply the result of Problem 10, page 359, to find $s_n = \frac{\sin \frac{n}{2} \sin \frac{n+1}{2}}{\sin \frac{1}{2}}$ and, hence, $|s_n| \le \frac{1}{\sin \frac{1}{2}}$.

11. Do the following items:

    (a) Show that at least one of the numbers $|\sin n|$, $|\sin(n+1)|$ or $|\sin(n+2)|$ is always greater than or equal to $\frac{1}{2}$.
    (b) Conclude that the series $\sum_{k \ge 1} \frac{|\sin k|}{k}$ diverges.
    (c) Show that $\int_0^{+\infty} \frac{\sin t}{t} \, dt$ is not absolutely convergent.

12. If $p_1 = 2 < p_2 = 3 < p_3 = 5 < \cdots$ is the sequence of prime numbers, a theorem of Euler states that the series $\sum_{k \geq 1} \frac{1}{p_k}$ diverges.[12] In order to show this, do the following items:

(a) Show that $(1 - x)^{-1} \leq e^{2x}$ for $0 \leq x \leq \frac{1}{2}$.
(b) For $n \in \mathbb{N}$, let $p_1 < p_2 < \cdots < p_l$ be the prime numbers less that or equal to n. Show that

$$\sum_{k=1}^{n} \frac{1}{k} \leq \prod_{k=1}^{l} \left(1 + \frac{1}{p_k} + \frac{1}{p_k^2} + \cdots \right).$$

(c) Use the results of (a) and (b) to prove that $\sum_{k=1}^{n} \frac{1}{k} \leq e^{\sum_{k=1}^{l} \frac{2}{p_k}}$.
(d) Conclude that $\sum_{k \geq 1} \frac{1}{p_k}$ diverges.

10.10 Two Important Applications

We recall that our initial motivation for the study of the integral was geometric, namely, to define and compute the area of the region of the cartesian plane under the graph of a given nonnegative (integrable) function. We could also use the concept of integral to deal with the following geometric problems:

(I) How to define and compute the *length* of the graph of $f : [a, b] \to \mathbb{R}$?
(II) In case f is positive, how to define and compute the *volume* of the *solid of revolution* generated by the rotation of the region under the graph of f around the horizontal axis?
(III) Yet in the case of a positive f, how could one define and compute the *area* of the *surface of revolution* generated by the rotation of the graph of f around the horizontal axis?

In this section we analyse Problem (I), leaving the analysis of Problems (II) and (III) to [4]. As a byproduct of our discussion, we will present a thorough definition and computation of the length of a circle.

Then, let $f : [a, b] \to \mathbb{R}$ be continuous. Fix (cf. Fig. 10.11) an interval $[c, d] \subset (a, b)$, a partition $P = \{c = x_0 < x_1 < \cdots < x_k = d\}$ of $[c, d]$ and let $\Delta x_j = x_j - x_{j-1}$ for $1 \leq j \leq k$.

For $|P|$ sufficiently small, it is a reasonable guess to think of the line segment joining points $(x_{j-1}, f(x_{j-1}))$ and $(x_j, f(x_j))$ as a good approximation for the portion of the graph of f situated between the vertical lines $x = x_{j-1}$ and $x = x_j$. If this is so, it is also reasonable to assume that the length ℓ_j of such a segment should serve as a good approximation for the *length* of this portion of the graph of f, whichever way we define it.

---

[12] For a sligthly different proof, see Chap. 9 of [5].

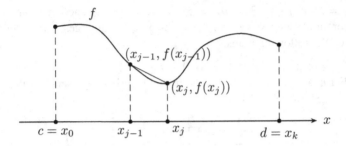

Fig. 10.11 Approximating the length of a graph

This being said, the formula for the distance between two points of the cartesian plane gives

$$
\begin{aligned}
\ell_j &= \sqrt{(x_j - x_{j-1})^2 + (f(x_j) - f(x_{j-1}))^2} \\
&= \sqrt{1 + \left(\frac{f(x_j) - f(x_{j-1})}{x_j - x_{j-1}} \right)^2} \Delta x_j.
\end{aligned}
$$

Now, if f is continuously differentiable in (a, b), then Lagrange's MVT guarantees the existence of $\xi_j \in (x_{j-1}, x_j)$ such that $\frac{f(x_j) - f(x_{j-1})}{x_j - x_{j-1}} = f'(\xi_j)$; this way,

$$
\ell_j \cong \sqrt{1 + f'(\xi_j)^2} \Delta x_j = \sqrt{1 + f'(\xi_j)^2}(x_j - x_{j-1}).
$$

Adding up the above expression for ℓ_j for $1 \leq j \leq k$, we conclude that

$$
\sum_{j=1}^{k} \sqrt{1 + f'(\xi_j)^2}(x_j - x_{j-1}) \tag{10.70}
$$

would be a reasonable approximation for what we would think of as the *length* of the graph of the restriction of f to the interval $[c, d]$. Moreover, we could equally hope that such an approximation would get better and better as $|P| \to 0$.

On the other hand, since (10.70) coincides with the Riemann sum

$$
\Sigma(\sqrt{1 + (f')^2}; P; \xi),
$$

it follows from Riemann's Theorem 10.7 that

$$
\sum_{j=1}^{k} \sqrt{1 + f'(\xi_j)^2} \Delta x_j \longrightarrow \int_{c}^{d} \sqrt{1 + f'(x)^2} dx
$$

as $|P| \to 0$.

The discussion above allows us to conclude that $\int_c^d \sqrt{1 + f'(x)^2} dx$ is the only reasonable way of defining the length of the graph of the restriction of f to the interval $[c, d]$. If we now notice that the union of the graphs of such restrictions (as $[c, d]$ varies over all bounded and closed subintervals of (a, b)) coincides with the graph of f but the points $(a, f(a))$ and $(b, f(b))$, we are naturally led to state the following

Definition 10.79 Let $f : [a, b] \to \mathbb{R}$ be continuous in $[a, b]$ and continuously differentiable in (a, b). If the improper integral

$$\ell = \int_a^b \sqrt{1 + f'(x)^2} dx, \tag{10.71}$$

does exist, then we say that the graph of f is **rectifiable** and has **length** ℓ.

Yet with respect to the former definition, recall that

$$\int_a^b \sqrt{1 + f'(x)^2} dx = \lim_{\epsilon \to 0} \int_{a+\epsilon}^{b-\epsilon} \sqrt{1 + f'(x)^2} dx.$$

Therefore, if f is continuously differentiable in $[a, b]$, then Problem 11, page 380, assures that (10.71) holds in the ordinary sense (i.e., not as an improper integral).

Example 10.80 Let Γ be a circle of center O and radius R, and choose a cartesian system of coordinates such that $O(0, 0)$. If $f : [-R, R] \to \mathbb{R}$ is given by $f(x) = \sqrt{R^2 - x^2}$, then the graph of f is the semicircle of Γ situated on the upper half-plane of the cartesian plane. Note that f is continuous in $[-R, R]$ and continuously differentiable in $(-R, R)$. Hence, according to (10.71), the length of Γ equals twice the value of the improper integral

$$\int_{-R}^{R} \sqrt{1 + f'(x)^2} dx,$$

provided it exists.

To see that this is indeed the case, first note that

$$\sqrt{1 + f'(x)^2} = \sqrt{1 + \left(\frac{-x}{\sqrt{R^2 - x^2}}\right)^2} = \frac{R}{\sqrt{R^2 - x^2}}.$$

Now, performing the trigonometric substitution $x = R \cos t$ and observing that $-R + \epsilon \le x \le R - \epsilon \Leftrightarrow \arccos(-1 + \frac{\epsilon}{R}) \le t \le \arccos(1 - \frac{\epsilon}{R})$, we get

$$\int_{-R+\epsilon}^{R-\epsilon} \sqrt{1+f'(x)^2}dx = \int_{-R+\epsilon}^{R-\epsilon} \frac{R}{\sqrt{R^2-x^2}}dx$$

$$= \int_{\arccos(-1+\frac{\epsilon}{R})}^{\arccos(1-\frac{\epsilon}{R})} \frac{R}{\sqrt{R^2-R^2\cos^2 t}} \cdot (-R\sin t)dt$$

$$= -\int_{\arccos(-1+\frac{\epsilon}{R})}^{\arccos(1-\frac{\epsilon}{R})} R\,dt$$

$$= R\left(\arccos\left(-1+\frac{\epsilon}{R}\right) - \arccos\left(1-\frac{\epsilon}{R}\right)\right).$$

Finally, since the arccos function arccos : $[-1,1] \to [0,\pi]$ is continuous, we obtain

$$\lim_{\epsilon \to 0} \int_{-R+\epsilon}^{R-\epsilon} \sqrt{1+f'(x)^2}dx = \lim_{\epsilon \to 0} R\left(\arccos\left(-1+\frac{\epsilon}{R}\right) - \arccos\left(1-\frac{\epsilon}{R}\right)\right)$$

$$= R\left(\arccos(-1) - \arccos(1)\right) = R\pi.$$

Finally, since essentially the same argument holds for $-f$, we conclude (as expected) that the length of Γ equals $2\pi R$.

In spite of the former example, the integral in (10.71) is generally difficult (or even *impossible*—cf. Problem 2) to compute exactly. Let's see an illustrative example.

Example 10.81 Given $b > 0$, let's compute the length ℓ of the portion of the parabola $y = x^2$ situated between the points $(0,0)$ and (b,b^2).

With $f(x) = x^2$ in (10.71) we get

$$\ell = \int_0^b \sqrt{1+4x^2}dx.$$

In order to compute such an integral, we apply the trigonometric substitution $x = \frac{1}{2}\tan t$. Since $1 + \tan^2 t = \sec^2 t$, $\frac{d}{dt}\tan t = \sec^2 t$ and $0 \le x \le b \Leftrightarrow 0 \le t \le \arctan(2b)$, it follows from the formula of integration by substitution that

$$\ell = \int_0^{\arctan(2b)} \sqrt{1+\tan^2 t} \cdot \sec^2 t\,dt = \int_0^{\arctan(2b)} \sec^3 t\,dt.$$

Now, item (c) of Problem 15, page 412, gives

$$\int \sec^3 t\,dt = \frac{1}{2}(\sec t \cdot \tan t - \log|\sec t - \tan t|) + C.$$

If we set $\alpha = \arctan(2b)$, then $\tan \alpha = 2b$ and, hence, $\sec \alpha = \sqrt{1 + \tan^2 \alpha} = \sqrt{1 + 4b^2}$. Therefore,

$$
\ell = \left(\frac{1}{2}(\sec t \cdot \tan t - \log | \sec t - \tan t |) + C \right) \Big|_0^\alpha
$$

$$
= b\sqrt{1 + 4b^2} - \frac{1}{2} \log(\sqrt{1 + 4b^2} - 2b).
$$

We now turn to a completely different topic, using Proposition 10.73 to introduce and discuss some of the amazing properties of the *Gamma function*.

The **Gamma function** is the function $\Gamma : (0, +\infty) \to (0, +\infty)$ defined by

$$
\Gamma(x) = \int_0^{+\infty} e^{-t} t^{x-1} \, dt.
$$

Since

$$
\int_0^{+\infty} e^{-t} t^{x-1} \, dt = \int_0^1 e^{-t} t^{x-1} \, dt + \int_1^{+\infty} e^{-t} t^{x-1} \, dt,
$$

in order to see that $\Gamma(x)$ is well defined it suffices to establish the convergence of both improper integrals at the right hand side above.

To what is left to do, initially observe that, since $x > 0$, Problem 3, page 435, guarantees the convergence of $\int_0^1 t^{x-1} \, dt$. Since $0 < e^{-t} t^{x-1} < t^{x-1}$ for $0 < t \leq 1$, Remark 10.74 (modified in the obvious way to be applied to the interval $(0, 1]$) assures the convergence of $\int_0^1 e^{-t} t^{x-1} \, dt$.

On the other hand, for a fixed $x > 0$, we claim that $e^{-t} t^{x-1} \leq C e^{-t/2}$ for $t \geq 1$, where C is a positive constant that only depends on x. Indeed, letting $n > x + 1$ be a natural number, it follows from Theorem 10.54 that

$$
\lim_{t \to +\infty} \frac{t^n}{e^{t/2}} = 0.
$$

Therefore, there exists $C > 0$ such that $\frac{t^n}{e^{t/2}} \leq C$ for every $t \geq 1$. However, if this is so, then

$$
e^{-t} t^{x-1} \leq e^{-t} t^n \leq C e^{-t/2} \tag{10.72}
$$

for every $t \geq 1$, as we wished to show.

Finally, note that an obvious variation of Example 10.69 guarantees the convergence of the improper integral $\int_1^{+\infty} e^{-t/2} \, dt$. Hence, inequality (10.72) allows us to apply Proposition 10.73 (with $A = 1, f(t) = e^{-t} t^{x-1}$ and $g(t) = C e^{-t/2}$) to conclude that $\int_1^{+\infty} e^{-t} t^{x-1} \, dt$ does converge.

As a quick look through the classics [3] and [17] makes it clear, the Gamma function has many important applications in Mathematics, notably in Mathematical

Physics and Analytic Number Theory. Roughly, one can say that this is due to the fact that it extends the notion of factorial of natural numbers to the positive reals.

In order to check the above claim, let's start by noting that Example 10.69 gives

$$\Gamma(1) = \int_0^{+\infty} e^{-t}\,dt = 1.$$

On the other hand, for $x > 0$ the integration by parts formula furnishes

$$\Gamma(x) = \lim_{s \to 0} \int_s^1 e^{-t} t^{x-1}\,dt + \lim_{s \to +\infty} \int_1^s e^{-t} t^{x-1}\,dt$$

$$= \lim_{s \to 0} \left(\frac{1}{x} e^{-t} t^x \Big|_{t=s}^{t=1} + \frac{1}{x} \int_s^1 e^{-t} t^x\,dt \right)$$

$$+ \lim_{s \to +\infty} \left(\frac{1}{x} e^{-t} t^x \Big|_{t=1}^{t=s} + \frac{1}{x} \int_1^s e^{-t} t^x\,dt \right)$$

$$= \frac{1}{x} \int_0^1 e^{-t} t^x\,dt + \frac{1}{x} \int_1^{+\infty} e^{-t} t^x\,dt + \lim_{s \to +\infty} \frac{1}{x} e^{-s} s^x.$$

Now, according to Theorem 10.54, we have $\lim_{s \to +\infty} \frac{1}{x} e^{-s} s^x = 0$. Therefore, it follows from the above computations that

$$\Gamma(x) = \frac{1}{x} \int_0^1 e^{-t} t^x\,dt + \frac{1}{x} \int_1^{+\infty} e^{-t} t^x\,dt = \frac{1}{x} \int_0^{+\infty} e^{-t} t^x\,dt = \frac{1}{x} \Gamma(x+1).$$

Then,

$$\Gamma(x + 1) = x\,\Gamma(x)$$

for every $x > 0$, and an easy induction starting from $\Gamma(1) = 1$ gives $\Gamma(n) = (n-1)!$ for every $n \in \mathbb{N}$.

We now show that Γ is continuous and $\log \Gamma : (0, +\infty) \to \mathbb{R}$ is convex in $(0, +\infty)$. Actually, since

$$\Gamma = e^{\log \Gamma}$$

in view of Proposition 9.68 it suffices to establish the second claim above.

The convexity of $\log \Gamma$ amounts to showing that

$$\log \Gamma((1 - s)x + sy) \le (1 - s) \log \Gamma(x) + s \log \Gamma(y),$$

for all $x, y > 0$ and $0 < s < 1$. Letting $s = \frac{1}{q}$ and $1 - s = \frac{1}{p}$, we have $p, q > 0$ such that $\frac{1}{p} + \frac{1}{q} = 1$. Then, the fact that $\log$ is an increasing function that takes products into sums shows that the validity of the above inequality is equivalent to that of

$$\Gamma\left(\frac{x}{p} + \frac{y}{q}\right) \le \Gamma(x)^{1/p}\Gamma(y)^{1/q}, \tag{10.73}$$

for all $x, y > 0$ and all $p, q > 0$ such that $\frac{1}{p} + \frac{1}{q} = 1$.

In order to establish (10.73), we first observe that

$$\Gamma\left(\frac{x}{p} + \frac{y}{q}\right) = \int_0^{+\infty} e^{-t}t^{\frac{x}{p}+\frac{y}{q}-1}\,dt = \int_0^{+\infty}\left(e^{-t/p}t^{\frac{x-1}{p}} \cdot e^{-t/q}t^{\frac{y-1}{p}}\right)dt.$$

Therefore, letting $f(t) = e^{-t/p}t^{\frac{x-1}{p}}$, $g(t) = e^{-t/q}t^{\frac{y-1}{q}}$ and applying Hölder's inequality (cf. Problem 8, page 436), we get

$$\Gamma\left(\frac{x}{p} + \frac{y}{q}\right) = \int_0^{+\infty} f(t)g(t)dt \le \left(\int_0^{+\infty} f(t)^p\,dt\right)^{1/p}\left(\int_0^{+\infty} g(t)^q\,dt\right)^{1/q}$$

$$= \left(\int_0^{+\infty} e^{-t}t^{x-1}\,dt\right)^{1/p}\left(\int_0^{+\infty} e^{-t}t^{y-1}\,dt\right)^{1/q}$$

$$= \Gamma(x)^{1/p}\Gamma(y)^{1/q}.$$

Remark 10.82 One can actually show that Γ is infinitely differentiable, and that the expression for its $k-$th derivative can be computed by differentiating under the integral that defines Γ. In particular, we must have

$$\Gamma'(x) = \int_0^{+\infty} e^{-t}t^{x-1}\log t\,dt.$$

The convergence of the improper integral at the right hand side above is the object of Problem 5, and the proof of the continuous differentiability of the Gamma function, together with the above formula for $\Gamma'(x)$, will be the object of Problems 18 and 19, page 471.

Problems: Section 10.10

1. For $x_0 \ne 0$, compute the length of the portion of the catenary situated between the points of abscissas 0 and x_0.
2. Let $\mathcal{E}$ be an ellipse of major axis AA' and minor axis BB', with $\overline{AA'} = 2a$ and $\overline{BB'} = 2b$. If we choose a cartesian system in which $A'(a, 0)$ and $B'(0, b)$, it is well known that $\mathcal{E}$ has equation $\frac{x^2}{a^2} + \frac{y^2}{b^2} = 1$. In this respect, do the following items:

(a) If $f : [-a, a] \to \mathbb{R}$ is given by $f(x) = \frac{b}{a}\sqrt{a^2 - x^2}$, show that the graph of f is the portion of $\mathcal{E}$ situated in the upper halfplane. Then, conclude that the length of $\mathcal{E}$ is given by twice the value of the improper integral

$$\int_{-a}^{a} \sqrt{1 + \frac{b^2 x^2}{a^2(a^2 - x^2)}}\, dx.$$

(b) Let $c = \sqrt{a^2 - b^2}$ and $x = a\cos t$, with $0 \le t \le \pi$. Use the formula of integration by substitution to show that the length of $\mathcal{E}$ is given by

$$2b \int_{0}^{\pi} \sqrt{1 + \left(\frac{c}{b}\right)^2 \sin^2 t}\, dt. \tag{10.74}$$

The definite integral of item (b) cannot be computed exactly. More generally, it is possible to show (albeit this is well beyond the scope of these notes) that the indefinite integral

$$\int \sqrt{1 + \kappa \sin^2 t}\, dt, \tag{10.75}$$

where κ is a positive real, cannot be explicitly computed in terms of *elementary functions*.[13] Thanks to (10.74), one says that (10.75) is an **elliptic function** or a **highly transcendental function**. Such functions were brought into prominence by the German mathematician of the XIX century Carl Gustav Jacob Jacobi, and for this reason (10.75) is also known as **Jacobi's function**.

3. We are given in the plane a straight line r and a circle Γ of radius 1, such that r and Γ are tangent to each other. The **cycloid** generated by Γ (cf. Fig. 10.12) is the curve described by a point P on Γ as Γ rolls along r without sliding. In this respect, do the following items:

(a) Fix a position of Γ and let O denote its point of tangency with r at this position. Then, choose a cartesian system xOy having r as horizontal axis

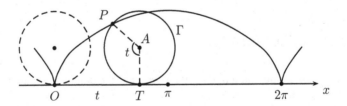

Fig. 10.12 The cycloid generated by Γ

---

[13] A proof of this fact can be found in [12].

and such that Γ is in the upper half-plane. Show that, after Γ rolls in the positive direction (with respect to the horizontal axis) by t radians starting from O, this point will be at position $P(x(t), y(t))$, with $x(t) = t - \sin t$ and $y(t) = 1 - \cos t$.

(b) Note that function $t \mapsto x(t)$, from $[0, 2\pi]$ into itself, is a differentiable bijection with inverse $x \mapsto t(x)$ differentiable in $(0, 2\pi)$. Therefore, if $f : [0, 2\pi] \to \mathbb{R}$ is the composite $f(x) = y(t(x))$, the graph of f is the portion of the cycloid situated in the vertical strip of the cartesian plane defined by the inequalities $0 \le x \le 2\pi$. Such a portion is called a **step** of the cycloid. Compute the length of it.

4. It is possible to show that $\int_0^{+\infty} e^{-s^2} ds = \frac{\sqrt{\pi}}{2}$ (see, for instance, Chap. 8 of [20]). Use this fact to compute $\Gamma(\frac{1}{2}) = \sqrt{\pi}$.

5. Show that, for $x > 0$, the integral $\int_0^{+\infty} e^{-t} t^{x-1} \log t \, dt$ is absolutely convergent.

6. Let $k > 1$ and $n_1, n_2 \dots, n_k$ be naturals such that $k \mid (n_1 + n_2 + \cdots + n_k)$. Prove that

$$\left(\frac{n_1 + n_2 + \cdots + n_k}{k} \right)! \ge \sqrt[k]{n_1! n_2! \dots n_k!}.$$

Then, if k is odd, conclude that

$$\left(\frac{k+1}{2} \right)! \ge \sqrt[k]{1! 2! \dots k!}.$$

Chapter 11
Series of Functions

This last chapter presents the adequate notion of convergence for sequences and series of functions, with an emphasis on the study of *power series*. We start by examining a power series naturally attached to an infinitely differentiable function, called its *Taylor series*. Then, in Sects. 11.2 and 11.3 we discuss the basic results on uniformly convergent series of functions and on power series. Finally, in Sect. 11.4, we present some applications of the theory to sequences defined by linear recurrence relations.

11.1 Taylor Series

This section begins by introducing another important generalization of Lagrange's MVT, namely, **Taylor formula[1] with Lagrange remainder**. In order to properly state it, recall (cf. Sect. 9.3) that a function $f : [a, b] \to \mathbb{R}$ is *n times continuously differentiable* in $[a, b]$ if f is n times differentiable in $[a, b]$ and $f^{(n)} : [a, b] \to \mathbb{R}$ is continuous.

Theorem 11.1 (Taylor) *Let I be an interval and $f : I \to \mathbb{R}$ be a function n times differentiable in I. Given distinct $x_0, x \in I$, there exists c between x_0 and x such that*

$$f(x) = \sum_{k=0}^{n-1} \frac{f^{(k)}(x_0)}{k!}(x - x_0)^k + \frac{f^{(n)}(c)}{n!}(x - x_0)^n. \tag{11.1}$$

---

[1] After Brook Taylor, English mathematician of the XVIII century.

© Springer International Publishing AG 2017
A. Caminha Muniz Neto, *An Excursion through Elementary Mathematics, Volume I*,
Problem Books in Mathematics, DOI 10.1007/978-3-319-53871-6_11

Proof Firstly, assume that $x_0 < x$ and let $g : [x_0, x] \to \mathbb{R}$ be given by

$$g(t) = f(x) - \sum_{k=0}^{n-1} \frac{f^{(k)}(t)}{k!}(x - t)^k - \frac{\alpha}{n!}(x - t)^n,$$

with $\alpha \in \mathbb{R}$ chosen in such a way that $g(x_0) = 0$ (evidently, such a choice is always possible).

Since f is n times differentiable in I and $[x_0, x] \subset I$, several applications of the results of Problem 3, page 300, guarantee that g is differentiable in $[x_0, x]$. Since $g(x_0) = g(x) = 0$, Rôlle's theorem assures the existence of $c \in (x_0, x)$ such that $g'(c) = 0$. On the other hand, a simple computation furnishes

$$g'(t) = \frac{\alpha - f^{(n)}(t)}{(n-1)!}(x - t)^{n-1},$$

so that for $c \in (x_0, x)$ we have $g'(c) = 0$ if and only if $\alpha = f^{(n)}(c)$. Hence,

$$0 = g(x_0) = f(x) - \sum_{k=0}^{n-1} \frac{f^{(k)}(x_0)}{k!}(x - x_0)^k - \frac{f^{(n)}(c)}{n!}(x - x_0)^n,$$

as we wished to show.

Now, let $x < x_0$ and $J = x_0 + x - I = \{x_0 + x - t; \ t \in I\}$. Then J is an interval and $I = x_0 + x - J$, so that we can define $h : J \to \mathbb{R}$ by $h(s) = f(x_0 + x - s)$. Corollary 9.42 assures that h is n times differentiable in J, with $h^{(k)}(s) = (-1)^k f^{(k)}(x_0 + x - s)$ for $0 \le k \le n$. Hence, by applying (11.1) to h and to $s_0 < s$ in J, we obtain

$$h(s) = \sum_{k=0}^{n-1} \frac{h^{(k)}(s_0)}{k!}(s - s_0)^k + \frac{h^{(n)}(c)}{n!}(s - s_0)^n$$

for some $c \in (s_0, s)$. In turn, this gives

$$f(x_0 + x - s) = \sum_{k=0}^{n-1} \frac{(-1)^k f^{(k)}(x_0 + x - s_0)}{k!}(s - s_0)^k + \frac{(-1)^n f^{(n)}(c)}{n!}(s - s_0)^n$$

for some $c \in (s_0, s)$. Finally, letting $s_0 = x$ and $s = x_0$ (so that $s_0 < s$), we get (11.1) for $x < x_0$.   □

If $f : I \to \mathbb{R}$ is a twice differentiable function, then, for distinct $x_0, x \in I$, Taylor's formula with Lagrange remainder guarantees that

$$f(x) = f(x_0) + f'(x_0)(x - x_0) + \frac{f''(c)}{2}(x - x_0)^2 \tag{11.2}$$

for some c between x_0 and x.

Fig. 11.1 The trapezium rule

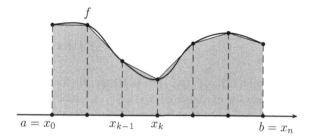

$$a = x_0 \qquad x_{k-1} \quad x_k \qquad b = x_n$$

In what follows, given a positive and twice differentiable function $f : [a, b] \to \mathbb{R}$, we use the above relation to estimate the difference between the area of the region $\mathcal{R}$ under the graph of f and that of the polygonal approximation of $\mathcal{R}$ by the union of the gray trapezoids of Fig. 11.1. For this reason, the coming result is known in mathematical literature as the **trapezium rule**.

Example 11.2 Let $f : [a, b] \to \mathbb{R}$ be a twice differentiable (but not necessarily positive) function. If $\{a = x_0 < x_1 < \cdots < x_n = b\}$ is a uniform partition of interval $[a, b]$, then

$$\left| \int_a^b f(x)\,dx - \sum_{k=1}^n \frac{1}{2}(f(x_k) + f(x_{k-1}))\left(\frac{b-a}{n}\right) \right| \leq \frac{(b-a)^3}{4n^2} \cdot \sup_{[a,b]} |f''|.$$

Proof For $1 \leq k \leq n$, let $g_k : [x_{k-1}, x_k] \to \mathbb{R}$ be given by

$$g_k(x) = \int_{x_{k-1}}^x f(t)\,dt - \frac{1}{2}(f(x) + f(x_{k-1}))(x - x_{k-1}).$$

Letting Δ denote the expression at the right hand side of the stated inequality, the additivity of the integral, together with the fact that $x_k - x_{k-1} = \frac{b-a}{n}$ for $1 \leq k \leq n$, gives

$$\Delta = \left| \sum_{k=1}^n \left(\int_{x_{k-1}}^{x_k} f(x)\,dx - \frac{1}{2}(f(x_k) + f(x_{k-1}))(x_k - x_{k-1}) \right) \right|$$

$$= \left| \sum_{k=1}^n g_k(x_k) \right| \leq \sum_{k=1}^n |g_k(x_k)|, \tag{11.3}$$

where we used triangle inequality in the last passage above.

Now, since f is twice differentiable, the same happens with g_k. Moreover, $g_k(x_{k-1}) = 0$ and easy computations furnish

$$g_k'(x) = f(x) - \frac{1}{2}f'(x)(x - x_{k-1}) - \frac{1}{2}(f(x) + f(x_{k-1}))$$

and

$$g_k''(x) = -\frac{1}{2}f''(x)(x - x_{k-1}),$$

so that $g_k'(x_{k-1}) = g_k''(x_{k-1}) = 0$. Hence, Taylor's formula with Lagrange remainder guarantees the existence of $c_k \in (x_{k-1}, x_k)$ such that

$$g_k(x_k) = g_k(x_{k-1}) + g_k'(x_{k-1})(x_k - x_{k-1}) + \frac{g_k''(c_k)}{2}(x_k - x_{k-1})^2$$

$$= -\frac{1}{4}f''(c_k)(c_k - x_{k-1})(x_k - x_{k-1})^2.$$

However, since $x_k - x_{k-1} = \frac{b-a}{n}$, it follows from the above computations that

$$|g_k(x_k)| = \frac{1}{4}|f''(c_k)||c_k - x_{k-1}|(x_k - x_{k-1})^2$$

$$\leq \frac{1}{4}|f''(c_k)||x_k - x_{k-1}|^3 = |f''(c_k)|\frac{(b-a)^3}{4n^3}.$$

Finally, by substituting this last expression in estimate (11.3) for Δ, we get

$$\Delta \leq \sum_{k=1}^{n} |g_k(x_k)| \leq \sum_{k=1}^{n} |f''(c_k)|\frac{(b-a)^3}{4n^3} = \frac{(b-a)^3}{4n^3}\sum_{k=1}^{n} |f''(c_k)|$$

$$\leq \frac{(b-a)^3}{4n^3} \cdot n \sup_{[a,b]} |f''| = \frac{(b-a)^3}{4n^2} \cdot \sup_{[a,b]} |f''|.$$

$\square$

In words, the result of the previous example gives a numerical approximation, with controlled error, for the computation of an integral. As the discussion of the computation of the length of an ellipse shows (cf. Problem 2, page 443), sometimes this is the best one can get. In this respect, see also Problem 3 of this section.

For more on the use of Calculus methods to get useful numerical approximations, see Chap. 15 of [2] or Part B of [7].

Back to the development of the theory, let $f : I \to \mathbb{R}$ be infinitely differentiable. For $x_0 \in I$, we say that

$$\sum_{k \geq 0} \frac{f^{(k)}(x_0)}{k!}(x - x_0)^k$$

is the **Taylor series** of f at x_0.

The coming result gives a necessary and sufficient condition for the Taylor series of f at x_0 to converge to f in I.

Proposition 11.3 *Let I be an interval and $f : I \to \mathbb{R}$ be an infinitely differentiable function. If there exists a constant $C \geq 0$ such that $|f^{(n)}(x)| \leq C^n$ for every $n \geq 1$ and every $x \in I$, then, given $x_0 \in I$, we have*

$$f(x) = \sum_{k \geq 0} \frac{f^{(k)}(x_0)}{k!}(x - x_0)^k$$

for every $x \in I$.

Proof Fixed $x \in I$ and $n \in \mathbb{N}$, it follows from Taylor formula with Lagrange remainder that

$$f(x) = \sum_{k=0}^{n-1} \frac{f^{(k)}(x_0)}{k!}(x - x_0)^k + \frac{f^{(n)}(c)}{n!}(x - x_0)^n \qquad (11.4)$$

for some c between x_0 and x. Now,

$$\left| \frac{f^{(n)}(c)}{n!}(x - x_0)^n \right| \leq \frac{(C|x - x_0|)^n}{n!} \xrightarrow{n} 0,$$

where we used the result of Problema 3, page 218, in the last passage above. Therefore, it suffices to let $n \to +\infty$ in (11.4) to get the desired result. □

Example 11.4 Since $|\sin^{(2j)} x| = |\sin x| \leq 1$ and $|\sin^{(2j-1)} x| = |\cos x| \leq 1$, the previous proposition gives

$$\sin x = \sum_{k \geq 0} \frac{\sin^{(k)} 0}{k!} x^k = \sum_{j \geq 1} \frac{(-1)^{j-1}}{(2j-1)!} x^{2j-1}$$

for every $x \in \mathbb{R}$. Analogously,

$$\cos x = \sum_{j \geq 0} \frac{(-1)^j}{(2j)!} x^{2j},$$

for every $x \in \mathbb{R}$. In Sect. 11.5, these series expansions will be the departure points for the rigorous construction of the sine and cosine functions.

We now apply Taylor formula with Lagrange remainder to show that the Taylor series of the exponential function converges to such a function along the whole real line.

Theorem 11.5 *For $x \in \mathbb{R}$, one has*

$$e^x = \sum_{k \geq 0} \frac{1}{k!} x^k. \qquad (11.5)$$

Proof For a fixed $x \in \mathbb{R}$, it follows from Theorem 11.1 (with $x_0 = 0$) that

$$e^x = \sum_{k=0}^{n-1} \frac{x^k}{k!} + e^c \cdot \frac{x^n}{n!}, \tag{11.6}$$

for some c between 0 and x. Now, using the fact that the exponential function is increasing, together with the result of Problem 3, page 218, we obtain

$$\left| \frac{e^c}{n!} x^n \right| = e^c \cdot \frac{|x|^n}{n!} \leq \max\{e^0, e^x\} \cdot \frac{|x|^n}{n!} \xrightarrow{n} 0.$$

Hence, the squeezing principle gives $\lim_{n \to +\infty} e^c \cdot \frac{x^n}{n!} = 0$. Finally, letting $n \to +\infty$ in (11.6), we get the desired result. $\qquad \square$

Note that the previous result largely generalizes Example 7.41. For future reference, we observe that changing x by ax in (11.5), we obtain

$$e^{ax} = \sum_{k \geq 0} \frac{a^k}{k!} x^k \tag{11.7}$$

for every $x \in \mathbb{R}$.

Sometimes, estimates of the error $\frac{f^{(n)}(c)}{n!}(x - x_0)^n$ in (11.1) as simple as those we did above do not suffice to guarantee that the Taylor series of an infinitely differentiable function converges to such a function at all points it could do. We now look at such an example.

Example 11.6 Function $x \mapsto \log(1 + x)$ is defined in the whole interval $(-1, 1)$ and is infinitely differentiable there. Letting $f(x) = \log(1 + x)$ for $|x| < 1$, it's immediate to verify that $f^{(k)}(x) = \frac{(-1)^{k-1}(k-1)!}{(1+x)^k}$ for every $k \in \mathbb{N}$. Hence, $f^{(k)}(0) = (-1)^{k-1}(k - 1)!$, and it immediately follows that the Taylor series of $\log(1 + x)$ centered at 0 is given by

$$\sum_{k \geq 1} \frac{(-1)^{k-1}}{k} x^k. \tag{11.8}$$

Since $\left| \frac{(-1)^{k-1}}{k} x^k \right| \leq |x|^k$ and $\sum_{k \geq 1} |x|^k$ converges when $|x| < 1$, the comparison test for series assures that (11.8) converges whenever $x \in (-1, 1)$. Nevertheless, if we try to estimate the error in (11.1) as we did before, we will only be able to conclude that

$$\log(1 + x) = \sum_{k \geq 1} \frac{(-1)^{k-1}}{k} x^k$$

for $x \in [-\frac{1}{2}, 1)$. Indeed, it follows from (11.1) that, for $0 < |x| < 1$, we have

$$\log(1 + x) = \sum_{k=0}^{n-1} \frac{(-1)^{k-1}}{k} x^k + \frac{(-1)^{n-1}}{n(1 + c)^n} \cdot x^n$$

for some c between 0 and x.

If $0 < x < 1$, then $0 < c < x < 1$ and we get

$$\left| \frac{(-1)^{n-1}}{n(1 + c)^n} \cdot x^n \right| \le x^n \xrightarrow{n} 0.$$

If $-1 < x < 0$, then $-1 < x < c < 0$, so that $1 + c > 1 + x > 0$. Therefore,

$$\left| \frac{(-1)^{n-1}}{n(1 + c)^n} \cdot x^n \right| = \frac{1}{n} \cdot \frac{|x|^n}{(1 + c)^n} \le \frac{1}{n} \cdot \frac{|x|^n}{(1 + x)^n} = \frac{1}{n} \left(\frac{|x|}{1 + x} \right)^n.$$

For $-1 < x < 0$, one easily concludes that $\frac{|x|}{1+x} < 1 \Leftrightarrow -\frac{1}{2} \le x < 0$; yet in this interval, the last estimate above gives

$$\left| \frac{(-1)^{n-1}}{n(1 + c)^n} \cdot x^n \right| \le \frac{1}{n} \xrightarrow{n} 0.$$

On the other hand, for $-1 < x < -\frac{1}{2}$ we have $\frac{|x|}{1+x} > 1$, so that

$$\frac{1}{n} \left(\frac{|x|}{1 + x} \right)^n \xrightarrow{n} +\infty;$$

hence, in the interval $\left(-1, -\frac{1}{2} \right)$ we are not able to conclude that $\left| \frac{(-1)^{n-1}}{n(1+c)^n} \cdot x^n \right| \to 0$ when $n \to +\infty$.

We shall remedy situations like that of the previous example in Sect. 11.3. For the time being, let us show that it may well happen that the Taylor series of an infinitely differentiable function $f : \mathbb{R} \to \mathbb{R}$, centered at some $x_0 \in \mathbb{R}$, does not converge to f in any open interval centered at x_0.

Example 11.7 Let $f : \mathbb{R} \to \mathbb{R}$ be given by

$$f(x) = \begin{cases} e^{-1/x}, & \text{if } x > 0 \\ 0, & \text{if } x \le 0 \end{cases}.$$

Then, f is infinitely differentiable in $\mathbb{R}$, with $f^{(k)}(0) = 0$ for every $k \in \mathbb{Z}_+$. In particular, the Taylor series of f centered at 0 vanishes identically.

Proof Evidently, $\lim_{x\to 0+} e^{-1/x} = 0$, so that f is continuous at 0, thus, in the whole real line. Also evidently, f is infinitely differentiable in $(0, +\infty)$ and in $(-\infty, 0)$, with $f^{(n)}(x) = 0$ if $x < 0$ and $n \in \mathbb{Z}_+$.

We claim that, for $n \in \mathbb{Z}_+$, there exists a polynomial p_n such that

$$f^{(n)}(x) = p_n(1/x)e^{-1/x} \tag{11.9}$$

for every $x > 0$. Let's check this by induction on $n \geq 0$, the case $n = 0$ being obvious. By induction hypothesis, assume that for some integer $k \geq 0$ there exists a polynomial p_k such that $f^{(k)}(x) = p_k(1/x)e^{-1/x}$ for every $x > 0$. Then, the chain rule gives, for $x > 0$,

$$f^{(k+1)}(x) = \frac{1}{x^2}\left(p_k\left(\frac{1}{x}\right) - p_k'\left(\frac{1}{x}\right)\right)e^{-1/x} = p_{k+1}(1/x)e^{-1/x},$$

with $p_{k+1}(x) = x^2(p_k(x) - p_k'(x))$, so that p_{k+1} is again a polynomial.

Now that we have (11.9), let's make one more induction to prove that f is infinitely differentiable at 0, with $f^{(n)}(0) = 0$ for every $n \in \mathbb{Z}_+$. By induction hypothesis, assume that f is k times differentiable at 0, with $f^{(k)}(0) = 0$. Since f is infinitely differentiable in $\mathbb{R} \setminus \{0\}$, in order to show that f is $k+1$ times differentiable at 0, with $f^{(k+1)}(0) = 0$, it suffices to show that $f^{(k+1)}(0) = 0$. To this end, note that for $h > 0$ we have

$$\frac{f^{(k)}(h) - f^{(k)}(0)}{h} = \frac{p_k(1/h)e^{-1/h}}{h} = q_k(1/h)e^{-1/h},$$

where $q_k(x) = xp_k(x)$, a polynomial; then, again by Theorem 10.54, we have

$$\lim_{h\to 0+} \frac{f^{(k)}(h) - f^{(k)}(0)}{h} = \lim_{h\to 0+} \frac{q_k(1/h)}{e^{1/h}} = \lim_{y\to +\infty} \frac{q_k(y)}{e^y} = 0.$$

On the other hand, for $h < 0$ we have $\lim_{h\to 0-} \frac{f^{(k)}(h) - f^{(k)}(0)}{h} = 0$, since $f^{(k)}(h) = f^{(k)}(0) = 0$. Therefore, $f^{(k+1)}(0)$ exists and equals 0. $\qquad\square$

Problems: Section 11.1

1. Let I be an open interval and $f : I \to \mathbb{R}$ be convex and twice differentiable in I. If $x_0 \in I$ and r denotes the tangent to the graph of f at the point $(x_0, f(x_0))$, show that no point on the graph of f lies under r.

2. Let I be an open interval, $f : I \to \mathbb{R}$ be n times continuously differentiable and $x_0 \in I$ be such that $f'(x_0) = f''(x_0) = \cdots = f^{(n-1)}(x_0) = 0$. If n is even and

$f^{(n)}(x_0) > 0$ (resp. $f^{(n)}(x_0) < 0$), show that x_0 is a point of *strict* local minimum (resp. maximum) for f, i.e.,

$$x \in I \setminus \{x_0\} \text{ sufficiently close to } x_0 \Rightarrow f(x) > f(x_0) \text{ (resp.,} f(x) < f(x_0)).$$

3. Let $\ell(\mathcal{E})$ be the length of the ellipse $\mathcal{E}$ of major axis $2a$, minor axis $2b$ and focal distance $2c$. Show that, for every $n \in \mathbb{N}$, one has

$$\left| \ell(\mathcal{E}) - \frac{2\pi b}{n} \left(1 + \sum_{j=1}^{n-1} \sqrt{1 + \kappa^2 \sin^2\left(\frac{j\pi}{n}\right)} \right) \right| \leq \frac{2\kappa(\kappa + 1)b\pi^3}{n^2},$$

where $\kappa = \frac{c}{b}$.

4. Obtain the Taylor series of the functions sinh and cosh. In each case, show that it converges to the corresponding function in the whole real line.

5. Let I be an interval and $f : I \to \mathbb{R}$ be n times differentiable and such that $f^{(n)}$ is constant. Prove that f either vanishes identically or is a polynomial of degree at most n. More precisely, if $x, x_0 \in I$, show that

$$f(x) = \sum_{j=0}^{n} \frac{f^{(j)}(x_0)}{j!} (x - x_0)^j.$$

6. Use (11.6) to give another proof of Theorem 10.54.

The next problem revisits the analysis of **Newton's method** for numerical approximations of roots of differentiable functions (cf. Problem 9, page 300).

7. Let $f : [a, b] \to \mathbb{R}$ be continuous in $[a, b]$ and twice continuously differentiable in (a, b), with $f', f'' > 0$ in (a, b). Assume that $f(a) < 0 < f(b)$, and let α be the only root of f in the interval $[a, b]$. Do the following items:

   (a) If $\beta \in (\alpha, b]$ and $\gamma = \beta - \frac{f(\beta)}{f'(\beta)}$, show that $\alpha \leq \gamma < \beta$.

   (b) If $(a_n)_{n \geq 1}$ is such that $a_1 \in (\alpha, b]$ and $a_{n+1} = a_n - \frac{f(a_n)}{f'(a_n)}$ for every $n \geq 1$, show that $(a_n)_{n \geq 1}$ converges and $a_n \to \alpha$ as $n \to +\infty$.

   (c) Refine the analysis of item (b) in the following way:

   (i) Show that there exists $\xi_n \in (\alpha, a_n)$ such that $f(a_n) = f'(a_n)(a_n - \alpha) - \frac{1}{2}f''(\xi_n)(a_n - \alpha)^2$.

   (ii) Conclude that $0 \leq a_{n+1} - \alpha = \frac{f''(\xi_n)}{2f'(a_n)}(a_n - \alpha)^2 \leq \frac{f''(\xi_n)}{2f'(\xi_n)}(a_n - \alpha)^2$.

   (iii) Suppose that $a < c < d < b$ satisfy $d - c < 1$ and $f(c) < 0 < f(d)$. If $\lambda = \max_{[c,d]} \frac{f''}{2f'}$ and we start with $a_1 = d$, use (ii) to successively conclude that $0 \leq a_{n+1} - \alpha \leq \lambda(a_n - \alpha)^2$ and $0 \leq a_n - \alpha \leq \lambda^n(d-c)^{2^n}$ for every $n \geq 1$.

Yet with respect to the previous problem, note that Problem 10, page 219, guarantees that $\lambda^n(d - c)^{2^n} \to 0$ as $n \to +\infty$. Hence, the result of item (iii) above estimates the error with which a_n approaches α. Note also that we can get results similar to those of the previous problem by assuming that $f' > 0$ and

$f'' < 0$ in (a, b); it suffices to start with $a_1 \in [a, \alpha)$. The same is true if we assume that $f' < 0$ and $f'' < 0$ (resp. $f'' > 0$) in (a, b); it suffices to start with $a_1 \in (\alpha, b]$ (resp. $a_1 \in [a, \alpha)$).

8. The polynomial $f(x) = x^3 - 2x - 5$ is such that $f(2) = -1$ and $f(\frac{5}{2}) = \frac{9}{2}$. Therefore, the IVT guarantees the existence of a root $\alpha \in (2, \frac{5}{2})$ for f. Apply the results of the previous problem to estimate α with five correct decimal places.

11.2   Series of Functions

The material on this section is a prelude to the study of power series and extends, to sequences and series of functions, some concepts and results of Sects. 7.2 and 7.4. We start by defining the concept of limit of a *sequence of functions*.

In all that follows, unless stated otherwise, I denotes an interval.

Definition 11.8 For each $n \in \mathbb{N}$, let a function $f_n : I \rightarrow \mathbb{R}$ be given. If $\lim_{n \to +\infty} f_n(x)$ exists for each $x \in I$, we define the **pointwise limit** of the sequence of functions $(f_n)_{n \geq 1}$ as the function $f : I \rightarrow \mathbb{R}$ such that

$$f(x) = \lim_{n \to +\infty} f_n(x) \tag{11.10}$$

for every $x \in I$. In this case, we also say that $(f_n)_{n \geq 1}$ is a **pointwise convergent** sequence of functions, or that $(f_n)_{n \geq 1}$ **converges pointwise** to f.

The pointwise limit of a sequence $f_n : I \rightarrow \mathbb{R}$ of functions, if exists, is unique. Indeed, if $(f_n)_{n \geq 1}$ converges pointwise to $f : I \rightarrow \mathbb{R}$, it follows from (11.10) that for each $x \in I$ the sequence of real numbers $(f_n(x))_{n \geq 1}$ converges to $f(x)$; hence, the uniqueness of the limit of a convergent sequence of reals (cf. Proposition 7.14) assures that there exists only one possible value for $f(x)$.

Example 11.9 We collect here two examples illustrating the fact that the pointwise limit of a pointwise convergent sequence of functions is not necessarily a *well behaved* function.

(a) For each $n \in \mathbb{N}$, let $f_n : [0, 1] \rightarrow \mathbb{R}$ be such that $f_n(x) = x^n$ for every $x \in [0, 1]$. It follows from Example 7.12 that the pointwise limit of $(f_n)_{n \geq 1}$ exists and is the function $f : [0, 1] \rightarrow \mathbb{R}$ such that

$$f(x) = \begin{cases} 0, & \text{if } x \in [0, 1) \\ 1, & \text{if } x = 1 \end{cases}.$$

In particular, this example shows that the (pointwise) limit of a pointwise convergent sequence of continuous (actually, even infinitely differentiable) functions can be discontinuous.

(b) For each $n \in \mathbb{N}$, let $f_n : [0, 1] \to \mathbb{R}$ be given by $f_n(x) = nx(1 - x^2)^n$ for every $x \in [0, 1]$. Since $0 < x < 1 \Rightarrow 0 < 1 - x^2 < 1$, Problem 4, page 218, assures that $\lim_{n \to +\infty} nx(1 - x^2)^n = 0$ if $0 < x < 1$. Also, since $f_n(0) = f_n(1) = 0$ for every $n \in \mathbb{N}$, we conclude that the pointwise limit of the sequence $(f_n)_{n \geq 1}$ exists and equals the function $f : [0, 1] \to \mathbb{R}$ that vanishes identically in $[0, 1]$; in particular, f is integrable, with $\int_0^1 f(x)dx = 0$. Now, note that the FTC gives

$$\int_0^1 f_n(x)dx = -\frac{n(1 - x^2)^{n+1}}{2(n + 1)} \Big|_{x=0}^{x=1} = \frac{n}{2(n + 1)}.$$

Hence, $\lim_{n \to +\infty} \int_0^1 f_n(x)dx = \lim_{n \to +\infty} \frac{n}{2(n+1)} = \frac{1}{2}$, so that

$$\lim_{n \to +\infty} \int_0^1 f_n(x)dx \neq \int_0^1 \lim_{n \to +\infty} f_n(x)dx.$$

We now introduce a notion of convergence for sequences of functions which is stronger than that of pointwise convergence. As we shall see right after the coming definition, under such more restrictive notion of convergence pathologies like those of the previous will not take place.

Definition 11.10 A sequence $(f_n)_{n \geq 1}$ of functions $f_n : I \to \mathbb{R}$ **converges uniformly** for a function $f : I \to \mathbb{R}$ if the following condition is satisfied: given $\epsilon > 0$, there exists $n_0 \in \mathbb{N}$ such that

$$n \geq n_0 \Rightarrow |f_n(x) - f(x)| < \epsilon, \ \forall x \in I. \tag{11.11}$$

In words, a sequence $(f_n)_{n \geq 1}$ of real functions defined in I converges uniformly to $f : I \to \mathbb{R}$ if the choice of a *sufficiently large* index n ($n \geq n_0$, in the notations of the former definition) makes $|f_n(x) - f(x)| < \epsilon$ for every $x \in I$.

Example 11.11 For $n \in \mathbb{N}$, let $f_n : \mathbb{R} \to \mathbb{R}$ be given by $f_n(x) = (\frac{x}{1+x^2})^n$. Then, $(f_n)_{n \geq 1}$ converges uniformly to the function that vanishes identically in $\mathbb{R}$.

Proof Since $\frac{|x|}{1+x^2} \leq \frac{1}{2}$ for every $x \in \mathbb{R}$, we have $|f_n(x) - 0| = |f_n(x)| \leq \frac{1}{2^n}$ for every $x \in \mathbb{R}$. In view of this inequality and given $\epsilon > 0$, choose $n_0 \in \mathbb{N}$ such that $\frac{1}{2^n} < \epsilon$ for $n \geq n_0$. Then, for $n \geq n_0$ we have $|f_n(x) - 0| < \epsilon$ for every $x \in \mathbb{R}$, as wished. $\square$

The coming corollary is an immediate consequence of Definition 11.10.

Corollary 11.12 *If a sequence $(f_n)_{n \geq 1}$ of functions $f_n : I \to \mathbb{R}$ converges uniformly to $f : I \to \mathbb{R}$, then $(f_n)_{n \geq 1}$ converges pointwise to f.*

The concept of uniform convergence has the following *geometric interpretation*: since

$$|f_n(x) - f(x)| < \epsilon \Leftrightarrow f(x) - \epsilon < f_n(x) < f(x) + \epsilon,$$

Fig. 11.2 Geometric
interpretation of uniform
converge

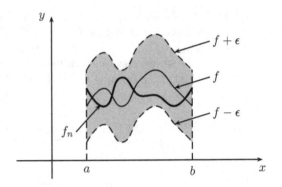

we conclude that $(f_n)_{n \geq 1}$ converges uniformly to $f : I \to \mathbb{R}$ if, given arbitrarily
$\epsilon > 0$, there exists $n_0 \in \mathbb{N}$ such that, for $n \geq n_0$, the graph of f_n is contained in
the *strip of the plane of width 2ϵ centered in the graph of f* (the gray region of
Fig. 11.2).

The actual importance of the concept of uniform convergence lies in the coming
Convergence Theorems 11.13 and 11.14. In particular, comparing such results with
the sequences of functions collected in Example 11.9, we conclude that the converse
of Corollary 11.12 is not necessarily true. In other words, we conclude that there
exist sequences of functions $f_n : I \to \mathbb{R}$ such that $(f_n)_{n \geq 1}$ converges pointwise, but
not uniformly, to some function $f : I \to \mathbb{R}$.

Theorem 11.13 *If a sequence $(f_n)_{n \geq 1}$ of continuous functions $f_n : I \to \mathbb{R}$ converges
uniformly to $f : I \to \mathbb{R}$, then f is continuous.*

Proof Given $x_0 \in I$ and $n \in \mathbb{N}$, the triangle inequality gives

$$|f(x) - f(x_0)| \leq |f(x) - f_n(x)| + |f_n(x) - f_n(x_0)| + |f_n(x_0) - f(x_0)|.$$

Now, for a given $\epsilon > 0$, the uniform convergence of $(f_n)_{n \geq 1}$ to f guarantees the
existence of $n_0 \in \mathbb{N}$ such that $n \geq n_0 \Rightarrow |f_n(x) - f(x)| < \frac{\epsilon}{4}$ for every $x \in I$. Hence,
by writing the previous inequality for $n = n_0$ we get

$$|f(x) - f(x_0)| \leq 2 \cdot \frac{\epsilon}{4} + |f_{n_0}(x) - f_{n_0}(x_0)| = \frac{\epsilon}{2} + |f_{n_0}(x) - f_{n_0}(x_0)|.$$

On the other hand, the continuity of f_{n_0} assures the existence of $\delta > 0$ such that

$$x \in I, \ |x - x_0| < \delta \Rightarrow |f_{n_0}(x) - f_{n_0}(x_0)| < \frac{\epsilon}{2}.$$

Therefore, for $x \in I$ such that $|x - x_0| < \delta$, we have

$$|f(x) - f(x_0)| \leq \frac{\epsilon}{2} + |f_{n_0}(x) - f_{n_0}(x_0)| < \frac{\epsilon}{2} + \frac{\epsilon}{2} = \epsilon,$$

so that f is continuous at x_0.

Finally, since x_0 was chosen arbitrarily in I, we conclude that f is continuous in I.

$\square$

Theorem 11.14 *Let* $(f_n)_{n\geq 1}$ *be a sequence of continuous functions* $f_n : [a, b] \to \mathbb{R}$, *converging uniformly to* $f : [a, b] \to \mathbb{R}$. *If* $g_n, g : [a, b] \to \mathbb{R}$ *are defined by*

$$g_n(x) = \int_a^x f_n(t)dt \quad and \quad g(x) = \int_a^x f(t)dt,$$

then $(g_n)_{n\geq 1}$ *converges uniformly to* g. *In particular,*

$$\int_a^b f(x)dx = \lim_{n\to+\infty} \int_a^b f_n(x)dx. \tag{11.12}$$

Proof For $x \in [a, b]$, it follows from the triangle inequality for integrals (cf. Proposition 10.14) that

$$|g_n(x) - g(x)| = \left| \int_a^x (f_n(t) - f(t))dt \right| \leq \int_a^x |f_n(t) - f(t)|dt.$$

Now, given $\epsilon > 0$, the uniform convergence of $(f_n)_{n\geq 1}$ to f gives $n_0 \in \mathbb{N}$ such that

$$n \geq n_0 \Rightarrow |f_n(t) - f(t)| \leq \frac{\epsilon}{b - a}$$

for every $t \in [a, b]$. Hence, for $n \geq n_0$ and $x \in [a, b]$, the above inequalities allow us to estimate

$$|g_n(x) - g(x)| \leq \int_a^x \frac{\epsilon}{b - a} dt = \frac{\epsilon}{b - a} \cdot (x - a) \leq \frac{\epsilon}{b - a} \cdot (b - a) = \epsilon.$$

For the second part, recall that uniform convergence implies pointwise convergence. Therefore, $g_n(b) \to g(b)$ as $n \to +\infty$, which is the same as (11.12). □

Problem 5 extends the above result to the realm of integrable functions. Actually, the following important remark holds true.

Remark 11.15 In the notations of the previous result, a much more general result is available. In order to state it properly, let $I \subset \mathbb{R}$ be an interval with endpoints α and β (in the sense of Sect. 10.9) and $f_n : I \to \mathbb{R}$ be given integrable functions (in the improper sense, if I or f is unbounded), with $(f_n)_{n\geq 1}$ converging *pointwise* to an integrable function $f : I \to \mathbb{R}$. If there exists an integrable function $F : I \to [0, +\infty)$ such that

$$|f_n(x)| \leq F(x), \ \forall \ x \in I, \ n \in \mathbb{N}, \tag{11.13}$$

then it is possible to show that

$$\int_\alpha^\beta f(x)dx = \lim_{n\to+\infty} \int_\alpha^\beta f_n(x)dx.$$

This is the content of **Lebesgue's dominated convergence theorem** (we abbreviate **DCT**), whose proof is beyond the scope of these notes. For the interested reader, we refer to [20] or [27].

Also with respect to Lebesgue's DCT, it is worth observing that, if $I = [a, b]$, it suffices to assume that there exists $L > 0$ such that

$$|f_n(x)| \leq L, \ \forall \ x \in I, \ n \in \mathbb{N}.$$

On the other hand, item (b) of Example 11.9 shows that such a condition is necessary for Lebesgue's DCT to hold true.

Finally, Problem 4 shows that proving (11.12) in a particular case and without the aid of Lebesgue's DCT can be a somewhat difficult task. Problem 18 asks you to prove a very important consequence of Lebesgue's DCT.

We continue our study of uniform convergence by presenting the famous **Weierstrass approximation theorem**, that states that every continuous function defined in a closed and bounded interval can be uniformly approximated by a sequence of polynomials.

Theorem 11.16 (Weierstrass) *Every continuous function $f : [a, b] \to \mathbb{R}$ is the uniform limit of a sequence of polynomial functions $p_n : [a, b] \to \mathbb{R}$.*

Before we jump into the proof, it is worth to do some heuristics to motivate it. For $n \in \mathbb{N}$, the binomial theorem gives

$$f(x) = f(x)\big(x + (1 - x)\big)^n = f(x) \sum_{k=0}^{n} \binom{n}{k} x^k (1 - x)^{n-k}$$

Letting $a_0 = a < a_1 < a_2 < \ldots < a_n = b$, with $a_k = a + \frac{k}{n}(b - a)$, the uniform continuity of f guarantees that, for large n, the values $f(x)$ for $x \in [a_{k-1}, a_k]$ do not differ too much from $f(a_k)$. Since every $x \in [a, b]$ belongs to one of the intervals $[a_{k-1}, a_k]$, we expect that

$$f(x) \cong \sum_{k=0}^{n} f(a_k) \binom{n}{k} x^k (1 - x)^{n-k}$$

for large n, and the right hand side is a polynomial function. This is precisely what we shall prove.

Proof To make the forthcoming computations a little simpler, assume first that $a = 0$ and $b = 1$, and let

$$p_n(x) = \sum_{k=0}^{n} f\left(\frac{k}{n}\right) \binom{n}{k} x^k (1 - x)^{n-k}.$$

Since f is uniformly continuous in $[0, 1]$, given $\epsilon > 0$ we can choose $\delta > 0$ such that $|f(x) - f(y)| < \epsilon$ whenever $x, y \in [0, 1]$ satisfy $|x - y| < \delta$. Letting $M = \max\{|f(t)|; t \in [0, 1]\}$, we have for a fixed $x \in [0, 1]$ that

$$|f(x) - p_n(x)| = \left| \sum_{k=0}^{n} \left(f(x) - f\left(\frac{k}{n}\right) \right) \binom{n}{k} x^k (1-x)^{n-k} \right|$$

$$\leq \sum_{\substack{0 \leq k \leq n \\ |x - \frac{k}{n}| < \delta}} \left| f(x) - f\left(\frac{k}{n}\right) \right| \binom{n}{k} x^k (1-x)^{n-k}$$

$$+ \sum_{\substack{0 \leq k \leq n \\ |x - \frac{k}{n}| \geq \delta}} \left| f(x) - f\left(\frac{k}{n}\right) \right| \binom{n}{k} x^k (1-x)^{n-k}$$

$$\leq \sum_{\substack{0 \leq k \leq n \\ |x - \frac{k}{n}| < \delta}} \epsilon \binom{n}{k} x^k (1-x)^{n-k} + \sum_{\substack{0 \leq k \leq n \\ |x - \frac{k}{n}| \geq \delta}} 2M \binom{n}{k} x^k (1-x)^{n-k}.$$

In the last line above, the first summand does not exceed

$$\sum_{k=0}^{n} \epsilon \binom{n}{k} x^k (1-x)^{n-k} = \epsilon (x + (1-x))^n = \epsilon.$$

On the other hand, the difficulty in estimating the second summand lies in estimating how many integers $0 \leq k \leq n$ satisfy the inequality $|x - \frac{k}{n}| \geq \delta$. We overcome this by inserting the factor $\frac{1}{\delta^2} \left(x - \frac{k}{n} \right)^2 \geq 1$ under the Σ sign to get

$$2M \sum_{\substack{0 \leq k \leq n \\ |x - \frac{k}{n}| \geq \delta}} \binom{n}{k} x^k (1-x)^{n-k} \leq \frac{2M}{\delta^2} \sum_{k=0}^{n} \left(x - \frac{k}{n} \right)^2 \binom{n}{k} x^k (1-x)^{n-k}.$$

Letting

$$S = \sum_{k=0}^{n} \left(x - \frac{k}{n} \right)^2 \binom{n}{k} x^k (1-x)^{n-k},$$

we proved that

$$|f(x) - p_n(x)| \leq \epsilon + \frac{2M}{\delta^2} S \tag{11.14}$$

for every $x \in [0, 1]$.

Substituting $\left(x - \frac{k}{n}\right)^2 = x^2 - \frac{2k}{n}x + \left(\frac{k}{n}\right)^2$ we compute

$$S = x^2 \sum_{k=0}^{n} \binom{n}{k} x^k (1-x)^{n-k} - 2x \sum_{k=0}^{n} \frac{k}{n}\binom{n}{k} x^k (1-x)^{n-k}$$

$$+ \sum_{k=0}^{n} \frac{k^2}{n^2}\binom{n}{k} x^k (1-x)^{n-k}$$

$$= x^2 - 2x^2 \sum_{k=1}^{n} \binom{n-1}{k-1} x^{k-1}(1-x)^{n-k} + \sum_{k=1}^{n} \frac{k}{n}\binom{n-1}{k-1} x^k (1-x)^{n-k}$$

$$= x^2 - 2x^2 + \sum_{k=1}^{n} \frac{k-1}{n}\binom{n-1}{k-1} x^k (1-x)^{n-k} + \sum_{k=1}^{n} \frac{1}{n}\binom{n-1}{k-1} x^k (1-x)^{n-k}$$

$$= -x^2 + \left(\frac{n-1}{n}\right) \sum_{k=1}^{n} \frac{k-1}{n-1}\binom{n-1}{k-1} x^k (1-x)^{n-k} + \frac{x}{n}$$

$$= -x^2 + \left(\frac{n-1}{n}\right) x^2 \sum_{k=2}^{n} \binom{n-2}{k-2} x^{k-2}(1-x)^{n-k} + \frac{x}{n}$$

$$= -x^2 + \left(\frac{n-1}{n}\right) x^2 + \frac{x}{n} = \frac{1}{n}(x - x^2),$$

so that $S \leq \frac{1}{4n}$ on the interval $[0, 1]$.

Therefore, back to (11.14) we get

$$|f(x) - p_n(x)| \leq \epsilon + \frac{M}{2n\delta^2},$$

for every $x \in [0, 1]$. Now, it suffices to see that this last expression is less than 2ϵ, provided $n > \frac{M}{2\epsilon\delta^2}$.

For the general case, let $f : [a, b] \to \mathbb{R}$ be a continuous function. Let $g : [0, 1] \to \mathbb{R}$ be given by $g(x) = f\big((1-x)a + xb\big)$, so that g is also continuous. What we did so far guarantees the existence of a sequence $(q_n)_{n \geq 1}$ of polynomial functions such that $q_n \xrightarrow{n} g$ uniformly on $[0, 1]$. If $p_n(y) = q_n\left(\frac{y-a}{b-a}\right)$, then p_n is obviously a polynomial function and, for $y \in [a, b]$,

$$|f(y) - p_n(y)| = \left| g\left(\frac{y-a}{b-a}\right) - q_n\left(\frac{y-a}{b-a}\right) \right|.$$

Thus, $p_n \xrightarrow{n} f$ uniformly on $[a, b]$. $\square$

Example 11.17 Let $f : [a, b] \to \mathbb{R}$ be a continuously differentiable function. Prove that there exists a sequence $(p_n)_{n\geq 1}$ of real polynomials such that $p_n \overset{n}{\longrightarrow} f$ and $p_n' \overset{n}{\longrightarrow} f'$ uniformly on $[a, b]$.

Proof By the Weierstrass approximation theorem, we can choose a sequence $(q_n)_{n\geq 1}$ of polynomial functions such that $q_n \overset{n}{\longrightarrow} f'$ uniformly on $[a, b]$. Letting

$$p_n(x) = \int_a^x q_n(t)dt + f(a)$$

for $x \in [a, b]$, we get a polynomial function satisfying $p_n(a) = f(a)$ and $p_n' = q_n$ for every $n \in \mathbb{N}$.

Hence, we can use the FTC again to write

$$f(x) - p_n(x) = \int_a^x f'(t)dt - \int_a^x q_n(t)dt = \int_a^x \big(f'(t) - q_n(t)\big)dt$$

for every $x \in [a, b]$, so that

$$|f(x) - p_n(x)| \leq \int_a^x |f'(t) - q_n(t)|dt \leq \int_a^b |f'(t) - q_n(t)|dt.$$

Since $q_n \to f'$ uniformly as $n \to +\infty$, given $\epsilon > 0$ we can find $n_0 \in \mathbb{N}$ such that $|f'(t) - q_n(t)| < \epsilon$ for every $t \in [a, b]$ and every $n > n_0$. Therefore, for $n > n_0$, we have

$$|f(x) - p_n(x)| \leq \int_a^b \epsilon\, dt = \epsilon(b - a)$$

for every $x \in [a, b]$.   □

We now turn our attention to the study of *series of functions*.

Definition 11.18 Given a sequence $(f_n)_{n\geq 1}$ of functions $f_n : I \to \mathbb{R}$, we define the **series of functions** $\sum_{k\geq 1} f_k$ as a shorthand for the sequence $(s_n)_{n\geq 1}$ of functions $s_n : I \to \mathbb{R}$, such that $s_n = \sum_{k=1}^n f_k$ for every $n \geq 1$.

In the notations of the former definition, we say that $\sum_{k\geq 1} f_k$ converges pointwise (resp. uniformly) in I to $f : I \to \mathbb{R}$ if the sequence $(s_n)_{n\geq 1}$ converges pointwise (resp. uniformly) to f. In this case, we write

$$f = \sum_{k\geq 1} f_k$$

and note that $f(x) = \sum_{k\geq 1} f_k(x)$ for every $x \in I$.

We shall generally apply Theorems 11.13 and 11.14 to uniformly convergent series of functions. In this sense, the coming corollary is of paramount importance for us. Note that, in words, its item (b) says that uniformly convergent series of continuous functions can be integrated *termwise*.

Corollary 11.19 *For each $n \geq 1$, let $f_n : [a, b] \to \mathbb{R}$ be a continuous function. If the series $\sum_{k \geq 1} f_k$ converges uniformly to $f : [a, b] \to \mathbb{R}$, then:*

(a) f *is continuous.*
(b) $\int_a^b \sum_{k \geq 1} f_k(x) dx = \sum_{k \geq 1} \int_a^b f_k(x) dx.$

Proof (a) If $s_n = \sum_{k=1}^n f_k$, then s_n is a finite sum of continuous functions, so that it is itself continuous. Since $s_n \to f$ uniformly, Theorem 11.13 guarantees that f is a continuous function.

(b) Since $s_n \to f$ uniformly, Theorem 11.14, together with the additivity of the integral, gives

$$\int_a^b f(x) dx = \lim_{n \to +\infty} \int_a^b s_n(x) dx = \lim_{n \to +\infty} \int_a^b \sum_{k=1}^n f_k(x) dx$$

$$= \lim_{n \to +\infty} \sum_{k=1}^n \int_a^b f_k(x) dx = \sum_{k \geq 1} \int_a^b f_k(x) dx.$$

$\square$

The former corollary will only be useful if we have an efficient way of finding out, in cases of interest, whether or not a given series of functions is uniformly convergent. The coming result, known in mathematical literature as **Weierstrass M-test**, provides a simple sufficient condition for the uniform convergence of a series of functions. Note that the "*M*" stands for *majorization*.

Theorem 11.20 (Weierstrass M-Test) *Let $\sum_{k \geq 1} f_k$ be a series of functions defined in an interval I and satisfying the following conditions:*

(a) *For each $k \geq 1$, there exists $M_k > 0$ such that $|f_k(x)| \leq M_k$ for every $x \in I$.*
(b) *The series $\sum_{k \geq 1} M_k$ converges.*

Then, the series of functions $\sum_{k \geq 1} f_k$ converges uniformly in I. In particular, if all of the f_k's are continuous in I, then so is $\sum_{k \geq 1} f_k$.

Proof Given $x \in I$, since $|f_k(x)| \leq M_k$ for every $n \geq 1$ and $\sum_{k \geq 1} M_k$ converges, the comparison test for series of real numbers (cf. Proposition 7.44) guarantees the absolute convergence of the series $\sum_{k \geq 1} f_k(x)$. Hence, we get a well defined function $f : I \to \mathbb{R}$ such that

$$f(x) = \sum_{k \geq 1} f_k(x), \ \forall \ x \in I.$$

In order to establish the uniform convergence of $\sum_{k \geq 1} f_k$, let $s_n = \sum_{k=1}^{n} f_k$. For $x \in I$ and $n \in \mathbb{N}$, we have

$$|f(x) - s_n(x)| = \left| \sum_{k>n} f_k(x) \right| \leq \sum_{k>n} |f_k(x)| \leq \sum_{k>n} M_k. \qquad (11.15)$$

Now, given $\epsilon > 0$, the convergence of the sequence $\left(\sum_{k=1}^{n} M_k \right)_{n \geq 1}$ to $\sum_{k \geq 1} M_k$ assures the existence of $n_0 \in \mathbb{N}$ such that

$$\left| \sum_{k \geq 1} M_k - \sum_{k=1}^{n} M_k \right| < \epsilon, \ \forall \ n \geq n_0.$$

Hence, $\left| \sum_{k>n} M_k \right| < \epsilon$ for every $n \geq n_0$.

Back to (11.15) we conclude that

$$|f(x) - s_n(x)| \leq \sum_{k>n} M_k < \epsilon$$

for $n \geq n_0$ and every $x \in I$. Therefore, $(s_n)_{n \geq 1}$ converges uniformly to f, and this is the same as saying that $\sum_{k \geq 1} f_k$ converges uniformly to f.

The last part follows at once from Theorem 11.13. $\qquad \square$

Example 11.21 The Weierstrass M-test can be used to show that the series of functions $\sum_{k \geq 1} \frac{1}{k^2} \sin(kx)$ converges uniformly in $\mathbb{R}$. Indeed, on the one hand, since $|\sin(kx)| \leq 1$ for every $k \geq 1$ and $x \in \mathbb{R}$, we get

$$\left| \frac{1}{k^2} \sin(kx) \right| \leq \frac{1}{k^2}$$

for every $k \geq 1$ and $x \in \mathbb{R}$. On the other, just note that the series $\sum_{k \geq 1} \frac{1}{k^2}$ is convergent.

Accidentally, the previous example also shows that, given a uniformly convergent series $\sum_{k \geq 1} f_k$ of differentiable functions $f_n : I \to \mathbb{R}$, the series $\sum_{k \geq 1} f_k'$ of their derivatives is not necessarily convergent, even pointwise. Indeed,

$$\sum_{k \geq 1} \frac{d}{dx} \left(\frac{1}{k^2} \sin(kx) \right) = \sum_{k \geq 1} \frac{1}{k} \cos(kx),$$

which doesn't converge pointwise in any real x of the form $x = 2\ell\pi$, with $\ell \in \mathbb{Z}$.

We finish this section by using the material developed here to give an example of a continuous function which is not differentiable at any point. For what comes next, the reader might find it helpful to read the statement of Example 8.23 again. Up to details, we follow the discussion in Sect. 9.7 of [9].

Example 11.22 For $y \in \mathbb{R}$, let $d(y)$ denote the distance from x to the nearest integer. If $f : \mathbb{R} \to \mathbb{R}$ is given by

$$f(x) = \sum_{k \geq 1} \left(\frac{9}{10}\right)^k d(10^k x),$$

then f is well defined and continuous, albeit not differentiable at any $x \in \mathbb{R}$.

Proof Example 8.23 guarantees the continuity of $x \mapsto d(10^k x)$, for every $k \geq 1$. On the other hand, since $0 \leq \left(\frac{9}{10}\right)^k d(10^k x) \leq \frac{1}{2}\left(\frac{9}{10}\right)^k$ and $\sum_{k \geq 1}\left(\frac{9}{10}\right)^k$ converges, the Weierstrass M-test assures the well definiteness and continuity of f.

Fix $x > 0$ (the case of $x \leq 0$ is analogous and will be left to the reader). We shall prove that f is not differentiable at x by constructing sequences $(a_n)_{n \geq 2}$ and $(b_n)_{n \geq 2}$ such that $a_n \leq x < b_n$ and $a_n, b_n \xrightarrow{n} x$, but

$$\left|\frac{f(b_n) - f(a_n)}{b_n - a_n}\right| \xrightarrow{n} +\infty.$$

This will contradict an obvious slight modification of Problem 8, page 300 (by allowing $a_n \leq x < b_n$, instead of $a_n < x < b_n$, there).

Given $x > 0$ and an integer $n \geq 2$, let $m_n \in \mathbb{N}$ be such that $m_n \leq 10^n x < m_n + 1$. If $a_n = 10^{-n} m_n$ and $b_n = 10^{-n}(m_n + 1)$, we obviously have $a_n \leq x < b_n$ and $b_n - a_n = 10^{-n}$. If $k \geq n$ is also integer, then $10^k a_n, 10^k b_n \in \mathbb{N}$, so that $\left|d(10^k b_n) - d(10^k a_n)\right| = 0$; on the other hand, if $k < n$, then Example 8.23 gives

$$\left|d(10^k b_n) - d(10^k a_n)\right| \leq 10^{k-n}. \tag{11.16}$$

Therefore,

$$f(b_n) - f(a_n) = \sum_{k=0}^{n-1} \left(\frac{9}{10}\right)^k \left(d(10^k b_n) - d(10^k a_n)\right). \tag{11.17}$$

Since $10^{n-1} a_n = \frac{m_n}{10}$ and $10^{n-1} b_n = \frac{m_n + 1}{10}$, by separately considering the cases $0 < x < 1$ and $x \geq 1$ one easily sees that $\left|d(10^{-1} b_n) - d(10^{-1} a_n)\right| = \frac{1}{10}$. Therefore, the triangle inequality in (11.17), together with (11.16), gives

$$|f(b_n) - f(a_n)| \geq \frac{1}{10}\left(\frac{9}{10}\right)^{n-1} - \left|\sum_{k=0}^{n-2}\left(\frac{9}{10}\right)^k (d(10^k b_n) - d(10^k a_n))\right|$$

$$\geq \frac{9^{n-1}}{10^n} - \sum_{k=0}^{n-2}\left(\frac{9}{10}\right)^k |d(10^k b_n) - d(10^k a_n)|$$

$$\geq \frac{9^{n-1}}{10^n} - \sum_{k=0}^{n-2}\left(\frac{9}{10}\right)^k \cdot 10^{k-n} = \frac{9^{n-1}}{10^n} - \frac{1}{10^n}\sum_{k=0}^{n-2} 9^k$$

$$= \frac{7 \cdot 9^{n-1} + 1}{8 \cdot 10^n} = \frac{7 \cdot 9^{n-1} + 1}{8}|b_n - a_n|.$$

Thus,

$$\left|\frac{f(b_n) - f(a_n)}{b_n - a_n}\right| \geq \frac{7 \cdot 9^{n-1} + 1}{8} \xrightarrow{n} +\infty.$$

$\square$

Problems: Section 11.2

1. Prove that the series of functions $\sum_{k\geq 1}\left(\frac{x}{1+x^2}\right)^k$ converges uniformly in $\mathbb{R}$ and compute its sum.

2. Prove that the Taylor series of the functions e^x, $\sin x$ and $\cos x$ converge uniformly to such functions in each interval $[-a, a]$, with $a > 0$.

3. For $n \in \mathbb{N}$, let $f_n : [0, 1] \to \mathbb{R}$ be given by $f_n(x) = nxe^{-nx}$. Prove that $(f_n)_{n\geq 1}$ converges pointwise, albeit not uniformly, 0. Also, show (without invoking Lebesgue's DCT) that $\int_0^1 f_n(x)\,dx \to 0$ as $n \to +\infty$.

4. For $n \geq 0$ integer, let $f_n : [0, 1] \to \mathbb{R}$ be given by $f_n(x) = x^n e^{-x}$. Do the following items:

   (a) Show that $(f_n)_{n\geq 0}$ converges pointwise to $f : [0, 1] \to \mathbb{R}$, with $f(x) = \begin{cases} 0, & \text{if } x < 1 \\ e^{-1}, & \text{if } x = 1 \end{cases}$.

   (b) Show that, for each integer $n \geq 0$, there exist natural numbers a_n and b_n such that $\int_0^1 f_n(x)\,dx = a_n - b_n e^{-1}$, with $a_0 = b_0 = 1$, $a_n = na_{n-1}$ and $b_n = nb_{n-1} + 1$ for every $n \in \mathbb{N}$.

   (c) Conclude that $a_n = n!$ and $b_n = n!\sum_{k=0}^n \frac{1}{k!}$ for every $n \in \mathbb{N}$.

   (d) Use the result of the previous items to show that $\int_0^1 f_n(x)\,dx \to 0$ as $n \to +\infty$.

5. Let $f_n : [a, b] \to \mathbb{R}$ be a sequence of integrable functions, converging uniformly to a function $f : [a, b] \to \mathbb{R}$. Show that f is integrable and

$$\int_a^b f(x)dx = \lim_{n \to +\infty} \int_a^b f_n(x)dx.$$

6. Let $(f_n)_{n \geq 1}$ be a sequence of differentiable functions $f_n : [a, b] \to \mathbb{R}$, such that $|f_n'(x)| \leq M_n$ for every $n \in \mathbb{N}$ and $x \in [a, b]$. If $\sum_{k \geq 1} M_k$ is a convergent series and there exists $x_0 \in [a, b]$ such that $\sum_{k \geq 1} f_k(x_0)$ converges absolutely, show that the series of functions $\sum_{k \geq 1} f_k$ converges uniformly in the interval $[a, b]$ to a differentiable function, such that

$$\frac{d}{dx} \sum_{k \geq 1} f_k(x) = \sum_{k \geq 1} f_k'(x)$$

for every $x \in [a, b]$.

7. (Putnam) Let $f : [a, b] \to \mathbb{R}$ be a continuous function such that $\int_a^b f(x)x^k dx = 0$ for every $k \in \mathbb{Z}_+$. Prove that f vanishes identically.

8. (Berkeley) Does there exist a continuous function $f : [0, 1] \to \mathbb{R}$ such that $\int_0^1 f(x)x dx = 1$ and $\int_0^1 f(x)x^k dx = 0$ for every nonnegative integer $k \neq 0$? Justify your answer.

9. (Berkeley) Let $\varphi_n : [0, 1] \to \mathbb{R}$ be a sequence of nonnegative continuous functions such that

$$\lim_{n \to +\infty} \int_0^1 x^k \varphi_n(x)dx$$

exists, for each $k \in \mathbb{Z}_+$. Prove that, for any given continuous function $f : [0, 1] \to \mathbb{R}$, the limit

$$\lim_{n \to +\infty} \int_0^1 f(x)\varphi_n(x)dx$$

also exists.

10. With respect to the series of functions $\sum_{k \geq 1} \frac{1}{k} \sin(kx)$, do the following items:

   (a) Use Abel's criterion (cf. Problem 18, page 243), together with the discussion of items (b) and (c) of Problem 20, page 244, to show that the given series converges pointwise in the interval $(0, 2\pi)$.

   (b) Revisit the proof of Abel's criterion, as sketched in the hints given to Problem 18, page 243, to show that Abel's identity guarantees that the convergence of item (a) is uniform in every interval of the form $[\delta, 2\pi - \delta]$, for $0 < \delta < \pi$.

11. Let $f : \mathbb{R} \to \mathbb{R}$ be a piecewise continuous function, periodic with period 2π. The **Fourier series**[2] of f is the series of functions

$$\frac{a_0(f)}{2} + \sum_{k \geq 1} (a_k(f) \cos(kx) + b_k(f) \sin(kx)),$$

where

$$a_k(f) = \frac{1}{\pi} \int_{-\pi}^{\pi} f(x) \cos(kx)dx \quad \text{and} \quad b_k(f) = \frac{1}{\pi} \int_{-\pi}^{\pi} f(x) \sin(kx)dx$$

for all k. In this respect, do the following items:

(a) If f is continuously differentiable and $k \geq 1$, prove that

$$a_k(f) = -\frac{1}{k} b_k(f') \quad \text{and} \quad b_k(f) = \frac{1}{k} a_k(f').$$

Then, conclude that $a_k(f), b_k(f) \to 0$ as $k \to +\infty$ (this is a special case of Riemann-Lebesgue's lemma, to be proved in greater generality in Problem 15).

(b) If f is twice continuously differentiable and $k \geq 1$, prove that

$$a_k(f) = -\frac{1}{k^2} a_k(f'') \quad \text{and} \quad b_k(f) = -\frac{1}{k^2} b_k(f'').$$

Then, conclude that the Fourier series of f converges uniformly in $\mathbb{R}$ and, hence, defines a continuous function in $\mathbb{R}$.

(c) Yet assuming f to be twice continuously differentiable, let $g : \mathbb{R} \to \mathbb{R}$ be the continuous function given by

$$g(x) = \frac{a_0(f)}{2} + \sum_{k \geq 1} (a_k(f) \cos(kx) + b_k(f) \sin(kx)).$$

Prove that $a_l(g) = a_l(f)$ for $l \geq 0$ and $b_l(g) = b_l(f)$ for $l \geq 1$.

For the next two problems, we say that a function $f : I \to \mathbb{R}$ is piecewise continuously differentiable if there exists reals $a_1 < a_2 < \ldots < a_n$ in I such that f is continuously differentiable in all of the intervals $(-\infty, a_1] \cap I$, $[a_i, a_{i+1}]$ for $1 \leq i < n$ and $[a_n, +\infty) \cap I$. We shall assume without proof the validity of **Fourier's convergence theorem**,[3] which states that if $f : \mathbb{R} \to \mathbb{R}$ is piecewise

---

[2]Fourier's studies on heat conduction, collected in his famous book *Théorie Analytique de la Chaleur*, laid the groundwork for modern Mathematical Physics, which in turn has greatly influenced the development of the theory of Partial Differential Equations.

[3]In spite of being known by the name of Fourier, such a result is actually due to Dirichlet.

continuous, piecewise continuously differentiable and periodic of period 2π, then its Fourier series is pointwise convergent in the whole real line, converging at $x_0 \in \mathbb{R}$ to

$$\frac{1}{2}\Big(\lim_{x \to x_0-} f(x) + \lim_{x \to x_0+} f(x) \Big).$$

For a proof of Fourier's convergence theorem, together with a discussion of several other interesting properties of Fourier series, see [10, 20] or [24].

12. Let $f : \mathbb{R} \to \mathbb{R}$ be periodic of period 2π and given by $f(x) = x^2$ in the interval $[-\pi, \pi]$.

(a) Compute its Fourier series.

(b) Use Fourier's theorem to show that $\sum_{k \geq 1} \frac{1}{k^2} = \frac{\pi^2}{6}$.

13. Let $f : \mathbb{R} \to \mathbb{R}$ be periodic of period 2π and such that

$$f(x) = \begin{cases} 0, & \text{if } -\pi \leq x < 0 \\ 1, & \text{if } 0 \leq x < \pi \end{cases}.$$

(a) Compute its Fourier series.

(b) Use Fourier's theorem to deduce **Leibniz formula**[4] for π:

$$\frac{\pi}{4} = 1 - \frac{1}{3} + \frac{1}{5} - \frac{1}{7} + \cdots .$$

14. (Miklós-Schweitzer—adapted) Let $p > 0$ be a real number and $f, g : \mathbb{R} \to \mathbb{R}$ be continuous functions, with g being periodic of period p.

(a) Prove that

$$\sum_{k=1}^{n} \int_{\frac{(k-1)p}{n}}^{\frac{kp}{n}} f\Big(\frac{kp}{n}\Big) g(nx)dx = \Big(\int_0^p g(y)dy \Big) \cdot \frac{1}{n} \sum_{k=1}^{n} f\Big(\frac{kp}{n}\Big)$$

$$\longrightarrow \frac{1}{p}\Big(\int_0^p f(x)dx \Big)\Big(\int_0^p g(y)dy \Big).$$

(b) Prove the following theorem of Fejér[5]:

$$\lim_{n \to +\infty} \int_0^p f(x)g(nx)dx = \frac{1}{p}\Big(\int_0^p f(x)dx \Big)\Big(\int_0^p g(x)dx \Big).$$

---

[4] We shall give a self-contained proof of such a formula in Problem 9, page 484.

[5] After Lipót Fejér, Hungarian mathematician of the XX century.

15. Prove the **Riemann-Lebesgue lemma**: let $f : \mathbb{R} \to \mathbb{R}$ be a continuous function, periodic with period 2π. If $a_k(f)$ and $b_k(f)$ denote the Fourier coefficients of f, as defined in Problem 11, prove that $a_k(f), b_k(f) \to 0$ as $k \to +\infty$.

For the next problem, let $f : \mathbb{R} \to \mathbb{R}$ be continuous and periodic with period 2π. For $n \in \mathbb{Z}_+$, let

$$S_n f(x) = \frac{a_0}{2} + \sum_{k=1}^{n} \left(a_k \cos(kx) + b_k \sin(kx) \right)$$

denote the **n-th partial sum** of the Fourier series of f and

$$\sigma_n f(x) = \frac{S_0 f(x) + S_1 f(x) + \cdots + S_n f(x)}{n+1}.$$

Another theorem of Fejér states that $\sigma_n f \xrightarrow{n} f$ uniformly on $\mathbb{R}$.

16. Let $f, g : \mathbb{R} \to \mathbb{R}$ be continuous and periodic with period 2π. If $a_k(f) = a_k(g)$ and $b_k(f) = b_k(g)$ for every k, prove that $f = g$.

17. (Berkeley) Find all continuous functions $f : \mathbb{R} \to \mathbb{R}$ such that $f(x) = f(x+1) = f(x + \sqrt{2})$ for every $x \in \mathbb{R}$.

18. * This problem establishes a fairly general version of **Leibniz' rule** of differentiation under the integral sign, assuming the validity of Lebesgue's DCT. To this end, let $I, J \subset \mathbb{R}$ be intervals and $f : I \times J \to \mathbb{R}$ be *continuous in each variable separately*.[6] Let α and β be the endpoints of J (in the sense of Sect. 10.9).

    (a) Assume that for every $x_0 \in I$ and every sequence $(x_n)_{n \geq 1}$ in I converging to x_0, there exists an integrable function $g : J \to \mathbb{R}$ such that $|f(x_n, t)| \leq g(t)$ for every $n \geq 1$ and every $t \in J$. Then, $t \mapsto f(x, t)$ is integrable for each $x \in I$. Moreover, if $F : I \to \mathbb{R}$ is given by

    $$F(x) = \int_{\alpha}^{\beta} f(x, t) dt,$$

    then F is continuous.

    (b) Let the function $x \mapsto f(x, t)$ be continuously differentiable for each $t \in J$, and let $\frac{\partial f}{\partial x}$ denote its derivative. Assume that for every $x_0 \in I$ and every sequence $(x_n)_{n \geq 1}$ in I converging to x_0, there exist integrable functions $g : J \to \mathbb{R}$ and $G : J \to \mathbb{R}$ such that $|f(x_n, t)| \leq g(t)$ and $\left| \frac{\partial f}{\partial x}(x_n, t) \right| \leq G(t)$ for every $n \geq 1$ and every $t \in J$. Then, $t \mapsto f(x, t)$ and $t \mapsto \frac{\partial f}{\partial x}(x, t)$ are integrable for each $x \in I$ and $F : I \to \mathbb{R}$, defined as in item (a), is continuously differentiable in I, with

    $$F'(x) = \int_{\alpha}^{\beta} \frac{\partial f}{\partial x}(x, t) dt. \tag{11.18}$$

---

[6]I.e., such that both $t \mapsto f(x, t)$ and $x \mapsto f(x, t)$ are continuous functions.

19. * Show that the Gamma function is continuously differentiable, with

$$\Gamma'(x) = \int_0^{+\infty} e^{-t} t^{x-1} \log t \, dt.$$

11.3 Power Series

In Sect. 11.1 we saw several examples of infinitely differentiable functions defined in open intervals and which coincide with their Taylor series in such intervals. The Taylor series of an infinitely differentiable function is a particular case of a **power series**, i.e., of a series of functions of the form

$$\sum_{k \geq 0} a_k (x - x_0)^k, \tag{11.19}$$

where $a_0, a_1, a_2, \ldots$ are given real numbers. In this case, as it happens with number series, we say that $a_n (x - x_0)^n$ is the **n-th term** of the power series and that a_n is the **n-th coefficient** of it.

In this section we develop the basic aspects of the theory of power series, starting with the following central result.

Theorem 11.23 *Given a power series $\sum_{k \geq 0} a_k (x - x_0)^k$, there exists $0 \leq R \leq +\infty$ such that the series:*

(a) *Converges absolutely in the interval $(x_0 - R, x_0 + R)$ and diverges in $\mathbb{R} \setminus [x_0 - R, x_0 + R]$.*
(b) *Converges uniformly in the interval $[x_0 - r, x_0 + r]$, $\forall \, 0 < r < R$.*

Proof Firstly, note that the series $\sum_{k \geq 0} a_k (x - x_0)^k$ converges absolutely in the interval $(x_0 - R, x_0 + R)$ if and only if the series $\sum_{k \geq 0} a_k x^k$ converges absolutely in the interval $(-R, R)$. Accordingly, $\sum_{k \geq 0} a_k (x - x_0)^k$ converges uniformly in $[x_0 - r, x_0 + r]$ if and only if $\sum_{k \geq 0} a_k x^k$ converges uniformly in $[-r, r]$. Therefore, we can assume that $x_0 = 0$.

Let's first deal with item (a).

Claim 4 if $\sum_{k \geq 0} a_k x^k$ converges at $x = \alpha \neq 0$, then it converges absolutely at any $x \in (-\alpha, \alpha)$.

Indeed, for such an x, we have

$$\sum_{k \geq 0} |a_k x^k| = \sum_{k \geq 0} |a_k \alpha^k| \left| \frac{x}{\alpha} \right|^k. \tag{11.20}$$

Since $\sum_{k \geq 0} a_k \alpha^k$ converges, Proposition 7.36 guarantees the existence of $n_0 \in \mathbb{N}$ such that $n \geq n_0 \Rightarrow |a_n \alpha^n| < 1$; therefore, it follows from (11.20) that $|a_k x^k| \left| \frac{x}{\alpha} \right|^k \leq$

$\left|\frac{x}{\alpha}\right|^k$ for $k \geq n_0$. Since the geometric series $\sum_{k \geq n_0} \left|\frac{x}{\alpha}\right|^k$ is convergent (for, $\left|\frac{x}{\alpha}\right| < 1$), the comparison test assures that the same holds true for $\sum_{k \geq n_0} |a_k x^k|$ and, thus, for $\sum_{k \geq 0} |a_k x^k|$.

Claim 5 if $\sum_{k \geq 0} a_k x^k$ diverges at $x = \beta \neq 0$, then it also diverges at any $x \in \mathbb{R}$ such that $|x| > |\beta|$.

For such an x, if $\sum_{k \geq 0} a_k x^k$ converged, the previous claim would assure the (absolute) convergence of the series $\sum_{k \geq 0} a_k \beta^k$, which is an absurd.

Claims 4 and 5 guarantee that, with respect to $\sum_{k \geq 0} a_k x^k$, one of the three following possibilities does happen: (i) it converges only at $x = 0$; (ii) it is absolutely convergent at any $x \in \mathbb{R}$; (iii) there exist $\alpha, \beta \neq 0$ such that $|\alpha| < |\beta|$ and the series converges absolutely when $|x| < \alpha$ and diverges when $|x| > |\beta|$.

If either (i) or (ii) happens, there is nothing left to do. If (iii) happens, let

$$R = \sup\{r > 0; \sum_{k \geq 0} a_k u^k \text{ converges absolutely when } |u| < r\}.$$

We claim that $\sum_{k \geq 0} a_k x^k$ converges absolutely when $|x| < R$ and diverges when $|x| > R$. To this end, let's look at two separate cases:

(i) If $|x| < R$, take r such that $|x| < r < R$ and $\sum_{k \geq 0} a_k u^k$ converges absolutely when $|u| < r$. Then, in particular $\sum_{k \geq 0} a_k x^k$ is absolutely convergent.

(ii) If $|x| > R$ and $\sum_{k \geq 0} a_k x^k$ converged, then, taking $\tilde{R}$ satisfying $R < \tilde{R} < |x|$, it would follow from Claim 4 that $\sum_{k \geq 0} a_k u^k$ converges absolutely when $|u| < \tilde{R}$. But this obviously contradicts the definition of R. Therefore, $\sum_{k \geq 0} a_k x^k$ diverges.

For item (b), given $0 < r < R$, it follows from what we did above that $\sum_{k \geq 0} |a_k r^k|$ converges. Hence, letting $M_k = |a_k r^k|$, we have, for $|x| \leq r$, that $|a_k x^k| \leq |a_k r^k| = M_k$. Therefore, Weierstrass M-test assures that $\sum_{k \geq 0} a_k x^k$ converges uniformly in the interval $[-r, r]$. □

In the notations of the previous result, we say that $0 \leq R \leq +\infty$ is the **radius of convergence** of the power series (11.19), and $(x_0 - R, x_0 + R)$ is its **interval of convergence**.

Problems 10 and 11 give a general formula for the radius of convergence in terms of the coefficients of a given power series. For the time being, the following consequence of the previous result allows us to easily compute it in a number of interesting examples. At this point, we suggest that the reader runs through the proof of the ratio test (cf. Proposition 7.51) once more, just to note that it remains valid if (in the corresponding notations) $l = +\infty$.

Corollary 11.24 *Let $(a_n)_{n \geq 0}$ be a sequence of nonzero real numbers. If there exists $0 \leq R \leq +\infty$ such that $\lim_{n \to +\infty} \left|\frac{a_n}{a_{n+1}}\right| = R$, then the power series $\sum_{k \geq 0} a_k (x - x_0)^k$ has radius of convergence equal to R.*

Proof Since

$$\left| \frac{a_{n+1}(x - x_0)^{n+1}}{a_n(x - x_0)^n} \right| = \frac{|x - x_0|}{\left| \frac{a_n}{a_{n+1}} \right|} \xrightarrow{n} \frac{|x - x_0|}{R},$$

the ratio test assures that $\sum_{k\geq 0} a_k(x - x_0)^k$ converges absolutely if $\frac{|x-x_0|}{R} < 1$ and diverges if $\frac{|x-x_0|}{R} > 1$. Since $\frac{|x-x_0|}{R} < 1$ if and only if $|x - x_0| < R$, the previous result guarantees that R is precisely the radius of convergence of the given power series.   □

Example 11.25

(a) The power series $\sum_{k\geq 1} \frac{1}{k}x^k$ has radius of convergence 1, for $\left| \frac{1/n}{1/(n+1)} \right| = \frac{n+1}{n} \to$ 1 as $n \to +\infty$.

(b) The series $\sum_{k\geq 1} k!x^k$ has radius of convergence 0, for $\left| \frac{n!}{(n+1)!} \right| = \frac{1}{n+1} \to 0$.

(c) The series $\sum_{k\geq 1} \frac{1}{k!}(x - 2)^k$ has radius of convergence $+\infty$, for $\left| \frac{1/n!}{1/(n+1)!} \right| = n + 1 \to +\infty$ as $n \to +\infty$. Actually, we already know from the material of Sect. 11.1 that $\sum_{k\geq 1} \frac{1}{k!}(x - 2)^k = e^{x-2}$ for every real x.

(d) The former corollary does not apply to the power series $\sum_{k\geq 1} \frac{1}{k!}x^{2^k}$, for it has infinitely many vanishing coefficients. See, however, Problem 2.

(e) Given $\alpha \neq 0$, the previous corollary assures that the power series $\sum_{k\geq 0}(\alpha x)^k$ has radius of convergence $\frac{1}{|\alpha|}$, for $\left| \frac{\alpha^n}{\alpha^{n+1}} \right| = \frac{1}{|\alpha|}$ for every $n \geq 0$. By Proposition 7.38, for $|x| < \frac{1}{|\alpha|}$ we have

$$\sum_{k\geq 0}(\alpha x)^k = \frac{1}{1 - \alpha x}. \tag{11.21}$$

We now collect another important consequence of Theorem 11.23, which will be of crucial importance for the proof of the subsequent result.

Proposition 11.26 *If the power series $\sum_{k\geq 0} a_k(x - x_0)^k$ has radius of convergence $R > 0$, then:*

(a) *The function $f : (x_0 - R, x_0 + R) \to \mathbb{R}$, given by $f(x) = \sum_{k\geq 0} a_k(x - x_0)^k$, is continuous.*

(b) *For every $x \in (x_0 - R, x_0 + R)$, we have $\int_{x_0}^x f(t)dt = \sum_{k\geq 0} \frac{a_k}{k+1}(x - x_0)^{k+1}$. In particular, the radius of convergence of the series at the right hand side above is greater than or equal to R.*

Proof Let $f_k : (x_0 - R, x_0 + R) \to \mathbb{R}$ be given by $f_k(x) = a_k(x - x_0)^k$. Theorem 11.23 gives $f = \sum_{k\geq 1} f_k$ in $(x_0 - R, x_0 + R)$, the convergence being uniform in $[x_0 - r, x_0 + r]$, for every $0 < r < R$. Since f_k is continuous for every k, items (a) and (b) follow immediately from items (a) and (b) of Corollary 11.19.   □

The coming theorem is the central result in the theory of power series.

Theorem 11.27 *If the power series $\sum_{k\geq0} a_k(x - x_0)^k$ has radius of convergence $R > 0$, then:*

(a) *The function $f : (x_0 - R, x_0 + R) \to \mathbb{R}$, given by $f(x) = \sum_{k\geq0} a_k(x - x_0)^k$, is infinitely differentiable.*
(b) *For every $n \in \mathbb{N}$, we have $f^{(n)}(x) = \sum_{k\geq n} \frac{k!}{(k-n)!} \cdot a_k(x - x_0)^{k-n}$ for every $x \in (x_0 - R, x_0 + R)$, and the series defining $f^{(n)}$ also has radius of convergence equal to R.*

Proof As in the proof of Theorem 11.23 we can assume, without any loss of generality, that $x_0 = 0$. We shall first prove that, letting $\tilde{R}$ be the radius of convergence of the power series $\sum_{k\geq1} ka_kx^{k-1}$, then $\tilde{R} = R$ and $f'(x) = \sum_{k\geq1} ka_kx^{k-1}$ for every $x \in (-R, R)$.

Note first that $|ka_kx^k| \geq |a_kx^k|$ for every integer $k \geq 1$ and every $x \in \mathbb{R}$. Therefore, if the power series $\sum_{k\geq1} ka_kx^{k-1}$ converges absolutely at some x, then the same holds true for the power series $\sum_{k\geq1} ka_kx^k$ and, thus, for $\sum_{k\geq0} a_kx^k$. Hence, $\tilde{R} \leq R$.

Now, for $0 < x < R$ and $0 < h < R - x$, we have

$$\frac{f(x + h) - f(x)}{h} = \frac{1}{h}\left(\sum_{k\geq1} a_k(x + h)^k - \sum_{k\geq1} a_kx^k\right) = \sum_{k\geq1} a_k\left(\frac{(x + h)^k - x^k}{h}\right).$$

On the other hand, Lagrange's MVT gives, for each $k \geq 1$, some $c_k \in (x, x + h)$ such that $\frac{(x+h)^k - x^k}{h} = kc_k^{k-1}$. Therefore,

$$\frac{f(x + h) - f(x)}{h} = \sum_{k\geq1} ka_kc_k^{k-1},$$

so that the power series $\sum_{k\geq1} ka_kc_k^{k-1}$ converges absolutely. Since $|ka_kx^{k-1}| \leq |ka_kc_k^{k-1}|$, the comparison test guarantees that $\sum_{k\geq0} ka_kx^{k-1}$ also converges absolutely. Analogously, such a series converges absolutely if $-R < x < 0$, so that $\tilde{R} \geq R$.

Our discussion up to this point assures that $g : (-R, R) \to \mathbb{R}$, given by $g(x) = \sum_{k\geq0} ka_kx^{k-1}$, is a well defined function. By item (b) of the previous proposition, we have

$$\int_0^x g(t)dt = \sum_{k\geq1} \frac{ka_k}{k}x^k = \sum_{k\geq1} a_kx^k = f(x) - a_0.$$

Hence, the FTC gives, for $|x| < R$,

$$f'(x) = g(x) = \sum_{k\geq0} ka_kx^{k-1}.$$

For what is left to do assume, by induction hypothesis, that we have already shown f to be m times differentiable in $(-R, R)$, with $f^{(m)}(x) = \sum_{k \geq m} \frac{k!}{(k-m)!} \cdot a_k(x - x_0)^{k-m}$; assume also that this last power series has radius of convergence equal to R. Then, by the first part, $f^{(m)}$ is a differentiable function, the power series defining $\left(f^{(m)}\right)' = f^{(m+1)}$ has radius of convergence R and

$$f^{(m+1)}(x) = \sum_{k \geq m+1} \frac{k!}{(k-m)!} \cdot (k-m)a_k(x-x_0)^{k-m-1}$$

$$= \sum_{k \geq m+1} \frac{k!}{(k-(m+1))!} \cdot a_k(x-x_0)^{k-(m+1)}.$$

This completes the inductive argument and, thus, the proof of (b).   □

In the notations of item (b) of the previous result, we observe that the power series expansion of $f^{(n)}$ is obtained by *termwise differentiating*, n times, the power series expansion of f. Indeed, letting $\frac{d^n}{dx^n}$ denote the n-th derivative of a function (if it exists), an immediate computation gives

$$\frac{d^n}{dx^n}(x-x_0)^k = \frac{k!}{(k-n)!}(x-x_0)^{k-n}.$$

We now collect two useful consequences of the previous theorem, the first of which refines the analysis of item (b) of Proposition 11.26.

Corollary 11.28 *If the power series $\sum_{k \geq 0} a_k(x - x_0)^k$ has radius of convergence $R > 0$, then the radius of convergence of the integrated power series $\sum_{k \geq 0} \frac{a_k}{k+1}(x - x_0)^{k+1}$ is also equal to R.*

Proof Since $\sum_{k \geq 0} a_k(x - x_0)^k = \frac{d}{dx} \sum_{k \geq 0} \frac{a_k}{k+1}(x - x_0)^{k+1}$, item (b) of the previous theorem guarantees that both of the given series have the same radius of convergence.   □

Corollary 11.29 *Assume that the power series $\sum_{k \geq 0} a_k(x - x_0)^k$ has radius of convergence $R > 0$, and let $f : (x_0 - R, x_0 + R) \to \mathbb{R}$ be given by $f(x) = \sum_{k \geq 0} a_k(x - x_0)^k$. Then:*

(a) $a_n = \frac{f^{(n)}(x_0)}{n!}$, for every $n \geq 0$.
(b) $\sum_{k \geq 0} a_k(x - x_0)^k$ is the Taylor series of f.

Proof Item (b) follows immediately from (a). For (a), item (b) of the previous result gives $f^{(n)}(x_0) = \frac{n!}{(n-n)!} \cdot a_n = n!a_n$; therefore, $a_n = \frac{f^{(n)}(x_0)}{n!}$.   □

We shall now see that the results presented so far allow us to reobtain, by means of a unified approach, some of the Taylor expansions discussed in Sect. 11.1. In this respect, see also Problem 3.

Example 11.30 Item (e) of Example 11.25 assures that $\frac{1}{1+x} = \sum_{k\geq0}(-1)^k x^k$, with radius of convergence equal to 1. Hence, item (b) of Proposition 11.26 gives, for $|x| < 1$,

$$\log(1 + x) = \int_0^x \frac{1}{1+t}\,dt = \sum_{k\geq0}\int_0^x (-1)^k t^k\,dt = \sum_{k\geq0}\frac{(-1)^k}{k+1}x^{k+1}. \qquad (11.22)$$

Example 11.31 Note that

$$\frac{1}{1+x^2} = 1 - x^2 + x^4 - x^6 + \cdots$$

for $|x| < 1$. Therefore, again by item (b) of Proposition 11.26, we have, for $|x| < 1$,

$$\arctan x = \int_0^x \frac{1}{1+t^2}\,dt = \sum_{k\geq0}\int_0^x (-1)^k t^{2k}\,dt$$

$$= x - \frac{x^3}{3} + \frac{x^5}{5} - \frac{x^7}{7} + \cdots. \qquad (11.23)$$

Problems 8 and 9 will show that (11.22) and (11.23) remain true for $x = 1$.

Example 11.32 It follows easily from Corollary 11.24 that the power series $\sum_{k\geq0}\frac{1}{k!}x^k$ converges in the whole real line. If $f(x) = \sum_{k\geq0}\frac{1}{k!}x^k$ for $x \in \mathbb{R}$, then item (b) of Theorem 11.27 furnishes

$$f'(x) = \sum_{k\geq1}\frac{1}{(k-1)!}x^{k-1} = \sum_{k\geq0}\frac{1}{k!}x^k = f(x).$$

Therefore,

$$\frac{d}{dx}(e^{-x}f(x)) = e^{-x}(-f(x) + f'(x)) = 0,$$

so that $x \mapsto e^{-x}f(x)$ is a constant function. Finally, since $f(0) = 1$, we get $e^{-x}f(x) = 1$ or, which is the same, $f(x) = e^x$.

We now apply Theorem 11.27 to show that, for any $\alpha \in \mathbb{R} \setminus \{0\}$ and $|x| < 1$, one can write $(1 + x)^\alpha$ as a convergent power series. To this end, given $\alpha \in \mathbb{R}$ and an integer $n \geq 0$, we define the **generalized binomial number** $\binom{\alpha}{n}$ by letting $\binom{\alpha}{0} = 1$ and, for $n \geq 1$,

$$\binom{\alpha}{n} = \frac{\alpha(\alpha-1)(\alpha-2)\ldots(\alpha-n+1)}{n!}. \qquad (11.24)$$

The following lemma establishes some useful properties of generalized binomial numbers. As in Sect. 4.2, the property of item (a) is also known as **Stifel's relation**.

Lemma 11.33 *Given $\alpha \in \mathbb{R}$ and $n \in \mathbb{N}$, we have:*

(a) $\binom{\alpha}{n} = \binom{\alpha-1}{n} + \binom{\alpha-1}{n-1}$.

(b) $\frac{n}{\alpha} \binom{\alpha}{n} = \binom{\alpha-1}{n-1}$ *for every $\alpha \neq 0$.*

(c) $\left| \binom{\alpha}{n} \right| \leq 1$ *when $|\alpha| \leq 1$.*

Proof Item (a) is an easy computation:

$$
\binom{\alpha}{n} - \binom{\alpha-1}{n} = \frac{1}{n!}\alpha(\alpha-1)(\alpha-2)\ldots(\alpha-n+1)
$$

$$
- \frac{1}{n!}(\alpha-1)(\alpha-2)\ldots(\alpha-n)
$$

$$
= \frac{1}{n!}(\alpha-1)(\alpha-2)\ldots(\alpha-n+1)(\alpha-(\alpha-n))
$$

$$
= \frac{1}{(n-1)!}(\alpha-1)(\alpha-2)\ldots(\alpha-n+1)
$$

$$
= \binom{\alpha-1}{n-1}.
$$

Item (b) follows immediately from (11.24).

Finally, for item (c), if $|\alpha| \leq 1$ then (11.24) and the triangle inequality give

$$
\left| \binom{\alpha}{n} \right| \leq \frac{|\alpha|(|\alpha|+1)(|\alpha|+2)\ldots(|\alpha|+n-1)}{n!} \leq \frac{1 \cdot 2 \cdot \ldots \cdot n}{n!} = 1.
$$

$\square$

The coming result is known in mathematical literature as the *binomial series theorem*, or simply as the **binomial theorem**, and is due to Newton. Notice that (11.25) generalizes (4.11), for $\binom{\alpha}{k} = 0$ if $\alpha \in \mathbb{N}$ and $k > \alpha$.

Theorem 11.34 (Newton) *For $\alpha \neq 0$ and $|x| < 1$, we have*

$$
(1+x)^\alpha = \sum_{k \geq 0} \binom{\alpha}{k} x^k. \tag{11.25}
$$

Proof Firstly, assume $0 < |\alpha| \leq 1$. Since

$$
\left| \frac{\binom{\alpha}{n}}{\binom{\alpha}{n+1}} \right| = \frac{n+1}{|\alpha-n|} \xrightarrow{n} 1,
$$

Corollary 11.24 assures that $\sum_{k\geq 0} \binom{\alpha}{k} x^k$ has radius of convergence equal to 1. Hence, by Theorem 11.27, the function $f : (-1, 1) \to \mathbb{R}$ given by $f(x) = \sum_{k\geq 0} \binom{\alpha}{n} x^k$ is differentiable. Also from that result,

$$f'(x) = \sum_{k\geq 1} k \binom{\alpha}{k} x^{k-1} = \sum_{k\geq 1} \alpha \binom{\alpha - 1}{k - 1} x^{k-1}, \tag{11.26}$$

where we used item (c) of the previous lemma in the last equality above. Therefore,

$$
\begin{aligned}
(1 + x)f'(x) &= \alpha(1 + x) \sum_{k\geq 1} \binom{\alpha - 1}{k - 1} x^{k-1} \\
&= \alpha \left(\sum_{k\geq 1} \binom{\alpha - 1}{k - 1} x^{k-1} + \sum_{k\geq 1} \binom{\alpha - 1}{k - 1} x^k \right) \\
&= \alpha \left(1 + \sum_{k\geq 2} \binom{\alpha - 1}{k - 1} x^{k-1} + \sum_{k\geq 2} \binom{\alpha - 1}{k - 2} x^{k-1} \right) \\
&= \alpha \left(1 + \sum_{k\geq 2} \binom{\alpha}{k - 1} x^{k-1} \right) = \alpha \sum_{k\geq 0} \binom{\alpha}{k} x^k \\
&= \alpha f(x).
\end{aligned}
$$

If $g(x) = (1 + x)^{-\alpha} f(x)$, then we have for $|x| < 1$ that

$$
\begin{aligned}
g'(x) &= -\alpha(1 + x)^{-\alpha-1} f(x) + (1 + x)^{-\alpha} f'(x) \\
&= (1 + x)^{-\alpha-1} \left(-\alpha f(x) + (1 + x)f'(x) \right) = 0.
\end{aligned}
$$

Therefore, g is constant in $(-1, 1)$. Since $g(0) = 1$, we get $(1 + x)^{-\alpha} f(x) = 1$ for $|x| < 1$, as wished.

For the general case, suppose that $(1 + x)^\alpha = \sum_{k\geq 0} \binom{\alpha}{k} x^k$ for some $\alpha \neq 0$ and every $x \in (-1, 1)$. Let us show that similar formulas hold true for $\alpha - 1$ and $\alpha + 1$ (and every $x \in (-1, 1)$):

(a) Theorem 11.27, together with item (b) of Lemma 11.33, gives, for $|x| < 1$,

$$
\begin{aligned}
(1 + x)^{\alpha-1} &= \frac{1}{\alpha} \cdot \frac{d}{dx} (1 + x)^\alpha = \frac{1}{\alpha} \sum_{k\geq 1} k \binom{\alpha}{k} x^{k-1} \\
&= \sum_{k\geq 1} \binom{\alpha - 1}{k - 1} x^{k-1} = \sum_{k\geq 0} \binom{\alpha - 1}{k} x^k.
\end{aligned}
$$

(b) Item (a) of lemma 11.33 gives us

$$(1+x)^{\alpha+1} = (1+x)\sum_{k\geq0}\binom{\alpha}{k}x^k = \sum_{k\geq0}\binom{\alpha}{k}x^k + \sum_{k\geq0}\binom{\alpha}{k}x^{k+1}$$

$$= 1 + \sum_{k\geq1}\left(\binom{\alpha}{k}+\binom{\alpha}{k-1}\right)x^k = 1 + \sum_{k\geq1}\binom{\alpha+1}{k}x^k$$

$$= \sum_{k\geq0}\binom{\alpha+1}{k}x^k.$$

By induction, we conclude that (11.25) is true for all $\alpha \neq 0$ and $|x| < 1$.   □

Corollary 11.35 *Given $\alpha, \beta \neq 0$, we have*

$$(1+\beta x)^{\alpha} = \sum_{k\geq0}\binom{\alpha}{k}(\beta x)^k$$

for every $x \in \mathbb{R}$ such that $|x| < \frac{1}{|\beta|}$.

Proof It suffices to apply (11.25) with βx instead of x, noticing that $|\beta x| < 1 \Leftrightarrow |x| < \frac{1}{|\beta|}$.   □

Example 11.36 As an application of the corollary above, note that for $|x| < \frac{1}{2}$ we have

$$(1-2x)^{-1/2} = \sum_{k\geq0}\binom{-1/2}{k}(-2x)^k,$$

with

$$\binom{-1/2}{n} = \frac{1}{n!}\left(-\frac{1}{2}\right)\left(-\frac{1}{2}-1\right)\cdots\left(-\frac{1}{2}-n+1\right)$$

$$= \frac{(-1)^n}{n!}\cdot\frac{1}{2}\cdot\frac{3}{2}\cdot\frac{5}{2}\cdots\frac{2n-1}{2}$$

$$= \frac{(-1)^n}{n!}\cdot\frac{(2n)!}{2^n\cdot2\cdot4\cdots(2n)}$$

$$= \frac{(-1)^n}{n!}\cdot\frac{(2n)!}{2^n\cdot2^n n!} = \frac{(-1)^n}{4^n}\binom{2n}{n}.$$

Hence, for $|x| < \frac{1}{2}$ we have

$$(1-2x)^{-1/2} = \sum_{k\geq0}\frac{1}{2^k}\binom{2k}{k}x^k.$$

We finish this section by presenting a result on the extension of a power series to the endpoints of its interval of convergence. To this end, we first need to discuss a lemma which is interesting in itself.

Lemma 11.37 *Let* $f(x) = \sum_{k\geq 0} a_k x^k$ *be defined in the interval* $(-R, R)$, *with* $a_k \geq 0$ *for every* $k \geq 0$. *If* $\sum_{k\geq 0} a_k R^k$ *converges, then*

$$\lim_{x\to R-} f(x) = \sum_{k\geq 0} a_k R^k.$$

Proof Since $a_k \geq 0$ for each $k \geq 0$, we have, for $0 < x < R$,

$$\sum_{k\geq 0} a_k R^k - f(x) = \sum_{k\geq 0} a_k (R^k - x^k) \geq 0.$$

On the other hand,

$$\sum_{k\geq 0} a_k R^k - f(x) = \sum_{k=0}^{n} a_k (R^k - x^k) + \sum_{k>n} a_k (R^k - x^k)$$

$$\leq \sum_{k=0}^{n} a_k (R^k - x^k) + \sum_{k>n} a_k R^k,$$

where we used again the fact that $a_k \geq 0$ to obtain the above inequality.

Now, given $\epsilon > 0$, the convergence of $\sum_{k\geq 0} a_k R^k$ assures the existence of $n_0 \in \mathbb{N}$ such that $\sum_{k>n} a_k R^k < \epsilon$ for $n > n_0$. Fix such an $n > n_0$. Since $\lim_{x\to R-} \sum_{k=0}^{n} a_k (R^k - x^k) = 0$, there exists $\delta > 0$ such that

$$x \in (R - \delta, R) \Rightarrow \sum_{k=0}^{n} a_k (R^k - x^k) < \epsilon.$$

With such choices and in view of the above computations, we conclude that

$$x \in (R - \delta, R) \Rightarrow 0 \leq \sum_{k=0}^{n} a_k R^k - f(x) < 2\epsilon.$$

$\square$

Remark 11.38 In the notations of the statement of the previous lemma, Problem 7 shows that the assumption on the sign of the a_k's can be dropped.

It is somewhat surprising that the converse of the previous lemma also holds true. Such a result is known in mathematical literature as a **tauberian theorem** after the work of Alfred Tauber, Austrian mathematician of the XIX and XX centuries. For the sake of simplifying the notation, we assume that $R = 1$, leaving to the reader the (easy!) task of dealing with the general case.

Theorem 11.39 *Let* $f(x) = \sum_{k\geq 0} a_k x^k$ *in the interval* $(-1, 1)$, *with* $a_k \geq 0$ *for every* $k \geq 0$. *If* $\lim_{x\to 1-} f(x)$ *does exist, then* $\sum_{k\geq 0} a_k$ *converges and* $\lim_{x\to 1-} f(x) = \sum_{k\geq 0} a_k$.

Proof The last part follows from the previous lemma. For what is left to do, note that, for $0 \leq x < 1$,

$$xf'(x) = \sum_{k\geq 1} k a_k x^k = \sum_{j\geq 0} \sum_{k=2^j}^{2^{j+1}-1} k a_k x^k$$

$$\geq \sum_{j\geq 0} \sum_{k=2^j}^{2^{j+1}-1} 2^j a_k x^{2^{j+1}-1}$$

$$= \sum_{j\geq 0} 2^j \Big(\sum_{k=2^j}^{2^{j+1}-1} a_k \Big) x^{2^{j+1}-1}.$$

In particular, for a fixed natural number l and $0 < x < 1$, we have

$$\int_0^x tf'(t)dt \geq \sum_{j=0}^{l} 2^j \Big(\sum_{k=2^j}^{2^{j+1}-1} a_k \Big) \int_0^x t^{2^{j+1}-1} dt$$

$$= \sum_{j=0}^{l} 2^j \Big(\sum_{k=2^j}^{2^{j+1}-1} a_k \Big) \frac{x^{2^{j+1}}}{2^{j+1}}$$

$$= \frac{1}{2} \sum_{j=0}^{l} \Big(\sum_{k=2^j}^{2^{j+1}-1} a_k \Big) x^{2^{j+1}}.$$

Now, let $a = \lim_{x\to 1-} f(x)$. Since f is increasing and nonnegative, we get

$$\int_0^x tf'(t)dt = xf(x) - \int_0^x f(t)dt \leq a.$$

Taking these two estimates together, we arrive at

$$\sum_{j=0}^{l} \Big(\sum_{k=2^j}^{2^{j+1}-1} a_k \Big) x^{2^{j+1}} \leq 2a$$

for every $0 \leq x < 1$, so that making $x \to 1-$ we find

$$\sum_{k=1}^{2^{l+1}-1} a_k = \sum_{j=0}^{l} \Big(\sum_{k=2^j}^{2^{j+1}-1} a_k \Big) \leq 2a.$$

Finally, since the sequence of partial sums of the series $\sum_{k\geq0} a_k$ is nondecreasing, this last inequality gives

$$\sum_{k\geq0} a_k \leq a_0 + 2a.$$

$\square$

Example 11.40 By applying the formula of Problem 11 for the radius of convergence, we conclude that $(-1, 1)$ is the interval of convergence of the power series $\sum_{j\geq0} x^{2^j}$. Alternatively, we can observe that, for $|x| < 1$,

$$0 \leq \sum_{j\geq0} x^{2^j} \leq \sum_{k\geq1} |x|^k < +\infty.$$

On the other hand, since the sequence of the coefficients of this series clearly diverges, the tauberian theorem assures that

$$\lim_{x\to1-} \sum_{j\geq1} x^{2^j} = +\infty.$$

For a more refined estimate on the growth of $\sum_{j\geq1} x^{2^j}$ as $x \to 1-$, see Problem 12.

Problems: Section 11.3

1. Compute the radii of convergence of the power series given below:

   (a) $\sum_{k\geq0} \frac{1}{2k+1} x^k$.
   (b) $\sum_{k\geq0} e^{-k^2} x^k$.
   (c) $\sum_{k\geq0} (-1)^k x^{2^k}$.

2. Compute the radius of convergence of the power series $\sum_{k\geq0} \frac{1}{k!} x^{2^k}$.
3. Use the approach of Example 11.32, together with the result of Example 9.48, to show that

   $$\sin x = \sum_{k\geq0} \frac{(-1)^{k-1}}{(2k-1)!} x^{2k-1} \quad \text{and} \quad \cos x = \sum_{k\geq0} \frac{(-1)^k}{(2k)!} x^{2k}.$$

4. * For $k \in \mathbb{N}$ and $|x| < 1$, prove that

   $$\frac{1}{(1-x)^k} = \sum_{n\geq0} \binom{k+n-1}{n} x^n.$$

5. Show that for $|x| < 1$ we have

$$\arcsin x = \sum_{k \geq 0} \frac{1}{4^k(2k+1)} \binom{2k}{k} x^{2k+1}.$$

6. The formula of Example 11.30 can be modified in order to compute $\log a$ for every $a > 0$. To this end, do the following items:

(a) Show that, for $x \in (-1, 1)$, one has

$$\frac{1}{2} \log \left(\frac{1+x}{1-x} \right) = x + \frac{x^3}{3} + \frac{x^5}{5} + \frac{x^7}{7} + \cdots .$$

(b) Show that $x \mapsto \frac{1+x}{1-x}$ defines a bijection from $(-1, 1)$ onto $(0, +\infty)$.
(c) Use the formula of item (a) to compute $\log 3$ with four correct decimal places.

7. * Let $f : (-R, R) \to \mathbb{R}$ be given by $f(x) = \sum_{k \geq 0} a_k x^k$. The purpose of this problem is to show that, if $\sum_{k \geq 0} a_k R^k$ converges, then

$$\lim_{x \to R-} f(x) = \sum_{k \geq 0} a_k R^k. \qquad (11.27)$$

To this end, do the following items:

(a) Let $r_n = \sum_{k \geq n} a_k R^k$ and, for $|x| < R$, let $y = \frac{x}{R}$. Show that, for natural numbers $n < m$,

$$\sum_{k=n}^{m} a_k x^k = \left(r_n y^n - r_{m+1} y^m \right) + \sum_{k=n+1}^{m} r_k \left(y^k - y^{k-1} \right).$$

(b) Show that, for natural numbers $n < m$,

$$\left| \sum_{k=n}^{m} a_k x^k \right| \leq |r_n| + |r_m| + \left(\sup_{k>n} |r_k| \right) (y^n - y^m).$$

(c) Show that $\sum_{k \geq 0} a_k x^k$ converges uniformly on the interval $[0, R]$. Then, conclude that (11.27) is true.

8. Use the result of the previous problem to show that

$$\log 2 = 1 - \frac{1}{2} + \frac{1}{3} - \frac{1}{4} + \cdots .$$

9. * Use the material of this section to give a self-contained proof of the Leibniz formula for π:

$$\frac{\pi}{4} = 1 - \frac{1}{3} + \frac{1}{5} - \frac{1}{7} + \cdots.$$

For the next two problems, we extend the notion of supremum by saying that a sequence $(a_k)_{k \geq 0}$ has supremum $+\infty$ if it is unbounded from above; in this case, we write

$$\sup\{a_k; \, k \geq 0\} = +\infty.$$

Accordingly, we extend the notion of infimum by saying that a sequence $(a_k)_{k \geq 0}$ has infimum $-\infty$ if it is unbounded from below; in this case, we write

$$\inf\{a_k; \, k \geq 0\} = -\infty.$$

10. Let $(a_k)_{k \geq 0}$ be a sequence of real numbers. Define its **limit superior**, denoted $\limsup a_k$, by letting

$$\limsup a_k = \lim_{j \to +\infty} \sup\{a_j, a_{j+1}, \ldots\}.$$

(a) Prove that $\limsup a_k$ is a well defined concept and that, if $s_j = \sup\{a_j, a_{j+1}, \ldots\} \in \mathbb{R} \cup \{+\infty\}$, then $\limsup a_k = \inf\{s_j; \, j \geq 0\} \in \mathbb{R} \cup \{\pm\infty\}$.

(b) If $\limsup a_k \in \mathbb{R}$, prove that it is the only real number M satisfying the two following conditions:

  i. Given $\epsilon > 0$, there exists infinitely many $n \in \mathbb{N}$ such that $a_n > M - \epsilon$.
  ii. Given $\epsilon > 0$, there exists at most finitely many $n \in \mathbb{N}$ such that $a_n \geq M + \epsilon$.

11. Let $\sum_{k \geq 0} a_k(x - x_0)^k$ be a power series with radius of convergence R. The purpose of this problem is to prove that

$$R = \frac{1}{\limsup \sqrt[k]{|a_k|}},$$

where the right hand side is to be interpreted as being equal to 0 if $\limsup \sqrt[k]{|a_k|} = +\infty$. To this end, assume without loss of generality that $x_0 = 0$, let R be given as above and do the following items:

(a) Fix $0 < r < R$ and take $r < s < R$. Show that there exists $k_0 \in \mathbb{N}$ such that $|a_k| < \frac{1}{s^k}$ for every $k \geq k_0$. Then, show that

$$|x| \leq r \Rightarrow \sum_{k \geq k_0} |a_k x^k| \leq \sum_{k \geq k_0} \left(\frac{r}{s}\right)^k < +\infty.$$

(b) Fix $|x| > R$ and show that there exist infinitely many indices $k \in \mathbb{N}$ such that $|a_k x^k| > 1$. Conclude that the series $\sum_{k \geq 0} a_k x^k$ diverges.

(c) Conclude that R, defined as above, is indeed the convergence radius of the given series.

12. As we saw in Example 11.40, the power series $\sum_{j \geq 1} x^{2^j}$ defines a differentiable function $f : (-1, 1) \to \mathbb{R}$. In this respect, prove that:

(a) $x f'(x) > \sum_{k \geq 2} x^k = \frac{x^2}{1-x}$, for $0 \leq x < 1$.

(b) $f(x) > -\log(1 - x) - x$, for $0 \leq x < 1$.

11.4   Some Applications

In this section, we briefly discuss the use of power series in Algebra and differential equations. For further applications of power series to Algebra and Combinatorics, we refer the reader to [5].

We begin with the following

Definition 11.41 The (**ordinary**) **generating function**[7] of a sequence of real numbers $(a_n)_{n \geq 0}$ is the power series

$$\sum_{k \geq 0} a_k x^k. \qquad (11.28)$$

The previous definition suggests that the main difference between the theory of power series and the *method* of generating function lies in a change of point of view. In the first case, we are primarily interested in examining the properties of the function f defined by the power series at the right hand side of (11.28); in the second (as we shall see next), we want to use the properties of f to infer conclusions about the terms of the sequence $(a_n)_{n \geq 0}$.

Let's illustrate this by revisiting Problem 20, page 220 with the aid of generating functions.

Example 11.42 (TT) The set of naturals is partitioned into m disjoint, infinite and nonconstant arithmetic progressions, of common ratios $d_1, d_2, \ldots, d_m$. Prove that

$$\frac{1}{d_1} + \frac{1}{d_2} + \cdots + \frac{1}{d_m} = 1.$$

Proof If $f(x) = \sum_{k \geq 1} x^k$ and a_i is the initial term of the i-th AP (that of common ratio d_i), then the given condition, together with item (e) of Example 11.25, gives

---

[7]In opposition to *exponential* generating functions, cf. Chap. 3 of [5], for instance.

for $|x| < 1$ that

$$f(x) = \sum_{i=1}^{m}(x^{a_i} + x^{a_i+d_i} + x^{a_i+2d_i} + \cdots) = \sum_{i=1}^{m}\frac{x^{a_i}}{1 - x^{d_i}}.$$

Since $1 - x^{d_i} = (1 - x)(1 + x + x^2 + \cdots + x^{d_i-1})$, multiplying both sides of the equality above by $1 - x$ we get

$$x = \sum_{i=1}^{m}\frac{x^{a_i}}{1 + x + x^2 + \cdots + x^{d_i-1}} \tag{11.29}$$

for $|x| < 1$. Now, both sides of this last equality define functions continuous in $[0, 1]$ and which coincide in $[0, 1)$; hence, such functions coincide for $x = 1$ too, so that, letting $x = 1$ in (11.29), we obtain $\sum_{i=1}^{m}\frac{1}{d_i} = 1$. □

The use of generating functions is particularly useful in the study of sequences $(a_n)_{n\geq 0}$ satisfying a given linear recurrence relation. Here, as in Sect. 3.2, we shall treat the cases of (certain) linear recurrence relations of orders 2 and 3, postponing the analysis of the general case to Chap. 20 of [5].

The idea is to consider the generating function $\sum_{k\geq 0}a_k x^k$ corresponding to the given sequence and, then, follow the various stages below, which comprise a sort receipt for several similar problems:

I. Use the initial terms of the sequence, as well as the recurrence relation it satisfies, to conclude that the given generating function converges at some interval of the form $(-r, r)$.
II. Again with the aid of the initial terms and the given recurrence relation, perform appropriate (generally algebraic) operations with the equality $f(x) = \sum_{k\geq 0}a_k x^k$ to get a *formula* for $f(x)$.
III. Develop the formula obtained in item II in Taylor series.
IV. Use the uniqueness of the power series representation of f, given by Corollary 11.29, to conclude that a_n equals the coefficient of x^n in the power series expansion of stage III (i.e., $a_n = \frac{f^{(n)}(0)}{n!}$).

In order to go through stage I, the following result is frequently useful.

Lemma 11.43 *Let $(a_n)_{n\geq 0}$ be a sequence of real numbers. If there exist positive reals c and M such that $|a_n| \leq cM^n$ for every $n \geq 0$, then the power series $\sum_{k\geq 0}a_k x^k$ converges in the interval $\left(-\frac{1}{M}, \frac{1}{M}\right)$.*

Proof Since $|a_n x^n| = |a_n||x|^n \leq cM^n|x|^n = c|Mx|^n$ and the geometric series $\sum_{k\geq 0}|Mx|^k$ converges when $|x| < \frac{1}{M}$ (cf. item (e) of Example 11.25), the comparison test for series guarantees that $\sum_{k\geq 0}a_k x^k$ converges when $|x| < \frac{1}{M}$. □

The following example implements stages I through IV above to get a positional formula for the n-th Fibonacci number.

Example 11.44 Let $(F_n)_{n\geq 1}$ be the Fibonacci sequence, so that $F_1 = 1, F_2 = 1$ and $F_{k+2} = F_{k+1} + F_k$ for every integer $k \geq 1$. Compute F_n as a function of n.

Solution Let $f(x) = \sum_{k\geq 1} F_k x^k$ be the generating function corresponding to the Fibonacci sequence, and let's run through the previously listed stages I to IV.

Step I: note first that an easy induction gives $F_n \leq 2^n$ for every $n \geq 1$. Hence, Lemma 11.43 guarantees that the power series $\sum_{k\geq 1} F_k x^k$ converges in the interval $\left(-\frac{1}{2}, \frac{1}{2}\right)$.

Step II: for $x \in \left(-\frac{1}{2}, \frac{1}{2}\right)$, we can write

$$f(x) = F_1 x + F_2 x^2 + \sum_{k\geq 3} F_k x^k$$

$$= x + x^2 + \sum_{k\geq 3}(F_{k-1} + F_{k-2})x^k$$

$$= x + x^2 + x\sum_{k\geq 3} F_{k-1}x^{k-1} + x^2\sum_{k\geq 3} F_{k-2}x^{k-2}$$

$$= x + x^2 + x(f(x) - F_1 x) + x^2 f(x)$$

$$= x + (x + x^2)f(x).$$

Then, for $x \in \left(-\frac{1}{2}, \frac{1}{2}\right)$ we have

$$f(x) = \frac{x}{1 - x - x^2}.$$

Step III: writing $1 - x - x^2 = (1 - \alpha x)(1 - \beta x)$, with $\alpha, \beta \in \mathbb{R}$, we have $\alpha + \beta = 1$, $\alpha\beta = -1$ and

$$f(x) = \frac{x}{1 - x - x^2} = \frac{x}{(1 - \alpha x)(1 - \beta x)}.$$

Imposing, with no loss of generality, that $\alpha > \beta$, we get $\alpha = \frac{1+\sqrt{5}}{2}$ and $\beta = \frac{1-\sqrt{5}}{2}$, so that $\alpha - \beta = \sqrt{5}$ and, thus,

$$f(x) = \frac{1}{\sqrt{5}}\left(\frac{1}{1 - \alpha x} - \frac{1}{1 - \beta x}\right).$$

Now, developing $\frac{1}{1-\alpha x}$ and $\frac{1}{1-\beta x}$ as geometric series, we obtain

$$f(x) = \frac{1}{\sqrt{5}}\left(\sum_{k\geq 0}(\alpha x)^k - \sum_{k\geq 0}(\beta x)^k\right) = \sum_{k\geq 1}\left(\frac{\alpha^k - \beta^k}{\sqrt{5}}\right)x^k,$$

as long as $|x| < \min \left\{ \frac{1}{2}, \frac{1}{|\alpha|}, \frac{1}{|\beta|} \right\} = \frac{1}{2}$. Then,

$$\sum_{k \geq 1} F_k x^k = \sum_{k \geq 1} \left(\frac{\alpha^k - \beta^k}{\sqrt{5}} \right) x^k,$$

for every $x \in \left(-\frac{1}{2}, \frac{1}{2} \right)$.

Step IV: finally, Corollary 11.29 assures that the power series expansion of f is unique, so that the last equality above, together with the initial definition of f, gives $F_n = \frac{\alpha^n - \beta^n}{\sqrt{5}}$ for every $n \geq 1$.

$\square$

Proceeding similarly to the above example, we now use generating functions to give another proof of Theorem 3.16 (see also Problem 6).

Theorem 11.45 *Given $u, v \in \mathbb{R}$, with $v \neq 0$, let $(a_n)_{n \geq 1}$ be such that $a_{k+2} = ua_{k+1} + va_k$ for every $k \geq 1$. Moreover, assume that the characteristic equation $x^2 - ux - v = 0$ has real roots α and β.*

(a) *If $\alpha \neq \beta$, then $a_n = A\alpha^{n-1} + B\beta^{n-1}$ for every $n \geq 1$, where A and B are the solutions of the linear system $\begin{cases} A + B = a_1 \\ \alpha A + \beta B = a_2 \end{cases}$.*

(b) *If $\alpha = \beta$, then $a_n = (A + Bn)\alpha^{n-1}$ for $n \geq 1$, where A and B are the solutions of the linear system $\begin{cases} A + B = a_1 \\ (A + 2B)\alpha = a_2 \end{cases}$.*

Proof We start by showing that there exists $q > 0$ such that $|a_n| \leq q^n$ for every $n \geq 1$. Indeed, assuming that $|a_k| \leq q^k$ and $|a_{k+1}| \leq q^{k+1}$, it follows from the triangle inequality that

$$|a_{k+2}| = |ua_{k+1} + va_k| \leq |u||a_{k+1}| + |v||a_k| \leq |u|q^{k+1} + |v|q^k,$$

so that $|a_{k+2}| \leq q^{k+2}$ provided $|u|q^{k+1} + |v|q^k \leq q^{k+2}$ or, which is the same, $|u|q + |v| \leq q^2$. Since the greatest root of the second degree equation $x^2 - |u|x - |v| = 0$ is $x_0 = \frac{1}{2} \left(|u| + \sqrt{u^2 + 4|v|} \right)$, we get $|u|q + |v| \leq q^2$ whenever $q > x_0$. Hence, $|a_n| \leq q^n$ for every $n \geq 1$, as long as $|a_1| \leq q$, $|a_2| \leq q^2$ and $q > x_0$, for which it suffices to choose $q > \max\{|a_1|, \sqrt{|a_2|}, x_0\}$.

Fix such a q, so that $|a_n| \leq q^n$ for every $n \geq 1$. Lemma 11.43 guarantees that the generating function $\sum_{k \geq 1} a_k x^k$ converges in the interval $\left(-\frac{1}{q}, \frac{1}{q} \right)$, thus defining $f : \left(-\frac{1}{q}, \frac{1}{q} \right) \to \mathbb{R}$ by $f(x) = \sum_{k \geq 1} a_k x^k$.

In order to get a simpler expression for $f(x)$, we argue as in the previous example, using the recurrence relation satisfied by the sequence $(a_k)_{k \geq 0}$ of the coefficients of f:

$$f(x) = \sum_{k\geq 1} a_k x^k = a_1 x + a_2 x^2 + \sum_{k\geq 3} a_k x^k$$

$$= a_1 x + a_2 x^2 + \sum_{k\geq 3}(u a_{k-1} + v a_{k-2}) x^k$$

$$= a_1 x + a_2 x^2 + u x \sum_{k\geq 3} a_{k-1} x^{k-1} + v x^2 \sum_{k\geq 3} a_{k-2} x^{k-2}$$

$$= a_1 x + a_2 x^2 + u x (f(x) - a_1 x) + v x^2 f(x).$$

Therefore, $(1 - ux - vx^2) f(x) = a_1 x + (a_2 - u a_1) x^2$ for every $x \in \left(-\frac{1}{q}, \frac{1}{q}\right)$.

Now, since α and β are the roots of $x^2 - ux - v = 0$, we get $\alpha + \beta = u$ and $\alpha\beta = -v$. Hence, $1 - ux - vx^2 = (1 - \alpha x)(1 - \beta x)$, so that

$$f(x) = \frac{a_1 x + (a_2 - u a_1) x^2}{(1 - \alpha x)(1 - \beta x)} \qquad (11.30)$$

for $|x| < \min\{\frac{1}{q}, \frac{1}{|\alpha|}, \frac{1}{|\beta|}\}$.

Let's look separately at cases (a) and (b):

(a) If $\alpha \neq \beta$, we *decompose* the right hand side of (11.30) *in partial fractions*,[8] i.e., we take $A, B \in \mathbb{R}$ such that $f(x) = \frac{Ax}{1-\alpha x} + \frac{Bx}{1-\beta x}$. Obviously, such A and B must satisfy $\begin{cases} A + B = a_1 \\ \beta A + \alpha B = u a_1 - a_2 \end{cases}$, and one promptly notes that such a linear system has a single solution, exactly because $\alpha \neq \beta$.

Expanding $\frac{1}{1-\alpha x}$ and $\frac{1}{1-\beta x}$ in geometric series, we get

$$f(x) = Ax \sum_{k\geq 0}(\alpha x)^k + Bx \sum_{k\geq 0}(\beta x)^k$$

$$= \sum_{k\geq 0}(A\alpha^k + B\beta^k) x^{k+1}$$

$$= \sum_{k\geq 1}(A\alpha^{k-1} + B\beta^{k-1}) x^k.$$

Finally, comparing this last expression with the fact that $f(x) = \sum_{k\geq 1} a_k x^k$, we obtain $a_k = A\alpha^{k-1} + B\beta^{k-1}$ for every $k \geq 1$.

(b) If $\alpha = \beta$, then $2\alpha = u$, $\alpha^2 = -v$ and $f(x) = \frac{a_1 x + (a_2 - u a_1) x^2}{(1-\alpha x)^2}$. This time we try to decompose the right hand side of (11.30) in partial fractions by writing

---

[8] A general theorem on the existence of partial fraction decomposition for quotients of polynomials will be seen in [5]. For the time being, we shall content ourselves to describe what such a theorem says in the simple cases we consider here.

$f(x) = \frac{Ax}{1-\alpha x} + \frac{Bx}{(1-\alpha x)^2}$, for some $A, B \in \mathbb{R}$. Obviously, such A and B must

satisfy $\begin{cases} A + B = a_1 \\ \alpha A = ua_1 - a_2 \end{cases}$, and again this linear system has a unique solution,

for $v \neq 0 \Rightarrow \alpha \neq 0$.

Expanding $\frac{1}{1-\alpha x}$ and $\frac{1}{(1-\alpha x)^2}$ i power series (for the second fraction with the aid of Problem 4), we get

$$f(x) = Ax \sum_{k \geq 0} (\alpha x)^k + Bx \sum_{k \geq 0} (k+1)(\alpha x)^k$$

$$= \sum_{k \geq 0} (A + B(k+1))\alpha^k x^{k+1}$$

$$= \sum_{k \geq 1} (A + Bk)\alpha^{k-1} x^k.$$

Finally, arguing as in the previous case we get $a_k = (A + Bk)\alpha^{k-1}$ for every integer $k \geq 1$. $\qquad\square$

Generating functions can also be used to deal with linear recurrence relations with *nonconstant* coefficients. To present such an example, we first need the following result.

Proposition 11.46 *If* $f, g : (x_0 - R, x_0 + R) \to \mathbb{R}$ *have power series expansions* $f(x) = \sum_{k \geq 0} a_k(x - x_0)^k$ *and* $g(x) = \sum_{l \geq 0} b_l(x - x_0)^l$, *then the function* $fg : (x_0 - R, x_0 + R) \to \mathbb{R}$ *has power series expansion* $(fg)(x) = \sum_{n \geq 0} c_n(x - x_0)^n$, *with* $c_n = \sum_{k+l=n} a_k b_l = \sum_{k=0}^{n} a_k b_{n-k}$.

Proof Theorem 11.23 assures that the power series defining f and g converge absolutely in the interval $(x_0 - R, x_0 + R)$. Hence, by Theorem 7.53 and for every $x \in (x_0 - R, x_0 + R)$, we have

$$f(x)g(x) = \left(\sum_{k \geq 0} a_k(x - x_0)^k \right) \left(\sum_{l \geq 0} b_l(x - x_0)^l \right)$$

$$= \sum_{n \geq 0} \left(\sum_{k+l=n} a_k(x - x_0)^k \cdot b_l(x - x_0)^l \right)$$

$$= \sum_{n \geq 0} \left(\sum_{k+l=n} a_k b_l \right) (x - x_0)^n$$

$$= \sum_{n \geq 0} c_n(x - x_0)^n.$$

Finally, note that Theorem 7.53 guarantees that the convergence of this last series is also absolute in the interval $(x_0 - R, x_0 + R)$.   □

Example 11.47 Let $(a_n)_{n \geq 0}$ be the sequence such that $a_0 = 1$, $a_1 = -1$ and

$$a_k = -\frac{a_{k-1}}{k} + 2a_{k-2}$$

for every integer $k \geq 2$. Compute a_n as a function of n.

Solution Denoting by $f(x) = \sum_{k \geq 0} a_k x^k$ the generating function of the given sequence, let's once again run through the previously described stages I to IV.

Step I: once more we shall try to apply the comparison test for series, in the form of Lemma 11.43. Assuming that $|a_{k-2}| \leq \alpha^{k-2}$, $|a_{k-1}| \leq \alpha^{k-1}$, the triangle inequality gives

$$|a_k| \leq \frac{|a_{k-1}|}{k} + 2|a_{k-2}| \leq \frac{\alpha^{k-1}}{k} + 2\alpha^{k-2};$$

hence, we shall have $|a_k| \leq \alpha^k$ provided $\frac{\alpha^{k-1}}{k} + 2\alpha^{k-2} \leq \alpha^k$ or, which is the same,

$$\frac{\alpha}{k} + 2 \leq \alpha^2.$$

Since such an inequality is true for $\alpha = 2$ and every $k \geq 2$, and $|a_0| \leq 2^0$, $|a_1| \leq 2^1$, it follows by induction that $|a_n| \leq 2^n$ for every integer $n \geq 1$. Hence, the above mentioned lemma, together with the theory of power series, guarantees that f is infinitely differentiable in the whole interval $\left(-\frac{1}{2}, \frac{1}{2}\right)$.

Step II: writing the given recurrence relation as $ka_k = -a_{k-1} + 2ka_{k-2}$ for $k \geq 2$, we get

$$f'(x) = \sum_{k \geq 1} ka_k x^{k-1} = a_1 + \sum_{k \geq 2} ka_k x^{k-1} = a_1 + \sum_{k \geq 2}(-a_{k-1} + 2ka_{k-2})x^{k-1}$$

$$= a_1 - \sum_{k \geq 2} a_{k-1}x^{k-1} + 2x\sum_{k \geq 2} ka_{k-2}x^{k-2}$$

$$= a_1 - (f(x) - a_0) + 2x\left(\sum_{k \geq 2}(k-2)a_{k-2}x^{k-2} + 2\sum_{k \geq 2} a_{k-2}x^{k-2}\right)$$

$$= a_1 + a_0 - f(x) + 2x(f'(x) + 2f(x)).$$

Taking into account that $a_1 + a_0 = 0$, we get $f'(x) = (4x - 1)f(x) + 2xf'(x)$ or, which is the same,

$$(2x - 1)f'(x) = -(4x - 1)f(x).$$

In order to solve such a *differential equation*, first note that f is positive in the interval $(-r, r)$, for some $0 < r \leq \frac{1}{2}$ (for, $f(0) = a_0 = 1 > 0$ and f continuous imply f positive in some neighborhood of 0). Therefore, for $|x| < r$ we can write

$$\frac{f'(x)}{f(x)} = -\frac{4x - 1}{2x - 1} = -2 - \frac{1}{2x - 1}$$

so that, for $|x| < r \leq \frac{1}{2}$,

$$\log f(x) = \log f(t)\Big|_{t=0}^{t=x} = \int_0^x \frac{f'(t)}{f(t)} dt$$

$$= -\int_0^x \left(2 + \frac{1}{2t - 1}\right) dt$$

$$= -2x - \frac{1}{2} \log(1 - 2x).$$

Thus, for $|x| < r \leq \frac{1}{2}$ we have

$$f(x) = e^{-2x}(1 - 2x)^{-1/2}. \tag{11.31}$$

Step III: note that the power series expansion of e^{-2x} is given by (11.7), with $a = -2$, and holds in the whole real line:

$$e^{-2x} = \sum_{k \geq 0} \frac{(-2)^k}{k!} x^k.$$

Hence, it follows from Example 11.36 and Proposition 11.46 that, for f given as in (11.31) and $|x| < r \leq \frac{1}{2}$, we have

$$f(x) = \left(\sum_{k \geq 0} \frac{(-2)^k}{k!} x^k\right)\left(\sum_{l \geq 0} \frac{1}{2^l}\binom{2l}{l} x^l\right)$$

$$= \sum_{n \geq 0}\left(\sum_{k+l=n} \frac{(-2)^k}{k!} \cdot \frac{1}{2^l}\binom{2l}{l}\right) x^n$$

$$= \sum_{n \geq 0}\left(\sum_{l=0}^{n} \frac{(-2)^{n-l}}{(n - l)!} \cdot \frac{1}{2^l}\binom{2l}{l}\right) x^n$$

$$= \sum_{n \geq 0}\left(2^n \sum_{l=0}^{n} \frac{(-1)^{n-l}}{4^l(n - l)!}\binom{2l}{l}\right) x^n.$$

Step IV: comparing the last expression above with $f(x) = \sum_{n\geq0} a_n x^n$, it follows once more from the uniqueness of power series expansions that

$$a_n = 2^n \sum_{l=0}^{n} \frac{(-1)^{n-l}}{4^l(n-l)!} \binom{2l}{l}.$$

□

All of the above results can be seen as sorts of *discrete versions* of the following result on ordinary differential equations:

Theorem 11.48 *Let $x_0 \in \mathbb{R}$, $R > 0$ and $p, q : (x_0-R, x_0+R) \to \mathbb{R}$ be given by their Taylor series centered at x_0. For $\alpha, \beta \in \mathbb{R}$, there exist a unique twice differentiable function $f : (x_0 - R, x_0 + R) \to \mathbb{R}$ such that*

$$\begin{cases} f'' + pf' + qf = 0 \\ f(x_0) = \alpha, f'(x_0) = \beta \end{cases}. \tag{11.32}$$

Moreover, f is actually infinitely differentiable and given by its Taylor series centered at x_0.

Interesting (and important) applications of the above result can be found, for instance, in Chap. 6 of [2], or in Chap. VI of [19]. We now discuss a particular case of it, referring to Problem 11.48 for a proof of the general case. See, also, Problem 4, page 510.

Given real numbers a and b and a twice differentiable function $f : \mathbb{R} \to \mathbb{R}$ such that

$$f'' + af' + bf = 0, \tag{11.33}$$

we say that the second degree equation

$$\lambda^2 + a\lambda + b = 0 \tag{11.34}$$

is the **characteristic equation** of (11.33). The discriminant $\Delta = a^2 - 4b$ of (11.34) is said to be the **discriminant** of (11.33).

Theorem 11.49 *Let a and b be given real numbers and $f : \mathbb{R} \to \mathbb{R}$ be a twice differentiable function satisfying (11.33). In the notations of the discussion above, there exist real constants A and B such that*

$$f(x) = Af_1(x) + Bf_2(x)$$

for every real x, where:

(a) $f_1(x) = e^{-\frac{ax}{2}}$ and $f_2(x) = xe^{-\frac{ax}{2}}$ if $\Delta = 0$.
(b) $f_1(x) = e^{\alpha x}$ and $f_2(x) = e^{\beta x}$ if $\Delta > 0$ and α and β are the roots of (11.34).
(c) $f_1(x) = e^{-\frac{ax}{2}} \cos\left(\frac{\sqrt{-\Delta}x}{2}\right)$ and $f_2(x) = e^{-\frac{ax}{2}} \sin\left(\frac{\sqrt{-\Delta}x}{2}\right)$ if $\Delta < 0$.

In particular, for given values of $f(0)$ and $f'(0)$, (11.33) has exactly one solution.

Proof First of all, writing $f'' = -af' - bf$ it is immediate that any solution of (11.33) is infinitely differentiable.

Claim 1 f is given by its Taylor series in the whole real line.

Fix $M > 0$ and choose $C > 0$ such that $|f| \leq C$ and $|f'| \leq C^2$ in $[-M, M]$. Since $f^{(k+2)} + af^{(k+1)} + bf^{(k)} = 0$ for every $k \in \mathbb{Z}_+$, we get

$$|f^{(k+2)}| \leq |a||f^{(k+1)}| + |b||f^{(k)}|.$$

Thus, if $|f^{(j)}| \leq C^{j+1}$ in $[-M, M]$ for $0 \leq j \leq k+1$, then

$$|f^{(k+2)}| \leq |a|C^{k+1} + |b|C^k \leq C^{k+2},$$

provided $C^2 \geq |a|C + |b|$. Therefore, if we choose such a C from the beginning, we conclude that

$$|f^{(n)}| \leq C^{n+1} \text{ in } [-M, M], \ \forall \ n \geq 0.$$

Now, for $x \in [-M, M]$, Taylor's formula gives

$$f(x) = \sum_{k=0}^{n-1} \frac{f^{(k)}(0)}{k!} x^k + \frac{f^{(n)}(c)}{n!} x^n, \tag{11.35}$$

for some $c \in [-M, M]$. However, since

$$\left| \frac{f^{(n)}(c)}{n!} x^n \right| \leq \frac{C^{n+1}}{n!} M^n = C \cdot \frac{(CM)^n}{n!} \xrightarrow{n} 0,$$

letting $n \to +\infty$ in (11.35) we conclude that $f(x) = \sum_{k \geq 0} \frac{f^{(k)}(0)}{k!} x^k$ in $[-M, M]$. Actually, since $M > 0$ was arbitrarily chosen, this holds for every $x \in \mathbb{R}$.

Claim 2 for given values of $f(0)$ and $f'(0)$, there is at most one solution for (11.33). (This claim will also establish the uniqueness part of the theorem.)

For simplicity, write $f(x) = \sum_{k \geq 0} \frac{a_k}{k!} x^k$. It follows from Theorem 11.27 that

$$f'(x) = \sum_{k \geq 1} \frac{a_k}{(k-1)!} x^{k-1} = \sum_{k \geq 0} \frac{a_{k+1}}{k!} x^k$$

and, analogously, $f''(x) = \sum_{k \geq 0} \frac{a_{k+2}}{k!} x^k$. Substituting these expressions into $f'' + af' + bf = 0$, we find

$$\sum_{k \geq 0} (a_{k+2} + aa_{k+1} + ba_k) \frac{x^k}{k!} = 0,$$

and the uniqueness of power series expansions gives $a_{k+2} + aa_{k+1} + ba_k = 0$ for every $k \geq 0$. However, since $a_0 = f(0)$ and $a_1 = f'(0)$, the sequence $(a_k)_{k \geq 0}$, and hence the function f, is completely determined.

Finally, to finish the proof it suffices to check, in each of the cases (a), (b) and (c), that: (i) the given functions f_1 and f_2 satisfy (11.33); (ii) for any given values for $f(0)$ and $f'(0)$, one is able to find real values for A and B such that

$$Af_1(0) + Bf_2(0) = f(0)$$
$$Af_1'(0) + Bf_2'(0) = f'(0)$$

Once this has been done, the uniqueness of solution, as established in Claim 2, allows us to conclude that $f = Af_1 + Bf_2$.

Checking items (i) and (ii) in each of (a), (b) and (c) is actually quite simple. For item (c), for instance, item (ii) amounts to showing that the system of equations

$$A = f(0), \quad -\frac{a}{2}A + \frac{\sqrt{-\Delta}}{2}B = f'(0)$$

always has a solution, which is immediate. As for (i), one just needs to compute f_i' and f_i'' for $i = 1, 2$, then showing that $f_i'' + af_i' + bf_i = 0$. We leave this as an exercise to the reader.   □

Remark 11.50 Note that the prescription of the values of $f(0)$ and $f'(0)$, instead of $f(x_0)$ and $f'(x_0)$ for some $x_0 \in \mathbb{R}$, is irrelevant. Indeed, letting $g(x) = f(x + x_0)$, we have $g(0) = f(x_0)$, $g'(0) = f'(x_0)$ and $g'' + ag' + bg = 0$, so that we can apply the previous result to find g and then find $f(x) = g(x - x_0)$.

Example 11.51 Find all differentiable functions $f, g : \mathbb{R} \to \mathbb{R}$ such that $f(0) = 0$, $g(0) = 1, f' = f + g$ and $g' = 2f$.

Solution Since $f' = f + g$, which is a differentiable function, we conclude that f is twice differentiable. Then $f'' = f' + g' = f' + 2f$ or, which is the same,

$$f'' - f' - 2f = 0.$$

Now, the second degree equation $\lambda^2 - \lambda - 2 = 0$ has real roots 2 and -1, so that the previous theorem gives $f(x) = Ae^{2x} + Be^{-x}$ for every real x. Since $f(0) = 0$ and $f'(0) = f(0) + g(0) = 1$, we get $A + B = f(0) = 0$ and $2A - B = f'(0) = 1$.

Solving such a system of equations, it follows that $A = \frac{1}{3}$ and $B = -\frac{1}{3}$, so that $f(x) = \frac{1}{3}(e^{2x} - e^{-x})$ and $g(x) = f'(x) - f(x) = \frac{2}{3}(e^{2x} + 2e^{-x})$.   □

Problems: Section 11.4

1. Compute the number of nonnegative integer solutions of the equation

$$a_1 + a_2 + \cdots + a_k = m,$$

where k and m are given natural numbers.

2. Compute the number of nonnegative integer solutions of the equation $a_1 + a_2 + a_3 + a_4 = 20$, such that $a_1 \geq 2$ and $a_3 \leq 7$.

3. Generalize the result of Proposition 11.46, showing that if $f : (x_0 - R, x_0 + R) \to \mathbb{R}$ is given by $f(x) = \sum_{n \geq 0} a_n (x - x_0)^n$, then for $k \in \mathbb{N}$ we have $f(x)^k = \sum_{n \geq 0} c_n (x - x_0)^n$, where

$$c_n = \sum a_{i_1} a_{i_2} \ldots a_{i_k}$$

and the above sum extends over all k-tuples $(i_1, \ldots, i_k)$ of nonnegative integers satisfying $i_1 + \cdots + i_k = n$.

4. In the notations of the discussion of Example 11.42, prove that

$$\frac{a_1}{d_1} + \frac{a_2}{d_2} + \cdots + \frac{a_m}{d_m} = \frac{m+1}{2}.$$

5. Use generating functions to find a_n as a function of n, where $(a_n)_{n \geq 1}$ is given by $a_1 = 2$ and $a_{k+1} = a_k + (k+1)$ for $k \geq 1$.

6. Use the methods of this section to give another proof to Theorem 3.19.

7. The sequence $(a_n)_{n \geq 0}$ is given by $a_0 = 1$ and $a_{n+1} = 2a_n + n$ for $n \geq 0$. In order to compute a_n as a function of n, do the following items:

   (a) If $a_n \leq \alpha^n$, show that $a_{n+1} \leq \alpha^{n+1}$ as long as $\alpha^n (\alpha - 2) \geq n$; then, conclude that $a_n \leq 3^n$ for every $n \geq 0$.

   (b) Show that the generating function of $(a_n)_{n \geq 0}$ converges in the interval $\left(-\frac{1}{3}, \frac{1}{3}\right)$ and is given by $f(x) = \frac{1 - 2x + 2x^2}{(1-x)^2(1-2x)}$.

   (c) Find real constants A, B and C such that

$$\frac{1 - 2x + 2x^2}{(1-x)^2(1-2x)} = \frac{A}{(1-x)^2} + \frac{B}{1-x} + \frac{C}{1-2x}.$$

   (d) Expand each of the functions of the right hand side above in power series to conclude that $a_n = 2^{n+1} - (n+1)$ for $n \geq 0$.

8. Let $(a_n)_{n \geq 0}$ be given by $a_0 = 1$, $a_1 = -3$, $a_2 = 5$ and $a_{k+3} = a_{k+2} + \frac{1}{2}a_{k+1} - \frac{1}{2}a_k$ for every integer $k \geq 0$. The purpose of this problem is to show that $(a_n)_{n \geq 0}$ converges and to compute the corresponding limit. To this end, do the following items:

(a) Show that $|a_n| \leq 5^n$ for every integer $n \geq 0$. Then, conclude that the radius of convergence of $\sum_{k \geq 0} a_k x^k$ is at least $\frac{1}{5}$.

(b) If $f : (-\frac{1}{5}, \frac{1}{5}) \to \mathbb{R}$ is given by $f(x) = \sum_{k \geq 0} a_k x^k$, use the given recurrence relation to get $f(x) = \frac{15x^2 - 8x + 2}{(x-1)(x^2-2)}$.

(c) Writing $f(x) = \frac{A}{x-1} + \frac{B}{x-\sqrt{2}} + \frac{C}{x+\sqrt{2}}$, show that $A = -9$. Then, expand $\frac{1}{1-x}$, $\frac{1}{1-\frac{x}{\sqrt{2}}}$ in $\frac{1}{1+\frac{x}{\sqrt{2}}}$ power series to get

$$f(x) = \sum_{k \geq 0} \left(-9 - \frac{B + (-1)^{k-1}C}{(\sqrt{2})^{k+1}} \right) x^k.$$

(d) Conclude that $a_n = -9 - \frac{B + (-1)^{n-1}C}{(\sqrt{2})^{n+1}}$ for $n \geq 0$, and hence that $a_n \xrightarrow{n} -9$.

9. (Putnam) Let $u, v, w : \mathbb{R} \to \mathbb{R}$ be given by the power series expansions

$$u(x) = 1 + \frac{x^3}{3!} + \frac{x^6}{6!} + \frac{x^9}{9!} + \cdots$$

$$v(x) = x + \frac{x^4}{4!} + \frac{x^7}{7!} + \frac{x^{10}}{10!} + \cdots$$

$$w(x) = \frac{x^2}{2!} + \frac{x^5}{5!} + \frac{x^8}{8!} + \frac{x^{11}}{11!} + \cdots$$

Prove that, for every real x, one has

$$u(x)^3 + v(x)^3 + w(x)^3 = 3u(x)v(x)w(x) + 1.$$

10. The purpose of this problem is to give a proof of Theorem 11.48. To this end, let

$$p(x) = \sum_{k \geq 0} b_k (x - x_0)^k \text{ and } q(x) = \sum_{k \geq 0} c_k (x - x_0)^k$$

in the interval $(x_0 - R, x_0 + R)$ and do the following items:

(a) If $f(x) = \sum_{k \geq 0} a_k (x - x_0)^k$ in the interval $(x_0 - R, x_0 + R)$ and f satisfies $f'' + pf' + qf = 0$, show that

$$(k+2)(k+1)a_{k+2} + \sum_{j=0}^{k} \left((j+1)b_{k-j}a_{j+1} + c_{k-j}a_j \right) = 0$$

for every $k \geq 0$. Then, conclude that $a_2, a_3, a_4, \ldots$ are uniquely determined by $a_0 = f(x_0)$, $a_1 = f'(x_0)$, p and q.

(b) Let $a_0 = \alpha$, $a_1 = \beta$ and $0 < r < R$ be given. If $|x - x_0| < r$, $\lambda = \frac{1}{r}|x - x_0|$ and $A_k = |a_k(x - x_0)^k|$ for $k \geq 0$, show that

$$(k + 2)(k + 1)A_{k+2} \leq \sum_{j=0}^{k}(j + 1)|b_{k-j}|r^{k-j+1}\lambda^{k-j}A_{j+1}$$

$$+ \sum_{j=0}^{k}|c_{k-j}|r^{k-j}\lambda^{k-j+2}A_j.$$

(c) Let $B, C > 0$ such that $|b_k|r^k \leq B$ and $|c_k|r^k \leq C$ for every $k \geq 0$. If $M = \max\{Br, C\}$, show that

$$(k + 2)(k + 1)A_{k+2} \leq M \sum_{j=0}^{k+1}(j + 1)\lambda^{k+1-j}A_j.$$

(d) Let $(\tilde{A}_k)_{k \geq 0}$ be defined by $\tilde{A}_0 = A_0$, $\tilde{A}_1 = A_1$ and

$$(k + 2)(k + 1)\tilde{A}_{k+2} = M \sum_{j=0}^{k+1}(j + 1)\lambda^{k+1-j}\tilde{A}_j.$$

Show that $A_k \leq \tilde{A}_k$ for every $k \geq 0$.

(e) If $\tilde{A}_0 \neq 0$ or $\tilde{A}_1 \neq 0$, show that $\tilde{A}_k > 0$ for every $k \geq 2$. Then, show that

$$\frac{\tilde{A}_{k+2}}{\tilde{A}_{k+1}} = \frac{\lambda k(k + 1) + M(k + 2)}{(k + 2)(k + 1)} \xrightarrow{k} \lambda < 1.$$

(f) Conclude that $\sum_{k \geq 0} \tilde{A}_k$ converges, and that this implies the convergence of $\sum_{k \geq 0} A_k$. Then, note that this is the same as saying that $\sum_{k \geq 0} a_k(x - x_0)^k$ is absolutely convergent in $(x_0 - r, x_0 + r)$. Finally, show that f, given as in (a), is well defined and does satisfy (11.32).

11.5 A Glimpse on Analytic Functions

In this section, we show that a function defined by a power series is *analytic*, i.e., is given by a convergent power series around each point of its interval of convergence. We also study the set of zeros of an analytic function and use the results we get to give purely analytic derivations of the properties of the sine and cosine functions.

Our purpose in doing so is to show the reader that all of the arguments we have done so far concerning these functions are actually independent of any considerations on the geometry of circles. As we anticipated in Remark 9.12 and now reinforce in a slightly different way, this is not a simple pedantism, for:

(i) the proof we have given for the formula $\sin' x = \cos x$ relied upon the fundamental trigonometric limit (cf. Lemma 9.11);
(ii) in turn, the proof of that result made use of the formula for the area of a circular sector;
(iii) then, in Example 10.43 we computed the area of a circle of radius R by using the change of variables formula to reduce it to a simple application of the FTC, which relies upon the fact that $\sin' x = \cos x$.

Hence, if we are not able to free the properties of the sine and cosine functions from the geometry of the circle, we will be forced to conclude that all of the arguments related to items (i), (ii) and (iii) above—and, therefore, to a large part of this book—are totally fallacious.

We begin our presentation with the following

Definition 11.52 Let $I \subset \mathbb{R}$ be an open interval. A function $f : I \to \mathbb{R}$ is **analytic** if, for every $x_0 \in I$, there exists $R > 0$ and a power series $\sum_{k \geq 0} a_k x^k$ such that $(x_0 - R, x_0 + R) \subset I$ and $f(x) = \sum_{k \geq 0} a_k x^k$ for every $x \in (x_0 - R, x_0 + R)$.

As we pointed out in the beginning of this section, a function defined by a power series in an open interval $(x_0 - R, x_0 + R)$ (possibly with $R = +\infty$) is analytic. In order to prove this, we need the following result on *double series* which is interesting in itself.

Proposition 11.53 *For each $j, k \in \mathbb{Z}_+$, let $a_{jk} \in \mathbb{R}$ be given, such that $\sum_{k \geq 0} |a_{jk}|$ converges for each $j \geq 0$ and $\sum_{j \geq 0} \sum_{k \geq 0} |a_{jk}|$ also converges. Then:*

(a) $\sum_{k \geq 0} a_{jk}$ converges for each $j \geq 0$ and $\sum_{j \geq 0} a_{jk}$ converges for each $k \geq 0$.
(b) $\sum_{j \geq 0} \sum_{k \geq 0} a_{jk}$ and $\sum_{k \geq 0} \sum_{j \geq 0} a_{jk}$ converge and

$$\sum_{j \geq 0} \sum_{k \geq 0} a_{jk} = \sum_{k \geq 0} \sum_{j \geq 0} a_{jk}. \tag{11.36}$$

Proof First of all, given $m, n \in \mathbb{N}$, our hypothesis gives

$$\sum_{k=0}^{n} \sum_{j=0}^{m} |a_{jk}| = \sum_{j=0}^{m} \sum_{k=0}^{n} |a_{jk}| \leq \sum_{j=0}^{m} \sum_{k \geq 0} |a_{jk}| \leq \sum_{j \geq 0} \sum_{k \geq 0} |a_{jk}| < +\infty.$$

Hence, letting $m \to +\infty$ we get

$$\sum_{k=0}^{n} \sum_{j \geq 0} |a_{jk}| \leq \sum_{j \geq 0} \sum_{k \geq 0} |a_{jk}| < +\infty,$$

for every $n \in \mathbb{N}$. Therefore, $\sum_{j \geq 0} |a_{jk}| < +\infty$ for every $k \geq 0$ and, letting $n \to +\infty$ in the above inequality, we get

$$\sum_{k \geq 0} \sum_{j \geq 0} |a_{jk}| \leq \sum_{j \geq 0} \sum_{k \geq 0} |a_{jk}| < +\infty.$$

Thus, $\sum_{k \geq 0} \sum_{j \geq 0} |a_{jk}|$ also converges.

Now we can prove (a) and (b), except for (11.36): since $\sum_{j \geq 0} |a_{jk}|$ converges for every $k \geq 0$, it is also true that $\sum_{j \geq 0} a_{jk}$ converges for every $k \geq 0$, with $\left| \sum_{j \geq 0} a_{jk} \right| \leq \sum_{j \geq 0} |a_{jk}|$. Summing this inequality over $k \geq 0$, we get

$$\sum_{k \geq 0} \left| \sum_{j \geq 0} a_{jk} \right| \leq \sum_{k \geq 0} \sum_{j \geq 0} |a_{jk}| < +\infty.$$

Therefore, $\sum_{k \geq 0} \sum_{j \geq 0} a_{jk}$ is absolutely convergent, hence convergent. Arguing in an entirely analogous way, we conclude that both of the series $\sum_{k \geq 0} a_{jk}$ and $\sum_{k \geq 0} \sum_{j \geq 0} a_{jk}$ are (absolutely) convergent.

In order to prove (11.36), let $\Delta = \left| \sum_{j \geq 0} \sum_{k \geq 0} a_{jk} - \sum_{k \geq 0} \sum_{j \geq 0} a_{jk} \right|$. Then, for $m, n \in \mathbb{N}$, we have

$$\Delta = \left| \sum_{j=0}^{m} \sum_{k \geq 0} a_{jk} + \sum_{j > m} \sum_{k \geq 0} a_{jk} - \sum_{k=0}^{n} \sum_{j \geq 0} a_{jk} - \sum_{k > n} \sum_{j \geq 0} a_{jk} \right|$$

$$\leq \left| \sum_{j=0}^{m} \sum_{k \geq 0} a_{jk} - \sum_{k=0}^{n} \sum_{j \geq 0} a_{jk} \right| + \left| \sum_{j > m} \sum_{k \geq 0} a_{jk} \right| + \left| \sum_{k > n} \sum_{j \geq 0} a_{jk} \right|.$$

Let $\epsilon > 0$ be given. Since $\sum_{j \geq 0} \sum_{k \geq 0} |a_{jk}|$ and $\sum_{k \geq 0} \sum_{j \geq 0} |a_{jk}|$ converge, we can choose $m_0, n_0 \in \mathbb{N}$ such that

$$m \geq m_0 \Rightarrow \sum_{j > m} \sum_{k \geq 0} |a_{jk}| < \epsilon \quad \text{and} \quad n \geq n_0 \Rightarrow \sum_{k > n} \sum_{j \geq 0} |a_{jk}| < \epsilon.$$

In turn, this gives $\left| \sum_{j > m} \sum_{k \geq 0} a_{jk} \right| < \epsilon$ and $\left| \sum_{k > n} \sum_{j \geq 0} a_{jk} \right| < \epsilon$, so that, with $m \geq m_0$ and $n \geq n_0$, we get in the above estimate for Δ

$$\Delta \le \left| \sum_{j=0}^{m} \sum_{k\ge 0}^{} a_{jk} - \sum_{k=0}^{n} \sum_{j\ge 0}^{} a_{jk} \right| + 2\epsilon = \left| \sum_{k\ge 0}^{} \sum_{j=0}^{m} a_{jk} - \sum_{j\ge 0}^{} \sum_{k=0}^{n} a_{jk} \right| + 2\epsilon$$

$$\le \left| \sum_{k=0}^{n} \sum_{j=0}^{m} a_{jk} - \sum_{j=0}^{m} \sum_{k=0}^{n} a_{jk} \right| + \left| \sum_{k>n}^{} \sum_{j=0}^{m} a_{jk} \right| + \left| \sum_{j>m}^{} \sum_{k=0}^{n} a_{jk} \right| + 2\epsilon$$

$$\le \sum_{k>n}^{} \sum_{j=0}^{m} |a_{jk}| + \sum_{j>m}^{} \sum_{k=0}^{n} |a_{jk}| + 2\epsilon \le \sum_{k>n}^{} \sum_{j\ge 0}^{} |a_{jk}| + \sum_{j>m}^{} \sum_{k\ge 0}^{} |a_{jk}| + 2\epsilon < 4\epsilon.$$

□

We are now in position to prove the following

Theorem 11.54 *If $f : (x_0 - R, x_0 + R) \to \mathbb{R}$ is given by the power series $f(x) = \sum_{k\ge 0} a_k(x-x_0)^k$, then f is analytic. More precisely, for a given $y_0 \in (x_0 - R, x_0 + R)$, there exist $b_0, b_1, b_2, \ldots \in \mathbb{R}$ such that $f(x) = \sum_{k\ge 0} b_k(x - y_0)^k$ in $(y_0 - r, y_0 + r)$, where $r = R - |y_0 - x_0| > 0$.*

Proof First of all, note that $(y_0 - r, y_0 + r) \subset (x_0 - R, x_0 + R)$. Now, for $x \in (y_0 - r, y_0 + r)$, we can write

$$f(x) = \sum_{k\ge 0} a_k(x - x_0)^k = \sum_{k\ge 0} a_k\big((x - y_0) + (y_0 - x_0)\big)^k$$

$$= \sum_{k\ge 0} a_k \sum_{j=0}^{k} \binom{k}{j} (y_0 - x_0)^{k-j}(x - y_0)^j.$$

If we can change the order of the sums in the last expression above, we will get

$$f(x) = \sum_{j\ge 0} \left(\sum_{k\ge j} a_k \binom{k}{j} (y_0 - x_0)^{k-j} \right)(x - y_0)^j.$$

Letting $b_j = \sum_{k\ge j} a_k \binom{k}{j}(y_0 - x_0)^{k-j}$, we will have

$$f(x) = \sum_{j\ge 0} b_j(x - y_0)^j,$$

thus finishing the proof.

Therefore, we are left to justifying the equality

$$\sum_{k\ge 0} a_k \sum_{j=0}^{k} \binom{k}{j} (y_0 - x_0)^{k-j}(x - y_0)^j = \sum_{j\ge 0} \left(\sum_{k\ge j} a_k \binom{k}{j} (y_0 - x_0)^{k-j} \right)(x - y_0)^j.$$

According to the last proposition, this will be true provided

$$\sum_{k\geq 0}\sum_{j=0}^{k}|a_k|\binom{k}{j}|y_0 - x_0|^{k-j}|x - y_0|^j < +\infty.$$

However, this expression equals

$$\sum_{k\geq 0}|a_k|\big(|y_0 - x_0| + |x - y_0|\big)^k,$$

which is finite, for $|y_0 - x_0| + |x - y_0| < |y_0 - x_0| + r = R$ and $\sum_{k\geq 0} a_k u^k$ converges absolutely for $u \in (-R, R)$.   □

We show next that the set of zeros of an analytic function $f : I \to \mathbb{R}$ do not *accumulate* on I. To this end, we first need a formal definition of this concept.

Definition 11.55 Given $X \subset \mathbb{R}$ and $a \in \mathbb{R}$, we say that a is an **accumulation point** of X if there exists a sequence $(a_n)_{n\geq 1}$ of pairwise distinct elements of X such that $a_n \xrightarrow{n} a$.

Theorem 11.56 *Let $I \subset \mathbb{R}$ be an open interval, $f : I \to \mathbb{R}$ be an analytic function and $\mathcal{Z} = \{x \in I; f(x) = 0\}$. The following conditions are equivalent:*

(a) $\mathcal{Z}$ has no accumulation point in I.
(b) There exists $x_0 \in I$ such that $f^{(k)}(x_0) = 0$ for every integer $k \geq 0$.
(c) f vanishes identically.

Proof (a) $\Rightarrow$ (b): let $x_0 \in I$ be an accumulation point of $\mathcal{Z}$, say $x_0 = \lim_{n\to+\infty} x_n$, where $(x_n)_{n\geq 1}$ is a sequence of pairwise distinct elements of $\mathcal{Z}$. Since $f(x_n) = 0$ for every $n \geq 1$, the continuity of f gives $f(x_0) = 0$.

By contradiction, assume that there exists an integer $k_0 \geq 1$ such that $f^{(k_0)}(x_0) \neq 0$. Without loss of generality, we can assume that k_0 is minimum with such a property. By the analyticity of f, we can choose $R > 0$ such that $(x_0 - R, x_0 + R) \subset I$ and, in such an interval,

$$f(x) = \sum_{k\geq k_0}\frac{f^{(k)}(x_0)}{k!}(x - x_0)^k = (x - x_0)^{k_0}\sum_{k\geq k_0}\frac{f^{(k)}(x_0)}{k!}(x - x_0)^{k-k_0}.$$

If $g(x) = \sum_{k\geq k_0}\frac{f^{(k)}(x_0)}{k!}(x - x_0)^{k-k_0}$, then the series which defines g converges in $(x_0 - R, x_0 + R) \setminus \{x_0\}$ (for $g(x) = \frac{f(x)}{(x-x_0)^{k_0}}$ in such an interval). Since it obviously converges at x_0, we conclude that g is continuous in $(x_0 - R, x_0 + R)$, with $g(x_0) = \frac{f^{(k_0)}(x_0)}{k_0!} \neq 0$. By the sign preserving lemma for continuous functions, there exists $0 < \delta \leq R$ such that $g \neq 0$ on $(x_0 - \delta, x_0 + \delta)$. Since $f(x) = g(x)(x - x_0)^{k_0}$ in this last interval, we conclude that $f \neq 0$ in $(x_0 - \delta, x_0 + \delta) \setminus \{x_0\}$. But this contradicts the fact that $x_n \in (x_0 - \delta, x_0 + \delta) \setminus \{x_0\}$ for every sufficiently large n.

(b) $\Rightarrow$ (c): let $x_0 \in I$ be such that $f^{(k)}(x_0) = 0$ for every integer $k \geq 0$, and let R be the radius of convergence of the power series defining f around x_0. Then, for $x \in (x_0 - R, x_0 + R)$ we have

$$f(x) = \sum_{k \geq k_0} \frac{f^{(k)}(x_0)}{k!}(x - x_0)^k = 0,$$

i.e., f vanishes identically in $(x_0 - R, x_0 + R)$.

Let's use this fact to show f vanishes identically in $I \cap [x_0, +\infty)$. By contradiction, assume that there would exist $a \in I \cap [x_0, +\infty)$ such that $f(a) \neq 0$. If

$$A = \{x \in I \cap [x_0, +\infty); f(x) \neq 0\},$$

then $A \neq \emptyset$ (since $a \in A$) and $x_0 + R$ is a lower bound for A; therefore, A has an infimum x_0', which obviously satisfies $x_0 + R \leq x_0' \leq a$. Hence, $x_0' \in I$ and, since x_0' is the greatest lower bound of A, we conclude that $[x_0, x_0') \cap A = \emptyset$. This is the same as saying that f vanishes identically on $[x_0, x_0')$, hence on $[x_0, x_0']$. In turn, this gives that $f^{(k)}(x_0') = 0$ for every integer $k \geq 0$. Now, an argument analogous to the one in the previous paragraph guarantees the existence of $R' > 0$ such that $f(x) = 0$ for every $x \in (x_0' - R', x_0' + R')$. Therefore, x_0 could not be the infimum of A, which is an absurd.

Finally, by the same token one can prove that f vanishes identically in $I \cap (-\infty, x_0]$, so that f vanishes identically in I.

(c) $\Rightarrow$ (a): obvious! $\square$

We now collect two straightforward consequences of the previous result which will be useful for what comes next.

Corollary 11.57 *Let $I \subset \mathbb{R}$ be an open interval and $f, g : I \to \mathbb{R}$ be analytic functions. If $f^{(j)}(x_0) = g^{(j)}(x_0)$ for some $x_0 \in I$ and every integer $j \geq 0$, then $f = g$.*

Proof Letting $h = f - g$, we get $h^{(j)}(x_0) = 0$ for every integer $j \geq 0$. Since h is also analytic on I, the previous result guarantees that h vanishes identically on I, and this is the same as having $f = g$. $\square$

Corollary 11.58 *Let $f : (-R, R) \to \mathbb{R}$ be an analytic function, possibly with $R = +\infty$. If f vanishes somewhere in $(0, R)$ but doesn't vanish identically, then f has a smallest positive zero. More precisely, there exists $a \in (0, R)$ such that $f(a) = 0$ but f doesn't change sign on $(0, a)$.*

Proof Otherwise, there would exist a sequence $(x_n)_{n \geq 1}$ of zeros of f in $(0, R)$, with $x_1 > x_2 > \cdots$. If $x_0 = \lim_{n \to +\infty} x_n$, then x_0 would be an accumulation point of the set of zeros of f in $[0, R)$, hence in its interval of convergence. By the previous result, f would vanish identically, which is a contradiction. $\square$

We are now in position to undertake a purely analytical development of Trigonometry. To this end, first notice that if $\sin$ and $\cos$ are to satisfy $\sin' = \cos$

and $\cos' = -\sin$, then they are infinitely differentiable and ultimately must be given by their Taylor series centered at 0, as in Example 11.4.

The key to free Trigonometry from Euclidean Geometry is to turn this around, *defining* sin and cos by means of those power series. Therefore, we start by letting

$$f(x) = \sum_{j \geq 1} \frac{(-1)^{j-1}}{(2j-1)!} x^{2j-1} \quad \text{and} \quad g(x) = \sum_{j \geq 0} \frac{(-1)^j}{(2j)!} x^{2j} \tag{11.37}$$

for every $x \in \mathbb{R}$.

The analytic foundations of Trigonometry are the content of the four coming results.

Proposition 11.59 *With notations as above, we have that:*

(a) f and g are well defined and analytical, with f odd and g even.
(b) $f^{(4k)} = f$, $f^{(4k+1)} = g$, $f^{(4k+2)} = -f$, $f^{(4k+3)} = -g$ and $g^{(4k)} = g$, $g^{(4k+1)} = -f$, $g^{(4k+2)} = -g$, $g^{(4k+3)} = f$, for every integer $k \geq 0$.
(c) f and g are nonconstant and such that $f(x)^2 + g(x)^2 = 1$ for every $x \in \mathbb{R}$.
(d) $f(x+y) = f(x)g(y) + f(y)g(x)$ and $f(x-y) = f(x)g(y) - f(y)g(x)$ for every $x, y \in \mathbb{R}$. In particular, $f(2x) = 2f(x)g(x)$ for every $x \in \mathbb{R}$.

Proof (a) The convergence of $\sum_{k \geq 0} \frac{|x|^k}{k!}$, together with the comparison test for series, assures the absolute convergence, in the whole real line, of both power series in (11.37). Then, Theorem 11.54 shows that they are analytical functions. Now, the oddness of f and the evenness of g follow from their very definitions.
(b) Theorem 11.27 gives

$$f'(x) = \sum_{j \geq 1} \frac{(-1)^{j-1}}{(2j-2)!} x^{2j-2} = \sum_{i \geq 0} \frac{(-1)^i}{(2i)!} x^{2i} = g(x);$$

analogously, $g' = -f$. Now, $f'' = g' = -f$, $g'' = (-f)' = -g$ and the given relations easily follow by induction.
(c) It follows from (b) that $(f^2 + g^2)' = 2ff' + 2gg' = 2fg + 2g(-f) = 0$. Therefore, $f^2 + g^2$ is constant and, since $f(0) = 0$ and $g(0) = 1$, we get $f(x)^2 + g(x)^2 = 1$ for every $x \in \mathbb{R}$. Also, $f'(0) = g(0) = 1$ and $g''(0) = -g(0) = -1$ assure that f and g are nonconstant.
(d) For a fixed $y \in \mathbb{R}$, let $h(x) = f(x+y)$ and $l(x) = f(x)g(y) + f(y)g(x)$. By the chain rule, $h, l : \mathbb{R} \to \mathbb{R}$ are infinitely differentiable. Moreover, by item (b) and the chain rule again, these functions satisfy

$$\begin{cases} h'' + h = 0 \text{ on } \mathbb{R} \\ h(0) = f(y), \ h'(0) = f'(y) \end{cases} \quad \text{and} \quad \begin{cases} l'' + l = 0 \text{ on } \mathbb{R} \\ l(0) = f(y), \ l'(0) = g(y) \end{cases}$$

Since $f'(y) = g(y)$, the uniqueness part of the proof of Theorem 11.49 (which did not depend on any properties of sin and cos) guarantees that $h = l$.

Now, the first formula, together with the fact that f is odd and g is even, gives

$$f(x-y) = f(x)g(-y) + f(-y)g(x) = f(x)g(y) - f(y)g(x).$$

Finally, the expression for $f(2x)$ also follows from the first formula, once we let $y = x$. □

Proposition 11.60 *The function f has a smallest positive zero.*

Proof Start by observing that, with item (b) of the above proposition at our disposal, the results of Problem 4, page 325 remain true for f and g in place of sin and cos, respectively, so that

$$x - \frac{x^3}{3!} \le f(x) \le x \quad \text{and} \quad g(x) \ge 1 - \frac{x^2}{2!}$$

for every $x \ge 0$. In turn, this gives

$$f(x) \le x - \frac{x^3}{3!} + \frac{x^5}{5!} \quad \text{and} \quad g(x) \le 1 - \frac{x^2}{2!} + \frac{x^4}{4!}$$

for every $x \ge 0$. For instance, for $x \ge 0$ one has

$$\frac{d}{dx}\left(g(x) - 1 + \frac{x^2}{2!} - \frac{x^4}{4!}\right) = -f(x) + x - \frac{x^3}{3!} \le 0,$$

so that $g(x) - 1 + \frac{x^2}{2!} - \frac{x^4}{4!} \le g(0) - 1 = 0$.

Now, the maximum of $x - \frac{x^3}{3!}$ for $x \ge 0$ is attained at $x = \sqrt{2}$ (check it!). Therefore, it follows from the above that

$$f(\sqrt{2}) \ge \sqrt{2} - \frac{\sqrt{2}^3}{3!} = \frac{2\sqrt{2}}{3}.$$

On the other hand, the minimum of $1 - \frac{x^2}{2!} + \frac{x^4}{4!}$ for $x \ge 0$ is attained at $x = \sqrt{10}$ (check it too!). Since $f(x) \le x\left(1 - \frac{x^2}{2!} + \frac{x^4}{4!}\right)$ for $x \ge 0$, this gives

$$f(\sqrt{10}) \le \sqrt{10}\left(1 - \frac{\sqrt{10}^2}{2!} + \frac{\sqrt{10}^4}{4!}\right) = \frac{\sqrt{10}}{6}.$$

Since $\frac{2\sqrt{2}}{3} > \frac{\sqrt{10}}{6}$, the IVT (applied to the intervals $[0, \sqrt{2}]$ and $[\sqrt{2}, \sqrt{10}]$) assures the existence of $0 < \alpha < \sqrt{2} < \beta < \sqrt{10}$ such that $f(\alpha) = f(\beta) = \frac{\sqrt{10}}{6}$. Then, the MVT of Lagrange gives $\gamma \in (\alpha, \beta)$ such that $f'(\gamma) = 0$.

It follows from the last part of item (d) of the previous proposition that

$$f(2\gamma) = 2f(\gamma)g(\gamma) = 2f(\gamma)f'(\gamma) = 0,$$

so that f has a positive root. Since f doesn't vanish identically, Corollary 11.58 shows that it has a smallest positive zero, as wished.   □

From now on, we let π denote the smallest positive zero of f. The last part of item (d) of Proposition 11.59 gives

$$0 = f(\pi) = 2f\left(\frac{\pi}{2}\right)g\left(\frac{\pi}{2}\right),$$

and since f is positive on $(0, \pi)$ this shows that $g\left(\frac{\pi}{2}\right) = 0$. Item (c) of Proposition 11.59 allows us to write

$$0 < f\left(\frac{\pi}{2}\right) = \pm\sqrt{1 - g\left(\frac{\pi}{2}\right)^2} = \pm 1,$$

so that $f\left(\frac{\pi}{2}\right) = 1$. Then, item (d) of Proposition 11.59 gives

$$f\left(\frac{\pi}{2} - x\right) = f\left(\frac{\pi}{2}\right)g(x) - f(x)g\left(\frac{\pi}{2}\right) = g(x). \tag{11.38}$$

We are finally in position to go one step further.

Proposition 11.61 *With notations as above, we have that:*

(a) $g : [0, \pi] \to [-1, 1]$ *is a decreasing bijection with continuous inverse.*
(b) $g^{-1} : (-1, 1) \to (0, \pi)$ *is differentiable.*
(c) $\mathrm{Im}(f) = \mathrm{Im}(g) = [-1, 1].$
(d) $f(x + \pi) = -f(x)$ *and* $g(x + \pi) = -g(x).$
(e) f *and* g *are periodic, with period* $2\pi.$

Proof (a) Since $g' = -f < 0$ in $(0, \pi)$, we conclude that g is decreasing in $[0, \pi]$. Since

$$1 = g(0) > g(\pi) = \pm\sqrt{1 - f(\pi)^2} = \pm 1,$$

we conclude that $g(\pi) = -1$. The rest follows from Theorem 8.35.
(b) This follows directly from Theorem 9.28. Indeed, since $f \neq 0$ in $(0, \pi)$, we get for $x \in (0, \pi)$ that $g'(x) = -f(x) \neq 0$.
(c) Item (c) of Proposition 11.59 gives $\mathrm{Im}(g) \subset [-1, 1]$. Then, item (a) above gives $\mathrm{Im}(g) = [-1, 1]$, from where (11.38) gives $\mathrm{Im}(f) = [-1, 1]$.
(d) An application of item (d) of Proposition 11.59, together with $g(\pi) = -1$, gives

$$f(x + \pi) = f(x)g(\pi) + f(\pi)g(x) = -f(x).$$

Therefore,

$$f\left(\frac{3\pi}{2}\right) = f\left(\frac{\pi}{2} + \pi\right) = -f\left(\frac{\pi}{2}\right) = -1$$

and, hence,

$$g\left(\frac{3\pi}{2}\right) = \pm\sqrt{1 - f\left(\frac{3\pi}{2}\right)^2} = 0.$$

In turn, (11.38) gives

$$g(x + \pi) = f\left(-x - \frac{\pi}{2}\right) = -f\left(x + \frac{\pi}{2}\right)$$
$$= -f(x)g\left(\frac{\pi}{2}\right) - f\left(\frac{\pi}{2}\right)g(x) = -g(x).$$

(e) Firstly, two applications of (d) give

$$f(x + 2\pi) = f\left((x + \pi) + \pi\right) = -f(x + \pi) = f(x)$$

and, analogously, $g(x + 2\pi) = g(x)$.

Now, let $p > 0$ be such that $f(x + p) = f(x)$ for every $x \in \mathbb{R}$. Then $f(p) = f(0) = 0$, so that $p \geq \pi$ (for, π is the smallest positive zero of f). Since $f(x + \pi) = -f(x)$ and f is not identically zero, we also have $p \neq \pi$. If $\pi < p < 2\pi$, then $0 < p - \pi < \pi$ and

$$f(p - \pi) = f(p)g(\pi) - f(\pi)g(p) = 0,$$

so that $p - \pi \geq \pi$. Therefore, $p \geq 2\pi$, and 2π is the smallest $p > 0$ such that $f(x + p) = f(x)$ for every $x \in \mathbb{R}$.

Finally, we leave to the reader the (analogous) task of checking that 2π is the smallest $p > 0$ such that $g(x + p) = g(x)$ for every $x \in \mathbb{R}$.   □

Our final result shows that f and g parametrize the unit circle of the cartesian plane.

Proposition 11.62 *For every point (x, y) in the unit circle $x^2 + y^2 = 1$ of the cartesian plane, there exists a unique $\theta \in [0, 2\pi)$ such that $x = g(\theta)$ and $y = f(\theta)$. Moreover, if $y_0 \geq 0$, then θ can be chosen in the interval $[0, \pi]$. Also, as θ varies from 0 to 2π, the point $(g(\theta), f(\theta))$ turns around the circle exactly once.*

Proof For the existence part, let (x_0, y_0) belong to the unit circle, so that $x_0^2 + y_0^2 = 1$. Since g is periodic of period 2π and $\text{Im}(g) = [-1, 1]$, there exists $\alpha \in [0, 2\pi)$ such that $x_0 = g(\alpha)$. Therefore,

$$y_0^2 = 1 - x_0^2 = 1 - g(\alpha)^2 = f(\alpha)^2,$$

so that $y_0 = \pm f(\alpha)$. If $y_0 = f(\alpha)$, let $\theta = \alpha$. If $y_0 = -f(\alpha)$, then the evenness g, the oddness of f and their periodicity give

$$\begin{aligned} x_0 &= g(\alpha) = g(-\alpha) = g(2\pi - \alpha) \\ y_0 &= -f(\alpha) = f(-\alpha) = f(2\pi - \alpha) \end{aligned},$$

and it suffices to let $\theta = 2\pi - \alpha$.

For the uniqueness part, assume, for the sake of contradiction, that there exist $0 \leq \theta_1 < \theta_2 < 2\pi$ such that $(x_0, y_0) = (g(\theta_1), f(\theta_1)) = (g(\theta_2), f(\theta_2))$. Then,

$$f(\theta_2 - \theta_1) = f(\theta_2)g(\theta_1) - f(\theta_1)g(\theta_2) = 0,$$

with $0 < \theta_2 - \theta_1 < 2\pi$. Since π is the smallest positive zero of f and the next one is 2π, this gives $\theta_2 - \theta_1 = \pi$. However, if this is so, then

$$x_0 = g(\theta_2) = g(\theta_1 + \pi) = -g(\theta_1) = -x_0$$
$$y_0 = f(\theta_2) = f(\theta_1 + \pi) = -f(\theta_1) = -y_0 ,$$

so that $x_0 = y_0 = 0$. This is a contradiction to the fact that (x_0, y_0) belongs to the unit circle.

We leave the rest as an (easy) exercise to the reader.   □

At this point, the reader might perhaps want to look again at Example 10.80 to realize that, in view of the four propositions above, that result gives a genuine proof of the fact that the unit circle has length 2π.

On the one hand, this validates the whole construction of the trigonometric functions, as usually done in elementary school Trigonometry. On the other, Leibniz formula for π (cf. Problem 9, page 484) provides a purely analytical way of getting numerical approximations for π, which can be used to get the usual estimate $\pi \cong 3.14159$, with five correct decimal places.

Problems: Section 11.5

1. Prove that, for every integer $k \geq 1$,

$$\sum_{j=0}^{2k} \frac{(-1)^{j-1}}{(2j-1)!} x^{2j-1} \leq \sin x \leq \sum_{j=0}^{2k+1} \frac{(-1)^{j-1}}{(2j-1)!} x^{2j-1}$$

and

$$\sum_{j=0}^{2k-1} \frac{(-1)^j}{(2j)!} x^{2j} \leq \cos x \leq \sum_{j=0}^{2k} \frac{(-1)^j}{(2j)!} x^{2j}.$$

2. One can prove that if $I, J \subset \mathbb{R}$ are open intervals and $g : I \to J$ and $f : J \to \mathbb{R}$ are analytic, then so is $f \circ g : I \to \mathbb{R}$. Use this fact to show that $\log : (0, +\infty) \to \mathbb{R}$ is analytic and to find its Taylor series expansion around $x_0 > 0$.

3. Prove the following theorem of S. Bernstein[9]: let $I \subset \mathbb{R}$ be an open interval, $f : I \to \mathbb{R}$ be infinitely differentiable and such that $f^{(n)}(x) \geq 0$ for every $x \in I$ and every sufficiently large natural n. Then, f is analytic.

4. Let $I \subset \mathbb{R}$ be an open interval and $p, q : I \to \mathbb{R}$ be analytic functions. The purpose of this problem is to show that, for $\alpha, \beta \in \mathbb{R}$ and $x_0 \in I$, there exists a unique analytic function $f : I \to \mathbb{R}$ such that

$$\begin{cases} f'' + pf' + qf = 0 \\ f(x_0) = \alpha, f'(x_0) = \beta \end{cases}. \tag{11.39}$$

To this end, do the following items:

(a) Show that (11.39) has at most one solution.
(b) Show that there exist some open interval $J \subset I$, containing x_0, and an analytic function $f_J : J \to \mathbb{R}$ satisfying (11.39).
(c) If J is as in (b) and $J \neq I$, show that there exist an interval $J' \subset I$, properly containing J, and an analytic function $f_{J'} : J' \to \mathbb{R}$ satisfying (11.39) and such that $f_{J'} = f_J$ along J.
(d) Finish the proof of the problem.

---

[9] After S. N. Bernstein, Russian mathematician of the XX century.

Bibliography

1. S. Abbott, *Understanding Analysis* (Springer, New York, 2001)
2. T. Apostol, *Calculus*, vol. 2 (Wiley, New York, 1967)
3. T. Apostol, *Introduction to Analytic Number Theory* (Springer, New York, 1976)
4. A. Caminha, *An Excursion Through Elementary Mathematics II - Euclidean Geometry* (Springer, New York, 2018)
5. A. Caminha, *An Excursion Through Elementary Mathematics III - Discrete Mathematics and Polynomial Algebra* (Springer, New York, 2018)
6. L.W. Cohen, G. Ehrlich, *The Structure of the Real Number System* (Literary Licensing, Whitefish, 2012)
7. K. Davidson, A. Donsig, *Real Analysis with Real Applications* (Prentice Hall, Inc., Upper Saddle River, 2002)
8. M.P. do Carmo, *Differential Geometry of Curves and Surfaces* (Prentice-Hall, Inc., Englewood Cliffs, 1976)
9. D.G. de Figueiredo, *Análise I* (in Portuguese) (LTC, Rio de Janeiro, 1996)
10. D.G. de Figueiredo, *Análise de Fourier e Equações Diferenciais Parciais* (in Portuguese) (IMPA, Rio de Janeiro, 2012)
11. D.G. de Figueiredo, *Números Irracionais e Transcendentes* (in Portuguese) (Brazilian Mathematical Society, Rio de Janeiro, 2002)
12. O. Foster, *Lectures on Riemann Surfaces* (Springer, New York, 1981)
13. C. Goffman, *Real Functions* (Rinehart & Company, Inc., New York, 1953)
14. C.R. Hadlock, *Field Theory and Its Classical Problems* (MAA, Washington, 1978)
15. P.R. Halmos, *Naive Set Theory* (Springer, New York, 1974)
16. D. Hilbert, *Foundations of Geometry* (Open Court Publ. Co., Peru, 1999)
17. H. Hochstadt, *The Functions of Mathematical Physics* (Dover, Mineola, 1986)
18. K. Hoffman, R. Kunze, *Linear Algebra* (Prentice-Hall, Boston, 1971)
19. D. Kreider, R. Küller, D. Ostberg, F. Perkins, *An Introduction to Linear Analysis* (Addison-Wesley Publ. Co., Boston, 1966)
20. W. Rudin, *Principles of Mathematical Analysis* (McGraw-Hill, Inc, New York, 1976)
21. E. Scheinerman, *Mathematics, a Discrete Introduction* (Cengage Learning, Boston, 2012)
22. G.F. Simmons, *Calculus Gems. Brief Lives and Memorable Mathematics*, 3rd edn. (McGraw-Hill, Inc., New York, 1992)
23. G.F. Simmons, *Differential Equations with Applications and Historical Notes*, 3rd edn. (Chapman and Hall/CRC, Boca Raton, 2016)
24. E. Stein, R. Shakarchi, *Fourier Analysis, an Introduction* (Princeton University Press, Princeton, 2003)
25. I. Stewart, *Why Beauty Is Truth: A History of Symmetry* (Basic Books, New York, 2008)

© Springer International Publishing AG 2017   511
A. Caminha Muniz Neto, *An Excursion through Elementary Mathematics, Volume I*,
Problem Books in Mathematics, DOI 10.1007/978-3-319-53871-6

26. M.B.W. Tent, *The Prince of Mathematics: Carl Friedrich Gauss* (A.K. Peters Ltd, Wellesley, 2006)
27. R. Wheeden, A. Zygmund, *Measure and Integral: An Introduction to Real Analysis* (Chapman & Hall, New York, 1977)

Appendix A
Glossary

Problems tagged with a country's name refer to any round of the corresponding national mathematical olympiad. For example a problem tagged "Brazil" means that it appeared in some round of some edition of the Brazilian Mathematical Olympiad. Problems proposed in other mathematical competitions, or which appeared in mathematical journals, are tagged with a specific set of initials, as listed below:

1. **APMO**: Asian-Pacific Mathematical Olympiad.
2. **Austrian-Polish**: Austrian-Polish Mathematical Olympiad.
3. **BMO**: Balkan Mathematical Olympiad.
4. **Berkeley**: Berkeley Preliminary Examination.
5. **Baltic Way**: Baltic Way Mathematical Contest.
6. **Crux**: Crux Mathematicorum, a mathematical journal of the Canadian Mathematical Society.
7. **EKMC**: Eötvös-Kürschák Mathematics Competition (Hungary).
8. **IMO**: International Mathematical Olympiad.
9. **IMO shortlist**: problem proposed to the IMO, though not used.
10. **Israel-Hungary**: Binational Mathematical Competition Israel-Hungary.
11. **Miklós-Schweitzer**: The Miklós-Schweitzer Mathematics Competition (Hungary).
12. **NMC**: Nordic Mathematical Contest.
13. **OCM**: State of Ceará Mathematical Olympiad.
14. **OCS**: South Cone Mathematical Olympiad.
15. **OBMU**: Brazilian Mathematical Olympiad for University Students.
16. **OIM**: Iberoamerican Mathematical Olympiad.
17. **OIM shortlist**: problem proposed to the OIM, though not used.
18. **OIMU**: Iberoamerican Mathematical Olympiad for University Students.
19. **Putnam**: The William Lowell Mathematics Competition.
20. **TT**: The Tournament of the Towns.

© Springer International Publishing AG 2017   513
A. Caminha Muniz Neto, *An Excursion through Elementary Mathematics, Volume I*,
Problem Books in Mathematics, DOI 10.1007/978-3-319-53871-6

Appendix B
Hints and Solutions

Section 1.1

1. Write $\frac{a}{b} = \frac{c}{d} = r$, thus obtaining $a = br$, $c = dr$ and, then, $a \pm c = r(b \pm d)$.

2. Let $x = 0.a_1a_2a_3\ldots$ and suppose that the sequence $(a_1, a_2, a_3, \ldots)$ is periodic from some point on, as in (1.3), say. If $y \in \mathbb{N}$ is the integer with decimal representation $b_1b_2\ldots b_p$, conclude that $10^{l+p}x = y + 10^l x$, so that $x = \frac{y}{10^{l+p}-10^p}$, a rational number. Conversely, let $x = \frac{a}{b}$, with $a, b \in \mathbb{N}$ and $0 < a \leq b$. If $y_k \in \mathbb{N}$ is the integer with decimal representation $a_1a_2\ldots a_k$, use the division algorithm to conclude that $10^k a = by_k + r_k$, with $0 \leq r_k < b$. Then, use the fact that there is only a finite number of possibilities for r_k to assure the existence of natural numbers l and p such that $r_{l+p} = r_l$. From this point on, conclude that the list $(a_1, a_2, a_3, \ldots)$ is as in (1.3), with $b_1 = a_{l+1}$, $b_2 = a_{l+2}$, ..., $b_p = a_{l+p}$.

3. Adapt, to the present case, the proof of the uniqueness of additive inverses, changing $+$ by $\cdot$ and 0 by 1.

4. Start by observing that $0 = 0 \cdot a = (1 + (-1))a = 1 \cdot a + (-1)a = a + (-1)a$; then, use the uniqueness of additive inverses.

5. Among the given decimal representations, the only one that does not correspond to a rational number is that of item (d); in order to prove this claim, observe that, in the given decimal representation, there will be arbitrarily long sequences of zeros. In order to write the numbers of items (a), (b) and (c) as irreducible fractions, follow the steps suggested in the hint to Problem 2.

© Springer International Publishing AG 2017
A. Caminha Muniz Neto, *An Excursion through Elementary Mathematics, Volume I,*
Problem Books in Mathematics, DOI 10.1007/978-3-319-53871-6

Section 1.2

1. For items (b) and (c), use (7') and (a); item (g) follows from items (b) and (c).
2. with equality if and only if $a = b = c = 0$. Adapt the proof of Corollary 1.5.
3. For item (a), observe that $(rs)^n = \underbrace{(rs)\cdots(rs)}_{n} = \underbrace{(r\cdots r)}_{n} \cdot \underbrace{(s\cdots s)}_{n} = r^n s^n$. For the other items, apply analogous arguments.
4. Consider separately the cases $m < n$, $m = n$ and $m > n$. In each case, make judicious use of the property of item (b) of the last problem.
5. If $b = 2^k \cdot 5^l$, with $k, l \in \mathbb{Z}_+$, and $n = \max\{k, l\}$, then $\frac{a}{b} = \frac{a \cdot 2^{n-k} \cdot 5^{n-l}}{10^n}$.
6. If $n = 2k$, with $k \in \mathbb{N}$, write $x^n = (x^2)^k$ and, then, apply item (g) of Proposition 1.2. The case of an odd n can be treated similarly.
7. Start by showing that $\frac{1}{2} - \frac{1}{3} + \frac{1}{4} - \frac{1}{5} > \frac{1}{5}$.
8. Compute the difference $\frac{a+1}{b+1} - \frac{a+2}{b+2}$.
9. Letting S denote the sum of the ten given numbers, show that $4S$ can be written as the sum of ten real numbers, each of which equals the sum of four of the ten given numbers.
10. Argue as in Example 1.4, using the fact that $31 < 32 = 2^5$ and $17 > 16 = 2^4$.
11. For items (a) and (b), begin by observing that $a^n < b^n$ if and only if $\left(\frac{a}{b}\right)^n < 1$; then, apply the result of Corollary 1.3. For item (c), begin by noticing that $a^n + b^n < (a+b)^n$ if and only if $\left(\frac{a}{a+b}\right)^n + \left(\frac{b}{a+b}\right)^n < 1$; then, apply the result of Corollary 1.3.
12. Use the fact that $a^3 < c \cdot a^2$ and $b^3 < c \cdot b^2$ and, then, apply Pythagoras' theorem.
13. The given inequality is equivalent to $\left(\frac{1}{7}\right)^n + \left(\frac{4}{7}\right)^n + \left(\frac{9}{7}\right)^n \geq 2$. This last one holds trivially for $n = 1$ and $n = 2$; for $n \geq 3$, apply item (b) of Corollary 1.3 to conclude that $\left(\frac{9}{7}\right)^n \geq \left(\frac{9}{7}\right)^3 > 2$.
14. Start by observing that, if $a \geq 4$, then $0 < \frac{1}{a} + \frac{1}{b} + \frac{1}{c} \leq \frac{1}{4} + \frac{1}{4} + \frac{1}{4} < 1$, so that $\frac{1}{a} + \frac{1}{b} + \frac{1}{c}$ cannot be an integer.
15. If $a > 5$ is an integer, show that $4(a-4) > a$; from this, conclude that it is not worth to have summands greater that 5. Then, by performing similar changes of summands, discard all of those which are equal to 4 or 5. Finally, show that is more advantageous to have more summands equal to 3 than summands equal to 2.
16. If we let a be the common leftmost digit, show that there exist nonnegative integers k and l such that $a \cdot 10^k < 2^n < (a+1) \cdot 10^k$ and $a \cdot 10^l < 5^n < (a+1) \cdot 10^l$; then, multiply these inequalities.
17. Suppose that all three inequalities are true; the second one can be written as $(a-d)(c-b) < 0$, so that the first one gives $a - d < 0 < c - b$. Now, use the third inequality to get $ad(c-b) < bc(a-d)$ and, then, arrive at a contradiction.

Section 1.3

2. For item (a), for instance, show that $(\sqrt[n]{xy})^n = xy$ and $(\sqrt[n]{x}\sqrt[n]{y})^n = xy$; then, use the definition of the n-th root.
3. By *contraposition*,[1] show that, if $b \neq 0$, then r is rational.
4. Reduce this problem to the previous one.
5. Argue by contraposition.
6. If $ab \neq 0$, compute $(a + b\sqrt{2})^2 = (-c\sqrt{3})^2$ to conclude that $\sqrt{2}$ should be rational, which is an absurd. Therefore, $ab = 0$ and, thus, $a = 0$ or $b = 0$. Then, apply the result of Problem 3.
7. By the sake of contradiction, write $\sqrt{2} = \frac{a}{b}$, with a and b relatively prime natural numbers, so that $2b^2 = a^2$. Then, successively show that a is even and b is even, thus reaching a contradiction.
8. Adapt, to the present case, the hint given to the previous problem.

Section 2.1

1. For item (a), it suffices to develop the right hand side to get

$$(x - y)(x + y) = x(x + y) - y(x + y)$$
$$= (x^2 + xy) - (xy + y^2)$$
$$= x^2 - y^2.$$

In what concerns (b), we have

$$(x \pm y)^2 = (x \pm y)(x \pm y) = x(x \pm y) \pm y(x \pm y)$$
$$= (x^2 \pm xy) \pm (xy \pm y^2) = x^2 \pm 2xy + y^2.$$

Finally, item (c) is also obtained by expanding the right hand side, and we leave this to the reader.

---

[1] The standard ways to prove a proposition of the form $A \Rightarrow B$ (i.e., *If A, then B*) are either *directly*, by *contraposition* or by *contradiction*. In the first case, we assume the validity of assertion A and deduce the validity of assertion B *directly*; in the second case, we assume that assertion B is *false* and deduce, directly, that assertion A is also false; finally, in the third case, we assume that assertion A is true and assertion B is false and, from this, directly deduce a contradiction (i.e., deduce that an assertion that is obviously false should be true, which is something impossible to happen). For a detailed discussion on the fundamentals of Logic and methods of proof, we refer the reader to [21].

2. It suffices to notice that

$$\frac{m}{np} + \frac{n}{mp} + \frac{p}{mn} = \frac{m^2 + n^2 + p^2}{mnp}$$

$$= \frac{(m + n + p)^2 - 2(mn + mp + np)}{mnp}.$$

3. The statement of the problem, together with item (a) of Proposition 2.1, gives
us

$$\left(\frac{b}{a}\right)^2 = \left(\frac{1-b}{1-a}\right)^2 \Leftrightarrow (b(1-a))^2 - (a(1-b))^2 = 0$$

$$\Leftrightarrow (b(1-a) - a(1-b))(b(1-a) + a(1-b)) = 0$$

$$\Leftrightarrow (b-a)(a+b-2ab) = 0.$$

However, since $a \neq b$, it follows that $a + b = 2ab$. Hence, we finally arrive at

$$\frac{1}{a} + \frac{1}{b} = \frac{a+b}{ab} = \frac{2ab}{ab} = 2.$$

4. Successively applying items (b) and (a) of Proposition 2.1, we get

$$\frac{1 - \left(\frac{x}{y}\right)^{-2}}{(\sqrt{x} - \sqrt{y})^2 + 2\sqrt{xy}} = \frac{1 - \left(\frac{y}{x}\right)^2}{(x - 2\sqrt{xy} + y) + 2\sqrt{xy}}$$

$$= \frac{x^2 - y^2}{x^2(x+y)} = \frac{(x-y)(x+y)}{x^2(x+y)} = \frac{x-y}{x^2}.$$

5. Applying item (a) of Proposition 2.1, we obtain

$$\frac{(x^3 + y^3 + z^3)^2 - (x^3 - y^3 - z^3)^2}{y+z} =$$

$$= \frac{[(x^3 + y^3 + z^3) - (x^3 - y^3 - z^3)][(x^3 + y^3 + z^3) + (x^3 - y^3 - z^3)]}{y+z}$$

$$= \frac{2(y^3 + z^3) \cdot 2x^3}{y+z}$$

$$= \frac{2(y+z)(y^2 - yz + z^2) \cdot 2x^3}{y+z}$$

$$= 4x^3(y^2 - yz + z^2).$$

6. Since $ab = 1 \Leftrightarrow \frac{1}{a} = b \Leftrightarrow \frac{1}{b} = a$, we have

$$\frac{\left(a - \frac{1}{a}\right)\left(b + \frac{1}{b}\right)}{a^2 - b^2} = \frac{(a-b)(b+a)}{a^2 - b^2} = \frac{a^2 - b^2}{a^2 - b^2} = 1,$$

where we have used item (a) of Proposition 2.1 in the next to last equality above.

7. We have $(y - x)(y + x) = 19^2$, with $y - x$ and $y + x$ integers such that $0 < y - x < y + x$. Therefore, the only possible choice is to have $y - x = 1$ and $y + x = 361$, so that $y = 181$ and $x = 180$.

8. A judicious application of item (b) of Proposition 2.1 gives

$$a^4 + b^4 = (a^2 + b^2)^2 - 2a^2b^2 = [(a+b)^2 - 2ab]^2 - 2(ab)^2$$
$$= (m^2 - 2n)^2 - 2n^2 = m^4 - 4m^2n + 2n^2.$$

9. Item (c) of Proposition 2.1 gives

$$a^6 + b^6 = (a^2)^3 + (b^2)^3 = (a^2 + b^2)(a^4 - a^2b^2 + b^4)$$
$$= (a^4 + b^4) - a^2b^2 = (a^2 + b^2)^2 - 3a^2b^2$$
$$= 1 - 3(ab)^2,$$

so that $\frac{1 - 3(ab)^2}{a^6 + b^6} = \frac{a^6 + b^6}{a^6 + b^6} = 1$.

10. Start by observing that

$$(ac + bd)^2 + (ad - bc)^2 =$$
$$= (a^2c^2 + 2acbd + b^2d^2) + (a^2d^2 - 2adbc + b^2c^2)$$
$$= a^2(c^2 + d^2) + b^2(d^2 + c^2) = (a^2 + b^2)(c^2 + d^2).$$

11. Adding 1 to both sides of the given equation, we get

$$11^2 = 121 = x + y + xy + 1 = (x + 1)(y + 1).$$

However, since $x + 1, y + 1 > 1$, the only possible choice is to have $x + 1 = y + 1 = 11$.

12. For item (a), it follows from item (a) of Proposition 2.1, with $\sqrt{x}$ and $\sqrt{y}$ in place of x and y, respectively, that

$$\frac{1}{\sqrt{x} \pm \sqrt{y}} = \frac{\sqrt{x} \mp \sqrt{y}}{(\sqrt{x} \pm \sqrt{y})(\sqrt{x} \mp \sqrt{y})}$$
$$= \frac{\sqrt{x} \mp \sqrt{y}}{(\sqrt{x})^2 - (\sqrt{y})^2} = \frac{\sqrt{x} \mp \sqrt{y}}{x - y}.$$

As for (b), we apply item (c) of Proposition 2.1, with $\sqrt[3]{x}$ and $\sqrt[3]{y}$ in place of x and y, respectively, and observe that $(\sqrt[3]{x})^2 = \sqrt[3]{x^2}$ and $(\sqrt[3]{x})^3 = x$ (and analogously for y). This way, we get

$$
\frac{1}{\sqrt[3]{x} \pm \sqrt[3]{y}} = \frac{\sqrt[3]{x^2} \mp \sqrt[3]{xy} + \sqrt[3]{y^2}}{(\sqrt[3]{x} \pm \sqrt[3]{y})(\sqrt[3]{x^2} \mp \sqrt[3]{xy} + \sqrt[3]{y^2})}
$$
$$
= \frac{\sqrt[3]{x^2} \mp \sqrt[3]{xy} + \sqrt[3]{y^2}}{x - y}.
$$

Finally, item (c) follows immediately from (b).

13. Item (a) of the previous problem gives

$$
2\left(\sqrt{n+1} - \sqrt{n}\right) = 2\frac{(n+1) - n}{\sqrt{n+1} + \sqrt{n}} = \frac{2}{\sqrt{n+1} + \sqrt{n}} < \frac{2}{2\sqrt{n}} = \frac{1}{\sqrt{n}}.
$$

The other inequality can be proved in a similar way.

14. Applying twice item (a) of Problem 12, we obtain

$$
\frac{1}{2 + \sqrt{2} + \sqrt{3}} = \frac{2 + \sqrt{2} - \sqrt{3}}{(2 + \sqrt{2})^2 - (\sqrt{3})^2} = \frac{2 + \sqrt{2} - \sqrt{3}}{3 + 4\sqrt{2}}
$$
$$
= \frac{(2 + \sqrt{2} - \sqrt{3})(3 - 4\sqrt{2})}{3^2 - (4\sqrt{2})^2}
$$
$$
= -\frac{1}{23}(2 + \sqrt{2} - \sqrt{3})(3 - 4\sqrt{2}).
$$

15. First of all, note that, for all real x, we have $x^3 + 3 = x^3 + (\sqrt[3]{3})^3 = (x + \sqrt[3]{3})(x^2 - x\sqrt[3]{3} + \sqrt[3]{9})$. Therefore, making $x = \sqrt{2}$, we get

$$
2\sqrt{2} + 3 = (\sqrt{2} + \sqrt[3]{3})(2 - \sqrt{2} \cdot \sqrt[3]{3} + \sqrt[3]{9})
$$

and, hence,

$$
\frac{1}{\sqrt{2} + \sqrt[3]{3}} = \frac{2 - \sqrt{2} \cdot \sqrt[3]{3} + \sqrt[3]{9}}{3 + 2\sqrt{2}}
$$
$$
= \frac{2 - \sqrt{2} \cdot \sqrt[3]{3} + \sqrt[3]{9}}{3 + 2\sqrt{2}} \cdot \frac{3 - 2\sqrt{2}}{3 - 2\sqrt{2}}
$$
$$
= (3 - 2\sqrt{2})(2 - \sqrt{2} \cdot \sqrt[3]{3} + \sqrt[3]{9}).
$$

16. It suffices to write $y + z = -x$, $x + z = -y$ and $x + y = -z$ to realize that each of $y + z$, $x + z$ and $x + y$ is also nonzero. Now, making use of these equalities in (a), we obtain

$$\frac{x^2}{(y+z)^2} + \frac{y^2}{(x+z)^2} + \frac{z^2}{(x+y)^2} = \frac{x^2}{(-x)^2} + \frac{y^2}{(-y)^2} + \frac{z^2}{(-z)^2} = 3.$$

Analogously, $\frac{x^3}{(y+z)^3} + \frac{y^3}{(x+z)^3} + \frac{z^3}{(x+y)^3} = -3$.

17. Applying item (a) of Proposition 2.1 several times, we get

$$
\begin{aligned}
a^{64} - b^{64} &= (a^{32} + b^{32})(a^{32} - b^{32}) \\
&= (a^{32} + b^{32})(a^{16} + b^{16})(a^{16} - b^{16}) \\
&= (a^{32} + b^{32})(a^{16} + b^{16})(a^8 + b^8)(a^8 - b^8) \\
&= (a^{32} + b^{32})(a^{16} + b^{16})(a^8 + b^8)(a^4 + b^4)(a^4 - b^4) \\
&= (a^{32} + b^{32})(a^{16} + b^{16}) \ldots (a^2 + b^2)(a^2 - b^2) \\
&= (a^{32} + b^{32})(a^{16} + b^{16}) \ldots (a^2 + b^2)(a + b)(a - b),
\end{aligned}
$$

so that

$$\frac{a^{64} - b^{64}}{(a+b)(a^2+b^2)(a^4+b^4)(a^8+b^8)(a^{16}+b^{16})} = (a^{32} + b^{32})(a+b).$$

18. For item (a), it suffices to see that

$$
\begin{aligned}
(a - b)(a^{n-1} + a^{n-2}b + a^{n-3}b^2 + \cdots + ab^{n-2} + b^{n-1}) &= \\
&= a(a^{n-1} + a^{n-2}b + a^{n-3}b^2 + \cdots + ab^{n-2} + b^{n-1}) \\
&\quad - b(a^{n-1} + a^{n-2}b + a^{n-3}b^2 + \cdots + ab^{n-2} + b^{n-1}) \\
&= (a^n + a^{n-1}b + a^{n-2}b^2 + \cdots + a^2b^{n-2} + ab^{n-1}) \\
&\quad - (a^{n-1}b + a^{n-2}b^2 + a^{n-3}b^3 + \cdots + ab^{n-1} + b^n) \\
&= a^n - b^n.
\end{aligned}
$$

With respect to (b), observe that the changes of sign make sense precisely because n is odd:

$$
\begin{aligned}
(a + b)(a^{n-1} - a^{n-2}b + a^{n-3}b^2 - \cdots - ab^{n-2} + b^{n-1}) &= \\
&= a(a^{n-1} - a^{n-2}b + a^{n-3}b^2 - \cdots - ab^{n-2} + b^{n-1}) \\
&\quad + b(a^{n-1} - a^{n-2}b + a^{n-3}b^2 - \cdots - ab^{n-2} + b^{n-1}) \\
&= (a^n - a^{n-1}b + a^{n-2}b^2 - \cdots - a^2b^{n-2} + ab^{n-1}) \\
&\quad + (a^{n-1}b - a^{n-2}b^2 + a^{n-3}b^3 - \cdots - ab^{n-1} + b^n) \\
&= a^n + b^n.
\end{aligned}
$$

19. More generally, let us factorise $x^{4n} + 4y^{4n}$, where $n \in \mathbb{N}$, inspiring ourselves in the formula for $(a+b)^2$:

$$
\begin{aligned}
x^{4n} + 4y^{4n} &= (x^{2n})^2 + (2y^{2n})^2 \\
&= [(x^{2n})^2 + (2y^{2n})^2 + 2x^{2n} \cdot 2y^{2n}] - 2x^{2n} \cdot 2y^{2n} \\
&= (x^{2n} + 2y^{2n})^2 - (2x^n y^n)^2 \\
&= (x^{2n} + 2y^{2n} + 2x^n y^n)(x^{2n} + 2y^{2n} - 2x^n y^n).
\end{aligned}
$$

20. It follows from Example 2.7 that

$$(a+b+c)^3 = a^3 + b^3 + c^3 + 3(a+b)(a+c)(b+c).$$

Now, since $(a+b) + (a+c) + (b+c) = 2(a+b+c)$, which is an even integer, we conclude that at least one of $a+b$, $a+c$ or $b+c$ is also even, so that $3(a+b)(a+c)(b+c)$ is a multiple of 6. Therefore,

$$
\begin{aligned}
6 \mid (a+b+c) &\Leftrightarrow 6 \mid (a+b+c)^3 \\
&\Leftrightarrow 6 \mid [a^3 + b^3 + c^3 + 3(a+b)(a+c)(b+c)] \\
&\Leftrightarrow 6 \mid (a^3 + b^3 + c^3).
\end{aligned}
$$

21. We present two different solutions. For the first one, just note that

$$
\begin{aligned}
a+b+c = 0 &\Rightarrow (a+b)^3 = (-c)^3 \\
&\Rightarrow a^3 + b^3 + 3ab(a+b) = -c^3 \\
&\Rightarrow a^3 + b^3 + 3ab(-c) = -c^3 \\
&\Rightarrow a^3 + b^3 + c^3 - 3abc = 0.
\end{aligned}
$$

Another possibility is to use the result of Example 2.7:

$$
\begin{aligned}
a+b+c = 0 &\Rightarrow (a+b+c)^3 = 0 \\
&\Rightarrow a^3 + b^3 + c^3 + 3(a+b)(a+c)(b+c) = 0.
\end{aligned}
$$

Now, from $a+b+c = 0$ we get $a+b = -c$, $a+c = -b$, $b+c = -a$ and, then,

$$3(a+b)(a+c)(b+c) = 3(-c)(-b)(-a) = -3abc.$$

22. Since $a + \sqrt{a^2 - b} \geq a - \sqrt{a^2 - b}$, both sides of the equality we wish to prove are nonnegative real numbers. Hence, it suffices to prove that their squares are equal, i.e., that

$$a \pm \sqrt{b} = \left(\frac{a + \sqrt{a^2 - b}}{2} \right) + \left(\frac{a - \sqrt{a^2 - b}}{2} \right)$$

$$\pm 2 \sqrt{\left(\frac{a + \sqrt{a^2 - b}}{2} \right) \left(\frac{a - \sqrt{a^2 - b}}{2} \right)}$$

$$= a \pm 2 \sqrt{\frac{a^2 - (\sqrt{a^2 - b})^2}{4}}.$$

In view of the above, this is pretty clear.

23. By the sake of contradiction, suppose that such x, y and z did exist. Then, writing $\frac{1}{x+y+z} - \frac{1}{x} = \frac{1}{y} + \frac{1}{z}$ and performing the additions on both sides, we would easily conclude that $x(x + y + z) = -yz$, or, which is the same, that $x^2 = -(xy + xz + yz)$. Similar reasoning would give us $x^2 = y^2 = z^2 = -(xy + xz + yz)$. Now, from $x^2 = y^2$ we would get $x = y$ or $x = -y$; however, since $x + y \neq 0$, we should have $x = y$. Analogously, we should also have $x = z$, so that the given equality would reduce to $\frac{1}{3x} = \frac{3}{x}$. This is obviously impossible.

24. Making $x = \frac{1}{b-c}$, $y = \frac{1}{c-a}$ and $z = \frac{1}{a-b}$, we have

$$xy + xz + yz = \frac{1}{(b-c)(c-a)} + \frac{1}{(b-c)(a-b)} + \frac{1}{(c-a)(a-b)}$$

$$= \frac{1}{(b-c)(c-a)(a-b)} [(a-b) + (c-a) + (b-c)] = 0.$$

Now, if x, y, z are real numbers such that $xy + xz + yz = 0$, then we have $x^2 + y^2 + z^2 = (x + y + z)^2$. Therefore, in our case we have

$$\frac{1}{(b-c)^2} + \frac{1}{(c-a)^2} + \frac{1}{(a-b)^2} = \left(\frac{1}{b-c} + \frac{1}{c-a} + \frac{1}{a-b} \right)^2.$$

Finally, just note that $\frac{1}{b-c} + \frac{1}{c-a} + \frac{1}{a-b}$ is a rational number.

25. Initially, note that

$$(\sqrt[3]{a} + \sqrt[3]{b})^3 = a + b + 3\sqrt[3]{ab}(\sqrt[3]{a} + \sqrt[3]{b})$$

and, hence,

$$\sqrt[3]{ab} = \frac{(\sqrt[3]{a} + \sqrt[3]{b})^3 - (a + b)}{3(\sqrt[3]{a} + \sqrt[3]{b})} \in \mathbb{Q}.$$

From this, we conclude that

$$\sqrt[3]{a} - \sqrt[3]{b} = \frac{a-b}{(\sqrt[3]{a})^2 + \sqrt[3]{ab} + (\sqrt[3]{b})^2} = \frac{a-b}{(\sqrt[3]{a} + \sqrt[3]{b})^2 - \sqrt[3]{ab}} \in \mathbb{Q}.$$

Finally,

$$\sqrt[3]{a} = \frac{1}{2}[(\sqrt[3]{a} + \sqrt[3]{b}) + (\sqrt[3]{a} - \sqrt[3]{b})] \in \mathbb{Q}$$

and, analogously, $\sqrt[3]{b} \in \mathbb{Q}$.

26. Raise both members of the equality $a + c + d = -(b + e + f)$ to the third power and apply the formula of Example 2.7 to get

$$a^3 + c^3 + d^3 + 3(a+c)(a+d)(c+d) = -(b^3 + e^3 + f^3) - 3(b+e)(b+f)(e+f).$$

Now, since $a^3 + b^3 + c^3 + d^3 + e^3 + f^3 = 0$, it follows from the above equality that

$$(a + c)(a + d)(c + d) = -(b + e)(b + f)(e + f).$$

Analogously, starting from $a + e + f = -(b + c + d)$ and arguing as above, we find that

$$(a + e)(a + f)(e + f) = -(b + c)(b + d)(c + d).$$

Finally, multiplying the two relations thus obtained and cancelling out the common factor $(c + d)(e + f)$, we arrive at the desired relation.

27. Start by observing that

$$b^3 < b^3 + 6ab + 1 \le b^3 + 6b^2 + 1 < b^3 + 6b^2 + 12b + 8 = (b + 2)^3.$$

Therefore, if $b^3 + 6ab + 1$ is a perfect cube, one has

$$b^3 + 6ab + 1 = (b + 1)^3 = b^3 + 3b^2 + 3b + 1,$$

so that $2a = b + 1$. Substituting $b = 2a - 1$ into $a^3 + 6ab + 1$, we conclude that $a^3 + 6a(2a - 1) + 1 = a^3 + 12a^2 - 6a + 1$ is a perfect cube. However, it is immediate to verify that

$$a^3 < a^3 + 12a^2 - 6a + 1 < (a + 4)^3,$$

so that the only possibilities are

$$a^3 + 12a^2 - 6a + 1 = (a + 1)^3, \ (a + 2)^3 \text{ or } (a + 3)^3.$$

These possibilities give, respectively, the equations $9a^2a - 9a = 0$, $6a^2 - 18a = 7$ and $3a^2 - 33a = 26$. The first one gives $a = 1$; the second has no integer solutions, for the left hand side is even, while the right hand one is odd; the third also has no integer solutions, for the left hand side is a multiple of 3, while the right hand one is not.

Section 2.2

1. For the first inequality, interpret $|x - a|$ as the distance from x to a in the real line (alternatively, apply the definition of modulus, separately considering the cases $x - a \geq 0$ and $x - a < 0$). For the other three inequalities, adapt the hint given to the analysis of the first one.
2. Separately analyse the cases (i) $x, y \geq 0$, (ii) $x \geq 0 > y$, (iii) $y \geq 0 > x$ and (iv) $x, y < 0$, showing that equality occurs in all of them.
3. For item (a), consider the cases $x \geq 0$ and $x < 0$. For item (b), argue as in Example 2.9.
4. Separately analyse cases $x < 0$, $0 < x < 1$ and $x > 1$. The solution set is $(-\infty, 0) \cup (1, +\infty)$.
5. Since $|y| \geq y$ and $|-y| = |y|$ for all real y, we have $|x - a| \geq x - a$ and $|x - b| = |b - x| \geq b - x$. Therefore, in order for the equation to have a real root x, we must necessarily have

$$c = |x - a| + |x - b| \geq (x - a) + (b - x) = b - a.$$

Thus, a necessary condition for the existence of solutions is that $c \geq b - a$. In other words, if $c < b - a$, then the given equation will have no solutions. Let us see what happens if $c \geq b - a$ (i.e., let us see whether this condition also suffices to the existence of solutions). Since every real number x satisfies one of $x \leq a$, $a < x \leq b$ or $x > b$, we shall analyse these cases separately:

- $x \leq a$: then, $x - a \leq 0$ and $x - b \leq 0$ (why?), so that the given equation reduces to $-(x - a) - (x - b) = c$, whose root is $x = \frac{1}{2}(a + b - c)$. Verify that such a root indeed satisfies the condition $x \leq a$.
- $a < x \leq b$: the given equation reduces to $(x - a) - (b - x) = c$, or $b - a = c$. If such an equality is true, then every real number x satisfying $a < x \leq b$ will be a solution of the equation; if it is false, then the equation will have no solutions x satisfying the inequalities $a < x \leq b$.
- $x > b$: arguing as in the previous cases, we easily arrive at the solution $x = \frac{1}{2}(a + b + c)$, which does satisfy condition $x > b$.

6. It suffices to see that

$$\left| \frac{r+2}{r+1} - \sqrt{2} \right| = \left| \frac{1}{r+1} + 1 - \sqrt{2} \right| = \left| \frac{1}{r+1} - \frac{1}{\sqrt{2}+1} \right|$$

$$= \frac{|\sqrt{2}-r|}{(r+1)(\sqrt{2}+1)} < \frac{1}{2}|r - \sqrt{2}|,$$

for $r \geq 0$ and $\sqrt{2} > 1$.

7. For item (a), compute $\frac{y}{1+y} - x1 + x$. For item (b), start by using the result of (a), together with the triangle inequality $|a+b| \leq |a|+|b|$, to get $\frac{|a+b|}{1+|a+b|} \leq \frac{|a|+|b|}{1+|a|+|b|}$.

8. For the first part, adapt the proof of Example 2.13. It may help you to notice that

$$((n+1)+(n+2)+\cdots+2n)-(1+2+\cdots+n) = n^2+(1+2+\cdots+n)-(1+2+\cdots+n) = n^2.$$

For the second part, show that every $x \in (n, n+1)$ is a root of the equation.

Section 2.3

1. If $x = \alpha$ is a root of the given equation, then $\alpha^2 + b|\alpha| + c = 0$. But, then, we also have $(-\alpha)^2 + b|-\alpha| + c = \alpha^2 + b|\alpha| + c = 0$, so that $x = -\alpha$ is also a root of the equation. Therefore, since $\alpha + (-\alpha) = 0$, the sum of the roots has to be equal to 0.

2. For item (a), squaring both sides of $\sqrt{x+2} = 10-x$ we get $x+2 = (10-x)^2$, or $x^2 - 21x + 98 = 0$. Since $7 + 14 = 21$ and $7 \cdot 14 = 98$, the roots of this last equation are $x = 7$ and $x = 14$. However, only $x = 7$ satisfies the original equation, for, substituting $x = 14$ in it, we find $\sqrt{16} = -4$, which is an absurd. In what concerns (b), we initially observe that, for the square roots to have a meaning in $\mathbb{R}$, we must have $-\frac{3}{2} \leq x \leq \frac{1}{3}$; on the other hand,

$$\sqrt{x+10} = \sqrt{2x+3} + \sqrt{1-3x} \Leftrightarrow$$

$$\Leftrightarrow x + 10 = (\sqrt{2x+3} + \sqrt{1-3x})^2$$

$$\Leftrightarrow x + 10 = 4 - x + 2\sqrt{(2x+3)(1-3x)}$$

$$\Leftrightarrow x + 3 = \sqrt{(2x+3)(1-3x)}$$

$$\Leftrightarrow (x+3)^2 = (2x+3)(1-3x)$$

$$\Leftrightarrow 7x^2 + 13x + 6 = 0,$$

so that $x = -1$ or $x = -\frac{6}{7}$. Since both of these values belong to the interval $[-\frac{3}{2}, \frac{1}{3}]$, we conclude that they are the solutions of the given equation. Finally, for item (c), we note that the substitution of variables $y = x^2 + 18x$ transform the given equação into $y + 30 = 2\sqrt{y + 45}$. Squaring both sides, we easily arrive at $y^2 + 56y + 720 = 0$, an equation whose roots are $y = -36$ and $y = -20$. From these, only $y = -20$ serves us, for $y + 30 = 2\sqrt{y + 45}$ gives $y + 30 \geq 0$ and $y + 45 \geq 0$, i.e., $y \geq -30$. Therefore, $x^2 + 18x + 20 = 0$ and, then, $x = -9 \pm \sqrt{61}$.

3. Start by observing that, in order for $\sqrt{2x - 1}$ to have a meaning in the set of reals, we must have $x \geq \frac{1}{2}$. By a similar reason, it must be the case that $x \geq \sqrt{2x - 1}$. However, since

$$x \geq \sqrt{2x - 1} \Leftrightarrow x^2 \geq 2x - 1 \Leftrightarrow x^2 - 2x + 1 \geq 0,$$

which is always true, the conditions for the existence of the square roots in the given equation reduce to $x \geq \frac{1}{2}$. Now,

$$A^2 = \left(\sqrt{x + \sqrt{2x - 1}} + \sqrt{x - \sqrt{2x - 1}} \right)^2$$

$$= 2x + 2\sqrt{(x + \sqrt{2x - 1})(x - \sqrt{2x - 1})}$$

$$= 2x + 2\sqrt{x^2 - (2x - 1)} = 2x + 2|x - 1|.$$

We shall consider here only item (b), leaving the other two items as exercises to the reader. If $A = 1$, we have to solve the equation $2x + 2|x - 1| = 1$, under the condition $x \geq \frac{1}{2}$. If $x \geq 1$, that equation reduces to $2x + 2(x - 1) = 1$, so that $x = \frac{3}{4}$ (which does not satisfy the condition $x \geq 1$). If $\frac{1}{2} \leq x < 1$, the equation reduces to $2x + 2(1 - x) = 1$, an equality which is false for every $x \in \left(\frac{1}{2}, 1 \right)$. Therefore, there are no solutions in item (b).

4. Let $ax^2 + bx + c = 0$ be the equation of the first quiz, $a'x^2 + bx + c = 0$ be that of the second one and $ax^2 + bx + c' = 0$ be that of the third one. Since the roots of the second quiz are 2 and 3, we have $-\frac{b}{a'} = 5$ and $\frac{c}{a'} = 6$, so that $-\frac{b}{c} = \frac{5}{6}$; on the other hand, since the roots of the third quiz are 2 and -7, we have $-\frac{b}{a} = -5$. Therefore, the second degree equation of the first quiz has the form $ax^2 + 5ax - 6a = 0$, and, thus, has the same roots as those of $x^2 + 5x - 6 = 0$, i.e., -6 and 1.

5. It follows from Proposition 2.16 that $a + b = -a$ and $ab = b$. The second equality gives $a = 1$, and substituting this value into the first equality we get $b = -2$.

6. Again by Proposition 2.16 (applied to both equations), we have $-a = \alpha^2 + \beta^2 = (\alpha + \beta)^2 - 2\alpha\beta = 13 - 2 \cdot 9 = -5$ and $b = \alpha^2\beta^2 = (\alpha\beta)^2 = 9^2 = 81$. Therefore, $a + b = 86$.

7. The simplest possibility is $x^2 - Sx + P = 0$, where

$$S = u^3 + v^3 = (u + v)(u^2 - uv + v^2)$$
$$= -[(u + v)^2 - 3uv]$$
$$= -[(-1)^2 - 3(-1)] = -4$$

and $P = u^3 v^3 = (uv)^3 = (-1)^3 = -1$.

8. A possibility is the equation $x^2 - S'x + P' = 0$, where $S' = (\alpha S + P) + (\beta S + P) = (\alpha + \beta)S + 2P = S^2 + 2P$ and $P' = (\alpha S + P)(\beta S + P) = \alpha\beta S^2 + (\alpha + \beta)SP + P^2 = 2S^2 P + P^2$. Here, we used the relations $\alpha + \beta = S$ and $\alpha\beta = P$.

9. If $\alpha = 7 + 4\sqrt{3}$ and $\beta = 7 - 4\sqrt{3}$, then $\alpha + \beta = 14$ and $\alpha\beta = 1$, so that α and β are the roots of the second degree equation $x^2 - 14x + 1 = 0$. Hence, $\alpha^2 = 14\alpha - 1$ and $\beta^2 = 14\beta - 1$ and, starting from such equalities, we get

$$\alpha^{k+2} + \beta^{k+2} = 14(\alpha^{k+1} + \beta^{k+1}) - (\alpha^k + \beta^k).$$

Now, make k successively equal to 0, 1, 2 and 3 to compute, also successively, $\alpha^2 + \beta^2, \alpha^3 + \beta^3, \alpha^4 + \beta^4$ and $\alpha^5 + \beta^5$.

10. Substituting x by α, we get the equality $\alpha^2 = \alpha + 1$ and, from it, $\alpha^4 = (\alpha^2)^2 = (\alpha + 1)^2 = \alpha^2 + 2\alpha + 1 = 3\alpha + 2$ and $\alpha^5 = \alpha \cdot \alpha^4 = \alpha(3\alpha + 2) = 3\alpha^2 + 2\alpha = 3(\alpha + 1) + 2\alpha = 5\alpha + 3$. Hence, $\alpha^5 - 5\alpha = 3$.

11. Since the roots of the equation must be integers whose product equals 5, we conclude that they must be (i) 1 and 5; or (ii) -1 and -5. In the first case $m = -(1 + 5) = -6$, whereas, in the second, $m = -((-1) + (-5)) = 6$.

12. The given equation is equivalent to $x^2 - (b + c + 2a^2)x + [bc + a^2(b + c)] = 0$. In order for the roots of this last equation to be real and distinct, it suffices to show that $\Delta > 0$, where Δ is its discriminant. In fact, since $a \neq 0$, we have

$$\Delta = (b + c + 2a^2)^2 - 4[bc + a^2(b + c)]$$
$$= (b + c)^2 + 4a^4 - 4bc = (b - c)^2 + 4a^4 > 0.$$

13. Making $a = x - \frac{1}{x}$, we get $x^2 - ax - 1 = 0$ (1) and, hence, $\frac{1}{x} = x - a$. Therefore, the original equation becomes $x = \sqrt{a} + \sqrt{1 - (x - a)}$ or, which is the same, $x - \sqrt{a} = \sqrt{1 + a - x}$ (2). Squaring both sides of (2), we get $x^2 + a - 2\sqrt{ax} = 1 + a - x$, or $x^2 - (2\sqrt{a} - 1)x - 1 = 0$ (3). Subtracting (3) from (1), it follows that $(2\sqrt{a} - 1 - a)x = 0$. Now, since $x \neq 0$, it must be that $2\sqrt{a} - 1 - a = 0$, from which $a = 1$. Then, $x - \frac{1}{x} = 1$, so that $x = \frac{1 \pm \sqrt{5}}{2}$. However, (2) gives $x - 1 = x - \sqrt{a} = \sqrt{1 + a - x} \geq 0$, so that $x = \frac{1 + \sqrt{5}}{2}$.

14. If α is a common root of both equations, then $a\alpha^3 - \alpha^2 - \alpha - (a + 1) = 0$ ($*$) and $a\alpha^2 - \alpha - (a + 1) = 0$. Multiply the second equality by α to get $a\alpha^3 - \alpha^2 - (a + 1)\alpha = 0$, and subtract this result from ($*$) to arrive at $(a + 1)\alpha - \alpha - (a + 1) = 0$,

so that $\alpha = 1 + \frac{1}{a}$. Therefore, if the two given equations are to have a common root, this must be equal to $1 + \frac{1}{a}$. It now suffices to verify that $1 + \frac{1}{a}$ actually is a root of both equations, and we leave this simple task to the reader.

15. For item (a), it suffices to see that

$$x^3 - 3x^2 + 5x = (x^3 - 3x^2 + 3x - 1) + (2x - 2) + 3$$
$$= (x - 1)^3 + 2(x - 1) + 3.$$

Now, it follows from (a) that the equalities in the statement of the problem can be rewritten as

$$(x - 1)^3 + 2(x - 1) + 3 = 1 \text{ and } (y - 1)^3 + 2(y - 1) + 3 = 5;$$

adding these, we arrive at

$$(x - 1)^3 + (y - 1)^3 + 2(x + y - 2) = 0.$$

In the above relation, substitute the factorisation

$$(x - 1)^3 + (y - 1)^3 = (x + y - 2)[(x - 1)^2 - (x - 1)(y - 1) + (y - 1)^2]$$

to get

$$(x + y - 2)[(x - 1)^2 - (x - 1)(y - 1) + (y - 1)^2 + 2] = 0.$$

Now, there are two possibilities: $x + y - 2 = 0$ or $(x - 1)^2 - (x - 1)(y - 1) + (y - 1)^2 + 2 = 0$. In order to show that the second one doesn't happen, make $x - 1 = a$ and $y - 1 = b$ and observe that

$$a^2 - ab + b^2 + 2 = a^2 - ab + \frac{b^2}{4} + \frac{3b^2}{4} + 2 = \left(a - \frac{b}{2}\right)^2 + \frac{3b^2}{4} + 2 > 0.$$

17. Compare the coefficients at both sides of the equality

$$x^3 + ax^2 + bx + c = (x - \alpha)(x^2 + b'x + c')$$

to get the equalities $b' - \alpha = a$, $c' - \alpha b' = b$ and $-\alpha c' = c$. Now, use the first and third relations to get $b' = a + \alpha$ and $c' = -\frac{c}{\alpha}$, showing, then, that these values for b' and c' also satisfy the relation $c' - \alpha b' = b$ (at this step you shall need to use the fact that $\alpha^3 + a\alpha^2 + b\alpha + c = 0$.

18. For item (a), let $a = \sqrt[3]{2 + \sqrt{5}}$ and $b = \sqrt[3]{2 - \sqrt{5}}$, so that $\alpha = a + b$. Then, use the fact that

$$\alpha^3 = a^3 + b^3 + 3ab(a + b) = a^3 + b^3 + 3ab\alpha,$$

together with the relation $ab = -1$. For (b), use the result of the previous problem to show that $x^3 + 3x - 4 = (x - 1)(x^2 + x + 4)$, so that $x = 1$ is the only real root of $x^3 + 3x - 4 = 0$.

19. Use the result of Problem 21 to compare the coefficients of $a_3x^3 + a_2x^2 + a_1x + a_0 = 0$ and $a_3(x - x_1)(x - x_1)(x - x_3)$.

20. In order to compute the value of the first sum of powers, use relations (2.21), together with the identity $\alpha^2 + \beta^2 + \gamma^2 = (\alpha + \beta + \gamma)^2 - 2(\alpha\beta + \alpha\gamma + \beta\gamma)$; for the other two sums of powers, start by observing that, since α is a root of the equation, we have $\alpha^3 = 3\alpha - 1$, and analogously for β and γ. Then add the left and right hand sides of these three identities.

21. Expand the products in $(y + d)^3 + a(y + d)^2 + b(y + d) + c$ and impose that the coefficient of y^2 equals 0.

23. It suffices to see that

$$x^3 + \frac{1}{x^3} = \left(x + \frac{1}{x}\right)\left(x^2 - 1 + \frac{1}{x^2}\right)$$
$$= \left(x + \frac{1}{x}\right)\left(\left(x + \frac{1}{x}\right)^2 - 3\right).$$

24. Just note that

$$x^4 + \frac{1}{x^4} = \left(x^2 + \frac{1}{x^2}\right)^2 - 2 = \left(\left(x + \frac{1}{x}\right)^2 - 2\right)^2 - 2.$$

25. First, note that the equality $x^2 - x - 1 = 0$ is equivalent to $x - \frac{1}{x} = 1$. Then, use the identity

$$x^3 - \frac{1}{x^3} = \left(x - \frac{1}{x}\right)\left(x^2 + 1 + \frac{1}{x^2}\right)$$
$$= \left(x - \frac{1}{x}\right)\left(\left(x - \frac{1}{x}\right)^2 + 3\right).$$

26. The equality $x^2 - 4x + 1 = 0$ is equivalent to $x + \frac{1}{x} = 4$. Now, use the expression for $x^3 + \frac{1}{x^3}$, deduced in the hint to Problem 23, and observe that $x^6 + \frac{1}{x^6} = \left(x^3 + \frac{1}{x^3}\right)^2 - 2$.

28. Initially, consider the cases $n = 2$, 4 and 6; then, adapt the reasoning of these particular cases to the general one.)

29. For item (a), see the hint to Problem 23. For item (b), argue as was done in the text for biquadratic equations.

30. For item (a), let's consider two cases: (i) if $a, b > 0$ then $c + d = -a < 0$ and $cd = b > 0$, so that $c, d < 0$; but then $ef = d < 0$, so that $e > 0 > f$. (ii) If $a, b < 0$, then, in the notations of (i), $c > 0 > d$. Therefore, starting

with the second equation, if necessary, we can assume from the beginning that $a > 0 > b$. Then, letting $c \geq d$ be the roots of $x^2 + ax + b = 0$, we get $c + d = -a < 0$ and $cd = b < 0$, so that $c > 0 > d$. For (b), we have $\Delta = m^2 - 4n$ and $\Delta' = p^2 - 4q$. Compute

$$\Delta' = \left(\frac{-m + \sqrt{\Delta}}{2}\right)^2 - 4\left(\frac{-m - \sqrt{\Delta}}{2}\right)$$

$$= \frac{1}{4}\left(m^2 - 2m\sqrt{\Delta} + \Delta + 8m + 8\sqrt{\Delta}\right)$$

and notice that, since $m > 0$,

$$\Delta' < \Delta \Leftrightarrow m^2 - 2m\left(\sqrt{\Delta} - 4\right) + \left(8\sqrt{\Delta} - 3\Delta\right)$$

$$\Leftrightarrow -\sqrt{\Delta} < m < 3\sqrt{\Delta} - 8 \Leftrightarrow m < 3\sqrt{\Delta} - 8$$

$$\Leftrightarrow (m + 8)^2 < 9(m^2 - 4n)$$

$$\Leftrightarrow 2(m - 1)^2 > 9(n + 2).$$

For (c), note that the process cannot continue indefinitely if the inequality $\Delta' < \Delta$ always holds. Then, show that at some point we reach a situation in which $n = -1$ or $m = 1, n = -2$. If $n = -1$, conclude that the roots of $x^2 + mx - 1 = 0$ should be 1 and -1, so that m should be equal to 0, which is not the case. Then, we are left to $m = 1$ and $n = -2$. In this case, argue backwards to show that $a = 1, b = -2$.

Section 2.4

1. Execute the elimination algorithm in each of the items above.
2. Apply the elimination algorithm, in the way done in Example 2.26.
3. Apply the variable substitution $a = \frac{1}{x}, b = \frac{1}{y}, c = \frac{1}{z}$ to transform the given system into a linear system of three equations in three unknowns.
4. Show that $x_j = a_{1j}b_1 + a_{2j}b_2 + a_{3j}b_3$, for $1 \leq j \leq 3$.
5. For item (a), it suffices to see that $x_1 = x_2 = \cdots = x_n = 0$ always is a solution. In what concerns (b), suppose that $x_1 = \alpha_1, x_2 = \alpha_2, \ldots, x_n = \alpha_n$ and $x_1 = \beta_1, x_2 = \beta_2, \ldots, x_n = \beta_n$ are two distinct solutions of (2.25). If $t \in \mathbb{R}$ is arbitrary, show that $x_i = t\alpha_i + (1 - t)\beta_i$, for $1 \leq i \leq n$, is also a solution. Finally, for item (c), assume that the system has only one solution when $b_1 = b_2 = \cdots = b_m = 0$, so that this solution is $x_1 = x_2 = \cdots = x_n = 0$. Let $b_1, b_2, \ldots, b_m$ be arbitrary real numbers, and $x_1 = \alpha_1, x_2 = \alpha_2, \ldots, x_n = \alpha_n$ and $x_1 = \beta_1, x_2 = \beta_2, \ldots, x_n = \beta_n$ be two solutions of (2.25). Then, it is immediate to check that $x_1 = \alpha_1 - \beta_1, x_2 = \alpha_2 - \beta_2, \ldots, x_n = \alpha_n - \beta_n$ is a solution of (2.25) when $b_1 = b_2 = \cdots = b_m = 0$, so that our assumption gives $\alpha_i - \beta_i = 0$, for $1 \leq i \leq n$.

Sections 2.5

1. Equations $x + yz = 2$ and $y + xz = 2$ give $(x - y)(1 - z) = 0$, so that either $x = y$ or $z = 1$. Do the same for the other two pairs of equations, and compare the different possibilities that arise.

2. Letting $z = 1/(x + y)$, we obtain the second degree system

$$\begin{cases} z + x = a + 1 \\ zx = a \cdot 1 \end{cases}.$$

Hence, there are two possibilities: $(z, x) = (a, 1)$ or $(z, x) = (1, a)$. Those give rise to the systems

$$\begin{cases} x + y = a^{-1} \\ x = 1 \end{cases} \quad \text{and} \quad \begin{cases} x + y = 1 \\ x = a^{-1} \end{cases},$$

both of which can be immediately solved.

3. Set $u = \frac{x+y}{xy}$ and $v = \frac{x-y}{xy}$, and show that the given system is equivalent to the second degree system

$$\begin{cases} u + v = 5 \\ uv = 6 \end{cases},$$

so that $u = 2$ and $v = 3$, or vice-versa.

4. Add the three given equations and write the result as

$$(ax_1^2 + (b - 1)x_1 + c) + (ax_2^2 + (b - 1)x_2 + c) + (ax_3^2 + (b - 1)x_3 + c) = 0.$$

Then, apply the result of Lemma 2.14 to the second degree trinomial $ax^2 + (b - 1)x + c$ and analyse separately itens (a) and (b).

5. Suppose, without loss of generality, that $x \geq y$. Then, use the equations of the system to conclude that $x = y = z$. Finally, note that $x = 1$ is a root of the third degree polynomial equation $x^3 - 2x + 1 = 0$, and apply the result of Problem 21, page 43.

6. Write the left hand side as a sum of squares and apply Lemma 2.31, with $n = 3$.

7. Substituting $y = \frac{1}{4}(a - z - 3x)$ into the first equation, we get

$$x^2 + \frac{1}{16}(a - z - 3x)^2 = 4z$$

or, which is the same,

$$25x^2 - 6x(a - z) + (a - z)^2 - 64z = 0.$$

For this last equation to have a single solution, the discriminant of the second degree trinomial (in x) of the left hand side must be equal to 0, i.e., we should have

$$36(a - z)^2 = 100[(a - z)^2 - 64z].$$

or, which is the same, $(a - z)^2 = 100z$. If this is so, then (from the second degree equation in x above) $x = \frac{3}{25}(a - z)$ and (from the second equation of the system) $y = \frac{4}{25}(a - z)$. Therefore, the system has a single solution if and only if the equation (in z) $(a - z)^2 = 100z$ has a single solution. Since it is equivalent to $z^2 - 2z(a + 50) + a^2 = 0$, its discriminant must also be equal to 0, i.e., we should have $(a + 50)^2 = a^2$. Thus, $a = -25$.

8. Start by writing

$$x + \frac{3}{x - 1} - 2\sqrt{x + 2} = (x - 1) + 1 + \frac{3}{x - 1} - 2\sqrt{x + 2}$$

$$= (x - 1) + \frac{x + 2}{x - 1} - 2\sqrt{x + 2}$$

$$= \left(\sqrt{x - 1} - \sqrt{\frac{x + 2}{x - 1}}\right)^2$$

Then, apply Lemma 2.31 to conclude that it suffices to solve the system of equations

$$\sqrt{x - 1} - \sqrt{\frac{x + 2}{x - 1}} = \sqrt{y - 1} - \sqrt{\frac{y + 2}{y - 1}} = \sqrt{z - 1} - \sqrt{\frac{z + 2}{z - 1}} = 0,$$

so that $x = y = z = \frac{3 + \sqrt{13}}{2}$.

9. Letting $a = \sqrt[3]{x + 5}$ and $b = \sqrt[3]{4 - x}$, we get $a + b = 3$ and $a^3 + b^3 = 9$. Hence

$$9 = a^3 + b^3 = a^3 + (3 - a)^3 = 27 - 27a + 9a^2$$

or, which is the same, $a^2 - 3a + 2 = 0$. Therefore, $a = 1$ or 2, from where $x = -4$ or $x = 3$.

10. Argue as in the previous problem, letting $y = \sqrt{5 - x}$ to transform the given equation into a system of equation in x and y.

11. First of all, observe that

$$x^2 + \frac{x^2}{(x+1)^2} = x^2 + \left(1 - \frac{1}{x+1}\right)^2$$

$$= x^2 + 1 + \frac{1}{(x+1)^2} - \frac{2}{x+1}$$

$$= (x^2 + 2x + 1) + \frac{1}{(x+1)^2} - 2x - \frac{2}{x+1}$$

$$= (x+1)^2 + \frac{1}{(x+1)^2} - 2(x+1) - \frac{2}{x+1} + 2.$$

Now, make the variable substitution $y = x + 1 + \frac{1}{x+1}$ to get

$$(x+1)^2 + \frac{1}{(x+1)^2} = y^2 - 2,$$

so that the given equation simplifies to the second degree equation $y^2 - 2y - 3 = 0$, whose roots are 3 and -1. Then, we have

$$x + 1 + \frac{1}{x+1} = 3 \quad \text{or} \quad -1,$$

and each of these possibilities gives rise to a second degree equation in x.

12. Firstly, observe that, if $x = 0$, then $y = z = 0$ (and analogously if $y = 0$ or $z = 0$). Therefore, we can assume that $xyz \neq 0$. Setting $u = \frac{1}{x}$, $v = \frac{1}{y}$ and $w = \frac{1}{z}$, we get

$$v = \frac{1}{y} = \frac{1 + 4x^2}{4x^2} = \frac{1}{4x^2} + 1 = \frac{u^2}{4} + 1.$$

Transforming the other two equations in analogous ways, we get the system below, which is equivalent to the original one:

$$u^2 + 4 = 4v, \quad v^2 + 4 = 4w, \quad w^2 + 4 = 4u.$$

Adding these three equations, we arrive at the equality

$$(u - 2)^2 + (v - 2)^2 + (w - 2)^2 = 0,$$

and Lemma 2.31 gives $u = v = w = 2$ and, hence, $x = y = z = \frac{1}{2}$. Thus, the solutions of the system are $x = y = z = 0$ or $x = y = z = \frac{1}{2}$.

13. Add the equations of the system to conclude that it suffices to solve the system of equations

$$\begin{cases} a+b+c = 3d \\ b+c+d = 3a \\ c+d+a = 3b \\ d+a+b = 3c \end{cases}.$$

Then, note that $a = b = c = d = \frac{1}{4}(a+b+c+d)$.

14. Write the given equation as $(E_1 - F_1) + (E_2 - F_2) + \cdots + (E_n - F_n) = 0$, observing that $E_i - F_i \geq 0$ for $1 \leq i \leq n$.

15. For $u \in \mathbb{R}$, note that

$$4u^2 - u^4 = 4 - (4 - 4u^2 + u^4) = 4 - (2 - u^2)^2.$$

Hence,

$$6 = \sqrt{4x^2 - x^4 - 3} + \sqrt{4y^2 - y^4} + \sqrt{4z^2 - z^4 + 5}$$

$$= \sqrt{1 - (2 - x^2)^2} + \sqrt{4 - (2 - y^2)^2} + \sqrt{9 - (2 - x^2)^2}$$

$$\leq \sqrt{1} + \sqrt{4} + \sqrt{9} = 1 + 2 + 3,$$

and it follows from the previous problem that $2 - x^2 = 2 - y^2 = 2 - z^2 = 0$, i.e., $x = \pm\sqrt{2}, y = \pm\sqrt{2}, z = \pm\sqrt{2}$.

16. If $x \geq 4$, write $\frac{5x}{4} - \frac{4}{x} = x + \left(\frac{x}{4} - \frac{4}{x}\right)$ to conclude that $z \geq 4$; then, argue similarly with the second equation to conclude that $y \geq 4$. Then, either $x, y, z \geq 4$ or $x, y, z \leq 4$. In either case, add the three equations of the system to get

$$\frac{4}{x} + \frac{4}{y} + \frac{4}{z} = \frac{x}{4} + \frac{y}{4} + \frac{z}{4},$$

and use the result of Problem 14.

17. For item (a), write $x + \frac{2}{x} - 2\sqrt{2} = \frac{1}{x} \cdot (x - \sqrt{2})^2$. For item (b), start by observing that $x + \frac{2}{x} = 2y$ gives that x and y are both positive or both negative; analogously, the same is true for y and z. Therefore, either x, y and z are all positive or all negative. Now, note that (x, y, z) is a solution of the system if and only if $(-x, -y, -z)$ is also a solution, so that we can restrict to the case $x, y, z > 0$. Now, the result of item (a) gives

$$2y = x + \frac{2}{x} \geq 2\sqrt{2},$$

so that $y \geq \sqrt{2}$; analogously, $x, z \geq \sqrt{2}$. Adding the three equations of the system, we get

$$x + y + z = \frac{2}{x} + \frac{2}{y} + \frac{2}{z}.$$

Finally, note that $u \geq \frac{2}{u}$ for $u \geq \sqrt{2}$, and use the result of Problem 14.

Section 3.1

2. For item (a), let $a_n = n$, for odd $n \geq 1$, and $a_n = (-1)^{n-1}n$, for even $n \geq 1$. For (b), let $a_n = \frac{n}{n+1}$, for $n \geq 1$. Finally, for item (c), let $a_n = n$, for odd $n \geq 1$, and $a_n = \frac{1}{n}$ for even $n \geq 1$.

3. It suffices to note that $b_{k+1} = \frac{1}{a_{k+1}} = \frac{3a_k+1}{a_k} = 3 + \frac{1}{a_k} = 3 + b_k$.

4. For item (a), we can write $a_1 = a_2 = a_3 = 1$ and $a_k = a_{k-1} + a_{k-2} + a_{k-3}$, for $k \geq 4$. For item (b), we can write $a_1 = 1$ and $a_{k+1} = 2^{a_k}$, for $k \geq 1$.

5. Observe that the entries in the n-th line form and AP of initial term n, common difference 2 and $2n - 1$ terms.

6. Write $\underbrace{11 \ldots 1}_{n}$ as a sum of powers of 10, then use the formula for the sum of the terms of a GP.

7. Use the result of the previous problem.

8. It suffices to show that $b_{k+1} - b_k$ does not depend on k; the common value we shall obtain will be the common difference of the AP $(b_k)_{k \geq 1}$. By definition, we have

$$\begin{aligned} b_{k+1} - b_k &= (a_{k+2}^2 - a_{k+1}^2) - (a_{k+1}^2 - a_k^2) \\ &= (a_{k+2} - a_{k+1})(a_{k+2} + a_{k+1}) - (a_{k+1} - a_k)(a_{k+1} + a_k) \\ &= r(a_{k+2} + a_{k+1}) - r(a_{k+1} + a_k) \\ &= r(a_{k+2} - a_k) = r \cdot 2r = 2r^2. \end{aligned}$$

9. Apply the formula for the general term of an AP to $a_p + a_q$ and $a_u + a_v$.

10. Let $a_1 = rm$, with $m \in \mathbb{Z}_+$. By the formula for the general term of AP's, we have

$$a_k + a_l = 2a_1 + (k + l - 1)r = a_1 + (m + k + l - 1)r = a_{m+k+l}.$$

11. The formula for the general term of AP's gives us the relations $\alpha = a_p = a_1 + (p-1)r$, $\beta = a_q = a_1 + (q-1)r$ and $a_{p+q} = a_1 + (p+q-1)r$, where r is the common difference of the AP. By viewing the first two relations above as a linear system of equations in a_1 and r, express them in terms of α, β, p and q; then, substitute these expressions for a_1 and r into the formula for a_{p+q}.

12. Letting $2n - 1$ denote the smallest of such integers and k their quantity, the formula for the general term of an AP assures that the biggest of the numbers equals $(2n - 1) + 2(k - 1) = 2n + 2k - 3$. Therefore, their sum is equal to $\frac{1}{2}[(2n - 1) + (2n + 2k - 3)]k = 7^3$, so that $(2n + k - 2)k = 7^3$. Now, since $2n + k - 2 \geq k > 1$, the only possibility is $2n + k - 2 = 7^2$ and $k = 7$, so that $n = 22$.

13. Let r be the common difference of the AP and let $a_2 = m$, so that $r \in \mathbb{N}$ and $m > 1$. Then, $a_{mk+2} = a_2 + ((mk + 2) - 2)r = m + mkr = m(1 + kr)$, which is clearly a composite number.

14. Item (b) follows immediately from (a). For item (a), separately consider the cases n even and n odd.

15. By contradiction, suppose that $a_m = \sqrt{2}$, $a_n = \sqrt{3}$ and $a_p = \sqrt{5}$, where $(a_k)_{k \geq 1}$ is an AP and m, n and p are pairwise distinct natural numbers. Then, use the formula for the general term of AP's to compute $\frac{n-m}{p-m}$.

16. Apply, to the left hand side, the formula for the general term of GP's.

17. Adapt, to the present case, the solution given to the Example 3.14.

18. Adapt, to the present case, the solution given to the Example 3.14.

19. Suppose that there exists such an AP $(a_k)_{k \geq 1}$, of common difference r, i.e., such that $\frac{a_1}{r} \notin \mathbb{Q}$ but $a_n^2 = a_m a_p$, for some distinct indices m, n and p. Use the formula for the general term of an AP to reach a contradiction.

20. Show that the sequence $(b_k)_{k \geq 1}$, such that $b_k = a_k - 1$, is a GP.

22. Suppose that $(a_k)_{k \geq 1}$ is a GP of common ratio q, such that $a_m = 2$, $a_n = 3$ and $a_p = 5$. Then, $a_1 q^{m-1} = 2$, $a_1 q^{n-1} = 3$ and $a_1 q^{p-1} = 5$, so that $q^{m-n} = \frac{2}{3}$ and $q^{m-p} = \frac{2}{5}$. Therefore,

$$\left(\frac{2}{3}\right)^{m-p} = q^{(m-n)(m-p)} = \left(\frac{2}{5}\right)^{m-n}$$

or, which is the same, $2^{n-p} \cdot 5^{m-n} = 3^{m-p}$. Now, get a contradiction.

Section 3.2

1. Apply the result of Theorem 3.16.

2. Apply the result of Theorem 3.16.

3. Start by observing that $a_{k+2} - a_{k+1} = (2a_{k+1} - 1) - (2a_k - 1) = 2a_{k+1} - 2a_k$ and, hence, that $a_{k+2} - 3a_{k+1} + 2a_k = 0$, for every integer $k \geq 1$. Now, apply the result of Theorem 3.16.

4. Use the fact that $x^2 + rx + s = (x - \alpha)^2$.

7. For item (a), apply Theorem 3.16. For item (b), use the formula of (a). Finally, for (c), use the formulas of item (a) and (3.10).

8. For item (a), the given recurrence relation furnishes

$$4b_{k+1}^2 = 4(1 + 24a_{k+1}) = 4 + 6(1 + 4a_k + \sqrt{1 + 24a_n})$$

$$= 4 + 6\left(1 + \frac{1}{6}(b_k^2 - 1) + b_k\right)$$

$$= b_k^2 + 6b_k + 9 = (b_k + 3)^2.$$

For item (b), it follows from (a) that $2b_{n+2} - b_{n+1} = 3 = 2b_{n+1} - b_n$ and, thus, $2b_{n+2} - 3b_{n+1} + b_n = 0$. Since $b_1 = 5$ and $b_2 = \sqrt{1 + 24a_2} = 4$, Theorem 3.16 guarantees that $b_n = 3 + \frac{4}{2^n}$. At last, for item (c), it follows from items (a) and (b) that

$$a_n = \frac{1}{24}(b_n^2 - 1) = \frac{1}{24}\left(\left(3 + \frac{4}{2^n}\right)^2 - 1\right)$$

$$= \frac{1}{24}\left(2 + \frac{4}{2^n}\right)\left(4 + \frac{4}{2^n}\right)$$

$$= \frac{1}{3}\left(1 + \frac{1}{2^{n-1}}\right)\left(1 + \frac{1}{2^n}\right).$$

Sections 3.3

1. If $(a_k)_{k\geq 1}$ is a second order AP, then $(a_{k+1} - a_k)_{k\geq 1}$ is a non constant AP, so that there exists $r \neq 0$ for which $(a_{k+2} - a_{k+1}) - (a_{k+1} - a_k) = r$, for every $k \geq 1$. In particular, for every $k \geq 1$, we have $a_{k+2} - 2a_{k+1} + a_k \neq 0$ and

$$(a_{k+3} - a_{k+2}) - (a_{k+2} - a_{k+1}) = (a_{k+2} - a_{k+1}) - (a_{k+1} - a_k).$$

The converse can be prove in an analogous way.
2. Write $a_{k+1} - a_k = 3k - 1$ and use telescoping sums.
3. Write $a_{k+1} - a_k = 8k$ and use telescoping sums.
4. Observe that $\frac{1}{(k-1)k} = \frac{1}{k-1} - \frac{1}{k}$ and use telescoping sums.
5. The previous problem is a particular case of the present one.
6. Start by noting that $\frac{1}{k^2} < \frac{1}{(k-1)k}$ for every integer $k > 1$; then, apply the result of Problem 4.
7. Adapt to the present case the hint to Problem 4, writing

$$\frac{1}{(4k - 1)(4k + 3)} = \frac{1}{4}\left(\frac{1}{4k - 1} - \frac{1}{4k + 3}\right).$$

8. For item (a), it suffices to observe that $(2k + 1)^3 - (2k - 1)^3 = 24k^2 + 2 = 16k^2 + 8k^2 + 2 = 16k^2 + (2k - 1)^2 + (2k + 1)^2$. For item (b), start by using the formula for telescoping sums to write

$$(2n + 1)^3 - 2 = \sum_{k=2}^{n}[(2k + 1)^3 - (2k - 1)^3] + (3^3 - 2)$$

$$= \sum_{k=2}^{n}[(4k)^2 + (2k + 1)^2 + (2k - 1)^2] + 25$$

$$= \sum_{k=2}^{n}[(4k)^2 + (2k + 1)^2 + (2k - 1)^2] + 4^2 + 3^2.$$

Then, observe that the last expression above is a sum of $(n - 1) \cdot 3 + 2 = 3n - 1$ perfect squares.

9. Note that $\frac{k}{(k+1)!} = \frac{(k+1)-1}{(k+1)!} = \frac{1}{k!} - \frac{1}{(k+1)!}$ and use telescoping sums.

10. For item (a), observe that $(k+1)^2 + k^2 + k^2(k+1)^2 = k^4 + 2k^3 + 3k^2 + 2k + 1 = (k^2 + k + 1)^2$. In what concerns (b), it follows from (a) that

$$\sqrt{\frac{1}{k^2} + \frac{1}{(k + 1)^2} + 1} = \sqrt{\frac{(k^2 + k + 1)^2}{k^2(k + 1)^2}} = \frac{k^2 + k + 1}{k(k + 1)}.$$

It now suffices to write

$$\frac{k^2 + k + 1}{k(k + 1)} = 1 + \frac{1}{k(k + 1)} = 1 + \frac{1}{k} - \frac{1}{k + 1}$$

and use the formula for telescoping sums or the result of Problem 4.

11. Write $\frac{F_{k+1}}{F_k F_{k+2}} = \frac{F_{k+2} - F_k}{F_k F_{k+2}} = \frac{1}{F_k} - \frac{1}{F_{k+2}}$ and, then, use the formula for telescoping sums.

12. Rationalise the fraction $\frac{1}{\sqrt{a_k} + \sqrt{a_{k+1}}}$—see Problem 12, page 25—and, then, use the formula for telescoping sums.

13. For item (a), just note that $x^4 + x^2 + 1 = (x^4 + 2x^2 + 1) - x^2 = (x^2 + 1)^2 - x^2 = (x^2 + 1 - x)(x^2 + 1 + x)$. For (b), use the result of (a) to write

$$\frac{k}{k^4 + k^2 + 1} = \frac{k}{(k^2 + k + 1)(k^2 - k + 1)}$$

$$= \frac{1}{2}\left(\frac{1}{k^2 - k + 1} - \frac{1}{k^2 + k + 1}\right).$$

Then, use the formula for telescoping sums.

14. Rationalise the fraction under the $\sum$ notation—see Problem 12, page 25—and, then, use the formula for telescoping sums.

15. Adapt, to the present case, the hint given to the previous problem.
16. Initially, observe that

$$\prod_{j=2}^{n}\left(1-\frac{1}{j^2}\right) = \prod_{j=2}^{n}\left(1-\frac{1}{j}\right)\prod_{j=2}^{n}\left(1+\frac{1}{j}\right)$$

$$= \prod_{j=2}^{n}\left(\frac{j-1}{j}\right)\prod_{j=2}^{n}\left(\frac{j+1}{j}\right).$$

Now, take $a_k = k$ for $k \geq 1$. It follows from Proposition 3.29 that

$$\prod_{j=2}^{n}\left(1-\frac{1}{j^2}\right) = \prod_{j=2}^{n}\left(\frac{a_{j-1}}{a_j}\right)\prod_{j=2}^{n}\left(\frac{a_{j+1}}{a_j}\right)$$

$$= \frac{a_1}{a_n} \cdot \frac{a_{n+1}}{a_2} = \frac{1}{n} \cdot \frac{n+1}{2} = \frac{n+1}{2n}.$$

17. If S is the sum we wish to compute, then

$$S = \sum_{k=0}^{101} \frac{x_k^3}{(1-x_k)^3 + x_k^3}.$$

Since $1 - x_k = 1 - \frac{k}{101} = x_{101-k}$, we get

$$S = \sum_{k=0}^{101} \frac{x_k^3}{x_{101-k}^3 + x_k^3}$$

and can write

$$2S = \sum_{k=0}^{101} \frac{x_k^3}{x_{101-k}^3 + x_k^3} + \sum_{k=0}^{101} \frac{x_k^3}{x_{101-k}^3 + x_k^3}$$

$$= \sum_{k=0}^{101} \frac{x_k^3}{x_{101-k}^3 + x_k^3} + \sum_{k=0}^{101} \frac{x_{101-k}^3}{x_{101-k}^3 + x_k^3}$$

$$= \sum_{k=0}^{101} \frac{x_k^3 + x_{101-k}^3}{x_{101-k}^3 + x_k^3} = \sum_{k=0}^{101} 1 = 102.$$

Hence, $S = 51$.

18. Start by factorising both the numerator and denominator of the fraction under the $\prod$ notation. If $a_k = k^2 - k + 1$, show that $a_{k+1} = k^2 + k + 1$ and, then, use the formula for telescoping products.

19. We first show that n must be even. To this end, suppose that $A = A_1 \cup \cdots \cup A_n$, with $A_1, \ldots, A_n$ satisfying the conditions of the problem. For $1 \le k \le n$, let $x_k \in A_k$ be the element of A_k that equals the arithmetic mean of the other three. Then $4x_k = \sum_{x \in A_k} x$, and it follows that

$$4 \sum_{k=1}^{n} x_k = \sum_{k=1}^{n} \sum_{x \in A_k} x = \sum_{x \in A} x = 1 + 2 + \cdots + 4n = 2n(4n + 1).$$

Hence, $n(4n + 1)$ must be even and, since $4n + 1$ is odd, it follows that n must be even. Now, let $n = 2k$, with $k \in \mathbb{N}$, and write

$$\{1, 2, 3, \ldots, 8k\} = A_1 \cup \cdots \cup A_{2k},$$

where

$$A_{2j-1} = 8(j-1) + \{1, 3, 4, 8\} \quad \text{and} \quad A_{2j} = 8(j-1) + \{2, 5, 6, 7\}.$$

(Here, for $X \subset \mathbb{R}$, we define the set $X + t$ by $X + t = \{x + t; \ x \in X\}$). Since $3 \cdot 4 = 1 + 3 + 8$ and $3 \cdot 5 = 2 + 6 + 7$, it is immediate to verify that the sets A_i, defined as above, satisfy the conditions of the problem.

20. By the result of Problem 13, page 26, we have

$$\left(\sqrt{n+1} - \sqrt{n} \right) \sqrt{n} = \sqrt{n^2 + n} - \sqrt{n^2}$$

$$< \sqrt{\left(n + \frac{1}{2} \right)^2 - \sqrt{n^2}} = \frac{1}{2}$$

and, thus, $2(\sqrt{n+1} - \sqrt{n}) < \frac{1}{\sqrt{n}}$. Analogously, we get $\frac{1}{\sqrt{n}} < 2(\sqrt{n} - \sqrt{n-1})$. Taking these two inequalities together, we get

$$\sum_{k=2}^{10000} \left(\sqrt{k+1} - \sqrt{k} \right) < \sum_{k=2}^{10000} \frac{1}{\sqrt{k}} < \sum_{k=2}^{10000} \left(\sqrt{k} - \sqrt{k-1} \right).$$

However,

$$\sum_{k=2}^{10000} \left(\sqrt{k+1} - \sqrt{k} \right) = 2\sqrt{10001} - 2\sqrt{2}$$

and

$$\sum_{k=2}^{10000} \left(\sqrt{k} - \sqrt{k-1} \right) = 2 \left(\sqrt{10000} - 1 \right) = 198.$$

Hence, if we show that $2\sqrt{10001} - 2\sqrt{2} > 197$, it will come that the greatest integer which is less than or equal to S equals 198. To what is left to do, just see that

$$2\sqrt{10001} > 2\sqrt{2} + 197 \Leftrightarrow \left(2\sqrt{10001}\right)^2 > \left(2\sqrt{2} + 197\right)^2$$

$$\Leftrightarrow 40004 > 788\sqrt{2} + 38817$$

$$\Leftrightarrow 788\sqrt{2} < 1187 \Leftrightarrow 788^2 \cdot 2 < 1187^2$$

$$\Leftrightarrow 1241888 < 1408969,$$

which is indeed true.

21. For item (a), factorising $(k + 1)^{p+1} - k^{p+1}$ with the aid of the result of Problem 18, page 26, we get

$$(k + 1)^{p+1} - k^{p+1} = \sum_{j=1}^{p+1} (k + 1)^{p+1-j} k^{j-1}$$

$$< \sum_{j=1}^{p+1} (k + 1)^{p+1-j} (k + 1)^{j-1}$$

$$= (p + 1)(k + 1)^p.$$

Analogously,

$$(k + 1)^{p+1} - k^{p+1} = \sum_{j=1}^{p+1} (k + 1)^{p+1-j} k^{j-1}$$

$$> \sum_{j=1}^{p+1} k^{p+1-j} k^{j-1} = (p + 1)k^p.$$

For item (b), it follows from (a) and from the formula for telescoping sums that

$$\sum_{k=1}^{n-1} k^p < \sum_{k=1}^{n-1} \frac{(k + 1)^{p+1} - k^{p+1}}{p + 1} = \frac{n^{p+1} - 1}{p + 1} < \frac{n^{p+1}}{p + 1}$$

and

$$\sum_{k=1}^{n} k^p > \sum_{k=1}^{n} \frac{k^{p+1} - (k - 1)^{p+1}}{p + 1} = \frac{n^{p+1}}{p + 1}.$$

Section 4.1

1. It follows from the induction hypothesis that

$$1 + 2 + \cdots + k + (k+1) = \frac{k(k+1)}{2} + (k+1) = \frac{(k+1)(k+2)}{2}.$$

2. It follows from the induction hypothesis that

$$\sum_{j=1}^{k} j^3 = \left(\frac{k(k+1)}{2}\right)^2 + (k+1)^3 = \left(\frac{(k+1)(k+2)}{2}\right)^2.$$

3. It follows from the induction hypothesis that

$$1 - \frac{1}{2} + \frac{1}{3} - \cdots + \frac{1}{2k-1} - \frac{1}{2k} + \frac{1}{2k+1} =$$

$$= \frac{1}{k} + \frac{1}{k+1} + \cdots + \frac{1}{2k-1} - \frac{1}{2k} + \frac{1}{2k+1}$$

$$= \frac{1}{k+1} + \cdots + \frac{1}{2k-1} + \frac{1}{2k} + \frac{1}{2k+1}.$$

4. By induction hypothesis, we have

$$(k+1) + h(1) + h(2) + h(3) + \cdots + h(k-1) + h(k) = 1 + kh(k) + h(k),$$

Therefore, it suffices to prove that $1 + (k+1)h(k) = (k+1)h(k+1)$ or, which is the same, that $(k+1)(h(k+1) - h(k)) = 1$. However,

$$(k+1)(h(k+1) - h(k)) = (k+1) \cdot \frac{1}{k+1} = 1.$$

5. It follows from the induction hypothesis that

$$\sum_{j=1}^{k} j(j+1) = \frac{1}{3}(k-1)k(k+1) + k(k+1)$$

$$= \frac{1}{3}k(k+1)(k+2).$$

6. It follows from the induction hypothesis that

$$\sum_{j=1}^{k+1} (2j-1)^2 = \frac{1}{6}(2k-1)2k(2k+1) + (2k+1)^2$$

$$= \frac{1}{6}(2k+1)2(k+1)(2k+3).$$

8. In trying to appeal to the induction hypothesis, we observe that $4^{k+1} + 15(k+1) - 1 = (4^k + 15k - 1) + 3(4^k + 5)$. Then, since $4^k + 15k - 1$ is a multiple of 9 (by that induction hypothesis), it suffices to show that $4^k + 5$ is a multiple of 3. To this end, make another proof by induction: $4^{l+1} + 5 = (4^l + 5) + 3 \cdot 4^l$; since $4^l + 5$ (by the new induction hypothesis) and $3 \cdot 4^l$ are multiples of 3, it follows that $4^{l+1} + 5$ is also a multiple of 3.

9. Appealing to the induction hypothesis, we observe that $(k + 1)^3 - (k + 1) = (k^3 - k) + 3(k^2 + k)$. Now, since $k^3 - k$ (by that induction hypothesis) and $3(k^2 + k)$ are multiples of 3, it comes that $(k + 1)^3 - (k + 1)$ is also a multiple of 3.

10. If $4^{3^{k-1}} - 1 = 3^k q$, for some $q \in \mathbb{N}$, then

$$4^{3^k} - 1 = (4^{3^{k-1}})^3 - 1 = (4^{3^{k-1}} - 1)[(4^{3^{k-1}})^2 + 4^{3^{k-1}} + 1]$$

$$= 3^k q[((4^{3^{k-1}})^2 - 1) + (4^{3^{k-1}} - 1) + 3]$$

$$= 3^k q[(4^{3^{k-1}} - 1)(4^{3^{k-1}} + 1) + 3^k q + 3]$$

$$= 3^k q[3^k q(4^{3^{k-1}} + 1) + 3^k q + 3]$$

$$= 3^{k+1} q[3^{k-1} q(4^{3^{k-1}} + 1) + 3^{k-1} q + 1].$$

11. Suppose we have n_1 ways of choosing an object of type 1, n_2 ways of choosing an object of type 2, ..., n_k ways of choosing an object of type k, n_{k+1} ways of choosing an object of type $k + 1$. In order to choose an object of each of these $k + 1$ types, start by choosing an object of each of the types from 1 to k. By induction hypothesis, there are exactly $n_1 n_2 \ldots n_k$ possible distinct choices. On the other hand, for each of these choices, we have n_{k+1} ways of choosing an object of type $k + 1$. Therefore, the number of ways of choosing an object of each of the types from 1 to $k + 1$ is equal to

$$\underbrace{n_1 n_2 \ldots n_k + n_1 n_2 \ldots n_k + \cdots + n_1 n_2 \ldots n_k}_{n_{k+1} \text{ times}} = n_1 n_2 \ldots n_k n_{k+1}.$$

12. By induction hypothesis, suppose that any set with k elements has exactly 2^k subsets. If $A = \{a_1, a_2, \ldots, a_k, a_{k+1}\}$ is a set with $k + 1$ elements. It has two distinct kinds of subsets: those which are subsets of $B = \{a_1, a_2, \ldots, a_k\}$ and those which contain a_{k+1}. By the induction hypothesis, A has exactly 2^k subsets of the first type. On the other hand, the subsets of the second type are of the form $B' \cup \{a_{k+1}\}$, where B' is a subset of B. Again by induction hypothesis, there are also 2^k subsets of this second type. Hence, A has $2^k + 2^k = 2^{k+1}$ subsets.

13. To the first case, by using the induction hypothesis (strong induction!), we get $u_{k+2} = -r u_{k+1} - s u_k = -r a_{k+1} - s a_k = a_{k+2}$. The second case is completely analogous.

14. By induction hypothesis, suppose that $(a_1, a_2, \ldots, a_k)$ is an AP of common difference r. Then,

$$\frac{k}{a_1 a_{k+1}} = \sum_{j=1}^{k} \frac{1}{a_j a_{j+1}} = \sum_{j=1}^{k-1} \frac{1}{a_j a_{j+1}} + \frac{1}{a_k a_{k+1}} = \frac{k-1}{a_1 a_k} + \frac{1}{a_k a_{k+1}}$$

and, hence, $k a_k = (k-1) a_{k+1} + a_1$. However, $a_k = a_1 + (k-1)r$, so that

$$(k-1) a_{k+1} = k(a_1 + (k-1)r) - a_1 = (k-1)(a_1 + kr).$$

Therefore, $a_{k+1} = a_1 + kr$, and $(a_1, a_2, \ldots, a_k, a_{k+1})$ is also an AP.

15. For item (a), the induction hypothesis gives

$$a_{k+1} = a_k^2 - a_k + 1 = a_k(a_k - 1) + 1$$
$$= a_k \cdot (a_1 \ldots a_{k-1}) + 1 = a_1 \ldots a_{k-1} a_k + 1.$$

For (b), it follows from the induction hypothesis and the result of (a) that

$$\sum_{j=1}^{k+1} \frac{1}{a_j} = \sum_{j=1}^{k} \frac{1}{a_j} + \frac{1}{a_{k+1}} = 2 - \frac{1}{a_1 a_2 \ldots a_k} + \frac{1}{a_{k+1}}$$

$$= 2 - \frac{a_{k+1} - a_1 a_2 \ldots a_k}{a_1 a_2 \ldots a_k a_{k+1}} = 2 - \frac{1}{a_1 a_2 \ldots a_k a_{k+1}}.$$

16. By induction hypothesis, we have $2^{2^{k+1}} = (2^{2^k})^2 > k^{2k}$, so that it suffices to show that $k^{2k} \geq (k+1)^{k+1}$ or, which is the same, $\left(\frac{k^2}{k+1}\right)^k \geq k+1$. Such an inequality is true for every integer $k \geq 3$, for we have

$$\left(\frac{k^2}{k+1}\right)^k = \left(k - 1 + \frac{1}{k+1}\right)^k > (k-1)^k \geq (k-1)^3 \geq k+1.$$

17. For $n \in \mathbb{Z}$, let $a_n = x^n + x^{-n}$. We first show that $a_n \in \mathbb{Z}$ for every $n \in \mathbb{N}$. The hypothesis of the problem gives $a_1 \in \mathbb{Z}$, so that

$$a_2 = x^2 + \frac{1}{x^2} = \left(x + \frac{1}{x}\right)^2 - 2 = a_1^2 - 2 \in \mathbb{Z}.$$

Now suppose, by induction hypothesis, that $a_1, a_2, \ldots, a_k \in \mathbb{Z}$, for a certain integer $k \geq 2$. Then,

$$a_{k+1} = x^{k+1} + \frac{1}{x^{k+1}}$$

$$= \left(x^k + \frac{1}{x^k}\right)\left(x + \frac{1}{x}\right) - \left(x^{k-1} + \frac{1}{x^{k-1}}\right)$$

$$= a_k a_1 - a_{k-1} \in \mathbb{Z},$$

and the induction step is complete. To what was left to show, it suffices to see that $a_0 = 2$ and, if $n < 0$ is an integer, then $a_n = a_{-n} \in \mathbb{Z}$.

18. It follows from induction hypothesis that

$$x_{k+1}^2 = (x_k^3 - 3x_k)^2 = x_k^6 - 6x_k^4 + 9x_k^2$$
$$= (y_k + 2)^3 - 6(y_k + 2)^2 + 9(y_k + 2)$$
$$= y_k^3 - 3y_k + 2 = y_{k+1} + 2.$$

19. It suffices to prove that the quotient $\frac{x_{n+1}+x_{n-1}}{x_n}$ does not depend on n. This is immediate by induction, for

$$\frac{x_{n+2} + x_n}{x_{n+1}} - \frac{x_{n+1} + x_{n-1}}{x_n} = \frac{x_{n+2}x_n + x_n^2 - x_{n+1}^2 - x_{n-1}x_{n+1}}{x_{n+1}x_n}$$

$$= \frac{(x_{n+1}^2 - 1) + x_n^2 - x_{n+1}^2 - (x_n^2 - 1)}{x_{n+1}x_n} = 0.$$

20. For item (a) we have, by induction hypothesis, $F_1 + F_2 + \cdots + F_k + F_{k+1} = (F_{k+2} - 1) + F_{k+1} = F_{k+3} - 1$. For (b) (and also by induction hypothesis),

$$F_1^2 + F_2^2 + \cdots + F_k^2 + F_{k+1}^2 = F_k F_{k+1} + F_{k+1}^2 = F_{k+1} F_{k+2}.$$

Items (c), (d) and (e) can be dealt with in an analogous way.

21. We make induction on $n \geq 1$. For $n = 1$, the relation in the statement of the problem is simply the recursive definition of F_{m+1}. In general, it follows from the induction hypothesis that

$$F_{m+k+1} = F_{m+k} + F_{m+k-1}$$
$$= (F_m F_{k+1} + F_{m-1} F_k) + (F_m F_k + F_{m-1} F_{k-1})$$
$$= F_m (F_{k+1} + F_k) + F_{m-1} (F_k + F_{k-1})$$
$$= F_m F_{k+1} + F_{m-1} F_{k+1}.$$

22. In both items, make induction on n, for a fixed $m \geq 1$. For the initial cases $2F_{m+1} = F_m + L_m$ and $2L_{m+1} = 5F_m + L_m$, you will also have to make inductive arguments.

23. For (a), observe that $F_1 = 1$, $F_2 = 1$, $F_3 = 2$, $F_4 = 3$, $F_5 = 5$, $F_6 = 8$, $F_7 = 13$, $F_8 = 21$, $F_9 = 34$, $F_{10} = 55$, $F_{11} = 89$, $F_{12} = 144$, $F_{13} = 233$, $F_{14} = 377$. Therefore, $F_1 = 1^2$, $F_{12} = 12^2$ and $F_{13} > 13^2$, $F_{14} > 14^2$. On the other hand, if $F_k > k^2$ and $F_{k+1} > (k+1)^2$, then $F_{k+2} = F_k + F_{k+1} > k^2 + (k+1)^2 > (k+2)^2$ for every $k > 3$. Hence, we get by induction that $F_n > n^2$ for every $n \geq 13$. For the first part of item (b), note that α is a root of the second degree equation $x^2 - x - 1 = 0$, so that $\alpha^2 = \alpha + 1$. Thus, we have by induction that

$$\alpha^{k+1} = \alpha^k \cdot \alpha = (F_k \alpha + F_{k-1})\alpha = F_k \alpha^2 + F_{k-1}\alpha$$
$$= F_k(\alpha + 1) + F_{k-1}\alpha = (F_k + F_{k-1})\alpha + F_k$$
$$= F_{k+1}\alpha + F_k,$$

Now, note that $\alpha^n - n^2\alpha = F_n\alpha + F_{n-1} - n^2\alpha = (F_n - n^2)\alpha + F_{n-1}$. Since α is irrational, it follows from this equality that $\alpha^n - n^2\alpha$ will be an integer if and only if $F_n - n^2 = 0$. Hence, the solution of ii. follows from that of the item (a).

24. Use induction to construct a subsequence $(a_{n_k})_{k \geq 1}$ of $(a_n)_{n \geq 1}$ such that $a_{n_k} > n_k, a_1, a_2, \ldots, a_{n_{k-1}}$.

25. If $a_j = j$ for $1 \leq j \leq k$, then we get from

$$a_1^3 + a_2^3 + \cdots + a_k^3 + a_{k+1}^3 = (a_1 + a_2 + \cdots + a_k + a_{k+1})^2$$

that

$$1^3 + 2^3 + \cdots + k^3 + a_{k+1}^3 = (1 + 2 + \cdots + k + a_{k+1})^2.$$

According to the result of Problem 2, this last relation gives

$$\frac{1}{4}[k(k+1)]^2 + a_{k+1}^3 = \left(\frac{1}{2}k(k+1) + a_{k+1}\right)^2.$$

Performing the computations at the right hand side, we arrive at $a_{k+1}^3 = a_{k+1}^2 + k(k+1)a_{k+1}$ or, which is the same, $a_{k+1}^2 - a_{k+1} - k(k+1) = 0$. Since $a_{k+1} > 0$, it immediately follows that $a_{k+1} = k + 1$.

26. If $x_k > \sqrt{a}$, then

$$x_{k+1} - \sqrt{a} = \frac{1}{2}\left(x_k - 2\sqrt{a} + \frac{a}{x_k}\right) = \frac{1}{2x_k}(x_k - \sqrt{a})^2 > 0.$$

To the other inequality, we start by observing that, if $x \geq y > \sqrt{a}$, then $x + \frac{a}{x} \geq y + \frac{a}{y}$ (*); in fact,

$$\left(x+\frac{a}{x}\right)-\left(y+\frac{a}{y}\right)=(x-y)+a\left(\frac{1}{x}-\frac{1}{y}\right)=\frac{(x-y)}{xy}(xy-a)\geq 0,$$

for $xy\leq\sqrt{a}^2=a$. Now, if $x_k\leq\sqrt{a}+\frac{1}{2^{k-1}}$, then, from (*), we get

$$x_{k+1}=\frac{1}{2}\left(x_k+\frac{a}{x_k}\right)\leq\frac{1}{2}\left(\sqrt{a}+\frac{1}{2^{k-1}}+\frac{a}{\sqrt{a}+\frac{1}{2^{k-1}}}\right)$$

$$<\frac{1}{2}\left(\sqrt{a}+\frac{1}{2^{k-1}}+\frac{a}{\sqrt{a}}\right)=\sqrt{a}+\frac{1}{2^k}.$$

27. If $k\geq 6$ and $a_1,a_2,\ldots,a_k$ are positive integers such that

$$\frac{1}{a_1^2}+\frac{1}{a_2^2}+\cdots+\frac{1}{a_k^2}=1,$$

then

$$\frac{1}{4}+\frac{1}{4}+\frac{1}{4}+\frac{1}{(2a_1)^2}+\frac{1}{(2a_2)^2}+\cdots+\frac{1}{(2a_k)^2}=\frac{3}{4}+\frac{1}{4}=1.$$

Hence, by a simple variant of the strong form of the principle of mathematical induction, it suffices to consider the cases $k=6$, $k=7$ and $k=8$. Before we look at them, note that

$$\frac{1}{9}+\frac{1}{9}+\frac{1}{36}=\frac{1}{4}\quad\text{and}\quad\frac{1}{4}+\frac{1}{4}+\frac{1}{4}+\frac{1}{4}=1.$$

Therefore, for $k=6$, we have $\frac{1}{4}+\frac{1}{4}+\frac{1}{4}+\frac{1}{9}+\frac{1}{9}+\frac{1}{36}=1$; for $k=7$, we have $\frac{1}{16}+\frac{1}{16}+\frac{1}{16}+\frac{1}{16}+\frac{1}{9}+\frac{1}{9}+\frac{1}{36}=1$; for $k=8$, we have $\frac{1}{4}+\frac{1}{4}+\frac{1}{9}+\frac{1}{9}+\frac{1}{36}+\frac{1}{9}+\frac{1}{9}+\frac{1}{36}=1$.

28. Argue as in the proof of Example 4.12.

29. For the existence of a representation for n, if $(k+1)!\leq n<(k+2)!$, let a_{k+1} be the greatest natural number such that $a_{k+1}(k+1)!\leq n<(a_{k+1}+1)(k+1)!$. Then, the inequalities $a_{k+1}(k+1)!\leq n<(k+2)!=(k+2)\cdot(k+1)!$ guarantee that $a_{k+1}\leq k+1$; on the other hand, since

$$n-a_{k+1}(k+1)!<(a_{k+1}+1)(k+1)!-a_{k+1}(k+1)!=(k+1)!,$$

it follows from the induction hypothesis that

$$n-a_{k+1}(k+1)!=a_1\cdot 1!+a_2\cdot 2!+\cdots+a_k\cdot k!,$$

for certain integers $a_1,a_2,\ldots,a_k$, such that $0\leq a_j\leq j$, for $1\leq j\leq k$. For the uniqueness of the representation of n, suppose we have established it for every natural n such that $n\leq(k+1)!$. Take $n\in\mathbb{N}$ such that $(k+1)!\leq n<(k+2)!$, and assume that

$$n = a_1 \cdot 1! + a_2 \cdot 2! + \cdots + a_{k+1} \cdot (k+1)! = b_1 \cdot 1! + b_2 \cdot 2! + \cdots + b_{k+1} \cdot (k+1)!,$$

with $0 \le a_j, b_j \le j$ for every $j \ge 1$. Then, by the result of Example 3.26, we get

$$\begin{aligned}
a_{k+1} \cdot (k+1)! &\le n = b_1 \cdot 1! + b_2 \cdot 2! + \cdots + b_{k+1} \cdot (k+1)! \\
&\le 1 \cdot 1! + 2 \cdot 2! + \cdots + k \cdot k! + b_{k+1} \cdot (k+1)! \\
&= (b_{k+1} + 1) \cdot (k+1)! - 1 < (b_{k+1} + 1) \cdot (k+1)!,
\end{aligned}$$

$$(B.1)$$

so that $a_{k+1} \le b_{k+1}$. Analogously, $b_{k+1} \le a_{k+1}$ and, thus, $a_{k+1} = b_{k+1}$. Therefore, we have

$$n - a_{k+1} \cdot (k+1)! = a_1 \cdot 1! + a_2 \cdot 2! + \cdots + a_k \cdot k! = b_1 \cdot 1! + b_2 \cdot 2! + \cdots + b_k \cdot k!,$$

with (again by the result of Example 3.26)

$$n - a_{k+1} \cdot (k+1)! \le 1 \cdot 1! + 2 \cdot 2! + \cdots + k \cdot k! = (k+1)! - 1.$$

It thus follows from the induction hypothesis that $a_j = b_j$, for $1 \le j \le k$.

30. For the existence of a representation as asked by the problem assume, as induction hypothesis, that for some natural $n \ge 3$ every $m \le F_n$ can be uniquely written as asked. Now, take a natural m such that $F_n < m \le F_{n+1}$. If $m = F_{n+1}$, there is nothing to do. Otherwise, $0 < m - F_n < F_{n+1} - F_n = F_{n-1} < F_n$. Therefore, by induction hypothesis, there exist an integer $r \ge 1$ and nonconsecutive indices $1 < t_1 < \cdots < t_r < n$ such that $m - F_n = F_{t_1} + \cdots + F_{t_r}$. Then, $m = F_{t_1} + \cdots + F_{t_r} + F_n$, and we claim that $n - t_r > 1$. In fact, if $n - t_r = 1$, we would have $m \ge F_n + F_{t_r} = F_n + F_{n-1} = F_{n+1}$, thus contradicting the fact that $m < F_{n+1}$. For the uniqueness of representation assume, as induction hypothesis, that for some natural $n \ge 3$ every natural $m \le F_n$ can be uniquely represented as asked. Now, take $m \in \mathbb{N}$ such that $F_n < m \le F_{n+1}$, and consider two cases separately:

(a) $m = F_{n+1}$: there are two subcases. First, suppose that $F_{n+1} = F_{t_1} + F_{t_2} + \cdots + F_{t_r}$, with $r \ge 1$, $1 < t_1 < t_2 < \cdots < t_r < n$ and $t_{j+1} - t_j > 1$ for $1 \le j < r$. Then,

$$\begin{aligned}
F_{n+1} &= F_{t_1} + F_{t_2} + \cdots + F_{t_r} < F_{t_1-1} + F_{t_1} + F_{t_2} + \cdots + F_{t_r} \\
&= F_{t_1+1} + F_{t_2} + \cdots + F_{t_r} < F_{t_2-1} + F_{t_2} + F_{t_3} + \cdots + F_{t_r} \\
&= F_{t_2+1} + F_{t_3} + \cdots + F_{t_r} < F_{t_3-1} + F_{t_3} + \cdots + F_{t_r} \\
&= \cdots \\
&< F_{t_r+1} \le F_n < F_{n+1},
\end{aligned}$$

which is an absurd. Now, if $F_{n+1} = F_{t_1} + \cdots + F_{t_r} + F_n$, then

$$F_{t_1} + \cdots + F_{t_r} = F_{n+1} - F_n = F_{n-1}$$

and, by the uniqueness of the representation of F_{n-1}, we would conclude that any other representation of F_{n+1} would be $F_{n+1} = F_n + F_{n-1}$, which does not satisfy the conditions of the statement of the problem.

(b) $F_n < m < F_{n+1}$: let us show that F_n is necessarily one of the summands of the representation. In fact, if $m = F_{t_1} + \cdots + F_{t_r}$, with $t_r < n$, then a reasoning analogous to that of the first subcase of item (a) would give us the absurd conclusion $m \leq F_n$. Therefore, for a representation of m as in the statement of the problem, we should surely have $m = F_{t_1} + \cdots + F_{t_r} + F_n$, with $t_r < n-1$. Thus, since $m - F_n < F_{n+1} - F_n = F_{n-1}$, the uniqueness of the representation of m follows from the uniqueness of the representation of $m - F_n$.

Section 4.2

1. It suffices to note that $\binom{2n}{n} = \frac{2n}{n}\binom{2n-1}{n-1} = 2\binom{2n-1}{n-1}$.
2. Adapt, to the present case, the idea of the solution of Example 4.18.
3. Compare $\binom{n}{k}$ to $\binom{n}{k+1}$ using the definition of $\binom{n}{j}$.
4. Make a proof by induction on n.
5. Apply Stiefel's relation to the numerator of the fraction and, then, use telescoping sums.
6. Make a proof by induction, applying Stiefel's relation in the induction step.
7. Let a_n be defined by the expression at the right hand side of the equality we wish to establish. Show that $a_1 = 1$, $a_2 = 1$ and (using Stiefel's relation) $a_{n+2} = a_{n+1} + a_n$, for every integer $n \geq 1$.
8. Make induction on n; more precisely, show that, for each $n \in \mathbb{N}$, the set $\{1, 2, \ldots, n\}$ has $\binom{n}{k}$ subsets of k elements each, for every $0 \leq k \leq n$. For the induction step, note that each of the subsets of k elements of $\{1, 2, \ldots, n+1\}$ is of one of two possible types: those not containing $n+1$ – thus, being subsets of k elements of $\{1, 2, \ldots, n\}$ – and those containing $n+1$ – thus, being equal to sets of the type $A \cup \{n+1\}$, where A is a subset of $k-1$ elements of $\{1, 2, \ldots, n\}$.

Section 4.3

1. Use Newton's binomial formula to expand $4^n = (1 + 3)^n$.

2. It follows from the previous problem that $A = 4^n$. Moreover, by an analogous reasoning to that of the previous problem, we get $B = 12^{n-1}$. Therefore, $\frac{3^8}{4} = \frac{B}{A} = \frac{12^{n-1}}{4^n} = \frac{3^{n-1}}{4}$, so that $n = 9$.

3. Compute the first three terms in the expansion of $(1, 1)^n = \left(1 + \frac{1}{10}\right)^n$.

4. The sum we wish to compute equals

$$\frac{1}{11!}\left(\frac{11!}{1!10!} + \frac{11!}{3!8!} + \frac{11!}{5!6!} + \frac{11!}{7!4!} + \frac{11!}{9!2!} + \frac{11!}{11!0!}\right) =$$

$$= \frac{1}{11!}\left(\binom{11}{1} + \binom{11}{3} + \binom{11}{5} + \binom{11}{7} + \binom{11}{9} + \binom{11}{11}\right)$$

$$= \frac{1}{11!} \cdot 2^{11-1}.$$

5. The general term of such expansion is $\binom{65}{k}\frac{1}{3^k}$. Now, observe that

$$\binom{65}{k}\frac{1}{3^k} < \binom{65}{k+1}\frac{1}{3^{k+1}} \Leftrightarrow \frac{1}{65-k} < \frac{1}{k+1}\cdot\frac{1}{3} \Leftrightarrow 65 - k > 3(k+1) \Leftrightarrow k \le 15.$$

Analogously, $\binom{65}{k}\frac{1}{3^k} > \binom{65}{k+1}\frac{1}{3^{k+1}} \Leftrightarrow k \ge 16$, so that the maximal term is $\binom{65}{16}\frac{1}{3^{16}}$.

6. The given condition is equivalent to $a^2 + b^2 + ab = c^2 + d^2 + cd$. On the other hand, it follows from the binomial expansion that $a^4 + b^4 + (a+b)^4 = c^4 + d^4 + (c+d)^4$ if and only if $a^4 + b^4 + 2a^3b + 3a^2b^2 + 2ab^3 = c^4 + d^4 + 2c^3d + 3c^2d^2 + 2cd^3$. Finally, observe that $a^4 + b^4 + 2a^3b + 3a^2b^2 + 2ab^3 = (a^2 + b^2 + ab)^2$ and, analogously, $c^4 + d^4 + 2c^3d + 3c^2d^2 + 2cd^3 = (c^2 + d^2 + cd)^2$.

7. Use the result of Problem 3, page 103—with $2n$ in place of n—together with the theorem on the lines of the Pascal triangle.

8. Adapt, to the present case, the idea of the solution of Example 4.27.

9. It suffices to see that

$$\sum_{j=0}^{n}\binom{n}{j}ja^j = \sum_{j=1}^{n}an\binom{n-1}{j-1}a^{j-1} = an(1+a)^{n-1}$$

and

$$\sum_{j=0}^{n}\binom{n}{j}j^2 = \sum_{j=1}^{n}n\binom{n-1}{j-1}j = \sum_{j=1}^{n}n\binom{n-1}{j-1}(j-1) + \sum_{j=1}^{n}n\binom{n-1}{j-1}$$

$$= \sum_{j=2}^{n}n(n-1)\binom{n-2}{j-2} + n\sum_{j=1}^{n}\binom{n-1}{j-1}$$

$$= n(n-1)\cdot 2^{n-2} + n\cdot 2^{n-1}.$$

10. Use the formula for the general term of an AP, together with the result of the previous problem.

11. Write $\frac{1}{q} = 1 + a$, with $a > 0$, and use the fact that $(1 + a)^n \geq 1 + na$.

12. Making $\alpha = \sqrt{a} + \sqrt[n]{b}$, we get $(\alpha - \sqrt{a})^n = b$. Now, arguing as in the proof of the Example 4.23, we conclude that

$$(\alpha - \sqrt{a})^n = (\alpha^n + a_{n-2}\alpha^{n-2} + a_{n-4}\alpha^{n-4} + \cdots) +$$
$$+ (a_{n-1}\alpha^{n-1} + a_{n-3}\alpha^{n-3} + \cdots)\sqrt{a},$$

for certain integers $a_0, a_1, a_2, \ldots, a_{n-2}, a_{n-1}$. Hence,

$$[(\alpha^n + a_{n-2}\alpha^{n-2} + a_{n-4}\alpha^{n-4} + \cdots) - b]^2 =$$
$$= a(a_{n-1}\alpha^{n-1} + a_{n-3}\alpha^{n-3} + \cdots)^2.$$

so that α is the root of the polynomial equation of degree $2n$

$$[(x^n + a_{n-2}x^{n-2} + a_{n-4}x^{n-4} + \cdots) - b]^2 -$$
$$- a(a_{n-1}x^{n-1} + a_{n-3}x^{n-3} + \cdots)^2 = 0.$$

13. First of all, let's find out which is the (constant) value of the given expression. To this end, substituting $x = 2$ and $y = z = -1$ (observe that $x + y + z = 0$), we get $x^5 + y^5 + z^5 = 30$ and $xyz(xy + yz + zx) = -6$, so that $\frac{x^5+y^5+z^5}{xyz(xy+yz+zx)} = -5$. Therefore, we ought to prove that $\frac{x^5+y^5+z^5}{xyz(xy+yz+zx)} = -5$, for all nonzero reals x, y and z such that $x + y + z = 0$. To this end, note that

$$x^5 + y^5 + z^5 = x^5 + y^5 + (-x-y)^5 = x^5 + y^5 - (x+y)^5$$
$$= x^5 + y^5 - (x^5 + 5x^4y + 10x^3y^2 + 10x^2y^3 + 5xy^4 + y^5)$$
$$= -(5x^4y + 5xy^4 + 10x^3y^2 + 10x^2y^3)$$
$$= -5xy[(x^3 + y^3) + 2xy(x+y)]$$
$$= -5xy(x+y)[(x^2 - xy + y^2) + 2xy]$$
$$= -5xy(-z)[(x^2 + 2xy + y^2) - xy]$$
$$= 5xyz[(x+y)(x+y) - xy] = 5xyz[(x+y)(-z) - xy]$$
$$= -5xyz(xy + yz + zx).$$

14. Applying Newton's expansion formula twice, we get

$$(x + y + z)^n = \sum_{l=0}^{n} \binom{n}{l} (x + y)^{n-l} z^l$$

$$= \sum_{l=0}^{n} \binom{n}{l} \sum_{k=0}^{n-l} \binom{n-l}{k} x^{n-l-k} y^k z^l$$

$$= \sum_{l=0}^{n} \sum_{k=0}^{n-l} \binom{n}{l} \binom{n-l}{k} x^{n-l-k} y^k z^l.$$

Now, observe that $\binom{n}{l}\binom{n-l}{k} = \frac{n!}{(n-k-l)!k!l!}$, so that it suffices to make $j = n - k - l$ to get the formula in the statement of the problem.

16. The general term in the expansion of $\left(1 + x + \frac{6}{x}\right)^{10}$ is $\binom{10}{j,k,l} 1^j x^k \left(\frac{1}{x}\right)^l = \binom{10}{j,k,l} x^{k-l}$, with $j + k + l = 10$. Therefore, we should have $k + l = 10 - j$ and $k - l = 0$, so that $k = l = 5 - \frac{j}{2}$ and, hence, j must be even. Now, make j successively equal to 0, 2, 4, 6, 8 and 10 and add all of the corresponding summands $\binom{10}{j,5-\frac{j}{2},5-\frac{j}{2}}$.

17. Use the trinomial expansion formula to get $\sum_{j+k+l=n} \binom{n}{j,k,l} = (1+1+1)^n$ and $\sum_{j+k+l=n} (-1)^l \binom{n}{j,k,l} = (1+1-1)^n$.

18. Make induction on n. To the induction step you will need to use Stiefel's relation, together with the fact that $F_{k+2} = F_{k+1} + F_k$, for every $k \geq 1$.

19. Let's show that $\sum_{k=0}^{995} (-1)^k \frac{1991}{1991-k} \binom{1991-k}{k} = 1$. To this end, start by writing

$$\sum_{k=0}^{995} (-1)^k \frac{1991}{1991-k} \binom{1991-k}{k} =$$

$$= \sum_{k=0}^{995} (-1)^k \binom{1991-k}{k} + \sum_{k=0}^{995} (-1)^k \frac{k}{1991-k} \binom{1991-k}{k}$$

$$= \sum_{k=0}^{995} (-1)^k \binom{1991-k}{k} + \sum_{k=0}^{995} (-1)^k \binom{1990-k}{k-1}.$$

Now, for $n \in \mathbb{N}$, define $S_n = \sum_{k\geq 0} (-1)^k \binom{n-k}{k}$ by setting $\binom{n-k}{k} = 0$ for $k > n - k$, so that $S_0 = S_1 = 1$. Use the columns' theorem to get $\sum_{m=0}^{n-2} S_m = \sum_{k\geq 0} (-1)^k \binom{n-1-k}{k+1} = 1 - S_n$ or, which is the same, $S_n = 1 - \sum_{m=0}^{n-2} S_m$. From this last relation, prove that $S_{n+1} = S_n - S_{n-1}$ for every $n \in \mathbb{Z}_+$, and compute

$S_2 = 0$, $S_3 = S_4 = -1$, $S_5 = 0$, $S_6 = S_7 = 1$. Finally, show that $S_m = S_n$ whenever 6 divides $m - n$, and observe that

$$\sum_{k=0}^{995}(-1)^k\frac{1991}{1991-k}\binom{1991-k}{k} = S_{1991} - S_{1989}$$

$$= S_5 - S_3 = 0 - (-1) = 1.$$

Section 5.1

2. First of all, note that

$$\left(\frac{a^3}{b} - a^2\right) + \left(\frac{b^3}{a} - b^2\right) = \frac{a^2(a-b)}{b} + \frac{b^2(b-a)}{a}$$

$$= (a-b)\left(\frac{a^2}{b} - a^2\right) = (a-b)\left(\frac{a^3 - b^3}{ab}\right)$$

$$= \frac{1}{ab}(a-b)^2(a^2 + ab + b^2).$$

Therefore, it suffices to show that $a^2 + ab + b^2 \geq 0$. To this end, write

$$a^2 + ab + b^2 = \left(a + \frac{b}{2}\right)^2 + \frac{3b^2}{4} > 0.$$

Of course, equality holds if and only if $a = b$.

3. For $x \neq a, b, c$, to solve the given equation is the same as to solve the second degree equation

$$(x - b)(x - c) + (x - a)(x - c) + (x - a)(x - b) = 0.$$

Now, observe that the discriminant of such an equation equals

$$4(a + b + c)^2 - 12(ab + ac + bc) = 4[(a^2 + b^2 + c^2) - (ab + ac + bc)],$$

which is a strictly positive number.

4. Write the left hand side as $a(a + b + c) + bc$. Then, apply inequality (5.2).

5. Firstly, show that $a^3 + b^3 \geq (a + b)ab$. Then, deduce that $\frac{1}{a^3+b^3+abc} \leq \frac{c}{abc(a+b+c)}$, getting analogous inequalities for the other two summands of the left hand side. Finally, add the three inequalities thus obtained.

6. Apply Ravi's transformation to conclude that is sufficient to prove that $(y + z)(x + z)(x + y) \geq 8xyz$, for every positive real numbers x, y and z. To this end, use inequality (5.2) three times.

7. Apply Ravi's transformation to rewrite the left hand side as

$$\frac{1}{2}\left(\frac{y}{x}+\frac{z}{x}+\frac{x}{y}+\frac{z}{y}+\frac{y}{z}+\frac{x}{z}\right).$$

Then, use the inequality between the arithmetic and geometric means for six numbers.

8. By adequately grouping in pairs the summands at the left hand side and applying inequality (5.7), between the arithmetic and harmonic means, we get

$$\frac{a+c}{a+b}+\frac{c+a}{c+d}=(a+c)\left(\frac{1}{a+b}+\frac{1}{c+d}\right)\geq(a+c)\cdot\frac{4}{(a+b)+(c+d)}$$

and

$$\frac{b+d}{b+c}+\frac{d+b}{d+a}=(b+d)\left(\frac{1}{b+c}+\frac{1}{a+d}\right)\geq(b+d)\cdot\frac{4}{(b+c)+(a+d)}.$$

Now, it suffices to add the two inequalities above.

9. The inequality we wish to prove is equivalent to

$$1+x+x^2+\cdots+x^{2n-1}+x^{2n}\geq(2n+1)x^n$$

or, which is the same, to

$$(1+x^{2n})+(x+x^{2n-1})+\cdots+(x^{n-1}+x^{n+1})\geq 2nx^n.$$

In order to prove this last one, it suffices to apply inequality (5.1) n times, adding the n inequalities thus obtained.

10. It follows from (5.4) that

$$3a^4+b^4=a^4+a^4+a^4+b^4\geq 4\sqrt[4]{a^4\cdot a^4\cdot a^4\cdot b^4}=4|a^3b|\geq 4a^3b.$$

11. We've already shown that $a^2+b^2+c^2\geq ab+bc+ca$, for every positive reals a, b and c. Multiply both sides of this inequality by $a+b+c$ to get the desired inequality.

12. Since $(ab+bc+ca)^2=(ab)^2+(bc)^2+(ca)^2+2abc(a+b+c)$, it suffices to show that

$$(ab)^2+(bc)^2+(ca)^2\geq abc(a+b+c).$$

Making $x=ab$, $y=bc$ and $z=ca$ in (5.3), we get

$$\begin{aligned}(ab)^2+(bc)^2+(ca)^2&=x^2+y^2+z^2\\&\geq xy+yz+zx\\&=abc(a+b+c).\end{aligned}$$

13. Apply inequality (5.3) once more.
14. Expand the right hand side and use the algebraic identity (2.5). After performing the obvious cancellations, apply the inequality of Example 2.3 three times. Then, use inequality (5.5).
15. Let $S = a^4(1 + b^4) + b^4(1 + c^4) + c^4(1 + a^4)$. Then, apply the inequality between the arithmetic and geometric means for two and three numbers, to get

$$S \geq a^4 \cdot 2b^2 + b^4 \cdot 2c^2 + c^4 \cdot 2a^2$$

$$\geq 3\sqrt[3]{(2a^4b^2)(2b^4c^2)(2c^4a^2)}$$

$$= 6\sqrt[3]{a^6b^6c^6}$$

$$= 6a^2b^2c^2.$$

16. Apply the inequality between the arithmetic and geometric means.
17. Change a_i by $-a_i$ and, then, apply the inequality between the arithmetic and geometric means.
18. For item (a), it follows from the inequality between the arithmetic and geometric means that

$$n! = 1 \cdot 2 \cdots n \leq \left(\frac{1 + 2 + \cdots + n}{n}\right)^n = \left(\frac{n + 1}{2}\right)^n.$$

For item (b), argue as in (a), applying the inequality between the arithmetic and geometric means to the numbers $1^2, 2^2, \ldots, n^2$ and, then, using the result of Example 4.18.
19. For $1 \leq k \leq m - 1$, apply the inequality between the arithmetic and geometric means to $x + k = x + \underbrace{1 + \cdots + 1}_{k}$. Then, multiply the $m - 1$ inequalities thus obtained.
20. Since $\sum_{j=1}^{n}(S - x_j) = (n - 1)S$, the inequality in the statement of the problem is equivalent to

$$\left(\sum_{j=1}^{n}(S - x_j)\right)\left(\sum_{j=1}^{n}\frac{1}{S - x_j}\right) \geq n^2.$$

Then, it suffices to apply inequality (5.7).
21. For item (a), just expand the product of the left hand side. For (b), apart from the hint given in the statement of the problem, apply the result of Problem 8, page 104 (with $k = 2$), to conclude that the sum at the right hand side of item (a) has exactly $\frac{1}{2}n(n - 1)$ summands.

22. First of all, note that

$$k_1 k_2 \ldots k_n \left(\frac{1}{k_1} + \frac{1}{k_2} + \cdots + \frac{1}{k_n} \right) = p_1 + p_2 + \cdots + p_n,$$

with $p_i = k_1 \ldots k_{i-1} \widehat{k_i} k_{i+1} \ldots k_n$, the hat over k_i indicating that the product contains all of $k_1, k_2, \ldots, k_n$, with the exception of k_i. Then, we can write

$$\frac{a_1^{k_1}}{k_1} + \frac{a_2^{k_2}}{k_2} + \cdots + \frac{a_n^{k_n}}{k_n} = \frac{p_1 a_1^{k_1} + p_2 a_2^{k_2} + \cdots + p_n a_n^{k_n}}{k_1 k_2 \ldots k_n}.$$

Now, expand the summands of the numerator in such a way that each $a_i^{k_i}$ appears p_i times and apply the inequality between the arithmetic and geometric means to get

$$\frac{1}{k_1 k_2 \ldots k_n} [\underbrace{(a_1^{k_1} + \cdots + a_1^{k_1})}_{p_1 \text{ vezes}} + \cdots + \underbrace{(a_n^{k_n} + \cdots + a_n^{k_n})}_{p_n \text{ times}} \geq$$

$$\geq \left((a_1)^{k_1})^{p_1} \ldots (a_n)^{k_n})^{p_n} \right)^{\frac{1}{k_1 k_2 \ldots k_n}}$$

$$= \prod_{i=1}^{n} a_i^{\frac{p_i k_i}{k_1 k_2 \ldots k_n}} = a_1 a_2 \ldots a_n.$$

23. Make $a_0 = 0$ and apply (5.7) to get

$$\frac{1}{a_j - a_{j-1}} + \cdots + \frac{1}{a_1 - a_0} \geq \frac{j^2}{a_j}.$$

Then, add the above inequalities for $1 \leq j \leq n$ and group equal terms to get the desired inequality. Finally, conclude that equality holds if and only if the sequence $(a_k)_{k \geq 1}$ is an AP.

24. Substituting a, b and c respectively by $6x^6$, $6y^6$ and $6z^6$, show that it suffices to prove that

$$7x^{12} + 12x^6 y^6 + 7y^6 z^6 + 9y^{12} + 9x^6 z^6 \geq$$

$$\geq 2x^3 y^9 + 6x^9 y^3 + 6x^2 y^8 z^2 + 12x^5 y^5 z^2 + 6x^4 y^4 z^4 + 6xy^7 z^4 + 6x^8 y^2 z^2.$$

To this end, write the left hand side expression as a sum of seven other expressions, chosen in such a way that, when applying the inequality between the arithmetic and geometric means to each one of them, we get the seven summands of the right hand side. For instance:

$$2x^6 z^6 + 2x^6 z^6 + 2x^{12} + 2y^{12} + 2y^{12} + 2x^6 y^6 \geq 12x^5 y^5 z^2.$$

Section 5.2

1. By Cauchy' inequality, we have $12 = 3x + 4y \leq \sqrt{3^2 + 4^2} \sqrt{x^2 + y^2} = 5\sqrt{x^2 + y^2}$. Therefore, $x^2 + y^2 \geq \frac{144}{25}$, with equality holding if and only if $\frac{x}{3} = \frac{y}{4}$. However, since $3x + 4y = 12$, equality holds if and only if $x = \frac{36}{25}$ and $y = \frac{48}{25}$.

2. Adapt the discussion of the case $n = 3$, presented in the text.

3. Making $b_1 = b_2 = \cdots = b_n = 1$ in Cauchy's inequality, we get

$$a_1 + a_2 + \cdots + a_n \leq \sqrt{a_1^2 + a_2^2 + \cdots + a_n^2} \sqrt{n},$$

with equality if and only if there exists a nonzero real number λ such that $\frac{a_1}{1} = \frac{a_2}{1} = \cdots = \frac{a_n}{1} = \lambda$, i.e., if and only if $a_1, a_2, \ldots, a_n$ are all equal. Dividing by n both sides of the above inequality, we get the decided inequality.

4. Apply the inequality between the quadratic and arithmetic means (see last problem) to the numerator of the fraction at each summand above. Then, add the results thus obtained.

5. We want to prove that

$$\left(\sum_{j=1}^{n} x_j \right) \left(\sum_{j=1}^{n} \frac{1}{x_j} \right) \geq n^2,$$

for all positive reals $x_1, x_2, \ldots, x_n$. To this end, make $a_j = \sqrt{x_j}$, $b_j = \frac{1}{\sqrt{x_j}}$ and apply Cauchy's inequality.

6. Multiply both sides by $a + b + c + d$ and apply Cauchy's inequality for $n = 4$, with $a_1 = \sqrt{a}$, $a_2 = \sqrt{b}$, $a_3 = \sqrt{c}$, $a_4 = \sqrt{d}$ and $b_1 = \frac{1}{\sqrt{a}}$, $b_2 = \frac{1}{\sqrt{b}}$, $b_3 = \frac{2}{\sqrt{c}}$, $b_4 = \frac{4}{\sqrt{d}}$. Alternatively, apply (5.7), writing $c = \frac{c}{2} + \frac{c}{2}$, $\frac{4}{c} = \frac{1}{c/2} + \frac{1}{c/2}$, $d = \frac{d}{4} + \frac{d}{4} + \frac{d}{4} + \frac{d}{4}$ and $\frac{16}{d} = \frac{1}{d/4} + \frac{1}{d/4} + \frac{1}{d/4} + \frac{1}{d/4}$.

7. For the left hand inequality, let $S = \sqrt{2x + 3} + \sqrt{2y + 3} + \sqrt{2z + 3}$. Applying the inequality between the arithmetic and geometric means for three numbers, we get

$$S \geq 3\sqrt[6]{(2x + 3)(2y + 3)(2z + 3)}$$

$$= 3\sqrt[6]{8xyz + 12(xy + xz + yz) + 18(x + y + z) + 27}$$

$$> 3\sqrt[6]{18 \cdot 3 + 27} = 3\sqrt[3]{9}.$$

For the right hand inequality, apply the inequality between the quadratic and arithmetic means for three numbers (cf. Problem 3), with $a_1 = \sqrt{2x + 3}$, $a_2 = \sqrt{2y + 3}$ and $a_3 = \sqrt{2z + 3}$, and use the fact that $x + y + z = 3$.

8. Apply the inequality between the quadratic and arithmetic means (cf. Problem 3), with $a_k = \sqrt{\binom{n}{k}}$. Then, use the result of item (a) of Corollary 4.25.

9. By Cauchy's inequality, we have

$$\left(\frac{x^2}{y^2} + \frac{y^2}{z^2} + \frac{z^2}{x^2}\right)\left(\frac{y^2}{z^2} + \frac{z^2}{x^2} + \frac{x^2}{y^2}\right) \geq \left(\frac{x}{y} \cdot \frac{y}{z} + \frac{y}{z} \cdot \frac{z}{x} + \frac{z}{x} \cdot \frac{x}{y}\right)^2$$

$$= \left(\frac{y}{x} + \frac{z}{y} + \frac{x}{z}\right)^2.$$

10. Since

$$\frac{a_k^2}{a_k + b_k} = \frac{a_k^2 - b_k^2 + b_k^2}{a_k + b_k} = (a_k - b_k) + \frac{b_k^2}{a_k + b_k},$$

we get

$$\sum_{k=1}^{n} \frac{a_k^2}{a_k + b_k} = \sum_{k=1}^{n} (a_k - b_k) + \sum_{k=1}^{n} \frac{b_k^2}{a_k + b_k} = \sum_{k=1}^{n} \frac{b_k^2}{a_k + b_k}.$$

Hence,

$$\sum_{k=1}^{n} \frac{a_k^2}{a_k + b_k} \geq \frac{1}{2} \sum_{k=1}^{n} a_k \Leftrightarrow \sum_{k=1}^{n} \frac{a_k^2 + b_k^2}{a_k + b_k} \geq \frac{1}{2} \sum_{k=1}^{n} (a_k + b_k),$$

and it suffices to show that $\frac{x^2+y^2}{x+y} \geq \frac{x+y}{2}$, for all positive reals x and y. This follows immediately from the inequality between the quadratic and arithmetic means (cf. Problem 3).

11. Start by letting $a_{n+1} = a_1$. Now, for $1 \leq j \leq n$, use Cauchy's inequality to write

$$(1 + a_j)^2 = \left(1 \cdot 1 + \sqrt{a_{j+1}} \cdot \frac{a_j}{\sqrt{a_{j+1}}}\right)^2 \leq (1 + a_{j+1})\left(1 + \frac{a_j^2}{a_{j+1}}\right),$$

so that

$$1 + \frac{a_j^2}{a_{j+1}} \geq \frac{(1 + a_j)^2}{1 + a_{j+1}}.$$

Then, multiply the n inequalities obtained in this way to reach the desired one.

12. Perform Ravi's transformation (according to the discussion that precedes the statement of Problem 6, page 120) to show that the given inequality is equivalent to

$$\sqrt{2x} + \sqrt{2y} + \sqrt{2z} \le \sqrt{y+z} + \sqrt{x+z} + \sqrt{x+y},$$

for positive reals x, y and z. Then, use the inequality between the quadratic and arithmetic means (cf. Problem 3) to show that $\sqrt{\frac{y+z}{2}} \ge \frac{\sqrt{y}+\sqrt{z}}{2}$, with equality if and only if $y = z$; analogously, get $\sqrt{\frac{x+z}{2}} \ge \frac{\sqrt{x}+\sqrt{z}}{2}$ and $\sqrt{\frac{x+y}{2}} \ge \frac{\sqrt{x}+\sqrt{y}}{2}$, and add these three inequalities to get the desired one.

13. Use Cauchy's inequality to get

$$\left(\sum_{k=1}^{n} a_k b_k c_k\right)^2 \le \left(\sum_{k=1}^{n} (a_k b_k)^2\right)\left(\sum_{k=1}^{n} c_k^2\right).$$

and

$$\sum_{k=1}^{n} a_k^2 b_k^2 \le \sqrt{\left(\sum_{k=1}^{n} a_k^4\right)\left(\sum_{k=1}^{n} b_k^4\right)}$$

Now, check that the inequality between the quadratic and arithmetic means (cf. Problem 3) gives

$$\sum_{k=1}^{n} c_k^2 \le \sqrt{n\left(\sum_{k=1}^{n} c_k^4\right)}.$$

Finally, it suffices to combine both inequalities above.

14. Make $x = \frac{1}{a}, y = \frac{1}{b}, z = \frac{1}{c}$ and, then, apply Cauchy's inequality to get

$$((y+z) + (x+z) + (x+y))\left(\frac{x^2}{y+z} + \frac{y^2}{x+z} + \frac{z^2}{x+y}\right) \ge (x+y+z)^2.$$

Finally, use the inequality between the arithmetic and geometric means to conclude the proof.

Section 5.3

1. Start by observing that

$$\frac{\left(1 + \frac{1}{n+1}\right)^{n+1}}{\left(1 + \frac{1}{n}\right)^n} = \left(1 + \frac{1}{n+1}\right)\left(1 - \frac{1}{(n+1)^2}\right)^n.$$

Then, apply Bernoulli's inequality.

2. Write $a^m = m^m \left(1 + \frac{a-m}{m}\right)^m$ and apply Bernoulli's inequality. Do the same with a^n and, then, add the results.

3. The desired inequality is equivalent to

$$\frac{a^6}{b^2c^2} + \frac{b^6}{a^2c^2} + \frac{c^6}{a^2b^2} \geq ab + ac + bc.$$

Suppose, without loss of generality, that $a \leq b \leq c$. Then, $a^6 \leq b^6 \leq c^6$ and $\frac{1}{b^2c^2} \leq \frac{1}{a^2c^2} \leq \frac{1}{a^2b^2}$. Applying in succession Chebyshev's inequality and the inequality between the arithmetic and geometric means, we get

$$\frac{a^6}{b^2c^2} + \frac{b^6}{a^2c^2} + \frac{c^6}{a^2b^2} \geq \frac{1}{3}(a^6 + b^6 + c^6)\left(\frac{1}{b^2c^2} + \frac{1}{a^2c^2} + \frac{1}{a^2b^2}\right)$$

$$\geq a^2b^2c^2\left(\frac{1}{b^2c^2} + \frac{1}{a^2c^2} + \frac{1}{a^2b^2}\right)$$

$$= a^2 + b^2 + c^2.$$

Finally, since we already know that $a^2 + b^2 + c^2 \geq ab + ac + bd$, there is nothing left to do.

4. Write $x_1^4 + x_2^4 + \cdots + x_{1994}^4 = x_1^3 \cdot x_1 + x_2^3 \cdot x_2 + \cdots + x_{1994}^3 \cdot x_{1994}$ and, then, apply Chebyshev's inequality to get

$$x_1^4 + x_2^4 + \cdots + x_{1994}^4 \geq \frac{1}{1994}(x_1^3 + x_2^3 + \cdots + x_{1994}^3)(x_1 + x_2 + \cdots + x_{1994}).$$

Now, use the equations of the system.

5. Apply Chebyshev's inequality to each summand of the left hand side.

6. Let's show that, for every $n > 1$, one has

$$\sum_{i=1}^{n} \lambda_i a_i b_i - \left(\sum_{i=1}^{n} \lambda_i a_i\right)\left(\sum_{i=1}^{n} \lambda_i b_i\right) \geq 0.$$

To this end, start by observing that the left hand side equals

$$\sum_{i=1}^{n} \lambda_i(\lambda_1 + \cdots + \widehat{\lambda_i} + \cdots + \lambda_n)a_i b_i - \sum_{\substack{i,j=1 \\ i \neq j}}^{n} \lambda_i\lambda_j a_i b_j, \qquad (B.2)$$

where $\widehat{\lambda_i}$ indicates that the summand λ_i is omitted. Then, notice that, to each pair (i, j) with $i < j$, the sum

$$\lambda_i\lambda_j a_i b_i + \lambda_i\lambda_j a_j b_j - \lambda_i\lambda_j a_i b_j - \lambda_i\lambda_j a_j b_i = -\lambda_i\lambda_j(a_i - a_j)(b_i - b_j)$$

appears in (B.2) exactly once, which means that

$$\sum_{i=1}^{n} \lambda_i a_i b_i - \left(\sum_{i=1}^{n} \lambda_i a_i\right)\left(\sum_{i=1}^{n} \lambda_i b_i\right) \lambda_i \lambda_j (a_i - a_j)(b_i - b_j) \geq 0.$$

Equality holds when all of the a_i's or all of the b_i's are equal. Finally, the usual Chebyshev's inequality corresponds to the particular case in which $\lambda_i = \frac{1}{n}$, for $1 \leq i \leq n$.

7. Apply Chebyshev's inequality to conclude that the expression at the left hand side is greater than or equal to

$$\frac{1}{4}(a^3 + b^3 + c^3 + d^3)\left(\frac{1}{b+c+d} + \frac{1}{a+c+d} + \frac{1}{a+b+d} + \frac{1}{a+b+c}\right).$$

Then, apply the result of Corollary 5.20, together with (5.7), to conclude that the last expression above is greater than or equal to

$$\left(\frac{a+b+c+d}{4}\right)^3 \cdot \frac{4^2}{3(a+b+c+d)} = \frac{(a+b+c+d)^2}{12}.$$

Finally, observe that the condition given in the statement of the problem is equivalent to $(a+c)(b+d) = 1$, and apply the inequality between the arithmetic and geometric means.

8. By symmetry we can suppose, without loss of generality, that $a \geq b \geq c$. This way, we get $\frac{1}{b+c} \geq \frac{1}{a+c} \geq \frac{1}{a+b}$. Applying Chebyshev's inequality and (5.7), we obtain

$$\frac{a^n}{b+c} + \frac{b^n}{a+c} + \frac{c^n}{a+b} \geq \frac{1}{3}\left(\frac{1}{b+c} + \frac{1}{a+c} + \frac{1}{a+b}\right)(a^n + b^n + c^n)$$

$$\geq \frac{3}{2(a+b+c)}(a^n + b^n + c^n).$$

Also by Chebyshev's inequality, it follows that

$$a^n + b^n + c^n \geq \frac{1}{3}(a+b+c)(a^{n-1} + b^{n-1} + c^{n-1}),$$

which completes the proof.

9. Suppose, with no loss of generalidade, that $x \geq y \geq z$. Use Chebyshev's inequality to conclude that the expression in the statement of the problem is greater than or equal to

$$\left(\frac{x^3 + y^3 + z^3}{3}\right) \cdot \frac{(x+1) + (y+1) + (z+1)}{(x+1)(y+1)(z+1)}.$$

Apply Chebyshev's inequality to the first factor and the inequality between the arithmetic and geometric means to the second factor to conclude that the last expression above is greater than or equal to $\frac{3t^3}{(t+1)^2}$, where $t = \frac{1}{3}(x + y + z)$. Finally, use again the inequality between the arithmetic and geometric means to conclude that $t \geq 1$, and prove that $t \geq s \Rightarrow \frac{3t^3}{(t+1)^2} \geq \frac{3s^3}{(s+1)^2}$.

10. In order to get the first inequality, apply Chebyshev's inequality to the first sum, followed by the result of Corollary 5.8. For the second inequality, use the inequality between the arithmetic and geometric means.

11. Suppose, with no loss of generalidade, that $a_1 \leq a_2 \leq \cdots \leq a_{2n}$. Arrange the given numbers in n pairs $(b_1, c_1), (b_2, c_2), \ldots, (b_n, c_n)$, with $b_1 \leq b_2 \leq \cdots \leq b_n$ and $b_j \leq c_j$, for each j. We wish to maximize

$$S = b_1 c_1 + b_2 c_2 + \cdots + b_n c_n.$$

To this end, we will show that $c_1 \leq c_2 \leq \cdots \leq c_n$. Indeed, if the sequence (c_i) isn't nondecreasing, there will exist indices $i > j$ such that $c_i \leq c_j$. In this case, we change the positions of c_i and c_j in S, after which the new sum will be

$$S' = S - b_i c_i - b_j c_j + b_i c_j + b_j c_i = S + (b_i - b_j)(c_j - c_i) \geq S.$$

Hence, the sum will be maximal when $c_1 \leq c_2 \leq \cdots \leq c_n$. Finally, observe that, for all indices $i < j$, we have $b_i \leq c_i \leq c_j$. Suppose that, for some pair of indices $i < j$, we would have $b_j \leq c_i$. Change c_i by b_j, so that the new sum is

$$S'' = S - b_i c_i - b_j c_j + b_i b_j + c_i c_j = S + (b_i - c_j)(b_j - c_i) \geq S.$$

Therefore, we must have $b_i \leq c_i \leq b_j \leq c_j$ whenever $i < j$. In general, we must have

$$b_1 \leq c_1 \leq b_2 \leq c_2 \leq \cdots \leq b_n \leq c_n,$$

so that the maximal sum is $a_1 a_2 + a_3 a_4 + \cdots + a_{2n-1} a_{2n}$.

12. Adapt, to the present case, the idea of the proof of rearrangement's inequality.

13. For item (a), it suffices to see that

$$a(x + y - a) - xy = (ax - xy) + a(y - a) = -x(y - a) + a(y - a)$$
$$= (y - a)(a - x) \geq 0,$$

for $y \geq a \geq x$. In what concerns (b), we want to prove that, given an integer $n > 1$ and positive reals $x_1, \ldots, x_n$, one always has

$$\text{AM}(x_1, \ldots, x_n) \geq \text{GM}(x_1, \ldots, x_n),$$

where AM and GM denote the arithmetic and geometric means, respectively. To this end we can suppose, without loss of generality, that $x_1 \leq \cdots \leq x_n$, with $x_1 < x_n$. If $a = \mathrm{AM}(x_1, \ldots, x_n)$, then $x_1 < a < x_n$. Similarly to the proof of item (a), we obtain $a(x_1 + x_n - a) > x_1 x_n$, so that

$$\mathrm{AM}(x_1 + x_n - a, x_2, \ldots, x_{n-1}, a) = \mathrm{AM}(x_1, x_2, \ldots, x_{n-1}, x_n)$$

and

$$\mathrm{GM}(x_1 + x_n - a, x_2, \ldots, x_{n-1}, a) > \mathrm{GM}(x_1, x_2, \ldots, x_{n-1}, x_n).$$

If the numbers $x_1 + x_n - a, x_2, \ldots, x_{n-1}, a$ are all equal, we will have

$$\mathrm{AM}(x_1 + x_n - a, x_2, \ldots, x_{n-1}, a) = \mathrm{GM}(x_1 + x_n - a, x_2, \ldots, x_{n-1}, a)$$

and, therefore

$$\mathrm{AM}(x_1, x_2, \ldots, x_{n-1}, a) > \mathrm{GM}(x_1, x_2, \ldots, x_{n-1}, a).$$

Otherwise, we order those n numbers as $y_1 \leq \cdots \leq y_n$, with $y_1 < y_n$. Since $\mathrm{AM}(y_1, \ldots, y_n) = a$, it follows that $y_1 < a < y_n$. Now, changing y_1 per $y_1 + y_n - a$ and y_n per a we obtain, as above,

$$\mathrm{AM}(y_1 + y_n - a, y_2, \ldots, y_{n-1}, a) = \mathrm{AM}(y_1, y_2, \ldots, y_{n-1}, y_n)$$

and

$$\mathrm{GM}(y_1 + y_n - a, y_2, \ldots, y_{n-1}, a) > \mathrm{GM}(y_1, y_2, \ldots, y_{n-1}, y_n).$$

Among the numbers $y_1 + y_n - a, y_2, \ldots, y_{n-1}, a$ we now have at least two which are equal to a. If the numbers $y_1 + y_n - a, y_2, \ldots, y_{n-1}, a$ are all equal, we will have

$$\mathrm{AM}(y_1 + y_n - a, y_2, \ldots, y_{n-1}, a) = \mathrm{GM}(y_1 + y_n - a, y_2, \ldots, y_{n-1}, a)$$

and, hence,

$$\mathrm{AM}(y_1, y_2, \ldots, y_{n-1}, y_n) = \mathrm{GM}(y_1, y_2, \ldots, y_{n-1}, y_n).$$

Otherwise, we operate a third exchange of numbers, as described above. Observe that this algorithm ends after a finite number of steps, and when we reach this point all numbers in our list will be equal to a. In particular, the corresponding arithmetic and geometric means of this las set of (n equal) numbers will be equal. However, since each performed operation preserves arithmetic means and increases geometric means, it follows that, at the beginning, we had

$$AM\,(x_1, x_2, \ldots, x_{n-1}, a) > GM\,(x_1, x_2, \ldots, x_{n-1}, a).$$

14. Make a proof by induction on $n > 1$. For the induction step, it suffices to prove that

$$\left(1 + \frac{a_n^2}{a_{n+1}}\right)\left(1 + \frac{a_{n+1}^2}{a_1}\right) \geq \left(1 + \frac{a_n^2}{a_1}\right)(1 + a_{n+1})$$

or, equivalently, that

$$a_{n+1}^2 \cdot \frac{1}{a_1} + a_n^2 \cdot \frac{1}{a_{n+1}} \geq a_{n+1} + \frac{a_n^2}{a_1}.$$

Finally, observe that such an inequality is true by an immediate application of the rearrangement's inequality.

15. (By Prof. Emanuel Carneiro) Let's make induction on n, leaving the initial case $n = 3$ as an exercise. More precisely, let's show that, in order to maximize the left hand side expression, one of the x_i's must be equal to 0, and that this allows us to invoke the induction hypothesis. To this end, let

$$E(x_1, x_2, \ldots, x_{n-1}, x_n) = x_1^2 x_2 + x_2^2 x_3 + x_3^2 x_4 + \cdots + x_n^2 x_1.$$

and suppose that the expression attains its greatest value[2] for some sequence $(x_1, x_2, \ldots, x_n)$ such that none of the x_i's is equal to 0. Substituting x_{n-1} by 0 and x_1 by $x_1 + x_{n-1}$, we obtain a new expression, which is less than or equal to the original one, i.e., which is such that

$$0 \geq E(x_1 + x_{n-1}, x_2, \ldots, 0, x_n) - E(x_1, x_2, \ldots, x_{n-1}, x_n)$$

$$= 2x_1 x_2 x_{n-1} + x_2 x_{n-1}^2 + x_{n-1} x_n^2 - x_{n-2}^2 x_{n-1} - x_{n-1}^2 x_n.$$

Upon dividing this last expression by x_{n-1}, we get

$$2x_1 x_2 + x_2 x_{n-1} + x_n^2 - x_{n-2}^2 - x_{n-1} x_n \leq 0.$$

Analogously, performing the exchange operations

$$x_{i-1} \mapsto 0 \quad \text{and} \quad x_{i+1} \mapsto x_{i+1} + x_{i-1},$$

[2]Here, we're tacitly assuming that there exists a sequence $(x_1, x_2, \ldots, x_n)$ such that $x_1 + x_2 + \cdots + x_n = 1$ and $E(x_1, x_2, \ldots, x_n)$ is maximal. Although this could be rigorously proved, such a proof is beyond the scope of these notes, so that we shall not present it here.

for $1 \leq i \leq n$ (with $x_0 = x_n$ and $x_{n+1} = x_1$), we get the inequalities

$$2x_{i+1}x_{i+2} + x_{i+2}x_{i-1} + x_i^2 - x_{i-2}^2 - x_{i-1}x_i \leq 0,$$

for $1 \leq i \leq n$. Adding all of these, we arrive at the inequality

$$\sum_{i=1}^{n} x_i(x_{i+1} + x_{i+3}) \leq 0,$$

which is an absurd. Therefore, if $(x_1, x_2, \ldots, x_n)$ maximizes $E(x_1, x_2, \ldots, x_n)$, then at least one of the x_i's must be equal to 0. Suppose, with no loss of generality, that $x_n = 0$. Then,

$$\begin{aligned}
E_{max} &= E(x_1, x_2, \ldots, x_{n-1}, 0) \\
&= x_1^2 x_2 + x_2^2 x_3 + x_3^2 x_4 + \cdots + x_{n-2}^2 x_{n-1} \\
&\leq x_1^2 x_2 + x_2^2 x_3 + x_3^2 x_4 + \cdots + x_{n-2}^2 x_{n-1} + x_{n-1}^2 x_1 \\
&\leq \frac{4}{27},
\end{aligned}$$

where, in the last inequality, we applied the induction hypothesis.

17. Make $\lambda_i = \frac{b_i}{a_i}$, for $1 \leq j \leq n$, and conclude that

$$b_1 + \cdots + b_n \geq a_1 + \cdots + a_n \Leftrightarrow a_1(\lambda_1 - 1) + a_2(\lambda_2 - 1) + \cdots + a_n(\lambda_n - 1) \geq 0.$$

Then, apply Abel's inequality to get

$$\sum_{j=1}^{n} a_j(\lambda_j - 1) \geq a_1 \cdot \min \{\lambda_1 - 1, \lambda_1 + \lambda_2 - 2, \ldots, \lambda_1 + \cdots + \lambda_n - n\}.$$

Finally, use the condition given in the statement of the problem, together with the inequality between the arithmetic and geometric means to show that $\lambda_1 + \cdots + \lambda_k \geq k$, for $1 \leq k \leq n$.

Section 6.1

1. We must have $x - 1 \geq 0$ and $3 - x > 0$, so that $x \in [1, 3)$.

2. We must have $x \geq 0$, $3 - \sqrt{x} \geq 0$, $\frac{3}{2} - \sqrt{3 - \sqrt{x}} \geq 0$ and $\frac{1}{2} - \sqrt{\frac{3}{2} - \sqrt{3 - \sqrt{x}}}$. The maximal domain of f is the set formed by the intersection of the solution sets of these inequalities.

3. For item (a), just see that $f(1) = f\left(\frac{1}{1}\right) = \frac{|1^2-1^2|}{1^2+1^2} = 0, f(10) = f\left(\frac{10}{1}\right) = \frac{|10^2-1^2|}{10^2+1^2} = \frac{99}{101}$ and $f\left(\frac{24}{36}\right) = f\left(\frac{2}{3}\right) = \frac{|2^2-3^2|}{2^2+3^2} = \frac{5}{13}$. For item (b), if $|a^2-b^2| = 55$ and $a^2 + b^2 = 73$, then $(a, b) = (8, 3)$ or $(3, 8)$, so that $f\left(\frac{3}{8}\right) = f\left(\frac{8}{3}\right) = \frac{55}{73}$. If $|a^2 - b^2| = 32$ and $a^2 + b^2 = 257$, then $2a^2 = 289$ or $2a^2 = 225$ and, hence, $a \notin \mathbb{N}$. However, setting $a^2 - b^2 = 32k$ and $a^2 + b^2 = 257k$, with $k \in \mathbb{N}$, we get $2a^2 = 289k$ and $2b^2 = 225k$, so that $k = 2$ gives $a = 17$ and $b = 15$ (analogously, by letting $b^2 - a^2 = 32k$, we obtain $a = 15$ and $b = 17$). Therefore, $f\left(\frac{15}{17}\right) = f\left(\frac{17}{15}\right) = \frac{32}{257}$. Finally, if $a^2-b^2 = 101k$ and $a^2+b^2 = 89k$, for some $k \in \mathbb{N}$, then $a^2 = 95k$ and $b^2 = -6k$, so that $(ab)^2 = -570k^2$; if $b^2 - a^2 = 101k$ and $a^2 + b^2 = 89k$, we conclude, analogously, that $(ab)^2 = -570k^2$. In any case we reach a contradiction, for, $(ab)^2 > 0 > -570k^2$.

5. Start by making $x = 1$ and $y = \sqrt{2}$ in the given relation, to compute $f(1+\sqrt{2})$.

6. One has to prove that $f(a_{k+1}) = f(a_k) + f(r)$, for every $k \geq 1$. To this end, let $x = a_k$ and $y = r$, so that $x + y = a_{k+1}$.

7. Adapt, to the present case, the hint given to the previous problem.

8. In the case of $\frac{f}{g}$, notice that one has to shrink the common domain X of f and g, to avoid those $x \in X$ for which $g(x) = 0$.

9. The image is the set $\mathbb{Z}$ of integers.

10. Start by showing that $\{x\} \in [0, 1)$, for every $x \in \mathbb{R}$. Then, show that the image of $\{\cdot\}$ is precisely the interval $[0, 1)$.

11. If a_n denotes the n-th natural number which is not a perfect square, then there exist positive integers s and t, such that $1 \leq s \leq 2t$ and $a_n = t^2+s$; in particular, t^2 is the greatest perfect square which is less than or equal to a_n. Since there are exactly a_n integers from 1 to a_n, exactly t of which are perfect squares (these being $1^2, 2^2, \ldots, t^2$), it follows that a_n is the $(a_n - t)$-th non perfect square. Hence, $n = a_n - t = t^2 + s - t$, so that

$$t = \sqrt{n - s + \frac{1}{4}} + \frac{1}{2}.$$

Now, since $s > 0$, we have

$$a_n = t^2 + s = n + t = n + \sqrt{n - s + \frac{1}{4}} + \frac{1}{2} < n + \sqrt{n} + \frac{1}{2}.$$

On the other hand, by using the fact that $s \leq 2t$, we get, as above, $a_n > n + \sqrt{n} - \frac{1}{2}$. Therefore,

$$\left(n + \sqrt{n} + \frac{1}{2}\right) - 1 < a_n < n + \sqrt{n} + \frac{1}{2},$$

and, thus, $a_n = \lfloor n + \sqrt{n} + \frac{1}{2} \rfloor$.

12. For the first part of item (a), make $x = y = 0$; for the second, make $y = -x$. For item (b), use (a). For item (c), consider initially the case of $m \in \mathbb{N}$, by induction; then, use (a) to get the case $m < 0$. Finally, apply the result of (c) to get (d) and the results of (c) and (d) to get (e).

Section 6.2

1. Show that it is equal to $(0, +\infty)$.
2. Start by recalling that $x + \frac{1}{x} \geq 2$, for every real $x > 0$, with equality if and only if $x = 1$. Then, use this fact to conclude that the image is $\mathbb{R} \setminus (-2, 2)$.
3. Let's analyse the case in which f increases in $(-\infty, a] \cap I$ and decreases in $[a, +\infty) \cap I$ (the other case is completely analogous). For $x_0 \in I$ such that $x_0 < a$, we have $f(x_0) > f(a)$, so that x_0 is not a minimum point of f. Accordingly, for $x_0 \in I$ such that $x_0 > a$, we have $f(x_0) > f(a)$ and, as before, x_0 is not a minimum point of f. Hence, a is the only minimum point of f.
4. For $y_0 \in Y$, take $x_0 \in X$ such that $f(x_0) = y_0$. Then, by the definition of the function $f + c$, we have $(f + c)(x_0) = f(x_0) + c = y_0 + c$, so that $y_0 + c \in \text{Im}\,(f + c)$. Since $y_0 + c$ is a generic element of $Y + c$, it follows that $Y + c \subset \text{Im}\,(f + c)$. Now, if $y_1 \in \text{Im}\,(f + c)$, then, by definition, there exists $x_1 \in X$ such that $(f + c)(x_1) = y_1$ or, which is the same, $f(x_1) + c = y_1$. Then, $f(x_1) = y_1 - c$, and hence $y_1 - c \in \text{Im}\,(f) = Y$. This way, $y_1 = (y_1 - c) + c \in Y + c$ and, since this is valid for every $y_1 \in \text{Im}\,(f + c)$, we get $\text{Im}\,(f + c) \subset Y + c$. Therefore, $\text{Im}\,(f + c) = Y + c$.
5. Adapt, to the present case, the hint given for the previous problem.
6. Let's consider the case $a > 0$, the case $a < 0$ being totally analogous. For $y \in \mathbb{R}$, we have $f(x) = y$ if and only if $\left(x + \frac{b}{2a}\right)^2 - \frac{\Delta}{4a^2} = \frac{y}{a}$ or, which is the same,

$$\left(x + \frac{b}{2a}\right)^2 = \frac{y}{a} + \frac{\Delta}{4a^2} = \frac{4ay + \Delta}{4a^2}.$$

Since $\left(x + \frac{b}{2a}\right)^2 \geq 0$, the last equation above has a solution if and only if $4ay + \Delta \geq 0$. However, since $a > 0$, such a condition is equivalent to $y \geq -\frac{\Delta}{4a}$. Hence, there exists $x \in \mathbb{R}$ for which $f(x) = y$ if and only if $y \geq -\frac{\Delta}{4a}$, so that $\text{Im}\,(f) = [-\frac{\Delta}{4a}, +\infty)$.
7. Use the canonical form f (see the statement of the previous problem) to get, in the case $a > 0$,

$$f(x) < 0 \Leftrightarrow a\left[\left(x + \frac{b}{2a}\right)^2 - \frac{\Delta}{4a^2}\right] < 0 \Leftrightarrow \left(x + \frac{b}{2a}\right)^2 - \frac{\Delta}{4a^2} < 0$$

$$\Leftrightarrow \left(x + \frac{b}{2a}\right)^2 < \frac{\Delta}{4a^2} \Leftrightarrow \left|x + \frac{b}{2a}\right| < \frac{\sqrt{\Delta}}{2a} \Leftrightarrow -\frac{\sqrt{\Delta}}{2a} < x + \frac{b}{2a} < \frac{\sqrt{\Delta}}{2a}$$

$$\Leftrightarrow \frac{-b - \sqrt{\Delta}}{2a} < x < \frac{\sqrt{-b + \Delta}}{2a}.$$

8. In order to show that $\Delta > 0$, use the canonical form of f to get

$$af(x_0) = a^2\left[\left(x_0 + \frac{b}{2a}\right)^2 - \frac{\Delta}{4a^2}\right],$$

so that $af(x_0) < 0$ if and only if $\left(x_0 + \frac{b}{2a}\right)^2 - \frac{\Delta}{4a^2} < 0$. Now, since $\left(x_0 + \frac{b}{2a}\right)^2 \geq 0$, we must have $0 \leq \left(x_0 + \frac{b}{2a}\right)^2 < \frac{\Delta}{4a^2}$ and, hence, $\Delta > 0$. For the second part, use the result of the previous problem.

9. If a rectangle of perimeter $2p$ has dimensions x and y, then $x + y = p$. Show that its area depends on x according to the quadratic function $f(x) = -x^2 + px$, and apply the result of Proposition 6.25.

10. Letting l be the width and h be the height of the truck (measured in meters), it is immediate that the load volume of each truck is $18lh\,\mathrm{m}^3$. Hence, we have to maximize the product lh, subjected to the condition that the trucks can enter the tunnel using the correct lanes. Since the cross section of the trucks are $l \times h$ rectangles, the common length d of their diagonals should satisfy $d \leq 5m$. However, by Pythagoras' Theorem, we have $d = \sqrt{l^2 + h^2}$ and, hence,

$$lh = l\sqrt{d^2 - l^2} \leq l\sqrt{25 - l^2} = \sqrt{25l^2 - l^4}.$$

Setting $x = l^2$, we conclude that it suffices to maximize the quadratic function $f(x) = 25x - x^2$, with $0 < x < 5$.

11. We must find the smallest positive real number a such that $f(x) \leq a$, for every $x \in \mathbb{R}$; equivalently, this is the same as finding the smallest positive real number a such that $ax^2 - 5x + (a + 1) \geq 0$, for every $x \in \mathbb{R}$.

12. For even n, say $n = 2k$, using the triangle inequality we get

$$f(x) = (|x - \alpha_1| + \cdots + |x - \alpha_k|) + (|\alpha_{k+1} - x| + \cdots + |\alpha_{2k} - x|)$$

$$\geq \sum_{j=1}^{k}(x - \alpha_j) + \sum_{j=k+1}^{2k}(\alpha_j - x) = \sum_{j=k+1}^{2k}\alpha_j - \sum_{j=1}^{k}\alpha_j.$$

Now, show that every $x \in (\alpha_k, \alpha_{k+1})$ satisfies $f(x) = \sum_{j=k+1}^{2k}\alpha_j - \sum_{j=1}^{k}\alpha_j$. Finally, the case of an odd n can be dealt with in an analogous way.

13. For item (a), apply the inequality between the arithmetic and geometric means to the denominator of the defining formula for $f(x)$.For item (b), write $f(x) = 2 + \frac{5x}{x^2+1}$ and, then, proceed as in (a). Finally, for item (c), apply the inequality between the arithmetic and geometric means to get

$$x^3(1-x^3)^3 = \frac{1}{3} \cdot 3x^3(1-x^3)^3$$

$$\leq \frac{1}{3}\left(\frac{3x^3 + (1-x^3) + (1-x^3) + (1-x^3)}{4}\right)^4.$$

14. For item (a), write $\frac{(x+10)(x+2)}{x+1} = \frac{((x+1)+9)((x+1)+1)}{x+1} = (x+1) + \frac{9}{x+1} + 10$; for item (b), write $\frac{a}{x} = \frac{a}{2x} + \frac{a}{2x}$; for (c), write $x^3 = \frac{x^3}{2} + \frac{x^3}{2}$; finally, for (d), write $6x = 3x + 3x$.

15. Start by writing $x^4 + y^4 = (x^2 + y^2)^2 - 2(xy)^2 \leq (2 + xy)^2 - 2(xy)^2$, and look at this last expression as a quadratic function of $z = xy$.

16. First of all, prove that the function $f : \mathbb{R} \to \mathbb{R}$, given by $f(x) = x^3 + 2x$, is increasing. Then, observe that the equations of the given system can be written as $y = f(x)$, $z = f(y)$, $t = f(z)$ and $x = f(t)$. Finally, suppose that $x \geq y$ and use fact that f is increasing to conclude that $x \geq y \geq z \geq t \geq x$.

17. For fixed $m, n \in \mathbb{N}$, let $I_k = [(kmn+1)n, (kmn+1)m]$. Since $((k+1)mn+1)n < (kmn + 1)m$ for $k > \frac{mn^2+n-m}{mn(m-n)}$, there exists $k_0 \in \mathbb{N}$ such that

$$\bigcup_{k \geq k_0} I_k = [(k_0mn + 1)n, +\infty).$$

Since f has infinitely many strangulation points, there exist natural numbers $k \geq k_0$ and p such that p is a strangulation point of f and $p \in I_k$. Then, $f((kmn+1)n) < f(p) < f((kmn + 1)m)$. However, $kmn + 1$ is relatively prime to m and n, so that $f(kmn + 1) + f(n) < f(kmn + 1) + f(m)$ and, hence, $f(n) < f(m)$.

Section 6.3

1. For every $z \in \mathbb{R}$, we have $|z| > z$ if and only if $z < 0$. Hence, we want to find the set of $x \in \mathbb{R}$ for which $g(f(x)) < 0$. Make $y = f(x)$, and apply the result of Problem 7, page 161, to conclude that $-\frac{1}{2} < y < \frac{1}{2}$. Now, recall that $y = f(x)$ to show that $-\frac{1}{2} < x - \frac{7}{2} < \frac{1}{2}$.

2. Start by observing that $(f \circ g)(x) = f(g(x)) = 2g(x) + 7$ and, hence, that $2g(x) + 7 = x^2 - 2x + 3$.

3. Make $g(x) = y$ to obtain $x = \frac{y+3}{2}$ and, then, $f(y) = f(g(x)) = 2x^2 - 4x + 1 = 2\left(\frac{y+3}{2}\right)^2 - 4\left(\frac{y+3}{2}\right) + 1$.

4. Since $f \circ g$ and $g \circ g$ are functions from $\mathbb{R}$ to itself, we have $f \circ g = g \circ g \Leftrightarrow$ $(f \circ g)(x) = (g \circ f)(x) \Leftrightarrow f(g(x)) = g(f(x))$. Now, it suffices to compute $f(g(x)) = (ac)x + (ad + b)$, $g(f(x)) = (ac)x + (bc + d)$ and compare the results.

5. For $x, f(x) \neq -b$, we have

$$f(f(x)) = \frac{f(x) + a}{f(x) + b} = \frac{\frac{x+a}{x+b} + a}{\frac{x+a}{x+b} + b} = \frac{(a + 1)x + (ab + a)}{(b + 1)x + (a + b^2)}.$$

From this, conclude that we must have $\frac{(a+1)x+(ab+a)}{(b+1)x+(a+b^2)} = x$ for every $x \in \mathbb{R}$, except for at most four values of x. Hence, except for such values of x, we have $(a + 1)x + (ab + a) = (b + 1)x^2 + (a + b^2)x$. Finish by showing that $b = -1$.

7. Let's analyse the case in which f and g are increasing, leaving the analysis of the other two cases to the reader. Given $x_1 < x_2$ in I, it follows from the fact that f is increasing that $f(x_1) < f(x_2)$. Now, since $f(x_1), f(x_2) \in J$ and g is also increasing, we have $g(f(x_1)) < g(f(x_2))$.

8. Make $x = \frac{u}{v}$ in both given relations.

9. Verify that F is injective by showing that $F(x_1) = F(x_2) \Rightarrow x_1 = x_2$. Show that the surjectivity of F reduces to its very definition.

10. Start by assuming that $f = g + h$, with $g, h : X \to \mathbb{R}$ such that g is even and h is odd. Then, use the definitions of even and odd function to conclude that $g(x) = \frac{1}{2}(f(x) + f(-x))$ and $h(x)\frac{1}{2}(f(x) - f(-x))$.

11. Initially, show that $f(1) = 0$. Then, make $a = 1$ and $b = -1$ to compute $f(-1)$. Finally, make $b = -a$.

12. It suffices to compute $(f \circ f)(-x) = f(f(-x)) = f(-f(x)) = -f(f(x)) = -(f \circ f)(x)$, where, in the second and third equalities, we used the fact that f is odd.

13. Take $f(x) = -g(x)$ for $x \leq 0$ and $f(x) = -x$ for $x \geq 0$.

14. Start by writing

$$f(x) = \frac{4x^2}{x + 1} + k = \frac{4(x^2 - 1)}{x + 1} + \frac{4}{x + 1} + k$$

$$= 4(x - 1) + \frac{4}{x + 1} + k = 4\left(x + 1 + \frac{1}{x + 1}\right) + (k - 8).$$

Hence, letting $g(x) = x + 1 + \frac{1}{x+1}$, we have $f(x) = 4g(x) + (k - 8)$. Now, Problem 2, together with the fact that g is an odd function, implies that $\text{Im}(g) = \mathbb{R}\backslash(-2, 2)$, so that (by the result of Problem 5) $\text{Im}(4g) = \mathbb{R}\backslash(-8, 8)$. Finally, Problem 4 guarantees that $\text{Im}(f) = \mathbb{R} \setminus (k - 16, k)$, and we must have $k - 16 = -L$ and $k = L$.

15. For item (a), start by showing that every $x \in \mathbb{R}$ can be written as $x = kp + \alpha$, with $k \in \mathbb{Z}$ and $\alpha \in [0, p)$. Then, make induction on k to

show that $f(x) = f(x - kp)$, for every $k \in \mathbb{Z}$. For item (b), first observe that $g\left(x + \frac{p}{|a|}\right) = f(ax \pm p) = f(ax) = g(x)$, for every $x \in \mathbb{R}$; subsequently, note that if $p' > 0$ satisfies $g(x+p') = g(x)$ for every $x \in \mathbb{R}$, then $f(ax+ap') = f(ax)$ for every $x \in \mathbb{R}$. Finally, use this fact to deduce that $|a|p' = |ap'| \geq p$.

14. In order to show that f is odd, start by observing that $f(20 + x) = f(10 + (10 + x)) = f(10 - (10 + x)) = f(-x)$ and, analogously, $f(20 - x) = f(x)$. Then, use the fact that $f(20 + x) = -f(20 - x)$. For the second part, compute $f(40 + x) = f(20 + (20 + x)) = -f(20 + x) = f(x)$.

17. Making $g(x) = f(x) + \frac{1}{2}$, show that $g(x + a) = \sqrt{\frac{1}{4} - g(x)^2}$ and, hence, that $g(x + 2a) = g(x)$, for every $x \in \mathbb{R}$.

18. If $X = \{x \in \mathbb{Z}; x > 100\}$, then $\text{Im}(f) = f(X) \cup f(\mathbb{Z} \setminus X)$. On the one hand, $f(X) = \{y \in \mathbb{Z}; y > 90\}$. Now, if $90 \leq x \leq 100$, then $101 \leq x + 11 \leq 111$ and $f(x) = f(f(x + 11)) = f(x + 1)$. Therefore, $f(90) = f(91) = \cdots = f(100) = f(f(111)) = f(101) = 91$. Use strong induction to show that $f(90-x) = 91$ for every integer $x \geq 1$. For the induction step, if $f(90 - x) = 91$ for every $x < n$, observe that, for $x = n$, one has $f(90 - n) = f(f(101 - n)) = f(91) = 91$. Finally, conclude that $\text{Im}(f) = \{91, 92, 93, \ldots\}$.

19. From $f(f(n)) = 4n + 1$, we get $f(4n + 1) = f(f(f(n))) = 4f(n) + 1$. On the other hand, it's easy to prove by induction, with the aid of items (a) and (b), that $f(2^k) = 2^{k+1} - 1$. Consequently, again by item (c), we obtain $f(2^{k+1} - 1) = f(f(2^k)) = 4 \cdot 2^k + 1$. Due to these facts, and since $1993 = 4 \cdot 498 + 1$, we successively compute $f(1993) = 4f(498) + 1$, $f(498) = 2f(249) + 1$, $f(249) = f(4 \cdot 62 + 1) = 4f(62) + 1$, $f(62) = 2f(31) + 1$ and $f(31) = f(2^5 - 1) = 4 \cdot 2^4 + 1 = 65$.

20. Write $\mathbb{N} = A_1 \cup A_2 \cup A_3 \cup \ldots$, with $A_1, A_2, \ldots$ pairwise disjoint infinite sets. Then, for each $n \in \mathbb{N}$, set $f(x) = n$ whenever $x \in A_n$.

21. For item (a), use the assumptions on A and B to construct an injection from $A \times B$ to $\mathbb{N} \times \mathbb{N}$. For item (b), take $f : \mathbb{Z} \times \mathbb{N} \to \mathbb{Q}$ such that $f(m, n) = \frac{m}{n}$. The first part of (c) now follows from (b) and Lemma 6.43. As for the second part, let $g : \mathbb{Q} \to \mathbb{N}$ be defined by letting $g(r) = \min\{n \in \mathbb{N}; f(n) = r\}$.

22. For item (a), let $B_1 = A_1$ and, once we have defined $B_1, \ldots, B_{k-1}$, for some natural k, let $B_k = A_k \setminus \left(\cup_{j=1}^{k-1} A_j\right)$. For (b), since $B_k \subset A_k$, we conclude that B_k is finite or countably infinite. If $f_k : B_k \to \mathbb{N}$ is injective, let $f : \cup_{k \geq 1} B_k \to \mathbb{N} \times \mathbb{N}$ be given by $f(x) = (f_k(x), k)$ if $x \in B_k$ and verify that f is injective. Finally, (c) follows from (b) as a much easier particular case, arguing by contradiction and taking $A_1 = \mathbb{I}$ and $A_2 = \mathbb{Q}$.

23. Suppose, for the sake of contradiction, that $\mathcal{F} = \{A_1, A_2, A_3, \ldots\}$. Define $A = \{x_1, x_2, x_3, \ldots\} \subset \mathbb{N}$ by choosing $x_1 \notin A_1$, then $x_2 \notin A_2$ such that $x_2 > x_1 + 1$, then $x_3 \notin A_3$ such that $x_3 > x_2 + 1$ etc. Since $x_k \notin A_k$, we have $A \neq A_k$. On the other hand since $x_1 < x_2 < x_3 < \cdots$, we conclude that A is infinite and $\mathbb{N} \setminus A$ contains the infinite set $\{x_1 + 1, x_2 + 1, x_3 + 1, \ldots\}$, hence is also infinite.

Section 6.4

1. For item (a), use Proposition 6.39. For item (b), suppose that $g_1, g_2 : Y \to X$ satisfy $g_i \circ f = \text{Id}_X$ and $f \circ g_i = \text{Id}_Y$, for $i = 1, 2$. Use Proposition 6.34 to get $g_1 = g_1 \circ \text{Id}_Y = g_1 \circ (f \circ g_2) = (g_1 \circ f) \circ g_2 = \text{Id}_X \circ g_2 = g_2$.

2. Use several times the result of Proposition 6.34, together with the fact that $f^{-1} \circ f = \text{Id}_X, f \circ f^{-1} = \text{Id}_Y, g \circ g^{-1} = \text{Id}_Z$ and $g^{-1} \circ g = \text{Id}_Y$.

3. Apply the result of Example 6.48.

4. See the hint to the next problem.

5. For the first part, note that the equation $\frac{ax+b}{cx+d} = y$ has, for every $y \neq \frac{a}{c}$, the single solution $x = \frac{-dy+b}{cy-a}$.

6. For the first part, it suffices to see that, for $0 \leq x_1 < x_2$, we have $x_2 - x_1 > 0$; hence, item (a) of Problem 18, page 26, gives

$$x_2^n - x_1^n = (x_2 - x_1)(x_1^{n-1} + x_1^{n-2}x_2 + \cdots + x_2^{n-1}) > 0,$$

so that $f(x_1) < f(x_2)$ and f is increasing, thus injective. For the second part, let $y \in [0, +\infty)$ be given. Since we are assuming that $\text{Im}(f) = [0, +\infty)$, there exists $x \in [0, +\infty)$ such that $y = x^n$; then, by the definition of the n-th root of y, we get $x = \sqrt[n]{y}$, so that $f^{-1}(y) = \sqrt[n]{y}$.

7. Setting $f(x) = 2x$, we have $f^{-1}(x) = \frac{x}{2}$, so that $(f + f^{-1})(x) = \frac{5x}{2}$ and $(f - f^{-1})(x) = \frac{3x}{2}$. These last two functions are bijections from $\mathbb{R}$ to itself.

8. Suppose that f is increasing (the case of a decreasing f is entirely analogous). Given $y_1 < y_2$ in J, take the elements $x_1, x_2 \in I$ such that $f(x_1) = y_1$ and $f(x_2) = y_2$. If $x_1 = x_2$, then $y_1 = f(x_1) = f(x_2) = y_2$, which is not the case; if $x_2 < x_1$, it follows from the fact that f is increasing that $y_2 = f(x_2) < f(x_1) = y_1$, which is not the case either. Therefore, $x_1 < x_2$ or, which is the same, $f^{-1}(y_1) < f^{-1}(y_2)$. Since the elements $y_1 < y_2$ of J were chosen arbitrarily, this assures that f^{-1} is also increasing.

9. Show that, for fixed $a, b \in \mathbb{R}$, the system of equations $x^3 = a$ and $x - f(y) = b$ admits the single solution $x = \sqrt[3]{a}$ and $y = f^{-1}(\sqrt[3]{a} - b)$.

10. Start by showing that it suffices to prove that any two functions in G commute in respect to the operation of composition of functions. To this end, use items (a) and (b), together with the fact that the only function $h \in G$ of the form $h(x) = x + a$, for some $a \in \mathbb{R}$, is $h = \text{Id}_{\mathbb{R}}$.

11. Set $g = f^{-1}$ and start by showing that what is needed to be proved is equivalent to the existence of positive integers $x < y < z$, in arithmetic progression and such that $g(x) < g(y) < g(z)$; then, show that we can suppose that $x = 1$. To what is left to do, fix $t \in \mathbb{R}$ and show that one cannot have $g(t) > g(2t - 1) > g(4t - 3) > g(8t - 7) > \cdots > g(1)$.

Section 6.5

1. If there existed $x_1 < x_2$ in I such that $f(x_1) = g(x_1)$ and $f(x_2) = g(x_2)$, we would have $f(x_1) > f(x_2) = g(x_2) > g(x_1) = f(x_1)$.
2. Initially, note that $x = 4$ is a solution. Then, look at both sides of the given equation as functions from $(0, +\infty)$ to $\mathbb{R}$ and apply the result of the former problem.
3. Letting $g(x) = f(x) - f(0)$ in the given relation, conclude that it suffices to consider the case in which $f(0) = 0$. Under this additional hypothesis, make $y = 0$ to get $f\left(\frac{x}{2}\right) = \frac{f(x)}{2}$, for every $x \in \mathbb{Q}$. Then, use the given relation to conclude that $f(x + y) = f(x) + f(y)$, for all $x, y \in \mathbb{Q}$, so that (according to the first part of the solution of Example 6.56) $f(x) = f(1)x$, for every $x \in \mathbb{Q}$.
4. Use the given relation to prove by induction that $f(n) = f(1)$, for every $n \in \mathbb{N}$; then, consider the case $n < 0$ by making $y = -x$.
5. Make $x = y = z = 0$ to get $f(0) = \frac{1}{2}$; then, obtain $f(1) = \frac{1}{2}$ by an analogous reasoning. Make $y = z = 1$ to conclude that $f(x) \geq \frac{1}{2}$, for every $x \in \mathbb{R}$. Finally, make an analogous substitution to prove that $f(x) \leq \frac{1}{2}$, for every $x \in \mathbb{R}$.
6. In each of the intervals $[n, n + f(n)]$ and $[f(n), 2f(n)]$ there are $f(n) + 1$ naturals. Use the increasing character of f, together with the given relation, to conclude that $f(n + k) = f(n) + k$, for every $1 \leq k \leq f(n)$. Now, fix natural numbers k and n, with $n > k$. Use the fact that $f(n) \geq n$ to show that $f(n) > k$ and, by what was done above, that $f(n) = n - 1 + f(1)$.
7. Set $x = a = 0$ in (a) to get $f(0) = 0$. Then, letting $x = 0$ in (a), show that $f(a) = a$ for every $a \in \mathbb{Z}$. Now taking $x \in [0, 1)$, use the fact that $f(x) \in \mathbb{Z}$ and $f(f(x)) = 0$ to conclude that $f(x) = 0$. Finally, for a general $x \in \mathbb{R}$, change x by $\{x\}$ (the fractional part of x) and make $a = \lfloor x \rfloor$ in (a) to obtain $f(x) = \lfloor x \rfloor$.
8. Start by computing $f(x + f(y + f(0)))$ in two different ways to get $f(0) = 0$. Then, deduce that $f(f(y)) = -y$ for every $y \in \mathbb{Z}$, so that f is bijective. From this last relation, compute $f(f(f(x)))$ in two different ways to conclude that f is odd. Change x by $f(x)$ in the given relation to obtain $f(f(x) + f(y)) = f(f(x + y))$ and, then, $f(x) + f(y) = f(x + y)$, for all $x, y \in \mathbb{Z}$. Finally, use induction to get $f(x) = f(1)x$, for every $x \in \mathbb{Z}$, thus arriving at the contradiction $-1 = f(f(1)) = f(1)f(1) \geq 0$.
9. Set $k = 0$ to conclude that $f(1) = 2$; then, use induction to show that $f(n) = n + 1$, for every $n \in \mathbb{N}$. Now, use the given relation to show that, if $f(-1) = a$, then $a < 0$ or $a > 0$ lead to contradictions. Finally, make another induction to conclude that $f(-n) = -n + 1$, for every $n \in \mathbb{N}$.
10. First of all, show that we can suppose that f has no fixed points. Then, write $A = B \cup C$, where B and C are disjoint sets such that $f(x) > x$ for every $x \in B$ and $f(x) < x$ for every $x \in C$. If B has l elements and a is the common value of $|f(x_j) - x_j|$, conclude that $0 = \sum_{j=1}^{k}(f(x_j) - x_j) = (2l - k)a$ and, hence, that $a = 0$, which is a contradiction.

11. Set $x = y = 0$ in the given relation to obtain $f(0) = 2f(0)f(a)$ and, hence, $f(a) = \frac{1}{2}$. Set just $y = 0$ to get $f(x) = f(a - x)$ for every $x \in \mathbb{R}$, so that $f(-x) = f(a + x)$, also for every $x \in \mathbb{R}$. Substitute $y = a$ in the given relation to get

$$f(-x) = f(x + a) = f(x)f(0) + f(a)f(a - x) = \frac{1}{2}f(x) + \frac{1}{2}f(x) = f(x),$$

for every $x \in \mathbb{R}$. Now, let $y = -x$ in the given relation to find

$$\frac{1}{2} = f(0) = f(x)f(a + x) + f(-x)f(a - x)$$
$$= f(x)f(-x) + f(-x)f(x) = 2f(x)^2.$$

Therefore, if we show that $f(x) \geq 0$ for every $x \in \mathbb{R}$, it will follow from the last relation above that $f(x) = \frac{1}{2}$ for every $x \in \mathbb{R}$. To what is left to do, it suffices to make $y = x$ in the given relation, to obtain

$$f(2x) = f(x)f(a - x) + f(x)f(a - x) = 2f(x)^2$$

and, hence, $f(x) = 2f\left(\frac{x}{2}\right)^2 \geq 0$, for every $x \in \mathbb{R}$.

12. Set $x = y = 0$ to get $f(0) = 1$. Then, use induction to prove that $f(nx) = f(x)^n$, for all $x \in \mathbb{Q}$ and $n \in \mathbb{N}$; therefore, use this relation to show that $f(x) = f(1)^x$, for every $x \in \mathbb{Q}_+^*$. Now, letting $y = -x$, show that $f(-x) = f(x)^{-1}$ and, hence, $f(x) = f(1)^x$ for every $x \in \mathbb{Q}$. Finally, use the fact that the codomain of f is the set of positive rationals to conclude that $f(1) = 1$.

13. Making $x + y = a$ and $x - y = b$, show that $f(a) + f(b) = 2f\left(\frac{a+b}{2}\right)$, for all $a, b \in [0, 1]$. Setting $x = y$, show that $f(2x) = 2f(x)$ for every $x \in [0, 1]$, and conclude that $f(a) + f(b) = f(a + b)$ for all $a, b \in [0, 1]$. From this point, show that $f(x) = x$ for every $x \in [0, 1] \cap \mathbb{Q}$. To what is left to do, reason as in the passage from $\mathbb{Q}$ to $\mathbb{R}$ in Example 6.56.

14. For item (a), start by making $m = 0$ in the given relation to get $f(f(n)) = f(0)^2 + n$; then, conclude that f is a bijection. Take $k \in \mathbb{Z}$ for which $f(k) = 0$, and let $l = f(0)$. We then have $l = f(0) = f(f(k)) = f(0)^2 + k = l^2 + k$, whereas, letting $m = k$ and $n = 0$ in the stated relation, we get $k^2 + l = k$; hence, $k = l = 0$. In what concerns (b), make $m = 1$ and $n = 0$ in the given relation to obtain $f(1) = 1$; then, deduce that $f(f(n) + 1) = n + 1$ and, hence, that $f(n) = n$ for every nonnegative integer n. Finally, extend the arguments above to the negative integers, thus showing that $f(n) = n$ for every $n \in \mathbb{Z}$.

15. Let $x = y = 1$ to conclude that $f(1) = 1$. Then, successively show that $f(f(xy)) = x^2 y^2 f(xy) = f(f(x)f(y))$, $f(xy) = f(x)f(y)$ and $f(x^2) = f(x)^2$, for every $x, y \in \mathbb{N}$. Now, if $f(x) < x^2$ for some $x \in \mathbb{N}$, write $f(x)^3 = f(x^3) > f(xf(x)) = x^2 f(x)^2$ to arrive at $f(x) > x^2$, which is a contradiction; analogously, show that one cannot have $f(x) > x^2$ for some $x \in \mathbb{N}$, so that $f(x) = x^2$ for every $x \in \mathbb{N}$.

16. Write $x - 2$ in place of x e 2 in place of y in (a) to conclude that $f(x) = 0$ for $x \geq 2$. Now, let $x = y = 0$ in (a) to get, from (b), that $f(0) = 1$. Finally, for $0 < x < 2$, write $2 - x$ in place of y in (a) to obtain $f(x) \geq \frac{2}{2-x}$; on the other hand, show that $f\left(x + \frac{2}{f(x)}\right) = f(2)f(x) = 0$ and, hence, that $f(x) \leq \frac{2}{2-x}$ for each one of those values of x.

17. Firstly, notice from (b) that f has at most one fixed point in each one of the intervals $(-1, 0)$ and $(0, +\infty)$. If there exists a fixed point x_0 of f such that $-1 < x_0 < 0$, let $x = y = x_0$ in (a) to conclude that $x_0^2 + 2x_0$ is another fixed point of f and, hence, that $x_0^2 + 2x_0 = x_0$, which is a contradiction. Argue analogously to conclude that f has no fixed points in the interval $(0, +\infty)$. Finally, set $x = y$ in (a) to conclude that $f(x) = -\frac{x}{x+1}$ for every $x \in S$.

18. Start by observing that, if such an f there exist, then $f(F_k) = F_{k+1}$ for every $k \in \mathbb{N}$, where $(F_n)_{n \geq 1}$ is the Fibonacci sequence. Then, use Zeckendorf's theorem (cf. Problem 30, page 98) to show that one possibility for f is to have

$$f(F_{i_1} + F_{i_2} + \cdots + F_{i_j}) = F_{i_1+1} + F_{i_2+1} + \cdots + F_{i_j+1},$$

for natural numbers j and $1 \leq i_1 < i_2 < \cdots < i_j$.

19. If f is such a function, set $y = 0$ in the given relation to get $f(z) = (c + 1)z$ for every $z \in \text{Im}(f)$, where $c = f(0)$. Then, use this fact to deduce that $cf(x+y) = f(x)f(y) - xy$ for all $x, y \in \mathbb{R}$. In order to show that $c = 0$, suppose that $c \neq 0$ and conclude that $0 \notin \text{Im}(f)$. Now, make $x = c$ and $y = -c$ in the last relation above to obtain $f(c)f(-c) = 0$ and, then, arrive at a contradiction. Therefore, we have got $f(x)f(y) = xy$, for all $x, y \in \mathbb{R}$. In order to finish, take $y_0 \in \text{Im}(f)$ for which $f(y_0) \neq 0$; then, $f(y_0) = (c + 1)y_0 = y_0$ and, hence, $f(x)y_0 = xy_0$, for every $x \in \mathbb{R}$.

20. From (a), notice that $f(x + k) = f(x) + k$, for every $k \in \mathbb{N}$. Now, take $m, n \in \mathbb{N}$ and use the fact that $\frac{(n^2+m)^3 - n^3}{n^3} \in \mathbb{N}$ to compute

$$f\left(\frac{m^3}{n^3} + \frac{(n^3 + m)^3 - n^3}{n^3}\right)$$

in two different ways, thus obtaining the equality

$$(a + n^2)^3 = a^3 + (n^6 + 3n^3 m + 3m^2),$$

where $a = f\left(\frac{m}{n}\right)$. Then, conclude that $f(x) = x$, for every $x \in \mathbb{Q}_+^*$.

Section 6.6

1. For the second part, show that the x-coordinates of the intersection points of the graphs are the solutions of the equation $f(x) = x$. In our case, the only such solution is $x = \frac{5+\sqrt{5}}{2}$.

2. If $|f(x)| \leq M$ for every $x \in I$ and $(x_0, y_0) \in G_f$, then $|y_0| = |f(x_0)| \leq M$, so that $-M \leq y_0 \leq M$. Hence, (x_0, y_0) belongs to the horizontal strip of the cartesian plane bounded by the parallel lines $y = -M$ and $y = M$.

3. Since the bisector of the odd quadrants of the cartesian plane is the set of points (x, y) for which $x = y$, we conclude that $x_0 \in I$ is a fixed point of f if and only if $f(x_0) = x_0$, i.e., if and only if $(x_0, x_0) \in G_f$. Therefore, the fixed points of f are exactly the abscissas of the points where the graph of f intersects that bisector. In Fig. B.1, if f is increasing for $x < x_1$ and decreasing for $x > x_3$, its fixed points are x_1, x_2 and x_3, which, in turn, are the abscissas of the points A, B and C, respectively.

4. Such a point (x, y) satisfies $x \in I$ and $y = f(x) = g(x)$. Therefore, it suffices to solve, for $x \in I$, the equation $f(x) = g(x)$.

5. Let's prove item (a), the proof of (b) being totally analogous. If f is even, then $f(x) = f(-x)$ for every $x \in I$. Hence, for $x \in I$, we have

$$(x, y) \in G_f \Leftrightarrow y = f(x) \Leftrightarrow y = f(-x) \Leftrightarrow (-x, y) \in G_f,$$

However, since the points (x, y) and $(-x, y)$ are symmetric with respect to the vertical axis, we conclude that the same is true of G_f.

6. Functions f_2 and f_4 are always nonnegative and vanish only at $x = 0$. Also, it is clear that they are even functions, so that (by the result of the previous problem) their graphs are symmetric with respect to the vertical axis of the cartesian system. On the other hand, as $|x|$ increases to $+\infty$, it's evident that

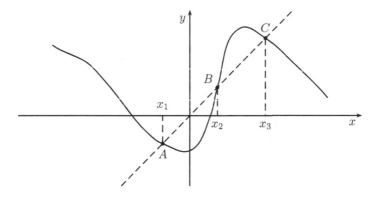

Fig. B.1 Fixed points of a function f

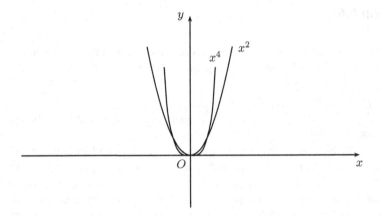

Fig. B.2 Graphs of f_2 and f_4

the values of f_2 and f_4 become bigger and bigger, eventually surpassing any predefined value. Finally,

$$|x| < 1 \Rightarrow x^4 < x^2, \quad |x| > 1 \Rightarrow x^4 > x^2 \text{ and } |x| = 1 \Rightarrow x^4 = x^2 = 1,$$

justifying the fact that the graph of f_4 is situated below (resp. above) the graph of f_2 in the interval $(-1, 1)$ (resp. outside the interval $[-1, 1]$). See Fig. B.2.

Functions f_3 and f_5 are positive for $x > 0$, negative for $x < 0$ and vanish at $x = 0$. Also, since they are odd functions, their graphs are symmetric with respect to the origin of the cartesian plane. Now, as $|x|$ increases, it is clear that the values of $|f_3|$ and $|f_5|$ become bigger and bigger, eventually surpassing any predefined value. Finally,

$$|x| < 1 \Rightarrow |x^5| < |x^3|, \quad |x| > 1 \Rightarrow |x^5| > |x^3| \text{ and } |x| = 1 \Rightarrow |x^5| = |x^3|,$$

which justifies the fact that, in the interval $(-1, 1)$ (resp. outside the interval $[-1, 1]$), the graph of f_5 is closer to (resp. further away from) the horizontal axis than the graph of f_3. See Fig. B.3.

7. Note that f is the inverse of the function $g : \mathbb{R} \to \mathbb{R}$ such that $g(x) = x^3$, for every $x \in \mathbb{R}$. Then, apply the result of Proposition 6.63.

8. Verify that the graph of the integer part function is the union of the sets $[n, n + 1) \times \{n\}$, when n varies in $\mathbb{Z}$.

9. For item (a), start by observing that, for $x \in [0, p)$, the periodicity of f guarantees that $f(x + kp) = f(x)$, for every $k \in \mathbb{Z}$. Hence, $(x, y) \in G_f \Leftrightarrow y = f(x) \Leftrightarrow y = f(x + kp) \Leftrightarrow (x + kp, y) \in G_f$. Then, letting $\mathcal{F} = G_f \cap ([0, p) \times \mathbb{R})$, we have $G_f = \bigcup_{k \in \mathbb{Z}} (\mathcal{F} + k)$, where $\mathcal{F} + k$ denotes the translation of $\mathcal{F}$, of k units, in the direction of the vertical axis. For item (b), note that $\{x\} = x$ for $x \in [0, 1)$. Since $\{\cdot\}$ is periodic of period 1, it follows from (a) that Fig. B.4 sketches the graph of $\{\cdot\}$.

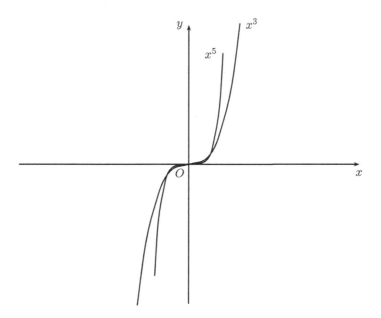

Fig. B.3 Graphs of f_3 and f_5

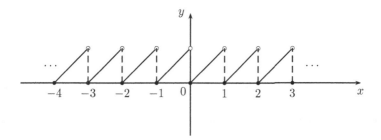

Fig. B.4 Graph of $x \mapsto \{x\}$

10. For the item (a), it suffices to show that (x, y) is on the graph of f if and only if $(x - a, y)$ is on the graph of g. Indeed, if (x, y) belongs to the graph of f, then $f(x) = y$ and, hence, $g(x - a) = f((x - a) + a) = y$, so that $(x - a, y)$ belongs to the graph of g; moreover, the converse statement can be established in an analogous way. For item (b), it suffices to show that (x, y) is on the graph of f if and only if $(x, y + a)$ is on the graph of g. Actually, if (x, y) belongs to the graph of f, then $f(x) = y$ and, hence, $g(x) = f(x) + a = y + a$, so that $(x, y + a)$ belongs to the graph of g; as in the previous case, the converse statement can be proved analogously. The other items can be similarly dealt with.

11. At a first glance, it may seem that we cannot use the results of the previous problem, for, the domain of the function under scrutiny is not the set $\mathbb{R}$ of real numbers and its expression doesn't match any of the ones considered there. Nevertheless, since

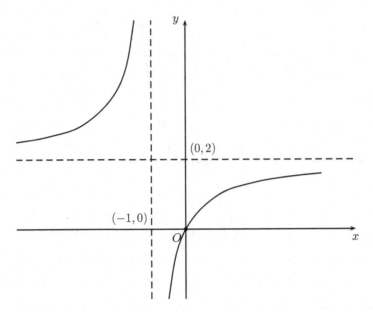

Fig. B.5 Graph of $x \mapsto \frac{2x}{x+1}$

$$\frac{2x}{x+1} = \frac{2x+2-2}{x+1} = 2 - \frac{2}{x+1},$$

we can reason in a way analogous to that of the previous problem, thus sketching the graph of f in the following way: firstly, we sketch the graph of $x \mapsto \frac{1}{x}$; then, we translate it one unit to the left, thus obtaining the graph of $x \mapsto \frac{1}{x+1}$; secondly, we stretch the previous graph in the vertical direction, by a factor 2, thus obtaining the graph of $x \mapsto \frac{2}{x+1}$; reflecting the result along the horizontal axis, we obtain the graph of $x \mapsto -\frac{2}{x+1}$; finally, if we vertically translate the reflected graph two units above, we obtain the graph $x \mapsto \frac{2x}{x+1}$. The final result is shown in Fig. B.5.

12. For the first part, show that, with the graph of f at our disposal, we can get the graph of g by *reflecting*, along the horizontal axis, the portion of the graph of f situated below such axis. For items (a) and (c), apply this procedure to the graphs of the functions $f(x) = x^2 - 4, f(x) = \frac{1}{x+1}$ and $f(x) = x^2 - |x+2| + 2$, respectively. Also notice that, according to Problem 10, the graph of $f(x) = \frac{1}{x+1}$ can be obtained by translating the graph of $x \mapsto \frac{1}{x}$ one unit to the left. In what concerns the graph of $f(x) = x^2 - |x+2| + 2$, take a separate look at the cases $x \leq -2$ and $x > -2$, observing that $|x+2|$ is respectively equal to $-x-2$ and $x+2$.

13. Apply a (clockwise) rotation of $\frac{\pi}{4}$ radians to the original cartesian system (this is equivalent to a counterclockwise rotation, of $\frac{\pi}{4}$ radians, of the hyperbola of

equation $x^2 - y^2 = 4$). By standard analytic geometry (see, for instance, Chapter 6 of [4]), we obtain the curve of equation $\left(\frac{x-y}{\sqrt{2}}\right)^2 - \left(\frac{x+y}{\sqrt{2}}\right)^2 = 2\sqrt{2}$ or, which is the same, $xy = 1$.

Section 6.7

1. For the first part of item (a), let $A = (1,0)$ and P be the point on the trigonometric circle such that $\overset{\frown}{AP} = x$ (of course, in the counterclockwise direction). Since $\sin x$ is the ordinate of P, if $x_1, x_2 \in \left[-\frac{\pi}{2}, \frac{\pi}{2}\right]$, with $x_1 < x_2$, we clearly have $\sin x_1 < \sin x_2$, so that the restriction of the sine function to the interval $\left[-\frac{\pi}{2}, \frac{\pi}{2}\right]$ is increasing and, as such, injective. Since this restriction has image $[-1, 1]$, the arcsin function is well defined and, by Problem 8, page 176, is also increasing. The first part of item (b) can be dealt with in a similar way, observing that $\cos x$ is the abscissa of P. Finally, the first part of item (c) follows from the fact that, for $x \in \left(-\frac{\pi}{2}, \frac{\pi}{2}\right)$, the real number $\tan x$ is the ordinate of the intersection point of the half-line $\overrightarrow{OA}$ with the vertical line passing through A.

2. For item (a), observe that $\cot x = -\tan\left(x + \frac{\pi}{2}\right)$ and use the results of Problema 10, page 194. For items (b) to (d), use the result of Proposition 6.63.

3. For items (a) and (b), we refer the reader to the discussion of Example 6.68, more precisely to Eq. (6.15); as for item (c), we suggest the reader to recall items (a) and (e) of Problem 10, page 194.

4. Argue as in Example 6.68 to get $f(x) = \sqrt{a^2 + b^2} \cos(\lambda x - \alpha)$, where $\cos \alpha = \frac{a}{\sqrt{a^2+b^2}}$ and $\sin \alpha = \frac{b}{\sqrt{a^2+b^2}}$. Now, show that f ha period $\frac{2\pi}{|\lambda|}$.

5. For item (a), start by using trigonometric identities to get $f(x) = -2\sin^2 x + 2\sin x + 1$. Then, observe that the quadratic function $y \mapsto -2y^2 + 2y + 1$ attains its maximum value at $y = \frac{1}{2}$, increases on $\left[-1, \frac{1}{2}\right]$ and decreases on $\left[\frac{1}{2}, 1\right]$. Item (b) follows from the expression obtained above for $f(x)$, together with the fact that the sine function has period 2π. Finally, for item (c), sketch the graphs of $x \mapsto 2\sin x$ and $x \mapsto \cos(2x)$ on the interval $[0, 2\pi]$ and in a single cartesian coordinate system; then, add corresponding ordinates of the points of these two graphs to get the desired sketch of the graph of f.

6. Use the result of Problem 4, page 193.

7. Perform the change of variable $x = \sqrt{5}\cos\theta$, with $0 \le \theta \le \pi$. Then, use the discussion of Example 6.68 to conclude that $f(x) = 25\cos(\theta - \alpha)$, with α satisfying $\cos \alpha = \frac{3}{5}$ and $\sin \alpha = \frac{4}{5}$.

8. Suppose, by the sake of contradiction, that f is periodic, of period $\tau > 0$. Then, we must have $f(\tau) = f(0) = 2$. Use this equality, together with the fact that the cosine of every real number is at most 1, to get $\cos \tau = \cos(\alpha\tau) = 1$. Conclude that there must exist (nonzero) integers k and l for which $\tau = 2k\pi$ and $\alpha\tau = 2l\pi$, and arrive at a contradiction.

9. Initially, show that the equality $f(x + 3\pi) = f(x)$ implies the equality

$$(-1)^n \sin\left(\frac{5x}{n} + \frac{15\pi}{n}\right) = \sin\left(\frac{5x}{n}\right),$$

whenever $nx \neq \frac{\pi}{2} + k\pi$, for every $k \in \mathbb{Z}$. Then, if n is odd, for instance, show that the last equality implies $\sin\left(\frac{5x}{n} + \frac{15\pi}{n}\right) \cos\left(\frac{15\pi}{2n}\right) = 0$, for every $x \in \mathbb{R}$ such that $nx \neq \frac{\pi}{2} + k\pi$, for every $k \in \mathbb{Z}$. Finally, conclude that n divides 15.

10. Suppose that the given function is periodic of period $p > 0$. Compute $f(x+p)$ with the aid of the standard trigonometric identities for the sine of the sum of two arcs and, then, take a careful look at the equality $f(x+p) = f(x)$.

Section 7.1

1. For item (a), observe that $\left(\frac{1}{\sqrt[n]{1/x}}\right)^n = \frac{1}{1/x} = x$. For (b), argue in an analogous way. Item (c) now follows immediately from the results of Problems 6 and 8, page 176. Finally, for item (d), Problem 6, page 176, gives $\sqrt[m]{a} > \sqrt[n]{a}$ if and only if $a^n > a^m$; the rest follows from Corollary 1.3.

2. For item (a) suppose, by the sake of contradiction, that $a > 0$. Then, by the Archimedian property of $\mathbb{N}$, we can choose $n \in \mathbb{N}$ such that $n > \frac{1}{a}$. This is equivalent to $a < \frac{1}{n}$, which is an absurd. For item (b), choose $n \in \mathbb{N}$ such that $n > \frac{c-b}{a}$.

3. For item (b), let $r_1, \ldots, r_n \in (a, b) \cap \mathbb{Q}$, with $r_1, \ldots, r_n$ pairwise distinct. If $r = \min\{r_1, \ldots, r_n\}$, then $r \in (a, b) \cap \mathbb{Q}$, and item (a) gives $a < \frac{a+r}{2} < r$. Finally, argue similarly, using the second part of (a), to get infinitely many irrational numbers in the interval (a, b).

4. For item (a), suppose that we already have the truth of the result for $a \geq 0$. If $b \leq 0$, then $-b \geq 0$ and, by item (a), there existe $r \in \mathbb{Q}$ and $\alpha \in \mathbb{R} \setminus \mathbb{Q}$ such that $r, \alpha \in (-b, -a)$. Hence, $-r, -\alpha \in (a, b)$, with $-r \in \mathbb{Q}$ and $-\alpha \in \mathbb{R} \setminus \mathbb{Q}$. If $a \leq 0 < b$, show that it suffices to apply, to the interval $(0, b)$, the result we assumed to be known. For item (b), use the Archimedian property of natural numbers to get $n \in \mathbb{N}$ such that $n > \frac{\sqrt{2}}{b-a}$. Finally, in what concerns (c), starts by using the Archimedian property to guarantee the existence of $m \in \mathbb{N}$ for which $\frac{m}{n} > b$ (resp. $\frac{m\sqrt{2}}{n} > b$); continue by showing that, if m is the least possible natural number fullfilling this requirement, then $m > 1$ and $\frac{(m-1)\sqrt{2}}{n} \in (a, b)$.

5. Adapt the proof of items (b) and (c) of the previous problem to the present case.

6. Use the result of Example 7.2, together with Theorem 7.4.

7. Use the result of item (b) of Problem 4.

8. For $a = k^2$ and $b = k^2 + 1$, we have $|\sqrt{a} - \sqrt{b}| = \frac{|a-b|}{\sqrt{a}+\sqrt{b}} = \frac{1}{\sqrt{k^2+1}+\sqrt{k^2}} < \frac{1}{2k}$. Since every element of X is positive, it follows that $\inf(X) = 0$.

9. Observe that, for $k \in \mathbb{N}$, every number of each of the forms $\frac{1}{3^k}$ and $1 - \frac{1}{3^k}$ belong to C. Now, use the fact that $\frac{1}{3^k} < \frac{1}{k}$, for every $k \in \mathbb{N}$.

10. Adapt, to the present case, the proof of Proposition 7.6.
11. Use the result of Proposition 7.6, together with the result of Problem 10, to establish the existence of $x_n \in X$ and $y_n \in Y$ such that $\alpha - \frac{1}{2n} < x_n < \alpha$ and $\alpha < y_n < \alpha + \frac{1}{2n}$.
12. For item (a), show that, if α (resp. β) is an upper bound for X (resp. for cX), then $c\alpha$ (resp. $\frac{\beta}{c}$) is an upper bound for cX (resp. for X). The other cases can be dealt with similarly.
13. For item (a), start by showing that, if α and β are upper bounds for X and Y, respectively, then $\alpha + \beta$ is an upper bound for $X + Y$; then, conclude that $\sup(X + Y) \leq \sup X + \sup Y$. Show next that, if $\sup(X + Y) - \sup X < \sup Y$, then Y would have an upper bound less than $\sup Y$.
15. The given conditions imply that R_n has vertices $(0,0)$, $(a_n, 0)$, $(0, b_n)$ and (a_n, b_n), for some integers a_n and b_n. Passing to a subsequence, if necessary, we can assume without loss of generality that $a_n, b_n > 0$. Show that one can assume, also without loss of generality, that $a_1 \leq a_2 \leq a_3 \leq \cdots$. Then, show that one cannot have $b_1 > b_2 > b_3 > \cdots$.
16. Since $|f(x)| \leq 1$ for every $x \in \mathbb{R}$, we can let $L = \sup\{|f(x)|;\ x \in \mathbb{R}\}$. Suppose that $\sup\{|g(x)|;\ x \in \mathbb{R}\} > 1$ and apply triangle inequality to the relation given in the statement of the problem to reach a contradiction.

Section 7.2

1. Show that, if $a > b$, then we would have $a_n > \frac{a+b}{2} > b_n$ for every sufficiently large n.
2. Write $|\sqrt{a_n} - \sqrt{a}| = \frac{|a_n - a|}{\sqrt{a_n} + \sqrt{a}} < \frac{1}{\sqrt{a}}|a_n - a|$; then, use the convergence of $(a_n)_{n \geq 1}$ to a.
3. For $n > 16a^2$, show that $\frac{|a|^n}{n!} < \frac{|a|^n}{(\frac{n}{2})^{\frac{n}{2}}} < \left(\frac{2|a|}{\sqrt{n}}\right)^n < \frac{1}{2^n}$.
4. Firstly, note that

$$\sqrt[n]{\frac{n^k}{|a|^n}} = \frac{(\sqrt[n]{n})^k}{|a|} \rightarrow \frac{1}{|a|} < 1$$

as $n \rightarrow +\infty$. Hence, fixed a real number α such that $\frac{1}{|a|} < \alpha < 1$, there exists $n_0 \in \mathbb{N}$ for which $n > n_0 \Rightarrow \sqrt[n]{\frac{n^k}{|a|^n}} < \alpha$, or $\frac{n^k}{|a|^n} < \alpha^n$. However, since $\alpha^n \rightarrow 0$ as $n \rightarrow +\infty$, the same happens with $\frac{n^k}{a^n}$. Alternatively, let $|a| = 1 + \alpha$, with $\alpha > 0$, and note that, for $n > k$,

$$|a|^n = (1 + \alpha)^n = \sum_{j=0}^{n} \binom{n}{j}\alpha^j > \binom{n}{k+1}\alpha^{k+1}$$

$$= \frac{n(n-1)\ldots(n-k)}{(k+1)!} \cdot \alpha^{k+1}.$$

Hence, for $n > k$, we have

$$\frac{n^k}{|a|^n} < \frac{(k+1)!}{\alpha^{k+1}} \cdot \frac{n^k}{n(n-1)\dots(n-k)}$$

$$= \frac{(k+1)!}{\alpha^{k+1}} \cdot \frac{1}{n(1-\frac{1}{n})\dots(1-\frac{k}{n})},$$

with the last expression above tending to 0 as $n \to +\infty$.

5. Write $l = (1 - t_n)l + t_n l$ and use the triangle inequality to get $|c_n - l| \le (1 - t_n)|a_n - l| + t_n|b_n - l|$. Then, show that $|a_n - l|, |b_n - l| < \epsilon$ implies $|c_n - l| < \epsilon$.

6. First of all, note that $|b_n - l| \le \max\{|a_n - l|, |c_n - l|\}$, for every $n \in \mathbb{N}$. Hence, given $\epsilon > 0$ and taking $n_0 \in \mathbb{N}$ such that $n > n_0 \Rightarrow |a_n - l|, |c_n - l| < \epsilon$, we have $|b_n - l| < \epsilon$ for $n > n_0$, so that $b_n \xrightarrow{n} l$.

7. For item (a), write $\frac{n\sqrt{n}}{n^2+1} = \frac{1}{\sqrt{n}+\frac{1}{n\sqrt{n}}} < \frac{1}{\sqrt{n}}$. For (b), write

$$\sqrt{n^2 + an + b} - n = \frac{an + b}{\sqrt{n^2 + an + b} + n} = \frac{a + \frac{b}{n}}{\sqrt{1 + \frac{a}{n} + \frac{b}{n^2}} + 1}.$$

For item (c), observe that, if $a_n = \sqrt[n]{1 + q^n}$, then $a_n > 1$ and, by item (a) of Problem 18, page 26,

$$0 < a_n - 1 = \frac{a_n^n - 1}{a_n^{n-1} + a_n^{n-2} + \dots + a_n + 1} < \frac{q^n}{n}.$$

Finally, for (d), write $\sqrt[n]{a^n + b^n} = a\sqrt[n]{1 + (\frac{b}{a})^n}$ and use the result of the previous item.

8. Adapt, to the present case, the reasoning that led to the proof of Proposition 7.18.

9. For the first part, write $q = 1 + \alpha$, with $\alpha > 0$, and note that $a_n = (1 + \alpha)^n \ge 1 + n\alpha$, so that $a_n > M$ for $n > \frac{M-1}{\alpha}$.

10. Let $b = \frac{1}{q}$, so that $b > 1$. Use the first part of the previous problem to get $k \in \mathbb{N}$ such that $b^k > a$. Then, show that $2^n > kn$ for every sufficiently large n and, from this, conclude that $a^n q^{2^n} = (\frac{a}{b^k})^n \cdot \frac{1}{b^{2^n - kn}} \to 0$ when $n \to +\infty$.

11. First, show that $|a_m - a_n| \le \frac{2mk}{m^2+k^2} + \frac{2nk}{n^2+k^2}$; then, make $k \to +\infty$.

12. Observe that $a_{n+1} = \sqrt{1 + a_n}$, for every $n \ge 1$. Then, successively conclude that $a_{n+1}^2 - a_n^2 = a_n - a_{n-1}$ and that $(a_n)_{n\ge 1}$ is increasing. Next, note that $a_{n+1}^2 = 1 + a_n$ implies $a_{n+1} = \frac{1 + a_n}{a_{n+1}} < \frac{1 + a_{n+1}}{a_{n+1}} < 2$, and apply Bolzano-Weierstrass theorem to guarantee the existence of $l = \lim_{n\to+\infty} a_n$. Now, make $n \to +\infty$ in the recurrence relation $a_{n+1} = \sqrt{1 + a_n}$ and use the result of item (a) of Problem 2, together with item (b) of Proposition 7.18, to get the equation $l = \sqrt{1 + l}$.

13. Setting $\alpha = \min\{a_1, a_2\}$ and $\beta = \max\{a_1, a_2\}$, conclude that $(a_n)_{n\geq 1}$ is bounded (to this end, separately consider the cases $0 < \alpha \leq \beta \leq 4, 4 \leq \alpha \leq \beta$ and $0 < \alpha \leq 4 \leq \beta$). Now, if $a_1 \leq a_2 \leq a_3$ or $a_1 \geq a_2 \geq a_3$, write $a_{k+2} - a_{k+1} = \sqrt{a_{k+1}} - \sqrt{a_{k-1}}$ to conclude that the given sequence is monotonic, thus convergent; make $k \to +\infty$ in the recurrence relation satisfied by the sequence to conclude that $a_k \xrightarrow{k} 4$. If $a_1 \leq a_3 \leq a_2$ or $a_1 \geq a_3 \geq a_2$, conclude, as above, that the subsequences of the terms of even and odd indices are convergent, say, $a_{2k-1} \xrightarrow{k} c$ and $a_{2k} \xrightarrow{k} d$; then, make $k \to +\infty$ in the given recurrence to conclude that $d = \sqrt{c} + \sqrt{d}$ and $c = \sqrt{c} + \sqrt{d}$, so that $c = d = 4$.

14. Use the fact that $t_n = -t_0 - t_1 - \cdots - t_{n-1} = 0$ to write

$$a_k = t_0(\sqrt{k} - \sqrt{k+n}) + t_1(\sqrt{k+1} - \sqrt{k+n}) + \cdots$$
$$+ t_{n-1}(\sqrt{k+n-1} - \sqrt{k+n})$$

$$= -\frac{t_0}{\sqrt{k} + \sqrt{k+n}} - \frac{t_1}{\sqrt{k+1} + \sqrt{k+n}} - \cdots$$
$$- \frac{t_{n-1}}{\sqrt{k+n-1} + \sqrt{k+n}}.$$

Then, make $k \to +\infty$.

15. Perform the trigonometric substitution $x_n = 2\cos y_n$, with $y_n \in [0, \frac{\pi}{2}]$, and use some Trigonometry to conclude that $y_{n+1} \geq 2y_n$, for every $n \geq 1$. Then, show that $y_n \leq \frac{y_{n+2k}}{2^k} \leq \frac{\pi}{2^{k+1}}$ for every $k \geq 1$, so that $y_n = 0$ for every $n \geq 1$.

16. First of all, use the given condition to show that $|(a_{n+1}-a_n)-(a_{k+1}-a_k)| < \frac{2}{n}$. Then, make $n \to +\infty$ to conclude that there exists $l = \lim_{n\to+\infty}(a_{n+1} - a_n)$, and show that $a_{k+1} - a_k = l$, for every $k \in \mathbb{N}$.

17. Use the definition of a_n to conclude that $a_{k-1} + 1 = \frac{2k}{k+1}a_k$, for every integer $k \geq 2$. Then, show that

$$a_{k-1} - a_k < \frac{2(k+1)^2}{(k+1)(k+2)}(a_k - a_{k+1})$$

and conclude that the sequence $(a_n)_{n\geq 1}$ is decreasing for $n > 2$. Finally, make $k \to +\infty$ in the equality $a_{k-1} + 1 = \frac{2k}{k+1}a_k$ to conclude that $a_n \to 1$ when $n \to +\infty$.

18. Successively show that $a_n^2 = k + a_{n-1}$ and $a_{n+1}^2 - a_n^2 = a_n - a_{n-1}$; then, conclude that $(a_n)_{n\geq 1}$ increases. Now, use the fact that $a_n^2 < k + a_n$ to get $a_n < \frac{1}{2}(1 + \sqrt{4k+1})$ for every $n \geq 1$, so that the given sequence is convergent. Make $n \to +\infty$ in $a_n^2 = k + a_{n-1}$ to find out $a_n \xrightarrow{n} \frac{1}{2}(1 + \sqrt{4k+1})$. From here, items (b) and (c) are relatively easy.

19. First of all, use the inequality between the arithmetic and geometric means of two numbers to conclude that $a_{k+1} \geq \sqrt{a}$, for every $k \in \mathbb{N}$. Now, make $k \to +\infty$ in the given recurrence relation to conclude that, if $(a_n)_{n\geq 1}$ converges, then its limit equals $\sqrt{a}$. To what is left to do, use the triangle inequality to obtain

$$\begin{aligned}
|a_{k+1} - a_k| &= \frac{1}{2}\left|a_k - a_{k-1} + \frac{a}{a_k} - \frac{a}{a_{k-1}}\right| \\
&= \frac{1}{2}|a_k - a_{k-1}| \cdot \left|1 - \frac{a}{a_k a_{k-1}}\right| \\
&\leq \frac{1}{2}|a_k - a_{k-1}|,
\end{aligned}$$

for each $k > 2$. Finally, establish the convergence of $(a_n)_{n\geq 1}$ from the result of Example 7.28.

20. Let $a_1, a_2, \ldots, a_n$ be the initial terms of the AP's, and say that the i-th AP is that of initial term a_i and common ratio d_i. Choose $n \in \mathbb{N}$ greater than all of the a_i's, and let k_i be the number of terms of the i-th AP belonging to the set $\{1, 2, \ldots, n\}$. Then, on the one hand, $k_1 + k_2 + \cdots + k_m = n$; on the other, $a_i + (k_i - 1)d_i \leq n < a_i + k_i d_i$. Use these relations to show that

$$1 + \frac{S - m}{n} < \sum_{i=1}^{m} \frac{1}{d_i} < 1 + \frac{S}{n},$$

where $S = \left(\sum_{i=1}^{m} \frac{a_i}{d_i}\right)$. Then, make $n \to +\infty$.

21. Suppose such a function does exist, let $k > 1$ be an integer and $x_j = \frac{j}{k}$, for $0 \leq j \leq k$. Given a real number $x \in [x_j, x_{j+1}]$, show that $d(f(x), f(x_j)) \leq \frac{c}{k^{\alpha+1/2}}$. Letting C_j denote the circle in the plane with center $f(x_j)$ and radius $\frac{c}{k^{\alpha+1/2}}$, conclude that $x \in [x_j, x_{j+1}] \Rightarrow f(x) \in C_j$. Then, use this fact to show that $1 \leq \sum_{j=0}^{k-1} A(C_j) = \frac{\pi c^2}{k^{2\alpha}}$. Finally, use the fact that $k > 1$ can be chosen arbitrarily and $\alpha > 0$ to reach a contradiction.

22. Start by proving that $a_{n+1} \geq a_n + 1$ and, hence, that $a_n \geq n$ for every $n \geq 1$. Now, use telescoping products to show that, for a fixed $p \in \mathbb{N}$, we have

$$1 < \frac{a_{n+p}}{a_n} < \prod_{j=0}^{p-1}\left(1 + \frac{1}{\sqrt{a_{n+j}}}\right)$$

and, then,

$$1 < \frac{a_{n+p}}{a_n} < \prod_{j=0}^{p-1} \frac{a_{n+j+1}}{a_{n+j}} < \prod_{j=0}^{p-1}\left(1 + \frac{1}{\sqrt{a_{n+j}}}\right).$$

Continue by showing that

$$\lim_{n \to +\infty} \left(1 + \frac{1}{\sqrt{a_n}}\right)\left(1 + \frac{1}{\sqrt{a_{n+1}}}\right) \cdots \left(1 + \frac{1}{\sqrt{a_{n+p-1}}}\right) = 1.$$

Now, if $y = 1$, conclude that what is asked to be proved follows from the limit above. If $y < 1$, conclude from the limit above that we can choose $n_p \in \mathbb{N}$ such that $n > n_p \Rightarrow \frac{a_n}{a_{n+p}} > y$. Use the fact that $a_k \geq k$ for every k to show that we can choose a natural number n_0 such that $\frac{1}{1+\sqrt{a_{n_0}+1}} < y - x$. Then, note that $n > n_0$ and p natural furnish

$$\frac{a_n}{a_{n+p}} - \frac{a_n}{a_{n+p} + \sqrt{a_{n+p}}} < y - x.$$

Fix $k > n_0, n_p$, so that $\frac{a_k}{a_{k+p}} > y$; establish the existence of a natural number $p_0 > p$ for which

$$\frac{a_k}{a_{k+p_0}} \geq y > \frac{a_k}{a_{k+p_0+1}}.$$

Finally, suppose that $\frac{a_k}{a_{k+p_0+1}} \leq x$ and arrive at a contradiction.

22. Let $c_0, c_1, c_2, \ldots$ be defined by $c_0 = 1$ and

$$c_n = \left(\frac{a_{n-1}}{1 + a_n}\right) c_{n-1}, \quad \forall \, n \geq 1.$$

Rewriting such a relation as $c_n = a_{n-1}c_{n-1} - a_n c_n$, we get the telescoping sum

$$c_1 + c_2 + \cdots + c_n = a_0 - a_n c_n. \tag{B.3}$$

On the other hand, the assertion of the problem is equivalent to $\frac{c_n}{c_{n-1}} < 2^{-1/n}$ for infinitely many values of $n \in \mathbb{N}$. Arguing by contradiction, suppose that there exists a natural N such that the opposite inequality is true for every $n \geq N$. Then, for $n > N$ we have

$$c_n \geq c_N \cdot 2^{-\left(\frac{1}{N+1} + \frac{1}{N+2} + \cdots + \frac{1}{n}\right)} = C \cdot 2^{-\left(1 + \frac{1}{2} + \cdots + \frac{1}{n}\right)},$$

where $C = c_N \cdot 2^{\left(1 + \frac{1}{2} + \cdots + \frac{1}{n}\right)}$ is a positive constant. If $2^{k-1} \leq n < 2^k$, then

$$\sum_{j=1}^{n} \frac{1}{j} \leq 1 + \left(\frac{1}{2} + \frac{1}{3}\right) + \left(\frac{1}{4} + \cdots + \frac{1}{7}\right) + \cdots \left(\frac{1}{2^{k-1}} + \cdots + \frac{1}{2^k - 1}\right)$$

$$\leq 1 + 1 + 1 + \cdots + 1 = k,$$

so that $c_n \geq C \cdot 2^{-k}$ for $2^{k-1} \leq n < 2^k$. Let $r \in \mathbb{N}$ be such that $2^{r-1} \leq N < 2^r$, and let $m > r$. Then,

$$
\begin{aligned}
c_{2^r} + c_{2^r+1} + \cdots + c_{2^m-1} &= \\
&= (c_{2^r} + \cdots + c_{2^{r+1}-1}) + (c_{2^{r+1}} + \cdots + c_{2^{r+2}-1}) \\
&\quad + \cdots + (c_{2^{m-1}} + \cdots + c_{2^m-1}) \\
&\geq C \cdot (2^r \cdot 2^{-r-1} + 2^{r+1} \cdot 2^{-r-2} + \cdots + 2^{m-1} \cdot 2^{-m}) \\
&= \frac{C \cdot (m-r)}{2},
\end{aligned}
$$

thus showing that the sum of the c_n's can be taken arbitrarily large. However, (B.3) guarantees that this sum cannot exceed a_0. This contradiction finishes the proof.

Section 7.3

2. For (a), note that there are more numbers of the form $\{j\alpha\}$ than intervals of the form $\left[\frac{k}{p}, \frac{k+1}{p}\right)$. For (b), take $1 \leq s < t \leq p+1$ such that $\{s\alpha\}, \{t\alpha\} \in \left[\frac{k}{p}, \frac{k+1}{p}\right)$, write $\{s\alpha\} = s\alpha - \lfloor s\alpha \rfloor, \{t\alpha\} = t\alpha - \lfloor t\alpha \rfloor$ and observe that $0 < |\{s\alpha\} - \{t\alpha\}| < \frac{1}{p}$.
3. Use Corollary 7.33 to inductively construct a sequence $(a_k)_{k\geq 1}$ such that $a_k = m_k + n_k\alpha$, for integers m_k and n_k satisfying item (a), and such that $|a_k - l| < \frac{1}{k}$, for every $k \geq 1$.
4. For item (c), choose A and B in π such that $O \notin \overleftrightarrow{AB}$, $\alpha \in \mathbb{R} \setminus \mathbb{Q}$ and make

$$
X = \{(m_1 + n_1\alpha)\overrightarrow{OA} + (m_2 + n_2\alpha)\overrightarrow{OB}; \ m_1, m_2, n_1, n_2 \in \mathbb{Z}\}.
$$

For item iii., conclude that there exist distinct lines r and s through O, such that $X \cap r, X \cap s \neq \emptyset$ and $X \cap r$ is not dense in r. Show that one can reconstruct X from $X \cap r$ and $X \cap s$, apply Kronecker's lemma and make $d = s$.
5. For item (b), write $n = \frac{n}{\alpha} + \frac{n}{\beta}$ and take integer and fractionary parts. For (c), argue by contradiction, using the fact that α and β are irrational to show that one cannot have either $\left\{\frac{n}{\alpha}\right\} = 1 - \frac{1}{\alpha}$ or $\left\{\frac{n}{\beta}\right\} = 1 - \frac{1}{\beta}$.

Section 7.4

1. Use que fact that $\frac{1}{a_{n+1}} = \frac{1}{a_n} - \frac{1}{a_{n+1}}$, for every $n \in \mathbb{N}$.
2. Use the result of Problem 5, page 84, to show that $\sum_{k=1}^{n-1} \frac{1}{a_k a_{k+1}} = \frac{n-1}{a_1 a_n}$, for every integer $n \geq 1$. Then, make $n \to +\infty$, noting that $a_n = a_1 + (n-1)r$.

3. Make $S_n = \sum_{k=1}^{n} \frac{2k-1}{a^k}$ and show that

$$(a-1)S_n = aS_n - S_n = a + 2\left(\frac{1}{a} + \frac{1}{a^2} + \cdots + \frac{1}{a^{n-1}}\right) - \frac{2n-1}{a^n}$$

$$= a + \frac{2a}{a-1}\left(\frac{1}{a} - \frac{1}{a^n}\right) - \frac{2n-1}{a^n}.$$

Now, use the results of Example 7.12 and of Problem 4, page 218, to conclude that $S_n \xrightarrow{n} \frac{a}{a-1} + \frac{2}{(a-1)^2}$.

4. Note that $\sqrt{k} + \sqrt{k^2 - 1} < \sqrt{2k}$, so that $\sum_{k\geq 1} \frac{1}{\sqrt{k+\sqrt{k^2-1}}} > \sum_{k\geq 1} \frac{1}{\sqrt{2k}}$. Since $\frac{1}{\sqrt{n}} \geq \frac{1}{n}$ for $n \geq 1$, use the comparison test to show that the given series diverges.

5. Show that $n^3 - 1000n^2 > (n - 500)^3$ for every sufficiently large natural n. Then, use the comparison test, together with the convergence of the series $\sum_{k>500} \frac{1}{(k-500)^{3/2}}$.

6. Let $r > 0$ be the common ratio of the AP. For item (a), we have $\frac{1}{a_k} = \frac{1}{a_1+(k-1)r} > \frac{1}{2(k-1)r}$, provided $k > \frac{a_1}{r} + 1$. For item (b), we have

$$\frac{1}{a_{2^k}} = \frac{1}{a_1 + (2^k - 1)r} < \frac{1}{(2^k - 1)r} \leq \frac{1}{2^{k-1}r}.$$

Now, apply the comparison test.

7. Start by showing that

$$\sum_{x\in A} \frac{1}{x} < \left(\sum_{a=0}^{N} \frac{1}{2^a}\right)\left(\sum_{b=0}^{N} \frac{1}{3^b}\right)\left(\sum_{c=0}^{N} \frac{1}{5^c}\right),$$

for some natural number N. Then, make $N \to +\infty$ and use the formula for the sum of a geometric series.

9. Use the comparison test to show that the series $\sum_{j\geq 1} \frac{a_j}{10^j}$ converges. Now, letting x denote its sum, note that, for $k \geq n$, we have

$$x - \sum_{j=1}^{n} \frac{a_j}{10^j} = \sum_{j\geq n+1} \frac{a_j}{10^j} \leq \sum_{j\geq n+1} \frac{9}{10^j} = \frac{1}{10^n}.$$

10. For the first part, suppose that, for some $x \in (0, 1)$, we have $x = 0.a_1a_2a_3 \ldots = 0.b_1b_2b_3 \ldots$, with $a_n \neq 0$ for infinitely many values of n and, accordingly, $b_n \neq 0$ for infinitely many values of n. Then, $\frac{a_1}{10} < \sum_{k\geq 1} \frac{a_k}{10^k} = \sum_{k\geq 1} \frac{b_k}{10^k} \leq \frac{b_1}{10} + \sum_{k\geq 2} \frac{9}{10^k} = \frac{b_1+1}{10}$, so that $a_1 \leq b_1$. Analogously, show that $a_1 \geq b_1$, so that $a_1 = b_1$. Then, argue similarly to establish, by induction on n, that $a_n = b_n$, for every $n \geq 1$. In what concerns the second part, define $f(x) = (y, z)$, with

$y = 0.a_1a_3a_5 \ldots$ and $z = 0.a_2a_4a_6 \ldots$ if $x = 0.a_1a_2a_3 \ldots$, with $a_n \neq 0$ for infinitely many values of n. Then, use the result of the first part and the former problem to show that f is well defined and is a surjection.

11. Use the fact that $a_k^2 + \frac{1}{k^{2\alpha}} \geq \frac{2|a_k|}{k^\alpha}$, for every $k \in \mathbb{N}$; then, apply the result of Proposition 7.39, together with the comparison test and Proposition 7.48.

12. Use the fact that $\sqrt{a_k a_{k+1}} \leq \frac{1}{2}(a_k + a_{k+1})$ for every $k \in \mathbb{N}$, together with the comparison test.

13. Imposing that $F_j \geq \alpha^j$, for every $j \leq k+1$, deduce that $F_{k+2} \geq \alpha^{k+2}$ if $\alpha + 1 \geq \alpha^2$. Use this *a priori* estimate to show that $F_n \geq \alpha^n$ for every $n \geq 3$, where $\alpha = \min\{\sqrt[3]{2}, \sqrt[4]{3}, \frac{1+\sqrt{5}}{2}\}$. Finally, note that $\alpha > 1$ and apply the comparison test.

14. For item (a) (the proof of item (b) is analogous), let $q = \frac{l+1}{2} \in (0, 1)$ and take $n_0 \in \mathbb{N}$ such that $\sqrt[n]{a_n} < q$ for $n > n_0$ or, which is the same, $a_n < q^n$ para $n > n_0$. Then, adapt the reasoning used at the proof of the ratio test.

15. Apply Theorem 7.53.

16. Start by observing that the series $\sum_{k \geq 0} \frac{(-1)^k}{k!}$ is absolutely convergent. Then, let $a = \sum_{k \geq 0} \frac{(-1)^k}{k!}$ and use the fact that $e = \sum_{k \geq 0} \frac{1}{k!}$, together with Theorem 7.53, to get $ae = 1$.

17. For item (a), let $s_n = \sum_{k=1}^{n} a_k$ and $t_n = \sum_{k=1}^{n} a_{\varphi(k)}$, so that $s_1 \leq s_2 \leq s_3 \leq \xrightarrow{n} s$, where $s = \sup\{s_n; n \geq 1\}$, and analogously for $(t_n)_{n \geq 1}$. Given $n \in \mathbb{N}$, prove that there exists $m \in \mathbb{N}$ such that $s_n \leq t_m$; then, make $n \to +\infty$ to conclude that $s \leq t$. Change the roles of the two sequences to get the reverse inequality. For item (b), note that $|a_n| = a_n^+ + a_n^-$, so that $a_n^+ = \frac{1}{2}(|a_n| + a_n)$ and $a_n^- = \frac{1}{2}(|a_n| - a_n)$. For (c), apply the result of (b) to both $\sum_{k \geq 1} a_k^+$ and $\sum_{k \geq 1} a_k^-$.

18. By the definition of convergent series, it suffices to show that the sequence $(x_n)_{n \geq 1}$, given for $n \geq 1$ by $x_n = a_1 b_1 + a_2 b_2 + \cdots + a_n b_n$, is convergent. As we know, this is the same as showing that it is a Cauchy sequence. So, let m and n be natural numbers, with $m > n$. Let $M > 0$ be such that $|s_k| < M$ for every $k \geq 1$. It follows from Abel's identity (5.19) (with the roles of a_i and b_i exchanged) and from triangle inequality that

$$|x_m - x_n| = \left| \sum_{i=n}^{m-1} s_i(b_i b_{i+1}) + s_m b_m - s_n b_n \right|$$

$$\leq \sum_{i=n}^{m-1} |s_i|(b_i b_{i+1}) + |s_m| b_m + |s_n| b_n$$

$$= M(b_n - b_m) + M b_m + M b_n$$

$$= 2M b_n \xrightarrow{n} 0.$$

Hence, $(x_n)_{n \geq 1}$ is indeed a Cauchy sequence.

19. In the notations of the former problem, change a_k by $(-1)^k$ and b_k by a_k.

20. For item (a), observe that $2\sin(a+jh)\sin\frac{h}{2} = \cos(a+(j-1)h)-\cos(a+jh)$.
Now, sum from $j=0$ to $j=k$ and transform the result in a product. For item
(b), note that it suffices to show that the sequence $(s_n)_{n\geq1}$, given for $n\geq1$ by
$s_n = \sum_{k=1}^{n}\sin k$, is bounded. To this end, apply the result of item (a) to show
that $s_n = \frac{\sin\frac{n}{2}\sin\frac{n+1}{2}}{\sin\frac{1}{2}}$ and, hence, that $|s_n|\leq\frac{1}{\sin\frac{1}{2}}$.

21. Letting $\alpha_n = A_n\widehat{O}A_{n+1}$, we have $\sin\alpha_n = \frac{1}{\sqrt{n}}$. Now, use the unit circle to show
that $\sin\alpha < \alpha$ for every $\alpha\in\left(0,\frac{\pi}{2}\right)$. Finally, use the divergence of $\sum_{k\geq1}\frac{1}{\sqrt{k}}$ to
show that $\sum_{k\geq1}\alpha_k$ also diverges.

22. Letting $\alpha_n = 2\beta_n$, it suffices to show that $\beta_1+\beta_2+\beta_3+\cdots = \frac{\pi}{4}$. Equivalently,
letting $b_n = \tan(\beta_1+\cdots+\beta_n)$ for $n\geq1$, it suffices to show that que $b_n\xrightarrow{n}1$.
Use some Trigonometry to show that $b_{n+1} = \frac{2(n+1)^2b_n+1}{2(n+1)^2-b_n}$, for every $n\in\mathbb{N}$.
Then, use induction to prove that $b_n = \frac{n}{n+1}$, for every $n\geq1$.

Section 8.1

1. For the second part of item (a) note that, if $a < 0 < b$, then the image of the
interval $[a,b]$ by f is the union $\left(-\infty,\frac{1}{a}\right)\cup\{0\}\cup\left(\frac{1}{b},+\infty\right)$. For item (b), start
by computing the image of the interval $\left[-\frac{1}{2},0\right]$ by f.

2. Suppose that g is continuous, and let $g(x_0) = c$. For $x\in I\setminus\{x_0\}$, it follows
from triangle inequality that $|c-f(x_0)|\leq |c-g(x)|+|g(x)-f(x_0)| = |g(x_0)-g(x)|+|f(x)-f(x_0)|$. Now, use the fact that f and g are continuous to conclude
that $|c-f(x_0)| < \epsilon$, for every $\epsilon > 0$.

3. For items (a) and (b), use the result of the previous problem. For items (c) and
(d), use the definition of continuity to conclude that there exists no value of c
that turns the given function into a continuous one – it may help to sketch the
graph of the function.

4. Start by observing that the set $(x_0-\delta, x_0+\delta)\times(f(x_0)-\epsilon, f(x_0)+\epsilon)$ is an open
(i.e., with its boundary removed) rectangle centered at $P_0(x_0,f(x_0))$ and with
sides parallel to the axis. Then, recall that every disk centered at P_0 contains an
open rectangle centered at P_0 and with sides parallel to the axis, and vice-versa.

5. Note that, in D, the tangent function is the quotient of the continuous functions
sin and cos.

6. Use the chain rule for continuous functions.

7. Use the chain rule for continuous functions.

8. For item (a) adapt, to the present case, the proof of the continuity of the square
root function at $x_0 = 0$. For (b), make $y = \sqrt[n]{x}$, $y_0 = \sqrt[n]{x_0}$ and apply item (a) of
Problem 18, page 26, to get

$$|y-y_0| = \frac{|y^n-y_0^n|}{y^{n-1}+y^{n-2}y_0+\cdots+y_0^{n-1}}.$$

Finally, for item (c), use the result of (b) to adapt, to the present case, the proof of the continuity of the square root function at x_0.

9. Use the result of the previous problem, the chain rule and Example 8.10.

10. Use the chain rule to show that f is continuous at every $x_0 \neq 0$. Then, use the definition of continuity, together with the fact that $|x \sin \frac{1}{x}| \leq |x|$, to show that f is continuous at 0.

11. Fixed $x_0 \in [0, 1]$, take $\epsilon = \frac{1}{2}$ and assume, by contradiction, that f is continuous at x_0. By the definition of continuity, there exists $\delta > 0$ such that $x \in [0, 1]$ and $|x - x_0| < \delta$ imply $|f(x) - f(x_0)| < \frac{1}{2}$. Now, separately consider the cases $x_0 \in \mathbb{Q}$ and $x_0 \notin \mathbb{Q}$, and use the result of Problem 4, page 206, to arrive at a contradiction.

12. In order to show that f is discontinuous at every rational number of the interval $[0, 1]$, reason as in the hint to the previous problem, using the result of Problem 4, page 206. If $x_0 \in [0, 1]$ is an irrational number and $n_0 \in \mathbb{N}$, show that there exists $\delta > 0$ such that every fractional representation of a rational number of the interval $(x_0 - \delta, x_0 + \delta)$ has denominator greater than n_0; then, conclude that $|f(x) - f(x_0)| < \frac{1}{n_0}$, for $x \in [0, 1] \cap (x_0 - \delta, x_0 + \delta)$.

13. The first part follows from the fact that $f(x_0)$ is an upper bound for the set on the left hand side, whereas is a lower bound for the set on the right hand side. For the second part, use the definition of continuity, together with Proposition 7.8.

14. That f is well defined follows from Problem 17, page 243, together with the fact that the geometric series $\sum_{n \geq 1} \frac{1}{2^n}$ converges. For item (b), show that $f(x) - f(x_m) \geq \frac{1}{2^m}$ if $x > x_m$, so that each x_m is a point of discontinuity of f. Now, let be given $x \notin D$ and $\epsilon > 0$. Take $m_0 \in \mathbb{N}$ such that $\frac{1}{2^m} < \epsilon$ whenever $m > m_0$; then, there exists $\delta > 0$ such that $\{x_1, \ldots, x_{m_0}\} \cap (x - \delta, x) = \emptyset$. In other words, $x_m \in (x - \delta, x) \Rightarrow m > m_0$, and this gives that, for $y \in (x - \delta, x)$,

$$f(x) - f(y) = \sum_{y \leq x_n < x} \frac{1}{2^n} \leq \sum_{x - \delta < x_m < x} \frac{1}{2^m} \leq \sum_{m \geq m_0 + 1} \frac{1}{2^m} = \frac{1}{2^{m_0}} < \epsilon.$$

Analogously, show that $f(y) - f(x) < \epsilon$ if $y \in (x, x + \delta)$ and $\delta > 0$ is sufficiently small.

Section 8.2

1. First of all, recall that Problem 12, page 152, gives $f(x) = f(1)x$ for every $x \in \mathbb{Q}$. Then, fix an irrational number $x_0 \in \mathbb{R}$ and use the result of Problem 4, page 206, to find a sequence $(a_n)_{n \geq 1}$ in $\mathbb{Q}$ satisfying $a_n \xrightarrow{n} x_0$. Finally, apply Theorem 8.16.

2. Start by using the second hypothesis to show that f is injective; to this end, take distinct $x, y \in \mathbb{R}$ and consider the divergent sequence $(a_n)_{n \geq 1}$ such that $a_{2k} = x$ and $a_{2k-1} = y$, for every $k \geq 1$. Finally, use the hypotheses on f to show that f^{-1} transforms convergent sequences into convergent sequences.

3. Given $\epsilon \leq 0$ and letting $c > 0$ be the Lipschitz constant of f, show that we can take $\delta = \frac{\epsilon}{c}$ in the definition of uniform continuity.

4. Note that, for every $\delta > 0$, there exist $x, y \in \mathbb{R}$ such that $|x - y| < \delta$ but $|f(x) - f(y)| \geq 1$; for instance, taking $x > 0$ and $y = x + \frac{\delta}{2}$, it follows from the binomial formula that $|f(x) - f(y)| = \left(x + \frac{\delta}{2}\right)^n - x^n \geq n \cdot \frac{\delta}{2} \cdot x^{n-1} \geq \frac{n\delta}{2}x$, and it suffices to take a sufficiently large x.

5. If f was uniformly continuous, to the given $\epsilon > 0$ there would correspond some $\delta > 0$ as in (8.4). Now, since $|a_n - b_n| \xrightarrow{n} 0$, we could choose $n \in \mathbb{N}$ satisfying $|a_n - b_n| < \delta$; hence, we should have $|f(a_n) - f(b_n)| < \epsilon$, which is a contradiction.

6. Use the result of the previous problem.

7. For item (c), consider $g : [a, b] \to \mathbb{R}$ given by $g(x) = \frac{1}{M-f(x)}$, for every $x \in [a, b]$.

8. It suffices to let $f(x) = \frac{1}{(x-a)(x-b)}$, for $x \in (a, b)$.

9. Initially, observe that the problem of computing the distance from P to G_f is equivalent to the problem of minimizing the continuous function $g : (a, b) \to \mathbb{R}$, given by $g(x) = \sqrt{(x - x_0)^2 + (f(x) - y_0)^2}$. Now, let $Q(x_0, f(x_0))$ and, for $0 < \delta < b - a$, let $R_\delta = (a + \delta, f(a + \delta))$ and $S_\delta = (b - \delta, f(b - \delta))$. Use the stated condition to show that there exists $\delta_0 > 0$ such that, for $0 < \delta < \delta_0$, we have $\overline{R_\delta P}, \overline{S_\delta P} > \overline{PQ}$. Then, conclude that it suffices to minimize g in the interval $[a + \delta_0, b - \delta_0]$. Finally, apply Weierstrass theorem to finish the proof.

10. For item (a), apply the triangle inequality to get

$$\frac{f(x) - a_0}{x^n} \geq a_n - \left| \frac{a_{n-1}}{x} + \cdots + \frac{a_1}{x^{n-1}} \right| \geq a_n - \frac{|a_{n-1}|}{|x|} - \cdots - \frac{|a_1|}{|x|^{n-1}}.$$

For (b), note that $|x| \geq 1$ implies $|x| \leq |x|^2 \leq \cdots \leq |x|^{n-1}$. Item (c) follows from (b), taking $A = \max\left\{1, \frac{1}{a_n} \sum_{j=1}^{n-1} |a_j|\right\}$ and observing that $x^n > 0$ for $x \neq 0$, since n is even. Finally, for item (d), use the result of item (c), together with Theorem 8.26 and $a_0 = f(0)$.

11. Item (a) follows immediately from (8.8). In what concerns (c), on the one hand we have

$$f(\alpha) = f(\lim_{n \to +\infty} x_n) = \lim_{n \to +\infty} f(x_n) = \lim_{n \to +\infty} x_{n+1} = \alpha;$$

on the other, if $\beta \in \mathbb{R}$ is such that $f(\beta) = \beta$, show that the inequalities $|\alpha - \beta| = |f(\alpha) - f(\beta)| \leq c|\alpha - \beta|$, together with the fact that $0 < c < 1$, give $\alpha = \beta$.

12. Setting $f(x) = \frac{1}{2}(x + \sqrt{x^2 + 1})$, we clearly have $f(x) < x$ for every real x. Moreover,

$$f(x) - f(y) = \frac{1}{2}\left| x - y + \frac{x^2 - y^2}{\sqrt{x^2 + 1} + \sqrt{y^2 + 1}} \right|$$

$$= \frac{1}{2}|x - y|\left| 1 + \frac{x + y}{\sqrt{x^2 + 1} + \sqrt{y^2 + 1}} \right|$$

$$\leq \frac{1}{2}|x - y|\left(1 + \frac{|x| + |y|}{\sqrt{x^2 + 1} + \sqrt{y^2 + 1}} \right)$$

$$< |x - y|.$$

13. Use the sign-preserving lemma, together with Problem 5, page 206, and Theorem 8.16.

14. Show that the stated condition gives $f(0) = f(m + n\alpha)$, for all $m, n \in \mathbb{Z}$. Then, use Corollary 7.32, together with Theorem 8.16, to prove that f is constant.

Section 8.3

1. We could only apply the IVT if the domain of f contained the interval $[0, 2]$, which is not the case.

2. Apply the IVT in each case. Sketching the graphs of the involved functions may help.

3. Let $y = ax + b$ be a non vertical line, so that $a \neq 0$. It suffices to show that the function $g : \mathbb{R} \to \mathbb{R}$, given by $g(x) = x^3 \sin x - ax - b$, has infinitely many zeros. To what is left to do, suppose that $a > 0$ (the case $a < 0$ can be handled similarly); show that, for every sufficiently large $k \in \mathbb{N}$, we have $g(2k\pi) < 0 < g(2k\pi + \frac{\pi}{2})$.

4. Look at the function $f : [0, 1] \to [0, 1]$ such that $f(x) = x + (1-x)^2 \sin \pi(n-2)x$.

5. Adapt the argument in the proof of Bolzano's theorem. More precisely, suppose $I = [a, b]$ (the other cases can be dealt with analogously), fix $\alpha \in X$ and let $A = \{x \in [\alpha, b]; [\alpha, x] \subset X\}$, so that $A \neq \emptyset$. If $c = \sup A$, take a sequence $(a_n)_{n \geq 1}$ in A such that $a_n \to c$ and use condition (ii) to show that $c \in A$. If $c < b$, use condition (i) to choose $0 < \delta < b - c$ (corresponding to $x_0 = c$) to conclude that $c + \frac{\delta}{2} \in A$, which is a contradiction. Finally, make an analogous reasoning to show that $[a, x_0] \subset X$.

6. Since $x = 0$ is not a solution, we can write the given equation as $f(x) = 0$, where $f : (0, +\infty) \to \mathbb{R}$ is the function given by $f(x) = \sum_{i=1}^{n} \sqrt{\frac{1}{x^2} + \frac{a_i}{x}} - n$. Thus, we need to show that f has exactly one positive zero. To this end, since the function $x \mapsto \sqrt{\frac{1}{x^2} + \frac{a_i}{x}}$ is continuous and decreasing, and since f is a finite sum of functions of this kind, we conclude that f is continuous and decreasing too, which forces it to have at most one positive zero. On the other hand, $f(1) =$

$\sum_{i=1}^{n} \sqrt{1+a_i} - n > 0$; also, $\sqrt{\frac{1}{x^2} + \frac{a_i}{x}} < 1$ if $x > \max\{\sqrt{2}, 2a_i\}$, so that $f(x) < 0$ if $x > \max\{\sqrt{2}, 2a_1, \dots, 2a_n\}$. Now, apply the IVT.

7. Let $f : [0, 1] \to \mathbb{R}$ be given by $f(x) = \frac{1}{n} \sum_{i=1}^{n} |x - x_i|$. It suffices to guarantee the existence of $x \in [0, 1]$ for which $f(x) = \frac{1}{2}$. To this end, note that

$$f(0) = \frac{1}{n} \sum_{i=1}^{n} x_i \quad \text{and} \quad f(1) = 1 - \frac{1}{n} \sum_{i=1}^{n} x_i,$$

so that $f(0) + f(1) = 1$. If $f(0) = f(1) = \frac{1}{2}$, there is nothing to do. Otherwise, we can suppose, without loss of generality, that $f(0) < \frac{1}{2} < f(1)$. Since f is continuous, the IVT finishes the task.

8. Firstly, suppose that $f(0) = 0$. Making $x = 0$ in the given relation, we get $f(1) = f(0)f(2) + f(1) = 0$. Now making $x = 1$, we get $f(2) = f(1)f(3) + f(2) = 0$. This way, an easy induction gives $f(n) = 0$ for every positive integer n. Suppose, then, that $f(0) > 0$. It follows from $f(0)f(2) + f(1) = 0$ that $f(1)$ and $f(2)$ must have opposite signs. But then, the IVT guarantees the existence of a real number $a \in (1, 2)$ for which $f(a) = 0$. Arguing as in the case $f(0) = 0$, conclude that $f(a + n) = 0$ for every positive integer n.

9. Make $x = 1000$ and, then, use the IVT to conclude that the image of f contains the interval $[\frac{1}{999}, 999]$. Use the IVT once more to guarantee the existence of $x_0 \in \mathbb{R}$ such that $f(x_0) = 500$. Finally, make $x = x_0$ in the given relation.

10. First of all, show that if $n \in \mathbb{N}$, then $a = \frac{1}{n}$ is one possible value. Then, if $a \neq \frac{1}{n}$ for every $n \in \mathbb{N}$, construct a continuous function $f : [0, 1] \to \mathbb{R}$ such that $f(x + a) \neq f(a)$ for every $x \in [0, 1 - a]$.

11. Use the given relation to arrive at a contradiction to the IVT.

12. For item (a), the fact that f is a bijection follows from Example 6.40; then, Theorem 8.35 guarantees that f is increasing or decreasing. For item (b), note first that, from (a), we have $f(0) = 0$ e $f(1) = 1$. If there exists $x \in (0, 1)$ such that $f(x) < x$, then the fact that f is increasing, together with the hypothesis of the problem, gives $x = f(f(x)) < f(x)$, which is a contradiction. Analogously, we cannot have $f(x) > x$, so that $f(x) = x$ is the only possibility. Finally, for (c), note first that, by item (a), we have $f(0) = 1$ and $f(1) = 0$. Then, fix $a \in (0, 1)$ and define f letting $f(a) = a$ and the restrictions of f to both intervals $[0, a]$ and $[a, 1]$ being affine functions. It's immediate to check that $f(f(x)) = x$ for every $x \in [0, 1]$.

13. Example 8.31 shows that g has a fixed point x_0. If $x_n = g(x_{n-1})$ for $n \geq 1$, show that the sequence $(x_n)_{n \geq 1}$ is nondecreasing and such that $f(x_n) = x_n$, for every $n \geq 1$. Letting $x_n \xrightarrow{n} a$, use Theorem 8.16 to show that $f(a) = g(a) = a$.

14. Consider the auxiliary function $g : \mathbb{R} \to \mathbb{R}$ given by $g(x) = \sum_{j=1}^{k} f(x + jr)$, where $r > 0$ is to be chosen. Then, if $f(x_0) < 0 < f(x_1)$, apply the sign-preserving lemma to g to show that r can be chosen in such a way that $g(x_0) < 0 < g(x_1)$. Finally, invoke the IVT.

15. Start by observing that, for $x \in (0,1)$, the function $g(x) = x(f(x) - f(0))$ measures the area of the closed rectangle having $(0, f(0))$ and $(x, f(x))$ as endpoints of a diagonal and sides parallel to the coordinate axis. Hence, there exists $0 < x_1 < 1$ such that this rectangle has area $\frac{1}{n^2}$. Arguing in an analogous way, we find rectangles $R_1, \ldots, R_k$, each of which having area $\frac{1}{n^2}$ and sides parallel to the coordinate axis, such that $R_1 \cup \ldots \cup R_k$ contains the graph of f and is contained in the closed square $[0,1] \times [0,1]$ of the cartesian plane. Finally, use Cauchy-Schwarz inequality to show that $k \leq n$.

16. Suppose, by the sake of contradiction, that there exist $x_0, y_0 \in \mathbb{R}$ for which $f(x_0) < f(y_0)$; moreover, suppose that $y_0 \in (x_0, x_0 + f(x_0))$ (the other cases can be treated analogously). Use the fact that $f(x_0 + f(x_0)) = f(x_0)$ to show that it is possible to choose $n \in \mathbb{N}$ and $c \in \mathbb{R}$ such that the line r of equation $x + ny = c$ leaves the point $(y_0, f(y_0))$ in a semiplane opposite to that of the points $(x_0, f(x_0))$ and $(x_0 + f(x_0), f(x_0 + f(x_0))) = (x_0, f(x_0))$. Now, show the IVT guarantees the existence of real numbers $a \in (x_0, y_0)$ and $b \in (y_0, x_0 + f(x_0))$ such that the points $A(a, f(a))$ and $B(b, f(b))$ are on the line r, i.e., such that $a + nf(a) = c$ and $b + nf(b) = c$. Finally, conclude that $f(c) = f(a + nf(a)) = f(a)$ and $f(c) = f(b + nf(b)) = f(b)$; hence, $f(a) = f(b)$ and, then, $a = b$.

17. Making $x = y = 0$ in the given relation, we get $f(0) + g(0) = h(0)$. Thus, letting $f_1(x) = f(x) - f(0)$, $g_1(x) = g(x) - g(0)$ and $h_1(x) = h(x) - h(0)$, we have

$$f_1(x + y^3) + g_1(x^3 + y) = h_1(xy),$$

with $f_1(0) = g_1(0) = h_1(0) = 0$. Hence, we can start by supposing that $f(0) = g(0) = h(0) = 0$. Making $y = -x^3$ in the given relation, we get $g(x - x^9) = h(-x^4)$ for every real x. Making $x = -y^3$, we get $f(-y^9 + y) = h(-y^4)$ for every $y \in \mathbb{R}$; therefore, it follows that $f(x - x^9) = g(x - x^9)$. However, since the image of the polynomial function $x \mapsto x - x^9$ is the whole set of reals (see Example 8.28), we have $f(x) = g(x)$ for every $x \in \mathbb{R}$. Then, the relation at the statement of the problem reduces to

$$f(x + y^3) + f(x^3 + y) = h(xy). \tag{B.4}$$

Performing the substitutions $y = -x^3$ and $x = -y^3$ in (B.4) and taking into account that $f(0) = 0$, we obtain, respectively,

$$f(x - x^9) = h(-x^4) \quad \text{and} \quad f(-y^9 + y) = h(-y^4),$$

so that $f(x - x^9) = f(-x^9 + x)$. Using once more the fact that the image of the polynomial function $x \mapsto x - x^9$ is $\mathbb{R}$, it follows that f is an even function, i.e., that $f(x) = f(-x)$ for every $x \in \mathbb{R}$. Now, letting $y = 0$ in (B.4), we get $f(x) + f(x^3) = 0$, so that $f(x) = -f(x^3)$. Back to (B.4), this relation gives us

$$f(x^3 + y) - f((x + y^3)^3) = h(xy).$$

However, the fact that f is even implies, then,

$$f(x^3 + y) - f(-(x + y^3)^3) = h(xy).$$

Therefore, if $a \in \mathbb{R}$ is such that the system of equations

$$\begin{cases} xy = a \\ x^3 + y = -(x + y^3)^3 \end{cases},$$

has at least one real solution, then $h(a) = 0$. Let us prove that, for every $a \leq 0$, this is indeed the case: if $a = 0$, we already have $h(a) = 0$. If $a \neq 0$, the the second equation of the system is equivalent to $y^9 + 3y^6x + 3x^2y^3 + y + 2x^3 = 0$. Writing $x = \frac{y}{a}$, we conclude that it suffices to guarantee the existence of a real number y such that

$$y^{12} + 3ay^8 + (3a^2 + 1)y^4 + 2a^3 = 0.$$

To this end, let $p(y) = y^{12} + 3ay^8 + (3a^2 + 1)y^4 + 2a^3$. If $a < 0$, then $p(0) = 2a^3 < 0$ and $p(-a) > 0$, so that the IVT guarantees that p has at least one real root. To finish, not that $f(x - x^9) = h(-x^4)$, so that $f(x - x^9) = 0$ for every $x \in \mathbb{R}$. Invoking once more the surjectivity of f, it follows that $f = 0$. Therefore, (B.4) gives $h = 0$. Then, the functions f, g and h that satisfy the stated conditions are the constant functions $f = f(0)$, $g = g(0)$ and $h = h(0)$, such that $f(0) + g(0) = h(0)$.

Section 9.2

1. The idea is to adapt, to the present case, the proof of the uniqueness of limits of sequences, given in Proposition 7.14. More precisely, suppose that L and M are distinct reals and that $f(x)$ converges simultaneously to L and M when $x \to x_0$. Take $\epsilon = \frac{1}{2}|L - M| > 0$ and $\delta > 0$ such that $x \in I$ and $0 < |x - x_0| < \delta$ give $|f(x) - L| < \epsilon$ and $|f(x) - M| < \epsilon$. Then, use the triangle inequality to arrive at a contradiction.

2. Suppose we had $L > M$, and let $\epsilon = \frac{L-M}{2} > 0$. Take $\delta > 0$ such that $x \in I$ and $0 < |x - x_0| < \delta$ give $|f(x) - L| < \epsilon$ and $|g(x) - M| < \epsilon$. Use the triangle inequality to arrive at a contradiction.

3. Make induction on n. The initial case $n = 2$ is exactly the content of items (a) and (b) of Proposition 9.7.

4. For right-handed limits, for instance, copy the statement and proof of Proposition 9.7, changing, whenever convenient, $\lim_{x \to x_0}$ by $\lim_{x \to x_0+}$ and $0 < |x - x_0| < \delta$ (resp. $0 < |x - x_0| < \delta_1, \delta_2$) by $x_0 < x < x_0 + \delta$ (resp. $x_0 + \delta_1, x_0 + \delta_2$).

5. Implication $\Rightarrow$ and the last part follow from the fact that each one of $x_0 - \delta < x < x_0$ and $x_0 < x < x_0 + \delta$ $0 < |x - x_0| < \delta$ do imply $0 < |x - x_0| < \delta$. Conversely, suppose $\lim_{x \to x_0+} f(x) = \lim_{x \to x_0-} f(x) = L$, and let $\epsilon > 0$ be given. Take $\delta_1, \delta_2 > 0$ such that each of $x_0 - \delta_1 < x < x_0$ and $x_0 < x < x_0 + \delta$ imply $|f(x) - L| < \epsilon$. Letting $\delta = \min\{\delta_1, \delta_2\} > 0$, conclude that $0 < |x - x_0| < \delta$ implies $|f(x) - L| < \epsilon$.

6. Suppose that $\lim_{x \to x_0} f(x) = L > 0$ and $\lim_{x \to x_0} g(x) = +\infty$ (the other cases are completely analogous), and let $M > 0$ be given. By the sign-preserving lemma, there exists $\delta_1 > 0$ such that $x \in I$, $0 < |x - x_0| < \delta_1 \Rightarrow f(x) > \frac{L}{2}$; by the definition of infinite limit, there exists $\delta_2 > 0$ such that $x \in I$, $0 < |x - x_0| < \delta_1 \Rightarrow g(x) > \frac{2M}{L}$. Then, let $\delta = \min\{\delta_1, \delta_2\}$ and conclude that for $x \in I$ satisfying $0 < |x - x_0| < \delta$ we have $f(x)g(x) > M$. For what is left, start by observing that the analogous result for limits at infinity (at $+\infty$, for instance) says that if $\lim_{x \to +\infty} f(x) = L > 0$ and $\lim_{x \to +\infty} g(x) = +\infty$, then $\lim_{x \to +\infty} f(x)g(x) = \pm\infty$. For the proof of this fact, copy the former one, changing $0 < |x - x_0| < \delta$ by $x > A$ whenever convenient.

7. Suppose first that $a, b, L, M \in \mathbb{R}$ (the cases in which $a = -\infty$, $b = +\infty$, $L = -\infty$ or $M = +\infty$ are totally analogous). Given $\epsilon > 0$, let $\delta > 0$ be such that $a < x < a + \delta$ and $b - \delta < y < b$ respectively imply $f(x) < L + \epsilon$ and $g(x) > M - \epsilon$. Then, apply the IVT to conclude that $[L + \epsilon, M - \epsilon] \subset \text{Im}(f)$.

8. In all items one has to use the fundamental trigonometric limit, together with some algebra. For items (a) and (b), multiply both the numerator and denominator by $1 + \cos x$; for item (c), write $\frac{\sin(2x)}{\sin(3x)} = \frac{2}{3} \cdot \frac{\sin(2x)}{2x} \cdot \frac{3x}{\sin(3x)}$; for (d), multiply both the numerator and denominator by $\frac{\pi}{2}$; for (e), use (d) and the fact that $\cos\left(\frac{\pi}{2}x\right) = \sin\left(\frac{\pi}{2} - \frac{\pi x}{2}\right)$; finally, for (f), apply the identity $1 - y = \frac{1 - y^3}{1 + y + y^2}$ with $y = \sqrt[3]{\cos x}$ and, then, the result of item (b).

9. For the first part adapt, to the present case, the proof of Corollary 9.9. In order to compute the limits, use the result of the first part, together with the fact that $|\sin x|, |\cos x| \leq 1$ for all $x \in \mathbb{R}$.

10. Given $\epsilon > 0$, let $\delta > 0$ be such that $x \in I$ and $0 < |x - x_0| < \delta$ imply $|f(x) - L| < \epsilon$. Take $n_0 \in \mathbb{N}$ such that $n > n_0 \Rightarrow |a_n - x_0| < \delta$. Then, for $n > n_0$ we have $0 < |a_n - x_0| < \delta$ (since $a_n \neq x_0$ for every $n \geq 1$), so that $|f(a_n) - L| < \epsilon$. For the second part, compute $f(a_n)$ and $f(b_n)$, where $a_n = \frac{1}{2n\pi}$ and $b_n = \frac{1}{2n\pi + \pi/2}$ for $n \geq 1$.

11. For the first case, if $\lim_{x \to +\infty}(f(x) - (ax + b)) = 0$, then

$$0 = \lim_{x \to +\infty}\left(\frac{f(x) - (ax + b)}{x}\right) = \lim_{x \to +\infty}\left(\frac{f(x)}{x} - a - \frac{b}{x}\right) = \lim_{x \to +\infty}\frac{f(x)}{x} - a.$$

and

$$0 = \lim_{x \to +\infty}(f(x) - (ax + b)) = \lim_{x \to +\infty}(f(x) - ax) - b.$$

The second case (i.e., that of $f : (-\infty, A) \to \mathbb{R}$) is entirely analogous the the first one, so will be left to the reader.

12. In item (a), if $y = \alpha x + \beta$ is an oblique asymptote when $x \to +\infty$, then $\alpha = \lim_{x \to +\infty} \frac{f(x)}{x} = \lim_{x \to +\infty} \left(1 + \frac{1}{x^2}\right) = 1$ and $\beta = \lim_{x \to +\infty}(f(x) - \alpha x) = \lim_{x \to +\infty} \frac{1}{x} = 0$. Analogously, $y = x$ is an oblique asymptote when $x \to -\infty$. In item (b), if $y = \alpha x + \beta$ is an oblique asymptote when $x \to +\infty$, then $\alpha = \lim_{x \to +\infty} \frac{f(x)}{x} = \lim_{x \to +\infty} \frac{b}{a}\sqrt{1 - \frac{a^2}{x^2}} = \frac{b}{a}$ and

$$\beta = \lim_{x \to +\infty}(f(x) - \alpha x) = \lim_{x \to +\infty} \frac{b}{a}\left(\sqrt{x^2 - a^2} - x\right)$$

$$= -\lim_{x \to +\infty} \frac{b}{a}\left(\frac{a^2}{\sqrt{x^2 - a^2} + x}\right) = 0.$$

Analogously, $y = -\frac{b}{a}x$ is an oblique asymptote when $x \to -\infty$. For item (c), observe that $\lim_{x \to +\infty} x \sin \frac{1}{x} = 1$. Finally, in all of these items, we leave to the reader the analysis of the existence of horizontal or vertical asymptotes.

13. Prior to making $x \to +\infty$, write successively

$$\sqrt{x + \sqrt{x + \sqrt{x}}} - \sqrt{x} = \frac{x + \sqrt{x + \sqrt{x}} - x}{\sqrt{x + \sqrt{x + \sqrt{x}}} + \sqrt{x}} = \frac{\sqrt{1 + \frac{1}{\sqrt{x}}}}{\sqrt{1 + \sqrt{\frac{1}{x} + \frac{1}{\sqrt{x^3}}}} + 1}.$$

14. We can suppose $n > 2$, so that $\frac{f(nx)}{f(x)} > \frac{f(2x)}{f(x)}$. On the other hand, if $k \in \mathbb{N}$ is such that $2^k > n$, then $\frac{f(nx)}{f(x)} < \frac{f(2^k x)}{f(x)} = \prod_{j=1}^{k-1} \frac{f(2^{j+1}x)}{f(2^j x)}$. Hence,

$$\frac{f(2x)}{f(x)} < \frac{f(nx)}{f(x)} < \prod_{j=1}^{k-1} \frac{f(2^{j+1}x)}{f(2^j x)},$$

and it suffices to make $x \to +\infty$ in the above inequalities and invoke the squeezing principle to conclude the proof.

15. For the second part of item (a), use induction. For item (b) and for the first part of (c) use some Trigonometry, together with the fact that $a_n = 2^n\sqrt{2 - b_n}$. Finally, for the second part of (c), compute θ_1, write $a_n = 2\theta_1 \cdot \frac{\sin(\theta_1/2^n)}{\theta_1/2^n}$ and apply the fundamental trigonometric limit.

16. Start by letting $x = 1$ in (a) to get $f(f(y)) = yf(1)$ for every $y > 0$. Then, use such a relation to show that f is injective and, thus, that $f(1) = 1$ and $f(f(x)) = x$ for every $x > 0$. If $a > 0$ is a fixed point of f, show that $\frac{1}{a}$ also is, so that we can assume $a \geq 1$. Conclude that a^k is a fixed point of f for every $k \in \mathbb{N}$, and use (b) to show that $a = 1$. Finally, make $x = y$ in (a) to get $f(x) = \frac{1}{x}$ for every $x > 0$.

Section 9.3

1. For item (a), observe that $\frac{g(x)-g(x_0)}{x-x_0} = c \cdot \frac{f(x)-f(x_0)}{x-x_0}$. For item (b), $\frac{h(x)-h(x_0)}{x-x_0} = \frac{f(x)-f(x_0)}{x-x_0}$.

2. For x in the domain of g, we have $\frac{g(x)-g\left(\frac{x_0}{c}\right)}{x-\frac{x_0}{c}} = c \cdot \frac{f(cx)-f(x_0)}{cx-x_0}$. Now, making $y = y(x) = cx$, we have $y \to x_0$ if $x \to \frac{x_0}{c}$, so that

$$\lim_{x \to \frac{x_0}{c}} \frac{g(x) - g\left(\frac{x_0}{c}\right)}{x - \frac{x_0}{c}} = c \lim_{y \to x_0} \frac{f(y) - f(x_0)}{y - x_0} = cf'(x_0).$$

3. Make induction on k.
4. For arcsin, take $x \in (-1, 1)$ and let $y \in \left(-\frac{\pi}{2}, \frac{\pi}{2}\right)$ be such that $\sin y = x$. Since $\sin' y = \cos y \neq 0$, it follows from Theorem 9.28 (with the roles of x and y changed) that arcsin is differentiable at x, with

$$\text{arcsin}' x = \frac{1}{\sin' y} = \frac{1}{\cos y} = \frac{1}{\sqrt{1 - \sin^2 y}} = \frac{1}{\sqrt{1 - x^2}}.$$

   The reasoning for arccos is completely analogous and will be left to the reader.
5. Use the results of the three previous problems.
6. For the second part of (a), note that $f'(1) = n$ and $g'(1) = m$, so that $f'(1) = g'(1)$ only if $m = n$. For item (b), we have $f(x_0) = f^{-1}(x_0) = y_0$ and $f'(x_0) = (f^{-1})'(x_0)$ (since the tangent lines to the graphs of f and f^{-1} at A must coincide). Now, apply Theorem 9.28 to conclude that $f'(x_0)f'(y_0) = 1$.
7. Express Newton' quotient of $x \mapsto f(x)$ at x_0 in terms of Newton's quotient of f at x_0.
8. Start by writing

$$\frac{f(b_n) - f(a_n)}{b_n - a_n} = \frac{f(b_n) - f(x_0)}{b_n - x_0} \cdot \frac{b_n - x_0}{b_n - a_n} + \frac{f(x_0) - f(a_n)}{x_0 - a_n} \cdot \frac{x_0 - a_n}{b_n - a_n}.$$

Then, use the triangle inequality to get

$$\left| \frac{f(b_n) - f(a_n)}{b_n - a_n} - L \right| \leq \left| \frac{f(b_n) - f(x_0)}{b_n - x_0} - L \right| \cdot \frac{b_n - x_0}{b_n - a_n} +$$

$$+ \left| \frac{f(x_0) - f(a_n)}{x_0 - a_n} - L \right| \cdot \frac{x_0 - a_n}{b_n - a_n}$$

$$\leq \left| \frac{f(b_n) - f(x_0)}{b_n - x_0} - L \right| + \left| \frac{f(x_0) - f(a_n)}{x_0 - a_n} - L \right|.$$

9. For item (a), note that the tangent line to the graph of f at $(a_n, f(a_n))$ has equation $y - f(a_n) = f'(a_n)(x - a_n)$. For (b), make $f(x) = x^2 - a$ and use the result of item (b) of Problem 1 to conclude that $a_{n+1} = a_n - \left(\frac{a_n^2 - a}{2a_n} \right) = \frac{1}{2} \left(a_n + \frac{a}{a_n} \right)$.

10. For item (b), write Newton's quotient of f in x_0 as

$$\frac{\sqrt{R^2 - x^2} - \sqrt{R^2 - x_0^2}}{x - x_0} = \frac{1}{x - x_0} \cdot \frac{(R^2 - x^2) - (R^2 - x_0^2)}{\sqrt{R^2 - x^2} + \sqrt{R^2 - x_0^2}}$$

$$= -\frac{x + x_0}{\sqrt{R^2 - x^2} + \sqrt{R^2 - x_0^2}};$$

then, make $x \to x_0$. For (c), let r be the straight line that passes through A and is perpendicular to OA. Elementary Analytic Geometry (cf. Chap. 6 of [4], for instance) teaches us that the slope of r equals $-\frac{1}{m}$, where m is the slope of $\overleftrightarrow{OA}$. Since $m = \frac{f(x_0)}{x_0}$, it follows that the slope of r equals $-\frac{x_0}{f(x_0)}$. Now, (b) gives $-\frac{x_0}{f(x_0)} = f'(x_0)$, as we wished to show.

11. For the first part, since (x_0, y_0) belongs to the graph of f, we must have $y_0 = f(x_0)$. Now, since r is tangent to the graph of f at (x_0, y_0), its slope must be equal to $f'(x_0)$; on the other hand, since r passes through (a, b), its slope must be also equal to $\frac{y_0 - b}{x_0 - a}$. For what is left to do, if a straight line passes through $(1, -1)$ and is tangent to the given parabola at (x_0, y_0), it follows from the first part and from Problem 1 that

$$\begin{cases} y_0 = \frac{x_0^2}{4} \\ y_0 - (-1) = \frac{x_0}{2}(x_0 - 1) \end{cases}.$$

Hence, $\frac{x_0^2}{4} + 1 = \frac{x_0}{2}(x_0 - 1)$, so that $x_0 = 1 \pm \sqrt{5}$ and $y_0 = \frac{3 \pm \sqrt{5}}{2}$.

12. Choose a cartesian coordinate system such that $F(0, \frac{p}{2})$ and d is the straight line $y = -\frac{p}{2}$. Since $Q(x, y) \in \mathcal{P}$ if and only if $\overline{FQ} = \text{dist}(Q; d)$, a simple computation gives $y = \frac{x^2}{2p}$ as the equation of the parabola. Now, letting $P(x_0, -\frac{p}{2})$ and $Q(x, \frac{x^2}{2p})$, use the result of the previous problem to conclude that $\overleftrightarrow{PQ}$ is tangent to $\mathcal{P}$ if and only if

$$\frac{d}{dx} \left(\frac{1}{2p} x^2 \right) = \frac{\frac{x^2}{2p} - \left(-\frac{p}{2} \right)}{x - x_0}$$

or, which is the same (cf. Problem 1), $x^2 - 2x_0 x - p^2 = 0$. Letting α and β be the roots of this second degree equation, we have $A(\alpha, \frac{\alpha^2}{2p})$ e $B(\beta, \frac{\beta^2}{2p})$. On the other hand, the slope of $\overleftrightarrow{AF}$ is $m = \frac{\frac{\alpha^2}{2p} - \left(-\frac{p}{2} \right)}{\alpha - 0} = \frac{\alpha^2 - p^2}{2p\alpha}$; however, since

$\alpha^2 - 2x_0\alpha - p^2 = 0$, it comes that $m = \frac{2x_0\alpha}{2p\alpha} = \frac{x_0}{p}$. Analogously, the slope of $\overleftrightarrow{BF}$ also equals $\frac{x_0}{p}$, so that $\overleftrightarrow{AF} = \overleftrightarrow{BF}$, and A, B and F are collinear.

13. Since $\sqrt{k} \notin \mathbb{Q}$, we have $f(\sqrt{k}) = 0$. Now, letting x be an irrational number different from $\sqrt{k}$, we have $\frac{f(x)-f(\sqrt{k})}{x-\sqrt{k}} = 0$. It thus suffices to show that

$$\lim_{\substack{x \to \sqrt{k} \\ x \in \mathbb{Q}}} \frac{f(x) - f(\sqrt{k})}{x - \sqrt{k}} = 0.$$

To this end, let $x = \frac{p}{q}$, with $p, q \in \mathbb{N}$, be an irreducible fraction. We have

$$\left| \frac{f(x) - f(\sqrt{k})}{x - \sqrt{k}} \right| = \frac{q^{-3}}{|p/q - \sqrt{k}|} = \frac{1}{q^2|p - q\sqrt{k}|}$$

$$= \frac{p + q\sqrt{k}}{q^2|p^2 - q^2k|} \le \frac{p + q\sqrt{k}}{q^2} = \frac{1}{q}\left(\frac{p}{q} + \sqrt{k}\right).$$

Given $\epsilon > 0$, take $n \in \mathbb{N}$ such that $n > \frac{2\sqrt{k}+1}{\epsilon}$ and let

$$A = \{p/q;\, p, q \in \mathbb{N},\, 0 < q \le n\}.$$

Note that A is a finite set. Letting

$$\delta = \min\{1, \{|x - \sqrt{k}|;\, x \in A\}\},$$

we have $|\frac{p}{q} - \sqrt{k}| < \delta \Rightarrow q \ge n$, whereas $\frac{p}{q} < \sqrt{k} + 1$ implies

$$\left| \frac{f(x) - f(\sqrt{k})}{x - \sqrt{k}} \right| = \frac{1}{q}\left(\frac{p}{q} + \sqrt{k}\right) < \frac{2\sqrt{k} + 1}{n} < \epsilon.$$

Section 9.4

1. Apply the formulas for differentiation of sums, products and quotients, in conjunction with the computations of derivatives performed in this and in the previous section.

2. For $x \in (0, +\infty)$, let $h(x) = \sqrt[n]{x}$, so that h is differentiable by Example 9.29, with $h'(x) = \frac{1}{n}x^{\frac{1}{n}-1}$. Since $g(x) = (h \circ f)(x)$ and f is differentiable at x_0, the chain rule guarantees that g is differentiable at x_0, with $g'(x_0) = h'(f(x_0))f'(x_0) = \frac{1}{n}f(x_0)^{\frac{1}{n}-1}f'(x_0)$.

3. Apply the differentiation formulas for the sums, product and quotient of two functions, together with the derivative computations of this and the last section. For items (e) and (f), start by reviewing the results of Problem 4, page 300.

4. If $r = \frac{m}{n}$, with $m, n \in \mathbb{Z}^*$, then $f(x) = \sqrt[n]{x^m}$. For $x > 0$, let $g(x) = \sqrt[n]{x}$, so that g is differentiable (by Example 9.29). Since $x \mapsto x^m$ is also differentiable (for $x \neq 0$ if $m < 0$) and $f(x) = g(x^m)$, the chain rule guarantees the differentiability of f, with $f'(x) = g'(x^m) \cdot mx^{m-1}$. Then, again by Example 9.29, we have

$$f'(x) = \frac{1}{n} \cdot (x^m)^{1/n-1} \cdot mx^{m-1} = \frac{m}{n} \cdot x^{m(\frac{1}{n}-1)+m-1} = r \cdot x^{\frac{m}{n}-1} = rx^{r-1}.$$

5. For the first part, differentiate both sides of (9.22) and, then, divide both sides of the resulting equality by $f(x)$. For the second part, differentiate both sides of the equality $f(x) = a(x - \alpha_1)(x - \alpha_2) \ldots (x - \alpha_n)$ and, then, divide both sides of the resulting equality by $f(x)$.

6. Straightforward computations give $f'(x) = -7(x-3)^{-2}, f''(x) = 14(x-3)^{-3}$, $f^{(3)}(x) = -42(x-3)^{-4}, f^{(4)}(x) = 168(x-3)^{-5}$ and $f^{(5)}(x) = -840(x-3)^{-6}$. Hence, $f^{(5)}(0) = -840 \cdot 3^{-6} = -\frac{840}{729}$.

7. Set $f_1(x) = \sqrt{x+1}$ and, for an integer $n \geq 2$,

$$f_n(x) = \sqrt{x + \sqrt{x + \cdots + \sqrt{x + \sqrt{x+1}}}},$$

with n square root signs. For $n \geq 2$, we have $f_n(x)^2 = x + f_{n-1}(x)$. Hence, it follows from the chain rule that $2f_n(0)f_n'(0) = 1 + f_{n-1}'(0)$ or, which is the same, $2f_n'(0) = 1 + f_{n-1}'(0)$ (since $f_n(0) = 1$). Now, since $f_1'(0) = \frac{1}{2}$, we can use the previous recurrence relation nine times to get $f_{10}'(0) = \frac{1023}{1024}$. Alternatively, observing that $f_n'(0) - 1 = \frac{1}{2}(f_{n-1}'(0) - 1)$, we get $f_n'(0) - 1 = \frac{1}{2^{n-1}}(f_1'(0) - 1) = -\frac{1}{2^n}$ and, thus, $f_n'(0) = 1 - \frac{1}{2^n}$.

8. Make induction on k. For the induction step, differentiate $x \mapsto \frac{p_k(x)}{(1+x^2)^k}$ using the formula for differentiation of a quotient, together with the chain rule.

9. Suppose that the straight line $y = ax + b$ is tangent to the given parabolas at the points with abscissas α e β, respectively. Then, $\alpha^2 = a\alpha + b$ and $2 + (\beta - 3)^2 = a\beta + b$, while $a = \frac{d}{dx}x^2\big|_{x=\alpha} = 2\alpha$ and $a = \frac{d}{dx}x^2\big|_{x=\beta}(2 + (\beta - 3)^2) = 2(\beta - 3)$. Hence, $\alpha = \beta - 3$ and $b = \alpha^2 - 2\alpha \cdot \alpha = -\alpha^2, b = 2 + (\beta - 3)^2 - 2(\beta - 3) \cdot \beta = -\beta^2 + 11$, so that $\alpha^2 = \beta^2 - 11$. Solving the system of equations in α and β thus obtained, we get $\alpha = \frac{1}{3}$ and $\beta = \frac{10}{3}$. From this, we easily find $a = \frac{2}{3}$ and $b = -\frac{1}{9}$, so that the common tangent line is $y = \frac{2}{3}x - \frac{1}{9}$.

10. For item (b), solve the system formed by the equations $\frac{x^2}{a^2} + \frac{y^2}{b^2} = 1$ and $\frac{x^2}{a'^2} -$
$\frac{y^2}{b'^2} = 1$, concluding from (a) that it has exactly four distinct solutions. For
(c), analyse the case in which $P(x_0, y_0)$, with $x_0, y_0 > 0$ (the remaining cases
can be dealt with in analogous ways). Show (using either the chain rule, implicit
differentiation or directly the result of Problem 2) that the slopes of the tangents
to $\mathcal{E}$ and $\mathcal{H}$ at P are respectively equal to $-\frac{b^2}{a^2} \cdot \frac{x_0}{y_0}$ and $\frac{b'^2}{a'^2} \cdot \frac{x_0}{y_0}$. Finally, conclude
that it suffices to show that $\frac{b^2 b'^2}{a^2 a'^2} \cdot \frac{x_0^2}{y_0^2} = 1$, and use the result of (b) to verify that
such an equality is indeed true.

11. For item (b), take g to be equal to the sum of the functions $x \mapsto f(x - x_k)$, where
f is as in item (a).

12. Adapt, to the present case, the discussion of Example 9.41.

13. Firstly, observe that $a_1 + 2a_2 + \cdots + na_n = f'(0)$. Then, use the fact that
$|f(x)| \le |\sin x|$, together with $f(0) = 0$ and the fundamental trigonometric
limit, to get $|f'(0)| \le 1$.

14. Letting $p > 0$ be the period of f, show that $f^{(k)}(p) = f^{(k)}(0)$ for every integer
$k \ge 0$. Then, looking at these n equalities as a linear system of equations in the
unknowns $\cos(jp)$, $1 \le j \le n$, conclude that its only solution is $\cos(jp) = 1$ for
$1 \le j \le n$.

15. Firstly, verify that, for $x \in (a, b)$, the discussion that precedes Lemma 9.36
allows us to write

$$\frac{f(x)}{g(x)} = \frac{f'(a) + \frac{r(x)}{x-a}}{g'(a) + \frac{s(x)}{x-a}},$$

with $\lim_{x \to a} \frac{r(x)}{x-a}$ and $\lim_{x \to a} \frac{s(x)}{x-a} = 0$.

Section 9.5

1. Let $g, h : \mathbb{R} \to \mathbb{R}$ be given by $g(x) = f(x) - f(0) \cos x$ and $h(x) = g'(x) = f'(x) + f(0) \sin x$. Verify that $h'(x) = -g(x)$, and apply the result
of Example 9.48.

2. Let $g : \mathbb{R} \to \mathbb{R}$ be given by $g(x) = f(\frac{x}{\sqrt{\lambda}})$. Verify that $g''(x) + g(x) = 0$, and
apply the result of the previous problem.

3. If $g(x) = f''(x_0)(x - x_0)$, use Corollary 9.47 to conclude that $f' - g$ is constant and, hence, that $f'(x) = f'(x_0) + f''(x_0)(x - x_0)$ for every $x \in I$. Then, put $h(x) = f'(x_0)(x - x_0) + \frac{f''(x_0)}{2}(x - x_0)^2$ and use a similar reasoning.

4. Use the given inequality to conclude that f is differentiable, with $f'(x) = 0$ for every $x \in \mathbb{R}$. Then, use Corollary 9.47.

5. Suppose first that c is the left end of I. Apply Lagrange's MVT to an interval $[c, x] \subset I$ to conclude that $f'_+(c) = \lim_{x \to c^+} \frac{f(x) - f(c)}{x - c}$ exists and equals l. Then, do the same if c is the right end or an interior point of I.

6. Apply Rôlle's theorem to $\frac{a_0}{1}x + \frac{a_1}{2}x^2 + \cdots + \frac{a_n}{n+1}x^{n+1}$.

7. Use the IVT to conclude that $f(x) = x^2 - x\sin x - \cos x$ has at least two real roots, one in the interval $(-\frac{\pi}{2}, 0)$ and another in the interval $(0, \frac{\pi}{2})$. Then, notice that f' has a single zero, and apply Rôlle's theorem to conclude that $f(x) = 0$ cannot have a third real root.

8. Apply Lagrange's MVT.

9. In each of the items, apply l'Hôpital's rule for $\frac{0}{0}$.

10. Apply l'Hôpital's rule for $\frac{0}{0}$.

11. Apply l'Hôpital's rule for $\frac{0}{0}$ twice.

12. For item (a), use Lagrange's MVT twice to get $f(x) - (x-a)f'(x) = -f''(\xi)(x - a)^2$, for some $\xi \in (a, x)$; then, use the continuity of f'', together with the fact that $f''(a) \neq 0$, to guarantee the existence of $c \in (a, b)$ such that $f'' \neq 0$ in $[a, c)$. For (b), let $g(x) = \frac{(x-a)f(x)}{f(x) - (x-a)f'(x)}$ and use l'Hôpital's rule twice to get $\lim_{x \to a} g(x) = -\frac{2f'(a)}{f''(a)}$.

13. Apply the ordinary l'Hôpital's rule to $F, G : (0, \frac{1}{a}) \to \mathbb{R}$, given by $F(x) = f(\frac{1}{x})$ and $G(x) = g(\frac{1}{x})$.

14. Firstly, observe that $\frac{\frac{d}{dx}(xf(x))}{\frac{d}{dx}x} = \frac{d}{dx}(xf(x)) = f(x) + xf'(x)$. Now, our hypotheses and the extension of Proposition 9.7 to limits at infinity guarantee that $\lim_{x \to +\infty}(f(x) + xf'(x))$ does exist. Hence, by the version of l'Hôpital's rule given by Problem 13, we have

$$\lim_{x \to +\infty} f(x) = \lim_{x \to +\infty} \frac{xf(x)}{x} = \lim_{x \to +\infty} \frac{\frac{d}{dx}(xf(x))}{\frac{d}{dx}x} = \lim_{x \to +\infty} (f(x) + xf'(x))$$

$$= \lim_{x \to +\infty} f(x) + \lim_{x \to +\infty} (xf'(x)).$$

Therefore, $\lim_{x \to +\infty}(xf'(x)) = 0$.

15. Suppose, without loss of generality, that $\lim_{x \to a} g(x) = +\infty$. Given $\epsilon > 0$, we want to find $0 < \delta < b - a$ such that $a < x < a + \delta \Rightarrow \left| \frac{f(x)}{g(x)} - L \right| < \epsilon$.

Take $0 < \eta < b - a$ such that $a < x \leq a + \eta \Rightarrow \left| \frac{f'(x)}{g'(x)} - L \right| < \frac{\epsilon}{2}$. Letting $x_0 = a + \eta$ and taking $a < x < x_0$, show that one has $\frac{f(x) - f(x_0)}{g(x) - g(x_0)} = \frac{f'(c_x)}{g'(c_x)}$ for some $c_x \in (x, x_0)$. Then, conclude that $L - \frac{\epsilon}{2} < \frac{f(x) - f(x_0)}{g(x) - g(x_0)} < L + \frac{\epsilon}{2}$ and, hence, that

$$L - \frac{\epsilon}{2} - \left(L - \frac{\epsilon}{2}\right)\frac{g(x_0)}{g(x)} + \frac{f(x_0)}{g(x)} < \frac{f(x)}{g(x)} < L + \frac{\epsilon}{2} - \left(L + \frac{\epsilon}{2}\right)\frac{g(x_0)}{g(x)} + \frac{f(x_0)}{g(x)}.$$

Finally, use the fact that $g(x) \to +\infty$ as $x \to a$ to find $0 < \delta < \eta$ such that, for $a < x < a + \delta$,

$$\left|\left(L - \frac{\epsilon}{2}\right)\frac{g(x_0)}{g(x)} - \frac{f(x_0)}{g(x)}\right|, \left|\left(L + \frac{\epsilon}{2}\right)\frac{g(x_0)}{g(x)} - \frac{f(x_0)}{g(x)}\right| < \frac{\epsilon}{2}.$$

16. Use Lagrange's MVT, together with the result of Problem 11, page 263.

17. Take a sequence $(x_n)_{n\geq 1}$ in $I \setminus \{x_0\}$ such that $x_n \to x_0$. If $x_n > x_0$ (resp. $x_n < x_0$), then, applying Lagrange's MVT to the interval $[x_0, x_n]$ (resp. $[x_n, x_0]$), conclude that there exists $y_n \in (x_0, x_n)$ (resp. $y_n \in (x_n, x_0)$) such that

$$f'(y_n) = \frac{f(x_n) - f(x_0)}{x_n - x_0}. \tag{B.5}$$

Now, since $y_n \to x_0$ and $y_n \neq x_0$, we have $f'(y_n) \to L$. Therefore, it follows from (B.5) that $\frac{f(x_n)-f(x_0)}{x_n-x_0} \to L$. The above reasoning has shown that

$$x_n \to x_0 \Rightarrow \frac{f(x_n) - f(x_0)}{x_n - x_0} \to L.$$

Argue by contradiction to show that this suffices to prove that $\lim_{x\to x_0} \frac{f(x)-f(x_0)}{x-x_0}$ does exist and equals L.

18. If such an f exists, start by writing the given relation as $\frac{f(x+y)-f(x)}{y} \leq -\frac{f(x+y)}{f(x)}$. Conclude that $f'(x) \leq -1$ for every $x > 0$ and, then, use Lagrange's MVT to show that $f(x + 1) \leq f(x) - 1$ for every $x > 0$. Finally, use this last inequality to show that f should take negative values, which is not the case.

19. Start by using Lagrange's MVT to show that there exists $0 < \delta < 1$ for which $|f(x)| < |x|$ whenever $|x| < \delta$. Then, observe that $|f'(x)| \leq |x|$ for $|x| < \delta$, and use Lagrange's MVT once more to obtain $|f(x)| \leq |x|^2$ for $|x| < \delta$. Iterate this reasoning, thus showing (with the same $\delta > 0$) that $|f(x)| \leq |x|^n$ for $|x| < \delta$ and every $n \in \mathbb{N}$. From this, conclude that $f(x) = 0$ for $|x| < \delta$. Extend the argument above to show that, if $f(x_0) = 0$, then there exists $\delta > 0$ such that $f(x) = 0$ for $|x - x_0| < \delta$. Finally, invoke the result of Problem 5, page 272, to get $f(x) = 0$ for every $x \in \mathbb{R}$.

Section 9.6

1. Consider the quadratic $f(x) = ax^2 + bx + c$, with $a > 0$ (the case $a < 0$ is completely analogous). Since $f'(x) = 2ax + b$ and $a > 0$, we have $f'(x) < 0$

if and only if $x < -\frac{b}{2a}$ and $f'(x) > 0$ if and only if $x > -\frac{b}{2a}$. Therefore, f decreases in $(-\infty, -\frac{b}{2a}]$ and increases in $(-\frac{b}{2a}, +\infty)$, so that f attains a global minimum at $x = -\frac{b}{2a}$ (and only there), with $f(-\frac{b}{2a}) = -\frac{\Delta}{4a}$.

2. In all of the items, apply the first derivative test in conjunction with the study of the first variation. For instance, in item (a) note that $f'(x) = \frac{16-3x^2}{2\sqrt{x}(x^2+16)^2}$ for $x > 0$. Therefore, $x = \frac{4}{\sqrt{3}}$ is the only critical point of f, which increases in $[0, \frac{4}{\sqrt{3}}]$ and decreases in $[\frac{4}{\sqrt{3}}, +\infty)$. Hence, $x = \frac{4}{\sqrt{3}}$ is the only global maximum point of f, with $f(\frac{4}{\sqrt{3}}) = \frac{3}{32\sqrt[4]{3}}$. Now, observe that $f(0) = 0$ and $f(x) > 0$ for $x > 0$, so that $x = 0$ is the only global minimum point of f. The remaining items can be dealt with likewise.

3. First of all, note that f has no points of global maximum or minimum, since (cf. Example 9.17) $\lim_{x \to -\infty} f(x) = -\infty$ and $\lim_{x \to +\infty} f(x) = +\infty$. Now, we have $f'(x) = 3x^2 + 2ax + b$, with $\Delta = 4(a^2 - 3b)$. If $\Delta < 0$, then $f'(x) > 0$ for every $x \in \mathbb{R}$, so that f increases along all of the real line. If $\Delta = 0$, then $f'(x) = 0 \Leftrightarrow x = -\frac{a}{3}$ and $f'(x) > 0$ for $x \neq -\frac{a}{3}$; hence, $-\frac{a}{3}$ is the only critical point of f, which still increases along all of the real line. Finally, if $\Delta > 0$, then f' has exactly two real roots, $x_1 < x_2$; moreover, $f'(x) > 0$ for $x < x_1$ or $x > x_2$, while $f'(x) < 0$ for $x_1 < x < x_2$; therefore, f increases in $(-\infty, x_1]$, decreases in $[x_1, x_2]$ and increases again in $[x_2, +\infty)$. Thus, f attains a local maximum at x_1 and a local minimum at x_2.

4. First, let $f : [0, +\infty) \to \mathbb{R}$ be given by $f(x) = \sin x - x$. Then, $f'(x) = \cos x - 1 \leq 0$, so that f is nonincreasing in $[0, +\infty)$. Hence, for $x \geq 0$ we have $f(x) \leq f(0) = 0$. Now, let $f : [0, +\infty) \to \mathbb{R}$ be given by $f(x) = \cos x - 1 + \frac{x^2}{2}$. By the first part, we have $f'(x) = -\sin x + x \geq 0$ for every $x \geq 0$, so that f is nondecreasing in $[0, +\infty)$; thus, for $x \geq 0$ we have $f(x) \geq f(0) = 0$. Finally, let $f : [0, +\infty) \to \mathbb{R}$ be given by $f(x) = \sin x - x + \frac{x^3}{6}$ and note that $f'(x) = \cos x - 1 + \frac{x^2}{2} \geq 0$; this gives $f(x) \geq f(0) = 0$ for $x \geq 0$.

5. Verify that $APQ \sim DQR$, so that $\overline{DR} = x(2-x)$ if $\overline{AQ} = x$. Then, compute $\overline{PQ} = \sqrt{x^2+1}$ and $\overline{QR} = (2-x)\sqrt{x^2+1}$, so that PQR has area equal to $\frac{1}{2}\overline{PQ} \cdot \overline{QR} = \frac{1}{2}(2-x)(x^2+1)$. Finally, study the first variation of $f : [0, 2] \to \mathbb{R}$, given by $f(x) = (2-x)(x^2+1)$, to conclude that it attains a maximum only for $x = \frac{2+\sqrt{7}}{3}$.

6. Let $A(\alpha, f(\alpha))$ and $A'(x, f(x))$, so that (thanks to elementary Analytic Geometry) α is a point of global minimum for $d : (a, b) \to \mathbb{R}$ given by $d(x) = (x - x_0)^2 + (f(x) - y_0)^2$. By the test of first derivative, we have $0 = d'(\alpha) = 2(\alpha - x_0) + 2(f(\alpha) - y_0)f'(\alpha)$, so that either $\alpha = x_0$ and $f'(\alpha) = 0$ (since $P \notin G_f$), or $\alpha \neq x_0$ and $f'(\alpha)\left(\frac{f(\alpha)-y_0}{\alpha-x_0}\right) = -1$. Now, use the fact that the product of the slopes of two perpendicular lines, none of which being vertical, equals -1.

7. Just notice that, from the previous problem, $\overleftrightarrow{AB}$ is perpendicular to the tangents to the graph of f at A and to the graph of g at B.

8. If $A(a, a^2)$ and $B(b, 1 - (b - 3)^2)$, then $\overline{AB}^2 = (a - b)^2 + (a^2 - 1 + (b - 3)^2)^2$. Condition $f'(a) = g'(b)$ gives us $b = 3 - a$, so that $\overline{AB}^2 = 4a^4 - 12a + 10$; hence, we are left to minimize $h(x) = 4x^4 - 12x + 10$. The test of the first derivative gives $a = \sqrt[3]{3/4}$ as the only minimum point of h, and an easy computation gives $\overline{AB} = \sqrt{10 - 9\sqrt[3]{3/4}}$.

9. Let $\overline{AH} = h$, $\overline{BH} = l$, $H\widehat{A}B = \alpha$ and $B\widehat{C} = \theta$. Apply some Trigonometry to get $\tan\theta = \frac{\tan(\alpha+\theta)-\tan\alpha}{1+\tan(\alpha+\theta)\cdot\tan\alpha} = \frac{2hl}{h^2+3l^2}$. Now, if $f(l) = \frac{2hl}{h^2+3l^2}$ for $l > 0$, compute $f'(l) = \frac{2h(h^2-3l^2)}{(h^2+3l^2)^2}$ and conclude that the critical point $l = \frac{h}{\sqrt{3}}$ of f is a point of global maximum.

10. Let $\theta = B\widehat{A}D = A\widehat{D}C$, so that $0 < \theta < \frac{\pi}{2}$, $\overline{AD} = a + 2b\cos\theta$ and $h = b\sin\theta$, where h is the height of $ABCD$. Hence, $A(ABCD) = \frac{1}{2}(\overline{AD} + \overline{BC})h = (a + b\cos\theta)b\sin\theta$, and it suffices to maximize $f : (0, \frac{\pi}{2}) \to \mathbb{R}$ given by $f(\theta) = (a + b\cos\theta)\sin\theta$. To what is left to do, use some Trigonometry to get $f'(\theta) = 2b\cos^2\theta + a\cos\theta - b$ and $f''(\theta) = -2b\sin(2\theta) - ab\sin\theta$. Now, check that $f''(\theta) < 0$ for every $\theta \in (0, \frac{\pi}{2})$ and that $\theta = \arccos\left(\frac{2b}{a+\sqrt{a^2+8b^2}+a}\right)$ is the only critical point of f. Finally, apply Corollary 9.60 to $-f$.

11. In the notations of Example 9.61, we have $\overline{PX} \cdot \overline{PY} = \frac{\overline{OP}^2 \sin\alpha\sin\beta}{\sin\theta\sin(\alpha+\beta+\theta)}$. Since $\overline{OP}^2 \sin\alpha\sin\beta$ is constant, we conclude that $\overline{PX} \cdot \overline{PY}$ is minimum if and only if $f(\theta) = \sin\theta\sin(\alpha + \beta + \theta)$ is maximum, where $f : (0, \pi - \alpha - \beta) \to \mathbb{R}$. In turn, the first derivative test assures that if a global maximum point for f exists at all, it should be a critical point of f. However, since

$$f'(\theta) = \cos\theta\sin(\alpha + \beta + \theta) + \sin\theta\cos(\alpha + \beta + \theta) = \sin(\alpha + \beta + 2\theta),$$

it follows that θ is critical if and only if $\alpha + \beta + 2\theta = \pi$ or, which is the same, if and only if $\theta = \pi - \alpha - \beta - \theta$, i.e., if and only if OXY is isosceles with basis XY. It now suffices to show that, OXY being isosceles with basis XY (or, which is the same, letting $\theta = \frac{1}{2}(\pi - \alpha - \beta)$ – which is the only critical point of f), then f does attain its maximum possible value. To this end, just note that if $0 < \theta < \frac{1}{2}(\pi - \alpha - \beta)$ then $0 < \alpha + \beta + 2\theta < \pi$, so that $f'(\theta) = \sin(\alpha+\beta+2\theta) > 0$; on the other hand, if $\frac{1}{2}(\pi-\alpha-\beta) < \theta < \pi-\alpha-\beta$, then $\pi < \alpha + \beta + 2\theta < 2\pi$, so that $f'(\theta) = \sin(\alpha + \beta + 2\theta) < 0$. Therefore, it follows from Proposition 9.57 that f is increasing in $(0, \frac{1}{2}(\pi - \alpha - \beta))$ and decreasing in $(\frac{1}{2}(\pi - \alpha - \beta), \pi - \alpha - \beta)$, so that $\theta = \frac{1}{2}(\pi - \alpha - \beta)$ is, indeed, the only global maximum point of f.

12. In the notations of Fig. B.6, we shall prove that the position of P along r that minimizes $\overline{AP} + \overline{BP}$ is such that $\alpha = \beta$.

To this end, start by observing that, had we drawn the point P to the left of A, then we would have $\alpha > \frac{\pi}{2}$; accordingly, had we drawn P to the right of B, we would have $\beta > \frac{\pi}{2}$. Therefore, $\alpha, \beta \in (0, \pi)$.

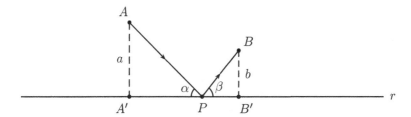

Fig. B.6 Minimizing $\overline{AP} + \overline{BP}$

If $P = A'$, Pythagoras' theorem gives $\overline{AP} + \overline{PB} = a + \sqrt{b^2 + c^2}$; if $P = B'$, we analogously have $\overline{AP} + \overline{PB} = b + \sqrt{a^2 + c^2}$. If $P \neq A', B'$, then, computing the sines of α and β in the right triangles APA' and BPB', we get (even if α or β is obtuse) $\overline{AP} + \overline{PB} = \frac{a}{\sin \alpha} + \frac{b}{\sin \beta}$. Also, it follows from $\overline{A'P} + \overline{B'P} = c$ that

$$\frac{a}{\tan \alpha} + \frac{b}{\tan \beta} = c, \tag{B.6}$$

and this is also valid even if α or β is obtuse.

Now, note that β continuously depends on α. On the other hand, a separate analysis of the cases of P to the left of A', P between A' and B' or P to the right of B', together with (B.6), show that β actually depends smoothly on α. Then, we just need to minimize $d(\alpha) = \frac{a}{\sin \alpha} + \frac{b}{\sin \beta}$, with β being given by (B.6).

Use the first derivative test to show that $d'(\alpha) = -\frac{a \cos \alpha}{\sin^2 \alpha} - \frac{b \cos \beta}{\sin^2 \beta} \cdot \beta'(\alpha)$, so that $d'(\alpha) = 0$ if and only if $\beta'(\alpha) = -\frac{a}{b} \cdot \frac{\cos \alpha \sin^2 \beta}{\cos \beta \sin^2 \alpha}$. On the other hand, differentiating (B.6) with respect to α we get $\frac{a}{\sin^2 \alpha} + \frac{b}{\sin^2 \beta} \cdot \beta'(\alpha) = 0$, so that $\beta'(\alpha) = -\frac{a}{b} \cdot \frac{\sin^2 \beta}{\sin^2 \alpha}$. Comparing both expressions above for $\beta'(\alpha)$, conclude that $d'(\alpha) = 0 \Leftrightarrow \frac{\cos \alpha}{\cos \beta} = 1 \Leftrightarrow \alpha = \beta$.

We are left to showing that the position of P at which $\alpha = \beta$ actually minimizes $\overline{AP} + \overline{PB}$. To this end, computing $d''(\alpha)$ (for a generic α) and recalling that $\alpha, \beta \in (0, \pi)$, we obtain $d''(\alpha) = \frac{a}{\sin^3 \alpha}(1 + \cos^2 \alpha) + \frac{b}{\sin^3 \beta}(1 + \cos^2 \beta) > 0$. Hence, Corollary 9.60 guarantees that the critical point of d is indeed a minimum global point, among all positions of P along r, as long as $P \neq A', B'$.

Finally, conclude that when $\alpha = \beta$, then $d(\alpha) = \sqrt{c^2 + (a + b)^2}$. Finally, observe that $\sqrt{c^2 + (a + b)^2} < a + \sqrt{b^2 + c^2}, b + \sqrt{a^2 + c^2}$ (which are the values of $\overline{AP} + \overline{PB}$ when $P = A'$ or B').

13. Fix a cartesian system for which $A(0, a)$ and $B(l, b)$, and let $P(x, 0)$, with $x \in \mathbb{R}$. Use basic Analytic Geometry to show that it suffices to prove that $f : \mathbb{R} \to \mathbb{R}$, given by $f(x) = (x^2 + a^2)((x - l)^2 + b^2)$, has a single minimum global point,

provided $l \leq \sqrt{2(a^2 + b^2)}$. Compute $f(x) = x^4 - 2lx^3 + (a^2 + b^2 + l^2)x^2 - 2a^2lx + a^2(b^2 + l^2), f'(x) = 4x^3 - 6lx^2 + 2(a^2 + b^2 + l^2)x - 2a^2l$ and $f''(x) = 12x^2 - 12lx + 2(a^2 + b^2 + l^2)$. Then, show that f' has a root in the interval $(0, l)$ and examine under what conditions we have $f''(x) \geq 0$ for every $x \in \mathbb{R}$.

14. Conclude that it suffices to show that $f(x) = \frac{(x+1)^{n+1}}{x^n}$, $x > 0$, attains its minimum value at $x = n$. To this end, study the first variation of f.

15. Firstly, show that $y = \frac{1}{2}\left(k(\sqrt{1 + x^2} - x) + \frac{1}{k}(\sqrt{1 + x^2} + x)\right)$. Then, study the first variation of $f : \mathbb{R} \to \mathbb{R}$ given by $f(x) = x + y$.

16. Use the inequality between the arithmetic and geometric means of two reals to show that the given expression is greater that or equal to $f(x + y)$, where $f(t) = t + \frac{2}{t} + \frac{2}{t^2}$ for $t > 0$. Then, study the first variation of f.

17. Letting $\alpha = \sin\frac{\widehat{A}}{2}$ and $\beta = \sin\frac{\widehat{B}}{2}$, show that it suffices to analyse the equality $f(\alpha) = f(\beta)$, where $f(x) = \frac{x^{23}}{(1-x^2)^{24}}$, for $x \in (0, 1)$. To this end, study the first variation of f.

18. Use the MVT to guarantee the existence of $c \in (a, b)$ such that $f'(c) = 1$. If $f(c) = c$, it suffices to apply the MVT twice. If $f(c) < c$, use the MVT to assure the existence of $\alpha \in (0, c)$ and $\beta \in (c, 1)$ such that $f'(\alpha) < 1$ and $f'(\beta) > 1$; then, apply Darboux's theorem. If $f(c) > c$, argue analogously.

19. Write $f(x) - f'(x) - f''(x) + f'''(x) = (f(x) - f'(x)) - \frac{d}{dx}(f'(x) - f''(x))$. Then, apply the following claim twice: if g is a polynomial function of positive degree, such that $g(x) - g'(x) \geq 0$ for every $x \in \mathbb{R}$, then $g(x) \geq 0$ for every $x \in \mathbb{R}$. In order to prove this claim, start by using the result of Example 9.17 to show that g has even degree and positive leading coefficient. Then, take (by the result of Problem 10, page 263) $x_0 \in \mathbb{R}$ such that g attains its minimum at x_0 and apply the first derivative test.

Section 9.7

1. For $a < b$ in I and $0 < t < 1$, it follows from the strictly convex character of g that $g((1 - t)a + tb) < (1 - t)g(a) + tg(b)$. Now, since f is increasing, we have $(f \circ g)((1 - t)a + tb) = f(g((1 - t)a + tb)) < f((1 - t)g(a) + tg(b))$. Finally, use the fact that f is also strictly convex to show that the same is true of $f \circ g$.

2. Firstly, Theorem 8.35 shows that f is either increasing or decreasing. Suppose that f is increasing (the other case in totally analogous). For $a < b$ in I and $0 < t < 1$, we have $f((1 - t)a + tb) < (1 - t)f(a) + tf(b)$. Let $\alpha = f(a)$ and $\beta = f(b)$, so that $a = f^{-1}(\alpha)$ and $b = f^{-1}(\beta)$. Applying f^{-1} to both sides of the above inequality and invoking the result of Problem 8, page 176, show that $(1 - t)f^{-1}(\alpha) + tf^{-1}(\beta) < f^{-1}((1 - t)\alpha + t\beta)$. From this, conclude that f^{-1} is strictly concave.

3. Use the definition of strictly convex (resp. concave) function, along the lines of the hints given to the two previous problems.

4. Write $f(x) = \frac{(x-1)+1}{\sqrt{1-x}} = -\sqrt{1-x} + \frac{1}{\sqrt{1-x}}$ and show that each one of the functions $t \mapsto -\sqrt{t}$ and $t \mapsto \frac{1}{\sqrt{t}}$, $t \in (0,1)$, is strictly convex. Then, use the result of the previous problem. Alternatively, compute $f''(x) = \frac{4-x}{4(1-x)^{5/2}}$ and solve the inequality $f''(x) > 0$ (resp. $f''(x) < 0$).

5. Since $f''(x) = 6x + 2a$, we have $f'' > 0$ in $(-\frac{a}{3}, +\infty)$ and $f'' < 0$ in $(-\infty, -\frac{a}{3})$. Hence, f is strictly convex in $(-\frac{a}{3}, +\infty)$ and strictly concave in $(-\infty, -\frac{a}{3})$, so that $x = -\frac{a}{3}$ is its only inflection point.

6. Computing derivative, we get $f''(x) = (2 - \frac{1}{x^2}) \sin \frac{1}{x} - \frac{2}{x} \cos \frac{1}{x}$. Now, if $n \in \mathbb{N}$ and $x_n = \frac{1}{n\pi}$, then $\frac{1}{x_n} = n\pi$, so that $f''(x_n) = -2n\pi \cos(n\pi) = (-1)^{n+1}2n\pi$. Therefore, $f''(x_k)f''(x_{k+1}) < 0$, and the IVT guarantees the existence of $y_n \in (x_{n+1}, x_n)$ such that $f''(y_n) = 0$ (notice that we need not invoke Darboux's Theorem 9.62, since f'' is continuous in $(0, \frac{1}{\sqrt{2}})$). Verify that y_n is an inflection point of f.

7. Note that g is also twice differentiable, with $g''(x) = 2f'(x) + xf''(x) \geq 0$. Then, apply item (a) of Theorem 9.69.

8. Suppose that f is convex. Give $x, y \in I$, let us first show that $f((1-t)x + ty)) \leq (1-t)f(x) + tf(y)$ (*) for every dyadic rational number $t \in [0,1]$ (cf. Problem 5, page 206); to this end, let's make induction on $k \geq 1$. For $k = 1$, we have the validity of (*) for $t = 0$, $\frac{1}{2}$ and 1. For $t = \frac{3}{4}$, applying the convexity of f twice we get

$$f\left(\frac{x+3y}{4}\right) = f\left(\frac{\frac{x+y}{2}+y}{2}\right) \leq \frac{f\left(\frac{x+y}{2}\right) + f(y)}{2}$$

$$\leq \frac{\frac{f(x)+f(y)}{2} + f(y)}{2} = \frac{1}{4}f(x) + \frac{3}{4}f(y),$$

as wished. Finally, for $t = \frac{1}{4}$ it suffices to interchange x and y in the reasoning above. Now, by the sake of induction hypothesis, let (*) be valid for a certain $k \in \mathbb{N}$, all $x, y \in I$ and every integer $0 \leq n \leq 2^k$. For $t = \frac{m}{2^{k+1}}$, where $0 \leq m \leq 2^{k+1}$ is an integer, let's distinguish two distinct subcases: (i) m is even, say $m = 2n$: then $t = \frac{n}{2^k}$, and the validity of (*) follows from the induction hypothesis. (ii) m is odd, say $m = 2n + 1$: then

$$t = \frac{m}{2^{k+1}} = \frac{1}{2}\left(\frac{m-1}{2^{k+1}} + \frac{m+1}{2^{k+1}}\right) = \frac{1}{2}\left(\frac{n}{2^k} + \frac{n+1}{2^k}\right).$$

Letting $s = \frac{n}{2^k}$ and $u = \frac{n+1}{2^k}$, we have $s, u \in [0,1]$ and $t = \frac{1}{2}(s+u)$. Hence, it follows from the convexity of f and the induction hypothesis that

$$f((1-t)x + ty) = f\left(\left(1 - \frac{s+u}{2}\right)x + \left(\frac{s+u}{2}\right)y\right)$$

$$= f\left(\frac{((1-s)x+sy) + ((1-u)x+uy)}{2}\right)$$

$$\leq \frac{f((1-s)x+sy) + f((1-u)x+uy)}{2}$$

$$\leq \frac{1}{2}[((1-s)f(x) + sf(y)) + ((1-u)f(x) + uf(y))]$$

$$= \left(1 - \frac{s+u}{2}\right)f(x) + \left(\frac{s+u}{2}\right)f(y)$$

$$= (1-t)f(x) + tf(y).$$

Now, given $x, y \in I$, the continuity of f assures that of $g : [0, 1] \to \mathbb{R}$, such that $g(t) = (1-t)f(x) + tf(y) - f((1-t)x+ty)$ for $t \in [0, 1]$. On the other hand, we have shown above that $g(t) \geq 0$ for every dyadic rational number $t \in [0, 1]$, so that the density of them in $[0, 1]$, with the help of Problem 13, page 264, assures that $g(t) \geq 0$ for every $t \in [0, 1]$.

9. Suppose f is strictly convex, fix $a < b$ in I and $0 < t < 1$. If $c = (1-t)a + tb$, then $a < c < b$, so that we can choose $\alpha \in (a, c)$ and $\beta \in (c, b)$ satisfying $c = \frac{\alpha+\beta}{2}$. Moreover, letting $\alpha = (1-s)a + sb$ and $\beta = (1-u)a + ub$, with $s, u \in (0, 1)$, show that $c = \frac{\alpha+\beta}{2} \Rightarrow t = \frac{s+u}{2}$. Now, use the given condition together with the definition of strict convexity and the result of the previous problem, to obtain $f((1-t)a + tb) < \frac{f(\alpha)+f(\beta)}{2} \leq (1-t)f(a) + tf(b)$.

10. Use the hypotheses on f to show that, given positive and distinct reals x and y, we have $\frac{xf(x)+yf(y)}{x+y} \geq f\left(\frac{x^2+y^2}{x+y}\right) \geq f\left(\frac{x+y}{2}\right)$, with at least one of these inequalities being a strict one. From this, deduce that $\frac{g(x)+g(y)}{2} > g\left(\frac{x+y}{2}\right)$ and apply the result of the previous problem.

11. For the function of item (a), use the inequality of item (b) of Example 5.1 to show that $f(\frac{a+b}{2}) \leq \frac{f(a)+f(b)}{2}$ for all $a, b > 0$, with equality if and only if $a = b$. For item (b), use induction on n, together with the fact that f is increasing, to show that $(\frac{a+b}{2})^n \leq \frac{a^n+b^n}{2}$ for all $a, b > 0$, with equality if and only if $a = b$. For (c), use Trigonometry to show that, on $(0, \pi)$, we have $\sin a + \sin b \leq 2\sin\left(\frac{a+b}{2}\right)$, with equality if and only if $a = b$. For item (d), argue as in (c), using Trigonometry to show that $2\tan(\frac{a+b}{2}) \leq \tan a + \tan b$ for $a, b \in (0, \frac{\pi}{2})$, with equality if and only if $a = b$.

12. If $x = x_0$, there is nothing to do. Otherwise, Lagrange's MVT (applied twice) gives a real c between x_0 and x such that $f(x) - f(x_0) = f'(c)(x - x_0)$, and d between c and x_0, such that $f'(c) - f'(x_0) = f''(d)(c - x_0)$. Hence,

$$f(x) - f(x_0) - f'(x_0)(x - x_0) = f'(c)(x - x_0) - f'(x_0)(x - x_0)$$
$$= (f'(c) - f'(x_0))(x - x_0)$$
$$= f''(d)(c - x_0)(x - x_0).$$

Now, since f is convex and twice differentiable, it follows from Corollary 9.70 that $f''(d) \geq 0$. Then, taking into account that $(c - x_0)(x - x_0) > 0$, the above computations give

$$f(x) - f(x_0) - f'(x_0)(x - x_0) \geq 0.$$

13. Suppose, to the contrary, that there exist natural numbers $n_1 < n_2 < n_3 < \cdots$ such that $(f(n_k))_{k \geq 1}$ is an arithmetic progression. Use the strict convexity and the increasing character of f to successively get, for $k > 1$,

$$f(n_k) > f\left(\frac{n_{k-1} + n_{k+1}}{2}\right)$$

and $2n_k > n_{k-1} + n_{k+1}$. Finally, arrive at a contradiction.

14. Firstly, note that $f : (0, +\infty) \to \mathbb{R}$, given by $f(x) = x^k$, is strictly convex. Now, apply Jensen's inequality (9.31).

15. For item (b), note first that (a) allows us to take O in the interior of $A_1 A_2 \ldots A_n$. In turn, letting $\alpha_i = A_i \widehat{O} A_{i+1}$ for $1 \leq i \leq n$ (with $A_{n+1} = A_1$), this gives $0 < \alpha_i < \pi$ for every i. Finally, adapt to the present case the reasoning of the solution of Example 9.78.

16. Letting $\alpha_i = A_i \widehat{O} A_{i+1}$ for $1 \leq i \leq n$ (with $A_{n+1} = A_1$), we have $0 < \alpha_i < \pi$. Now, show that the perimeter of $A_1 A_2 \ldots A_n$ equals $r \sum_{i=1}^n \tan \frac{\alpha_i}{2}$. Finally, observe that $\tan : \left(0, \frac{\pi}{2}\right) \to \mathbb{R}$ is strictly convex and apply (9.31).

17. Sine Law gives $a + b + c = 2R(\sin \widehat{A} + \sin \widehat{B} + \sin \widehat{C})$. Now, apply Jensen's inequality (9.31) to $\sin : (0, \pi) \to \mathbb{R}$.

18. For item (a), use the fact that $A(ABC) = A(ABP) + A(ACP) + A(BCP)$. For (b), show that $f : (0, h) \to \mathbb{R}$ given by $f(x) = \frac{h-x}{h+x}$ is strictly convex and apply Jensen's inequality to get $f(h) \geq \frac{3}{2}$, with equality if and only if P is the center of ABC.

19. For the first inequality, apply (9.31), together with the result of Problem 4, page 337. For the second, apply (9.31) again, this time using the fact that the square root function is strictly concave in $(0, +\infty)$.

Section 9.8

1. Since $\sin x = \cos(\frac{\pi}{2} - x)$, we have $\csc x = \sec(\frac{\pi}{2} - x)$. Now, use the discussion of Example 9.79, together with the results of Problem 10, page 194.

2. We sketch hints to some of the items, letting the other ones to the reader. For item (b), compute $f'(x) = 1 - \frac{1}{x^2}$ and $f''(x) = \frac{2}{x^3}$ to conclude that f decreases in $(-1, 1)$, increases in each of $(-\infty, -1)$ and $(1, +\infty)$, is strictly convex in $(0, +\infty)$ and strictly concave in $(-\infty, 0)$. Also, use the result of Problem 11, page 291, to conclude that $y = x$ is the only oblique asymptote of f. For item (f), compute $f'(x) = \frac{1-x^2}{(x^2+1)^2}$ and $f''(x) = \frac{2x(x^2-3)}{(x^2+1)^3}$ to conclude that $x = \pm 1$ are the only critical points of f, while $x = 0$ and $x = \pm\sqrt{3}$ are its only inflection points. Also, $f'(x) > 0$ if and only if $|x| < 1$, so that f increases along the interval $(-1, 1)$ and decreases along each of $(-\infty, -1)$ and $(1, +\infty)$; therefore, $x = -1$ is a local minimum point and $x = 1$ is a local maximum point of f. However, since $f(-1) = -\frac{1}{2}, f(1) = \frac{1}{2}$ and $\lim_{|x| \to +\infty} f(x) = 0$, it follows that $x = -1$ is the only global minimum point of f, while $x = 1$ is its only global maximum point. Now, at the inflection point $x = 0$, the tangent to the graph is (trigonometrically) at $\frac{\pi}{4}$ from the horizontal axis, while at the inflection points $x = \pm\sqrt{3}$ such an angle equals $\arctan\left(-\frac{1}{8}\right)$. Note that $f''(x) > 0$ if and only if $x \in (-\sqrt{3}, 0) \cup (\sqrt{3}, +\infty)$; therefore, f is strictly convex in each of these intervals. Accordingly, f is strictly concave in each of the intervals $(-\infty, -\sqrt{3})$ and $(0, \sqrt{3})$. Finally, the horizontal axis is the only horizontal asymptote of the graph. For item (h), compute $f'(x) = -\frac{3x^4+1}{(x^4-1)^2}$ and $f''(x) = \frac{12x^3(x^4+1)}{(x^4-1)^3}$ to conclude that f decreases along all of its domain, is strictly convex in each of the intervals $(-1, 0)$ and $(1, +\infty)$, and strictly concave in each of $(-\infty, -1)$ and $(0, 1)$. Also, $x = 0$ is its only inflection point, in which the tangent to the graph makes a trigonometric angle of $-\frac{\pi}{4}$ with the horizontal axis. Finally, this axis is the only horizontal asymptote of f, with $\lim_{x \to +\infty} f(x) = \lim_{x \to -\infty} f(x) = 0$; lines $x = -1$ and $x = 1$ are the only vertical asymptotes, with $\lim_{x \to (-1)-} f(x) = -\infty$, $\lim_{x \to (-1)+} f(x) = +\infty$, $\lim_{x \to 1-} f(x) = -\infty$ and $\lim_{x \to 1+} f(x) = +\infty$.

3. First of all, note that since $f'(x_0) = g'(x_0)$, line r is tangent to the graph of g (hence, to Γ) at A; thus, $\overleftrightarrow{OA} \perp r$. For item (a), if $f'(x_0) = 0$, then r is horizontal and, then, $\overleftrightarrow{OA}$ is vertical. If $f'(x_0) \neq 0$, basic Analytic Geometry assures that the slope of $\overleftrightarrow{OA}$ equals $-\frac{1}{f'(x_0)}$. For (c), it follows from $(x - \alpha)^2 + (g(x) - \beta)^2 = R^2$ and the chain rule that $2(x - \alpha) + 2(g(x) - \beta)g'(x) = 0$; therefore, $2 + 2g'(x)^2 + 2(g(x) - \beta)g''(x) = 0$. Now, substitute $x = x_0$ in the first equality and use the result of the second part of (a) to get the first equality of (c); then, substitute $x = x_0$ in the second equality and use the fact that $g(x_0) = f(x_0)$ and $g'(x_0) = f'(x_0)$ to get the second equality of (c). For item (e), observe that $R^2 = (x_0 - \alpha)^2 + (f(x_0) - \beta)^2$ and substitute the values of α and β computed at (d). Finally, for the second part of (f), notice that condition $f'(x_0) = g'(x_0)$ guarantees that $r \perp \overleftrightarrow{AO}$.

4. For item (a), note that f is even. For (b), compute $f'(x) = \sin\frac{1}{x} - \frac{1}{x}\cos\frac{1}{x}$. For (c), observe that $\sin y - y\cos y = 0$ implies $\cos y \neq 0$ and, then, $\tan y = y$. Now, use the IVT (sketching the graphs of $y \mapsto \tan y$ and $y \mapsto y$ may help) to guarantee

that the set of reals $y > 0$ such that $\tan y = y$ can be put in a sequence $(y_n)_{n\geq 1}$, satisfying the given conditions. For item (d), note that

$$2k\pi < y < y_{2k} \Rightarrow \tan y < y \Rightarrow \sin y - y\cos y < 0$$

and, hence, that $f'(x) < 0$ for $x \in (x_{2k}, \frac{1}{2k\pi})$; then, use a similar reasoning to show that $f'(x) > 0$ for $x \in (\frac{1}{2k\pi+\frac{\pi}{2}}, x_{2k})$, so that x_{2k} is a local maximum point of f. Then, argue analogously to conclude that x_{2k-1} is a local minimum point for f. For item (f), initially observe that, since $f'(x_n) = \sin\frac{1}{x_n} - \frac{1}{x_n}\cos\frac{1}{x_n} = 0$, we get $|f(x_n)| = |x_n \sin\frac{1}{x_n}| = |\cos\frac{1}{x_n}| = |\cos y_n|$; then, apply the last part of (c). For (f), conclude first that $f'(x) > 0$ for $x > x_1$; then, apply the fundamental trigonometric limit to get $\lim_{x\to+\infty} f(x) = \lim_{y\to 0+} \frac{\sin y}{y} = 1$. For (g), start by computing $f''(x) = -\frac{1}{x^3}\sin\frac{1}{x}$. Then, for $x > 0$, note that $f''(x) > 0$ (resp. $f''(x) < 0$) if and only if $\sin\frac{1}{x} < 0$ (resp. $\sin\frac{1}{x} > 0$, and solve such inequalities. Finally, for item (h), use the result of the previous items, also noticing that $|f(x)| \leq |x|$, with $|f(x)| = |x|$ if and only if $x = \frac{1}{n\pi+\frac{\pi}{2}}$ for some $n \in \mathbb{Z}$. Your drawing should be somewhat similar to the one shown in Fig. B.7).

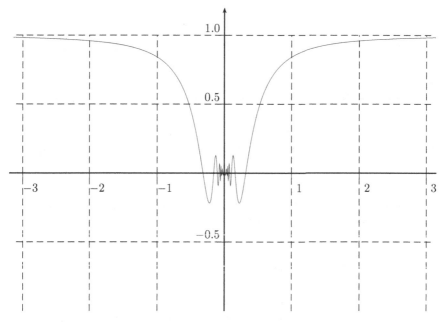

Fig. B.7 Sketching the graph of $x \mapsto x\sin\frac{1}{x}$

Section 10.2

1. Adapt the proof of $S(f; Q) \leq S(f; P)$ to this case. You shall need to use the fact that $\inf_{[x_{i-1}, x']} f$, $\inf_{[x', x_i]} f \geq \inf_{[x_{i-1}, x_i]} f$.
2. Take a partition P of $[a, b]$ and show that $s(f; P) = S(f; P) = c(b - a)$.
3. Read the proof of Theorem 10.5 again to check that $\lim_{k \to +\infty} (S(f; P_k) - s(f; P_k)) = 0$. Then, use the fact that f is integrable, together with the inequalities $s(f; P_k) \leq \int_a^b f(x)dx \leq S(f; P_k)$, to show that the limits $\lim_{k \to +\infty} S(f; P_k)$ and $\lim_{k \to +\infty} s(f; P_k)$) do exist and are equal to $\int_a^b f(x)dx$.
4. Letting $P_k = \{a = x_0 < x_1 < \cdots < x_k = b\}$ be a uniform partition, we have $\underline{A}(f; k) = s(f; P_k)$ and $\overline{A}(f; k) = S(f; P_k)$, so that (10.6) gives $s(f; P_k) < \frac{b^{n+1}}{n+1} \left(1 - \frac{1}{k}\right)^{n+1} < S(f; P_k)$. Now, make $k \to +\infty$.
5. For item (a), substitute $u = x^2$, $v = y^2$ and write down the expressions involved in terms of x and y. For item (b), let $u = x_j$ and $v = x_{j-1}$ in (a) and add the inequalities thus obtained, observing that $x_j - x_{j-1} = \frac{b-a}{k}$. Finally, for (c) let $k \to +\infty$ in the inequalities of (b) and use the result of Problema 3.
6. For $1 \leq k \leq n$, write $\frac{1}{\sqrt{n}} \cdot \frac{1}{\sqrt{k+n}} = \frac{1}{n} \cdot \frac{1}{\sqrt{\frac{k}{n}+1}}$. Then, if $f(x) = \frac{1}{\sqrt{x+1}}$ and $P_n = \{0, \frac{1}{n}, \frac{2}{n}, \ldots, \frac{n-1}{n}, 1\}$, note that $a_n = s(f; P_n)$. Finally, use the result of Problem 3.
7. For $1 \leq k \leq n$, let $n \cdot \frac{1}{n^2+k^2} = \frac{1}{n} \cdot \frac{1}{1+\left(\frac{k}{n}\right)^2}$. Then, take $f(x) = \frac{1}{1+x^2}$, $P_n = \{0, \frac{1}{n}, \frac{2}{n}, \ldots, \frac{n-1}{n}, 1\}$ and note that $a_n = s(f; P_n)$. Finally, use the result of Problem 3.
8. For the first part, use the given inequality to show that Cauchy's integrability criterion is satisfied. For the second part, conclude (also from the given inequality) that $\lim_{k \to +\infty} s(f; P_k)$ and $\lim_{k \to +\infty} S(f; P_k)$ exist and are equal. Then, use the fact that $s(f; P_k) \leq \int_a^b f(x)dx \leq S(f; P_k)$ to finish the proof.
9. For item (a), adapt (10.17) to the present case, noticing that, if $M_j = f(x_j')$ and $m_j = f(x_j'')$, then $M_j - m_j = f(x_j') - f(x_j'') \leq c|x_j' - x_j''| \leq c(x_j - x_{j-1}) = \frac{c(b-a)}{k}$. For item (b), start by showing that $\Sigma(f; P_k; \xi)$ and $\int_a^b f(x)dx$ both belong to the closed interval with endpoints $s(f; P_k)$ and $S(f; P_k)$.
10. For the first part of item (a), note that $2 \sin(a + jh) \sin \frac{h}{2} = \cos(a + (j-1)h) - \cos(a + jh)$, add such equalities from $j = 0$ to $j = k$ and transform the result thus obtained in product. For the second part, just note that $\cos x = \sin\left(\frac{\pi}{2} - x\right)$. For the first part of (b), Example 8.8 shows that the sine function is Lipschitz, with Lipschitz constant $c = 1$. Let $P_k = \{a = x_0 < x_1 < \cdots < x_k = b\}$ be a uniform partition of $[a, b]$ and (in the notations of the previous problem) take $\xi_{kj} = x_j$ for $1 \leq j \leq k$. Letting $h = \frac{b-a}{k} = x_j - x_{j-1}$ for $1 \leq j \leq k$, note that

$$\Sigma(\sin; P_k; \xi_k) = h \sum_{j=1}^{k} \sin(a + jh) = \frac{h \sin\left(a + \frac{(k-1)h}{2}\right) \sin \frac{(k+1)h}{2}}{\sin \frac{h}{2}} - h \sin a.$$

Then, apply the fundamental trigonometric limit, together with the result of item (b) of the previous problem, to conclude that

$$\int_a^b \sin x\, dx = \lim_{h\to 0}\left(\frac{h\sin\left(a + \frac{(k-1)h}{2}\right)\sin\frac{(k+1)h}{2}}{\sin\frac{h}{2}} - h\sin a\right)$$

$$= 2\lim_{h\to 0}\left[\frac{\frac{h}{2}}{\sin\frac{h}{2}}\cdot\sin\left(\frac{b+a}{2}-\frac{h}{2}\right)\sin\left(\frac{b-a}{2}+\frac{h}{2}\right)\right]$$

$$= 2\sin\left(\frac{b+a}{2}\right)\sin\left(\frac{b-a}{2}\right) = \cos a - \cos b.$$

For the second part, adapt the above argument.

11. Firstly, note that, by the density of the irrationals in the real line, we have $s(f;P) = 0$ for every partition P. Now, give $\epsilon > 0$, fix $n_0 \in \mathbb{N}$ such that $n_0 > \frac{2}{\epsilon}$. Then, if $A = \{x \in [0,1];\ x = \frac{m}{n},\ \text{with}\ m, n \in \mathbb{N}\ \text{and}\ n < n_0\}$, show that A is finite and use this fact to guarantee the existence of a partition P_0 of $[0,1]$, such that the sum of the lengths of the intervals of P_0 which contain some element of A is less than $\frac{\epsilon}{2}$. Finally, show that $S(f;P_0) < \epsilon$.

Section 10.3

1. Let $f(x) = x\sin(\pi x)$. If $P_n = \{0 = x_0 < x_1 < \cdots < x_n = 1\}$ is a uniform partition, show that $a_n = \pi\Sigma(f;P_n;\xi)$, where $\xi = (\xi_1,\ldots,\xi_n)$ is such that $\xi_j = x_j$, for $1 \le j \le n$. Now, apply Riemann's theorem.

2. More generally, let $A = \{x_1, x_2, x_3,\ldots\} \subset \mathbb{R}$ be countably infinite. Given $\epsilon > 0$, let $I_j = \left(x_j - \frac{\epsilon}{2^{j+2}}\right)$ for $j \ge 1$ and note that $\sum_{j\ge 1}|I_j| = \frac{\epsilon}{2} < \epsilon$.

3. Given $\epsilon > 0$ and $j \ge 1$, let $I_{1j}, I_{2j},\ldots$ be open intervals such that $A_j \subset \bigcup_{i\ge 1} I_{ij}$ and $\sum_{i\ge 1}|I_{ij}| < \frac{\epsilon}{2^{j+1}}$. Then, $\{I_{ij};\ i,j \ge 1\}$ is countably infinite, $\bigcup_{j\ge 1} A_j \subset \bigcup_{i,j\ge 1} I_{ij}$ and $\sum_{i,j\ge 1}|I_{ij}| < \frac{\epsilon}{2}$.

4. Let D_f, D_g and D_{fg} be the sets of points of discontinuity of f, g and D_{fg}. Show that $D_{fg} \subset D_f \cup D_g$ and, then, apply Lebesgue's theorem, together with the result of the previous problem.

5. Let $Q = \{0 = y_0 < y_1 < \cdots < y_l = 1\}$ be a partition of the codomain $[0,1]$ of f and, for $1 \le j < k$, let A_j be the inverse imagem of $[y_{j-1}, y_j)$; let also A_k be the inverse image of $[y_{k-1}, y_k]$. Show that $A_j = \emptyset$ for $1 < j < k$ and recall that, since $A_k = \mathbb{Q} \cap [0,1]$, we have $m(A_k) = 0$. Then, use (10.23) to conclude that $\int_{[0,1]} f_Q = 0$.

6. Firstly, assume that f is increasing, and let $Q = \{0 = y_0 < y_1 < \cdots < y_l = L\}$ be a partition of $[0, L]$. For $1 \le j < k$, let A_j be the inverse image of $[y_{j-1}, y_j)$; let also A_k be the inverse image of $[y_{k-1}, y_k]$. If $x_j = f^{-1}(y_j)$ for $1 \le j \le k$, show

that $A_j = [x_{j-1}, x_j)$ for $1 \leq j < k$, and $A_k = [x_{k-1}, x_k]$. Then, if $P = \{a = x_0 < x_1 < \cdots < x_k = b\}$, show that $\int_{[a,b]} f_Q = s(f; P)$. Now, consider the case of a nondecreasing f, and observe that the case of a nonincreasing f is completely analogous.

7. Look at both integrals as the areas of regions under the graphs of f and f^{-1}, and recall that the graph of f^{-1} is symmetric to the graph of f with respect to the straight line $y = x$ of the cartesian plane. Then, reflect the region under the graph of f^{-1} along this line and show that the union of the region under the graph of f to this reflected one contains a rectangle with sides a and b.

8. Start by showing that, given $t \in [0, 1]$, the fact that f is nondecreasing guarantees that $A(R_f)$ is greater than or equal to the area of the rectangle of basis $[t, 1]$ and height $f(t)$, i.e., that

$$\int_0^1 f(x)dx \geq (1 - t)f(t) \geq f(t) - t.$$

Then, fix $s \in [0, 1]$ and let $t = g(s)$ in the above inequality to get $f(g(s)) - g(s) \leq \int_0^1 f(x)dx$. Finally, use the fact that $s \in [0, 1]$ was chosen arbitrarily to conclude that

$$\int_0^1 (f(g(s)) - g(s))ds \leq \int_0^1 f(x)dx.$$

9. Let $f : [a, b] \to \mathbb{R}$ be a bounded function for which $\lim_{|P| \to 0} \Sigma(f; P; \xi) = I$, with I not depending on the choice of the intermediate points ξ of P. In order to show that f is integrable, given $\epsilon > 0$ we need to find a partition P of $[a, b]$ such that $S(f; P) - s(f; P) < \epsilon$. To this end, start by taking P_0 such that $\Sigma(f; P_0; \xi) - I| < \frac{\epsilon}{3}$ for every choice of intermediate points ξ. If $P = \{a = x_0 < x_1 < \ldots < x_k = b\}$ and $\xi_j, \eta_j \in [x_{j-1}, x_j]$, show that

$$\sum_{j=1}^k |f(\xi_j) - f(\eta_j)|(x_j - x_{j-1}) < \frac{2\epsilon}{3}. \tag{B.7}$$

Now, let $M_j = \sup_{[x_{j-1}, x_j]} f$, $m_j = \inf_{[x_{j-1}, x_j]} f$ and choose sequences $(\xi_{jn})_{n \geq 1}$ and $(\eta_{jn})_{n \geq 1}$ in $[x_{j-1}, x_j]$ satisfying $f(\xi_{jn}) \xrightarrow{n} M_j$ and $f(\eta_{jn}) \xrightarrow{n} m_j$. Make $\xi_j = \xi_{jn}$ and $\eta_j = \eta_{jn}$ in (B.7) to show that $\sum_{j=1}^k (M_j - m_j)(x_j - x_{j-1}) \leq \frac{2\epsilon}{3}$.

Section 10.4

1. Modify the function of item (a) of Problem 11, page 311, taking, for instance $f(x) = (x - a)^{4/3} \sin \frac{1}{x-a}$, if $x \in (a, b]$.

2. Making $x_0 = a$ and $x_{k+1} = b$, Proposition 10.21 gives

$$\int_a^b f(x)dx = \sum_{j=1}^k \int_{x_{j-1}}^{x_j} f(x)dx.$$

Now, observe that, in this last expression, the sum of the summands $\int_{x_{j-1}}^{x_j} f(x)dx$ for which $f > 0$ in (x_{j-1}, x_j) corresponds to $A(R_f^+)$, whereas the sum of those summands for which $f < 0$ in (x_{j-1}, x_j) corresponds to $-A(R_f^-)$.

3. The hypothesis gives $g_1(x) - f_1(x) = g_2(x) - f_2(x)$ for every $x \in [a, b]$.

4. Use Proposition 10.19, together with Problem 2, page 357, and Lemma 10.17.

5. For item (a) and for $j \in \mathbb{Z}_+$, let $f_j : [j, j + 1] \to \mathbb{R}$ be the restriction of the fractional part function to the interval $[j, j + 1]$. If $P = \{0 = x_0 < x_1 < \cdots < x_k = 1\}$ is a partition of the interval $[0, 1]$ and $y_j = x_j + n$ for $0 \leq j \leq k$ (so that $Q = \{n = y_0 < y_1 < \cdots < y_k = n + 1\}$ is a partition of $[n, n + 1]$), show that $s(f_0; P) = s(f_n; Q)$ and $S(f_0; P) = S(f_n; Q)$. For item (b), use Proposition 10.19, together with Problem 4, page 358, and Lemma 10.17.

6. First of all, show that it suffices to consider the case $f = 0$. In this case, if there exists $x_0 \in [a, b]$ such that $g(x_0) > 0$, use Lemma 8.20 to show that $\int_a^b g(x)dx > 0$.

7. Conclude first that $-\int_a^b |f(x)|dx \leq \int_a^b f(x)dx \leq \int_a^b |f(x)|dx$. Then, finish the proof.

8. Adapt, to this more general case, the approach to the case of $f = g$ in $[a, b)$, as presented in the text.

9. For the first part, use the result of Problem 4, page 358, together with Proposition 10.21. For the second part, use the result of Proposition 10.13.

11. If $M > 0$ is such that $|f(x)| \leq M$ for every $x \in [a, b]$, then

$$\left| \int_a^b f(x)dx - \int_{a+\epsilon}^{b-\epsilon} f(x)dx \right| = \left| \int_a^{a+\epsilon} f(x)dx + \int_{b-\epsilon}^b f(x)dx \right|$$

$$\leq \left| \int_a^{a+\epsilon} f(x)dx \right| + \left| \int_{b-\epsilon}^b f(x)dx \right|$$

$$\leq \int_a^{a+\epsilon} |f(x)|dx + \int_{b-\epsilon}^b |f(x)|dx$$

$$\leq \int_a^{a+\epsilon} Mdx + \int_{b-\epsilon}^b Mdx = 2M\epsilon.$$

Therefore, $\lim_{\epsilon \to 0} \left(\int_a^b f(x)dx - \int_{a+\epsilon}^{b-\epsilon} f(x)dx \right) = 0$.

12. From $\int_a^b (f(x) - tg(x))^2 dx \geq 0$ for every $t \in \mathbb{R}$, conclude that $At^2 - 2Bt + C \geq 0$ for every $t \in \mathbb{R}$, where $A = \int_a^b g(x)^2 dx$, $B = \int_a^b f(x)g(x)dx$ and $C = \int_a^b f(x)^2 dx$. Then, note that $\Delta = 4(B^2 - AC) \leq 0$. For the last part, use the result of Problem 6.

13. First consider the case in which f is positive, and look at both $\left(\frac{f(a)+f(b)}{2}\right)(b-a)$ and $\int_a^b f(x)dx$ as the areas of certain regions in the cartesian plane. For the general case, note that changing f by $f+c$ amounts to adding c to both sides of the inequality in the statement of the problem.

Section 10.5

1. For item (a), use that $\cos^2 t = \frac{1}{2}(1+\cos(2t))$; for (b), write $\cos t = 1-2\sin^2\frac{t}{2}$; for item (c), integrate by parts.
2. For item (a), integrate by parts; for (b), integrate by parts twice.
3. For the first integral, if $m \neq n$ write $\sin(mx)\sin(nx) = \frac{1}{2}(\cos(m-n)x-\cos(m+n)x)$; if $m = n$, use the fact that $\sin^2 nx = \frac{1}{2}(1 - \cos(2nx))$. For the second integral, argue in an analogous way.
4. In both cases, use (10.38). Additionally, for item (a) note that $x \mapsto \frac{2}{3}x^{3/2}$ is a primitive for $x \mapsto \sqrt{x}$ in $[0, +\infty)$; for (b), note that $x \mapsto \tan x$ is a primitive for $x \mapsto \sec^2 x$ in $\left(-\frac{\pi}{2}, \frac{\pi}{2}\right)$.
5. Start by taking a primitive F for f and computing the given integral with the aid of the FTC.
6. If $F(x) = \int_x^{x+p} f(t)dt$, use the result of the previous problem to show that F' vanishes identically.
7. Let $f(x) = mx + n$, compute $\int_a^b (mx_n)dx$ with the aid of the FTC and, then, check that the result coincides with the ordinary formulas for the area of a right triangle (if f vanishes at $x = a$ or $x = b$) or for the area of a right trapezoid (if f is positive in $[a, b]$).
8. Letting F be one such primitive and $a < x < y < b$, use the interpretation of $\int_x^y f(t)\, dt$ as an area and the monotonicity of f to compare the differences $F(y) - F\left(\frac{x+y}{2}\right)$ and $F\left(\frac{x+y}{2}\right) - F(x)$. Then, apply the result of Problems 8 and 9, page 338.
9. Let $f(x) = (x+1)^n = \sum_{k=0}^n \binom{n}{k}x^k$, $x \in \mathbb{R}$. Compute $\int_0^x f(t)dt$ in two different ways to get

$$\sum_{k=0}^n \binom{n}{k}\frac{x^{k+1}}{k+1} = \frac{(x+1)^{n+1}-1}{n+1}.$$

Then, integrate both sides from 0 to 1 to get

$$\sum_{k=0}^n \frac{1}{(k+1)(k+2)}\binom{n}{k} = \frac{2^{n+2}-(n+2)}{n+1} - \frac{1}{n+2}.$$

10. Apply Lagrange's MVT to the primitive of f based at the point a.

11. First of all, use Corollary 10.33 to get

$$\int_a^b f(x)^2 dx = xf(x)^2 \Big|_{x=a}^{x=b} - \int_a^b 2xf(x)f'(x)dx = -\int_a^b 2xf(x)f'(x)dx.$$

Then, apply Cauchy inequality for integrals (cf. Problem 12, page 381) to $x \mapsto xf(x)$ and f'.

12. Firstly, use the computations in Example 10.29 to show that there exists a single $f_0 \in \mathcal{F}$ of the form $f_0(x) = A\cos x + B\sin x$, with $A, B \in \mathbb{R}$. Then, expand the integrand in the inequality $\int_0^\pi (f(x) - f_0(x))^2 dx \geq 0$ to get the desired result.

13. Since $0 \leq \cos x \leq 1$ for $x \in [0, \frac{\pi}{2}]$, we have $(\cos x)^{n+1} \leq (\cos x)^n$ for every integer $n \geq 0$. Hence, $\frac{I_{n+1}}{I_n} = \frac{\int_0^{\pi/2}(\cos x)^{n+1}dx}{\int_0^{\pi/2}(\cos x)^n dx} \leq 1$ for every $n \geq 0$. On the other hand, we saw in Example 10.35 that $I_{n+1} = \frac{n}{n+1}I_{n-1}$ for every $n \geq 1$. Hence, $1 \geq \frac{I_{n+1}}{I_n} = \frac{\frac{n}{n+1}I_{n-1}}{\frac{n-1}{n}I_{n-2}} = \frac{n^2}{n^2-1} \cdot \frac{I_{n-1}}{I_{n-2}} \geq \frac{I_{n-1}}{I_{n-2}}$, so that each one of the sequences $\left(\frac{I_{2k-1}}{I_{2k-2}}\right)_{k\geq 1}$ and $\left(\frac{I_{2k}}{I_{2k-1}}\right)_{k\geq 1}$ is nondecreasing and bounded from above, thus convergent. Letting ℓ_0 and ℓ_1 respectively denote their limits, we have $\ell_0, \ell_1 > 0$. Now, again by (a), observe that $\frac{I_{n+1}}{I_n} = \frac{\frac{n}{n+1}I_{n-1}}{I_n} = \frac{n}{n+1}\left(\frac{I_n}{I_{n-1}}\right)^{-1}$. Therefore, letting $n = 2k \to +\infty$ we get $\ell_1 = \ell_0^{-1}$ or, which is the same, $\ell_0\ell_1 = 1$. However, since $\frac{I_{n+1}}{I_n} \leq 1$, it easily follows that $\ell_0, \ell_1 \leq 1$. Thus, $1 = \ell_0\ell_1 \leq 1 \cdot 1 = 1$, which gives us $\ell_0 = \ell_1 = 1$. Finally, since both sequences $\left(\frac{I_{2k-1}}{I_{2k-2}}\right)_{k\geq 1}$ and $\left(\frac{I_{2k}}{I_{2k-1}}\right)_{k\geq 1}$ converge to 1, we conclude that $\left(\frac{I_n}{I_{n-1}}\right)_{n\geq 1}$ also converges to 1.

Section 10.6

1. For item (a), use the substitution $x = t - 1$, together with the fact that $\frac{1}{\sqrt{t}} = t^{-1/2} = \frac{d}{dt}\left(\frac{t^{1/2}}{1/2}\right)$; for (b), use the substitution $x = \sin t$; for item (c), use the substitution $x = \sqrt{t}$.

2. For (a), make the substitution $y = 1 + x^3$; for (b), integrate by parts and use the result of (a); for item (c), make the substitution $x = \sin t$ and, then, integrate by parts; finally, for (d), perform the substitution $x = t^2$ and, then, integrate by parts.

3. In both cases, write $\int_{-a}^a f(x)dx = \int_{-a}^0 f(x)dx + \int_0^a f(x)dx$ and, for the first integral, perform the substitution $x = -t$.

4. For the first integral, note that the sine function is odd and apply the result of the previous problem. For the second one, argue in a similar way.

5. Assume (without loss of generality) that $O(0, 0)$ is the center and $A(-a, 0)$ and $A'(a, 0)$ are the endpoints of the major axis of the ellipse. It's a well known fact

that, in this setting, the equation of the ellipse is given by $\frac{x^2}{a^2} + \frac{y^2}{b^2} = 1$. Letting
$f : [-a, a] \to \mathbb{R}$ be given by $f(x) = b\sqrt{1 - \frac{x^2}{a^2}}$, show that the area of the ellipse
equals $2 \int_{-a}^{a} f(x)dx$. Then, perform the trigonometric substitution $x = a\cos t$,
with $0 \le t \le \pi$, and compute the integral.

6. Suppose that g is crescente (the case of a decreasing g in completely analogous).
    Let $t_1 < t_2 < \cdots < t_k$ be the points of the interval $[g(c), g(d)]$ in which f is
    discontinuous, and $c \le s_1 < s_2 < \cdots < s_k \le d$ be such that $g(s_j) = t_j$ for $1 \le
    j \le k$. Let $f_j : [t_j, t_{j+1}] \to \mathbb{R}$ be such that $f_j = f$ in $(t_j, t_{j+1}), f_j(t_j) = \lim_{t \to t_j+} f(t)$
    and $f_j(t_{j+1}) = \lim_{t \to t_{j+1}-} f(t)$, so that f_j is continuous. Use Lemma 10.17,
    together with Theorem 10.37, to get $\int_{t_j}^{t_{j+1}} f(t)dt = \int_{s_j}^{s_{j+1}} f(g(s))g'(s)ds$ for
    $1 \le j < k$ and, analogously, $\int_{g(c)}^{t_1} f(t)dt = \int_c^{s_1} f(g(s))g'(s)ds$ and $\int_{t_k}^{g(d)} f(t)dt =
    \int_{s_k}^{d} f(g(s))g'(s)ds$. Finally, add all such equalities to get the desired result.

7. First of all, note that $f(a) = a$ and $f(b) = b$. Hence, integrating successively
    by substitution and parts, we get

$$\int_a^b f^{-1}(x)\, dx = \int_{f(a)}^{f(b)} f^{-1}(x)\, dx = \int_a^b f^{-1}(f(x))f'(x)\, dx$$

$$= \int_a^b xf'(x)\, dx = xf(x)\Big|_{x=a}^{x=b} - \int_a^b f(x)\, dx$$

$$= bf(b) - af(a) - \int_a^b f(x)\, dx$$

$$= b^2 - a^2 - \int_a^b f(x)\, dx.$$

8. Start by observing that, since p has degree n, it is reciprocal if and only if
    $p(\frac{1}{x}) = \frac{p(x)}{x^n}$ for infinitely many nonzero real values of x (here we are using the
    fact—to be proved in [5]—that two real polynomial functions which assume
    the same values for every $x \in A$, where $A \subset \mathbb{R}$ is infinite, are actually
    identical). Now, assume the equality of the statement holds for every $x > 1$.
    Then, differentiating both sides with the aid of Problem 5, page 391, gives
    $\frac{p(x)}{x^{n/2+1}} = -\frac{p(1/x)}{(1/x)^{n/2+1}}\left(-\frac{1}{x^2}\right)$, which is the same as $p(\frac{1}{x}) = \frac{p(x)}{x^n}$ for every $x > 1$.
    Conversely, assuming p to be reciprocal, the variable substitution $t \mapsto \frac{1}{t}$ gives

$$\int_1^x \frac{p(t)}{t^{n/2+1}}dt = \int_1^{1/x} \frac{p(1/t)}{(1/t)^{n/2+1}}\left(-\frac{1}{t^2}\right)dt = \int_{1/x}^1 \frac{p(t)}{t^n} \cdot t^{n/2-1}dt$$

$$= \int_{1/x}^1 \frac{p(t)}{t^{n/2+1}}.$$

9. Since $f' > 0$, we conclude that f is increasing. Therefore, $f(x) > 1$ for $x > 1$ and the FTC gives, also for $x > 1$,

$$f(x) - 1 = \int_0^x \frac{1}{t^2 + f(t)^2} dt < \int_0^x \frac{1}{t^2 + 1} dt = \arctan x - \frac{\pi}{4} < \frac{\pi}{4}.$$

Finally, since f is increasing and bounded above by $1 + \frac{\pi}{4}$, we are done.

10. The change of variables formula gives, together with the semi-invariance of f over g, gives

$$\int_{g(x)}^{g^{(2)}(x)} f(t)dt = \int_x^{g(x)} f(g(t))g'(t)dt = \int_x^{g(x)} f(t)dt.$$

Therefore, if $g^{(m)} = \mathrm{Id}_I$, then

$$0 = \int_x^x f(t)dt = \sum_{k=0}^{m-1} \int_{g^{(k)}x}^{g^{(k+1)}(x)} f(t)dt$$

$$= \sum_{k=0}^{m-1} \int_x^{g(x)} f(t)dt = m \int_x^{g(x)} f(t)dt.$$

Section 10.7

1. Since $y = \log_a x \Leftrightarrow x = a^y$, we have $\log_a x \cdot \log a = y \log a = \log a^y = \log x$ and, then, $\log_a x = \frac{\log x}{\log a}$; on the other hand, Theorem 9.28 furnishes $\log_a' x = \frac{1}{f_a'(y)} = \frac{1}{a^y \log a} = \frac{1}{x \log a}$. Now, item (a) follows from the first part of (10.60), for, $\log e = 1$. Item (b) follows from the second part of (10.60), for, $\log_a$ is increasing $\Leftrightarrow \log_a' x > 0$ for every $x > 0 \Leftrightarrow \frac{1}{x \log a} > 0$ for every $x > 0 \Leftrightarrow \log a > 0$ for every $x > 0 \Leftrightarrow a > 1$. Concerning (c) and (d), we have $\log_a(xy) = \frac{\log(xy)}{\log a} = \frac{\log x + \log y}{\log a} = \frac{\log x}{\log a} + \frac{\log y}{\log a} = \log_a x + \log_a y$ and $\log_a c = \frac{\log c}{\log a} = \frac{\log c}{\log b} \cdot \frac{\log b}{\log a} = \log_b c \cdot \log_a b$. Finally, (e) follows immediately from (d).

2. Taking natural logarithms, conclude that it suffices to compare $\frac{1}{e}$ and $\frac{\log \pi}{\pi}$. To this end, study the first variation of $f : (0, +\infty) \to \mathbb{R}$ given by $f(x) = \frac{\log x}{x}$, showing that $x = e$ is its only point of global maximum.

3. Taking natural logarithms, show that it suffices to find out the solutions $a, b \in \mathbb{N}$ of the equation $\frac{\log a}{a} = \frac{\log b}{b}$. To do so, proceed as in the hint given to the previous problem.

4. Apply the formula of Corollary 10.33.

5. Let $f(x) = x - \log(1 + x)$ and show that f is increasing. Then, write $f(x) = x\left(1 - \frac{\log(1+x)}{x}\right)$ and use L'Hôpital's rule to get $\lim_{x\to 0+} f(x) = 0$.

6. It suffices to observe that, since log is increasing and $\frac{x+y}{2} \geq \sqrt{xy}$ for every $x, y > 0$, we have $\log\left(\frac{x+y}{2}\right) \geq \log \sqrt{xy} = \frac{\log x + \log y}{2}$, with equality if and only if $x = y$.

7. Since $f''(x) = e^x + 6ax$, we wish to find all $a < 0$ such that $e^x + 6ax$ has a single real root. Arguing geometrically, conclude that it suffices to find all $a < 0$ for which the straight line $y = -6ax$ is tangent to the graph of $x \mapsto e^x$. To this end, use the result of Problem 11, page 301 (with $(a, b) = (0, 0)$), to conclude that, if (x_0, e^{x_0}) is the point of tangency, then $e^{x_0} = -6ax_0$ and $-6a = \frac{d}{dx}e^x\big|_{x=x_0} = e^{x_0}$.

8. Recall that $x^\alpha = e^{\alpha \log x}$ and apply the chain rule.

9. Adapt, to the present case, the hint given to the previous problem to get $h'(x) = (f(x)g'(x)\log f(x) + f'(x)g(x))f(x)^{g(x)-1}$.

10. Study the first variation of f to conclude that $x = \frac{a}{a+b}$ is its only critical point. Then, use Weierstrass' Theorem 8.26 to conclude that such a point is its only maximum point.

11. Let $a = b = \sqrt{2}$ and analyse the possibilities a^b rational and a^b irrational, observing, in this last case, that $(a^b)^b \in \mathbb{Q}$.

12. The continuity of the natural logarithm guarantees that it suffices to show that

$$\lim_{x \to +\infty} x \log\left(1 + \frac{a}{x}\right) = a$$

or, which is the same (letting $y = \frac{1}{x}$), that $\lim_{y \to 0+} \frac{\log(1+ay)}{y} = a$. To this end, use l'Hôpital's rule.

13. Since $\left(\frac{x+a}{x-a}\right)^x = e^{x\log\left(\frac{x+a}{x-a}\right)}$, we wish all $a \in \mathbb{R}$ such that $\lim_{x\to +\infty} x\log\left(\frac{x+a}{x-a}\right) = 1$ or, which is the same, $\lim_{x\to +\infty} x\log\left(\frac{1+\frac{a}{x}}{1-\frac{a}{x}}\right) = 1$. Letting $y = \frac{1}{x}$, conclude that we wish to find all $a > 0$ such that $\lim_{y\to 0+} \frac{\log\left(\frac{1+ay}{1-ay}\right)}{y} = 1$. Finally, apply l'Hôpital's rule.

14. Differentiate $g(x) = e^{-x}f(x)$ and then apply Proposition 9.57.

15. For (b), use the result of (a), together with the FTC. Finally, for item (c), we have $\int \sec^3 t \, dt = \int \sec t(1 + \tan^2 t)dt = \int \sec t \, dt + \int (\tan t \cdot \sec t)\tan t \, dt = -\log|\sec t - \tan t| + \int \sec' t \cdot \tan t \, dt = -\log|\sec t - \tan t| + \sec t \cdot \tan t - \int \sec t \cdot \tan' t \, dt$.

16. For the second part of (a) use the chain rule, together with the fact that $\frac{d}{dx}e^x = e^x$. For item (b), use that $\cosh'' x = \cosh x > 0$, $\cosh x = \frac{e^x + e^{-x}}{2} \geq \sqrt{e^x \cdot e^{-x}} = 1$ and $\lim_{x\to +\infty} \cosh x = +\infty$. For (c), show that $\sinh' x = \cosh x > 0$ and $\sinh'' x = \sinh x$, with $\sinh x > 0$ if and only if $x > 0$. In item (d), you should get something similar to Fig. B.8. For item (e), letting $k(x_0)$ denote the curvature we wish to compute and $f(x) = \cosh x$, conclude from Problem 3, page 344, that $k(x_0) = \frac{\cosh x_0}{(1+\sinh^2 x_0)^{3/2}} = \frac{1}{\cosh^2 x_0}$.

Fig. B.8 Graphs of the
hyperbolic sine and cosine

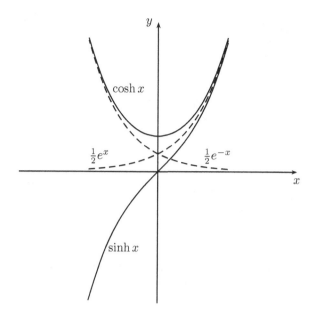

17. Since $x^2 - y^2 = 1 \Leftrightarrow y = \pm\sqrt{x^2 - 1}$, we have

$$A = \int_0^1 (\tanh t) x\, dx + \int_1^{\cosh t} ((\tanh t)x - \sqrt{x^2 - 1})dx$$

$$= \frac{1}{2}\tanh t + (\tanh t) \cdot \frac{1}{2}(\cosh^2 t - 1) - \int_0^t \sqrt{\cosh^2 s - 1}\,\sinh s\,ds$$

$$= \frac{1}{2}\sinh t \cdot \cosh t - \frac{1}{2}\int_0^t (\cosh(2s) - 1)ds = \frac{t}{2}.$$

18. Write $x^\lambda \log x = -\frac{\log y}{y^\lambda}$, with $y = \frac{1}{x}$. Then, adapt the proof of Theorem 10.51 to
     the present case to conclude that $\lim_{y \to +\infty} \frac{\log y}{y^\lambda} = 0$.
19. Rework the proof of Theorem 10.55 to get

$$\gamma - \left(\sum_{k=1}^n \frac{1}{k} - \log n\right) = \sum_{k \geq n}\left(\frac{1}{k} - \log\left(1 + \frac{1}{k}\right)\right)$$

$$< \sum_{k \geq n}\left(\frac{1}{k} - \frac{1}{k+1}\right) = \frac{1}{n}.$$

20. Firstly, note that $f(1) = \log(3e/8) > 0$, by Problem 8, page 242. Now, write
     $f(x) = \log\left(\frac{x+1/2}{x+1}\right) - \log\left(1 + \frac{1}{x}\right)^x + 1$ to get $\lim_{x \to +\infty} f(x) = 0$, and conclude
     that it suffices to show that $f'(x) < 0$ for $x \geq 1$. For what is left to do, compute

$f'(x) = \frac{1}{x+1/2} + \log\left(\frac{x}{x+1}\right), f''(x) = \frac{1}{x^2+x} - \frac{1}{x^2+x+1/4}$ and conclude that f' is increasing, with $\lim_{x\to+\infty} f'(x) = 0$.

21. Apply Stirling's formula.
22. Use Stirling's formula to get

$$\binom{(k+1)n}{n} \sim \frac{1}{\sqrt{2\pi n}}\sqrt{\frac{k+1}{k}}(k+1)^n\left(1+\frac{1}{k}\right)^{kn}.$$

Then, note that $k > \alpha n \Rightarrow k \to +\infty$ as $n \to +\infty$.

Section 10.8

1. Note that $a = 2$ and $a = 4$ are two roots. To show that there are exactly three real roots, start by sketching the graphs of $x \mapsto x^2$ and $x \mapsto 2^x$; then, use the IVT to show that there exists $x_0 \in (-2, 0)$ such that $x_0^2 = 2^{x_0}$. Finally, assume that x_0 is rational and derive a contradiction.

2. Firstly, note that $f(x) = \log\left(x + \sqrt{x^2 + 1}\right)$ is increasing in $\mathbb{R}$, so that the given system of equations can be written as

$$g(x) = y, \quad g(y) = z, \quad \text{and} \quad g(z) = x.$$

   Assume, without any loss of generality, that $x = \max\{x, y, z\}$. Then, successively conclude that $f(x) = \max\{f(x), f(y), f(z)\}$, $y \geq z, x$ and $y = x$. Arguing in an analogous way, show that $x = y = z$ and solve $\log(x + \sqrt{x^2 + 1}) = 0$.

3. By the sake of contradiction, differentiate the equality $p(x) = \log x$ and, then, use the results of Example 9.17.

4. Let $k \in \mathbb{Z}_+$ be such that $10^k \leq n < 10^{k+1}$ and conclude that $k = \lfloor \log_{10} n \rfloor$.

5. We want that

$$(a_1 a_2 \ldots a_m \underbrace{00 \ldots 0}_{l})_{10} \leq n^k \leq (a_1 a_2 \ldots a_m \underbrace{99 \ldots 9}_{l})_{10}.$$

   for some $n, l \in \mathbb{N}$. This is the same as $10^l p \leq n^k < 10^l(p + 1)$, where $p = (a_1 a_2 \ldots a_m)_{10}$. Take $l = kq$ and show that, in this case, the last condition above is equivalent to the existence of a natural number n such that $n \in [10^q \sqrt[k]{p}, 10^q \sqrt[k]{p+1})$, for which it suffices to have $10^q(\sqrt[k]{p+1} - \sqrt[k]{p}) > 1$ or, which is the same, $q > -\log_{10}(\sqrt[k]{p+1} - \sqrt[k]{p})$.

6. Start by showing the validity of the inequalities $\log_a(\log_a b) > \log_c(\log_a b)$ and $\log_b(\log_c a) > \log_c(\log_c a)$. Then, apply item (e) of Problem 1, page 411.

7. Use the given functional equation to show that, if $a_0 = 0$, $b_0 = 0$, $a_{n+1} = 2a_n + 1$ and $b_{n+1} = 3b_n + 5$, then $f(a_n) = b_n$ for every $n \geq 0$. Then, use the material of Sect. 3.2 to conclude that $f(2^n - 1) = \frac{5}{2}(3^n - 1)$ for every $n \geq 0$. Finally, solve for n the equation $2^n - 1 = x$.

8. Start by showing that $f(0) = 0$, 1 or 2. In the first two cases, show that f is constant. In the third case, show that $f(m) = 2m$ whenever m is a power of 2. Then, take $n < m$ and suppose that $f(m) = f(n)$; use induction to show that $f(m^{2^k}) = f(n^{2^k})$ and, to reach a contradiction, choose $a, b, c \in \mathbb{N}$ such that $\log_2 n < \frac{b}{2c} < \frac{a}{2c} < \log_2 m$.

9. Initially, use the formula of integration by parts to get

$$n \int_0^1 \frac{x^n}{x^n + a} dx = n \int_0^1 x \cdot \frac{nx^{n-1}}{x^n + a} dx$$

$$= \int_0^1 x \cdot \frac{d}{dx} \log(x^n + a) dx$$

$$= \log(a + 1) - \int_0^1 \log(x^n + a) dx.$$

Now, observe that

$$\log a \leq \int_0^1 \log(x^n + a) dx$$

$$= \int_0^1 \left(\log a + \log \left(\frac{x^n}{a} + 1 \right) \right) dx$$

$$\leq \log a + \int_0^1 \frac{x^n}{a} dx.$$

Finally, take together the two computations above to compute the desired limit.

10. Use the power means inequality.
11. Apply Jensen's inequality to the natural logarithm, as in Example 10.61.
12. Apply Jensen's inequality to $f(x) = x \log x$, $x > 0$.
13. Apply Jensen's inequality to $f(x) = \log \frac{1-x}{1+x}$, $0 < x < 1$.
14. Apply Jensen's inequality to $f(x) = \log \frac{\sin x}{x}$, $x \in (0, \pi)$.
15. If $a_i = e^{x_i}$, with $x_i \geq 0$, show that $f(x) = \frac{1}{e^x+1}$, $x \geq 0$, is strictly convex and apply Jensen's inequality.
16. Start by writing the given inequality as

$$\prod_{i=1}^n \frac{a_i}{1 - a_i} \leq \frac{1}{n^{n+1}} \cdot \frac{s}{1 - s},$$

where $s = a_1 + a_2 + \cdots + a_n$. Then, rewrite the inequality as

$$\sum_{i=1}^n \log \left(\frac{a_i}{1 - a_i} \right) \leq -(n + 1) \log n + \log \left(\frac{s}{1 - s} \right).$$

If $a_i < \frac{1}{2}$ for $1 \le i \le n$, apply Jensen's inequality to $f(x) = \log\left(\frac{x}{1-x}\right)$, $0 < x < \frac{1}{2}$; subsequently, show that

$$n \log\left(\frac{s}{n-s}\right) + (n+1) \log n \le \log\left(\frac{s}{1-s}\right).$$

17. Let $S = \left(\sum_{i=1}^{n}(a_i + b_i)^k\right)^{1/k}$ and write

$$S^k = \sum_{i=1}^{n} a_i(a_i + b_i)^{k-1} + \sum_{i=1}^{n} b_i(a_i + b_i)^{k-1}.$$

Then, choose $q > 0$ such that $\frac{1}{q} = 1 - \frac{1}{k}$ and apply Hölder's inequality to each of the sums above.

Section 10.9

1. For item (a), use the result of item (b) of Problem 15, page 412. For (b), apply the comparison test for improper integrals, together with the fact that $|e^{-t} \sin t| \le e^{-t}$ for every $t \in \mathbb{R}$. For item (c), note that $\frac{1}{\log t} \to 0$ when $t \to 0+$ and apply the comparison test when $t \to 1-$. Finally, item (d) follows from the fact that arctan is a primitive for $\frac{1}{1+t^2}$.

2. Apply Theorem 10.37 to compute $\int_{g(B)}^{g(x)} f(t)\, dt$ for $x > B$. Then, let $x \to +\infty$.

3. Adapt, to the present case, the argument of Example 10.70. Alternatively, apply the change of variables theorem for improper integrals (previous problem) with the variable substitution $t \mapsto \frac{1}{t}$, that maps $(0, 1]$ bijectively to $[1, +\infty)$.

4. For $a > \pi^2$, the change of variables formula, together with the formula for integration by parts, gives

$$\int_{\pi^2}^{a} \cos(x^2)\,dx = \int_{\pi}^{\sqrt{a}} t^{-1/2} \cos t\, dt$$

$$= t^{-1/2} \sin t \Big|_{\pi}^{\sqrt{a}} + \frac{1}{2} \int_{\pi}^{\sqrt{a}} t^{-3/2} \sin t\, dt$$

$$= \frac{\sin\sqrt{a}}{\sqrt[4]{a}} + \frac{1}{2} \int_{\pi}^{\sqrt{a}} \frac{\sin t}{t^{3/2}}\,dt.$$

Now, argue as in Example 10.75 to show that $\int_{\pi}^{+\infty} \frac{\sin t}{t^{3/2}}\,dt$ converges.

5. For the convergence of the integrals, note first that (cf. Problem 4, page 325) $\sin x \ge x - \frac{x^3}{6}$ for $x \ge 0$. Then, for $0 < x < \sqrt{6}$,

$$0 \geq \log(\sin x) \geq \log\left(x - \frac{x^3}{6}\right) = \log x + \log\left(1 - \frac{x^2}{6}\right).$$

Since both integrals $\int_0^\pi \log x \, dx$ and $\int_0^\pi \log\left(1 - \frac{x^2}{6}\right) dx$ converge, the same is true of $\int_0^\pi \log\left(x - \frac{x^3}{6}\right) dx$. Given $0 < \alpha < \beta < \sqrt{6}$, it follows from the above inequalities that

$$\left| \int_\alpha^\beta \log(\sin x) dx \right| \leq \left| \int_\alpha^\beta \log\left(x - \frac{x^3}{6}\right) dx \right|.$$

Therefore, $\int_0^\pi \log(\sin x) dx$ converges. Now, use the change of variables formula to show that

$$\int_0^\pi \log(\sin x) dx = \int_0^\pi \log|\cos x| dx,$$

so that $\int_0^\pi \log|\cos x| dx$ also converges. Finally, let $I = \int_0^\pi \log(\sin x) dx$ and using the change of variables formula again, we compute

$$2I = 2 \int_0^{2\pi} \log|\sin x| dx = 2 \int_0^\pi \log|\sin(2y)| dy$$

$$= 2 \int_0^\pi \log\left(2 \sin y |\cos y|\right) dy$$

$$= 2\pi \log 2 + 2I + 2 \int_0^\pi \log|\cos y| dy.$$

Then, $\int_0^\pi \log|\cos y| dy = -\pi \log 2$.

6. Apply the integral test to each of the functions $x \mapsto \frac{1}{(\log x)^\alpha}$, $x \mapsto \frac{1}{x(\log x)^\alpha}$ and $\frac{1}{x(\log x)(\log\log x)^\alpha}$.

7. First recall that, for some $A > 1$, we have $\log t < \sqrt{t}$ for $t > A$. Therefore, $\frac{\log t}{1+t^2} < \frac{\sqrt{t}}{1+t^2} < \frac{\sqrt{t}}{t^2} = t^{-3/2}$ for $t > A$, and $\int_A^{+\infty} t^{-3/2} dt$ converges. Now, take $x > 1$ and use the result of Problem 5, page 391, as well as Corollary 9.47, to show that $f : (0, +\infty) \to \mathbb{R}$, given by $f(x) = \int_{1/x}^x \frac{\log t}{1+t^2} dt$, is constant. Deduce that $\int_{1/x}^x \frac{\log t}{1+t^2} dt = 0$ and, then, that $\int_0^{1/A} \frac{\log t}{1+t^2} dt$ converges. Finally, compute $\int_0^{+\infty} \frac{\log t}{1+t^2} dt = 0$.

8. For item (a), apply Young's inequality (cf. Proposition 10.62) to $a = f(t)$ and $b = g(t)$ and then integrate the result on I. For (b), start by observing that we can assume that $\int_\alpha^\beta f(t)^p \, dt > 0$ and $\int_\alpha^\beta g(t)^q \, dt > 0$. Then, let $f_1 = f/A$ and $g_1 = g/B$, where $A = \left(\int_\alpha^\beta f(t)^p \, dt\right)^{1/p}$ and $B = \left(\int_\alpha^\beta g(t)^q \, dt\right)^{1/q}$. Finally, show that $\int_\alpha^\beta f_1(t)^p \, dt = \int_\alpha^\beta g_1(t)^q \, dt = 1$ and apply the result of item (a).

9. For item (a), on the one hand we have $f'(x) = -\frac{1}{x^2} - \frac{1}{x+1} + \frac{1}{x} = -\frac{1}{x^2} - \frac{1}{x(x+1)} < 0$, so that f is indeed decreasing; on the other, $\lim_{x \to +\infty} f(x) = \lim_{x \to +\infty} \left(\frac{1}{x} - \log\left(1 + \frac{1}{x}\right)\right) = -\log 1 = 0$. For (b), given $x > 1$ and using the result of Problem 4, page 412, together with Theorem 10.53, we get

$$\int_1^x f(t)dt = \int_1^x \left(\frac{1}{t} - \log(t+1) + \log t\right) dt$$

$$= \log t \Big|_{t=1}^{t=x} - ((t+1)\log(t+1) - (t+1)) \Big|_{t=1}^{t=x}$$

$$+ (t \log t - t) \Big|_{t=1}^{t=x}$$

$$= -\log\left(1 + \frac{1}{x}\right) - \log\left(1 + \frac{1}{x}\right)^x + 2\log 2 + 1$$

$$\xrightarrow{x \to +\infty} -\log 1 - \log e + 2\log 2 + 1 = 2\log 2.$$

10. For item (a) adapt, to the present case, the proof of the integral test.

11. For item (a), argue geometrically to show that if $2l\pi - \frac{\pi}{6} < n < n+1 < 2l\pi + \frac{\pi}{6}$, then $2l\pi + \frac{\pi}{6} < n+2 < 2l\pi + \frac{5\pi}{6}$, and if $2l\pi + \frac{5\pi}{6} < n < n+1 < 2l\pi + \frac{7\pi}{6}$, then $2l\pi + \frac{7\pi}{6} < n+2 < 2l\pi + \frac{11\pi}{6}$. For item (b), use (a) to conclude that $\frac{|\sin n|}{n} + \frac{|\sin(n+1)|}{n+1} + \frac{|\sin(n+2)|}{n+2} \geq \frac{1}{2(n+2)}$. Then, use the divergence of the harmonic series to show what is asked. Finally, for item (c), for each $k \in \mathbb{N}$ show that $\int_{k\pi}^{(k+1)\pi} \left|\frac{\sin t}{t}\right| dt \geq \int_{k\pi+\frac{\pi}{6}}^{(k+1)\pi - \frac{\pi}{6}} \left|\frac{\sin t}{t}\right| dt \geq \frac{1/2}{(k+1)\pi} \cdot \frac{2\pi}{6}$. Then, conclude what is asked again from the divergence of the harmonic series. (Why can't one use the integral test?)

12. For item (a), study the first variation of $f(x) = (1-x)e^{2x}$ in the interval $[0, \frac{1}{2}]$. For (b), use the Fundamental Theorem of Arithmetic (cf. introduction to Chap. 1). Item (c) follows from items (a) and (b), together with the fact that $1 + \frac{1}{p_k} + \frac{1}{p_k^2} + \cdots = \left(1 - \frac{1}{p_k}\right)^{-1}$. Finally, for item (d) use the divergence of the harmonic series.

Section 10.10

1. Setting $f(x) = \cosh x$ in (10.71), we have $\sqrt{1 + f'(x)^2} = \cosh x$, so that the desired length equals $\int_0^{x_0} \cosh x \, dx = \sinh x \Big|_{x=0}^{x=x_0} = \sinh x_0$.

2. Item (a) follows immediately from (10.71). For item (b), use the fact that $a^2 = b^2 + c^2$, together with the fundamental relation of Trigonometry to write

$$1 + \frac{b^2 x^2}{a^2(a^2 - x^2)} = 1 + \frac{b^2 \cos^2 t}{a^2 \sin^2 t} = \frac{a^2 \sin^2 t + b^2 \cos^2 t}{a^2 \sin^2 t}$$

$$= \frac{b^2 + c^2 \sin^2 t}{a^2 \sin^2 t} = \left(\frac{b}{a \sin t}\right)^2 \left(1 + \left(\frac{c}{b}\right)^2 \sin^2 t\right).$$

3. For item (a), observe (cf. Fig. 10.12) that after Γ has rolled to the right by t radians starting from O, the point of tangency between Γ and r will be $T(t, 0)$ and the center of Γ will be $A(t, 1)$. On the other hand, angle $\angle TOP$, measured in the counterclockwise sense, will be equal to t radians. Since $\widehat{BAP} = \frac{3\pi}{2} - t$ (in the counterclockwise sense), it's immediate that

$$x(t) = t + \cos\left(\frac{3\pi}{2} - t\right) = t - \sin t, \quad y(t) = 1 + \sin\left(\frac{3\pi}{2} - t\right) = 1 - \cos t.$$

For (b), first note that

$$\int_0^{2\pi} \sqrt{1 + f'(x)^2}\,dx = \int_0^{2\pi} \sqrt{1 + f'(x(t))^2}\,x'(t)\,dt = \int_0^{2\pi} \sqrt{1 + \frac{y'(t)^2}{x'(t)^2}}\,x'(t)\,dt$$

$$= \int_0^{2\pi} \sqrt{x'(t)^2 + y'(t)^2}\,dt = \sqrt{2} \int_0^{2\pi} \sqrt{1 - \cos t}\,dt.$$

For the computation of the last integral, see item (b) of Problem 1, page 397.

4. By definition, we have $\Gamma(\frac{1}{2}) = \int_0^{+\infty} e^{-t} t^{-1/2}\,dt$. Now, perform the substitution of variable $s = t^{1/2}$ and apply the result of Problem 2, page 435.

5. We start by using the result of Problem 18, page 413, to conclude that there exists $A > 0$ such that $t^{\frac{x}{2}} |\log t| \le A$ for $0 < t \le 1$. Then, observe that

$$\int_0^1 e^{-t} t^{x-1} |\log t|\,dt = \int_0^1 e^{-t} t^{\frac{x}{2}-1} \cdot t^{\frac{x}{2}} |\log t|\,dt < A \int_0^1 e^{-t} t^{\frac{x}{2}-1}\,dt.$$

Now, use the fact that $\log t < t$ for $t \ge 1$ to write

$$\int_1^{+\infty} e^{-t} t^{x-1} |\log t|\,dt = \int_1^{+\infty} e^{-t} t^{x-1} \log t\,dt < \int_1^{+\infty} e^{-t} t^x\,dt.$$

Finally, gather together the above estimates to obtain

$$\int_0^{+\infty} e^{-t} t^{x-1} |\log t|\,dt < A \int_0^1 e^{-t} t^{\frac{x}{2}-1}\,dt + \int_1^{+\infty} e^{-t} t^x\,dt$$

$$< A \int_0^{+\infty} e^{-t} t^{\frac{x}{2}-1}\,dt + \int_0^{+\infty} e^{-t} t^x\,dt$$

$$= A\Gamma\left(\frac{x}{2}\right) + \Gamma(x + 1).)$$

6. For the second part, let $n_j = j$ for $1 \leq j \leq k$. For the first part, use the fact that $\log \Gamma$ is convex, together with Jensen's inequality.

Section 11.1

1. Use version (11.2) of the Taylor's formula of f centered at x_0, together with the fact that $f'' \geq 0$.

2. Assume $f^{(n)}(x_0) > 0$ (the other case is entirely analogous). Since $f^{(n)}$ is continuous and I is open, we can take an open interval $J \subset I$, centered at x_0 and such that $f^{(n)}(x) > 0$ for every $x \in J$. For a fixed $x \in J \setminus \{x_0\}$, (11.1) guarantees the existence of a real c between x_0 and x such that $f(x) = f(x_0) + \frac{f^{(n)}(c)}{n!}(x - x_0)^n$. Conclude that $f(x) > f(x_0)$.

3. If $f(t) = \sqrt{1 + \kappa \sin^2 t}$ for $0 \leq t \leq \pi$, then $f''(t) = \frac{\kappa(1 - (\kappa + 2)\sin^2 t)}{(1 + \kappa \sin^2 t)^{3/2}}$. Now, look at (10.74) and substitute the formula for f in both sides of the inequality of Example 11.2, noticing that $|f''(t)| \leq \kappa(\kappa + 1)$ for $0 \leq t \leq \pi$.

4. The Taylor series of the exponential function, together with the definition of $\sinh x$, furnishes $\sinh x = \frac{1}{2}(e^x - e^{-x}) = \frac{1}{2}\sum_{k \geq 0} \frac{1 - (-1)^k}{k!} x^k = \sum_{j \geq 1} \frac{1}{(2j-1)!} x^{2j-1}$. Argue analogously to $\cosh x$.

5. Apply the result of Proposition 11.3, with $n + 1$ in place of n.

6. Let $p(x) = a_m x^m + a_{m-1} x^{m-1} + \cdots + a_1 x + a_0$ and assume $a_m > 0$ (the other case can be dealt with in an analogous way). By Example 9.17, we can take $A > 0$ such that $p(x) > 0$ for $x > A$. Then, for $x > A$, apply (11.6) with $n = m + 1$ to get the desired result.

7. For item (a), since $f(\beta), f'(\beta) > 0$, we clearly have $\gamma < \beta$; now, use the result of Problem 1 to conclude that $f(\gamma) \geq 0$, and hence $\gamma \geq \alpha$. For (b), use the result of (a) to show that $(a_n)_{n \geq 1}$ is decreasing and such that $a_n \in [\alpha, a_1]$ for $n \geq 1$, so that there exists $\ell \in [\alpha, a_1]$ satisfying $a_n \to \ell$ as $n \to +\infty$; then, let $n \to +\infty$ in the recurrence relation defining $(a_n)_{n \geq 1}$ to get $\ell = \alpha$. For (i), use (11.2) to get $f(\alpha) = f(a_n) + f'(a_n)(\alpha - a_n) + \frac{1}{2} f''(\xi_n)(\alpha - a_n)^2$ for some $\xi_n \in (\alpha, a_n)$. For the equality in (ii), use again the recurrence relation of $(a_n)_{n \geq 1}$; for the inequality, use the fact that f' is increasing. Finally, for the first part of item (iii), use the result of (ii); for the second part, make induction on n, noting that $0 \leq a_n - \alpha \leq d - c$.

8. Verify that $f', f'' > 0$ in $[2, \frac{5}{2}]$ and $\max_{[2, \frac{5}{2}]} \frac{f''}{2f'} = \frac{3}{5}$. In the notations of item (iii) of the previous problem, conclude that we can let $c = 2$ and $d = \frac{5}{2}$ to get
$$0 \leq a_n - \alpha \leq \left(\tfrac{3}{5}\right)^n \cdot \tfrac{1}{2^{2^n}} \text{ for every } n \geq 1, \text{ where } a_1 = \tfrac{5}{2} \text{ and } a_{n+1} = \frac{2a_n^3 + 5}{3a_n^2 - 2}. \text{ Finally,}$$
note that $\left(\tfrac{3}{5}\right)^4 \cdot \tfrac{1}{2^{24}} < 10^{-5}$.

Section 11.2

1. We saw in Example 11.11 that $(\frac{|x|}{1+x^2})^n \leq \frac{1}{2^n}$ for every $x \in \mathbb{R}$. Hence, letting $M_k = \frac{1}{2^k}$ and noting that $\sum_{k \geq 1} M_k < +\infty$, it follows from Weierstrass M-test that $\sum_{k \geq 1} (\frac{x}{1+x^2})^k$ converges uniformly in $\mathbb{R}$. Now, Proposition 7.38 gives

$$\sum_{k \geq 1} \left(\frac{x}{1 + x^2} \right)^k = \frac{\frac{x}{1+x^2}}{1 - \frac{x}{1+x^2}} = \frac{x}{x^2 - x + 1}.$$

2. In all cases we shall use Weierstrass M-test. For the Taylor series of the exponential function, we have $\left| \frac{1}{k!} x^k \right| \leq \frac{a^k}{k!}$ if $|x| \leq a$. Therefore, letting $M_k = \frac{a^k}{k!}$ for $k \geq 0$, we have $\sum_{k \geq 0} M_k = \sum_{k \geq 0} \frac{a^k}{k!} = e^a < +\infty$, and the M-test guarantees the uniform convergence of the series in the interval $[-a, a]$. For the sine function, we have $\left| \frac{(-1)^{j-1}}{(2j-1)!} x^{2j-1} \right| \leq \frac{a^{2j-1}}{(2j-1)!}$ if $|x| \leq a$. Letting $M_{2j} = 0$ for $j \geq 0$ and $M_{2j-1} = \frac{a^{2j-1}}{(2j-1)!}$ for $j \geq 1$, we get $\sum_{k \geq 0} M_k = \sum_{j \geq 1} M_{2j-1} = \sum_{j \geq 1} \frac{a^{2j-1}}{(2j-1)!} = \sinh a < +\infty$. Again, the M-test assures the uniform convergence of the Taylor series in the interval $[-a, a]$. Finally, for the Taylor series of the cosine function a similar argument is valid.

3. For the non uniform character of the convergence, show that the maximum value of f_n is e^{-1}, regardless of the chosen n. In order to compute $\int_0^1 f_n(x)\, dx$, integrate by parts.

4. For item (a), note that $x^n \xrightarrow{n} 0$ for $0 < x < 1$. For (b), integrate by parts to get $\int_0^1 f_n(x)\, dx = -e^{-1} + n \int_0^1 f_{n-1}(x)\, dx$ for every $n \in \mathbb{N}$; then, write $\int_0^1 f_n(x)\, dx = a_n - b_n e^{-1}$, $\int_0^1 f_{n-1}(x)\, dx = a_{n-1} - b_{n-1} e^{-1}$ and use the fact that e^{-1} is irrational. For item (c), make induction on n, using the recurrence relations found in (b). Finally, for (d), use the fact that $e = \sum_{k \geq 0} \frac{1}{k!}$.

5. First show that the definition of uniform convergence guarantees the existence of $C > 0$ and $n_0 \in \mathbb{N}$ such that $n \geq n_0 \Rightarrow |f_{n_0}(x)| \leq C$ for every $x \in [a, b]$. Then, conclude that $|f(x)| \leq C$ for every $x \in I$, so that f is bounded. To show that f is integrable let $\epsilon > 0$ be given and choose $n_0 \in \mathbb{N}$ such that $n \geq n_0$ implies $|f_n(x) - f(x)| < \epsilon$ for every $x \in I$. Fix such an n and invoke Theorem 10.4 to choose a partition $P_n = \{a = x_0 < x_1 < \cdots < x_k = b\}$ of the interval $[a, b]$ such that $S(f_n; P_n) - s(f_n; P_n) < \epsilon$. Noting that $m_j(f_n) \leq f_n(x) \leq M_j(f_n)$ for $x \in [x_{j-1}, x_j]$, conclude successively that $m_j(f_n) - \epsilon < f(x) < M_j(f_n) + \epsilon$ for $x \in [x_{j-1}, x_j]$, $m_j(f_n) - \epsilon \leq m_j(f) \leq M_j(f) \leq M_j(f_n) + \epsilon$ for $1 \leq j \leq k$ and $S(f; P_n) - s(f; P_n) < \epsilon(1 + 2(b - a))$. Finally, let $I_n = \int_a^b f_n(x)dx$, $I = \int_a^b f(x)dx$ and show that computations analogous to the ones above furnish $I_n - I < \epsilon(1 + (b - a))$ and $I_n - I > -\epsilon(1 + (b - a))$, so that $|I_n - I| \leq \epsilon(1 + (b - a))$. Alternatively, let D_n be the set of points of discontinuities of f_n and D that of f. Show that $D \subset \bigcup_{n \geq 1} D_n$ and apply the result of Problem 3, page 367, together with Lebesgue's Theorem 10.10, to show that f is integrable. Then, argue as in the proof of Theorem 11.14 to show that $\int_a^b f(x)dx = \lim_{n \to +\infty} \int_a^b f_n(x)dx$.

6. Start by using Lagrange's MVT to get $|f_k(x)| \leq |f_k(x) - f_k(x_0)| + |f_k(x_0)| \leq M_k|x - x_0| + |f_k(x_0)| \leq M_k(b - a) + |f_k(x_0)|$. Then, apply the M-test with $M'_k = M_k(b - a) + |f_k(x_0)|$ in place of M_k to show that $\sum_{k \geq 1} f_k$ converges uniformly in $[a, b]$ to a function $f : [a, b] \to \mathbb{R}$. The M-test also shows that $\sum_{k \geq 1} f'_k$ converges uniformly in $[a, b]$ to a function $g : [a, b] \to \mathbb{R}$, and we claim that $f' = g$. To this end, let $x_0, x \in [a, b]$, with $x \neq x_0$, and $S = \left|\frac{f(x) - f(x_0)}{x - x_0} - g(x_0)\right|$. With the aid of Corollary 11.19, write

$$S = \left|\sum_{k \geq 1} \frac{f_k(x) - f_k(x_0)}{x - x_0} - g(x_0)\right| = \left|\frac{1}{x - x_0} \sum_{k \geq 1} \int_{x_0}^{x} f'_k(t)dt - g(x_0)\right|$$

$$= \left|\frac{1}{x - x_0} \int_{x_0}^{x} \sum_{k \geq 1} f'_k(t)dt - g(x_0)\right| = \left|\frac{1}{x - x_0} \int_{x_0}^{x} (g(t) - g(x_0))dt\right|$$

$$\leq \frac{1}{|x - x_0|} \left|\int_{x_0}^{x} |g(t) - g(x_0)|dt\right|.$$

Now, use the continuity of g to show that this last expression can be made less than ϵ, provided $x \in [a, b]$ is chosen to be sufficiently close to x_0.

7. Use the given condition to conclude that $\int_a^b f(x)p(x)dx = 0$ for every real polynomial p. Then, use Weierstrass approximation theorem to find a sequence $(p_n)_{n \geq 1}$ of polynomials such that $p_n \xrightarrow{n} f$ uniformly on $[a, b]$. Finally, conclude from this fact that $\int_a^b f(x)^2 dx = 0$.

8. If such an f did exist, we would have $\int_0^1 f(x)p(x)dx = p'(0)$, for every polynomial function p. Now, if $F(x) = \int_0^x f(t)dt$, then $F' = f$ and $F(0) = F(1) = 0$. Integrating by parts, we get $\int_0^1 F(x)p'(x)dx = -p'(0)$, which is the same as $\int_0^1 F(x)q(x)dx = -q(0)$, for every polynomial function q. Then, follow the hint given to the previous problem to get

$$\int_0^1 F(x)^2 dx = \lim_{n \to +\infty} \int_0^1 F(x)q_n(x)dx = -\lim_{n \to +\infty} q_n(0) = -F(0) = 0,$$

so that F, and hence f, vanishes identically. But this contradicts the fact that $\int_0^1 f(x)xdx = 1$.

9. Given $\epsilon > 0$, use Weierstrass approximation theorem to choose a polynomial p so that $|f(x) - p(x)| < \epsilon$ for every $x \in [0, 1]$. Note that $\lim_{n \to +\infty} \int_0^1 p(x)\varphi_n(x)dx$ does exist does exist and estimate

$$\left|\int_0^1 f\varphi_m - \int_0^1 f\varphi_n\right| \leq \int_0^1 \epsilon(\varphi_m + \varphi_n) + \left|\int_0^1 p\varphi_m - \int_0^1 p\varphi_n\right|.$$

Now, use the fact that $n \mapsto \int_0^1 \varphi_n$ and $n \mapsto \int_0^1 p\varphi_n$ converge, together with the computations above, to show that $n \mapsto \int_0^1 f\varphi_n$ is a Cauchy sequence.

10. For item (a) and for $x \in (0, 2\pi)$, use the result of item (a) of Problem 10, page 359, to get

$$\sum_{k=1}^{n} \sin(kx) = \frac{\sin\left(\frac{(n-1)x}{2}\right) \sin\frac{(n+1)x}{2}}{\sin\frac{x}{2}}.$$

Then, conclude that $\left|\sum_{k=1}^{n} \sin(kx)\right| \leq \frac{1}{\sin\frac{x}{2}}$ and apply Abel's criterion. For item (b), note that for $0 < \delta < \pi$ and $x \in [\delta, 2\pi - \delta]$, one has $|\sin\frac{x}{2}| \geq \sin \delta$.

11. For the first parts of items (a) and (b), integrate by parts. For the second part of (a), use the triangle inequality for integrals to get $|a_k(f)|, |b_k(f)| \leq \frac{1}{2\pi} \int_0^{2\pi} |f(x)| dx$ for every $k \geq 1$. For the second part of (b), use again the triangle inequality for integrals, together with the M-test. Finally, for item (c) use Corollary 11.19, together with Problem 3, page 390.

12. If $a_k(f)$ and $b_k(f)$ are as in the statement of the previous problem, compute the integrals that define such numbers to get $a_0(f) = \frac{2\pi^2}{3}$, $a_k = \frac{4(-1)^k}{k^2}$ and $b_k = 0$ for every $k \in \mathbb{N}$. Then, use Fourier convergence theorem to write $f(x) = \frac{a_0(f)}{2} + \sum_{k\geq 1} a_k(f) \cos(kx)$. Finally, compute $f(\pi) = \pi^2$ with the aid of such a series.

13. Again in the notations of Problem 11, compute the appropriate integrals to find $a_0(f) = 1$, $a_n(f) = 0$ for $n \geq 1$, $b_{2k}(f) = 0$ for $k \geq 1$ and $b_{2k-1}(f) = \frac{2}{(2k-1)\pi}$ for $k \geq 1$. Then, evaluate the Fourier series of f at $\frac{\pi}{2}$, with the aid of Fourier's convergence theorem.

14. For item (a), apply the change of variables formula to g, followed by Riemann's Theorem 10.7. For (b), write

$$\int_0^p f(x)g(nx)dx = \sum_{k=1}^{n} \int_{\frac{(k-1)p}{n}}^{\frac{kp}{n}} f(x)g(nx)dx.$$

Then, given $\epsilon > 0$, use the uniform continuity of f to find $n_0 \in \mathbb{N}$ such that, for $n > n_0$,

$$\left|\sum_{k=1}^{n} \int_{\frac{(k-1)p}{n}}^{\frac{kp}{n}} \left(f(x) - f\left(\frac{kp}{n}\right)\right) g(nx)dx\right| \leq \epsilon \int_0^p |g(nx)|dx \leq \epsilon p \left(\max_{[0,p]} |g|\right).$$

15. First of all, it follows from Problem 6, page 391, that $a_k(f) = \frac{1}{\pi} \int_0^{2\pi} f(x) \cos(kx)dx$. Then, use the result of the previous problem. For $b_k(f)$, argue in a similar way.

16. Since $a_k(f) = a_k(g)$ and $b_k(f) = b_k(g)$ for every k, we have $S_k f = S_k g$ for every k, so that

$$\sigma_n f = \frac{1}{n+1} \sum_{k=0}^{n} S_k f(x) = \frac{1}{n+1} \sum_{k=0}^{n} S_k g(x) = \sigma_n g$$

for every $n \in \mathbb{N}$. Now, Fejér's theorem guarantees that $\sigma_n f \xrightarrow{n} f$ and $\sigma_n g \xrightarrow{n} g$ (even) uniformly on $\mathbb{R}$, so that $\sigma_n f = \sigma_n g$ implies $f = g$.

17. We are going to show that f is a constant function. To this end, first note that if $g(x) = f\left(\frac{x}{2\pi}\right)$, then g is continuous and periodic of period 2π, and f is constant if and only if g is so. Note also that $g(x) = g(x + 2\pi\sqrt{2})$ for every $x \in \mathbb{R}$. Now, let $\frac{a_0}{2} + \sum_{k\geq 1}(a_k\cos(kx) + b_k\sin(kx))$ be the Fourier series of g. Using the change of variables formula, together with the result of Problem 6, page 391, to get, for $k \geq 1$,

$$\pi a_k = \int_{-\pi}^{\pi} g(x)\cos(kx)dx = \int_{-\pi}^{\pi} g(x + 2\pi\sqrt{2})\cos(kx)dx$$

$$= \int_{-\pi-2\pi\sqrt{2}}^{\pi-2\pi\sqrt{2}} g(x)\cos\left(k(x - 2\pi\sqrt{2})\right)dx$$

$$= \int_{-\pi}^{\pi} g(x)\cos\left(k(x - 2\pi\sqrt{2})\right)dx$$

$$= \int_{-\pi}^{\pi} g(x)\left(\cos(kx)\cos(2\pi k\sqrt{2}) + \sin(kx)\sin(2\pi k\sqrt{2})\right)dx$$

$$= \pi a_k\cos(2\pi k\sqrt{2}) + \pi b_k\sin(2\pi k\sqrt{2}).$$

Analogously, show that $b_k = b_k\cos(2\pi k\sqrt{2}) - a_k\sin(2\pi k\sqrt{2})$, and solve the linear system in a_k and b_k thus obtained to get $a_k = b_k = 0$ for every $k \geq 1$. Finally, use the result of the previous problem to show that g is constant.

18. Since $f(x_n, t) \to f(x_0, t)$ and $\frac{\partial f}{\partial x}(x_n, t) \to \frac{\partial f}{\partial x}(x_0, t)$ as $n \to +\infty$, we have $|f(x_0, t)| \leq g(t)$ and $\left|\frac{\partial f}{\partial x}(x_0, t)\right| \leq G(t)$ for every $t \in J$. Then, the integrabilities of $t \mapsto f(x, t)$ and $t \mapsto \frac{\partial f}{\partial x}(x, t)$ (for each $x_0 \in I$) follow from the comparison test for improper integrals (Proposition 10.73). For the rest of item (a), fix $x_0 \in I$ and let $(x_n)_{n\geq 1}$ be a sequence of points in I, converging to x_0. Let $g_n(t) = f(x_n, t)$ and $g_0(t) = f(x_0, t)$, so that g_0 is the pointwise limit of g_n. Apply Lebesgue's DCT to the g_n's and g_0 to show that $F(x_n) \to F(x_0)$ as $n \to +\infty$. For the rest of (b), first note that if F' exists and is given by the right hand side of (11.18), then item (a) (applied to $\frac{\partial f}{\partial x}$, instead of f) guarantees that F' is continuous. Now, let $(x_n)_{n\geq 1}$ be a sequence of points in $I \setminus \{x_0\}$, converging to x_0. Lagrange's MVT assures the existence of ξ_n between x_n and x_0 such that

$$\frac{F(x_n) - F(x_0)}{x_n - x_0} = \int_{\alpha}^{\beta} \frac{f(x_n, t) - f(x_0, t)}{x_n - x_0}dt = \int_{\alpha}^{\beta} \frac{\partial f}{\partial x}(\xi_n, t)dt.$$

Let $G_n(t) = \frac{\partial f}{\partial x}(\xi_n, t)$, $G_0(t) = \frac{\partial f}{\partial x}(x_0, t)$, so that G_0 is the pointwise limit of G_n. Apply Lebesgue's DCT to the G_n's and G_0 to show that

$$\frac{F(x_n) - F(x_0)}{x_n - x_0} \xrightarrow{n} \int_{\alpha}^{\beta} \frac{\partial f}{\partial x}(x_0, t)dt.$$

18. Use the result of the previous problem, together with Problem 5, page 445. You may wish to use the following facts: if $(x_n)_{n\geq 1}$ is a sequence of positive reals that converges to the positive real x_0, then $(x_n)_{n\geq 1}$ is bounded; in turn, show that this guarantees the existence of integrable functions $g_0, G_0 : (0,1] \to \mathbb{R}$ and $g_1, G_1 : [1,+\infty) \to \mathbb{R}$ such that $e^{-t}t^{x_n-1} \leq g_0(t)$ and $e^{-t}t^{x_n-1}\log t \leq G_0(t)$ in $(0,1]$, whereas $e^{-t}t^{x_n-1} \leq g_1(t)$ and $e^{-t}t^{x_n-1}\log t \leq G_1(t)$ in $[1,+\infty)$.

Section 11.3

1. For items (a) and (b), use the result of Corollary 11.24. For item (c), note that the radius of convergence of the series is at most 1, since it doesn't converge when $x = 1$. Then, note that for $|x| < 1$ we have $\sum_{k\geq 0} x^{2^k} < +\infty$, so that $\sum_{k\geq 0}(-1)^k x^{2^k}$ is absolutely convergent.

2. For $|x| < 1$, note that $\sum_{k\geq 0}\frac{1}{k!}x^{2^k} < \sum_{k\geq 0}\frac{1}{k!} < +\infty$. On the other hand, for $|x| = a > 1$ and sufficiently large k,

$$\frac{x^{2^k}}{k!} = \frac{a^{2^k}}{k!} > \frac{a^{k^2}}{k!} > \frac{a^{k^2}}{k^k} = \left(\frac{a^k}{k}\right)^k \overset{k}{\longrightarrow} +\infty;$$

hence, the general term of the series doesn't go to 0 as k increases, and the series diverges. Therefore, its radius of convergence is 1.

3. If $f(x) = \sum_{k\geq 1}\frac{(-1)^{k-1}}{(2k-1)!}x^{2k-1}$ and $g(x) = \sum_{k\geq 0}\frac{(-1)^k}{(2k)!}x^{2k}$, then f and g are defined in the whole real line, and Theorem 11.27 gives $f'(x) = g(x)$ and $g'(x) = -f(x)$. However, since $f(0) = 0$ and $g(0) = 1$, it follows from Example 9.48 that $f(x) = \sin x$ and $g(x) = \cos x$.

4. Apply Theorem 11.27 $k-1$ times, starting from $\frac{1}{1-x} = \sum_{n\geq 0}x^n$. Alternatively, expand $(1-x)^{-k}$ with the aid of the binomial theorem.

5. Expand $(1-x^2)^{-1/2}$ in a power series with the aid of the binomial theorem. Then, use item (b) of Proposition 11.26, together with the result of Problem 4, page 300.

6. For item (a), observe that $\log\left(\frac{1+x}{1-x}\right) = \log(1+x) - \log(1-x)$ and apply the result of Example 11.30. For item (c), we have $\frac{1+x}{1-x} = 3$ for $x = \frac{1}{2}$. Substituting $x = \frac{1}{2}$ in the formula of item (a), note that $\frac{1}{2} + \frac{1}{2^3\cdot 3} + \frac{1}{2^5\cdot 5} + \frac{1}{2^7\cdot 7} + \frac{1}{2^9\cdot 9} + \frac{1}{2^{11}\cdot 11}$ give $\frac{1}{2}\log 3$ with four correct decimal places, since

$$\frac{1}{2^{13}\cdot 13} + \frac{1}{2^{15}\cdot 15} + \frac{1}{2^{17}\cdot 17} + \cdots < \frac{1}{13}\left(\frac{1}{2^{13}} + \frac{1}{2^{15}} + \frac{1}{2^{17}} + \cdots\right)$$

$$= \frac{1}{3\cdot 13\cdot 2^{11}} < \frac{1}{30\cdot 2000} = \frac{1}{60000}.$$

7. For item (a), adapt the idea of the proof of Abel's identity (5.19). More precisely, noting that $a_k = r_k - r_{k+1}$, we have

$$\sum_{k=n}^{m} a_k x^k = \sum_{k=n}^{m} a_k R^k y^k = \sum_{k=n}^{m} (r_k - r_{k+1}) y^k$$

$$= \sum_{k=n}^{m} r_k y^k - \sum_{k=n}^{m} r_{k+1} y^k = \sum_{k=n}^{m} r_k y^k - \sum_{k=n+1}^{m+1} r_k y^{k-1}$$

$$= r_n y^n + \sum_{k=n+1}^{m} r_k y^k - \sum_{k=n+1}^{m} r_k y^{k-1} - r_{m+1} y^m$$

$$= \left(r_n y^n - r_{m+1} y^m \right) + \sum_{k=n+1}^{m} r_k \left(y^k - y^{k-1} \right).$$

For item (b), since $|y| \le 1$ we get

$$\left| \sum_{k=n}^{m} a_k x^k \right| \le |r_n| + |r_m| + \sum_{k=n+1}^{m} |r_k| \left| y^k - y^{k-1} \right|$$

$$\le |r_n| + |r_m| + \sum_{k=n+1}^{m} \left(\sup_{k>n} |r_k| \right) \left(y^{k-1} - y^k \right)$$

$$= |r_n| + |r_m| + \left(\sup_{k>n} |r_k| \right) \left(y^n - y^m \right).$$

The first part of item (c) now follows from the fact that, on the one hand, $|y| \le 1$ gives

$$\left| \sum_{k=n}^{m} a_k x^k \right| \le |r_n| + |r_m| + \left(\sup_{k>n} |r_k| \right);$$

on the other, $r_n \xrightarrow{n} 0$. Finally, for the second part of (c), note that the first part guarantees that f extends continuously to $[0, R]$.

8. In Example 11.30, we saw that $\log(1 + x) = \sum_{k\ge0} \frac{(-1)^k}{k+1} x^{k+1}$ for $|x| < 1$. Since $\sum_{k\ge0} \frac{(-1)^k}{k+1}$ converges (cf. Example 7.50), it suffices to use the result of the previous problem.

9. Adapt the hint given to the previous problem, this time using (11.23). Alternatively, start by showing that

$$\frac{1}{1 + x^2} = 1 - x^2 + x^4 - x^6 + \cdots + (-1)^{n-1} x^{2n-2} + \frac{(-1)^n x^{2n}}{1 + x^2}$$

for every real x. Then, integrate both sides of the above inequality from 0 a 1, noticing that

$$\left| \int_0^1 \frac{(-1)^n x^{2n}}{1 + x^2} \, dx \right| \leq \int_0^1 \frac{x^{2n}}{1 + x^2} \, dx \leq \int_0^1 x^{2n} \, dx = \frac{1}{2n + 1};$$

finally, let $n \to +\infty$.

10. For item (a), just note that $s_0 \geq s_1 \geq s_j \geq \dots$, so that $\lim_{j \to +\infty} s_j = \inf\{s_j;$ $j \geq 0\} \in \mathbb{R} \cup \{\pm\infty\}$. For item (b), assume first that $\limsup a_k = M \in \mathbb{R}$. Then, in the above notations, $M = \inf\{s_j; j \geq 0\}$, so that, given $\epsilon > 0$, $M + \epsilon$ is no longer a lower bound for $\{s_j; j \geq 0\}$. In turn, this implies the existence of $j \in \mathbb{N}$ such that $s_j < M + \epsilon$, and thus $a_l < M + \epsilon$ for every $l \geq j$. This means that only $a_1, a_2, \dots, a_{l-1}$ can be greater than $M + \epsilon$. On the other hand, the fact that $M = \inf\{s_j; j \geq 0\}$ gives $\sup\{a_j, a_{j+1}, \dots\} = s_j \geq M$, and hence $M - \epsilon$ is not an upper bound for $\{a_j, a_{j+1}, \dots\}$, so that there exists $j_1 \geq j$ such that $a_{j_1} > M - \epsilon$. By repeating this argument with $s_{j_1+1} = \sup\{a_{j_1+1}, a_{j_1+2}, \dots\}$, we get $j_2 > j_1$ such that $a_{j_2} > M - \epsilon$. Then, proceeding by induction we construct a whole sequence $j_1 < j_2 < \dots$ of naturals such that $a_{j_l} > M - \epsilon$ for every $l \geq 1$. The converse of (b) can be established in a similar way.

11. For item (a), use item (b)ii. of the previous problem. For item (b), first show that $\limsup \sqrt[k]{|a_k x^k|} >$ and then use item (b)i. of the previous problem. Finally, since items (a) and (b) show that the series converges in $(-R, R)$ and diverges outside of $[-R, R]$, we conclude that R is its radius of convergence, thus establishing (c).

12. For item (a), Theorem 11.27 guarantees that $xf'(x) = \sum_{j \geq 1} 2^j x^{2^j}$ for $0 \leq x < 1$. Moreover, since

$$2^j x^{2^j} > x^{2^j} + x^{2^j+1} + x^{2^j+2} + \dots + x^{2^{j+1}-1},$$

we conclude that

$$xf'(x) > \sum_{j \geq 1} \left(x^{2^j} + x^{2^j+1} + x^{2^j+2} + \dots + x^{2^{j+1}-1} \right) = \sum_{k \geq 2} x^k = \frac{x^2}{1 - x}.$$

Integrating the inequality $f'(x) > \frac{x}{1-x}$ along the interval $[0, x]$, with $x < 1$, we get the desired inequality.

Section 11.4

1. Note that $a_1 + a_2 + \dots + a_k = m$ if and only if $x^{a_1} x^{a_2} \dots x^{a_k} = x^m$; hence, conclude that there are as many solutions $(a_1, a_2, \dots, a_k)$ of the given equation as ways of getting a summand x^m in

$$f(x) = \underbrace{(1 + x + x^2 + \cdots)(1 + x + x^2 + \cdots) \ldots (1 + x + x^2 + \cdots)}_{k}.$$

Next, show that $f(x) = \frac{1}{(1-x)^k} = \sum_{n \geq 0} \binom{k+n-1}{n} x^n$ for $|x| < 1$, where in the last equality we used the result of Problem 4, page 483, to get the power series expansion of $(1 - x)^{-k}$.

2. Proceeding as in the hint given to the previous problem, we want to compute the coefficient of x^{20} in

$$(1 + x + x^2 + \cdots)(x^2 + x^3 + \cdots)(1 + x + x^2 + \cdots)(1 + x + x^2 + \cdots + x^7).$$

Letting $f(x)$ denote such an expression when $|x| < 1$, conclude that $f(x) = (x^2 + x^3 + x^4 + \cdots + x^9) \cdot \frac{1}{(1-x)^3}$. Then, use once more the result of Problem 4, page 483.

3. Make induction on $k \geq 2$, the case $k = 2$ being given by Proposition 11.46.

4. Differentiate both sides of (11.29) and, then, make $x = 1$. Note that Theorem 11.27 assures we can do this.

5. Firstly, use induction to show that $a_n \leq n^2$ for $n \geq 2$. Secondly, letting $f(x) = \sum_{k \geq 1} a_k x^k$ and noticing that the radius of convergence of the power series $\sum_{k \geq 2} k^2 x^k$ is equal to 1, conclude that f is defined in the interval $(-1, 1)$. Now, use the given recurrence relation to show that $f(x) = 2x + xf(x) + xg'(x)$, with $g(x) = \frac{x}{1-x}$ for $|x| < 1$. From this point, get a closed formula for $f(x)$, valid for $|x| < 1$, expand such a formula in power series and find a_k in terms of k.

6. Adapt, to the present case, the proof given in this section to Theorem 11.45.

7. For item (b), compute

$$f(x) = 1 + \sum_{n \geq 1} a_n x^n = 1 + \sum_{n \geq 0} a_{n+1} x^{n+1}$$

$$= 1 + \sum_{n \geq 0} (2a_n + n) x^{n+1}$$

$$= 1 + 2xf(x) + \sum_{n \geq 0} (n + 2) x^{n+1} - 2 \sum_{n \geq 0} x^{n+1}$$

$$= 1 + 2xf(x) + g'(x) - \frac{2x}{1 - x},$$

with $g(x) = \frac{x^2}{1-x}$. For item (c), reduce both sides to a common denominator and, then, compare coefficients of the corresponding numerators. For item (d), use the result of Problem 4, page 483.

8. Adapt, to the present case, the hints given to the items of the previous problem.

9. Use identity (2.5) to get

$$e^{3x} = u^3 + v^3 + w^3 + 3(e^x - u)(e^x - v)(e^x - w)$$
$$= u^3 + v^3 + w^3 + e^x(uv + uw + vw) - uvw.$$

Let $f = uv + uw + vw$, $g = u^2 + v^2 + w^2$ and show that $f(0) = 0$, $g(0) = 1$, $f' = f + g$ and $g' = 2f$. Then, use the result of Example 11.51 to find f and finish the proof.

10. For the first part of item (a), use the results of of Theorem 11.27 and Proposition 11.46 to write $f'' + pf' + q$ as a power series. For (b), start by using the recurrence relation of (a) to get, with the aid of the triangle inequality,

$$(k+2)(k+1)|a_{k+2}| \le \sum_{j=0}^{k} \left((j+1)|b_{k-j}||a_{j+1}| + |c_{k-j}||a_j| \right).$$

Then, multiply both sides of the above inequality by $|x - x_0|^{k+2}$ to get

$$(k+2)(k+1)A_{k+2} \le \left(\sum_{j=0}^{k} (j+1)|b_{k-j}(x-x_0)^{k-j}|A_{j+1}| \right)|x - x_0|$$

$$+ \left(\sum_{j=0}^{k} |c_{k-j}(x-x_0)^{k-j}A_j| \right)|x - x_0|^2$$

$$\le \left(\sum_{j=0}^{k} (j+1)|b_{k-j}|r^{k-j}\left(\frac{|x-x_0|}{r} \right)^{k-j}|A_{j+1}| \right)r$$

$$+ \left(\sum_{j=0}^{k} |c_{k-j}|r^{k-j}\left(\frac{|x-x_0|}{r} \right)^{k-j+1} A_j \right)\frac{|x-x_0|}{r}.$$

Item (c) now follows directly from (b). Item (d) follows from (c) and induction on k. For second part of (e), use the recurrence relation of (d) to get

$$(k+2)(k+1)\tilde{A}_{k+2} - \lambda(k+1)k\tilde{A}_{k+1} = M(k+2)\tilde{A}_{k+1}.$$

Section 11.5

1. Make induction on k.
2. Write $\log x = \log(x_0 + (x-x_0)) = \log x_0 + \log\left(1 + \frac{x-x_0}{x_0}\right)$. Now, use the fact that $x \mapsto \log(1+x)$ is analytic on $(-1, 1)$ and $x \mapsto \frac{x-x_0}{x_0}$ maps $(0, 2x_0)$ into $(-1, 1)$ and is also analytic. For what is left to do, write $\log x = \log x_0 + \sum_{k\ge 1} \frac{(-1)^{k-1}}{kx_0^k}(x-x_0)^k$ in $(0, 2x_0)$.

3. Let $n_0 \in \mathbb{N}$ be such that $f^{(n)}(x) \geq 0$ for every $x \in I$ and $n \geq n_0$; then, $f^{(n)}$ is nondecreasing for every $n > n_0$. Now, given $x_0, x \in I$, the Taylor formula for f centered at x_0 gives

$$f(x) = \sum_{k=0}^{n-1} \frac{f^{(k)}(x_0)}{k!}(x-x_0)^k + \frac{f^{(n)}(c)}{n!}(x-x_0)^n,$$

for some c between x_0 and x. If $x > x_0$, show that

$$0 \leq f(x) - \sum_{k=0}^{n-1} \frac{f^{(k)}(x_0)}{k!}(x-x_0)^k \leq \frac{f^{(n)}(x)}{n!}(x-x_0)^n \xrightarrow{n} 0.$$

Argue analogously if $x < x_0$.

4. For item (a), show that $f(x_0), f'(x_0), f''(x_0), \ldots$ are uniquely determined by (11.39), and then apply Corollary 11.57. Item (b) is the content of Theorem 11.48. For item (c), let b be the right endpoint of J and suppose that b is not the right endpoint of I. Since p and q are analytic in I, there exists $r > 0$ such that $(b-r, b+r) \subset I$ and p and q are given in this interval by their Taylor series centered at b. If $a \in \left(b - \frac{r}{2}, b\right)$, use Theorem 11.54 to show that the Taylor series of p and q centered at a converge in the interval $(a-s, a+s)$, with $s = a - b + r$ and $a + s > b$. Let $g : (a-s, a+s) \to \mathbb{R}$ be the unique analytic solution of

$$\begin{cases} g'' + pg' + qg = 0 \\ g(a) = f_J(a), g'(a) = f'_J(a) \end{cases}.$$

Conclude that $g^{(k)}(a) = f_J^{(k)}(a)$ for every integer $k \geq 0$ and then that $g = f_J$ in $(a-s, b)$. Then, extend f_J to $J \cup (a-s, a+s)$. Finally, for (d), let J be the largest open interval contained in I such that (11.39) has a solution in J. Use item (c) to show that $J = I$.

Index

© Springer International Publishing AG 2017

A. Caminha Muniz Neto, *An Excursion through Elementary Mathematics, Volume I*,
Problem Books in Mathematics, DOI 10.1007/978-3-319-53871-6

Printed in the United States
By Bookmasters